T0350173

DIMENSION GROUPS AND DYNAMICAL SYSTEMS

This book is the first self-contained exposition of the fascinating link between dynamical systems and dimension groups. The authors explore the rich interplay between topological properties of dynamical systems and the algebraic structures associated with them, with an emphasis on symbolic systems, particularly substitution systems. It is recommended for anybody with an interest in topological and symbolic dynamics, automata theory or combinatorics on words.

Intended to serve as an introduction for graduate students and other newcomers to the field as well as a reference for established researchers, the book includes a thorough account of the background notions as well as detailed exposition – with full proofs – of the major results of the subject. A wealth of examples and exercises, with solutions, serve to build intuition, while the many open problems collected at the end provide jumping-off points for future research.

Fabien Durand is Full Professor in Mathematics at University of Picardie Jules Verne. His interests include topological dynamical systems and the relations with theoretical computer science. He is currently the president of the Société Mathématique de France.

Dominique Perrin is Emeritus Professor in Mathematics and Computer Science at Gustave Eiffel University. He is (co)author or editor of a number of books including *Profinite Semigroups and Symbolic Dynamics*, *Codes and Automata*, *Infinite Words*, and *Combinatorics on Words*, and (under the pseudonym Lothaire). He is a member of Academia Europea.

All the titles listed below can be obtained from good booksellers or from Cambridge University Press. For a complete series listing, visit www.cambridge.org/mathematics.

Dimension Groups and Dynamical Systems

Substitutions, Bratteli Diagrams and Cantor Systems

FABIEN DURAND
University of Picardie Jules Verne

DOMINIQUE PERRIN
Gustave Eiffel University

CAMBRIDGE
UNIVERSITY PRESS

CAMBRIDGE
UNIVERSITY PRESS

University Printing House, Cambridge CB2 8BS, United Kingdom

One Liberty Plaza, 20th Floor, New York, NY 10006, USA

477 Williamstown Road, Port Melbourne, VIC 3207, Australia

314–321, 3rd Floor, Plot 3, Splendor Forum, Jasola District Centre,
New Delhi – 110025, India

103 Penang Road, #05–06/07, Visioncrest Commercial, Singapore 238467

Cambridge University Press is part of the University of Cambridge.

It furthers the University's mission by disseminating knowledge in the pursuit of
education, learning, and research at the highest international levels of excellence.

www.cambridge.org
Information on this title: www.cambridge.org/9781108838689
DOI: 10.1017/9781108976039

© Fabien Durand and Dominique Perrin 2022

First published 2022

A catalogue record for this publication is available from the British Library.

Library of Congress Cataloging-in-Publication Data
Names: Durand, Fabien, 1969- author. | Perrin, Dominique, author.
Title: Dimension groups and dynamical systems : substitutions, Bratteli diagrams and
cantor systems / Fabien Durand, Dominique Perrin.
Description: Cambridge ; New York, NY : Cambridge University Press, 2022. |
Series: Cambridge studies in advanced mathematics ; 196 | Includes
bibliographical references and index.
Identifiers: LCCN 2021035082 (print) | LCCN 2021035083 (ebook) | ISBN
9781108838689 (hardback) | ISBN 9781108976039 (ebook)
Subjects: LCSH: Dynamics. | Dimension theory (Topology) | Group theory. |
BISAC: MATHEMATICS / General
Classification: LCC QA845 .D87 2022 (print) | LCC QA845 (ebook) | DDC
515/.39–dc23/eng/20211021
LC record available at https://lccn.loc.gov/2021035082
LC ebook record available at https://lccn.loc.gov/2021035083

ISBN 978-1-108-83868-9 Hardback

Contents

Introduction

In this monograph, we introduce the reader to the connection between topological dynamical systems and dimension groups. Let us first explain briefly each of these terms.

A topological dynamical system is a pair (X, T) of a compact metric space X and a continuous map T from X to itself. The system is minimal if the orbit of every point of X is dense. We will be mainly interested in minimal systems. We will often consider the case where X is a closed subset of the set of $A^{\mathbb{Z}}$ of infinite sequences over a finite alphabet A, and T is the shift on $A^{\mathbb{Z}}$. When X is, moreover, invariant by T, we obtain a topological dynamical system called a shift space.

A dimension group is an ordered abelian group having some additional specific properties. To every minimal shift space (and more generally to every minimal Cantor system), we will associate a dimension group in such a way that isomorphic systems have isomorphic dimension groups.

One of the main objects of this book is to describe various methods to compute these dimension groups. In this way, we will be able to distinguish topological dynamical systems, which can appear in many different forms, by comparing their dimension groups, which are easier to handle.

As a motivating example, consider the Fibonacci sequence, which is the sequence

$$x = abaababa \cdots$$

obtained by iterating indefinitely the substitution $a \mapsto ab$, $b \mapsto a$ starting with a. The sequence of iterates

$$a$$
$$ab$$
$$aba$$
$$abaab$$
$$\cdots$$

1

converges (in an obvious sense) to x. The shift space X formed of the sequences $y \in \{a, b\}^{\mathbb{Z}}$ with all its blocks appearing in x is a shift space called the Fibonacci shift. We will see that its dimension group is the discrete subgroup of $\mathbb{R}$ formed of the $x + y\alpha$ with $x, y \in \mathbb{Z}$ and where $\alpha = (1 + \sqrt{5})/2$ is the golden mean. We shall see how this is related to the fact that there is a unique invariant probability measure on X that is such that the probability of the set of sequences $y = (y_n)_{n \in \mathbb{Z}}$ such that $y_0 = a$ is $\alpha - 1$.

Dimension groups were first associated with a family of associative algebras called approximately finite, or AF-algebras. These algebras are themselves a class of C^*-algebras that are direct limits of finite dimensional algebras and were introduced by Ola Bratteli (1972). The algebra is built from a special kind of graph called a *Bratteli diagram*.

Dimension groups were introduced by George Elliott (1976) as a tool for classifying AF-algebras and he proved that the dimension group (together with an additional information called the scale) provides a complete algebraic invariant for these algebras.

The connection of these ideas with dynamical systems was first done by Wolfgang Krieger (1977) (see also Krieger (1980a)) who defined a dimension group for every shift of finite type. The link with Bratteli diagrams was done by Anatol Vershik (1982) who used a lexicographic order on paths of the Bratteli diagrams to define a topological dynamical system on the set of infinite paths of the graph. Later, Richard Herman, Ian Putnam and Christian Skau showed that every minimal system on a Cantor space is isomorphic to such system. As a consequence, a dimension group is attached to any minimal Cantor system and subsequent work by Thierry Giordano, Ian Putnam and Christian Skau (1995) showed that this group is related to the orbit structure of the system.

In this expository presentation, written after the unpublished notes by Bernard Host (1995) (see also Host (2000)), we present the basic elements of this theory, insisting on the computational and algorithmic aspects allowing one to effectively compute the dimension groups. The computation applies in particular to the case of substitution shifts, explicitly presented previously in Durand et al. (1999), in relation with Forrest (1997).

In the first chapter (Chapter 1) we present the basic notions of topological dynamical systems. Although such systems can be defined using a group or semigroup action, we restrict our attention to systems on which acts the group $\mathbb{Z}$ or the semigroup $\mathbb{N}$. We define recurrent systems and minimal dynamical systems (Section 1.1). Next, we introduce in Section 1.2 shift spaces, which are the basic systems we are interested in. We define return words and higher block shifts. In Section 1.4, we introduce substitution shifts. We define the notion of recognizable morphism and we prove the Theorem of Mossé (Theorem 1.4.35) asserting that any aperiodic primitive morphism is recognizable.

In the second chapter (Chapter 2), we shift to an algebraic and combinatorial environment. We first introduce, in Section 2.1, ordered groups (considering only abelian groups). We define several notions, as that of order unit and order ideal. We also define a simple ordered group as one with no nontrivial ideals. In Section 2.3 we define direct limits of ordered groups and we give examples of the computation of these ordered groups. In the last part of this section (Section 2.4), we finally define dimension groups. These groups are defined as direct limits of groups $\mathbb{Z}^n$ with the usual ordering. We prove the abstract characterization by Effros, Handelman and Shen Effros et al. (1980) using the property of Riesz interpolation.

In Chapter 3, we come to notions of cohomology defined in a Cantor system. We first introduce the notion of coboundary (Section 3.1) and prove in Section 3.2 the Gottshalk–Hedlund Theorem (Proposition 3.2.5) characterizing the continuous functions on a Cantor set that are coboundaries. We next define the ordered cohomology group $K^0(X, T)$ of a recurrent system (X, T) as the quotient of the group of integer valued continuous functions on X by the subgroup formed by coboundaries. In the next two sections (Sections 3.6 and 3.7), we consider the effect on the ordered cohomology group of applying a factor map or taking the system induced on a clopen set. In the second part of this chapter, beginning with Section 3.8, we define invariant probability measures on a Cantor system and recall that a substitutive shift defined by a primitive substitution has a unique invariant probability measure. We indicate a method to compute this measure. We show in Section 3.9 that there is a close connection between the cohomology group and the cone of invariant measures (Proposition 3.9.3). We use this connection to give a description of the dimension groups of Sturmian shifts (Theorem 3.9.3).

In Chapter 4, we introduce the fundamental tool of partitions in towers, or Kakutani–Rokhlin partitions. We prove the theorem of Herman, Putnam and Skau, which shows that any minimal Cantor system can be represented as the limit of a sequence of partitions in towers (Theorem 4.1.6). In Chapter 4 we come back to partition in towers. We first show how to associate an ordered group to a partition in towers (Section 4.2). Next, in Section 4.3, we use a sequence of partitions in towers to prove that the group $K^0(X, T)$ is, for any minimal dynamical system (X, T), a simple dimension group (Theorem 4.3.4). In the next sections, we present explicit methods to compute the dimension group of a minimal shift space. In Section 4.4, we use return words and in Section 4.5, we use Rauzy graphs. Finally, in Section 4.6, we show how to compute the dimension group of a substitutive shift, as exposed in Durand et al. (1999).

We introduce Bratteli diagrams in Chapter 5. We define the telescoping of a diagram. We define the dimension group of a Bratteli diagram and prove that it is a complete invariant for telescoping equivalence (Theorem 5.1.5). We next introduce ordered Bratteli diagrams and show that one may associate a dynamical system to every properly ordered Bratteli diagram. We prove the Bratteli–Vershik Model Theorem (Theorem 5.3.3) showing the completeness of the model for minimal Cantor systems. We next prove the Strong Orbit Equivalence Theorem (Theorem 5.5.1) showing that dimension groups are a complete invariant for strong orbit equivalence. We state (without proof) the related Orbit Equivalence Theorem (Theorem 5.5.3).

In Chapter 6, we focus on substitution shifts and their representations. We begin by considering odometers, which have BV-representations close to substitution shifts. We characterize, as a main result, the family of BV-systems associated with stationary Bratteli diagrams as the disjoint union of stationary odometers and substitution minimal systems (Theorem 6.2.1). We develop next the description of linearly recurrent shifts, which are characterized by their BV-representation (Theorem 6.3.5). We introduce in Section 6.4 the notion of an S-adic representation. The main result is an explicit description of the dimension group of a unimodular S-adic shift (Theorem 6.5.4). In the last section (Section 6.6), we consider the family of substitutive shifts, a natural generalization of substitution shifts. The main result is a characterization by a finiteness condition of substitutive sequences (Theorem 6.6.1).

Chapter 7 describes the class of dendric shifts, defined by a restrictive condition on the possible extensions of a word. This class is a simultaneous generalizations of several other classes of interest, such as Sturmian shifts or interval exchange shifts (introduced in the next chapter). The main result is the Return Theorem (Theorem 7.1.15), which states that the set of return words in a minimal dendric shift is a basis of the free GP on the alphabet. We use this result to describe the S-adic representation of dendric shifts and show that it can be defined using elementary automorphisms of the free group (Theorem 7.1.40). We illustrate these results by considering the class of Sturmian shifts (Section 7.2). The last part of the chapter is devoted to specular shifts, a class of eventually dendric shifts that plays a role in the next chapter, when we introduce linear involutions. The main result is a description of the dimension group of a specular shift (Theorem 7.3.40).

In Chapter 8, we introduce the notion of interval exchange transformation. We prove Keane's Theorem characterizing minimal interval exchanges (Theorem 8.1.2). We develop the notion of Rauzy induction and characterize the subintervals reached by iterating the transformation (Theorem 8.1.25). We generalize Rauzy induction to a two-sided version and characterize the intervals reached by this more general transformation (Theorem 8.2.2). We link these transformations with automorphisms of the free group (Theorem 8.2.14). We also relate these results with the theorem of Boshernizan and Carroll giving a finiteness condition on the systems induced by an interval exchange when the lengths of the intervals belong to a quadratic field (Theorem 8.3.2). In the last section (Section 8.4) we define linear involutions and show that the natural coding of a linear involution without connections is a specular shift (Theorem 8.4.9).

In the last chapter (Chapter 9) we give a brief introduction to the link between Bratteli diagrams and the vast subject of C^*-algebras. We define approximately finite algebras and show their relation to Bratteli diagrams. We relate simple Bratteli diagrams and simple AF algebras (Theorem 9.3.12). We prove Elliott's Theorem showing that AF algebras are characterized by their dimension groups (Theorem 9.3.21).

A point useful to mention is that each chapter ends with exercises that can be either illustrations of the results or proofs of some results stated in the chapter, or additional results. For each of them, a solution is provided. The style of writing for the solutions is often more concise than for a proof in the main text but is in general a full proof.

After the exercises, a section of notes concludes each chapter, giving the bibliographic references and also pointing to further results.

The book ends with a series of appendices. The first one (Appendix A) gives the solutions of the exercises proposed in the previous chapters.

A second appendix (Appendix B) is a guide to be used as a reference for notions from several domains of mathematics used in this book. There are also three appendices of a special kind. The first one (Appendix C) is a summary of the various systems (or classes of systems) discussed in the chapters. Next, Appendix D lists the many equivalent definitions of Sturmian shifts (we hope to be thanked by the readers for this). Finally, Appendix E gives a list of open problems in this field.

The book is written in such a way that it should be readable by a graduate student in mathematics or computer science. As a general rule (with a few exceptions, and notably in Chapter 9), complete proofs are given. Some chapters can be read independently of the others, although most of them rely on the introductory chapter (Chapter 1). It seems impossible to cover all chapters in one course, but a selection can be made, resulting in a significant content. One of the authors has recently taught with success the content of Chapters 1, 2 and 3.

Acknowledgments This book started as notes following a seminar held in Marne-la-Vallée in June 2016 and gathering Marie-Pierre Béal, Valérie Berthé, Francesco Dolce, Pavel Heller, Revekka Kyriakoglou, Julien Leroy,

Dominique Perrin and Giuseppina Rindone. The initial version followed closely the notes of Bernard Host (1995), trying to develop more explicitly some arguments. We have progressively enlarged it to its present version. We are deeply grateful to Bernard Host for letting us use his manuscript and encouraging us to continue to reach a more complete account of the field. We wish to thank Mike Boyle, Kenneth Davidson, Brian Marcus and Ian Putnam for fruitful discussions during the preparation of this text. We are also grateful to Paulina Cecchi, Francesco Dolce, Pavel Heller, Maryam Hosseini, Amir Khodayan Karim, Revekka Kyriakoglou, Sébastien Labbé, Christophe Reutenauer, Gwénaël Richomme and Wolfgang Steiner for reading the manuscript and finding many errors. Special thanks are due to Soren Eilers for reading closely Chapter 9 and to Christian Choffrut for his careful reading of most chapters. The manuscript was typeset using LaTeX for the text and TikZ for the figures. We acknowledge the use of many macros due to Jean Berstel.

1

Topological Dynamical Systems

We present in this chapter some definitions and basic properties concerning topological dynamical systems and symbolic dynamical systems. We define shift spaces and the important particular case of substitution shifts, obtained by iterating a substitution. We also prove some difficult results, including the Mossé recognizability theorem and the basic results on Sturmian sequences and their generalizations.

This long chapter aims at serving both as an introduction to the subject and also as a reference when reading forthcoming chapters. It should thus be considered both a tutorial and a memento.

We begin, in Section 1.1, with some general definitions concerning topological dynamical systems. The adjective topological is used to distinguish these systems, which are based on a topological space from dynamical systems based on a measurable space. We define the notion of recurrent and of minimal system.

In Section 1.2, we consider shift spaces and their language. In Sections 1.3, 1.4, 1.5 and 1.6, we present several particular types of symbolic systems, namely shifts of finite type, Sturmian shifts, substitution shifts and finally Toeplitz shifts.

1.1 Recurrent and Minimal Dynamical Systems

A *topological dynamical system* is a pair (X, T) where X is a *compact metric space* and $T: X \to X$ a continuous map.

Example 1.1.1 As a simple example, consider $X = [0, 1]$, which is metric and compact as a closed interval of the real line $\mathbb{R}$. Given $\alpha \in \mathbb{R}$, the transformation $T: x \mapsto (x + \alpha) \mod 1$ is a continuous map from X into X. The pair (X, T) is a topological dynamical system called a *rotation* of the interval $[0, 1]$.

7

Thus, in such a system, with each point x in the space X is associated a sequence $(x, T(x), T^2(x), \ldots)$ of points. It is convenient to imagine the action of T as the sequence of positions of the point x in the space X at discrete times $0, 1, 2, \ldots$. The effect of the hypothesis that X is compact is to guarantee that the sequence of these points will remain at a bounded distance of x.

When T is a homeomorphism, we say that the system (X, T) is *invertible*. Although we will meet, most of the time, invertible dynamical systems, we do not make this hypothesis systematically, mentioning each time when it is necessary. Note that, since X is assumed to be compact, if T is invertible, its inverse is continuous and thus T is a homeomorphism (Exercise 1.1).

Example 1.1.2 The system of Example 1.1.1 is not invertible since 1 is not in the image of T. If we consider, instead of $[0, 1]$, the *torus* $\mathbb{T} = \mathbb{R}/\mathbb{Z}$ in which 0 and 1 are identified, the transformation T is simply the translation $T_\alpha \colon x \mapsto x + \alpha$ and becomes a homeomorphism. The system $(\mathbb{T}, T_\alpha)$ is called the *rotation of angle α*.

For $x \in X$, we often denote Tx instead of $T(x)$. We also denote T^0 for the identity on X and for $n \geq 0$, $T^{-n}(x) = \{y \in X \mid T^n(y) = x\}$.

An important example is when X is a *Cantor space*, that is, a *totally disconnected* compact metric space without *isolated points*. We say then that (X, T) is a *Cantor dynamical system* or *Cantor system* (we shall come back shortly to Cantor spaces).

In particular, let A be a finite set called an *alphabet*. The set $A^\mathbb{Z}$ of all bi-infinite sequences endowed with the product topology is a compact space. Let d be the distance on $A^\mathbb{Z}$ defined for $x \neq y$ by $d(x, y) = 2^{-r(x,y)}$ with

$$r(x, y) = \min\{|n| \mid x_n \neq y_n, n \in \mathbb{Z}\},$$

which is actually an ultrametric distance (see the definition in Appendix B.4). The topology defined by this distance is the same as the product topology, and thus $A^\mathbb{Z}$ is a compact metric space. It is actually a Cantor space if $\mathrm{Card}(A) \geq 2$ (see below). Note that we denote by $\mathrm{Card}(A)$ the cardinality of the set A.

The *shift* transformation $S \colon A^\mathbb{Z} \to A^\mathbb{Z}$ is defined for $x = (x_n)_{n \in \mathbb{Z}}$ by $y = Sx$ where

$$y_n = x_{n+1}, \tag{1.1}$$

for all $n \in \mathbb{Z}$. The shift is obviously continuous and thus $(A^\mathbb{Z}, S)$ is a topological dynamical system. Note that S is not an isometry. Indeed, for $x, y \in A^\mathbb{Z}$, one has $d(x, y)/2 \leq d(Sx, Sy) \leq 2d(x, y)$ and both bounds can be reached.

As a variant, the set $A^{\mathbb{N}}$ of one-sided infinite words is also a topological space for the product topology, and this topology is defined by the metric analogous to the one above. It is also a Cantor space if $\mathrm{Card}(A) \geq 2$. The one-sided shift transformation is defined by Eq. (1.1) for $n \in \mathbb{N}$. It is not invertible as soon as the cardinality $\mathrm{Card}(A)$ of A is at least 2. Thus, $(A^{\mathbb{N}}, S)$ is an example of a noninvertible dynamical system on a Cantor space.

1.1.1 Recurrent Dynamical Systems

A point $x \in X$ in a topological dynamical system (X, T) is *recurrent* if for every open set U containing x, there is an $n \geq 1$ such that $T^n(x) \in U$ (and, in fact, then an infinity of such n; see Exercise 1.3).

A nonempty system (X, T) is *recurrent* (or *topologically transitive*) if for every pair of nonempty open sets U, V in X, there is an integer $n \geq 0$ such that $U \cap T^{-n}V \neq \emptyset$ (or equivalently $T^n U \cap V \neq \emptyset$; see Exercise 1.2).

The *positive orbit* of a point $x \in X$ in a dynamical system (X, T) is the set $\mathcal{O}_+(x) = \{T^n x \mid n \geq 0\}$. Its *orbit* is the set $\mathcal{O}(x) = \cup_{n \geq 0} T^{-n}(\mathcal{O}_+(x))$. Thus, we have also $\mathcal{O}(x) = \{y \in X \mid T^n x = T^m y \text{ for some } m, n \geq 0\}$ and when T is invertible, $\mathcal{O}(x) = \{T^n x \mid n \in \mathbb{Z}\}$. If a point has a dense positive orbit, it is recurrent.

Proposition 1.1.3 *The following conditions are equivalent for a topological dynamical system.*

(i) *(X, T) is recurrent.*

(ii) *There is a point in X with a dense positive orbit.*

The proof is left as Exercise 1.4.

A *morphism* of dynamical systems from (X, T) to (X', T') is a continuous map $\phi \colon X \to X'$ such that $\phi \circ T = T' \circ \phi$ (see the diagram below).

$$
\begin{array}{ccc}
X & \xrightarrow{\ T\ } & X \\
\downarrow{\scriptstyle \phi} & & \downarrow{\scriptstyle \phi} \\
X' & \xrightarrow{\ T'\ } & X'
\end{array}
$$

If ϕ is onto, it is called a *factor map* and (X', T') is called a *factor system* of (X, T).

A factor of a topological dynamical system inherits some of its dynamical properties. For instance, a factor of a recurrent system is recurrent.

When it is bijective, the morphism ϕ is called an *isomorphism* of dynamical systems, or also a *topological conjugacy* (or simply a *conjugacy*). Since

X is compact, the inverse is also continuous (Exercise 1.1) and thus ϕ is a homeomorphism.

Conjugate systems are indistinguishable concerning their dynamical properties. It can be very difficult to exhibit a conjugacy between dynamical systems, and much of what will follow in this book addresses this problem. In particular, we will be looking for *invariants*, that is, properties shared by conjugate systems and easier to determine than the conjugacy itself. Even more interesting are *complete invariants*, which characterize the conjugacy class. This applies to other equivalences than conjugacy, as we shall see. Formally, let θ be an equivalence on the class of dynamical systems, such as conjugacy. A map f assigning an element $f(X, T)$ to a dynamical system (X, T) is an invariant for θ if $(X, T) \equiv (X', T') \bmod \theta$ implies $f(X, T) = f(X', T')$. It is a complete invariant if the converse holds.

We will also be interested in the effective verification of properties of dynamical systems (such as conjugacy, for example). We say that a property is *decidable* if there is an algorithm that allows one to verify it (see Section 1.8 for a reference). Otherwise, it is called *undecidable*. Thus, decidability is desirable, although not the end of things since decidable properties can be very difficult to verify.

1.1.2 Minimal Dynamical System

Given a topological dynamical system (X, T), a subset Y of X is *stable* if $TY \subset Y$ (note that, for us, $\subset$ means $\subseteq$). The empty set and X are always (trivial) stable sets. As a stronger condition, the set $Y \subset X$ is *invariant* if $T^{-1}Y = Y$.

A nonempty topological dynamical system is *minimal* if X is the only closed and stable nonempty subset of X. A factor of a minimal dynamical system is minimal (Exercise 1.5).

Every nonempty dynamical system (X, T) contains a nonempty invariant subset Y such that (Y, T) is minimal (Exercise 1.6).

The following characterization of minimal systems is sometimes used for definition.

Proposition 1.1.4 *A topological dynamical system is minimal if and only if the positive orbit of every point is dense in X.*

Proof Assume first that (X, T) is minimal. For every $x \in X$, the closure of the positive orbit of x is a closed stable nonempty set, and thus it is equal to X. Conversely, let $Y \subset X$ be a closed stable nonempty set. For any $y \in Y$, the closure of the positive orbit of y is contained in Y. Since it is equal to X, we conclude that $Y = X$. ∎

It follows directly from Propositions 1.1.4 and 1.1.3 that a minimal system is recurrent.

The simplest example of a minimal dynamical system is a finite system (also called a *periodic system*), formed of a finite set $X = \{1, 2, \ldots, n\}$ on which acts a circular permutation T. A *periodic point* in a dynamical system (X, T) is a point $x \in X$ such that $T^n x = x$ for some $n \geq 1$. A dynamical system (X, T) is called *aperiodic* if it does not contain any periodic point. A minimal system is aperiodic if and only if it is infinite.

As a second example, we find the rotations of the circle.

Example 1.1.5 Consider the *unit circle* $\mathbb{S}^1 = \{z \in \mathbb{C} \mid |z| = 1\}$ and fix some $\lambda \in \mathbb{S}^1$. Let R_λ be the map defined by $R_\lambda(z) = \lambda z$. Then $(\mathbb{S}^1, R_\lambda)$ is a dynamical system. It is minimal if and only if λ is not a root of unity (Exercise 1.7). Otherwise, it is a disjoint union of periodic systems.

We have introduced above (Example 1.1.2) the *torus* $\mathbb{T}$ as the topological space $\mathbb{R}/\mathbb{Z}$. For $\alpha \in \mathbb{R}$, let $T_\alpha : \mathbb{T} \to \mathbb{T}$ be the map $T_\alpha(x) = x + \alpha$ and set $\lambda = \exp(2i\pi\alpha)$. The map $\phi : x \mapsto e^{2i\pi x}$ is a homeomorphism from $\mathbb{T}$ onto $\mathbb{S}^1$ and $R_\lambda \circ \phi = \phi \circ T_\alpha$ because

$$R_\lambda \circ \phi(x) = R_\lambda e^{2i\pi x} = e^{2i\pi x} e^{2i\pi\alpha} = \phi \circ T_\alpha(x).$$

Consequently, the topological dynamical systems $(\mathbb{S}^1, R_\lambda)$ and $(\mathbb{T}, T_\alpha)$ are isomorphic. The transformation R_α is called a *rotation* of angle α of the circle. The isomorphism is the map $\phi : x \mapsto \exp(2i\pi x)$ (see the diagram below).

$$
\begin{array}{ccc}
\mathbb{T} & \xrightarrow{\ T_\alpha\ } & \mathbb{T} \\
\downarrow{\scriptstyle \phi} & & \downarrow{\scriptstyle \phi} \\
\mathbb{S}^1 & \xrightarrow{\ R_\lambda\ } & \mathbb{S}^1
\end{array}
$$

1.1.3 Induced Systems

Let (X, T) be a minimal topological dynamical system and let U be a non-empty clopen subset of X. Since X is minimal, for every $x \in X$, there is an $n > 0$ such that $T^n x \in U$ and we can define the integer

$$n(x) = \inf\{n > 0 \mid T^n x \in U\}$$

called the *entrance time* of x in U.

Since U is clopen, the function $x \mapsto n(x)$ is continuous. Indeed, for each $n \geq 1$, the set of $x \in X$ such that $n(x) = n$ is $T^{-n}(U) \setminus \cup_{i=1}^{n-1} T^{-i}(U)$, which is open. Thus, the map $x \mapsto n(x)$ is locally constant. Consequently, the map $T_U : U \to U$ defined by

$$T_U(x) = T^{n(x)}(x)$$

is continuous, being locally a power of T. It is called the *induced transformation* on U, and (U, T_U) is called an *induced system* of (X, T) or also a *derivative* of (X, T). This system is minimal. Indeed, the orbit under T of every point x of U is dense in X, and thus its orbit under T_U is dense in U.

For $x \in U$, the integer $n(x)$ is called the *return time* to U. For $x \in X$, the function

$$m(x) = \begin{cases} n(x) & \text{if } x \notin U \\ 0 & \text{otherwise} \end{cases}$$

is called the *waiting time* to access U.

Note that the induced system can be defined even if (X, T) is not minimal, provided the clopen set U is such that the return time $n(x)$ is bounded on U.

The inverse operation can be described as follows. Let (X, T) be a topological dynamical system and let $f : X \to \mathbb{N}$ be a continuous function. Set

$$\hat{X} = \{(x, i) \mid x \in X, 0 \le i < f(x)\} \tag{1.2}$$

and define a map $\hat{T} : \hat{X} \to \hat{X}$ by

$$\hat{T}(x, i) = \begin{cases} (x, i + 1) & \text{if } i + 1 < f(x) \\ (Tx, 0) & \text{otherwise.} \end{cases}$$

Then $(\hat{X}, \hat{T})$ is a topological dynamical system, which is minimal if (X, T) is minimal (see Exercise 1.8). The map $x \mapsto (x, 0)$ identifies X to the system induced by $\hat{X}$ on $X \times \{0\}$. The space $\hat{X}$ defined above is said to be obtained from X by the *tower construction* relative to f, and the system $(\hat{X}, \hat{T})$ is called a *primitive* of X.

One can usefully figure out the result from the tower construction as a superposition of "floors" above X. The number of floors above $x \in X$ is $f(x)$. The transformation consists in accessing the point above until there is no floor above and then using the transformation defined on X (see Figure 1.1).

Example 1.1.6 Consider the system (X, T) with $X = [0, 1]$ and $T(x) = x + \alpha \bmod 1$ of Example 1.1.1 for $1/2 < \alpha < 1$. The transformation T_U induced by T on $U = [0, \alpha[$ is given by

$$T_U(x) = \begin{cases} x + 2\alpha - 1 & \text{if } 0 \le x < 1 - \alpha \\ x + \alpha - 1 & \text{otherwise.} \end{cases}$$

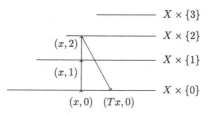

Figure 1.1 The tower construction.

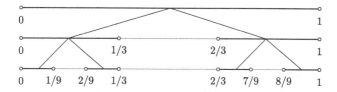

Figure 1.2 The sets C_0, C_1, C_2.

Thus, (U, T_U) is isomorphic to the system $(X, T_{2-(1/\alpha)})$ via the conjugacy $x \mapsto x/\alpha$.

1.1.4 Dynamical Systems on Cantor Spaces

We have defined a Cantor space as a totally disconnected compact metric space without isolated points. Recall that a topological space is called totally disconnected if every connected component is reduced to a point (see Appendix B.4).

Let us give classical examples of Cantor spaces. There is first the historical one, as defined originally by Cantor. The *Cantor set* is the subset C of the interval $[0, 1]$ defined by $C = \cap_n C_n$, where $C_0 = [0, 1]$ and (see Figure 1.2)

$$C_{n+1} = \frac{C_n}{3} \cup \left(\frac{2}{3} + \frac{C_n}{3} \right).$$

It is a Cantor space for the induced topology.

There is the abstract topological one that we used for the definition. It is a compact metric space, which is totally disconnected without isolated points or, equivalently, a topological space with a countable basis of the topology consisting of clopen sets and without isolated points (see Appendix B.4).

Next, there is the symbolic one that we have already seen, which consists in considering the set $A^{\mathbb{Z}}$ of two-sided infinite sequences on a finite alphabet A. It is compact as a product of finite sets. A basis of clopen sets is formed by the sets $[a_0, a_1, \ldots, a_n] = \{x \in A^{\mathbb{Z}} \mid x_0 = a_0, \ldots, x_n = a_n\}$ for $a_i \in A$ and $n \geq 0$. When $\text{Card}(A) \geq 2$, there is no isolated point. Thus, it is a Cantor set.

There is the algebraic one, the group $\mathbb{Z}_p$ of *p-adic integers*, where $p \geq 2$ is a prime number. It is the completion of $\mathbb{Z}$ for the *p-adic topology*, which is induced by the distance defined for $x \neq y$ by $d(x, y) = p^{-n}$ if p^n is the higher power of p dividing $x - y$. Every element can be uniquely represented as

$$x = x_0 + x_1 p + x_2 p^2 + \cdots$$

with $0 \leq x_i < p$, and thus $\mathbb{Z}_p$ is homeomorphic to the set of infinite sequences $(x_0, x_1, \ldots)$ with $x_i \in \mathbb{Z}/p\mathbb{Z}$. This shows that it is a Cantor space.

The transformation defined on $\mathbb{Z}_p$ by $T(x) = x + 1$ defines a minimal dynamical system $(\mathbb{Z}_p, T)$ called an *odometer* (we shall study odometers in more detail in Chapter 6).

From a topological point of view, all these topological spaces are the same as they are homeomorphic (see Section 1.8 for a reference to a proof).

A *Cantor system* is a dynamical system (X, T) where X is a T-stable Cantor space. The odometers are Cantor systems. A *symbolic system* is a dynamical system (X, T) where X is a T-stable closed subset of $A^{\mathbb{Z}}$. The transformation T need not be the shift (see the example of the odometer). Observe that symbolic systems are not necessarily Cantor systems, since closed subsets could have isolated points. But infinite recurrent symbolic systems are Cantor systems (Exercise 1.10).

The pair $(A^{\mathbb{Z}}, S)$ is called the *full shift* (on the alphabet A). If X is a closed shift-invariant subset of $A^{\mathbb{Z}}$, then the topological dynamical system (X, T), where T is the restriction of S to X, is called a *subshift* of the full shift on the alphabet A, or a *shift space*, or a *two-sided shift space*. Thus, a shift space is a symbolic system. For the full shifts on any finite alphabet and in general for shift spaces, the transformation will usually be denoted by S. Thus, we will often use the notation X instead of (X, S) for a shift space.

Similarly one can define the *one-sided full shift* as follows. We still denote by S the one-sided transformation, called the *one-sided shift*, which is defined for $x \in A^{\mathbb{N}}$ by $y = Sx$ if $y_n = x_{n+1}$, as in Eq. (1.1). The pair $(A^{\mathbb{N}}, S)$ is called the one-sided full shift on the alphabet A. If X is a closed subset of $A^{\mathbb{N}}$ such that $SX = X$, the pair (X, S) is called a *one-sided shift space*.

To every shift space (X, S) one may associate a one-sided shift space (Y, S), called its *associated one-sided shift space*, by considering the set Y of $y_0 y_1 \cdots$ such that $y_i = x_i$ $(i \geq 0)$ for some $x = (x_n)_{n \in \mathbb{Z}}$. The map $\theta \colon X \to Y$ defined by $\theta(x) = y$ is a surjective morphism.

Conversely, for every one-sided shift space (Y, S) there is a unique two-sided shift space (X, S) such that (Y, S) is associated with (X, S). Indeed, the set X of sequences $x \in A^{\mathbb{Z}}$ such that $x_n x_{n+1} \cdots$ belongs to Y for every $n \in \mathbb{Z}$ is closed and shift-invariant. It is clear that (X, S) is the unique shift space

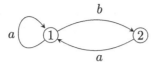

Figure 1.3 The golden mean shift.

such that $Y = \theta(X)$. We also say that X is the two-sided shift space *associated* with Y.

This shows that one-sided and two-sided shift spaces are closely related objects. For example, a one-sided shift space is minimal if and only if its associated two-sided shift space is minimal.

In general, in this book, we consider two-sided shift spaces rather than one-sided shift spaces because it is often convenient to have a transformation that is invertible. In general, by a shift space, we mean a two-sided shift space.

Example 1.1.7 The *golden mean shift* is the set X of two-sided sequences on $A = \{a, b\}$ with no consecutive b. Thus, X is the set of labels of two-sided infinite paths in the graph of Figure 1.3. It is recurrent, as one may easily verify. It is not minimal, since it contains the one-point set $\{a^{\mathbb{Z}}\}$, which is closed and shift-invariant.

Example 1.1.8 Let $\varphi: a \mapsto ab, b \mapsto a$ be the *Fibonacci morphism*. Since $\varphi(a)$ begins with a, any $\varphi^n(a)$ is a prefix of $\varphi^{n+1}(a)$. Let $x \in \{a, b\}^{\mathbb{N}}$ be the sequence x having all $\varphi^n(a)$ as prefixes. Thus,

$$x = abaababa \cdots .$$

It is known as the *Fibonacci sequence* (we use the name "Fibonacci sequence of numbers" for the well-known sequence $F_{n+1} = |\varphi^n(b)|$, given by $F_0 = 0$, $F_1 = 1$ and $F_{n+1} = F_n + F_{n+1}$). The subshift of $\{a, b\}^{\mathbb{N}}$, which is the closure of the orbit of x, is the *one-sided Fibonacci shift*. We will see that it is minimal (Example 1.4.19).

1.1.5 Measure-Theoretic Dynamical Systems

A *measure-theoretic dynamical system* is a triple (X, T, μ) where X is a compact metric space, μ is a Borel probability measure on X and $T: X \to X$ is a measurable map such that μ is *invariant*, that is, if $\mu(T^{-1}(U)) = \mu(U)$ for every Borel set $U \subset X$.

If (X, T) is a topological dynamical system and μ is an invariant probability measure on X, then (X, T, μ) is a measure-theoretic system.

Example 1.1.9 The rotation $T_\alpha : x \mapsto x + \alpha$ mod 1 on $X = [0, 1]$ is a continuous isometry and thus preserves the Lebesgue measure. Thus, (X, T_α, μ) is a measure-theoretic dynamical system with μ being the Lebesgue measure.

Although this book is devoted to topological dynamical systems, we will have several occasions to meet measure-theoretic ones (for example, with rotations in Section 1.5 or interval exchange transformations in Chapter 8).

A basic result concerning measure-theoretic dynamical systems is the *Poincaré Recurrence Theorem*, which we will use in Chapter 6.

Theorem 1.1.10 (Poincaré) *Let (X, T, μ) be a measure-theoretic dynamical system. For every Borel set $U \subset X$ such that $\mu(U) > 0$, the set of points $x \in U$ such that for some $N \geq 1$ one has $T^n(x) \notin U$ for all $n \geq N$ has measure 0.*

Proof Let V_N be the set of $x \in U$ such that $T^n(x) \notin U$ for all $n \geq N$. It is a Borel set because

$$V_N = U \cap \left(\cap_{n \geq N} T^{-n}(X \setminus U) \right).$$

Next, by definition of V_N, we have $T^{-n}(V_N) \cap V_N = \emptyset$ for all $n \geq N$, which implies that the sets $T^{-N}(V_N), T^{-2N}(V_N), \ldots$ are disjoint. Therefore,

$$\mu(X) \geq \mu \left(\bigcup_{n \geq 1} T^{-nN}(V_N) \right) = \sum_{n \geq 1} \mu(T^{-nN}(V_N)) = \sum_{n \geq 1} \mu(V_N).$$

Since $\mu(X) = 1$, this implies that $\mu(V_N) = 0$. Consequently, the set $V = \cup_{N \geq 1} V_N$ has measure 0 as $\mu(V) \leq \sum_{N \geq 1} \mu(V_N) = 0$. ∎

This result allows us to define induced transformations on every measurable subset $U \subset X$ since for almost every $x \in U$, the return time $n(x) = \inf\{n > 0 \mid T^n x \in U\}$ is finite.

Two measure-theoretic systems (X, T, μ) and (X, T', μ') are *isomorphic* if there are Borel subsets $X_1 \subset X, X'_1 \subset X'$ of measure 1 and a bimeasurable bijection $\varphi \colon X_1 \to X'_1$ such that $\varphi \circ T = T' \circ \varphi$.

Let $(X, T), (X', T')$ be topological dynamical systems and $\varphi \colon X \to X'$ be a conjugacy. If μ, μ' are invariant measures such that $\mu'(\varphi(U)) = \mu(U)$ for every Borel set $U \subset X$, then (X, T, μ) and (X', T', μ') are isomorphic. However, a measure-theoretic isomorphism need not be a conjugacy (we shall see an example in Section 1.5.3).

1.2 More on Shift Spaces

In this section, we develop in more detail the notions related to shift spaces and their language. We will see how the notions of recurrence and minimality can be expressed adequately for shift spaces. We will also introduce important notions like return words or Rauzy graphs.

1.2.1 Some Combinatorics on Words

Let A be a nonempty set called an *alphabet*. We will generally assume that the alphabet is finite. A *word* over A is an element of the *free monoid* generated by A, denoted by A^*. If $u = u_0 u_1 \cdots u_{n-1}$ (with $u_i \in A$, $0 \le i \le n - 1$) is a word, its *length* is n and is denoted by $|u|$. For $a \in A$, we denote by $|u|_a$ the number of occurrences of the letter a in u.

The *empty word* is denoted by ε. It is the unique word of length 0. The set of nonempty words over A, called the *free semigroup* on A, is denoted by A^+.

We consider the free monoid A^* as embedded in the free group on A (see Appendix B.2). Consequently, when $u = vw$, we also write $v = uw^{-1}$ and $w = v^{-1}u$.

A *factor* (also called a *subword* or a *block*) of a word u is a finite word y such that there exist two words v and w satisfying $u = vyw$. When v (resp. w) is the empty word, we say that y is a *prefix* (resp. *suffix*) of u. A factor y (resp. prefix, resp. suffix) of a word u is *proper* if $y \ne u$.

The *prefix order* is the partial order on words defined by $u \le v$ if u is a prefix of v. Similarly, the *suffix order* is defined by $u \le v$ if u is a sufix of v.

Two words u, v are *conjugate* if $u = rs$ and $v = sr$ for some words r, s or, equivalently, if v is obtained from u by a circular permutation of its letters. Conjugacy is an equivalence relation on words.

A word w is *primitive* if it is not a power of another word. Formally, w is primitive if $w = u^n$ implies $n = 1$. A primitive word of length n has n distinct conjugates (Exercise 1.13). Any nonempty word w can be written uniquely $w = u^n$ with u primitive and $n \ge 1$. The integer n is called the *exponent* of w.

A word $w = a_1 a_2 \cdots a_n$ with $a_i \in A$ has *period* p if $a_i = a_{i+p}$ for $1 \le i \le n - p$. In this case, w has a prefix of exponent $\lfloor n/p \rfloor$. Indeed, one has $w = u^q v$ where $n = pq + r$ with $0 \le r < p$, $|u| = p$ and $|v| = r$.

There is an important connection between the periods of a word and the overlap of its factors.

Proposition 1.2.1 *If a word y has two overlapping occurrences in u, that is, if $u = rys = r'ys'$ with $|r| < |r'| \le |r| + |y|$, then y has period $|r| - |r'|$.*

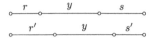

Figure 1.4 Two overlapping occurrences of y.

The proof is straightforward (see Figure 1.4).

A set of words on the alphabet A is also called a *language* on A. If U is a language on A, we denote by U^* the submonoid of A^* generated by U, that is,

$$U^* = \{w \in A^* \mid w = u_1 u_2 \cdots u_n, u_i \in U, n \geq 0\}. \tag{1.3}$$

When $U = \{u\}$, we denote simply u^* instead of $\{u\}^*$. Thus, $u^* = \{u^n \mid n \geq 0\}$. For $U, V \subset A^*$, we also denote

$$UV = \{uv \mid u \in U, v \in V\}. \tag{1.4}$$

If k, l are integers such that $0 \leq k \leq l < |u|$, we let $u_{[k,l]}$ denote the subword $u_k u_{k+1} \cdots u_l$ of u. We define $u_{[k,l+1)}$ to be $u_{[k,l]}$. If $l < k$, then $u_{[k,l]}$ is the empty word. If y is a factor of u, the *occurrences* of y in u are the integers i such that $u_{[i,i+|y|-1]} = y$. If y has an occurrence in u, we also say that y *occurs* in u.

Two occurrences $i < j$ of y in u are said to *overlap* if $j \leq i + |y| - 1$ (see Proposition 1.2.1).

The *reversal* of a word $u = u_0 u_1 \cdots u_n$, with $u_i \in A$, is the word $\tilde{u} = u_n \cdots u_1 u_0$. The reversal of a set U of words is the set $\tilde{U} = \{\tilde{u} \mid u \in U\}$.

The elements of $A^{\mathbb{K}}$, where $\mathbb{K}$ is equal to $\mathbb{N}$, $-\mathbb{N}$ or $\mathbb{Z}$, are called *sequences* or *infinite words*. When we need to precisely determine which kind of sequences we are dealing with, we sometimes say *right infinite sequence*, or *one-sided sequence* when $K = \mathbb{N}$, *left-infinite sequence* when $K = -\mathbb{N}$ and *two-sided sequence* (or also *bi-infinite sequence*) in the last case. For $x = (x_n)_{n \in \mathbb{Z}} \in A^{\mathbb{Z}}$, we let x^+ and x^- respectively denote the sequences $(x_n)_{n \geq 0}$ and $(x_n)_{n < 0}$. For $x \in A^{-\mathbb{N}}$ and $y \in A^{\mathbb{N}}$, we denote by $z = x \cdot y$ the two-sided sequence z such that $x = z^-$ and $y = z^+$. The notion of *factor* is naturally extended to sequences, as well as the notion of *prefix* when $\mathbb{K} = \mathbb{N}$.

The set of subwords of length n of x is written $\mathcal{L}_n(x)$, and the set of subwords of x, called the *language* of x, is denoted by $\mathcal{L}(x)$. We also denote by $\mathcal{L}_{\leq n}(x)$ the set of words of length at most n in $\mathcal{L}(x)$.

The *reversal* of a right-infinite sequence $x = x_0 x_1 \cdots$ is the left-infinite sequence $\tilde{x} = \cdots x_1 x_0$. The language of $\tilde{x}$ is the reversal of $\mathcal{L}(x)$.

For a finite word $u \in A^+$, we denote by u^ω the right infinite sequence $uuu \cdots$ and by u^∞ the two-sided infinite sequence $x = \cdots uuu \cdot uuu \cdots$, where the dot is placed to the left of x_0. In this way, we have $x^+ = u^\omega$. An integer $p \geq 1$ is a *period* of a sequence $x \in A^{\mathbb{N}}$ if $x_i = x_{i+p}$ for all $i \geq 0$. Clearly, the sequence x has period p if and only if $x = u^{\omega}$ with $p = |u|$.

Proposition 1.2.2 *Every sequence x has a unique minimal period and its multiples form the set of periods of x. Moreover, x has minimal period p if and only if $x = u^{\omega}$ with $p = |u|$ and u primitive.*

Proof The set of $p \in \mathbb{Z}$ such that $|p|$ is a period of x is a subgroup of $\mathbb{Z}$. Indeed, if p, q are periods of x with $p \leq q$, then $x_{i+q-p} = x_{(i+q-p)+p} = x_{i+q} = x_i$ for every $i \geq 0$ and thus $q - p$ is also a period of x. This shows that the set of periods of x coincides with the set of multiples of the minimal period. Next, the map $u \mapsto |u|$ sends the set of words u such that $x = u^{\omega}$ onto the set of periods of x. This proves the second statement. ∎

A refinement of this statement, called the Fine–Wilf Theorem, is given in Exercise 1.14. These notions carry easily to two-sided sequences.

The sequence $(p_n(x))_{n \geq 0}$ defined by $p_n(x) = \mathrm{Card}(\mathcal{L}_n(x))$ is the *factor complexity* (or *word complexity* or simply the *complexity*) of x. Note that $p_0(x) = 1$, that $p_n(x) \leq p_{n+1}(x)$ and that $p_{n+m}(x) \leq p_n(x)p_m(x)$ for all $n, m \geq 0$.

Example 1.2.3 Let x be the Fibonacci word (Example 1.1.8). We have $p_1(x) = 2$ since every letter a, b appears in x. Next $p_2(x) = 3$ since $\mathcal{L}_2(x) = \{aa, ab, ba\}$, as one may verify. We will see that, actually, one has $p_n(x) = n + 1$ for all $n \geq 1$ (see Section 1.5).

The sequence $x \in A^{\mathbb{N}}$ is *eventually periodic* if there exists a word u and a nonempty word v such that $x = uv^{\omega}$, where $v^{\omega} = vvv \cdots$. A sequence that is not eventually periodic is called *aperiodic*. It is *periodic* if u is the empty word.

A two-sided sequence $x \in A^{\mathbb{Z}}$ is *periodic* if $x = v^{\infty}$ for some $v \in A^+$. In this case, x^+ is periodic.

The following result is classical.

Theorem 1.2.4 (Morse, Hedlund) *Let x be a two-sided sequence. The following conditions are equivalent.*

(i) *For some $n \geq 1$, one has $p_n(x) \leq n$.*
(ii) *For some $n \geq 1$, one has $p_n(x) = p_{n+1}(x)$.*
(iii) *x is periodic.*

Morover, in this case, the least period of x is $\max p_n(x)$.

Proof (i) $\Rightarrow$ (ii). Since $p_n(x) \le p_{n+1}(x)$ for all $n \ge 0$, the hypothesis implies that $p_n(x) = p_{n+1}(x)$ for some $n \ge 0$.

(ii) $\Rightarrow$ (iii). For every $w \in \mathcal{L}_n(x)$, there is a unique letter $a \in A$ such that $wa \in \mathcal{L}_{n+1}(x)$. This implies that two consecutive occurrences of a word u of length n in x are separated by a fixed word depending only on u and thus that x is periodic.

(iii) $\Rightarrow$ (i) is obvious.

Let n be the least period of x. Since a primitive word of length n has n distinct conjugates, we have $p_n(x) = n$ and $p_m(x) = n$ for all $m \ge n$. ∎

Thus, by Proposition 1.2.4, either $p_n(x) \ge n + 1$ for all $n \ge 1$ or $p_n(x)$ is eventually constant. The case $p_n(x) = n + 1$ for all $n \ge 1$ corresponds to the Sturmian sequences (see below).

Note that for a one-sided sequence, the same result holds with condition (iii) replaced by the condition that x is eventually periodic. The proof is the same.

Proposition 1.2.5 *The following conditions are equivalent for $x \in A^{\mathbb{N}}$.*

(i) *Every $u \in \mathcal{L}(x)$ has at least two occurrences in x.*
(ii) *Every $u \in \mathcal{L}(x)$ has an infinite number of occurrences in x.*
(iii) *For every $u, w \in \mathcal{L}(x)$ there is $v \in \mathcal{L}(x)$ such that $uvw \in \mathcal{L}(x)$.*

Proof (i) $\Rightarrow$ (ii). Assume that u has a finite number of occurrences in x. Let v be the shortest prefix of x containing all these occurrences. Since v has a second occurrence in x, we have a contradiction.

(ii) $\Rightarrow$ (iii). Assume that $u = x_{[i,j)}$. Since w has an infinite number of occurrences in x, there is an index k larger than j such that $w = x_{[k,\ell)}$. Set $v = x_{[j,k)}$. Then $uvw = x_{[i,\ell)}$.

(iii) $\Rightarrow$ (i) is clear considering $u = w$. ∎

A word u is *recurrent* in $x \in A^{\mathbb{N}}$ if condition (ii) above is satisfied. The sequence x itself is called *recurrent* if one of the conditions is satisfied. Thus, a sequence $x \in A^{\mathbb{N}}$ is recurrent if and only if x is a recurrent point of the full shift. We could, of course, use Proposition 1.1.3 to prove Proposition 1.2.5. We also say that the language $\mathcal{L}(x)$ is *irreducible* if condition (iii) is satisfied.

Example 1.2.6 Let $A = \{a, b\}$ and let $x = abaaabbabb \cdots$ be the sequence formed of all words on A in *radix order* (that is, ordered first by length, then lexicographically). It is a recurrent sequence in which all words on A appear, that is, such that $\mathcal{L}(x) = A^*$. As a variant of this example, the *Champernowne sequence* is the sequence $x = 012345678910111213141516171819 20 \cdots$ formed of the decimal representation of all numbers in increasing order.

Proposition 1.2.7 *The following conditions are equivalent for a sequence* $x \in A^{\mathbb{N}}$.

(i) *Every* $u \in \mathcal{L}(x)$ *occurs infinitely often in* x *and the difference of two successive occurrences of* u *is bounded.*
(ii) *For every* $u \in \mathcal{L}(x)$, *there is an* $n \geq 1$ *such that* u *occurs in every word of* $\mathcal{L}_n(x)$.

Proof (i) $\Rightarrow$ (ii). Let k be the maximum of the differences between successive occurrences of u. Then u appears in every word of $\mathcal{L}_n(x)$ for $n = |u| + k$.

(ii) $\Rightarrow$ (i) is clear. ∎

A sequence $x \in A^{\mathbb{N}}$ is *uniformly recurrent* if one of these conditions holds. We also say in this case that $\mathcal{L}(x)$ is *uniformly recurrent*.

Example 1.2.8 The Fibonacci word x (Example 1.2.3) is uniformly recurrent. Indeed, let $u \in \mathcal{L}(x)$ and let n be the minimal integer such that u is a factor of $\varphi^n(a)$. Then $\varphi^{n+2}(a) = \varphi^n(a)\varphi^n(b)\varphi^n(a)$ and thus u has a second occurrence in $\varphi^{n+2}(a)$ at a bounded distance of the first one, which implies that it occurs infinitely often at a bounded distance.

A sequence $x \in A^{\mathbb{N}}$ is *linearly recurrent* with *constant* K if it is recurrent and the greatest difference between successive occurrences of u is bounded by $K|u|$.

Most of this terminology extends naturally to a two-sided infinite sequence. In particular, $x \in A^{\mathbb{Z}}$ is *recurrent* (resp. *uniformly recurrent*) if $\mathcal{L}(x)$ is recurrent (resp. uniformly recurrent). The same extension holds for linearly recurrent sequences.

Proposition 1.2.9 *Let x be a two-sided sequence that is linearly recurrent with constant K. Then*

1. *Every word of $\mathcal{L}_n(x)$ appears in every word of $\mathcal{L}_{(K+1)n-1}(x)$.*
2. *The factor complexity of x is at most Kn.*

Moreover, if x is not periodic, it is $(K + 1)$-power free, that is, for every nonempty word $u \in \mathcal{L}(x)$, $u^n \in \mathcal{L}(x)$ implies $n \leq K$.

Proof 1. Let $u \in \mathcal{L}_n(x)$. Since two successive occurrences of u differ by at most Kn, every word of $\mathcal{L}_{(K+1)n-1}$ contains u as a factor.

2. Set $p_n(x) = \mathrm{Card}(\mathcal{L}_n(x))$. A word of length $(K + 1)n - 1$ has at most Kn factors of length n. Thus, by Assertion 1, $p_n(x) \leq Kn$.

Assume now that u^{K+1} belongs to $\mathcal{L}(x)$. Set $n = |u|$. The word u^{K+1} has length $(K + 1)n$ and at most n factors of length n. By Assertion 1, this implies $p_n(x) \leq n$. By Theorem 1.2.4, this implies that x is periodic.   ∎

Clearly, a linearly recurrent sequence is uniformly recurrent, but the converse is not true (see Exercise 1.65).

Example 1.2.10 The Fibonacci word is linearly recurrent (Exercise 1.17).

1.2.2 The Language of a Shift Space

Let X be a shift space. The *language* of X is the set $\mathcal{L}(X)$ of subwords of elements belonging to X. The set $\mathcal{L}(X)$ is, of course, the union of the languages $\mathcal{L}(x)$ for $x \in X$. We also denote by $\mathcal{L}_n(X)$ the set of words of length n in $\mathcal{L}(X)$ and by $\mathcal{L}_{\leq n}(X)$ the set of those of length at most n.

The same notation can be used for the language of a one-sided shift space. The languages of a two-sided shift space and of its associated one-sided shift space are actually the same.

A set L of words on the alphabet A is *factorial* if it contains the factors of its elements. A word $u \in L$ is *extendable* in L if there are letters $a, b \in A$ such that aub belongs to L. The language L is said to be *extendable* if every $u \in L$ is extendable. The language of a shift space is factorial and extendable and, conversely, for every factorial extendable set L, there is a unique shift space X such that $\mathcal{L}(X) = L$ (Exercise 1.16).

For two words u, v such that $uv \in \mathcal{L}(X)$, the set

$$[u \cdot v]_X = \{x \in X \mid x_{[-|u|,|v|-1]} = uv\}$$

is nonempty. It is called the *cylinder* with basis (u, v). We set $[v]_X = [\varepsilon \cdot v]_X$, where ε is the empty word, or equivalently $[v]_X = \{x \in X \mid x_{[0,|v|-1]} = v\}$. Any cylinder is open, and every open set in a shift space is a union of cylinders. The clopen sets in X are the finite unions of cylinders.

For any sequence $x \in A^{\mathbb{Z}}$ there is a smallest shift space containing x called the *shift generated* by x and denoted $\Omega(x)$. It is the closure of the orbit of x. For example, if $x = u^{\infty}$ is a periodic sequence, the shift $\Omega(x)$ is periodic (that is, it is a periodic dynamical system).

The following property is sometimes taken for definition of shift spaces.

Proposition 1.2.11 *A set $X \subset A^{\mathbb{Z}}$ is a shift space if and only if there is a set $F \subset A^*$ of finite words such that X is the set X_F of infinite words without factor in F.*

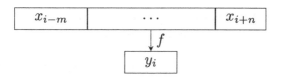

Figure 1.5 The sliding block code.

Proof Indeed, such a set is clearly closed and invariant by the shift. Conversely, let X be a shift space. Since X is closed, its complement $Y = A^{\mathbb{Z}} \setminus X$ is open. Thus, for every $y \in Y$, there is a cylinder $[u_y \cdot v_y]_{A^{\mathbb{Z}}}$ containing y and contained in Y. Set $F = \{u_y v_y \mid y \in Y\}$. Then $X_F \subset X$ since $X_F \cap Y = \emptyset$. Conversely, if $x \in X$ has a factor $u_y v_y$ in F, then $T^n(x)$ is in $[u_y \cdot v_y]_{A^{\mathbb{Z}}}$ for some $n \in \mathbb{Z}$, and this is a contradiction since X is shift-invariant. Thus, $X = X_F$. ∎

Let X and Y be shift spaces on alphabets A, B respectively. Given integers m, n such that $-m \le n$, let $f : \mathcal{L}_{m+n+1}(X) \to B$ be a map called a *block map*. We call m the *memory*, n the *anticipation* and $m + n + 1$ the *window size* of the block map f.

The *sliding block code* induced by f is the map $\varphi : X \to B^{\mathbb{Z}}$ defined, for all $x \in X$, by $y = \varphi(x)$ if (see Figure 1.5)

$$y_i = f(x_{i-m} \cdots x_{i+n}) \quad (i \in \mathbb{Z}).$$

When $\varphi(X)$ is included in Y, we write $\varphi : X \to Y$.

Theorem 1.2.12 (Curtis, Hedlund, Lyndon) *Let X and Y be shift spaces. A map $\varphi : X \to Y$ is a factor map if and only if it is a sliding block code from X onto Y.*

Proof A sliding block code is clearly continuous and commutes with the shifts, that is $\varphi \circ T = S \circ \varphi$. Thus, it is a morphism of dynamical systems. Conversely, let $\varphi : X \to Y$ be a morphism. For every $b \in B$ the set $[b]$ is clopen. The map φ being continuous, it is also the case for $\varphi^{-1}([b])$. The family $\{\varphi^{-1}([b]) \mid b \in B\}$ being a partition, the set B being finite and each $\varphi^{-1}([b])$ being a finite union of cylinders, there is an integer n such that $\varphi(x)_0$ depends only on $x_{[-n,n]}$. Set $f(x_{-n} \cdots x_n) = \varphi(x)_0$. Then φ is the sliding block code associated with f. ∎

The *factor complexity* (or *word complexity* or simply the *complexity*) of the shift space X is the sequence

$$p_n(X) = \text{Card}(\mathcal{L}_n(X)).$$

Observe that $p_0(X) = 1$ and that $p_n(X) \leq p_{n+1}(X)$. Indeed, for every $w \in \mathcal{L}_n(X)$, there is a letter $a \in A$ such that $wa \in \mathcal{L}(X)$.

Example 1.2.13 Let X be the golden mean shift (Example 1.1.7). The factor complexity of X is given by $p_n(X) = F_{n+1}$, where F_n is the *Fibonacci sequence* defined by $F_0 = 0, F_1 = 1$ and $F_{n+1} = F_n + F_{n-1}$ for $n \geq 1$. Indeed, the number of words in $\mathcal{L}_{n+1}(X)$ ending with a is equal to $p_n(X)$, and the number of those ending with b is $p_{n-1}(X)$ since the b has to be preceded by a. Thus $p_{n+1}(X) = p_n(X) + p_{n-1}(X)$.

The factor complexity of a shift is not invariant by conjugacy (since a conjugacy may change the alphabet) but its growth rate is an invariant (Exercise 1.18).

The following result is the counterpart for shift spaces of Theorem 1.2.4.

Theorem 1.2.14 (Morse, Hedlund) *Let X be a shift space. The following conditions are equivalent.*

(i) *For some $n \geq 1$, one has $p_n(X) \leq n$.*
(ii) *For some $n \geq 1$, one has $p_n(X) = p_{n+1}(X)$.*
(iii) *X is finite.*

Proof (i) $\Rightarrow$ (ii). Since $p_n(X) \leq p_{n+1}(X)$ for all $n \geq 0$, the hypothesis implies that $p_n(X) = p_{n+1}(X)$ for some $n \geq 0$.

(ii) $\Rightarrow$ (iii). By Theorem 1.2.4, every $x \in X$ is periodic and its period is bounded by $\max p_n(X)$.

(iii) $\Rightarrow$ (i). If X is finite, it is a finite union of orbits of periodic points of the form u^∞ for some word u. Since the set $\mathcal{L}_n(u^\infty)$ has at most $|u|$ elements, $p_n(u^\infty)$ is bounded and thus also $p_n(X)$. ∎

Thus, by Theorem 1.2.14, either $p_n(X) \geq n + 1$ for all $n \geq 1$ or $p_n(X)$ is eventually constant. The case $p_n(X) = n + 1$ for all $n \geq 1$ corresponds to the Sturmian shifts (see below).

A shift space X is *irreducible* if the language $\mathcal{L}(X)$ is irreducible, that is, if for every $u, v \in \mathcal{L}(X)$, there is a w such that $uwv \in \mathcal{L}(X)$.

Proposition 1.2.15 *A shift space is recurrent if and only if it is irreducible.*

Proof Assume first that X is recurrent and consider $u, v \in \mathcal{L}(X)$. Let $n > |u|$ be such that $S^{-n}([v]_X) \cap [u]_X \neq \emptyset$. Then every $x \in S^{-n}([v]_X) \cap [u]_X$ is in $[uwv]_X$ for some w. Thus, uwv belongs to $\mathcal{L}(X)$, showing that X is irreducible.

Conversely, assume that X is irreducible. Let U, V be open sets in X. We can find cylinder sets $[u]_X$ and $[v]_X$ that are respectively included in U and V. Since X is irreducible, there is some word w such that $uwv \in \mathcal{L}(X)$. Then $[uwv]_X$ is nonempty and in $S^{-n}V \cap U$ for $n = |uw|$. Thus, X is recurrent. ∎

A shift space is, of course, irreducible if and only if it is recurrent as a topological dynamical system. Thus, we could also have used Proposition 1.1.3 to prove Proposition 1.2.15.

Thus, for a shift space, the property of being recurrent can be expressed by a property of its language $\mathcal{L}(X)$. Likewise, we can translate the property of being minimal. A shift space X is *uniformly recurrent* if the language $\mathcal{L}(X)$ is uniformly recurrent, that is, if for every $u \in \mathcal{L}(X)$, there is an $n \geq 1$ such that u is a factor of every word in $\mathcal{L}_n(X)$.

Proposition 1.2.16 *The following conditions are equivalent for a shift space* X.

(i) *The shift space X is minimal.*
(ii) *The shift space X is uniformly recurrent.*
(iii) *Every $x \in X$ is uniformly recurrent and $\mathcal{L}(x) = \mathcal{L}(X)$.*

Proof (i) $\Rightarrow$ (ii). Let $u \in \mathcal{L}(X)$. Since $[u]_X$ is clopen and X is minimal, for every $x \in X$, the entrance time $n(x) = \min\{n > 0 \mid S^n x \in [u]_X\}$ exists and, being continuous, it is bounded. Set $n = \max n(x)$. Then every word w in $\mathcal{L}(X)$ of length $n + |u|$ has a factor u. Indeed, let $x \in X$ be such that $x \in [w]_X$. Then $T^i x \in [u]_X$ for some $i \leq n$ and thus u is a factor of w. This shows that X is uniformly recurrent.

(ii) $\Rightarrow$ (iii) is clear since every $u \in \mathcal{L}(X)$ appears in every word of $\mathcal{L}_n(x)$.

(iii) $\Rightarrow$ (i) is clear since the orbit of every $x \in X$ is dense. ∎

Again, we could have used Proposition 1.1.4 to prove Proposition 1.2.16.

1.2.3 Special Words

For $w \in \mathcal{L}(X)$, there is at least one letter $a \in A$ such that wa belongs to $\mathcal{L}(X)$ and symmetrically, at least one letter $a \in A$ such that aw belongs to $\mathcal{L}(X)$. The word w is called *right-special* if there are at least two letters $a \in A$ such that wa belongs to $\mathcal{L}(X)$. Symmetrically, w is *left-special* if there are at least two letters $a \in A$ such that aw belongs to $\mathcal{L}(X)$. It is *bispecial* if it is both left and right-special.

Special words are closely linked with the factor complexity of a shift space. Let indeed $p_n(X) = \text{Card}(\mathcal{L}_n(X))$ be the factor complexity of the shift space X on the alphabet A. Set for $n \geq 0$

$$s_n(X) = p_{n+1}(X) - p_n(X), \tag{1.5}$$
$$b_n(X) = s_{n+1}(X) - s_n(X). \tag{1.6}$$

For a word $w \in \mathcal{L}(X)$, let

$$\ell_X(w) = \mathrm{Card}\{a \in A \mid aw \in \mathcal{L}(X)\},$$
$$r_X(w) = \mathrm{Card}\{b \in A \mid wb \in \mathcal{L}(X)\},$$
$$e_X(w) = \mathrm{Card}\{(a,b) \in A \times A \mid awb \in \mathcal{L}(X)\}.$$

Thus, $\ell_X(w) > 1$ (resp. $r_X(w) > 1$) if and only if w is left-special (resp. right-special). Define also the *multiplicity* of $w \in \mathcal{L}(X)$ as

$$m_X(w) = e_X(w) - \ell_X(w) - r_X(w) + 1. \tag{1.7}$$

The word w is called *neutral* if $m_X(w) = 0$.

Proposition 1.2.17 *We have for all $n \geq 0$,*

$$s_n(X) = \sum_{w \in \mathcal{L}_n(X)} (\ell_X(w) - 1) = \sum_{w \in \mathcal{L}_n(X)} (r_X(w) - 1) \tag{1.8}$$

and

$$b_n(X) = \sum_{w \in \mathcal{L}_n(X)} m_X(w). \tag{1.9}$$

In particular, the number of left-special (resp. right-special) words of length n is bounded by $s_n(X)$.

Proof We have

$$\sum_{w \in \mathcal{L}_n(X)} (\ell_X(w) - 1) = \sum_{w \in \mathcal{L}_n(X)} \ell_X(w) - \mathrm{Card}(\mathcal{L}_n(X))$$
$$= \mathrm{Card}(\mathcal{L}_{n+1}(X)) - \mathrm{Card}(\mathcal{L}_n(X)) = p_{n+1}(X) - p_n(X)$$
$$= s_n(X)$$

with the same result for $\sum_{w \in \mathcal{L}_n(X)}(r_X(w) - 1)$. Next,

$$\sum_{w \in \mathcal{L}_n(X)} m_X(w) = \sum_{w \in \mathcal{L}_n(X)} (e_X(w) - \ell_X(w) - r_X(w) + 1)$$
$$= p_{n+2}(X) - 2p_{n+1}(X) + p_n(X) = s_{n+1}(X) - s_n(X)$$
$$= b_n(X). \qquad \blacksquare$$

Every infinite shift space X contains an infinity of right-special words (and also of left-special words). Indeed, if there is no right-special word of length n, we have $p_n(X) = p_{n+1}(X)$ and thus X is finite by Theorem 1.2.14. Moreover,

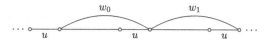

Figure 1.6 The factorization of x in right return words.

in a minimal nonperiodic shift X, every $w \in \mathcal{L}(X)$ is a prefix of a right-special word. Indeed, otherwise, there is a unique one-sided word x beginning with w in the one-sided shift associated with X, which is thus not minimal.

1.2.4 Return Words

Let X be a shift space. For $u \in \mathcal{L}(X)$ a *right return word* to u is a nonempty word w such that uw is in $\mathcal{L}(X)$, uw has u as a proper suffix and uw has no occurrence of a factor u which is not a prefix or a suffix. Thus, the return words to u allow us to factorize a sequence x as a sequence of return words to u (see Figure 1.6).

For example, in the golden mean shift, the word $w = aab$ is a right return word to $u = b$.

Symmetrically, a *left return word* to u is a word w such that wu is in $\mathcal{L}(X)$, wu has u as a proper prefix and wu has no other occurrence of a factor u.

We denote by $\mathcal{R}_X(u)$ (resp. $\mathcal{R}'_X(u)$) the set of right (resp. left) return words to u. The set $U = \mathcal{R}_X(u)$ is a *prefix code*, that is, no word in U is a prefix of another word of U. Equivalently, U is a prefix code if its elements are incomparable for the prefix order. Symmetrically, the set $U' = \mathcal{R}'_X(u)$ is a *suffix code*; that is, no word of U' is a suffix of another word of U.

For any prefix code U on an alphabet A, the set U^* satisfies the following property for every $r, s \in A^*$

$$r, rs \in U^* \Rightarrow s \in U^*. \tag{1.10}$$

The following property of return words is easy to verify.

Proposition 1.2.18 *Every word $w \in \mathcal{L}(X)$ that begins and ends with u has a unique factorization $w = uw_1 w_2 \cdots w_n$ with w_i in $\mathcal{R}_X(u)$ and $n \geq 0$.*

For example, if X is the golden mean shift (Example 1.1.7), we have $\mathcal{R}_X(b) = \{ab, aab, aaab, \ldots\}$ and $\mathcal{R}'_X(b) = \{ba, baa, \ldots\}$.

Clearly a recurrent shift space X is minimal if and only if $\mathcal{R}_X(w)$ is finite for every $w \in \mathcal{L}(X)$.

1.2.5 Rauzy Graphs

Let X be a shift space on the alphabet A. The *Rauzy graph* of X of order n, denoted $\Gamma_n(X)$, is the following labeled graph. The set of vertices of $\Gamma_n(X)$ is

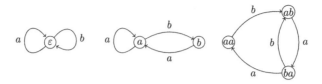

Figure 1.7 The Rauzy graphs of order $n = 1, 2, 3$ of the Fibonacci shift.

the set $\mathcal{L}_{n-1}(X)$ and the set of edges is the set $\mathcal{L}_n(X)$. The origin and end of the edge w are the words u, v such that $w = ua = bv$ with a, b in A. The label of the edge w is a.

Example 1.2.19 Let X be the Fibonacci shift. The Rauzy graphs of order $n = 1, 2, 3$ are represented in Figure 1.7 (with the edge from $w = ua$ labeled a).

Every infinite path $\cdots \overset{a_{i-1}}{\to} p_i \overset{a_i}{\to} p_{i+1} \overset{a_{i+1}}{\to} \cdots$ in $\Gamma_n(X)$ has a label, which is the sequence $(a_i)_{i \in \mathbb{Z}}$. The set of these labels is a shift space X_n. We have $X_1 \supset X_2 \supset \cdots \supset X_n \supset \cdots \supset X$ and $X = \cap_{n \geq 0} X_n$. Thus, the sequence X_n approximates X from above. A graph is *strongly connected* if for every pair of vertices v, w there is a path from v to w.

Proposition 1.2.20 *If a shift space X is recurrent, all the graphs $\Gamma_n(X)$ are strongly connected.*

Proof Assume that X is recurrent (or, equivalently, irreducible). If u, v belong to $\mathcal{L}_{n-1}(X)$, there is some $w \in \mathcal{L}(X)$ such that uwv belongs to $\mathcal{L}(X)$. But then there is a path labeled wv from u to v in $\Gamma_n(X)$. Thus, $\Gamma_n(X)$ is strongly connected. ∎

The converse of Proposition 1.2.20 is not true, as shown by the example of the shift space X such that $\mathcal{L}(X) = a^*b^* \cup b^*a^*$.

1.2.6 Higher Block Shifts

Let X be a shift space on the alphabet A and let $k \geq 1$ be an integer. Let $f : \mathcal{L}_k(X) \to A_k$ be a bijection from the set $\mathcal{L}_k(X)$ of blocks of length k of X onto an alphabet A_k. The map $\gamma_k : X \to A_k^{\mathbb{Z}}$ defined for $x \in X$ by $y = \gamma_k(x)$ if for every $n \in \mathbb{Z}$

$$y_n = f(x_n \cdots x_{n+k-1})$$

is the kth *higher block code* on X (see Figure 1.8). The set $X^{(k)} = \gamma_k(X)$ is a shift space on A_k, called the kth *higher block presentation* of X (one also uses the term of coding by *overlapping blocks* of length k).

Figure 1.8 The kth higher block code.

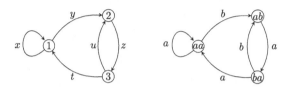

Figure 1.9 The third higher block coding of the golden mean shift X and the graph $\Gamma_3(X)$.

The higher block code is an isomorphism of dynamical systems and the inverse of γ_k is the map $y \mapsto x$ such that x_n is the first letter of $f^{-1}(y_n)$ for all n.

We sometimes, when no confusion arises, identify A_k and $\mathcal{L}_k(X)$ and write simply $y_0 y_1 \cdots = (x_0 x_1 \cdots x_{k-1})(x_1 x_2 \cdots x_k) \cdots$.

Example 1.2.21 Consider again the golden mean shift X (Example 1.1.7). We have $\mathcal{L}_3(X) = \{aaa, aab, aba, baa, bab\}$. Set $f: aaa \mapsto x, aab \mapsto y, aba \mapsto z, baa \mapsto t, bab \mapsto u$. The third higher block shift $X^{(3)}$ of X is the set of two-sided infinite paths in the graph of Figure 1.9 on the left (this graph is, up to the labeling, the Rauzy graph $\Gamma_3(X)$; see Figure 1.9 on the right).

Note that for $n \geq k$, the Rauzy graph $\Gamma_n(X^{(k)})$ is the same, up to the labels, as the graph $\Gamma_{n+k-1}(X)$. As an example, the graph of Figure 1.9 on the left can be identified either with $\Gamma_2(X^{(2)})$ or with $\Gamma_3(X)$ (see Figure 1.9 on the right).

1.3 Shifts of Finite Type

A shift space X is *of finite type* if $\mathcal{L}(X) = A^* \setminus A^* I A^*$ where $I \subset A^*$ is finite set. In other words, a two-sided infinite sequence x is in X if and only if it has no factor in the finite set I. The elements of I are called the *forbidden blocks* of X.

The class of shifts of finite type is closed under conjugacy (Exercise 1.21).

A well-known example is the golden mean shift X where X is the set of two-sided sequences on $A = \{a, b\}$ with no consecutive b (see Example 1.1.7). Thus, the set of forbidden blocks is $I = \{bb\}$, and X is the set of labels of two-sided infinite paths in the graph of Figure 1.3.

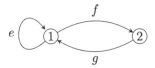

Figure 1.10 An edge shift.

As a more general example, for every shift space X and $n \geq 1$, the set X_n of labels of bi-infinite paths in the Rauzy graph $\Gamma_n(X)$ is a shift of finite type. Indeed, it is defined by the finite set of forbidden blocks, which is the set of words of length n that are not in $\mathcal{L}_n(X)$.

Given a finite graph $G = (V, E)$, the *edge shift* on G is the shift space X where $X \subset E^{\mathbb{Z}}$ is the set of bi-infinite paths in G and S is the shift on $E^{\mathbb{Z}}$. An edge shift is a shift of finite type, since it is defined by forbidden blocks of length 2. Moreover, if the graph G is strongly connected, the edge shift on G is irreducible (and thus recurrent).

Proposition 1.3.1 *Every shift of finite type is conjugate to an edge shift on some graph. The shift is recurrent if and only if the graph can be chosen strongly connected.*

The proof is left as an exercise (Exercise 1.22).

Example 1.3.2 The edge shift on the graph G represented in Figure 1.10 is conjugate to the golden mean shift by the 1-block map $e \mapsto a, f \mapsto b, g \mapsto a$.

1.4 Substitution Shifts

Let A and B be finite alphabets. By a *morphism* from A^* to B^* we mean a morphism of monoids; that is, a map $\sigma: A^* \rightarrow B^*$ such that $\sigma(\varepsilon) = \varepsilon$ and $\sigma(uv) = \sigma(u)\sigma(v)$ for all $u, v \in A^*$. When $A = B$, we say that σ is an *endomorphism*. When $\sigma(A) = B$, we say σ is a *letter-to-letter morphism*. Thus, letter-to-letter morphisms are onto.

We set $|\sigma| = \max_{a \in A} |\sigma(a)|$ and $\langle \sigma \rangle = \min_{a \in A} |\sigma(a)|$. The morphism σ is of *constant length* or *uniform* if $\langle \sigma \rangle = |\sigma|$, that is, if all $\sigma(a)$ have the same length n, called the *length* of σ. It is *growing* if $\lim_{n \to \infty} \langle \sigma^n \rangle = +\infty$ or, equivalently, if $|\sigma^n(a)| \to \infty$ for every $a \in A$.

We say that a morphism σ is *erasing* if there exists $b \in A$ such that $\sigma(b)$ is the empty word and *nonerasing* otherwise. A growing morphism is nonerasing. If σ is nonerasing, it induces, by infinite concatenation, a map from $A^{\mathbb{N}}$ to $B^{\mathbb{N}}$

defined for $x \in A^{\mathbb{N}}$ by

$$\sigma(x) = \sigma(x_0)\sigma(x_1)\cdots$$

and a map from $A^{\mathbb{Z}}$ to $B^{\mathbb{Z}}$ defined for $x \in A^{\mathbb{Z}}$ by

$$\sigma(x) = \cdots\sigma(x_{-1}) \cdot \sigma(x_0)\sigma(x_1)\cdots.$$

These maps are also denoted by σ. Both are continuous because, if x, y coincide on the first n letters, then so do $\sigma(x), \sigma(y)$.

If $\sigma : A^* \to C^*$ is a morphism, we denote $\ell(\sigma) = \sum_{a \in A}(|\sigma(a)| - 1)$.

Proposition 1.4.1 *If $\alpha : B^* \to C^*$ and $\beta : A^* \to B^*$ are such that $\sigma = \alpha \circ \beta$, and if every $b \in B$ appears in some $\beta(a)$ for $a \in A$, then*

$$\ell(\sigma) \geq \ell(\alpha) + \ell(\beta). \tag{1.11}$$

Proof We have

$$\ell(\sigma) - \ell(\beta) = \sum_{a \in A}(|\sigma(a)| - |\beta(a)|) = \sum_{a \in A}\sum_{b \in B}(|\alpha(b)||\beta(a)|_b - |\beta(a)|_b)$$

$$= \sum_{a \in A}\sum_{b \in B}(|\alpha(b)| - 1)||\beta(a)|_b = \sum_{b \in B}\left((|\alpha(b)| - 1)\sum_{a \in A}|\beta(a)|_b\right).$$

Since every b appears in some $\beta(a)$, every factor $\sum_{a \in A}|\beta(a)|_b$ is positive, whence the conclusion. $\blacksquare$

The *language* of the endomorphism $\sigma : A^* \to A^*$ is the set $\mathcal{L}(\sigma)$ of words occurring in some $\sigma^n(a)$, $a \in A$, $n \geq 0$. We denote by $X(\sigma)$ the set of sequences $y \in A^{\mathbb{Z}}$ whose subwords belong to $\mathcal{L}(\sigma)$. The set $X(\sigma)$ is clearly a shift space since it is closed and invariant. It is called the *shift generated* by σ. Similarly, the *one-sided shift generated* by σ is the set of sequences $x \in A^{\mathbb{N}}$ such that $\mathcal{L}(x) \subset \mathcal{L}(\sigma)$.

By definition $\mathcal{L}(\sigma^n)$ is included in $\mathcal{L}(\sigma)$ and $X(\sigma^n)$ in $X(\sigma)$ for all $n \geq 1$ (the equality does not always hold; see Exercise 1.27). Observe that, by compactness of $X(\sigma)$, the language $\mathcal{L}(\sigma)$ is finite if and only if $X(\sigma)$ is empty. Observe also that $\mathcal{L}(X(\sigma))$ is included in $\mathcal{L}(\sigma)$ but that the equality might not hold in general. For example, if σ is the morphism defined by $a \mapsto a$, $b \mapsto ba$, then $X(\sigma)$ is reduced to a^{∞} and thus b is not a factor of $X(\sigma)$. We will soon introduce a class of morphisms (the substitutions) where this kind of phenomenon cannot happen.

Example 1.4.2 The *Fibonacci morphism* (Example 1.1.8) is defined by $\varphi : a \mapsto ab, b \mapsto a$. The shift space associated with φ is called the *Fibonacci* *shift*. An example of a two-sided infinite sequence in $X(\varphi)$ is $z = \tilde{x} \cdot x$ where x is the Fibonacci word. Indeed, $\varphi^2(a) = aba$ ends with a and thus there is a unique left-infinite word y having all $\varphi^{2n}(a)$ as suffixes. Since $\varphi^2(a)$ is a palindrome, we have $y = \tilde{x}$. Then all factors of z are factors of some $\varphi^{2n}(aa)$ and thus, as aa is in $\mathcal{L}(\varphi)$, the sequence z belongs to $X(\varphi)$.

Example 1.4.3 The *Thue–Morse morphism* is the morphism defined by $\tau: a \mapsto ab, b \mapsto ba$. The associated shift space is the *Thue–Morse shift*.

A morphism $\sigma: A^* \to A^*$ is called *periodic* if its associated shift space is periodic and *aperiodic* if the associated shift space is aperiodic. When σ is periodic, its period can be bounded a priori, and thus this property is decidable for a morphism (Exercise 1.39).

Example 1.4.4 The shift space associated with the morphism $\varphi: a \mapsto aba, b \mapsto b$ is periodic. Indeed, $\varphi(ab) = (ab)^2$ and thus the set $X(\varphi)$ is formed of two points.

Let $\sigma: A^* \to A^*$ be a morphism and let $X = X(\sigma)$ be the two-sided shift defined by σ. A nonerasing morphism σ is a *substitution* if one has $\mathcal{L}(X) = \mathcal{L}(\sigma)$.

As an equivalent definition, σ is a substitution if and only if the language $\mathcal{L}(\sigma)$ is extendable. Indeed, this condition is necessary, and conversely, if it is satisfied, the shift space Y such that $\mathcal{L}(Y) = \mathcal{L}(\sigma)$ is clearly equal to X. Substitutions form a fairly general class of morphisms. They can fail to be growing (see the example of the Chacon binary morphism of Exercise 1.33). Conversely, not every growing morphism is a substitution (see Example 1.4.5). The property of being a substitution is easy to verify (Exercise 1.26). The examples of morphisms seen before as the Fibonacci or the Thue–Morse morphisms are all substitutions.

Note that if $\sigma: A^* \to A^*$ is a substitution, then $X(\sigma)$ is not empty since $\mathcal{L}(\sigma)$ always contains the alphabet. Consequently, there is a letter $a \in A$ such that $\lim_{n\to\infty} |\sigma^n(a)| = \infty$.

Example 1.4.5 The morphism $\sigma: a \to a, b \to b$ is not a substitution since $\mathcal{L}(\sigma) = \{a, b\}$ while $X = \emptyset$. As a less trivial example, let $\sigma: a \to b, b \to ab$. Then $\mathcal{L}(\sigma) = ab^* \cup b^*$, $X = b^\infty$ and $\mathcal{L}(X) = b^*$. Thus, σ is not a substitution.

1.4.1 Fixed Points

Fixed points of morphisms play an important role in this book. We first discuss one-sided fixed points.

Let $\sigma: A^* \to A^*$ be an endomorphism. If there exists a nonempty word $u \in \mathcal{L}(X)$ such that $\sigma(u)$ begins with u and if, moreover, $\lim_{n \to +\infty} |\sigma^n(u)| = +\infty$, then σ is said to be *right-prolongable on* u. We will first use this definition when u is a letter.

Suppose that σ is right-prolongable on $a \in A$. Since for all $n \in \mathbb{N}$, the word $\sigma^n(a)$ is a prefix of $\sigma^{n+1}(a)$, and because $|\sigma^n(a)|$ tends to infinity with n, there is a unique right infinite word denoted $\sigma^\omega(a)$, which has all $\sigma^n(a)$ as prefixes. Indeed, for every $i \geq 0$, the ith letter of this infinite word is the common ith letter of all words $\sigma^n(a)$, $n \in \mathbb{N}$, longer than i. Then $x = \sigma^\omega(a)$ is a *fixed point* of σ, which means by definition that $\sigma(x) = x$. Moreover, x belongs to the one-sided shift generated by σ since all factors of x are factors of some $\sigma^n(a)$.

By an *admissible one-sided fixed point* of σ, we mean a one-sided infinite sequence x that belongs to the one-sided shift generated by σ and such that $\sigma(x) = x$.

Observe that a morphism can have other fixed points, either finite or infinite. In fact, if σ is the morphism $a \mapsto a, b \mapsto ba$, then a is a finite fixed point since $\sigma(a) = a$ and a^ω is also a fixed point since $\sigma(a^\omega) = a^\omega$, but it is not admissible.

Proposition 1.4.6 *Every growing morphism* $\sigma: A^* \to A^*$ *has a power* $\tau = \sigma^n$ *that has an admissible one-sided fixed point* $\tau^\omega(a)$ *with* $a \in A$.

Proof Let $\sigma: A^* \to A^*$ be a growing morphism. Let $a \in A$. Since σ is growing, it is nonerasing and thus all $\sigma^n(a)$ are nonempty. Since A is finite, there are $n, p \geq 1$ such that $\sigma^n(a)$ and $\sigma^{n+p}(a)$ begin with the same letter, say b. Since σ is growing, $\lim_{k \to \infty} |\sigma^{kp}(b)| = \infty$. Thus, σ^p is right-prolongable on b and has an admissible one-sided fixed point. ∎

A sequence x that is an admissible one-sided fixed point of a substitution σ is said to be *purely substitutive* (with respect to σ). The subshift generated by x is then called a *substitution shift* and $\mathcal{L}(\sigma)$ a *substitution language*.

Example 1.4.7 The Thue–Morse morphism $\tau: a \mapsto ab, b \mapsto ba$ is prolongable on a and b. The fixed point $x = \lim \tau^n(a)$ is called the *Thue–Morse sequence*. One has

$$x = abbabaab \cdots .$$

One may show that $x_n = a$ if and only if the number of 1 in the binary expansion of n is even (Exercise 1.29).

If $x \in A^{\mathbb{N}}$ is purely substitutive (with respect to σ) and $\phi : A^* \to B^*$ is a letter-to-letter morphism, then $y = \phi(x)$ is said to be *substitutive* (with respect to (σ, ϕ)) and the shift space it generates, denoted $X(\sigma, \phi)$, is called a *substitutive shift*.

Admissible fixed points appear even for substitutions which are not growing.

Proposition 1.4.8 *A substitution $\sigma : A^* \to A^*$ generating a minimal shift space has a power $\tau = \sigma^n$ with an admissible fixed point $\tau^{\omega}(a)$ with $a \in A$.*

Proof Let A_i be the set of letters $a \in A$ such that $\lim |\sigma^n(a)| = \infty$. Set $A_f = A \setminus A_i$. Since σ is a substitution, A_i is not empty. The shift $X = X(\sigma)$ being minimal, every letter in A_i occurs in every $x \in X$ and the length of the words in $\mathcal{L}(X) \cap A_f^*$ is bounded.

Since the alphabet A is finite, there is a letter $a \in A_i$ such that $\sigma^n(a) = uav$ for some $n \geq 1$ and some words u, v with $u \in A_f^*$. The words $\sigma^n(u) \cdots uav \cdots \sigma^n(v)$ are all in $\mathcal{L}(X)$. If u is not empty, there are arbitrary long words in $\mathcal{L}(X) \cap A_f^*$, a contradiction. Thus, u is empty, σ^n is right prolongable on a and the property is proved. ∎

Proposition 1.4.8 is false if the shift is not minimal, as shown by the following example.

Example 1.4.9 Let $A = \{a, b\}$ and let $\sigma : A^* \to A^*$ be the substitution $a \mapsto a$, $b \mapsto aba$. Then no power of σ has an admissible fixed point of the form $\tau^{\omega}(a)$ with $a \in A$.

Two-sided fixed points of σ can be similarly defined. Assume that σ is right-prolongable on $u \in A^+$ and *left-prolongable* on $v \in A^+$, that is, $\sigma(v)$ ends with v and $\lim_{n \to +\infty} |\sigma^n(v)| = +\infty$. Let $x = \sigma^{\omega}(u)$ and let y be the left infinite sequence having all $\sigma^n(v)$ as suffixes. Let $z \in A^{\mathbb{Z}}$ be the two-sided sequence such that $x = z^+$ and $y = z^-$. Then z is a fixed point of σ denoted $\sigma^{\omega}(v \cdot u)$. It can happen that $\sigma^{\omega}(v \cdot u)$ does not belong to $X(\sigma)$. In fact, $\sigma^{\omega}(v \cdot u)$ belongs to $X(\sigma)$ if and only if vu belongs to $\mathcal{L}(\sigma)$.

An *admissible* two-sided fixed point of σ is a sequence $x \in X(\sigma)$ such that $\sigma(x) = x$. Thus, a fixed point of the form $\sigma^{\omega}(v \cdot u)$ with $vu \in \mathcal{L}(\sigma)$ is admissible.

When $z \in A^{\mathbb{Z}}$ is an admissible two-sided fixed point of σ, we say, as in the one-sided case, that z is a *purely substitutive* two-sided sequence and if

ϕ is a letter-to-letter morphism, we say that $x = \phi(z)$ is a *substitutive* two-sided sequence. Likewise, the shift generated by x, denoted $X(\sigma, \phi)$, is called a *substitutive shift*.

Proposition 1.4.10 *Every growing morphism* $\sigma : A^* \to A^*$ *has a power* τ *with an admissible two-sided fixed point* $\tau^\omega(a \cdot b)$ *with* $a, b \in A$ *and* $ab \in \mathcal{L}(X)$.

Proof Let $\sigma : A^* \to A^*$ be a growing morphism. Let $a, b \in A$ be such that $ab \in \mathcal{L}(\sigma)$. There are integers n, p such that simultaneously $\sigma^n(a), \sigma^{n+p}(a)$ end with the same letter c and $\sigma^n(b), \sigma^{n+p}(b)$ begin with the same letter d. This will also be true, with the same p but possibly different letters c, d, for $n + 1, n + 2, \ldots$ and thus we may also assume that p divides n. Thus, $\tau^\omega(c \cdot d)$ is a two-sided fixed point of $\tau = \sigma^p$. But cd is a factor of $\sigma^{n+p}(ab)$ and thus $cd \in \mathcal{L}(\tau)$, showing that $\tau^\omega(c.d)$ is an admissible two-sided fixed point of τ. ■

Example 1.4.11 Let $\varphi : a \mapsto ab, b \mapsto a$ be the Fibonacci morphism. Then $\psi = \varphi^2 : a \mapsto aba, b \mapsto ab$ is right prolongable on a and left prolongable on a and b. Since, moreover, ba is in $\mathcal{L}(\varphi)$, the two-sided infinite sequences $\psi^\omega(a \cdot a)$ and $\psi^\omega(b \cdot a)$ are admissible fixed points of φ^2.

Observe that a morphism σ could have nonadmissible fixed points.

Example 1.4.12 Let $\sigma : A^* \to A^*$, with $A = \{a, b, c\}$ be the morphism defined by $a \mapsto ab, b \mapsto ac, c \mapsto aa$. It can be checked that σ has a unique fixed point in $A^{\mathbb{N}}$ but no admissible fixed point in $A^{\mathbb{Z}}$, whereas σ^3 has three admissible fixed points : $\sigma^\omega(a \cdot a), \sigma^\omega(b \cdot a)$ and $\sigma^\omega(c \cdot a)$.

For morphisms that fail to be growing, we have the following result.

Proposition 1.4.13 *A substitution* σ *generating a minimal shift space has a power with an admissible fixed point.*

Proof Let A_i be the set of letters $a \in A$ such that $|\sigma^n(a)|$ tends to ∞ when n tends to ∞. Set $A_f = A \setminus A_i$. Since $X(\sigma)$ is minimal, the set A_i is nonempty. By Proposition 1.4.8, changing σ for one of its powers, there is a letter $a \in A$ such that the sequence $\sigma^\omega(a)$ is an admissible one-sided fixed point of σ. Since $X(\sigma)$ is minimal, there is some $b \in A_i$ such that bwa belongs to $\mathcal{L}(\sigma)$ for some word $w \in A_f^*$. There are integers $n, p \geq 1$ and a letter c such that $\sigma^n(b) = ucv$ with $v \in A_f^*$ and $\sigma^p(c)$ ends with c. Changing σ for one of its powers, we may assume that $\sigma(vw) = vw$. Then $\tau^\omega(cvw \cdot a)$ is an admissible two-sided fixed point of $\tau = \sigma^p$. ∎

The following example illustrates the case where there is no admissible fixed point $\sigma^\omega(a \cdot b)$ with $a, b \in A$.

Example 1.4.14 Let $A = \{a, b\}$ and let σ be the substitution $a \mapsto a, b \mapsto baabab$. Then $\sigma^\omega(ba \cdot b)$ is an admissible two-sided fixed point. It cannot be written $\sigma^\omega(a \cdot b)$ since $\sigma(a) = a$.

1.4.2 Primitive Morphisms

An endomorphism $\varphi \colon A^* \to A^*$ is said to be *primitive* if there is an $n \geq 1$ such that for every $a, b \in A$, the letter b appears in the word $\varphi^n(a)$.

In the following statement, we exclude the case of a one-letter alphabet for which $\sigma \colon a \to a$ is a primitive morphism that is neither growing nor a substitution.

Proposition 1.4.15 *If* $\sigma \colon A^* \to A^*$ *is a primitive morphism and A has at least two letters, then σ is a growing substitution.*

Proof Let $n \geq 1$ be such that every letter $b \in A$ occurs in $\sigma^n(a)$ for every $a \in A$. Then $|\sigma^n(a)| \geq \mathrm{Card}(A)$ and thus $|\sigma^{n+m}(a)| \geq \mathrm{Card}(A)^m$ for all $m \geq 1$. This shows that σ is growing.

Since σ is growing, there are letters a, b, c such that $abc \in \mathcal{L}(x)$. For large enough n, each of the three words $\sigma^n(a), \sigma^n(b), \sigma^n(c)$ contains every letter of A. This shows that every letter is extendable in $\mathcal{L}(\sigma)$ and thus, using Exercise 1.26, that σ is a substitution. ∎

The converse of Proposition 1.4.15 is not true since, for example, $a \mapsto aba, b \mapsto bb$ is a growing substitution, which is not primitive.

As a consequence of Proposition 1.4.15, the shift space $X(\sigma)$ generated by a primitive morphism σ on at least two letters is a substitution shift.

The following result is well known.

Proposition 1.4.16 *A primitive substitution shift is minimal.*

Proof Let $\sigma \colon A^* \to A^*$ be a primitive morphism. Let $n \geq 1$ be such that every $b \in A$ occurs in every $\sigma^n(a)$ for $a \in A$. By Proposition 1.2.16, it is enough to prove that $\mathcal{L}(\sigma)$ is uniformly recurrent. Let $u \in \mathcal{L}(\sigma)$. Let $b \in A$ and $m \geq 1$ be such that u is a factor of $\sigma^m(b)$. Then u is a factor of every $\sigma^{n+m}(a)$ and thus a factor of every word of $\mathcal{L}(\sigma)$ of length $2|\sigma|^{n+m}$. Therefore, $\mathcal{L}(\sigma)$ is uniformly recurrent. ∎

The converse of Proposition 1.4.16 is not true. A well-known example is the *Chacon binary morphism* $\sigma : 0 \mapsto 0010, 1 \mapsto 1$. This morphism is not primitive, but the corresponding shift space is minimal (see Exercise 1.33). The converse is true, however, for a morphism that is growing (Exercise 1.34).

When σ is primitive, we easily check that, for all $k \geq 1$, $\mathcal{L}(\sigma) = \mathcal{L}(\sigma^k)$ and that $\mathcal{L}(\sigma) = \mathcal{L}(\sigma^\omega(a))$ for any letter $a \in A$ on which σ is right-prolongable. In particular, we also have that $\Omega(x) = X(\sigma) = X(\sigma^k)$ for all $x \in X(\sigma)$ and $k \geq 1$. We say that $X(\sigma)$ is a *primitive substitution shift*. A sequence x is *primitive substitutive* if it is substitutive with respect to a primitive substitution. The subshift $\Omega(x)$ that it generates is then primitive substitutive. Since a letter-to-letter morphism is a morphism of dynamical systems, we have the following corollary of Proposition 1.4.16.

Corollary 1.4.17 *A primitive substitutive shift is minimal.*

The *composition matrix* of a morphism $\sigma : A^* \to B^*$ is the $B \times A$-matrix $M(\sigma)$ defined for every $a \in A$ and $b \in B$ by

$$M(\sigma)_{b,a} = |\sigma(a)|_b$$

where $|\sigma(a)|_b$ denotes the number of occurrences of the letter b in the word $\sigma(a)$. For a word w on the alphabet A, denote by $\ell(w)$ the vector $(|w|_a)_{a \in A}$. Then, one has more generally

$$\ell(\sigma(w)) = M(\sigma)\ell(w). \tag{1.12}$$

The transposed matrix of $M(\sigma)$ is called the *incidence matrix* of σ. Thus, on the composition matrix, the columns correspond to the words $\sigma(a)$ (in the sense that the column of index a is the vector $(|\sigma(a)|_b)_{b \in B}$) while, in the incidence matrix, this role is played by the rows.

Such a matrix gives, as we shall see, important information on the morphism. (although distinct morphisms may well have the same composition matrix). An important property is that $\sigma : A^* \to B^*$ and $\tau : B^* \to C^*$ are morphisms, then

$$M(\tau \circ \sigma) = M(\tau)M(\sigma). \tag{1.13}$$

Note that, with incidence matrices, the relative order of σ, τ is reversed since $M(\tau \circ \sigma)^t = M(\sigma)^t M(\tau)^t$.

The following mnemonic can be useful, since there is a risk of confusion between the composition matrix and the incidence matrix. The morphism is read on the Columns of the Composition matrix (note the initial letter C). Next, Composition matrices multiply in the same order as the Composition of the morphisms.

When $A = B$, the composition matrix is a square $A \times A$-matrix. It follows for (1.13) that for every $k \geq 0$, one has

$$M(\sigma^k) = M(\sigma)^k. \tag{1.14}$$

For example, the composition matrix of the substitution $\sigma \colon 0 \mapsto 01$, $1 \mapsto 00$ is

$$M(\sigma) = \begin{bmatrix} 1 & 2 \\ 1 & 0 \end{bmatrix}.$$

We easily derive from (1.14) the following statement.

Proposition 1.4.18 *The morphism σ is primitive if and only if the matrix $M(\sigma)$ is primitive.*

Example 1.4.19 The Fibonacci morphism is primitive. Indeed, $\varphi^2(a) = aba$ and $\varphi^2(b) = ab$ both contain a and b. Accordingly

$$M(\varphi) = \begin{bmatrix} 1 & 1 \\ 1 & 0 \end{bmatrix}, \quad M(\varphi)^2 = \begin{bmatrix} 2 & 1 \\ 1 & 1 \end{bmatrix}.$$

Thus, the Fibonacci shift is minimal.

Let $\varphi \colon A^* \to A^*$ be a primitive morphism. Then, the matrix $M = M(\sigma)$ is a primitive nonnegative matrix. Thus, we may use the Perron–Frobenius Theorem concerning the dominant eigenvalue λ_M of M (see Appendix B.3), also called, by abuse of language, the *dominant eigenvalue of φ*.

We illustrate the use of this theorem with the following example.

Example 1.4.20 Let $M = \begin{bmatrix} 1 & 1 \\ 1 & 0 \end{bmatrix}$. The eigenvalues of M are $\lambda = (1 + \sqrt{5})/2$ (called the *golden mean*) and $\hat{\lambda} = (1 - \sqrt{5})/2$. Since $\hat{\lambda} < \lambda$, we have $\lambda_M = \lambda$. As $\lambda^2 = \lambda + 1$, a left eigenvector corresponding to λ is $\begin{bmatrix} \lambda & 1 \end{bmatrix}$. The sequence (M^n/λ_M^n) tends to the matrix

$$\frac{1}{1 + \lambda^2} \begin{bmatrix} \lambda^2 & \lambda \\ \lambda & 1 \end{bmatrix} = \frac{1}{1 + \lambda^2} \begin{bmatrix} \lambda \\ 1 \end{bmatrix} \begin{bmatrix} \lambda & 1 \end{bmatrix}.$$

Example 1.4.21 Let φ be the Thue–Morse morphism. Then

$$M = \begin{bmatrix} 1 & 1 \\ 1 & 1 \end{bmatrix}.$$

The dominant eigenvalue is equal to 2 and the corresponding row eigenvector is $v = \begin{bmatrix} 1/2, & 1/2 \end{bmatrix}$.

Thus, if M is the incidence matrix of a primitive morphism, we have for any $a \in A$,

$$\lim_{n \to \infty} |\varphi^{n+1}(a)|/|\varphi^n(a)| = \lambda_M. \tag{1.15}$$

since $|\varphi^n(a)|/\lambda_M^n$ is the sum of coefficients of the row of index a of M^n/λ_M^n.

We now prove the following important property of primitive morphisms. For a morphism φ, we denote $\langle \varphi \rangle = \min_{a \in A} |\varphi(a)|$ and $|\varphi| = \max_{a \in A} |\varphi(a)|$.

Proposition 1.4.22 *For every primitive morphism φ, the sequence of quotients $(|\varphi^n|/\langle \varphi^n \rangle)_n$ is bounded.*

For this we need the following lemma (which will also be used again later).

Lemma 1.4.23 *Let M be a primitive matrix with dominant eigenvalue λ_M and positive left and right eigenvectors $x = (x_a)$, $y = (y_b)$ such that $\sum_{a \in A} x_a = 1$ and $\sum_{a \in A} x_a y_a = 1$. There are $c > 0$ and $\tau < \lambda_M$ such that for every $a, b \in A$ and $n \geq 1$, we have*

$$\left| |\varphi^n(a)|_b - \lambda_M^n y_a x_b \right| \leq c\tau^n, \tag{1.16}$$

$$\left| |\varphi^n(a)| - \lambda_M^n y_a \right| \leq c\,\mathrm{Card}(A)\tau^n, \tag{1.17}$$

$$\left| |\varphi^n(a)|_b - |\varphi^n(a)|x_b \right| \leq c(1 + \mathrm{Card}(A))\tau^n. \tag{1.18}$$

Proof The first equation results directly from (B.3.4). For the second one, we write, using the triangular inequality and (1.16)

$$\left| |\varphi^n(a)| - \lambda_M^n y_a \right| = |\sum_{b \in A} |\varphi^n(a)|_b - \lambda_M^n \sum_{b \in A} y_a x_b|$$

$$\leq \sum_{b \in A} \left| |\varphi^n(a)|_b - \lambda_M^n y_a x_b \right| \leq c\,\mathrm{Card}(A)\tau^n,$$

which is (1.17). Finally, using (1.16) and (1.17), we obtain, since $x_b \leq 1$,

$$\left| |\varphi^n(a)|_b - |\varphi^n(a)|x_b \right| = \left| |\varphi^n(a)|_b - \lambda_M^n y_a x_b + \lambda_M^n y_a x_b - |\varphi^n(a)|x_b \right|$$

$$\leq \left| |\varphi^n(a)|_b - \lambda_M^n y_a x_b \right| + x_b \left| |\varphi^n(a)| - \lambda_M^n y_a \right|$$

$$\leq c(1 + \mathrm{Card}(A))\tau^n,$$

which is (1.18). ∎

Proof of Proposition 1.4.22 Let M be the incidence matrix of φ. Equation (1.17) shows that for every $a \in A$, the sequence $(|\varphi^n(a)|/\lambda_M^n)_n$ converges to $y_a > 0$. This implies that for every $a, b \in A$, the sequence of quotients $(|\varphi^n(a)|/|\varphi^n(b)|)_n$ converges to y_a/y_b and thus proves the statement. ∎

We say that the shift space X is *linearly recurrent* (LR) (with constant $K \geq 0$) if it is minimal and if for all u in $\mathcal{L}(X)$ and for all right return words w to u in X we have $|w| \leq K|u|$.

Proposition 1.4.24 *A primitive substitution shift is linearly recurrent.*

Proof Let $\sigma \colon A^* \to A^*$ be a primitive substitution and let $X = X(\sigma)$ be the corresponding shift space. Set as usual $|\sigma| = \max_{a \in A} |\sigma(a)|$ and $\langle \sigma \rangle = \min_{a \in A} |\sigma(a)|$.

Since σ is primitive, it follows from Proposition 1.4.22 that there is a constant k such that

$$|\sigma^n| \leq k\langle \sigma^n \rangle \tag{1.19}$$

for all $n \geq 1$.

The substitution σ being primitive, the sequence $(\langle \sigma^n \rangle)_n$ is nondecreasing and unbounded. Thus, for any $w \in \mathcal{L}(X)$ there is an integer n such that $\langle \sigma^{n-1} \rangle < |w| \leq \langle \sigma^n \rangle$. The right inequality implies that w is a factor of $\sigma^n(ab)$ for some $a, b \in A$. Every return word to w is then a factor of a return word to $\sigma^n(ab)$. Let R be the maximal length of return words to words of length 2. Then, for every return word u to w, we have

$$|u| \leq R|\sigma^n| \leq Rk\langle \sigma^n \rangle \leq Rk|\sigma|\langle \sigma^{n-1} \rangle < Rk|\sigma||w|.$$

This shows that X is linearly recurrent with constant $K = kR|\sigma|$. ∎

We have seen an illustration of this result for the Fibonacci shift (Exercise 1.17). We add an illustration on the Thue–Morse shift.

Example 1.4.25 Let $\sigma \colon a \mapsto ab, b \mapsto ba$ be the Thue–Morse morphism. According to the above, since $k = 1$ for a constant-length morphism, and since the maximal length R of return words of length 2 is 6, the Thue–Morse shift is LR with $K = 12$.

Let us introduce a notion that will often come into play. Two morphisms $\sigma, \tau \colon A^* \to B^*$ are *conjugate* if there is a word $u \in B^*$ such that for every $a \in A$, one has $\sigma(a) = u\tau(a)u^{-1}$ (or equivalently $\sigma(a) = uv$, $\tau(a) = vu$ for some word v). Then $\sigma(w) = u\tau(w)u^{-1}$ holds for every $w \in A^*$. It is easy to verify that if $\sigma, \tau \colon A^* \to A^*$ are conjugate, then σ is primitive if and only if τ is primitive and, in this case, $X(\sigma) = X(\tau)$.

Example 1.4.26 The morphisms $\sigma : a \mapsto ab, b \mapsto a$ and $\tau : a \mapsto ba, b \mapsto a$ are conjugate. The shift $X(\sigma) = X(\tau)$ is the Fibonacci shift.

1.4.3 Circular Morphisms

We say that a set $U \subset A^+$ of nonempty words is a *code* if every word $w \in A^*$ admits at most one decomposition in a concatenation of elements of U. Thus, U is a code if the submonoid U^* generated by U is isomorphic to the free monoid on U. A prefix code is obviously a code.

A *coding morphism* for a code $U \subset A^+$ is a morphism $\varphi : B^* \to A^*$ whose restriction to B is a bijection onto U. Thus, in particular $\varphi(B) = U$ and $\operatorname{Card}(B) = \operatorname{Card}(U)$. Since U is a code, such a morphism is injective.

A submonoid M of A^* is *very pure* if for every $p, q \in A^*$

$$pq, qp \in M \Rightarrow p, q \in M. \tag{1.20}$$

The set $U \subset A^+$ is a *circular code* whenever it is a code and the submonoid U^* is very pure. A coding morphism for a circular code is called a *circular morphism*.

Thus, a morphism $\varphi : A^* \to B^*$ is circular if φ is a bijection from A onto a circular code.

One may visualize the definition of a circular code as follows. Imagine the word pq written on a circle (or infinitely repeated). Then pq and qp are two decompositions in words of U. Thus, the circular codes are such that the decomposition is unique on a circle (or, equivalently, on the infinite repetition $\cdots pqpq \cdots$).

Example 1.4.27 The set $U = \{0, 01\}$ is a circular code. On the contrary, $U' = \{010, 101\}$ is a code that is not circular since with $p = 0101$ and $q = 01$, one has $pq, qp \in U'^*$ although p, q do not belong to U'^*.

Proposition 1.4.28 *Let X be a shift space. For every $u \in \mathcal{L}(X)$, the set $\mathcal{R}_X(u)$ is a circular code.*

Proof The set $U = \mathcal{R}_X(u)$ is a prefix code and thus it is a code. Next, assume that pq, qp belong to U^*. Set $p = u_1 u_2 \cdots u_n s$ with $u_i \in U$ and s a proper prefix of some element of U. Let $k \geq 1$ be such that $(pq)^k$ is longer than u. Since $(pq)^k$ is in U^*, the word $u(pq)^k$ ends with u and thus $(pq)^k$ ends with u. Next, since $(qp)^{k+1} = q(pq)^k p$ also ends with u for the same reason, we obtain that up ends with u. Hence s belongs to U^* by (1.10). This forces $s = \varepsilon$ because U is a prefix code. Thus, p belongs to U^*, which implies that q is also in U^*. ∎

Circular codes have a property of unique decomposition of sequences.

Proposition 1.4.29 *Let* $U \subset A^+$ *be a finite code and let* $\varphi : B^* \to A^*$ *be a coding morphism for* U. *The following conditions are equivalent.*

(i) *The code* U *is circular.*
(ii) *For every* $x \in A^{\mathbb{Z}}$ *there is at most one pair* (k, y) *such that* $x = S^k \varphi(y)$ *with* $0 \le k < |\varphi(y_0)|$.

In particular, in this case, the map $\varphi \colon B^{\mathbb{Z}} \to A^{\mathbb{Z}}$ *is injective.*

The proof is left as an exercise (Exercise 1.43). Note that in the case where $U = \mathcal{R}_X(u)$, the proof is immediate since the occurrences of u in a sequence determine the factorization in words of U.

1.4.4 Recognizable Morphisms

Let $\varphi \colon A^* \to B^*$ be a nonerasing morphism. Let X be a shift space on the alphabet A and let Y be the closure under the shift of $\varphi(X)$. Then every $y \in Y$ has a representation as $y = S^k \varphi(x)$ with $x \in X$ and $0 \le k < |\varphi(x_0)|$. Thus, $Y = \{S^k \varphi(x) \mid x \in X, 0 \le k < |\varphi(x_0)|\}$. We say that φ is *recognizable* on X if every $y \in Y$ has only one such representation.

As an equivalent definition, consider the system obtained from X by the tower construction using the function $x \mapsto |\varphi(x_0)|$, already introduced in Section 1.1. It is the dynamical system (X^φ, T) where

$$X^\varphi = \{(x, i) \mid x \in X, \ 0 \le i < |\varphi(x_0)|\}$$

and

$$T(x, i) = \begin{cases} (x, i + 1) & \text{if } i + 1 < |\varphi(x_0)| \\ (Sx, 0) & \text{otherwise.} \end{cases}$$

The map $\widehat{\varphi} \colon (x, i) \mapsto S^i \varphi(x)$ is a morphism of dynamical systems from (X^φ, T) onto (Y, S). The morphism φ is recognizable on X if $\widehat{\varphi}^{-1}(\{y\})$ has only one element for every $y \in Y$. Thus, φ is recognizable on X if and only if $\widehat{\varphi}$ is injective (note that φ can be injective although $\widehat{\varphi}$, is not; see Exercise 1.44). Consequently $\hat{\varphi}$ is a homeomorphism. Since X is isomorphic to the system induced by (X^φ, T) on $X \times \{0\}$, we have proved the following useful statement.

Proposition 1.4.30 *If* φ *is recognizable on* X, *then*

1. The map $\hat{\varphi} : X^\varphi \to Y$ *is a homeomorphism.*

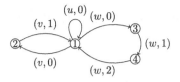

Figure 1.11 The shift space $\alpha(X^\varphi)$.

2. $\varphi(X)$ *is a clopen subset of* Y.
3. (X, T) *is isomorphic to the shift space induced by* Y *on* $\varphi(X)$.

Note that we may consider X^φ as a shift space on the alphabet

$$A^\varphi = \{(a, i) \mid a \in A, \ 0 \le i < |\varphi(a)|\}.$$

Indeed, there is a unique embedding α from $(A^{\mathbb{Z}})^\varphi$ into the full shift on A^φ such that $(\alpha(x, i))_0 = (x_0, i)$. The image of $(A^{\mathbb{Z}})^\varphi$ by α is actually a shift of finite type, as illustrated in the next example.

Example 1.4.31 Let X be the full shift on $\{u, v, w\}$. Consider the morphism $\varphi \colon u \mapsto a, v \mapsto ba, w \mapsto bba$. It is clearly recognizable on $\{u, v, w\}^{\mathbb{Z}}$. The shift space $\alpha(X^\varphi)$ is the edge shift on the graph G represented in Figure 1.11.

The following is a reformulation of Proposition 1.4.29.

Proposition 1.4.32 *A nonerasing morphism* $\varphi \colon B^* \to A^*$ *is recognizable on the full shift* $B^{\mathbb{Z}}$ *if and only if it is circular.*

As a positive example of application of Proposition 1.4.32, we have the case of the Fibonacci morphism.

Example 1.4.33 The Fibonacci morphism $\varphi \colon a \mapsto ab, b \mapsto a$ is recognizable on the full shift. Indeed, $\{a, ab\}$ is a circular code.

The negative example of the Thue–Morse morphism shows the necessity of the hypothesis that φ is circular in Proposition 1.4.29.

Example 1.4.34 Let $\tau \colon a \mapsto ab, b \mapsto ba$ be the Thue–Morse morphism. We have $(ab)^\infty = \tau(a^\infty) = S\tau(b^\infty)$. Thus, τ is not recognizable on $A^{\mathbb{Z}}$. But we will see below (Example 1.4.36) that τ is recognizable on $X(\tau)$.

The following important theorem will be used several times, in particular, in the representation of substitution shifts by sequences of partitions in towers in Chapter 4.

Theorem 1.4.35 (Mossé) *Every primitive aperiodic morphism φ is recognizable on $X(\varphi)$.*

The proof is given below.

Example 1.4.36 The Thue–Morse morphism $\tau\colon a \mapsto ab, b \mapsto ba$ is recognizable on the Thue–Morse shift $X = X(\tau)$. Indeed, let $y = S^k\tau(x)$ with $0 \le k < 2$. Every word in $\mathcal{L}(\tau)$ of length at least 5 contains aa or bb. Thus, y has an infinite number of occurrences of both aa and bb. Consider an $i > 1$ such that $y_i = y_{i+1} = a$. Then $y_{-k}\cdots y_i$ is in $\tau(A^*)$ and thus has even length. Consequently, i and k have opposite parities. This shows that k is unique and thus also x since τ has constant length and $ba \ne ab$.

The next example shows that the hypothesis that φ is aperiodic is necessary in Theorem 1.4.35.

Example 1.4.37 Let $\sigma\colon a \mapsto ab, b \mapsto ca, c \mapsto bc$. We have $\sigma(abc) = (abc)^2$ and thus $X(\sigma)$ is formed of the three shifts of $(abc)^\infty$. Therefore σ is primitive but periodic. Since

$$(abc)^\infty = \sigma(abc)^\infty = S\sigma(bca)^\infty$$

the morphism σ is not recognizable.

The proof of Theorem 1.4.35 uses the following concept. Let $\varphi\colon A^* \to B^*$ be a nonerasing morphism, let X be a shift space on the alphabet A, let Y be the orbit closure of $\varphi(X)$ under the shift. Let (p, q) be a pair of words on the alphabet B and (u, v) be a pair of words on the alphabet A such that $uv \in \mathcal{L}(X)$. We say that (p, q) is *parsable* for φ if there exists a pair (u, v) of words on A with $uv \in \mathcal{L}(X)$ such that

(i) p is a suffix of $\varphi(u)$ and
(ii) q is a prefix of $\varphi(v)$.

We also say in this case that (p, q) is (u, v)-parsable for φ.

Given a pair (p, q) that is (u, v)-parsable for φ, a word z on A such that pq is a factor of $\varphi(z)$ is *synchronized* with (p, q) if there is a factorization $z = yt$ such that (see Figure 1.12)

(i) p is a suffix of $\varphi(y)$,
(ii) q is a prefix of $\varphi(t)$,
(iii) the first letters of v, t are equal.

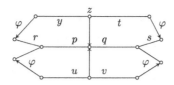

Figure 1.12 A synchronizing pair.

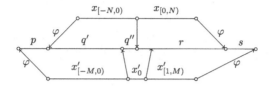

Figure 1.13 The intersection $\varphi(x) \cap S^{k-j}\varphi(x')$.

Finally, a pair (p, q) that is (u, v)-parsable for φ is *synchronizing* for φ if every $z \in \mathcal{L}(X)$ such that pq is a factor of $\varphi(z)$ is synchronized with (p, q).

Let $\varphi \colon A^* \to B^*$ be a morphism, let X be a shift space on A and let Y be the orbit closure of $\varphi(X)$ under the shift.

Proposition 1.4.38 *The morphism φ is recognizable on X if and only if there is an integer L such that every pair of words of length L, which is parsable for φ, is synchronizing for φ.*

Proof Let us first show that the condition is sufficient.

We have to prove that for $x, x' \in X$ such that $S^j\varphi(x) = S^k\varphi(x')$ with $0 \le j < |\varphi(x_0)|$ and $0 \le k < |\varphi(x'_0)|$, one has $j = k$ and $x_0 = x'_0$. We may suppose that $j \le k$.

By the hypothesis, for N large enough, the pair (q, r) with $q = \varphi(x_{[-N,0)})$ and $r = \varphi(x_{[0,N)})$ is synchronizing for φ.

We can also assume that $|\varphi(x_{[-N,0)})| \ge k - j$ and thus that $q = q'q''$ with $|q''| = k - j$ (see Figure 1.13). For $M \ge 1$ large enough, we have $\varphi(x'_{[-M,0)}) = pq'$ and $\varphi(x'_{[0,M)}) = q''rs$ for some words p and s (see Figure 1.13). Since (q, r) is synchronizing for φ, there is a prefix t of $x'_{[0,M)}$ such that $\varphi(t) = q''$ and x_0 is the first letter of $tx'_{[0,M)}$. Since $|q''| = k - j \le k < |\varphi(x'_0)|$, this forces $|t| = |q''| = 0$, $j = k$ and $x_0 = x'_0$.

Conversely, assume that φ is recognizable on X. Set $\psi = \hat{\varphi}^{-1}$. Since $\psi \colon Y \to X^\varphi$ is a conjugacy between shift spaces (we consider X^φ as a shift space on the alphabet A^φ), it is a sliding block code. Thus, there is an integer $L \ge 1$ such that the symbol $(\psi(y))_0$ depends only on $y_{[-L,L]}$.

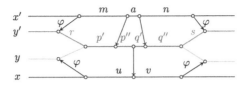

Figure 1.14 Proof that the condition is necessary.

Let (p, q) be a pair of words of length L which is (u, v)-parsable for φ, with $uv \in \mathcal{L}(X)$. Let $x \in X$ be such that u is a suffix of x^- and v is a prefix of x^+. Set $y = \varphi(x)$. Let $z \in \mathcal{L}(X)$ be such that $\varphi(z) = rpqs$. Let m, n, p', p'', q', q'' be words and $a \in A$ be a letter such that $z = man$ with (see Figure 1.14)

$$\varphi(m) = rp', \quad \varphi(a) = p''q', \quad \varphi(n) = q''s, \quad p = p'p'', \quad q = q'q''$$

and q' not empty.

Let $x' \in X$ be such that x'^- ends with m and x'^+ begins with an. Set $y' = S^{|p''|}\varphi(x')$ (in such a way that $\psi(y')_0 = (a, |p''|)$). Since $y_{[-L,L)} = y'_{[-L,L)} = pq$, it follows from the definition of L that $\psi(y)_0 = \psi(y')_0$. But $\psi(y)_0 = (b, 0)$, where b is the first letter of v while $\psi(y')_0 = (a, |p''|)$. We conclude that p'' is empty and $a = b$, which shows that the pair (p, q) is synchronizing for φ.   ∎

Proof of Theorem 1.4.35 Assume, by contradiction, that $\varphi \colon A^* \to A^*$ is not recognizable on $X = X(\varphi)$. By Proposition 1.4.38, this implies that for every ℓ, there is a pair (p, q) of words of length ℓ that is parsable for φ in $\mathcal{L}(X)$ but not synchronizing for φ.

Fix an integer $k \geq 1$, which will be chosen later. By the hypothesis, there is for every $n \geq 1$ a word r_n in $\mathcal{L}_{2k}(X)$ and a factorization $\varphi^{n-1}(r_n) = u_n v_n$ such that the pair (p_n, q_n) with $p_n = \varphi(u_n)$ and $q_n = \varphi(v_n)$ is not synchronizing for φ. Thus, there is for every $n \geq 1$ a word $z_n \in \mathcal{L}(X)$ such that pq is a factor of $\varphi(z_n)$ and

(i) $\varphi^{n-1}(z_n)$ is not synchronized with (p_n, q_n).

Moreover, choosing z_n of minimal length, we have

(ii) $z_n = a_n w_n b_n$ with $a_n, b_n \in A$ and $p_n q_n = s_n \varphi^n(w_n)t_n$ (see Figure 1.15).

Since $\varphi^n(r_n) = p_n q_n$ and $|r_n| = 2k$, we have

$$|p_n q_n| \leq 2k|\varphi^n|.$$

Moreover, since $p_n q_n = s_n \varphi^n(w_n)t_n$, we have

$$|p_n q_n| \geq |w_n|\langle\varphi^n\rangle.$$

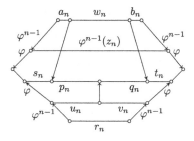

Figure 1.15 Proof of Mossé's Theorem.

Thus, we obtain

$$|w_n| \le 2k|\varphi^n|/\langle\varphi^n\rangle. \tag{1.21}$$

But, the morphism φ being primitive, by Proposition 1.4.22, the right-hand side of (1.21) is bounded independently of n. Thus, there is an infinite set E of integers n such that $r_n = r$, $z_n = z$, $a_n = a$, $b_n = b$ and $w_n = w$ for every $n \in E$. Hence we have the equalities

$$z = awb, \tag{1.22}$$

$$\varphi^n(r) = s_n\varphi^n(w)t_n. \tag{1.23}$$

Consider $n, m \in E$ with $n < m$. We have

$$\begin{aligned}
s_m\varphi^m(w)t_m = \varphi^m(r) &= \varphi^{m-n}(\varphi^n(r)) \\
&= \varphi^{m-n}(s_n)\varphi^m(w)\varphi^{m-n}(t_n).
\end{aligned}$$

Suppose that $s_m \ne \varphi^{m-n}(s_n)$ for arbitrary large values of k. Assume first that $|s_m| > |\varphi^{m-n}(s_n)|$. Then, by Proposition 1.2.1, the word $\varphi^m(w)$ is periodic of period $|s_m| - |\varphi^{m-n}(s_n)| \le |s_m|$. This implies that $\mathcal{L}(X)$ contains words of exponent larger than $\lfloor e \rfloor$ with

$$e = |\varphi^m(w)|/|s_m| \ge |p_m q_m|/|s_m| \ge 2k\langle\varphi^m\rangle/|\varphi^m|,$$

which tends to infinity with k. Thus, we may choose k to contradict Proposition 1.2.9 since, by Proposition 1.4.24, the shift space X is linearly recurrent.

In the case where $|s_m| < |\varphi^{m-n}(s_n)|$, we find that $\varphi^m(w)$ is periodic of period $|\varphi^{m-n}(s_n)| - |s_m| < |\varphi^{m-n}(s_n)|$ and thus that $\mathcal{L}(X)$ contains word of exponent larger than $\lfloor e \rfloor$ with

$$e = |\varphi^m(w)|/|\varphi^{m-n}(s_n)| \ge |p_m q_m|/|\varphi^m(a)| \ge 2k\langle\varphi^m\rangle/|\varphi^m|,$$

whence a contradiction again.

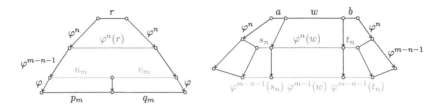

Figure 1.16 The case $s_m = \varphi^{m-n}(s_n)$.

Thus, $s_m = \varphi^{m-n}(s_n)$. Now we have by Eq. (1.23) $s_n\varphi^n(w)t_n = \varphi^n(r)$ and thus, applying φ^{m-n-1} to both sides (see Figure 1.16),

$$\varphi^{m-n-1}(s_n)\varphi^{m-1}(w)\varphi^{m-n-1}(t_n) = \varphi^{m-1}(r) = u_m v_m.$$

Let y, t be such that $\varphi^{m-1}(w) = yt$ with

$$u_m = \varphi^{m-n-1}(s_n)y, \quad v_m = t\varphi^{m-n-1}(t_n). \qquad (1.24)$$

Applying φ on both sides of (1.24), we obtain

$$p_m = \varphi^{m-n}(s_n)\varphi(y), \quad q_m = \varphi(t)\varphi^{m-n}(t_n).$$

Since $\varphi^{m-1}(z) = \varphi^{m-1}(a)yt\varphi^{m-1}(b)$, we find that

(i) p_m is a suffix of $\varphi^{m-1}(a)\varphi(y)$,
(ii) q_m is a prefix of $\varphi(t)\varphi^{m-1}(b)$,
(iii) the first letters of t and v_m are equal.

This shows that $\varphi^{m-1}(z)$ is synchronized with (p_m, q_m), a contradiction. ∎

The definition of recognizability for a primitive morphism can be formulated in terms of a fixed point of the morphism (see Exercise 1.45).

A striking feature of the Mossé Theorem is that it holds for noninjective morphisms. We illustrate this in the following example.

Example 1.4.39 The morphism $\varphi: a \mapsto abc, b \mapsto a, c \mapsto d, d \mapsto bcd$ is primitive and aperiodic. Thus, it is recognizable on $X(\varphi)$. It is not injective since $\varphi(ac) = \varphi(bd) = abcd$. However, one can check that the restriction of φ to $\mathcal{L}(X)$ is injective.

1.4.5 Block Presentations

We will need later to associate to a morphism φ and an integer $k \geq 1$ the k-block *presentation* of φ denoted φ_k.

Let $\varphi: A^* \to A^*$ be a nonerasing morphism and let X be its associated shift space. For $k \geq 1$, consider an alphabet A_k in one-to-one correspondence

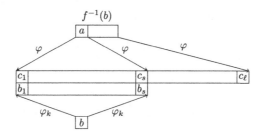

Figure 1.17 The morphism φ_k.

by $f: \mathcal{L}_k(X) \to A_k$ with the set $\mathcal{L}_k(X)$ of factors (or blocks) of length k of X. The map f extends naturally to a map, still denoted f, from $\mathcal{L}_{k+n}(X)$ to $\mathcal{L}_{n+1}(X^{(k)})$ defined for $n \geq 0$ by

$$f(a_1 a_2 \cdots a_{k+n}) = f(a_1 \cdots a_k) f(a_2 \cdots a_{k+1}) \cdots f(a_{n+1} \cdots a_{k+n}). \quad (1.25)$$

We define a morphism $\varphi_k: A_k^* \to A_k^*$ as follows. Let $b \in A_k$ and let a be the first letter of $f^{-1}(b)$ (see Figure 1.17). Set $s = |\varphi(a)|$. To compute $\varphi_k(b)$, we first compute the word $\varphi(f^{-1}(b)) = c_1 c_2 \cdots c_\ell$. Note that since φ is nonerasing, $\ell \geq |\varphi(a)| + k - 1 = s + k - 1$. We set $\varphi_k(b) = b_1 b_2 \cdots b_s$, where

$$b_1 = f(c_1 c_2 \cdots c_k), \ b_2 = f(c_2 c_3 \cdots c_{k+1}), \ldots, \ b_s = f(c_s \cdots c_{s+k-1}).$$

In other words, $\varphi_k(b)$ is the prefix of length $s = |\varphi(a)|$ of $f \circ \varphi \circ f^{-1}(b)$ where f is the map defined by (1.25) (see Figure 1.17).

Example 1.4.40 Let $\varphi: a \mapsto ab, b \mapsto a$ be the Fibonacci morphism. We have $\mathcal{L}_2(X) = \{aa, ab, ba\}$. Set $A_2 = \{x, y, z\}$ and let $f: aa \mapsto x, ab \mapsto y, ba \mapsto z$. Since $f \circ \varphi \circ f^{-1}(x) = f(\varphi(aa)) = f(abab) = yzy$, keeping the prefix of length $|\varphi(a)| = 2$, we have $\varphi_2(x) = yz$. Similarly, we have $\varphi_2(y) = yz$ and $\varphi_2(z) = x$.

Let $\pi: A_k^* \to A^*$ be the morphism defined by $\pi(b) = a$, where a is the first letter of $f^{-1}(b)$.

Then we have for each $n \geq 1$ the following commutative diagram, which expresses the fact that φ_k^n is the counterpart of φ^n for k-blocks.

$$
\begin{array}{ccc}
A_k^+ & \xrightarrow{\varphi_k^n} & A_k^+ \\
\downarrow{\pi} & & \downarrow{\pi} \\
A^+ & \xrightarrow{\varphi^n} & A^+
\end{array}
\qquad (1.26)
$$

Indeed, we have $\varphi \circ \pi(b) = \pi \circ \varphi_k(b)$ for every $b \in A_k$ by definition of φ_k. Since $\varphi\pi$ and $\pi\varphi_k$ are morphisms, this implies $\varphi \circ \pi = \pi \circ \varphi_k$ and thus

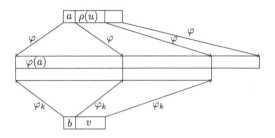

Figure 1.18 Comparing $\varphi_k(f(au))$ and $\varphi(au)$.

$\varphi^n \circ \pi = \pi \circ \varphi_k^n$ for all $n \geq 1$. This proves (1.26). In particular, since π is length preserving, we have

$$|\varphi_k^n(b)| = |\varphi^n(a)| \tag{1.27}$$

for $n \geq 1$ and $a = \pi(b)$.

In the following, we denote $u \leq v$ to express that the word u is a prefix of v.

Proposition 1.4.41 *We have for every $u \in \mathcal{L}(X)$ of length at least k,*

$$\varphi_k(f(u)) \leq f(\varphi(u)). \tag{1.28}$$

Proof For a word w of length at least n, we denote by $\text{Pref}_n(w)$ its prefix of length n and we set $\rho(u) = \text{Pref}_{|u|-k+1}(u)$ for u of length at least k. For $u \in \mathcal{L}(X)$, set $\ell(u) = |\varphi(\rho(u))|$. By definition, $\varphi_k(f(u))$ is, for every $u \in \mathcal{L}_k(X)$, the prefix of length $|\varphi(a)|$ of $f(\varphi(u))$, where a is the first letter of u. We prove that this property extends to longer words u and that, for every $u \in \mathcal{L}_{\geq k}(X)$, one has

$$\varphi_k(f(u)) = \text{Pref}_{\ell(u)} f(\varphi(u)). \tag{1.29}$$

Indeed, arguing by induction on the length of u, consider $a \in A$ and $u \in \mathcal{L}_{\geq k}(X)$. Set $f(au) = bv$ with $b \in A_k$ (see Figure 1.18). Then, since $\rho(au) = a\rho(u)$, and since $f^{-1}(b) = \text{Pref}_k(au)$,

$$\begin{aligned}
\text{Pref}_{\ell(au)} f(\varphi(au)) &= \text{Pref}_{|\varphi(a)|+\ell(u)} f(\varphi(au)) \\
&= \text{Pref}_{|\varphi(a)|} f(\varphi(au)) \, \text{Pref}_{\ell(u)} f(\varphi(u)) \\
&= \text{Pref}_{|\varphi(a)|} f(\varphi(f^{-1}(b))) \, \text{Pref}_{\ell(u)} f(\varphi(u)) \\
&= \varphi_k(b)\varphi_k(v) = \varphi_k(bv),
\end{aligned}$$

proving (1.29) and thus (1.28).   ∎

Proposition 1.4.42 *When φ is primitive, then for $k \geq 1$,*

1. the morphism φ_k is primitive,

2. *for every $b \in A_k$, $u = f^{(-1)}(b)$ and $n \geq 1$, $\varphi_k^n(b) \leq f(\varphi^n(u))$,*
3. *the shift space associated with φ_k is the kth higher block presentation of X.*

Proof 1. Since φ is primitive, there is an integer n such that for every $a \in A$, the word $\varphi^n(a)$ contains all the words of $\mathcal{L}_k(\varphi)$ as factors. Then for every $b \in A_k$, the word $\varphi_k^n(b)$ contains all letters of A_k. Thus, φ_k is primitive.

2. Since $\varphi_k(f(u)) \leq f(\varphi(u))$ for every $u \in \mathcal{L}_{\geq k}(X)$, we have also $\varphi_k^n(f(u)) \leq f(\varphi^n(u))$ for every $n \geq 1$.

3. This shows that for every $b \in A_k$ and $n \geq 1$, the word $\varphi_k^n(b)$ is in $\mathcal{L}(X^{(k)})$, which implies the conclusion since $X^{(k)}$ is minimal. $\blacksquare$

We denote by M_k the incidence matrix of the morphism φ_k. Thus, we have $M_k = M(\varphi_k)^t$.

Proposition 1.4.43 *All matrices M_k, for $k \geq 1$, have the same dominant eigenvalue.*

Proof Since $|\varphi_k^n(b)| = |\varphi^n(a)|$ for all $n \geq 1$ and $b \in A_k$ with $a = \pi(b)$ by (1.27), this follows from Eq. (1.15). $\blacksquare$

Actually, one can prove that all matrices M_k for $k \geq 2$ have the same nonzero eigenvalues (Exercise 3.31).

Example 1.4.44 Let $\varphi: a \mapsto ab, b \mapsto a$ be the Fibonacci morphism as in Example 1.4.40. Set $A_2 = \{x, y, z\}$ and $f : x \mapsto aa, y \mapsto ab, z \mapsto ba$. Then $\varphi_2: x \mapsto yz, y \mapsto yz, z \mapsto x$ as we have already seen. Thus,

$$M_2 = \begin{bmatrix} 0 & 1 & 1 \\ 0 & 1 & 1 \\ 1 & 0 & 0 \end{bmatrix}.$$

1.5 Sturmian and Arnoux–Rauzy Shifts

In this section, we introduce an important class of minimal shift spaces. We begin with the classical Sturmian shifts, which are on a binary alphabet, and we present next the episturmian and Arnoux–Rauzy shifts, which are a generalization of the former to an arbitrary finite alphabet.

1.5.1 Sturmian Shifts

A one-sided sequence s on the binary alphabet $\{0, 1\}$ is *Sturmian* if its word complexity is given by $p_n(s) = n + 1$ (we will say often, by a slight abuse of language, that its complexity is $n + 1$). Thus, by Proposition 1.2.14, Sturmian sequences have the minimal possible nonconstant word complexity.

Given two words $u, v \in \{0, 1\}^*$ their *balance* is $\delta(u, v) = ||u|_1 - |v|_1|$. A set of words $U \subset \{0, 1\}^*$ is *balanced* if for all $u, v \in U$,

$$|u| = |v| \Rightarrow \delta(u, v) \leq 1.$$

A sequence s is *balanced* if $\mathcal{L}(s)$ is balanced. The following statement gives an alternative definition of Sturmian sequences.

Proposition 1.5.1 *A one-sided sequence s is Sturmian if and only if it is balanced and aperiodic.*

The proof is left as Exercise 1.49.

A one-sided shift space X on a binary alphabet is called *Sturmian* if it is generated by a Sturmian sequence. It can be shown that a Sturmian shift is minimal (we will prove this later; see Eq. (1.34)).

A two-sided shift is Sturmian if its language is the language of a one-sided Sturmian sequence. Thus, the complexity of a Sturmian shift is $n + 1$ and conversely a minimal shift of complexity $n + 1$ is Sturmian. However, it should be noted that there exist nonminimal shifts of complexity $n + 1$, which are not Sturmian. An example is the shift space X such that $\mathcal{L}(X) = 0^* 1^*$ (recall from (1.3) and (1.4) that $0^* 1^* = \{0^n 1^m \mid n, m \geq 0\}$).

An element of a Sturmian shift is a *Sturmian sequence*. We shall see shortly an example of a Sturmian shift (Example 1.5.2).

The definition implies that Sturmian shifts are such that for every $n \geq 1$, there is exactly one right special word (and one left special word). Since a prefix of a left special word is left special, this implies that the left special words in a Sturmian shift are the prefixes of one right infinite sequence.

Example 1.5.2 The Fibonacci shift (see Example 1.4.2) is Sturmian (see Exercise 1.46).

As is well known, Sturmian shifts correspond to the coding of irrational rotations on the circle. Actually, for every real number α with $0 < \alpha \leq 1$ and $0 \leq \rho < 1$, let $s_{\alpha,\rho} = (s_n)_{n \in \mathbb{Z}}$ be the bi-infinite sequence defined by

$$s_n = \lfloor (n + 1)\alpha + \rho \rfloor - \lfloor n\alpha + \rho \rfloor. \tag{1.30}$$

Such a sequence is the *mechanical sequence* with *slope* α and *intercept* ρ. A mechanical sequence is *irrational* if α is irrational. It can be shown that an irrational mechanical sequence is Sturmian (Exercise 1.52). Conversely, every

Sturmian shift is generated by an irrational mechanical sequence with intercept $\rho = 0$ (we actually prove this statement below in Proposition 1.5.12).

For example, if $\alpha = (3 - \sqrt{5})/2$, then $s_0 s_1 \cdots = 001001 \cdots$. The closure of the orbit of s is the Fibonacci shift (see Example 1.5.15).

1.5.2 Episturmian Shifts

Sturmian shifts can be generalized to arbitrary alphabets as follows. A shift space X on an alphabet A is called *episturmian* if $\mathcal{L}(X)$ is closed under reversal and for every $n \geq 1$, there is at most one right-special word of length n.

It is called *strict episturmian* (or also *Arnoux–Rauzy*) if for each $n \geq 1$, there is a unique right-special word w of length n, which is, moreover, such that wa is in $\mathcal{L}(X)$ for every $a \in A$. Again, the definition applies to two-sided or one-sided shifts.

As an equivalent definition, a minimal shift X on the alphabet A is an Arnoux–Rauzy shift if for every n there is a unique left-special (resp. right-special) factor w of length n, which is, moreover, such that $wa \in \mathcal{L}(X)$ (resp. $aw \in \mathcal{L}(X)$) for every $a \in A$ (Exercise 1.59).

A Sturmian shift is strict episturmian. Indeed, if X is Sturmian, it can be shown that $\mathcal{L}(X)$ is closed under reversal (Exercise 1.50).

A one-sided infinite sequence s is called *episturmian* (resp. strict episturmian or Arnoux–Rauzy) if the subshift generated by s is episturmian (resp. strict episturmian). It is called *standard* if its left-special factors are prefixes of s. For every episturmian one-sided shift X, there is a unique standard infinite word s in X.

Example 1.5.3 The morphism $\varphi \colon a \mapsto ab, b \mapsto ac, c \mapsto a$ is called the *Tribonacci morphism*. The fixed point $s = \varphi^\omega(a)$ is called the *Tribonacci sequence*. It is a strict and standard episturmian sequence (Exercise 1.51). There are three two-sided infinite words z such that $z^+ = s$, namely $\varphi^{3\omega}(a \cdot a)$, $\varphi^{3\omega}(b \cdot a)$ and $\varphi^{3\omega}(c \cdot a)$.

For $a \in A$, denote by L_a the map defined for every $b \in A$ by

$$L_a(b) = \begin{cases} ab & \text{if } b \neq a \\ a & \text{otherwise.} \end{cases}$$

Each L_a is an automorphism of the free group on A (see Appendix B.2) since $L_a^{-1}(b) = a^{-1}b$ for $b \neq a$ and $L_a^{-1}(a) = a$ and these maps are called the *Rauzy automorphisms*. We define L_u for $u \in A^*$ by extending the map $a \mapsto L_a$ to a morphism $u \mapsto L_u$ from A^* into the group of automorphisms of the free group by $L_{ua} = L_u \circ L_a$.

A *palindrome* is a word equal to its reversal. It happens that palindromes play an essential role in Sturmian sequences, due to the fact that a bispecial factor of a Sturmian word, being the only one of its length, has to be a palindrome.

For a word w, the *palindromic closure* of w, denoted $w^{(+)}$, is the shortest palindrome that has w as a prefix. The *iterated palindromic closure* of a word w, denoted $\text{Pal}(w)$, is defined by $\text{Pal}(\varepsilon) = \varepsilon$ and $\text{Pal}(ua) = (\text{Pal}(u)a)^{(+)}$ for $u \in A^*$ and $a \in A$. The map $w \mapsto \text{Pal}(w)$ is called the *palindromization map*.

We may extend the palindromization map to one-sided infinite sequences. Indeed, since $\text{Pal}(u)$ is a prefix of $\text{Pal}(uv)$, there is for every $x \in A^{\mathbb{N}}$ a unique right infinite word s such that $\text{Pal}(u)$ is a prefix of s for every prefix u of x. We set $\text{Pal}(x) = s$. The map $x \mapsto \text{Pal}(x)$ is continuous.

An important property of the palindromization map is the *Justin Formula*. This formula expresses that, for every $u, v \in A^*$, one has

$$\text{Pal}(uv) = L_u(\text{Pal}(v))\,\text{Pal}(u). \tag{1.31}$$

The proof of Formula 1.31 is given as Exercise 1.57. One may verify that the palindromization map is the unique function $f : A^* \to A^*$ such that $f(uv) = L_u(f(v))f(u)$ for every $u, v \in A^*$.

As a consequence, for every $u \in A^*$ and $x \in A^{\mathbb{N}}$, one has

$$\text{Pal}(ux) = L_u(\text{Pal}(x)). \tag{1.32}$$

Moreover, one may verify that the palindromization map is the unique map $f : A^{\mathbb{N}} \to A^{\mathbb{N}}$ such that $f(ux) = L_u(f(x))$ for every $u \in A^*$ and $x \in A^{\mathbb{N}}$.

We will use the following result in Chapter 4.

Theorem 1.5.4 *A one-sided infinite sequence s on the alphabet A is standard episturmian if and only if there exists a one-sided sequence x on A such that $s = \text{Pal}(x)$. Moreover, the episturmian sequence s is strict if and only if every letter of A occurs infinitely often in x.*

The proof is given below. The one-sided infinite word x is called the *directive sequence* of the standard sequence s (or of the shift space generated by s).

Example 1.5.5 The standard sequence $s = \text{Pal}((01)^{\omega})$ is the Fibonacci word. Indeed, by Justin's Formula, one has $L_{01}(s) = s$, whence the result, since $L_{01} : 0 \mapsto 010, 1 \mapsto 01$ is the square of the Fibonacci morphism.

Note that since a Sturmian sequence is strict episturmian, the directive sequence of a Sturmian sequence has an infinite number of occurrences of 0 and 1.

The sequence s is actually in the limit set $\cap_{n\geq 0} L_{a_0 \cdots a_{n-1}} (A^{\mathbb{N}})$ obtained using the infinite sequence of morphisms

$$\ldots A^* \overset{L_{a_1}}{\to} A^* \overset{L_{a_0}}{\to} A^*. \tag{1.33}$$

In such a sequence, the infinite iteration of one morphism is replaced by the composition of an infinite sequence of morphisms, a feature that we will introduce later with S-adic representations of shift spaces (see Section 6.4).

If $x = x_0 x_1 \cdots$, the words $u_n = \text{Pal}(x_0 \cdots x_{n-1})$ are the palindromic prefixes of s. It can be shown that the set of left return words to u_n satisfies

$$\mathcal{R}'_X(u_n) \subset \{L_{x_0 \cdots x_{n-1}}(a) \mid a \in A\} \tag{1.34}$$

with equality when s is strict and that the sets of return words to other words have the same form up to conjugacy (Exercise 1.58). This formula implies, in particular, that the set of return words is finite and thus that an episturmian shift is minimal.

Formula (1.34) also shows the remarkable fact that, in a strict episturmian shift X, the set of return words to the words u_n (and, as a consequence, to every word in $\mathcal{L}(X)$) has a constant cardinality. We shall have more to say about this later, when we introduce dendric shifts (Chapter 7).

Example 1.5.6 The directive sequence of the Tribonacci sequence s (see Example 1.5.3) is $(abc)^\omega$. Indeed, we have by Justin's Formula $\text{Pal}((abc)^\omega) = L_{abc}(\text{Pal}(abc)^\omega)$, whence the result since $L_{abc} = \varphi^3$. The palindromic prefixes of s are

$$a, aba, abacaba, \ldots.$$

We have, for example, $\mathcal{R}'_X(abacaba) = \{L_{abc}(a), L_{abc}(b), L_{abc}(c)\}$, which is the set $\{abacaba, abacab, abac\}$.

We now give an example of an episturmian shift, which is not strict.

Example 1.5.7 We have $\text{Pal}(ab^\omega) = (ab)^\omega$, which is not Sturmian.

The proof of Theorem 1.5.4 uses the following notion. A one-sided infinite sequence s is *palindrome closed* if for every prefix u of s, the word $u^{(+)}$ is also a prefix of s. The following property is not surprising.

Proposition 1.5.8 *If $s = \text{Pal}(x)$ for some infinite sequence x, then s is palindrome closed.*

Proof Set $x = x_0x_1 \cdots$ and $u_n = \mathrm{Pal}(x_0 \cdots x_{n-1})$. Let u be a prefix of s. Then $|u_n| < u \le |u_{n+1}|$ for some n. Since $u^{(+)}$ is a palindrome with u_n as a prefix, we have $u^{(+)} = u_{n+1}$. Thus, s is palindrome closed. ∎

A factor of a word w is *unioccurrent* if it has exactly one occurrence in w. A word w is a *Justin word* if it has a palindromic suffix that is unioccurrent. This suffix is then also the longest palindromic suffix of w. As an example, every palindrome is a Justin word but $w = abca$ is not a Justin word.

A one-sided sequence s is *weakly palindrome closed* if, for every prefix u of s that is a Justin word, the word $u^{(+)}$ is a prefix of s.

Proposition 1.5.9 *Let s be a weakly palindrome closed sequence. Then every prefix of s is a Justin word. Consequently s is palindrome closed.*

Proof Let w be a prefix of s. We argue by induction on $n = |w|$. The property is true if $n = 1$ since every letter is a Justin word. Next, consider $w = va$ with $a \in A$. By induction hypothesis, v is a Justin word. Set $v = v_1v_2$ with v_2 palindrome unioccurrent in v. Then v_2 is also the longest palindromic suffix of v and $v^{(+)} = v_1v_2\tilde{v}_1$ (see Exercise 1.53). If v_1 is not empty, since s is weakly palindrome closed, $v^{(+)} = v_1v_2\tilde{v}_1$ is a prefix of s and thus v_1 ends with a. Then av_2a is a unioccurrent palindrome suffix of w and the property is true. Otherwise, v is a palindrome. If a does not occur in v, then a is a palindromic suffix of w unioccurrent in w. Otherwise, let ua be the prefix of v ending with a with $|u|$ minimal.

Claim 1: the word u is a palindrome. By induction hypothesis, we have $u = u_1u_2$ with u_2 a palindrome unioccurent in u. Again, if u_1 is not empty, since $u^{(+)} = u_1u_2\tilde{u}_1$ is a prefix of s, the word u_1 ends with a, a contradiction with the definition of u. This proves Claim 1.

Let qa be the longest prefix of v with q a palindrome. Since v is a palindrome, the word aqa is a palindromic suffix of w.

Claim 2: the palindrome aqa is unioccurrent in w. Otherwise, set $v = xaqay$ with aqa unioccurrent in $xaqa$.

Let us first prove that $z = xaq$ is a palindrome. Otherwise, we have by induction hypothesis $z = z_1z_2$ with z_2 a palindrome unioccurrent in z and z_1 nonempty. Since s is weakly palindrome closed, $z^{(+)} = z_1z_2\tilde{z}_1$ is a prefix of s and thus a is the last letter of z_1. But then az_2a and aqa are two palindromic unioccurrent suffixes of za, which forces $z_2 = q$. Since q is a prefix of z, this contradicts the unioccurrence of z_2. Thus, z is a palindrome.

But za is a prefix of v and $|z| > |q|$, which contradicts the definition of q. This proves Claim 2 and thus the proposition. ∎

The following result is the key of the proof of Theorem 1.5.4.

Proposition 1.5.10 *The following conditions are equivalent for a one-sided sequence s.*

(i) *s is palindrome closed.*

(ii) *s is weakly palindrome closed.*

(iii) *$s = \text{Pal}(x)$ for some one-sided sequence x.*

Proof (i)$\Leftrightarrow$ (ii) is clear by Proposition 1.5.9.

(ii)$\Rightarrow$ (iii) Let x_1 be the first letter of s and set $u_1 = x_1$. Assume that $x_0, \ldots, x_{n-1}$ and $u_i = \text{Pal}(x_0 \cdots x_{i-1})$ for $1 \leq i \leq n$ are already defined. Let x_n be the letter following u_n in s. By Proposition 1.5.9, $u = u_n x_n$ is a Justin word. Thus, since s is palindrome closed, $u^{(+)}$ is a prefix of s. This allows us to define $u_{n+1} = u^{(+)} = \text{Pal}(x_0 \cdots x_n)$. The word $x = x_0 x_1 \cdots$ build in this way is clearly such that $s = \text{Pal}(x)$.

(iii)$\Rightarrow$ (i) is Proposition 1.5.8. ∎

Proposition 1.5.11 *If $s = \text{Pal}(x)$ for some infinite word x, then all its left-special factors are prefixes of s.*

Proof Let u be the shortest left-special factor of s which is not a prefix of s. Set $u = va$ with $a \in A$. Then v is left-special and shorter than u. Thus, v is a prefix of s. Let $c \in A$ be such that vc is a prefix of s. Since $c \neq a$, v is right-special. Thus, $\tilde{v}$ is a prefix of s and consequently $v = \tilde{v}$. Since u is left-special, we have $bva, b'va \in \mathcal{L}(s)$ for two distinct letters b, b'.

Set $x = x_0 x_1 \cdots$ and $u_n = \text{Pal}(x_0 \cdots x_{n-1})$. Let n be such that u is a factor of u_{n+1} but not of u_n. Then bva or avb is a prefix of $x_n u_n$ (Exercise 1.60). Since $|u_n| > |v|$, vc is a prefix of u_n whence $c = a$ or $c = b$. Since $c \neq a$, we have $c = b$. In the same way, $c = b'$, whence $b = b'$, a contradiction. ∎

Proof of Theorem 1.5.4 Assume first that $s = \text{Pal}(x)$. By Proposition 1.5.11 its left-special factors are prefixes of s, and thus there is at most one of each length. Moreover, every factor u of s is a factor of some palindromic prefix and thus $\tilde{u}$ is also a factor of s. Thus, s is standard episturmian.

Conversely, consider a standard episturmian sequence s. Let us prove that s is weakly palindrome closed. Otherwise, there is a leftmost occurrence $s = uavw\tilde{v}bs'$ of a palindrome w with $a \neq b$. Then $vw\tilde{v}$ is left-special and thus a prefix of s. This contradicts the hypothesis that the leftmost occurrence of w appears after uav and proves the claim. By Proposition 1.5.10, this implies that $s = \text{Pal}(x)$ for some sequence x.

Let us prove now the last assertion. For this, it is enough to prove that for $a \in A$ and $n \geq 1$, one has $u_n a \in \mathcal{L}(s)$ if and only if a appears in $x_n x_{n+1} \cdots$.

If $a = x_m$ for some $m \geq n$, then $u_m a$ is a prefix of s and thus $u_n a \in \mathcal{L}(s)$ since u_n is a suffix of u_m. Thus, $u_n a \in \mathcal{L}(s)$.

Conversely, if $u_n a \in \mathcal{L}(s)$, we may assume that $a \neq x_n$ and thus that $u_n a$ is not a prefix of s. Let $b \in A$ be such that $b u_n a \in \mathcal{L}(s)$ and let m be such that $b u_n a$ is a factor of u_{m+1} but not of u_m. Then, by Exercise 1.60, either $b u_n a$ is a suffix of $u_m x_m$ or it is a prefix of $x_m u_m$. The first case implies $a = x_m$ and the conclusion. The second case is not possible. Indeed, since $|b u_n a| \leq |u_m x_m|$, we have $n < m$ and thus, $u_n a$ being a prefix of u_m, this forces $a = x_n$. ∎

1.5.3 Sturmian Sequences, Rotations and Continued Fractions

Let us come back to binary Sturmian sequences. We will use the traditional alphabet $A = \{0, 1\}$. To every standard Sturmian sequence $s \in \{0, 1\}^{\mathbb{N}}$ we associate a real number α, with $0 < \alpha \leq 1$, called its slope and defined as follows.

Let s be a standard Sturmian sequence and let x be its directive sequence. Since x has an infinite number of occurrences of 0 and 1, we can write

$$x = 0^{d_1} 1^{d_2} 0^{d_3} \cdots$$

with $d_1 \geq 0$ and $d_n \geq 1$ for $n \geq 2$. Define the *slope* of s as the irrational number

$$\pi(s) = [0; 1 + d_1, d_2, d_3, \ldots],$$

where $[a_0; a_1, a_2, \ldots]$ denotes the continued fraction with coefficients $a_0, a_1, \ldots$ (see Appendix B.1). The slope of s is actually the slope of the line in the plane approximating s when 0 is represented by a vector $(1, 0)$ and 1 by a vector $(1, 1)$ (see Figure 1.22).

Recall that we denote by $\mathbb{T} = \mathbb{R}/\mathbb{Z}$ the one-dimensional torus. For $0 < \alpha \leq 1$, let $T_\alpha \colon \mathbb{T} \to \mathbb{T}$ be the transformation defined by $T_\alpha(z) = z + \alpha$. The pair $(\mathbb{T}, T_\alpha)$ is a topological dynamical system called the *rotation* of angle α.

Proposition 1.5.12 *Let s be a standard Sturmian sequence, let X be the subshift generated by s and let α be the slope of s. The map $\gamma_\alpha \colon \mathbb{T} \to \{0, 1\}^{\mathbb{Z}}$, defined, for $z \in \mathbb{T}$, by $\gamma_\alpha(z) = t$ where*

$$t_n = \begin{cases} 0 & \text{if } T_\alpha^n(z) \in [0, 1 - \alpha) \\ 1 & \text{otherwise} \end{cases}$$

has the following properties.

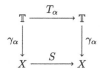

Figure 1.19 The map γ_α.

1. It is such that $s = \gamma_\alpha(\alpha)^+$. Thus, s is the mechanical sequence of slope α and intercept α.
2. It is an injective map from $\mathbb{T}$ to X.
3. It satisfies $\gamma_\alpha \circ T_\alpha = S \circ \gamma_\alpha$ (see Figure 1.19).

The map γ_α is called the *natural coding* of $(\mathbb{T}, T_\alpha)$. Note first that $\gamma_\alpha(z) = s_{\alpha,\rho}$ for $z = \lfloor \alpha + \rho \rfloor$, where $s_{\alpha,\rho}$ is the sequence defined by Eq. (1.30). Proposition 1.5.12 shows that every Sturmian shift is generated by such a sequence. It gives also the following corollary. The *slope* of a word $u \in \{0, 1\}^*$ is the rational number $\pi(u) = |u|_1/|u|$.

Corollary 1.5.13 *For every factor w of a standard Sturmian sequence s of slope α, the slope of w tends to α when $|w| \to \infty$.*

Indeed, by Proposition 1.5.12, we have $s = s_{\alpha,0}^+$, and thus the statement follows by Exercise 1.52.

Observe also that γ_α is not a conjugacy since it is neither continuous nor surjective. Indeed, it is not continuous since $\gamma_\alpha(0)_0 = 0$, while $\gamma_\alpha(z)_0 = 1$ for all $z \in (\alpha, 1)$ and thus for values of z arbitrary close to 0. Thus, γ_α is not continuous at zero. More precisely, there is a countable number of discontinuity points which form the orbit of 0. For each of these points z, the limits $\lim_{y \to z_-} \gamma_\alpha(y)$ and $\lim_{y \to z_+} \gamma_\alpha(y)$ exist but are distinct.

Next, the sequence $t = \lim_{z \to 1_-} \gamma_\alpha(z)$ is not in $\gamma_\alpha(\mathbb{T})$. Indeed, let $c_\alpha = \gamma_\alpha(\alpha)^+$. By definition of c_α and t, we have $t^+ = 1c_\alpha$. Moreover, if there exists some z with $t = \gamma_\alpha(z)$, then we should have $z = 0$, hence $\gamma_\alpha(0)_0 = 1$, a contradiction.

Although we will not use this, observe that the map γ_α is actually a measure-theoretic isomorphism since it is one-to-one except on a countable set, which is thus of measure 0.

The sequence c_α is called the *characteristic sequence* of slope α. Note that the left-special words in $\mathcal{L}(X)$ are the prefixes of c_α. Thus, c_α is a standard sequence.

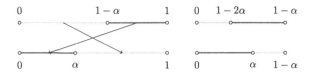

Figure 1.20 The action of T_α when $\alpha < 1/2$.

Moreover, since in a Sturmian shift X there is a unique left-special word of length n for all $n \in \mathbb{N}$, this implies that c_α is unique: if y^+ is standard for some $y \in X$, then $y^+ = c_\alpha$. We then say that X is the Sturmian shift of slope α.

To prove Proposition 1.5.12, we first prove the following lemma.

Lemma 1.5.14 *For every irrational α with $0 < \alpha \le 1$, the sequence $c_\alpha = \gamma_\alpha(\alpha)^+$ satisfies*

$$c_\alpha = \begin{cases} L_0(c_{\alpha/(1-\alpha)}) & \text{if } \alpha < 1/2 \\ L_1(c_{(1-\alpha)/\alpha}) & \text{otherwise.} \end{cases} \tag{1.35}$$

Proof Let $\alpha = [0; 1 + d_1, d_2, \ldots]$ be the continued fraction expansion of α. Assume that $d_1 > 0$ (or equivalently $\alpha < 1/2$). We have to prove that

$$c_\alpha = L_0(c_{\alpha/(1-\alpha)}). \tag{1.36}$$

We consider the transformation T_α as a map defined on $[0, 1)$ translating the semi intervals $[0, 1 - \alpha)$ and $[1 - \alpha, 1)$ as indicated in Figure 1.20 on the left. Consider the transformation T' induced by $T = T_\alpha$ on the semi-interval $[0, 1 - \alpha)$. It is defined by

$$T'(z) = \begin{cases} T(z) & \text{if } z \in [0, 1 - 2\alpha) \\ T^2(z) & \text{otherwise.} \end{cases}$$

The transformation T' is pictured on the right-hand side of Figure 1.20.

Let $\rho(z) \in \{0, 1\}^{\mathbb{Z}}$ be the natural coding of the T'-orbit of $z \in [0, 1 - \alpha)$ with the intervals $[0, 1 - 2\alpha)$, coded by 0, and $[1 - 2\alpha, 1 - \alpha)$ coded by 1 as we did to define γ_α. Then $\gamma_\alpha(z) = L_0(\rho(z))$. Moreover, normalizing the map T' to $[0, 1)$ it is the rotation of angle $\alpha/(1 - \alpha)$ and $\rho(z)$ is $\gamma_{\alpha/(1-\alpha)}(z/(1 - \alpha))$. Hence we have the relation

$$\gamma_\alpha(z) = L_0(\gamma_{\alpha/(1-\alpha)}(z/(1 - \alpha))),$$

which proves Eq. (1.36), taking $z = \alpha$.

If $\alpha > 1/2$, we consider instead the transformation T' induced by T_α on $[0, \alpha)$ (see Figure 1.21). Let $\rho(z) \in \{0, 1\}^{\mathbb{Z}}$ be the natural coding of the T'-orbit of $z \in [0, \alpha)$ with the intervals $[0, 1 - \alpha)$, coded by 0, and $[1 - \alpha, \alpha)$

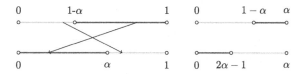

Figure 1.21 The action of T_α when $\alpha > 1/2$.

coded by 1. Then $\gamma_\alpha(z) = L_1(\rho(z))$. Moreover, normalizing the map T' to $[0, 1)$, it is the rotation of angle $\alpha/(1 - \alpha)$ and $\rho(z)$ is $\gamma_{\alpha/(1-\alpha)}(z/(1 - \alpha))$. Hence we have the relation

$$\gamma_\alpha(z) = L_1(\gamma_{\alpha/(1-\alpha)}(z/(1 - \alpha))),$$

which proves Eq. (1.35) taking $z = \alpha$.

We obtain this time a new rotation of angle $(1 - \alpha)/\alpha$, and for every $z \in [0, \alpha)$, we have $\gamma_\alpha(z) = L_1(\gamma_\alpha(R'(z)))$. Since $T'(z) = T_{(1-\alpha)/\alpha}(z/\alpha)$, the result follows also in this case. ∎

Proof of Proposition 1.5.12 For $y \in \{0, 1\}^{\mathbb{N}}$, denote $\alpha(y) = \pi(\text{Pal}(y))$. We have to prove that for all $u \in \{0, 1\}^*$ and $y \in \{0, 1\}^{\mathbb{N}}$ such that $x = uy$, we have

$$c_{\alpha(uy)} = L_u(c_{\alpha(y)}). \tag{1.37}$$

Indeed, by Justin's Formula (1.32), this implies that $c_{\alpha(x)} = \text{Pal}(x)$ and thus that $c_\alpha = s$.

Equation (1.37) is true for $u = \varepsilon$. Note that if $\alpha = \alpha(x)$ and $x = 0x'$, then $\alpha(x') = \alpha/(1 - \alpha)$. Arguing by induction on $|u|$, consider $u = 0v$. Then, by (1.35) and using the induction hypothesis, we have

$$c_{\alpha(uy)} = L_0(c_{\alpha(vy)}) = L_0(L_v(c_{\alpha(y)})) = L_u(c_{\alpha(y)}).$$

The case $u = 1v$ is similar.

To show that γ_α is injective, consider $z, z' \in \mathbb{T}$ with $0 \leq z < z' < 1$. Since α is irrational, the orbit of $1 - \alpha$ is dense and thus there is an integer $n \geq 0$ such that $z < T_\alpha^{-n}(1 - \alpha) < z'$. Then $T^n z < 1 - \alpha < T^n z'$ and thus $\gamma_\alpha(z)_n \neq \gamma_\alpha(z')_n$. Since T_α is minimal, the image of γ_α is the shift generated by s, that is, the Sturmian shift X.

Finally, the equality $\gamma_\alpha \circ T_\alpha = S \circ \gamma_\alpha$ is clear. ∎

The transformation used in the proof is called a *Rauzy induction* (we shall meet this notion again in Chapter 6 when we consider interval exchange transformations).

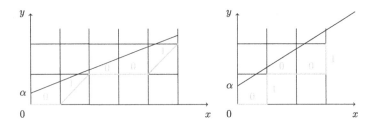

Figure 1.22 The mechanical sequence and the cutting sequence.

Let $s = (s_n)$ be the characteristic sequence of slope α. Using the representation of 0 by the vector $(1, 0)$ and 1 by $(1, 1)$, one associates to s a sequence (n, y_n) of integer points, which is an approximation of the line of equation $y = \alpha x + \alpha$ (see Figure 1.22 on the left). In fact, since s is the mechanical sequence of slope α and intercept α, it follows from Eq. (1.30) that for every $n \geq 0$, y_n is the integer part of $y = \alpha n + \alpha$.

Using instead the vector $(1, 0)$ to represent the symbol 1, one obtains a sequence of points (x'_n, y_n) called a *cutting sequence* of points. It is an approximation of the line $y = \beta x + \beta$ with $\beta = \alpha/(1 - \alpha)$ (see Figure 1.22 on the right). Indeed, the map $(x, y) \mapsto (x - y, y)$ sends the line of equation $y = \alpha x + \alpha$ to the line $y = \beta x + \beta$ and the point (n, y_n) to (x'_n, y_n). The sequence s is also called the *cutting sequence* of slope β and intercept β.

In the first representation, the slope of the line is the angle $0 < \alpha < 1$ of the rotation; in the second one it is β with $0 < \beta < \infty$. There is a risk of confusion between α and β induced by the name of slope for α, and Figure 1.22 helps to clarify it. For example, exchanging 0 and 1 in a Sturmian sequence changes α into $1 - \alpha$, while it changes β into $1/\beta$.

Example 1.5.15 The slope of the Fibonacci sequence s is $\alpha = \frac{3 - \sqrt{5}}{2}$. Indeed, its directive sequence is $0101\ldots$ (see Example 1.5.5) and thus $\alpha = [0, 2, 1, 1, \ldots]$. The first values of the mechanical sequence and of the cutting sequence are those indicated in Figure 1.22.

Theorem 1.5.16 *Let X be a Sturmian shift and let α be its slope. There exists a unique factor map $f : (X, S) \to (\mathbb{T}, T_\alpha)$ such that its restriction to $\gamma_\alpha(\mathbb{T})$ is the inverse of γ_α. Moreover, one has*

- *f is one-to-one on $X \setminus \gamma_\alpha^{-1}(\mathcal{O}_{T_\alpha}(0))$,*
- *$\mathrm{Card}(f^{-1}(\{T_\alpha^n(0)\})) = 2$ for all $n \in \mathbb{Z}$.*

Proof Set $s = \gamma_\alpha(\alpha)$ and $T = T_\alpha$. Then s is the standard Sturmian sequence generating X. Let $t \in X$. Since X is the shift generated by s, there is a sequence

(k_n) such that $t = \lim S^{k_n} s$. Set $z_n = T^{k_n} s$. Since an irrational rotation is minimal, for every $\varepsilon > 0$, there is an n such that every interval of size ε contains some $T^{-n}(0)$. Thus, there is $\eta > 0$ such that

$$d(\gamma_\alpha(z), \gamma_\alpha(z')) < \eta \Rightarrow |z - z'| < \varepsilon. \tag{1.38}$$

This implies that (z_n) is a Cauchy sequence and thus is convergent to some $z \in \mathbb{T}$. We define $f(t) = z$.

If $t = \gamma_\alpha(z)$, the sequence z_n above converges to z and thus $f(t) = z$. This shows that f is onto and equal to the inverse of γ_α on $\gamma_\alpha(\mathbb{T})$. It is continuous by (1.38).

If $z \notin \mathcal{O}_{T_\alpha}(0)$, then γ_α is continuous at z (this follows from the remarks after Proposition 1.5.12). Let $t \in X$ be such that $f(t) = z$. Let k_n be such that $t = \lim S^{k_n}(s)$. Then $z_n = T^{k_n}(\alpha)$ converges to z and

$$t = \lim S^{k_n}(s) = \lim \gamma_\alpha(z_n) = \gamma_\alpha(z)$$

showing that t is uniquely determined.

Finally, the cardinality of $f^{-1}(\{z\})$ is the same for every $z \in \mathcal{O}_{T_\alpha}(0)$. The set $f^{-1}(\{\gamma_\alpha(\alpha)\})$ is formed of two sequences $u \cdot s$ and $v \cdot s$ since, s being standard, all prefixes of s are left special. ∎

Note that the two elements t, t' of $f^{-1}(z)$ for $z \in \mathcal{O}_{T_\alpha}(0)$ are closely related. They are *right-asymptotic*, that is, such that $t_n = t'_n$ for all n large enough and also *left-asymptotic*, that is, such that $t_{-n} = t'_{-n}$ for all n large enough (see Exercise 1.66).

An interesting way of representing the link between a Sturmian sequence and its slope is to consider a binary tree with nodes labeled bijectively by the positive rationals and called the *Stern–Brocot tree* (see Figure 1.23).

The tree is built as follows. Every node can be identified with a word $u \in \{0, 1\}^*$. The label of u is the quotient $\lambda(u) = \pi(L_u(01)) = |L_u(01)|_1/$

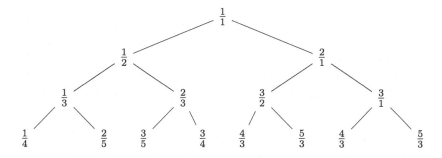

Figure 1.23 The Stern–Brocot tree.

$|L_u(01)|_0$. Thus, for example, since $L_{010}\colon 0 \rightarrow 010, 1 \rightarrow 01001$, the label of the node 010 is $3/5$. To compute the quotient $\lambda(u)$, one may, of course, use the composition matrix M_u of L_u. The numbers $\lambda(u)$ can also be computed directly on the tree (Exercise 1.67).

For every standard Sturmian sequence s, its directive word x can be identified with an infinite path in the Stern–Brocot tree starting at the root. The slope of s is, by Corollary 1.5.13, the limit of the slopes $\pi(L_u(01))$ for all prefixes u of x.

1.6 Toeplitz Shifts

A sequence $x = (x_n)_{n \in \mathbb{K}}$ with $\mathbb{K} = \mathbb{N}$ or $\mathbb{K} = \mathbb{Z}$ on the alphabet A satisfying for all $n \in \mathbb{K}$,

$$\exists p \geq 0, \forall k \geq 0, \ x_n = x_{n+kp}, \tag{1.39}$$

is called a *Toeplitz sequence*.

For such a two-sided sequence x and $p \geq 1$, we define

$$\mathrm{Per}_p(x) = \{n \in \mathbb{K} \mid x_n = x_{n+kp} \text{ for all } k \geq 0\}.$$

It is clear that x is a Toeplitz sequence if there exists a sequence $(p_n)_{n \geq 0}$ of positive integers such that $\mathbb{K} = \bigcup_{n \geq 0} \mathrm{Per}_{p_n}(x)$. Equivalently, x is a Toeplitz sequence if all finite blocks in x appear periodically. Hence, Toeplitz sequences are uniformly recurrent. We say that p_n is an *essential period* if for any $1 \leq p < p_n$, the sets $\mathrm{Per}_p(x)$ and $\mathrm{Per}_{p_n}(x)$ do not coincide. If the sequence $(p_n)_{n \geq 0}$ is formed by essential periods such that p_n divides p_{n+1} and if $\mathbb{K} = \bigcup_{n \geq 0} \mathrm{Per}_{p_n}(x)$, we call it a *periodic structure* of x. Clearly, if $(p_n)_{n \geq 0}$ is a periodic structure, then $(p_{i_n})_{n \geq 0}$ is also a periodic structure for any strictly increasing sequence of positive integers $(i_n)_{n \geq 0}$. Every Toeplitz sequence has a periodic structure.

A shift space X is a *Toeplitz shift* if it is generated by a Toeplitz sequence $x \in X$. Note that, unless it is periodic, a Toeplitz shift contains elements that are not Toeplitz sequences (Exercise 1.68). Note also that if x is a one-sided Toeplitz sequence, the two-sided shift X such that $\mathcal{L}(X) = \mathcal{L}(x)$ is a Toeplitz shift. Indeed, let (u_n, v_n) be the sequence of pairs of words of equal length and (p_n) be the sequence of integers such that

1. $u_n v_n$ is the prefix of length $2|u_n|$ of x,
2. $u_n v_n$ appears in x with period p_n,
3. $p_n | p_{n+1}$

defined as follows. Start with $u_0 = v_0 = \varepsilon$ and $p_0 = 1$. Assume that (u_n, v_n) and p_n are already constructed. Let $u_{n+1}v_{n+1}$ be the prefix of x of length $2p_n$ with $|u_{n+1}| = |v_{n+1}| = p_n$. Let p_{n+1} be a multiple of p_n such that $u_{n+1}v_{n+1}$ appears in x with period p_{n+1}. The unique two-sided sequence y in $\cap_n [u_n \cdot v_n]$ is a Toeplitz sequence and belongs to X.

Example 1.6.1 Let σ be the morphism $0 \mapsto 01, 1 \mapsto 00$. The substitution shift $X(\sigma)$ is a Toeplitz shift. Indeed, let us show that the admissible one-sided fixed point $x = \sigma^\omega(0)$ is a Toeplitz sequence. It is known as the *period-doubling sequence*, σ as the *period-doubling morphism* and $X(\sigma)$ as the *period-doubling shift*. First, all symbols x_{2n} of even index are equal to 0. This implies that for $k \in \mathbb{Z}$ the blocks $x_{[2k2^N, 2k2^N+2^N)}$ of length 2^N are all equal. Thus, for any $n \geq 1$, let N be such that $n < 2^N$. Then $x_n = x_{n+kp}$ for all $k \in \mathbb{Z}$ with $p = 2^{N+1}$. A period structure is $(2^n)_{n \geq 1}$.

More generally, a substitution σ of constant length n is said to have a *coincidence* at index k for $1 \leq k \leq n$ if the kth letter of every $\sigma(a)$ is the same.

Proposition 1.6.2 *A one-sided fixed point of a constant length substitution having a coincidence at index 1 is a Toeplitz sequence.*

The proof is similar to the one above. The general case of a coincidence at index k is proposed as Exercise 1.71.

There is a constructive way to obtain all Toeplitz sequences. Let A be a finite alphabet and $?$ a letter not in A (usually the symbol $?$ is referred as a "hole"). Let $x \in (A \cup \{?\})^{\mathbb{Z}}$. Given $x, y \in (A \cup \{?\})^{\mathbb{Z}}$, define $F_x(y)$ as the sequence obtained from x replacing consecutively all the $?$ by the symbols of y, where y_0 is placed in the first $?$ to the right of coordinate 0. In particular, if x has no holes, $F_x(y) = x$ for every $y \in (A \cup \{?\})^{\mathbb{Z}}$.

Now, consider a sequence of finite words $(w(n))_{n \geq 0}$ in $A \cup \{?\}$. For each n, let $y(n)$ be the periodic sequence $y(n) = w(n)^\infty$. We define the sequence $(z(n))_{n \geq 0}$ by $z(0) = y(0)$ and, for every $n \geq 1$,

$$z(n + 1) = F_{z(n)}(y(n + 1)). \tag{1.40}$$

It is not complicated to see that $z(n) = u_n^\infty$ for some word u_n of length $|w(1)||w(2)| \cdots |w(n)|$ and that the limit $z = \lim_{n \to \infty} z(n)$ is well defined as a sequence in $(A \cup \{?\})^{\mathbb{Z}}$. Moreover, if the $w(n)$ does not start or finish with a hole for infinitely many n, then the limit sequence belongs to $A^{\mathbb{Z}}$, that is, z has no holes. It is clear that z is a Toeplitz sequence.

The other way round, let $(p_n)_{n \geq 0}$ be a periodic structure of the Toeplitz sequence x. Then for every $n \geq 0$ we can define the *skeleton* of x at scale p_n by $z(n)_m$ equal to x_m if $m \in \mathrm{Per}_{p_n}(x)$ and to ? otherwise. Since $z(n)$ has period p_n and p_n divides p_{n+1}, there exists a periodic sequence $y(n)$ such that (1.40) holds.

We illustrate this constuction on the period-doubling sequence.

Example 1.6.3 Let x be the period-doubling sequence. Consider the periodic sequence $y = (10?0)^\infty$.

The iterated insertion of y in itself gives successively

$$10?0 \; 10?0 \; 10?0 \cdot 10?0 \; 10?0 \; 10?0 \; 10?0 \; 10?0$$
$$1000 \; 10?0 \; 1000 \cdot 1010 \; 1000 \; 10?0 \; 1000 \; 1010$$
$$1000 \; 1000 \; 1000 \cdot 1010 \; 1000 \; 1010 \; 1000 \; 1010,$$

which converges to a sequence that belongs to the shift generated by the period-doubling sequence.

1.7 Exercises

Section 1.1

1.1 Let $\phi \colon X \to Y$ be a continuous map between compact metric spaces X, Y. Show that if ϕ is bijective, its inverse is continuous.

1.2 Let (X, T) be a topological dynamical system. Show that for two non-empty sets $U, V \subset X$ and $n \geq 0$, one has $U \cap T^{-n}V \neq \emptyset$ if and only if $T^n U \cap V \neq \emptyset$.

1.3 Let (X, T) be a topological dynamical system. Show that if $x \in X$ is recurrent, for every open set U containing x, there are infinitely many $n > 0$ such that $T^n x$ belongs to U.

1.4 Prove Proposition 1.1.3. Hint: For (i)$\Rightarrow$ (ii), use the Baire category Theorem (Theorem B.4.2).

1.5 Show that a factor of a minimal system is minimal.

1.6 Show that every nonempty dynamical system (X, T) contains a nonempty invariant set Y such that (Y, T) is minimal.

1.7 Let T_α be the transformation $x \mapsto x + \alpha$ on the torus $\mathbb{T} = \mathbb{R}/\mathbb{Z}$. Show that the system (X, T_α) is minimal if and only if α is irrational.

1.8 Let (X, T) be a topological dynamical system and let $h: X \rightarrow \mathbb{N}$ be a continuous function. The set $X^h = \{(x, i) \mid 0 \leq i < h(x)\}$ is a compact metric space as a closed subset of the product of finitely many copies of X (indeed, since h is continuous on the compact space X, it is bounded). Moreover, the map

$$T^h(x, i) = \begin{cases} (x, i + 1) & \text{if } i + 1 < h(x) \\ (Tx, 0) & \text{otherwise} \end{cases}$$

is continuous. Thus, (X^h, T^h) is a topological dynamical system said to be obtained from X by the *tower construction* relative to h. Show that:

1. T^h is invertible if and only if T is invertible,
2. (X^h, T^h) is minimal if and only if (X, T) is minimal,
3. the system induced on $X \times \{0\}$ is isomorphic to (X, T).

1.9 Let $\phi: \{0, 1\}^{\mathbb{N}} \rightarrow [0, 1]$ be the map defined by $y = \phi(x)$ where $y = \sum_{n \geq 1} x_n 2^{-n}$. Thus, x is a binary expansion of y. Show that ϕ is a morphism from the full shift $(\{0, 1\}^{\mathbb{N}}, S)$ to the dynamical system $([0, 1], T)$ with $T(x) = 2x \bmod 1$.

1.10 Show that a recurrent infinite symbolic system is a Cantor system.

1.11 Let X be a compact metric space and let G be a group (or semigroup) with a continuous map $T_g: X \rightarrow X$ for every $g \in G$ such that $T_{gh} = T_g T_h$. The pair (X, G) is called a G-dynamical system. Show that a dynamical system (X, T) is a G-dynamical system with $G = \mathbb{N}$ and with $G = \mathbb{Z}$ if it is invertible.

1.12 Let (X, T) and (Y, H) be topological dynamical systems with H a semigroup and let $\varphi: x \rightarrow h_x$ be a continuous map from X into H (with the discrete topology on H). Show that the map $U: (x, y) \mapsto (Tx, T_{h_x} y)$ is continuous from $X \times Y$ to itself. The dynamical system $(X \times Y, U)$ is called the *skew product* of (X, T) and (Y, H). Show that the skew product of a shift space with a finite system is conjugate to a shift space.

Section 1.2

1.13 Show that a primitive word of length n has n distinct conjugates.

1.14 Prove the following property of periods of words, known as *Fine–Wilf Theorem*. Let $p, q \geq 1$ be integers and $d = \gcd(p, q)$. If a word w has periods p and q with $|w| \geq p + q - d$, then it has period d.

1.15 Let x be a one-sided infinite sequence. A factor of x is *conservative* if it is not right-special. Let $n \geq 1$ and let c be the number of conservative factors of x of length n. Show that if x has a factor of length $n + c$ whose factors of length n are all conservative, then x is eventually periodic.

1.16 Show that for every factorial extendable set L, there is a unique shift space X such that $\mathcal{L}(X) = L$.

1.17 Show that the Fibonacci sequence is linearly recurrent.

1.18 Let X, X' be two shift spaces. Show that if X' is a factor of X, there is an integer t such that $p_n(X') \leq p_{n+t}(X)$ for all $n \geq 0$. Show that if $r \geq 1$ and $s \geq 0$ are such that $p_n(X) \leq rn + s$ for all $n \geq 0$, there exists $s' \geq 0$ such that $p_n(X') \leq rn + s'$ for all $n \geq 0$.

1.19 Two points x, y of a shift space X are *right-asymptotic* if there are $i, j \in \mathbb{Z}$ such that $(T^i x)^+ = (T^j y)^+$. Show that this defines an equivalence on X. Its classes are the *right-asymptotic components* of X. A component is trivial if it is reduced to one orbit. Show that if $s_n(X) \leq k$, there are at most k non-trivial right-asymptotic components. Show that the number of right-asymptotic components is invariant by conjugacy.

Section 1.3

1.20 Show that a shift space X is of finite type if and only if there is an $n \geq 1$ such that every $v \in \mathcal{L}_n(X)$ satisfies

$$uv, vw \in \mathcal{L}(X) \Rightarrow uvw \in \mathcal{L}(X) \tag{1.41}$$

for every $u, w \in \mathcal{L}(X)$.

1.21 Show that the class of shifts of finite type is closed under conjugacy.

1.22 Prove Proposition 1.3.1.

1.23 Given a finite graph G with edges labeled by letters of an alphabet A, we denote by X_G the set of labels of infinite paths in G. A shift space X on the alphabet A is *sofic* if there is a finite labeled graph G, called a *presentation* of X, such that $X = X_G$. Show that sofic shifts are the factors of shifts of finite type.

1.24 A labeled graph G is *right-resolving* if the edges going out of the same vertex have different labels. A *right-resolving presentation* of a sofic shift is a right-resolving graph G such that $X = X_G$. Show that

1. every sofic shift has a right-resolving presentation,
2. every irreducible sofic shift has a unique minimal strongly connected right-resolving presentation.

Hint: consider the *follower sets* $F(u) = \{v \in \mathcal{L}(X) \mid uv \in \mathcal{L}(X)\}$ for $u \in \mathcal{L}(X)$. If $G = (Q, E)$ is a right-resolving presentation of X, denote $I(u) = \{q \in Q \mid$ there is a path ending at q labeled $u\}$ for $u \in \mathcal{L}(X)$. Show that $I(u) = I(v)$ implies $F(u) = F(v)$.

1.25 Two nonnegative integral square matrices M, N are *elementary equivalent* if there are nonnegative integral matrices U, V such that

$$M = UV, \quad N = VU.$$

The matrices are *strong shift equivalent* if there is a sequence $(M_1, M_2, \ldots, M_k)$ of matrices such that $M_1 = M, M_k = N$ and M_i is elementary equivalent to M_{i+1} for $1 \le i \le k - 1$.

Let M be a nonnegative $n \times n$-matrix. Denote by X_M the edge shift on a graph with adjacency matrix M. Show that if M, N are strong shift equivalent, then X_M, X_N are conjugate.

Section 1.4

1.26 Show that the following conditions are equivalent for a nonerasing morphism $\sigma : A^* \to A^*$.

(i) σ is a substitution.
(ii) The language $\mathcal{L}(\sigma)$ is extendable.
(iii) Every letter $a \in A$ is extendable in $\mathcal{L}(\sigma)$.

1.27 Let $\sigma : A^* \to A^*$ be a morphism. Show that if σ is primitive, then $\mathcal{L}(\sigma^n) = \mathcal{L}(\sigma)$ for every $n \ge 1$. Give an example of a morphism σ such that $\mathcal{L}(\sigma^2)$ is strictly contained in $\mathcal{L}(\sigma)$.

1.28 Let X be the Thue–Morse shift. Show that the complexity $p_n(X)$ is defined by $p_0(X) = 1, p_1(X) = 2, p_2(X) = 4$ and, for $n \ge 2$,

$$p_{n+1}(X) = \begin{cases} 4n - 2^k & \text{if } 2^k \le n < 2^k + 2^{k-1} \\ 4n - 2^k - 2\ell & \text{if } n = 2^k + 2^{k-1} + \ell, \ 0 \le \ell < 2^{k-1}. \end{cases}$$

1.29 Let x be the Thue–Morse sequence. Prove that $x_n = a$ if and only if the number of 1 in the binary expansion of n is even.

1.30 Show that the Thue–Morse shift has two nontrivial right-asymptotic components.

1.31 Let $\varepsilon_n \in \{-1, 1\}$ be the parity of the number of (possibly overlapping) factors 11 in the binary representation of n. The sequence $\varepsilon_0 \varepsilon_1 \varepsilon_2 \cdots$ is the binary *Rudin–Shapiro sequence*. Show that it is the image under the morphism $\phi \colon a \mapsto 1, b \mapsto 1, c \mapsto -1, d \mapsto -1$ of the fixed point x beginning with a of the morphism $\sigma \colon a \mapsto ab, b \mapsto ac, c \mapsto db, d \mapsto dc$, called the *Rudin–Shapiro morphism*.

1.32 Show that the complexity of the shift generated by the Rudin–Shapiro morphism, called the *Rudin–Shapiro shift*, is $p_X(n) = 8(n - 1)$ for $n \geq 2$. Show that, for $n \geq 8$, it is equal to the complexity of the binary Rudin–Shapiro sequence.

1.33 Let $\sigma \colon 0 \mapsto 0010, 1 \mapsto 1$ be the Chacon binary morphism. Show that σ is a substitution and that the *Chacon binary shift* generated by σ is minimal.

1.34 Let $\sigma \colon A^* \to A^*$ be a growing morphism. Show that if the shift space generated by σ is minimal, then σ is primitive.

1.35 Let σ be a morphism of constant length n. Show that n is the maximal eigenvalue of the matrix $M(\sigma)$.

1.36 Let $\sigma \colon A^* \to A^*$ be a morphism and let $M(\sigma)$ be its incidence matrix. Show that if the dominant eigenvalue of $M(\sigma)$ is a Pisot number and its characteristic polynomial is irreducible, then σ is primitive.

1.37 A morphism $\sigma \colon A^* \to C^*$ is *indecomposable* if it cannot be written $\sigma = \alpha \circ \beta$ with $\beta \colon A^* \to B^*, \alpha \colon B^* \to C^*$ and $\mathrm{Card}(B) < \mathrm{Card}(A)$. Prove that an indecomposable morphism defines an injective map from $A^{\mathbb{N}}$ to $B^{\mathbb{N}}$. More precisely, if $\sigma \colon A^* \to C^*$ is not injective on $A^{\mathbb{N}}$, there is a decomposition $\sigma = \alpha \circ \beta$ with $\alpha \colon B^* \to C^*$ and $\beta \colon A^* \to B^*$ such that α is injective on $B^{\mathbb{N}}$, $\mathrm{Card}(B) < \mathrm{Card}(A)$ and every $b \in B$ appears as the first letter of $\beta(a)$ for some $a \in A$.

1.38 Prove that if x is a periodic fixed point of a primitive indecomposable morphism, then every letter $a \in A$ can be followed by at most one letter in x and thus that the period of x is at most $\mathrm{Card}(A)$.

1.39 Prove that if a fixed point x of a primitive morphism φ is periodic, then the period of x is at most $|\varphi|^{\text{Card}(A)-1}$, where $|\varphi| = \max\{|\varphi(a)| \mid a \in A\}$. Conclude that the periodicity of fixed points of morphisms is decidable. Hint: Use Exercise 1.38 and the fact that if $\varphi = \alpha \circ \beta$ and if x is a periodic fixed point of φ, then $y = \beta(x)$ is a periodic fixed point of $\psi = \beta \circ \alpha$ and that the period of x is at most the period of y times $|\alpha| \leq |\varphi|$.

1.40 Prove that a submonoid M of A^* is generated by a code if and only if it satisfies

$$u, uv, vw, w \in M \Rightarrow v \in M \qquad (1.42)$$

for all $u, v, w \in A^*$.

1.41 Let $U \subset A^+$ be a finite set of words. The *flower automaton* of U is the following labeled graph $\mathcal{A}(U) = (Q, E)$. The set Q of vertices of $\mathcal{A}(U)$ is the set of pairs (u, v) of nonempty words such that $uv \in U$ plus the special vertex ω. For every u, a, v with $a \in A$ and $u, v \in A^*$ such that $uav \in U$, there is an edge e labeled a with

(i) $e : (u, av) \mapsto (ua, v)$ if $u, v \neq \varepsilon$,
(ii) $e : (\omega, av) \mapsto (a, v)$ if $u = \varepsilon, v \neq \varepsilon$,
(iii) $e : (u, a) \mapsto (ua, \omega)$ if $u \neq \varepsilon, v = \varepsilon$,
(iv) and finally $e : \omega \mapsto \omega$ if $u = v = \varepsilon$.

Show that the number of paths labeled w from ω to ω is equal to the number of factorizations of w in words of U. Show that U is

1. a code if and only if there is a unique path from p to q labeled w for every $p, q \in Q$ and $w \in A^*$,
2. a circular code if for every nonempty word w there is at most one $p \in Q$ such that there is a cycle labeled w from p to p.

1.42 Let U be a code. A pair (x, y) of words in U^* is *synchronizing* if for every $u, v \in U^*$, one has

$$uxyv \in U^* \Rightarrow ux, yv \in U^*.$$

Note that this definition is coherent with the definition of synchronizing pair given in Section 1.4. Consider the flower automaton $\mathcal{A}(U) = (Q, E)$. Let μ be the morphism from A^* into the monoid of $Q \times Q$-matrices with integer elements defined by

$$\mu(w)_{p,q} = \begin{cases} 1 & \text{if } p \xrightarrow{w} q \\ 0 & \text{otherwise.} \end{cases}$$

Show that a pair (x, y) of words in U^* is synchronizing if and only if $\mu(xy)_{p,q} = \mu(x)_{p,\omega}\mu(y)_{\omega,q}$ for all $p, q \in Q$ and thus $\mu(xy)$ has rank one.

Moreover, if $\mu(x)$, $\mu(y)$ have rank one, then (x, y) is synchronizing.

1.43 A code $U \subset A^+$ is said to have *finite synchronization delay* n (or to be *uniformly synchronized*) if there is an integer n such that every pair x, y of words in U^* of length at least n is synchronizing.

Show that the following conditions are equivalent for a finite code U on the alphabet A.

(i) U is a circular code,
(ii) U has finite synchronization delay,
(iii) every sequence in $A^{\mathbb{Z}}$ has at most one factorization in words of U, that is, for every $x \in A^{\mathbb{Z}}$ there is at most one pair (k, y) with $y \in B^{\mathbb{Z}}$ and $0 \le k < |\varphi(y_0)|$ such that $x = S^k\varphi(y)$.

Hint: for (i) $\Rightarrow$ (ii), use Exercise 1.42.

1.44 Let $\varphi: A^* \to B^*$ be a nonerasing morphism. Let $X = A^{\mathbb{Z}}$, $Y = B^{\mathbb{Z}}$ and $X^\varphi = \{(x, i) \mid x \in A^{\mathbb{Z}}, 0 \le i < |\varphi(x_0)|\}$ as in Section 1.4.4 and let $\varphi: X^\varphi \to B^{\mathbb{Z}}$ be defined by $\hat{\varphi}(x, i) = S^i\varphi(x)$. Show that φ is injective if and only if $\hat{\varphi}$ is *finite-to-one*, that is, for every $y \in Y$, the set $\hat{\varphi}^{-1}(y)$ is finite.

1.45 Let $\varphi: A^* \to A^*$ be a nonerasing morphism with an admissible fixed point $x \in A^{\mathbb{N}}$. Let $f: \mathbb{N} \to \mathbb{N}$ be defined by $f(n) = |\varphi(x_{[0,n)})|$. Show that φ is recognizable if and only if the following condition is satisfied. There is an integer $\ell > 0$ such that whenever $x_{[i-\ell,i+\ell)} = x_{[j-\ell,j+\ell)}$ and $i \in f(\mathbb{N})$, then $j \in f(\mathbb{N})$ and $x_{f^{-1}(i)} = x_{f^{-1}(j)}$.

Section 1.5

1.46 Show that the Fibonacci shift is Sturmian. Hint: show that the left-special words are the prefixes of the $\varphi^n(a)$.

1.47 Let U be a factorial set over $\{0, 1\}$. Show that if U is balanced, then $\text{Card}(U \cap \{0, 1\}^n) \le n + 1$.

1.48 Show that if x is unbalanced, there is a palindrome w such that $0w0$ and $1w1$ belong to $\mathcal{L}(x)$.

1.49 Prove Proposition 1.5.1. Hint: use Exercises 1.47 and 1.48.

1.50 Show that the set of factors of a Sturmian sequence is closed under reversal.

1.51 Prove that the Tribonacci sequence is a strict standard episturmian sequence. Hint: set $\psi(w) = \varphi(w)a$ and show that $\psi^n(a)$ is a palindrome.

1.52 Show that for α, ρ with $0 < \alpha < 1$ and $0 \leq \rho < 1$ with α irrational, the sequence $s = (s_n)_{n \geq 0}$ defined by $s_n = \lfloor (n+1)\alpha + \rho \rfloor - \lfloor n\alpha + \rho \rfloor$ is Sturmian. Moreover, define the *slope* of a word u on $\{0, 1\}$ as the real number $\pi(u) = |u|_1/|u|$. Show that for every factor u of s, the slope of u tends to α when $|u| \to \infty$ (justifying the name of slope for α, see Figure 1.22).

1.53 Show that for every word u, the palindromic closure of u is $u^{(+)} = uv^{-1}\tilde{u}$, where v is the longest palindrome suffix of u.

1.54 Prove that for every $w \in A^*$ and $a \in A$, one has

$$\mathrm{Pal}(wa) = \begin{cases} \mathrm{Pal}(w)a\,\mathrm{Pal}(w) & \text{if } |w|_a = 0 \\ \mathrm{Pal}(w)\,\mathrm{Pal}(w_1)^{-1}\,\mathrm{Pal}(w) & \text{if } w = w_1aw_2 \text{ with } |w_2|_a = 0. \end{cases}$$

1.55 Prove that if p is a palindrome and a is a letter, then $L_a(p)a$ is a palindrome.

1.56 Show that for $a \in A$ and $w \in A^*$, one has

$$\mathrm{Pal}(aw) = L_a(\mathrm{Pal}(w))a. \tag{1.43}$$

1.57 Prove Justin's Formula $\mathrm{Pal}(vw) = L_v(\mathrm{Pal}(w))\,\mathrm{Pal}(v)$.

1.58 Let $s = \mathrm{Pal}(x)$ be a standard episturmian sequence with directive sequence $x = x_0x_1 \cdots$ and let X be the shift generated by s.

Let $u_n = \mathrm{Pal}(x_0x_1 \cdots x_{n-1})$ for $n \geq 1$ be the palindromic prefixes of s. Show that one has

$$\mathcal{R}'_X(u_n) \subset \{L_{x_0 \cdots x_{n-1}}(a) \mid a \in A\} \tag{1.44}$$

with equality if s is strict. Show that for $w \in \mathcal{L}(s)$, one has $\mathcal{R}'_X(w) = y^{-1}\mathcal{R}'_X(u_n)y$, where y is the shortest word such that yw is a prefix of u_n for some $n \geq 1$. Conclude that in an Arnoux–Rauzy shift X, there are $\mathrm{Card}(A)$ return words for every $w \in \mathcal{L}(X)$.

1.59 Show that a minimal shift X on the alphabet A is strict episturmian (that is, Arnoux–Rauzy) if and only if for every $n \geq 1$, there is a unique right-special (resp. left-special) factor w of length n such that $wa \in L(X)$ (resp. $aw \in L(X)$) for every $a \in A$.

1.60 Let $s = \mathrm{Pal}(x)$ for some infinite sequence x. Set $x = x_0 x_1 \cdots$ and let $u_n = \mathrm{Pal}(x_0 \cdots x_{n-1})$. Let u be a factor of s and let $n \geq 0$ be such that u is a factor of u_{n+1} but not of u_n. Set $u_{n+1} = tw\tilde{t}$ where w is the unioccurrent palindrome suffix of $u_n x_n$. Show that

1. The leftmost occurrence of u in u_{n+1} is $u_{n+1} = yuy' = yzwz'y'$ for some words y, y', z, z'.
2. If $u = avb$ with v a palindrome prefix of s, then u is either a suffix of $u_n x_n$ or a prefix of $x_n u_n$.

1.61 Let $s = \mathrm{Pal}(x)$ be a standard strict episturmian sequence with directive sequence $x = a_0 a_1 \cdots$. Set $u_n = \mathrm{Pal}(a_0 \cdots a_{n-1})$. Show that $|u_n|/n$ tends to infinity with n. Hint: use Exercise 1.53 to show that $|u_{n+1}| - |u_n| \geq |u_n| - |u_{n-1}|$ with strict inequality if $a_n \neq a_{n-1}$.

1.62 Let $(d_1, d_2, \ldots)$ be a sequence of integers with $d_1 \geq 0$ and $d_n > 0$ for $n \geq 2$. The *standard sequence* with *directive sequence* $(d_1, d_2, \ldots)$ is the sequence (s_n) of words defined by $s_0 = 0$, $s_1 = 0^{d_1} 1$ and $s_n = s_{n-1}^{d_n} s_{n-2}$ for $n \geq 2$. Show that each s_n is a primitive word which is a prefix of the characteristic sequence of slope $\alpha = [0; 1 + d_1, d_2, \ldots]$.

1.63 Let $(s_n)_{n \geq 0}$ be the standard sequence of words with directive sequence $(d_1, d_2, \ldots)$ and let $\alpha = [0; 1 + d_1, d_2, \ldots]$. Show that for $n \geq 3$, the word $s_n^{1+d_{n+1}}$ is a prefix of the characteristic word c_α but not the word $s_n^{2+d_{n+1}}$.

1.64 A sequence x is said to be *d-power free* if for every nonempty word w, $w^n \in L(x)$ implies $n < d$. Show that if a Sturmian sequence of slope $\alpha = [a_0; a_1, a_2, \ldots]$ is d-power free for some d, the a_i are bounded (note that the converse is also true, see the Notes in Section 1.8). Hint : use Exercise 1.63.

1.65 Show that if the Sturmian sequence of slope $\alpha = [a_0; a_1, \ldots]$ is linearly recurrent, the a_i are bounded.

1.66 Let X be the Sturmian shift of slope α. Let $f : X \to \mathbb{T}$ be the factor map from X onto $(\mathbb{T}, T_\alpha)$ of Theorem 1.5.16. Show that the two sequences t, t' such that $f(t) = f(t') = 0$ are such that $t_n = t'_n$ for all $n \neq -1, 0$.

1.67 Show that the label $\lambda(u)$ of a node in the Stern–Brocot tree can be obtained as follows. The *mediant* of two fractions m/n and m'/n' is the fraction $(m+m')/(n+n')$. Show that the label of a node is the mediant of its nearest ancestors above and to the left m/n and above and to the right m'/n'. The ancestors of a node are those reachable by following the branches upwards. To give ancestors to the root, we give it a left ancestor $0/1$ and a right ancestor denoted $\infty = 1/0$.

Section 1.6

1.68 Show that a nonperiodic Toeplitz shift contains sequences that are not Toeplitz sequences.

1.69 Let $x \in \{0, 1\}^{\mathbb{N}}$ be the period-doubling sequence, which is the fixed point of the substitution $\sigma : 0 \mapsto 01, 1 \mapsto 00$. Show that $x_n = v_2(n + 1) \bmod 2$, where $v_2(m)$ is the number of 0 ending the binary representation of m.

1.70 Show that the period-doubling shift is a factor of the Thue–Morse shift. Hint: use the 2-block map

$$
\gamma(ab) = \begin{cases} b & \text{if } a = 1 \\ 1 - b & \text{otherwise.} \end{cases}
$$

1.71 Let σ be a primitive substitution of constant length n having a coincidence at index k. Then $X(\sigma)$ is a Toeplitz shift.

1.72 Show that the complexity of the period-doubling shift is such that $n \le p_n(X) \le 2n$.

1.73 Show that the 2-odometer is a factor of the period-doubling shift.

1.8 Notes

Dynamical systems occur in many contexts where systems depending on time appear. Such a system can be described by a set X of possible values and a map $t \mapsto x(t)$ giving the value of $x \in X$ at time t. The discretization of time gives rise to systems (X, T) in which $T(x)(n) = x(n + 1)$. In the context of mathematical physics, one associates to each particule in a system of N particules, a vector $x(t)$ recording its position and speed at time t. The map $T : X \to X$ defined by $T(x)(t) = x(t + 1)$ giving the values of these parameters at time $t + 1$ defines a dynamical system. Many other domains give rise to such systems, including linguistics, demography or biology. Natural questions are then: Will every element come back to a position close to its start? With which frequency will it visit a given subset?

The mathematical theory of dynamical systems turned into a theory called *ergodic theory*. The term "ergodic" was introduced by Boltzman, and the cornerstone of the theory is Birkhoff ergodic theorem (see Chapter 3). Topological dynamical systems arose as an independent object of study linked to the names of Birkhoff or Furstenberg. For a more detailed introduction to topological dynamical systems, see, for example, Brown (1976) or Petersen (1983).

1.8.1 Topological Dynamical Systems

Cantor spaces (Section 1.1.4) are a classical object in topology. See Willard (2004) for more details and, in particular, Theorem 30.3 of that book for a proof that all Cantor spaces are homeomorphic.

The mathematical definition of a decidable property (or problem), using Turing machines and due to Turing, can be found in any textbook on the Theory of Computation (see Aho et al. (1974) for example). Its adequacy to describe effectively computable properties is known as the *Church–Turing thesis*. It is worth noting that a famous undecidable problem is closely related to the subject of this book. Indeed, given two morphisms $\alpha, \beta: A^* \to B^*$, the existence of a word w such that $\alpha(w) = \beta(w)$ is an undecidable problem (called *Post correspondence problem*, see Aho et al. (1974)).

The minimality of irrational rotations (Exercise 1.7) is known as the (one-dimensional) Kronecker Theorem (see Hardy and Wright (2008)).

The original reference for Poincaré Recurrence Theorem (Theorem 1.1.10) is Poincaré (1890).

The skew product of dynamical systems (Exercise 1.12) is a classical construction (see Furstenberg (1981)). It is also called a *wreath product* (see Eilenberg (1976)).

1.8.2 Shift Spaces

Shift spaces are the basic object of symbolic dynamics. The classical reference on symbolic dynamics is Lind and Marcus (1995). Many classes of shift spaces (such as shifts of finite type) are described there in much more detail than we provide here. Shift spaces have also been called *symbolic flows* (see Furstenberg (1981) for example).

Section 1.2.1 is a brief introduction to combinatorics on words. For a more detailed exposition, see Lothaire (1997). The combinatorial properties of words are a source for many interesting algorithmic problems (see the recent

(Crochemore et al., 2021) for an appealing presentation of these problems). The original reference to the Curtis–Hedlund–Lyndon Theorem is Hedlund (1969).

The classical reference for the Morse–Hedlund Theorem (Theorem 1.2.14) is Morse and Hedlund (1938). The case of one-sided shifts is considered in Coven and Hedlund (1973). See also Lothaire (2002, Theorem 1.3.13), for example.

Proposition 1.2.17 is from Cassaigne (1997) (see also Berthé and Rigo (2010, Theorem 4.5.4)). As a remarkable result, it was shown by Cassaigne (1996) that if the factor complexity $p_n(X)$ of a shift space X is at most linear (as a function of n), then $s_n(X)$ is bounded.

The Champernowne sequence of Example 1.2.6 is from Champernowne (1933). It is usually referred to as the *Champernowne constant*, which is the real number 0.123456789101112....

The bound $K \leq 15$ given in Exercise 1.17 for the linear recurrence of the Fibonacci word is far from optimal. It is shown in Du et al. (2014), using the Walnut software (Mousavi, 2016) based on decidable properties of automatic sequences, that $K \leq 3$. Actually, one has more precisely $K \leq (3 + \sqrt{5})/2$, this bound being the best possible (Shallit, 2020).

Linearly recurrent sequences form an important class of sequences introduced in Durand et al. (1999). We shall study these sequences in more detail in Chapter 6.

1.8.3 Shifts of Finite Type

Our brief introduction to shifts of finite type (also called *topological Markov chains*) follows (Lind and Marcus, 1995). Sofic shifts (Exercise 1.23) were originally introduced by Benjamin Weiss (1973).

The notion of *right-resolving presentation* (Exercise 1.24) is close to the notion of *deterministic automaton*, which is classical in automata theory (see Berstel et al. (2009)). The unique right-resolving minimal presentation of a sofic shift (Exercise 1.24) is due to Fischer (1975), and it is often called its *Fischer cover*. It is closely related with the notion of *minimal automaton* of a language. It is the labeled graph with vertices the nonempty follower sets $F(u) = \{v \in A^* \mid uv \in L\}$ for $u \in A^*$ and edges $F(u) \xrightarrow{a} F(ua)$ (see Berstel et al. (2009), for example). A language is *recognizable* if its minimal automaton is finite. It follows from the definition of sofic shifts that a shift is sofic if and only if its language is a recognizable language.

The notion of strong shift equivalence (Exercise 1.25) was introduced by Robert Williams (1973). The converse of the statement of Exercise 1.25 is also true, and the equivalence is *Williams Classification Theorem* (Lind and Marcus, 1995, Theorem 7.2.7). The proof of the converse uses the *Decomposition Theorem*, which asserts that every conjugacy between shifts of finite type can be decomposed in elementary conjugacies called *input splits* and *output splits* and their inverses *input merges* and *output merges* (Lind and Marcus, 1995, Theorem 7.1.2). We will define ouput splits in Chapter 6. There is a close connection between strong shift equivalence and another notion called *shift equivalence,* which we will present in Chapter 2 (see Exercise 2.15).

1.8.4 Substitution Shifts

The substitution shifts of Section 1.4 form an important class of shifts, and a good part of this book is focused on this class. For a more detailed treatment, see the classical reference (Queffélec, 2010) or Berthé and Rigo (2010), where, in particular, Theorem 1.4.16 appears.

The terminology concerning substitution shifts admits some variations. What we call purely substitutive sequence is also called a *purely morphic sequence* (see Allouche and Shallit (2003), for example) or a substitutive sequence (as in Queffélec (2010)). Similarly, what we call a substitutive sequence is called a *morphic sequence* in Queffélec (2010), Rigo (2014), and Fogg (2002).

We warn the reader that our definition of a substitution and of a substitution shift differs from that used in Queffélec (2010), where a substitution is a morphism assumed to be growing and right-prolongable on some letter.

Substitutive sequences have also been considered as the result of a parallel rewriting system, often referred to as an *L-system* (see Rozenberg and Salomaa (1980)). The letter L is used as a tribute to Lindenmayer, who introduced these systems as a model for the development of biological organisms (Lindenmayer, 1968). In particular, a *D0L-system* G is given by a morphism $\varphi \colon A^* \to A^*$ and a word $w \in A^*$. The *language generated* by G is the set $L(G) = \{\varphi^n(w) \mid n \geq 1\}$.

The property of the Thue–Morse sequence stated in Exercise 1.29 shows that the Thue–Morse sequence is an *automatic sequence* (a sequence $x \in A^{\mathbb{N}}$ is k-automatic if for every $a \in A$, the set of expansions in base k of the indices n such that $x_n = a$ is a recognizable language). Automatic sequences form a class of substitutive sequences introduced by Cobham (1972) under the name of *uniform tag sequences*. This theory is developed in Allouche and Shallit (2003). Another example of automatic sequence is the period-doubling sequence of Example 1.6.1 (Exercise 1.69). Still another example is the Rudin-Shapiro sequence, also called the *Golay–Rudin–Shapiro sequence*

(Exercise 1.31). It is named after its independent invention by Golay and Shapiro in connection with extremal problems in analysis and in physics, and later by Rudin (1959) (see Allouche and Shallit (2003)).

The Chacon binary shift of Exercise 1.33 is a classical symbolic system (see Fogg (2002)) that we shall meet several times later.

Exercise 1.36 on morphisms with an incidence matrix having a domininant eigenvalue that is a Pisot number is from Canterini and Siegel (2001).

The definition of an indecomposable morphism is due to Ehrenfeucht and Rozenberg (1978). The property stated in Exercise 1.37 is originally due to Linna (1977) (see also Berstel et al. (1979) and Lothaire (2002, Chapter 6)). The decidability of periodicity of fixed points of morphisms (Exercises 1.38 and 1.39) is due to Pansiot (1986) and Harju and Linna (1986) independently (see also Kurka (2003)). It was extended by Durand (2013a) to the decidability of a more general question, with periodicity replaced by eventual periodicity and purely substitutive sequences replaced by substitutive sequences. See Crochemore et al. (2021, Problem 89) for an algorithmic description of this question.

The constant $K = 12$ for the linear recurrrence of the Thue–Morse shift given in Example 1.4.25 is not optimal. It is shown in Schaeffer and Shallit (2012) that the optimal bound is a computable rational number for every constant length substitution. Actually, the optimal bound for the Thue–Morse shift, computed using the Walnut sotware (Mousavi, 2016) is $K = 10$ (Shallit, 2020).

Codes and circular codes are described in detail in Berstel et al. (2009). A submonoid satisfying condition (1.42) is called *stable*. The notions of stable and very pure submonoids were originally introduced by Schützenberger (1955). The flower automaton (Exercise 1.41) is a particular case of *finite automaton*. The uniqueness of paths with given origin, end and label defines the so-called *unambigous automata*. The property of uniform synchronization of finite circular codes (Exercise 1.43) is due to Restivo (1975) (see also Berstel et al. (2009, Theorem 10.2.7)).

The notion of recognizability for morphisms was introduced initially by Martin (1973), and its status remained uncertain during many years. The definition of recognizability given here is from Bezuglyi et al. (2009), who have proved that any (primitive or not) substitution σ is recognizable on $X(\sigma)$ for aperiodic points (on this notion, see Chapter 6). The notion of recognizability for morphisms should not be confused with that of recognizable language (see the definition above).

The basic result on recognizability of morphisms is Mossé's Theorem (Theorem 1.4.35), which is from Mossé (1992, 1996). Our presentation follows Kyriakoglou (2019) where Proposition 1.4.38 is from. The idea of introducing synchronizing pairs to formulate recognizability was initiated by Cassaigne (1994) and pushed forward by Mignosi and Séébold (1993) and Klouda and Starosta (2019). The use of synchronizing pairs avoids relying heavily on a fixed point of the morphism, as the original formulation of the Theorem and subsequent proofs do. The proof of Theorem 1.4.35 presented here, however, follows essentially Kurka (2003).

The finite-to-one maps between shift spaces (and especially shifts of finite type) introduced in Exercise 1.44 are studied in Lind and Marcus (1995) (see also Ashley et al. (1993)).

Several software packages exist for computing with words and morphisms (we have already mentioned Walnut). A library of SageMath (2020) includes many useful functions to handle morphisms. For example, the command

```
phi=WordMorphism({0:[0,1],1:[0]})
phi.language(4)
```

will produce

```
word: 0010, word: 0100, word: 0101, word: 1001,
                    word: 1010,
```

which is the set $\{0010, 0100, 0101, 1001, 1010\}$ of words of length 4 of the language $\mathcal{L}(\varphi)$ where φ is the Fibonacci morphism $0 \mapsto 01, 1 \mapsto 0$.

1.8.5 Sturmian Shifts

The notion of Sturmian shift (and many ideas of symbolic dynamics including the term "symbolic dynamics" itself) was introduced by Morse and Hedlund (1938, 1940). An introduction can be found in Fogg (2002), Lothaire (2002) or Berthé and Rigo (2010). The proof of the equivalent definition of Sturmian sequences using balanced words (Exercise 1.49) is taken from Lothaire (2002).

Arnoux–Rauzy words are named after the paper (Arnoux and Rauzy, 1991), in which they are introduced as a generalization on more than two letters of Sturmian sequences. A reference for episturmian sequences is Droubay et al. (2001), where Theorem 1.5.4 is proved. We follow here the proof of Droubay et al. (2001) using palindrome closure and the notion of Justin word (called property *Ju* in Droubay et al. (2001)). The function Pal has been introduced by de Luca (1997) (see also Reutenauer (2019)). The Justin Formula (Eq. (1.32)) is from Justin and Vuillon (2000). The directive word of a standard episturmian sequence is called in Fogg (2002) the *additive coding sequence*.

Standard sequences (Exercise 1.62) are defined in Lothaire (2002), p. 75.

The statement of 1.64 (with a converse) is Lothaire (2002, Theorem 2.2.31) (see also Berstel (1999), where it is credited to Mignosi (1991). Powers in Sturmian sequences have been extensively studied (see Damanik and Lenz (2002, 2003)). We shall see in Chapter 7 a closely related statement concerning linearly recurrent sequences (Corollary 7.2.7).

The property of Sturmian sequences given in Exercise 1.66 is related to a property of factors of a Sturmian sequence that are consecutive for the lexicographic order (see Perrin and Restivo (2012)).

The Stern–Brocot tree is named after Moritz Stern (1858) and Achille Brocot (1860), who discovered it independently (see Graham et al. (1989)).

A summary of the equivalent characterizations of Sturmian shifts appears in Appendix D.

1.8.6 Toeplitz Shifts

Toeplitz sequences were introduced by Jacobs and Keane (1969) based on a construction of Toeplitz (1928). We refer to Downarowicz (2005) for a survey on Toeplitz shifts (see also Williams (1984); Jacobs and Keane (1969)). The coincidences in substitutions of constant length have been introduced in Dekking (1977/78).

2

Ordered Groups

We now introduce notions concerning abelian groups: ordered abelian groups and direct limits of abelian groups. This will allow us to define dimension groups, which are our main object of interest in this book.

In Section 2.1, we define abelian ordered groups and the notions of positive morphism or unit of an ordered group. In Section 2.3, we introduce direct limits of ordered groups, an essential notion for the following chapters. We come in Section 2.4 to the main focus of this book, that is, dimension groups, as direct limits of groups $\mathbb{Z}^d$ with the usual order. We prove the important theorem of Effros, Handelman and Shen characterizing dimension groups among abelian ordered groups (Theorem 2.4.3). The use of the term "dimension" in the name of dimension groups will be explained in the last chapter (Chapter 9), where the dimension groups are related to the dimensions of some algebras.

2.1 Ordered Abelian Groups

By an *ordered group* we mean an abelian group G with a partial order $\leq$ that is compatible with the group operations, that is, such that for all $g, h \in G$ with $g \leq h$, one has $g + k \leq h + k$ for every $k \in G$. As usual, we write $g < h$ if $g \leq h$ with $g \neq h$, and $g > h$ means $h < g$.

In an ordered group G, the *positive cone* is the set

$$G^+ = \{g \in G \mid g > 0\} \cup \{0\} = \{g \in G \mid g \geq 0\}.$$

It is a *submonoid* of G, that is, it contains 0 and satisfies $G^+ + G^+ \subset G^+$. Moreover, $G^+ \cap (-G^+) = \{0\}$. Indeed let $g \in G^+$. If $-g \in G^+$, then $g \geq 0$ and $0 \geq g$, which implies $g = 0$ since $\leq$ is an order relation.

The set G^+ is not a subgroup but, since G is abelian, the set $G^+ - G^+$ is a subgroup, which is itself an ordered group with the same positive cone as G.

Proposition 2.1.1 *For any pair (G, G^+) formed of an abelian group G and a subset G^+ of G such that*

$$G^+ + G^+ \subset G^+, \; G^+ \cap (-G^+) = \{0\},$$

the relation $g \leq h$ if $h - g \in G^+$ is a partial order compatible with the group operation and such that G^+ is the positive cone.

Proof The first condition on G^+ implies that the partial order $\leq$ is transitive and the second one that it is antisymmetric. For $g, h \in G$ such that $g \leq h$ and $k \in G$, one has $(h + k) - (g + k) = h - g$ and thus $g + k \leq h + k$. ∎

We will often denote an ordered group as a pair (G, G^+) where G^+ is the positive cone of G.

Example 2.1.2 The sets $\mathbb{R}^d$ and $\mathbb{Z}^d$ are, for $d \geq 1$, abelian groups for the componentwise addition

$$(x_1, x_2, \ldots, x_d) + (y_1, y_2, \ldots, y_d) = (x_1 + y_1, x_2 + y_2, \ldots, x_d + y_d),$$

with $\mathbf{0} = (0, \ldots, 0)$ as the neutral element. The pairs $(\mathbb{R}^d, \mathbb{R}_+^d)$ and $(\mathbb{Z}^d, \mathbb{Z}_+^d)$ with $\mathbb{R}_+$ (resp. $\mathbb{Z}_+$) formed of the nonnegative reals (resp. integers), are ordered groups. The corresponding partial order on $\mathbb{R}^d$ is called the *natural order*. It is defined by

$$(x_1, x_2, \ldots, x_d) \leq (y_1, y_2, \ldots, y_d)$$

if $x_i \leq y_i$ for $1 \leq i \leq d$. This partial order is a *lattice order*, which means that every pair x, y has a least upper bound, that is, an element z satisfying $x, y \leq z$ and such that $z \leq z'$ whenever $x, y \leq z'$.

In the next example, the order is a total order.

Example 2.1.3 Let $G = \mathbb{Z}^d$ ordered by the *lexicographic order* defined by $(x_1, x_2, \ldots, x_d) < (y_1, y_2, \ldots, y_d)$ if there is an index i with $1 \leq i \leq d$ such that $x_1 = y_1, \ldots, x_{i-1} = y_{i-1}$ and $x_i < y_i$. Then G^+ is the set

$$\{(x_1, \ldots, x_d) \mid x_1 = \ldots = x_{i-1} = 0 \text{ and } x_i > 0 \text{ for some } i \text{ with } 1 \leq i \leq d\} \cup \{0\}.$$

A *subgroup* of an ordered group (G, G^+) is a pair (H, H^+) where H is a subgroup of G and $H^+ = H \cap G^+$. Such a subgroup is itself an ordered group. Indeed, H^+ is clearly a submonoid and $H^+ \cap (-H^+) \subset G^+ \cap (-G^+)) = \{0\}$. In this way, the order on H is the restriction to $H \times H$ of the order on G.

An ordered group G is *directed* if for every $x, y \in G$ there is some $z \in G$ such that $x, y \leq z$. In other words, G is directed if every pair of elements has a common upper bound.

Example 2.1.4 Let $G = \mathbb{Z}^2$ with the positive cone $G^+ = \{(x_1, x_2) \mid x_1 > 0\} \cup \{(0, 0)\}$. It is a directed group.

In the next statement, we use the fact that if S is a submonoid of an abelian group G, then the set $S - S = \{s - t \mid s, t \in S\}$ is the subgroup generated by S (Exercise 2.1).

Proposition 2.1.5 *An ordered group G is directed if and only if it is generated by the positive cone, that is, if $G = G^+ - G^+$.*

Proof Assume first that G^+ generates G. For $x, y \in G$, there exist $z, t, u, v \in G^+$ such that $x = z - t$ and $y = u - v$. Then $w = z + u$ is such that $x, y \leq w$.

Conversely, for any $x \in G$, considering the pair $0, x$, there is some $y \in G$ such that $0, x \leq y$. Then $x = y - (y - x)$ belongs to $G^+ - G^+$. ∎

Note that a subgroup of a directed group need not be directed, as shown in the next example.

Example 2.1.6 Let $G = \mathbb{Z}^2$ with the positive cone $G^+ = \{(x_1, x_2) \mid x_1 > 0\} \cup \{(0, 0)\}$ as in Example 2.1.4. Then $H = \{0\} \times \mathbb{Z}$ is a subgroup of a directed group. But it is not directed since $H^+ = \{0\}$.

Let (G, G^+) and (H, H^+) be ordered groups. A morphism $\varphi \colon G \to H$ is *positive* if $\varphi(G^+)$ is a subset of H^+. Note that a morphism is positive if and only if it preserves the orders on G, H, that is, $g \leq g'$ implies $\varphi(g) \leq \varphi(g')$.

2.1.1 Ideals and Simple Ordered Groups

An *order ideal* J of an ordered group (G, G^+) is a subgroup J of G such that $J = J^+ - J^+$ (with $J^+ = J \cap G^+$) and such that $0 \leq a \leq b$ with $b \in J^+$ implies $a \in J$.

A *face* in G is a subset F of G^+, which is a submonoid and such that $0 \leq a \leq b$ with $b \in F$ implies $a \in F$.

Proposition 2.1.7 *Let $\mathcal{G} = (G, G^+)$ be an ordered group.*

1. *For every $g \in G^+$, the set*

$$[g] = \{h \in G \mid 0 \leq h \leq ng \text{ for some } n \geq 0\}$$

   is a face.

2. *For every face F, the subgroup $J = F - F$ satisfies $J \cap G^+ = F$ and is the smallest order ideal of G containing F.*

Proof 1. If h, k are in $[g]$, then $h \leq ng$ and $k \leq mg$ for some $n, m \geq 0$. Thus, $h+k \leq (n+m)g$, showing that $h+k$ belongs to $[g]$. Thus, $[g]$ is a submonoid. If $0 \leq h \leq k$ with $n \geq 0$ such that $k \leq ng$, then $0 \leq h \leq ng$ and thus h is in $[g]$. This shows that $[g]$ is a face.

2. The set $J = F - F$ is clearly a subgroup. The set $J^+ = J \cap G^+$ is equal to F. In fact, $F \subset J^+$ by definition. Conversely if $h \in J^+$, set $h = a - b$ with $a, b \in F$. Then $h \leq a$ implies $h \in F$ since F is a face. This shows that J is an ideal. Finally, if K is an ideal containing F, then $J \subset K$ since K is a subgroup. ∎

An ordered group is *simple* if it has no nonzero proper order ideals. Note that a simple group such that $G \neq \{0\}$ is directed since $G^+ - G^+$ is an ideal of (G, G^+), and thus $G^+ - G^+ = G$.

Proposition 2.1.8 *A subgroup of a simple ordered group is simple.*

Proof Let (H, H^+) be a subgroup of the simple group (G, G^+). Let J be a nonzero order ideal of H. Let k be a nonzero element of J^+. Then $F = \{h \in G \mid h \leq nk$ for some $n \geq 0\}$ is a face by Proposition 2.1.7. Thus, by Proposition 2.1.7 again, $K = F - F$ is an ideal and $K \cap G^+ = F$. Since G is simple, we have $K = G$ and also $F = G^+$. Thus, every h belonging to H^+ is in F and consequently in J, which shows that $J = H$. Therefore H is simple. ∎

Example 2.1.9 The ordered groups $\mathbb{Z}, \mathbb{Q}$ and $\mathbb{R}$ (with the natural order) are simple. On the contrary, the ordered group $(\mathbb{Z}^d, \mathbb{Z}_+^d)$ for $d \geq 2$ is not simple. Indeed, for $d = 2$, the set $\mathbb{Z} \times \{0\}$ is an order ideal of $\mathbb{Z}^2$.

By contrast, for every irrational α, the group $\mathbb{Z} + \alpha\mathbb{Z}$, with the order induced by $\mathbb{R}$, is simple since it is a subgroup of $\mathbb{R}$. We shall meet this simple ordered group several times. It is contained in the additive subgroup $\mathbb{Z}[\alpha]$ of $\mathbb{R}$ generated by the powers of α. When α is an *algebraic integer*, that is, such that $p(\alpha) = 0$ for some polynomial $p(x) = x^{k+1} + a_k x^k + \cdots + a_1 x + a_0$ with a_i in $\mathbb{Z}$, the group $\mathbb{Z}[\alpha]$ is generated by $1, \alpha, \ldots, \alpha^k$ and is thus finitely generated. For $k = 2$, the groups $\mathbb{Z} + \alpha\mathbb{Z}$ and $\mathbb{Z}[\alpha]$ coincide (see Appendix B.1).

An *order unit* of the ordered group G is a positive element u such that for every $g \in G^+$, there is an integer $n > 0$ such that $g < nu$. Equivalently, u is an order unit if the set $[u]$ defined in Proposition 2.1.7 is equal to G^+.

A *unital ordered group* is a triple $(G, G^+, 1_G)$ formed of an ordered group (G, G^+) and an order unit 1_G.

Note that if u is a unit of a directed ordered group G, then for every $g \in G$ there is an $n > 0$ such that $g < nu$. Indeed, since G is directed there is h such that $g, u \leq h$. Then $u \leq h$ implies $h \in G^+$. If $n > 0$ is such that $h < nu$, we have $g \leq h < nu$.

Proposition 2.1.10 *A directed ordered group (G, G^+) is simple if and only if every nonzero element of G^+ is an order unit.*

The proof is left as an exercise (Exercise 2.2).

Example 2.1.11 The triples $(\mathbb{R}^d, \mathbb{R}^d_+, 1)$ and $(\mathbb{Z}^d, \mathbb{Z}^d_+, 1)$ are unital ordered groups with order unit $1 = (1, 1, \dots, 1)$.

Let $\mathcal{G} = (G, G^+, 1_G)$ and $\mathcal{H} = (H, H^+, 1_H)$ be unital ordered groups. A *morphism* of a unital ordered group from $\mathcal{G}$ to $\mathcal{H}$ is a group morphism $\varphi \colon G \to H$, which is positive and such that $\varphi(1_G) = 1_H$.

A *subgroup* of a unital ordered group $\mathcal{G} = (G, G^+, u)$ is a unital ordered group $\mathcal{H} = (H, H^+, u)$ such that H is a subgroup of G containing u and $H^+ = H \cap G^+$.

2.1.2 Unperforated Ordered Groups

An ordered group (G, G^+) is *unperforated* if for every $g \in G$, and $n > 0$, if ng belongs to G^+, then g is in G^+. Otherwise, the group is *perforated*.

For example, $\mathbb{Z}^d$ with the natural order is unperforated.

Example 2.1.12 The group $G = \mathbb{Z}$ with positive cone the submonoid G^+ of $\mathbb{Z}$ generated by the set $\{2, 5\}$ is perforated since $9 = 3 \times 3$ belongs to G^+ whereas 3 does not.

A group G is *torsion-free* if for every $g \in G$ and $n > 0$, $ng = 0$ implies $g = 0$.

Proposition 2.1.13 *An unperforated group is torsion-free.*

Proof Suppose that $ng = 0$ for $g \in G$ and $n > 0$. Then $g \in G^+$ since G is unperforated. But $-g = (n - 1)g$ implies $-g \in G^+ \cap (-G^+)$ and thus $g = 0$. ∎

As an example, the group $G = \mathbb{Z} \times \mathbb{Z}/2\mathbb{Z}$ with $G^+ = \{(\alpha, \beta) \in G \mid \alpha > 0\} \cup \{(0, 0)\}$ is a perforated group. Indeed, $2(0, 1) = (0, 0)$, and thus G has torsion.

2.2 States

Let $\mathcal{G} = (G, G^+, \mathbf{1})$ be a unital ordered group. A *state* on $\mathcal{G}$ is a morphism of unital ordered groups from $\mathcal{G}$ to the unital ordered group $(\mathbb{R}, \mathbb{R}_+, 1)$. Thus, a group morphism $p \colon G \to \mathbb{R}$ is a state if $p(g) \geq 0$ for every $g \in G^+$ and $p(\mathbf{1}) = 1$.

Let $S(\mathcal{G})$ denote the set of states of $\mathcal{G}$. It is a convex set called the *state space* of $\mathcal{G}$. Indeed, let $p, q \in S(G)$ and let $t \in [0, 1]$. Then $r = tp + (1 - t)q$ is a morphism since the set of morphisms from a group to $\mathbb{R}$ forms a vector space. It is positive because for every $g \in G^+$, we have $r(g) \geq \min\{p(g), q(g)\} \geq 0$. Moreover, $r(\mathbf{1}) = tp(\mathbf{1}) + (1 - t)q(\mathbf{1}) = 1$. Thus, $tp + (1 - t)q$ belongs to $S(\mathcal{G})$, which shows that $S(\mathcal{G})$ is convex.

Example 2.2.1 Let $\mathcal{G}$ be the group $\mathbb{Z}^d$ with the usual order and $u = (1, \ldots, 1)$. The set $S(\mathcal{G})$ is the $d - 1$ simplex (see Appendix B.5 for the definition) formed of the maps $(\alpha_1, \ldots, \alpha_d) \mapsto p_1\alpha_1 + \cdots + p_d\alpha_d$ for $p_1, \ldots, p_d \geq 0$ of sum 1.

Example 2.2.2 Let λ be irrational. The unital ordered group $G = \mathbb{Z} + \lambda\mathbb{Z}$ (with the order induced by the order on $\mathbb{R}$, and order unit 1) is simple, as we have already seen (Example 2.1.9). There is a unique state that is the identity. Indeed, let p be a state on G. For every $x, y \in \mathbb{Z}$, as $p(1) = 1$, one classically obtains that $p(x) = x$, $p(y) = y$, and thus $p(x + \lambda y) = x + p(\lambda)y$. Moreover, if $x + \lambda y \geq 0$, then $x + p(\lambda)y \geq 0$. This implies $p(\lambda) = \lambda$ and shows that the unique state is the identity.

Example 2.2.3 Let G be a subgroup of $\mathbb{Q}$ with unit 1 and the order induced by $\mathbb{Q}$. Then G has a unique state, which is the identity, since for every state p and $m/n \in G$, we have $np(m/n) = p(m) = mp(1) = m$ and thus $p(m/n) = m/n$. This applies in particular when G is the group of *dyadic rationals* formed of all rational numbers m/n with n a power of 2.

Proposition 2.2.4 *Let* $\mathcal{G} = (G, G^+, \mathbf{1})$ *be a unital directed ordered group. The set* $S(\mathcal{G})$ *is convex and compact for the product topology on* $\mathbb{R}^G$.

Proof For every $g \in G$, we have $g = g' - g''$ with $g', g'' \in G^+$ since G is directed (Proposition 2.1.5). Let $n \geq 1$ be such that $g', g'' \leq n\mathbf{1}$. Then $|p(g)| \leq |p(g')| + |p(g'')| \leq 2n\mathbf{1}$ for every $p \in S(\mathcal{G})$. This implies that $p(g) \in [-n, n]$ for every $p \in S(\mathcal{G})$ and thus that $S(\mathcal{G})$ is compact for the product topology. ∎

We will now prove the following important result.

Theorem 2.2.5 *For every directed unital ordered group $\mathcal{G}$, the set of states on $\mathcal{G}$ is nonempty.*

We first prove the following lemmas.

Lemma 2.2.6 *Let $\mathcal{G} = (G, G^+, u)$ be a unital ordered group and let $\mathcal{H} = (H, H^+, u)$ be a unital ordered subgroup of $\mathcal{G}$. Let p be a state on $\mathcal{H}$. For $g \in G^+$, let*

$$\alpha(g) = \sup\{p(x)/m \mid x \in H, m > 0, x \le mg\}, \qquad (2.1)$$

$$\beta(g) = \inf\{p(y)/n \mid y \in H, n > 0, ng \le y\}. \qquad (2.2)$$

Then

1. $0 \le \alpha(g) \le \beta(g) < \infty$.
2. *If q is a state on $H + \mathbb{Z}g$ that extends p, then $\alpha(g) \le q(g) \le \beta(g)$.*
3. *If $\alpha(g) \le \gamma \le \beta(g)$, there is a state q on $(H + \mathbb{Z}g, H \cap (G^+ + \mathbb{Z}_+g), u)$ such that q extends p and $q(g) = \gamma$.*

Proof 1. Since $0 \le g$, we have by (2.1) with $x = 0$ and $m = 1$, that $\alpha(g) \ge p(0)/1 = 0$. Next, since u is an order unit, we have $g \le ku$ for some $k > 0$. Thus, using (2.2) with $y = ku$ and $n = 1$, we obtain $\beta(g) \le p(ku)/1 = k < \infty$.

Consider $x, y \in H$ and $m, n > 0$ such that $x \le mg$ and $ng \le y$. Then $nx \le mng \le my$, and consequently $p(x)/m \le p(y)/n$. Therefore $\alpha(g) \le \beta(g)$.

2. Let $x \in H$ and $m > 0$ be such that $x \le mg$. Then $p(x) = q(x) \le mq(g)$ and thus $\alpha(g) \le q(g)$. The proof that $q(g) \le \beta(g)$ is similar.

3. We first claim that if $z + kg \ge 0$ for some $z \in H$ and $k \in \mathbb{Z}$, then $p(z) + k\gamma \ge 0$. Indeed, if $k = 0$, then $z \ge 0$ and $p(z) + k\gamma = p(z) \ge 0$. If $k > 0$, then we have $-z \le kg$ with $-z \in H$, whence, by (2.1), $p(-z)/k \le \alpha(g) \le \gamma$ and so $p(z) + k\gamma \ge 0$. Finally, if $k < 0$, we have $-kg \le z$ with $z \in H$ and $-k > 0$, hence $\gamma \le \beta(g) \le p(z)/(-k)$ and so $p(z) + k\gamma \ge 0$. This proves the claim.

As a consequence of the claim, we obtain that if $z + kg = 0$ for some $z \in H$ and $k \in \mathbb{Z}$, then $p(z) + k\gamma = 0$. Indeed, we have both $z + kg \ge 0$, which implies $z + k\gamma \ge 0$, and $-z + (-k)g \ge 0$, which implies $-z + (-k)\gamma \ge 0$. As a consequence, the map $z + kg \mapsto p(z) + k\gamma$ induces a morphism q from $H + \mathbb{Z}g$ to $\mathbb{R}$. Moreover, the claim shows that q is positive. Since $q(g) = \gamma$ and $q(u) = p(u) = 1$, the proof is complete. ∎

Lemma 2.2.7 *Let $\mathcal{G} = (G, G^+, u)$ be a directed unital ordered group and let $\mathcal{H} = (H, H^+, u)$ be a subgroup of $\mathcal{G}$. Every state on $\mathcal{H}$ extends to a state on $\mathcal{G}$.*

Proof Consider the family of pairs $(\mathcal{K}, q)$ of a unital ordered group $\mathcal{K} = (K, K^+, u)$ such that K is a subgroup of G, which contains H with $K^+ = K \cap G^+$, and a state q on $\mathcal{K}$, which extends p. By Zorn's Lemma, this family has a maximal element $(K, q) \in \mathcal{K}$. Suppose that $K \neq G$. Since $\mathcal{G}$ is directed, G^+ generates G, which implies that there is some $g \in G^+ \setminus K$. By Lemma 2.2.6 there is a state q on $K + \mathbb{Z}g$ that extends p. But then $(K + \mathbb{Z}g, q)$ is in $\mathcal{K}$, a contradiction. We conclude that $K = G$ and that q is a state on $\mathcal{G}$ that extends p. ∎

We are now ready for the proof of Theorem 2.2.5.

Proof of Theorem 2.2.5 Set $\mathcal{G} = (G, G^+, u)$. The map $p \colon nu \mapsto n$ is a state on the unital ordered group $\mathcal{H} = (\mathbb{Z}u, \mathbb{Z}_+u, u)$, which is a unital ordered subgroup of $\mathcal{G}$. By Lemma 2.2.7, p extends to a state on G. ∎

Example 2.2.8 Let $\mathcal{G} = (G, G^+, u)$ with $G = \mathbb{Z}^2$, $G^+ = \{(x_1, x_2) \in G \mid x_1 > 0\} \cup \{(0, 0)\}$ and $u = (1, 0)$. There is a unique state that is the projection on the first component.

The following result shows that, for an unperforated simple ordered group, the order is determined by the set of states.

Proposition 2.2.9 *If* $\mathcal{G} = (G, G^+, u)$ *is an unperforated simple unital group, then* $G^+ = \{g \in G \mid p(g) > 0 \text{ for every } p \in S(\mathcal{G})\} \cup \{0\}$.

Proof Since G is simple, every nonzero element of G^+ is an order unit (Proposition 2.1.10). Thus, if g belongs to $G^+ \setminus \{0\}$, there is $n \geq 1$ such that $ng \geq u$. Then, for any $p \in S(\mathcal{G})$ (which is nonempty by Theorem 2.2.5), we have $p(ng) \geq p(u) = 1$ and thus $p(g) > 0$ since $p(ng) = np(g)$. Conversely, if $p(g) > 0$ for every $p \in S(\mathcal{G})$, then since $S(\mathcal{G})$ is closed by Proposition 2.2.4, there is $\varepsilon > 0$ such that $p(g) > \varepsilon$ for every $p \in S(\mathcal{G})$. Let $r, s > 0$ be such that $p(g) > r/s$ for every $p \in S(\mathcal{G})$. Since $p(sg - ru) \geq 0$ for every $p \in S(\mathcal{G})$ and since G is unperforated, we have $(sg - ru) + u \geq 0$ (Exercise 2.10). Thus, $sg \geq (r - 1)u \geq 0$, which implies $g \in G^+$ since G is unperforated. ∎

Example 2.2.10 Let $G = \mathbb{Z}^2$ with $G^+ = \{(x_1, x_2) \mid x_1 > 0\} \cup \{(0, 0)\}$ as in Example 2.2.8. There is a unique state $p \colon (x_1, x_2) \to x_1$. Thus, Proposition 2.2.9 is satisfied. By contrast, let $G = \mathbb{Z}^2$ with the lexicographic order, that is, with $G^+ = \{(x_1, x_2) \mid x_1 > 0 \text{ or } x_1 = 0 \text{ and } x_2 \geq 0\}$. There is only one state that is the projection on the first component. Thus, Proposition 2.2.9 does not hold. There is no contradiction since G is not simple.

2.2.1 Infinitesimals

Let (G, G^+, u) be an unperforated unital ordered group. We say that an element $g \in G$ is *infinitesimal* if $ng \leq u$ for every $n \in \mathbb{Z}$. It is easy to see that the definition does not depend on the choice of the order unit u (Exercise 2.4).

If $G = \mathbb{Z}^d$ with the usual order and unit $\mathbf{1} = (1, 1, \dots, 1)$, there are no non-zero infinitesimals. On the contrary, the following example exhibits nonzero infinitesimals.

Example 2.2.11 Let $G = \mathbb{Z}^2$ with $G^+ = \{(\alpha, \beta) \in G \mid \alpha > 0\} \cup \{(0, 0)\}$ and $u = (1, 0)$ as in Example 2.2.8. Any element $(0, \beta)$ with $\beta \in \mathbb{Z}$ is infinitesimal.

Let us introduce the following useful notation. For $\varepsilon \in \mathbb{Q}$, the inequality $g \leq \varepsilon u$ means that $qg \leq pu$ for some integers $p, q \geq 1$ such that $\varepsilon = p/q$.

Using this notation, one can give, as an equivalent definition, that g is infinitesimal if $-\varepsilon u \leq g \leq \varepsilon u$ for all $\varepsilon > 0$ in $\mathbb{Q}$ (Exercise 2.5).

Another equivalent definition is the following.

Proposition 2.2.12 *Let* $\mathcal{G} = (G, G^+, u)$ *be an unperforated simple unital group. An element* $g \in G$ *is infinitesimal if and only if* $p(g) = 0$ *for all* $p \in S(\mathcal{G})$.

Proof By Theorem 2.2.5, the set $S(\mathcal{G})$ of states is nonempty. Assume first that g is infinitesimal and let $p \in S(\mathcal{G})$. Since $-\varepsilon \leq p(g) \leq \varepsilon$ for every $\varepsilon \in \mathbb{Q}_+$, we conclude that $p(g) = 0$.

Conversely, suppose that $|p(g)| \geq 1/n$ for some trace p. Then $u - ng$ and $u + ng$ cannot be both in G^+. Thus, either $ng \leq u$ or $-ng \leq u$ is false, a contradiction. ∎

The collection of infinitesimal elements of G forms a subgroup, called the *infinitesimal subgroup of G*, which we denote by $\mathrm{Inf}(G)$.

The quotient group $G/\mathrm{Inf}(G)$ of a simple ordered group G is also a simple ordered group for the induced order, and the infinitesimal subgroup of $G/\mathrm{Inf}(G)$ is trivial (see Exercise 2.6). Furthermore, an order unit for G maps to an order unit for $G/\mathrm{Inf}(G)$. Moreover, the state spaces of G and $G/\mathrm{Inf}(G)$ are isomorphic.

It may be interesting to summarize the properties of some of the various orders on $\mathbb{Z}^2$ that we have considered in the examples. The first row of the table in Figure 2.1 concerns the group $\mathbb{Z}^2$ ordered by the first component. It is not directed since G^+ generates $\mathbb{Z} \times \{0\}$. The third row concerns the group of Example 2.1.3. The last one is the group of Example 2.1.6.

G^+	directed	total	simple
$\mathbb{Z}_+ \times \{0\}$	no	no	no
natural	yes	no	no
lexicographic	yes	yes	no
$(\mathbb{Z}_+ \setminus \{0\}) \times \mathbb{Z} \cup \{(0,0)\}$	yes	no	yes

Figure 2.1 The properties of various orders on $\mathbb{Z}^2$.

2.2.2 Image Subgroup

Let $\mathcal{G} = (G, G^+, u)$ be a unital ordered group. The *image subgroup* associated to $\mathcal{G}$ is the subgroup of $(\mathbb{R}, \mathbb{R}_+, 1)$ defined as

$$I(\mathcal{G}) = \bigcap_{p \in S(\mathcal{G})} p(G). \qquad (2.3)$$

When $S(\mathcal{G})$ consists of a unique trace, the group $I(\mathcal{G})$ is isomorphic to $\mathcal{G}/\operatorname{Inf}(\mathcal{G})$.

Example 2.2.13 Let $G = \mathbb{Z}^2$ with $G^+ = \{(\alpha, \beta) \in G \mid \alpha > 0\} \cup \{(0, 0\}$ as in Example 2.2.11. Then the image subgroup is $I(\mathcal{G}) = (\mathbb{R}, \mathbb{R}_+, 1)$.

2.3 Direct Limits

We now introduce the important notion of direct limit, which is central in this book. We will first formulate it for ordinary abelian groups and next for ordered ones.

2.3.1 Direct Limits of Abelian Groups

Let G_n be for each $n \geq 0$ an abelian group and let $i_{n+1,n} \colon G_n \to G_{n+1}$ be for every $n \geq 0$ a morphism. The sets

$$\Delta = \{(g_n)_{n\geq 0} \mid g_n \in G_n, g_{n+1} = i_{n+1,n}(g_n) \text{ for every } n \text{ large enough}\}$$

and

$$\Delta^0 = \{(g_n)_{n\geq 0} \mid g_n \in G_n, g_n = 0 \text{ for every } n \text{ large enough}\}$$

are subgroups of the direct product $\Pi_{n\geq 0} G_n$ and $\Delta^0 \subset \Delta$. Let G be the quotient group $G = \Delta/\Delta^0$ and $\pi \colon \Delta \to G$ be the natural projection. The group G, denoted $G = \varinjlim G_n$, is called the *direct limit* (or *inductive limit*) of the sequence $(G_n)_{n\geq 0}$ with the maps $i_{n+1,n}$. The maps $i_{n+1,n}$ and more generally the maps $i_{m,n} = i_{m,m-1} \circ \cdots \circ i_{n+1,n}$ for $n < m$ are called the *connecting morphisms*.

See Exercise 2.11 for an alternative definition of the direct limit as a quotient of the union of the G_n.

Given $g \in G_n$, all the sequences $(g_k)_{k \geq 0} \in \Pi_{k \geq 0} G_k$ such that

$$g_n = g \text{ and } g_{m+1} = i_{m+1,m}(g_m) \text{ for all } m \geq n$$

belong to Δ and have the same projection, denoted $i_n(g)$ in G. One easily checks that this defines a morphism i_n from G_n to G. It is called the *natural morphism* from G_n into G.

Note that $i_n \colon G_n \to G$ is such that

$$i_n = i_{n+1} \circ i_{n+1,n} \tag{2.4}$$

for every $n \geq 0$. Its kernel is

$$\ker(i_n) = \cup_{m \geq n} \ker(i_{m,m-1} \circ \cdots \circ i_{n+1,n}).$$

The group G is the union of the ranges of the i_n. Thus, for every $g \in G$, there exist an integer $n \geq 0$ and an element $g_n \in G_n$ such that $g = i_n(g_n)$.

Example 2.3.1 Consider the sequence $\mathbb{Z} \xrightarrow{2} \mathbb{Z} \xrightarrow{2} \mathbb{Z} \ldots$ where each map is the multiplication by 2. The direct limit G of this sequence can be identified with the group $\mathbb{Z}[1/2]$ of *dyadic rationals*, formed of all rational numbers p/q with q a power of 2. Indeed, Δ is formed of the sequences $(g_n)_{n \geq 0}$ such that for some $k \geq 1$, one has $g_{n+1} = 2g_n$ for every $n \geq k$. Consider the map $\pi \colon \Delta \to \mathbb{Z}[1/2]$ sending such a sequence on $2^{-k} g_k$. This map is a well-defined group morphism and its kernel is Δ^0. Thus, it induces an isomorphism from G onto $\mathbb{Z}[1/2]$. The natural morphism from $G_n = \mathbb{Z}$ to G is $i_n(g) = 2^{-n} g$.

Direct limits have the following property (called a *universal property*) expressing that the direct limit is, in some sense, the largest possible abelian group.

Proposition 2.3.2 *Let (G_n) be a sequence of abelian groups with connecting morphisms $i_{n+1,n} \colon G_n \to G_{n+1}$. For every abelian group H and every sequence (α_n) of morphisms from G_n to H such that $\alpha_n = \alpha_{n+1} \circ i_{n+1,n}$*

Figure 2.2 The universal property of the direct limit

(see Figure 2.2 on the left), there is a unique morphism φ from the direct limit $G = \lim\limits_{\rightarrow} G_n$ to H such that

$$\alpha_n = \varphi \circ i_n \tag{2.5}$$

for all $n \geq 0$ (see Figure 2.2 on the right).

The verification is left as Exercise 2.12.

2.3.2 Direct Limits of Ordered Groups

Now let $\mathcal{G}_n = (G_n, G_n^+, \mathbf{1}_n)$ be for each $n \geq 0$ a directed unital ordered group and let $i_{n+1,n} \colon G_n \to G_{n+1}$ be for every $n \geq 0$ a morphism of unital ordered groups. Let G be the direct limit of the sequence $(G_n)_{n\geq 0}$, let G^+ be the projection in G of the set

$$\Delta^+ = \{(g_n)_{n\geq 0} \mid g_n \in G_n^+ \text{ for every large enough } n\}$$

and let $\mathbf{1}$ be the projection in G of the sequence $(\mathbf{1}_n)_{n\geq 0} \in \Delta$.

Proposition 2.3.3 *The triple $\mathcal{G} = (G, G^+, \mathbf{1})$ is a directed unital ordered group, every natural morphism $i_n \colon \mathcal{G}_n \to \mathcal{G}$, $n \in \mathbb{N}$, is a morphism of unital ordered groups and G^+ is the union of the $i_n(G^+)$.*

Proof We first verify that G^+ satisfies the two conditions defining an ordered group. First, if g, g' belong to Δ^+, then g_n, g'_n belong to G_n^+ for every n large enough and thus, $g_n + g'_n$ also belong to G_n^+ for every n large enough. Thus, $g+g'$ belongs to Δ^+. This shows that $\Delta^+ + \Delta^+$ is included in Δ^+ and it implies that $G^+ + G^+$ is a subset of G^+. Similarly, the facts that $G = G^+ - G^+$ and $G^+ \cap (-G^+) = \{0\}$ follow from $\Delta = \Delta^+ - \Delta^+$ and $\Delta^+ \cap (-\Delta^+) = \Delta^0$. Thus, $\mathcal{G}$ is directed.

Finally, let us show that $\mathbf{1}$ is an order unit. Let $g = (g_n)_{n\geq 0} \in \Delta$. If $g_{n+1} = i_n(g_n)$ for some $n \geq 0$, and if $g_n \leq k\mathbf{1}_n$ for some $k \geq 1$, then, since $i_{n+1,n}$ is a morphism of ordered groups with order unit, we have $g_{n+1} \leq k\mathbf{1}_{n+1}$. Thus, there is a $k \geq 1$ such that $g_n \leq k\mathbf{1}_n$ for every n large enough. This implies that the projection of g in G is bounded by $k\mathbf{1}$. ∎

The triple $\mathcal{G}$ is called the *direct limit* (or *inductive limit*) of the sequence $(\mathcal{G}_n)$ (see Exercise 2.13 for an alternative definition).

The universal property of direct limits (Proposition 2.3.2) also holds for direct limits of ordered groups.

Example 2.3.4 Consider again the sequence $\mathbb{Z} \xrightarrow{2} \mathbb{Z} \xrightarrow{2} \cdots$ of Example 2.3.1 with $\mathbb{Z}$ ordered as usual. The direct limit of the corresponding sequence of ordered groups is the ordered group $(\mathbb{Z}[1/2], \mathbb{Z}_+[1/2], 1)$, where $\mathbb{Z}_+[1/2]$ is the set of nonnegative dyadic rationals.

2.3.3 Ordered Group of a Matrix

We can generalize Example 2.3.1 by considering $G_n = \mathbb{Z}^d$ for some integer $d \geq 1$, an integer $d \times d$-matrix M and the sequence of morphisms i_n being the multiplication by M on the elements of $\mathbb{Z}^d$ considered as column vectors.

Thus, we address the description of the direct limit of a sequence

$$\mathbb{Z}^d \xrightarrow{M} \mathbb{Z}^d \xrightarrow{M} \mathbb{Z}^d \dots$$

of groups all equal to $\mathbb{Z}^d$ with the same connecting morphisms.

Define the *eventual range* of M as

$$\mathcal{R}_M = \cap_{k \geq 1} M^k \mathbb{Q}^d$$

and the *eventual kernel* of M as

$$\mathcal{K}_M = \cup_{k \geq 1} \ker(M^k).$$

Note that

$$\mathbb{Q}^d = \mathcal{R}_M \oplus \mathcal{K}_M \tag{2.6}$$

and that the multiplication by M defines an automorphism of $\mathcal{R}_M$. Indeed, since

$$\dots \subset M^2 \mathbb{Q}^d \subset M \mathbb{Q}^d \subset \mathbb{R}^d \text{ and } \ker M \subset \ker M^2 \subset \dots$$

there is some $h \geq 0$ such that $\mathcal{R}_M = M^h \mathbb{Q}^d$ and $\mathcal{K}_M = \ker M^h$. Then $M\mathcal{R}_M = M^{h+1} \mathbb{Q}^d = M^h \mathbb{Q}^d = \mathcal{R}_M$ and thus the multiplication by M is an automorphism of $\mathcal{R}_M$. If x is in $\mathcal{R}_M$, then $Mx = 0$ implies $x = 0$. Thus, $\mathcal{R}_M \cap \mathcal{K}_M = \{0\}$. Next, for every $x \in \mathbb{Q}^d$ there is, since $M^h x$ belongs to $\mathcal{R}_M$, some $y \in \mathcal{R}_M$ such that $M^h x = M^h y$. Then $x = y + (x - y)$ belongs to $\mathcal{R}_M + \mathcal{K}_M$. This proves Eq. (2.6)

Let also

$$\Delta_M = \{v \in \mathcal{R}_M \mid M^k v \in \mathbb{Z}^d \text{ for some } k \geq 0\}. \tag{2.7}$$

The following result describes direct limits with identical connecting morphisms $i_{n+1,n}$ for all $n \geq 0$.

Proposition 2.3.5 *For every integer $d \times d$-matrix, the direct limit G of the sequence*

$$\mathbb{Z}^d \xrightarrow{M} \mathbb{Z}^d \xrightarrow{M} \mathbb{Z}^d \dots,$$

where each map is the multiplication by M, *is isomorphic to* Δ_M. *If, moreover, the matrix* M *is nonnegative, the triple* $(\Delta_M, \Delta_M^+, \mathbf{1}_M)$, *where*

$$\Delta_M^+ = \{v \in \mathcal{R}_M \mid M^k v \in \mathbb{Z}_+^d \text{ for every large enough } k \geq 0\}$$

and $\mathbf{1}_M$ *is the projection on* $\mathcal{R}_M$ *along* $\mathcal{K}_M$ *of the vector* $[1 \ 1 \ \dots \ 1]^t$, *is a unital ordered group. If* M *is primitive, the group* $(\Delta_M, \Delta_M^+, \mathbf{1}_M)$ *is simple.*

Proof Let

$$\Delta = \{(x_n)_{n \geq 0} \mid x_n \in \mathbb{Z}^d \text{ for all } n \geq 0 , x_{n+1} = M x_n \text{ for every } n \text{ large enough}\}$$

and

$$\Delta^0 = \{(x_n)_{n \geq 0} \in \Delta \mid x_n = 0 \text{ for every } n \text{ large enough}\}.$$

We have by definition $G = \Delta/\Delta^0$. Let $x \in \Delta$. We may assume, by choosing k large enough, that x_k is in $\mathcal{R}_M$ and $x_{n+1} = M x_n$ for every $n \geq k$. Since the multiplication by M is an automorphism of $\mathcal{R}_M$ there is a unique $y \in \mathcal{R}_M$ such that $M^k y = x_k$. The map $\pi : x \in \Delta \mapsto y \in \Delta_M$ is a well-defined group morphism and its kernel is Δ^0. Thus, π induces an isomorphism from G onto Δ_M. This proves the first statement.

Assume now that M is nonnegative. Let

$$\Delta^+ = \{(x_n)_{n \geq 0} \mid x_n \in \mathbb{Z}_+^d \text{ for every } n \text{ large enough}\}.$$

Since (G, G^+) is an ordered group by Proposition 2.3.3 and since $\pi(\Delta^+) = \Delta_M^+$, the group (Δ_M, Δ_M^+) is an ordered group.

Set $u = [1 \ 1 \ \dots \ 1]^t$ and let v be an element of Δ_M^+. Set $u = u' + w$ with $u' \in \mathcal{R}_M$ and $w \in \mathcal{K}_M$. There is an $n \geq 1$ such that $nu - v \in \mathbb{Q}_+^d$. Then $M^k(nu - v) = M^k(nu' - v)$ is in $\mathbb{Z}_+^d$ for k large enough. Consequently $nu' - v$ belongs to Δ_M^+. Thus, u' is an order unit.

Finally, assume that M is primitive. Let $u, v \in \Delta_M^+$ with u nonzero. There is an integer k such that M^k is strictly positive and $M^k u$, $M^k v$ are also strictly positive. The same argument as above shows that u is an order unit, proving the last statement by Proposition 2.1.10. ∎

The ordered group $(\Delta_M, \Delta_M^+, \mathbf{1}_M)$ is called the *ordered group of the matrix* M.

The following result is very useful.

Proposition 2.3.6 *Let* M *be a primitive matrix. Then*

$$\Delta_M^+ = \{v \in \Delta_M \mid z \cdot v > 0\} \cup \{0\} \tag{2.8}$$

where z *is a positive left eigenvector of* M *for the dominant eigenvalue.*

Proof Let λ be the dominant eigenvalue of M. Assume first that v is in $\Delta_M^+ \setminus \{0\}$. Then $M^k v$ belongs to $\mathbb{Z}_+^d$ for some $k \geq 0$. Since M is primitive, we may assume that all entries of $M^k v$ are positive. By assertion (iii) of the Perron–Frobenius (Theorem B.3.1), the vector $\lim_{n \to +\infty} \lambda^{-n} M^n v = (tz)v = t(z \cdot v)$ where t and z are respectively positive right and left eigenvectors of M relative to λ such that $z \cdot t = 1$. This implies that $z \cdot v$ is strictly positive.

Conversely, if z is strictly positive, the vector $\lim_{n \to +\infty} \lambda^{-n} M^n v$ has all its components positive, and thus there exists $n \geq 0$ such that $M^k v$ is an element of $\mathbb{Z}_+^d$. This shows that v is in Δ_M^+.   ∎

Example 2.3.7 Consider the primitive matrix $M = \begin{bmatrix} 1 & 1 \\ 1 & 0 \end{bmatrix}$. Since M is invertible, we have $\mathcal{R}_M = \mathbb{Q}^2$ and $\Delta_M = \mathbb{Z}^2$. Next, the dominant eigenvalue of M is $\lambda = (1 + \sqrt{5})/2$, and a corresponding row eigenvector is

$$z = \begin{bmatrix} \lambda & 1 \end{bmatrix}$$

(recall that $\lambda^2 = \lambda + 1$). Thus, one has $\Delta_M^+ = \{v \in \mathbb{Z}^2 \mid z \cdot v \geq 0\}$. The map $(\alpha, \beta) \mapsto \lambda \alpha + \beta$ is a positive isomorphism from (Δ_M, Δ_M^+) to the group of algebraic integers $\mathbb{Z}[\lambda] = \mathbb{Z} + \lambda \mathbb{Z}$. The order unit $\mathbf{1}_M = (1, 1)$ is mapped to $\lambda + 1$.

This shows that the direct limit of the sequence $\mathbb{Z} \xrightarrow{M} \mathbb{Z} \xrightarrow{M} \cdots$ is isomorphic to the group of algebraic integers $\mathbb{Z} + \lambda \mathbb{Z}$ with the order induced by the reals and $1 + \lambda$ as ordered unit. One can normalize the order unit to be 1 as follows. The map $x + \lambda y \to (x + \lambda y)/(1 + \lambda)$ is an isomorphism of groups, which sends $\mathbb{Z}[\lambda]$ to itself (because $\lambda^2 = \lambda + 1$ is invertible), and sends $1 + \lambda$ to 1. Thus, we find that the group $(\Delta_M, \Delta_M^+, \mathbf{1}_M)$ is isomorphic to $(\mathbb{Z}[\lambda], \mathbb{Z}[\lambda] \cap \mathbb{R}_+, 1)$.

We next give an example with a nonprimitive matrix.

Example 2.3.8 Consider now $M = \begin{bmatrix} 1 & 1 \\ 0 & 1 \end{bmatrix}$. We have again $\Delta_M = \mathbb{Z}^2$ but this time $\Delta_M^+ = \{(x, y) \mid y > 0\} \cup \{(x, 0) \mid x \geq 0\}$. Thus, we find that the ordered group of the matrix M is $\mathbb{Z}^2$ with the lexicographic order.

Note that there can be nonzero vectors v in Δ_M such that $z \cdot v = 0$ and thus that such a vector cannot be in Δ_M^+ (see Example 2.5.4).

Let $(H, H^+, \mathbf{1})$ be a unital ordered group and, for every $n \geq 1$, let

$$j_n : (G_n, G_n^+, \mathbf{1}_n) \to (H, H^+, \mathbf{1})$$

be a morphism such that $j_{n+1} \circ i_{n+1,n} = j_n$ for every n.

The following result will be used later (see Lemma 4.3.2).

Proposition 2.3.9 *There exists a unique morphism of unital ordered groups* $j \colon (G, G^+, 1_G) \to (H, H^+, 1_H)$ *such that* $j \circ i_n = j_n$, *for every* n, *where* $i_n \colon G_n \to G$ *is the natural morphism into* G *and* $j_n \colon G_n \to H$ *is defined above.*

It is surjective if $\cup_n j_n(G_n) = H$ *and it is injective if* $\ker(j_n) \subset \ker(i_n)$ *for every* n.

Proof Let $g = i_n(g_n) \in G$. Set $j(g) = j_n(g_n)$. Then, by the universal property of direct limits, j is a well-defined morphism from G into H. It is a morphism of unital ordered groups satisfying $j \circ i_n = j_n$.

Conversely, suppose $j \colon (G, G^+, 1_G) \to (H, H^+, 1_H)$ is a morphism of unital ordered groups such that $j \circ i_n = j_n$ for every n. Let g be an element of G. If $g = i_n(g_n)$, then $j(g) = j \circ i_n(g_n) = j_n(g_n)$. This shows the uniqueness.

The last assertions follow easily. ∎

2.4 Dimension Groups

A *dimension group* is a direct limit

$$\mathbb{Z}^{k_1} \overset{M_1}{\to} \mathbb{Z}^{k_2} \overset{M_2}{\to} \mathbb{Z}^{k_3} \cdots$$

of groups $\mathbb{Z}^{k_i}$, with $k_i \geq 1$, ordered in the usual way and with order unit $(1, \ldots, 1)$, with the morphisms defined by nonnegative matrices M_i. Thus, a dimension group is a unital ordered group.

The definition of dimension groups implies some properties of these groups that hold in any group $\mathbb{Z}^d$ with the natural order.

First of all, a dimension group (G, G^+) is unperforated. Let indeed g be a nonzero element of $G = \lim_{\to} \mathbb{Z}^{k_n}$ and assume that ng belongs to G^+ for some $n > 0$. Let $m \geq 1$ be such that $g = i_m(x)$ for $x \in \mathbb{Z}^{k_m}_+$. Then $nx > 0$ implies $x > 0$ and thus g belongs to G^+. Next, dimension groups satisfy the Riesz interpolation property that we introduce now.

2.4.1 Riesz Groups

An ordered group G satisfies the *Riesz interpolation property* if for any x_1, x_2, y_1, y_2 in G such that $x_1 \leq y_1, y_2$ and $x_2 \leq y_1, y_2$, there exists some $z \in G$ such that $x_1, x_2 \leq z \leq y_1, y_2$.

This property is equivalent to the *Riesz decomposition property*, requiring that given $x_1, x_2, y_1, y_2 \in G^+$ if $x_1 + x_2 = y_1 + y_2$, then there exists $z_{ij} \in G^+$ with $1 \leq i, j \leq 2$ such that $x_i = \sum_j z_{ij}$ and $y_j = \sum_i z_{ij}$ (Exercise 2.20).

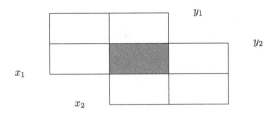

Figure 2.3 The Riesz interpolation property.

As a variant of the interpolation property, we have that for every $x, y_1, \ldots, y_k$ in G^+ such that $x \leq y_1 + \cdots + y_k$, there are $x_1, \ldots, x_k$ in G^+ such that $x = x_1 + \cdots + x_k$ and $x_i \leq y_i$ for $1 \leq i \leq k$ (see Exercise 2.21).

An ordered group G is said to be a *Riesz group* if it satisfies the Riesz interpolation property.

The groups $\mathbb{Z}$ and $\mathbb{R}$ clearly have the Riesz interpolation property as any totally ordered group. More generally, any lattice ordered group is a Riesz group. Next, we have the following more subtle example.

Example 2.4.1 Any dense subgroup of $\mathbb{R}^2$ is a Riesz group.

Indeed, let $x_1, x_2, y_1, y_2 \in \mathbb{R}^2$ with $x_1 \leq y_1, y_2$ and $x_2 \leq y_1, y_2$ as in Figure 2.3. The set of points z such that $x_1, x_2 \leq z \leq y_1, y_2$ is the central rectangle.

Every group $\mathbb{Z}^k$ with the natural order is a Riesz group. This implies that a direct limit of such groups is a Riesz group and therefore that a dimension group is a Riesz group.

Example 2.4.2 Let G be the quotient of $\mathbb{Z}^4$ by the subgroup generated by $(1, 1, -1, -1)$ and the order induced by the natural order. Denote by $[x]$ the projection on G of $x \in \mathbb{Z}^4$. The group G is not a Riesz group. Indeed, one has, with $x = (1, 0, 0, 0)$, $y = (0, 1, 0, 0)$, $z = (0, 0, 1, 0)$ and $t = (0, 0, 0, 1)$ the inequality $[x] \leq [z] + [t]$ since $[x] + [y] = [z] + [t]$. However, $[x]$ cannot be written as a sum of two positive smaller elements and thus the decomposition property fails to hold.

2.4.2 The Effros–Handelman–Shen Theorem

The following important theorem characterizes dimension groups among countable ordered groups.

Theorem 2.4.3 (Effros, Handelman and Shen) *A countable ordered group is a dimension group if and only if it is an unperforated directed Riesz group.*

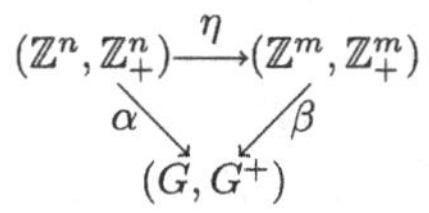

$$(\mathbb{Z}^n, \mathbb{Z}_+^n) \xrightarrow{\quad \eta \quad} (\mathbb{Z}^m, \mathbb{Z}_+^m)$$

$$\alpha \searrow \quad \swarrow \beta$$

$$(G, G^+)$$

Figure 2.4 The diagram of Lemma 2.4.4.

Theorem 2.4.3 gives a much easier way to verify that an ordered group is a dimension group than using the definition, since it does not require us to find an infinite sequence of morphisms. For example, it shows directly that any countable dense subgroup of $\mathbb{R}^2$ is a dimension group since it is an unperforated Riesz group (see Example 2.4.1).

The essential step of the proof is the following lemma. In the proof, we will find it convenient to identify the group $\mathbb{Z}^n$ with the *free abelian group* on a set A with n elements, denoted $\mathbb{Z}(A)$. The elements of $\mathbb{Z}(A)$ have the form $\sum_{a \in A} x_a a$ with $x_a \in \mathbb{Z}$. This amounts to identifying the set A with the canonical basis of $\mathbb{Z}^n$.

The ordered group $(\mathbb{Z}^n, \mathbb{Z}_+^n)$ is then identified with $(\mathbb{Z}(A), \mathbb{Z}(A)^+)$ where $\mathbb{Z}(A)^+$ is the set of sums $\sum_{a \in A} x_a a$ with $x_a \geq 0$.

Lemma 2.4.4 *Let (G, G^+) be an unperforated Riesz group and let $n \geq 1$. Let $\alpha : (\mathbb{Z}^n, \mathbb{Z}_+^n) \to (G, G^+)$ be a morphism and let $x \in \mathbb{Z}^n$ be such that $\alpha(x) = 0$. There is an integer $m \geq 1$, a surjective morphism $\eta : (\mathbb{Z}^n, \mathbb{Z}_+^n) \to (\mathbb{Z}^m, \mathbb{Z}_+^m)$ and a morphism $\beta : (\mathbb{Z}^m, \mathbb{Z}_+^m) \to (G, G^+)$ such that $\eta(x) = 0$ and $\beta \circ \eta = \alpha$ (see Figure 2.4).*

Proof We first remark that if we can prove the statement with a morphism $\eta \colon (\mathbb{Z}^n, \mathbb{Z}_+^n) \to (\mathbb{Z}^m, \mathbb{Z}_+^m)$ that is not surjective, we may replace $(\mathbb{Z}^m, \mathbb{Z}_+^m)$ by the ordered group $\eta(\mathbb{Z}^m, \mathbb{Z}_+^m)$, which is isomorphic to some $(\mathbb{Z}^{m'}, \mathbb{Z}_+^{m'})$ (by Exercise 2.3) and thus η becomes surjective.

We identify as above $\mathbb{Z}^n$ and $\mathbb{Z}(A)$. For $x = \sum_{a \in A} x_a a \in \mathbb{Z}(A)$, set $\|x\| = \max\{|x_a| \mid a \in A\}$ and $m(x) = \mathrm{Card}\{a \in A \mid |x_a| = \|x\|\}$. Let

$$A_+ = \{a \in A \mid x_a > 0\} \text{ and } A_- = \{a \in A \mid x_a < 0\}.$$

We will use an induction on the pairs $(\|x\|, m(x))$ ordered lexicographically. Suppose first that $\|x\| = 0$. Then $x = 0$ and there is nothing to prove.

Assume next that $\|x\| > 0$. Since $\alpha(x) = 0$, we may assume that A_+ and A_- are both nonempty. Changing, if necessary, x into $-x$, we may suppose that $\|x\| = \max\{x_a \mid a \in A_+\}$. Choose $a_0 \in A_+$ such that $x_{a_0} = \|x\|$. Since $\alpha(x) = 0$, we have

$$x_{a_0} \alpha(a_0) \leq \sum_{a \in A_+} x_a \alpha(a) = \sum_{a \in A_-} (-x_a) \alpha(a) \leq x_{a_0} \sum_{a \in A_-} \alpha(a).$$

Since G is unperforated, we derive that $\alpha(a_0) \leq \sum_{a \in A_-} \alpha(a)$. Since G is a Riesz group, there are some $g_a \in G^+$, for each $a \in A_-$, such that $\alpha(a_0) = \sum_{a \in A_-} g_a$ with $g_a \leq \alpha(a)$ for all $a \in A_-$.

Consider the set $B = (A \setminus \{a_0\}) \cup C$ where $C = \{a' \mid a \in A_-\}$ is a copy of A_-. We define two positive morphisms $\eta \colon \mathbb{Z}(A) \to \mathbb{Z}(B)$ and $\beta \colon \mathbb{Z}(B) \to G$ by

$$\eta(a) = \begin{cases} \sum_{a \in A_-} a' & \text{if } a = a_0 \\ a & \text{if } a \in A \setminus (A_- \cup \{a_0\}) \\ a + a' & \text{if } a \in A_- \end{cases}$$

and

$$\beta(b) = \begin{cases} \alpha(b) & \text{if } b \in A \setminus (A_- \cup \{a_0\}) \\ \alpha(b) - g_b & \text{if } b \in A_- \\ g_a & \text{if } b = a' \in C. \end{cases}$$

It is easy to verify that $\alpha = \beta \circ \eta$. Next, we claim that $y = \eta(x)$ is such that

$$(\|y\|, m(y)) < (\|x\|, m(x)).$$

Indeed, we have

$$y = \sum_{a \neq a_0} x_a a + \sum_{a \in A_-} (x_a + x_{a_0}) a'. \tag{2.9}$$

For every $a \in A_-$, we have $-x_{a_0} \leq x_a < 0$ and thus $0 \leq x_a + x_{a_0} < x_{a_0}$. This shows, by inspection of the right-hand side of Eq. (2.9), that $\|\eta(x)\| \leq \|x\|$. In the case of equality, we clearly have fewer terms with maximal absolute value since there is no term a_0. Thus, $m(y) < m(x)$.

This allows us to apply the induction hypothesis to the morphism $\beta \colon (\mathbb{Z}(B), \mathbb{Z}(B)^+) \to (G, G^+)$ and $y \in \mathbb{Z}(B)$. The solution is a pair of morphisms $\eta' \colon \mathbb{Z}(B) \to \mathbb{Z}(B')$ and $\beta' \colon \mathbb{Z}(B') \to G$ such that $\eta'(y) = 0$ with the diagram of Figure 2.5 being commutative. Since $\eta' \circ \eta(x) = 0$, the pair $(\eta' \circ \eta, \beta')$ is a solution. ∎

We prove a second lemma using iteratively the first one.

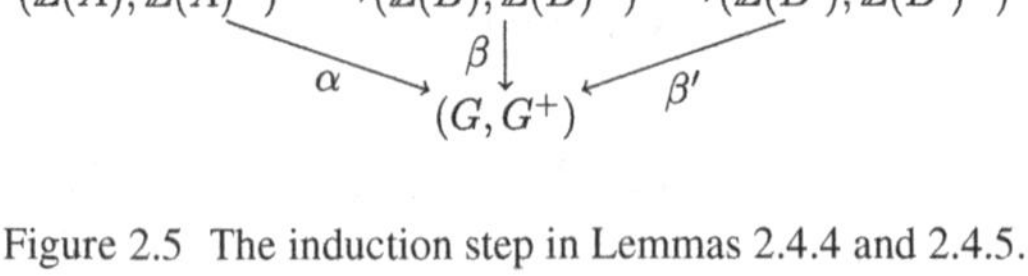

$$(\mathbb{Z}(A), \mathbb{Z}(A)^+) \xrightarrow{\eta} (\mathbb{Z}(B), \mathbb{Z}(B)^+) \xrightarrow{\eta'} (\mathbb{Z}(B'), \mathbb{Z}(B')^+)$$
$$\searrow^{\alpha} \quad \beta\downarrow \quad \swarrow_{\beta'}$$
$$(G, G^+)$$

Figure 2.5 The induction step in Lemmas 2.4.4 and 2.4.5.

Lemma 2.4.5 *Let (G, G^+) be an unperforated Riesz group and let $n \geq 1$. Let $\alpha \colon (\mathbb{Z}^n, \mathbb{Z}^n_+) \to (G, G^+)$ be a morphism. Then there is $m \geq 1$, a surjective morphism $\eta \colon (\mathbb{Z}^n, \mathbb{Z}^n_+) \to (\mathbb{Z}^m, \mathbb{Z}^m_+)$ and a morphism $\beta \colon (\mathbb{Z}^m, \mathbb{Z}^m_+) \to (G, G^+)$ such that $\ker \eta = \ker \alpha$ and $\beta \circ \eta = \alpha$ (see the diagram in Figure 2.4).*

Proof Since $\ker \alpha$ is a subgroup of $\mathbb{Z}^n$, it is finitely generated (see Exercise 2.3). Let $x_1, x_2, \ldots, x_k$ be a set of generators of $\ker \alpha$. We proceed by induction on k. If $k = 0$, there is nothing to prove. Otherwise, by Lemma 2.4.4, we find an integer $m \geq 1$, a surjective morphism $\eta \colon (\mathbb{Z}^n, \mathbb{Z}^n_+) \to (\mathbb{Z}^m, \mathbb{Z}^m_+)$ and a morphism $\beta \colon (\mathbb{Z}^m, \mathbb{Z}^m_+) \to (G, G^+)$ such that $\eta(x_1) = 0$ and $\alpha = \beta \circ \eta$. Then $\ker \beta$ is generated by $\eta(x_2), \ldots, \eta(x_k)$. By the induction hypothesis, there is an integer m', a surjective morphism $\eta' \colon (\mathbb{Z}^m, \mathbb{Z}^m_+) \to (\mathbb{Z}^{m'}, \mathbb{Z}^{m'}_+)$ and a morphism $\beta' \colon (\mathbb{Z}^{m'}, \mathbb{Z}^{m'}_+) \to (G, G^+)$ such that $\ker \eta' = \ker \beta$. Then $\alpha = \beta' \circ \eta' \circ \eta$ (see Figure 2.5) and $\ker \alpha = \ker \eta' \circ \eta$, whence the conclusion. ∎

We prove a third lemma. We will need only one direction of the equivalence but we state the full result.

Lemma 2.4.6 (Shen) *A countable directed ordered group (G, G^+) is a dimension group if and only if for every morphism $\alpha \colon (\mathbb{Z}^n, \mathbb{Z}^n_+) \to G$ there is an integer $m \geq 1$ and two morphisms $\eta \colon (\mathbb{Z}^n, \mathbb{Z}^n_+) \to (\mathbb{Z}^m, \mathbb{Z}^m_+)$ and $\beta \colon (\mathbb{Z}^m, \mathbb{Z}^m_+) \to (G, G^+)$ with η surjective such that $\alpha = \beta \circ \eta$ with $\ker \alpha = \ker \eta$.*

Proof Assume first that $(G, G^+) = \lim_{\to}(\mathbb{Z}^{k_n}, \mathbb{Z}^{k_n}_+)$. Let $\alpha \colon (\mathbb{Z}(A), \mathbb{Z}_+(A)) \to (G, G^+)$ be a morphism. Choosing n large enough, we can find in $\mathbb{Z}^{k_n}_+$ elements x_a such that $i_n(x_a) = \alpha(a)$ for all $a \in A$. Set $\eta(a) = x_a, a \in A$. Let $u_1, \ldots, u_k$ be a set of generators of $\ker \alpha$. Choosing n large enough, we will have $\eta(u_1) = \ldots = \eta(u_k) = 0$ and thus $\ker \alpha = \ker \eta$. Thus, the morphisms η and i_n are a solution.

Let us prove the converse. Since G is countable, G^+ is also countable. Let $S = \{g_0, g_1, \ldots\}$ with $g_n \in G^+$ be a set of generators of G and consider a set $A = \{a_0, a_1, \ldots\}$ in bijection with S.

We are going to build a sequence of finite sets $(A_0, B_0, A_1, B_1, \ldots)$ and morphisms $\alpha_n, \beta_n, \theta_n$ and η_n with $\ker \alpha_n = \ker \eta_n$ and η_n surjective, as in Figure 2.6. Set $A_0 = B_0 = \{a_0\}$ and define $\alpha_0(a_0) = \beta_0(a_0) = g_0$ while η_0 is the identity.

Assume that $A_n, B_n, \beta_n, \eta_n$ are already defined. Set $A_{n+1} = B_n \cup \{a_{n+1}\}$ and let θ_n be the natural inclusion of $\mathbb{Z}(B_n)$ into $\mathbb{Z}(A_{n+1})$.

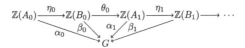

Figure 2.6 The construction of (A_i, B_i).

Define $\alpha_{n+1} \colon \mathbb{Z}(A_{n+1}) \to G$ by

$$\alpha_{n+1}(a) = \begin{cases} g_{n+1} & \text{if } a = a_{n+1} \\ \beta_n(a) & \text{otherwise.} \end{cases}$$

From Lemma 2.4.5 applied to the morphism α_{n+1}, we obtain B_{n+1}, η_{n+1} and β_{n+1}. Thus, the iteration can continue indefinitely (unless S is finite, in which case we stop).

Consider the sequence

$$\mathbb{Z}(B_0) \xrightarrow{\gamma_0} \mathbb{Z}(B_1) \xrightarrow{\gamma_1} \mathbb{Z}(B_2) \dots,$$

with $\gamma_n = \eta_{n+1} \circ \theta_n$, which is obtained by telescoping the top line of Figure 2.6. Let $(H, H^+) = \varinjlim(\mathbb{Z}(B_n), \mathbb{Z}_+(B_n))$ be its direct limit. In case S is finite, we take for H the last $\mathbb{Z}(B_n)$ instead of the direct limit.

Let $i_n \colon \mathbb{Z}(B_n) \to H$ be the natural morphism. By the universal property of direct limits, there is a morphism $h \colon (H, H^+) \to (G, G^+)$ such that $h \circ i_n = \beta_n$.

The morphism h is injective. Indeed, since $\alpha_n = \beta_n \circ \eta_n$ with $\ker \alpha_n = \ker \eta_n$, β_n is injective.

Finally, h is surjective. Indeed, let $g \in S$. Then $g = g_n$ for some n and thus $g = \alpha_n(a_n)$, which implies that g is in the image of h. Since G is directed, it is generated by G^+ and thus by S. Therefore the image of h is G. ∎

The proof of Theorem 2.4.3 is now reduced to a concluding sentence.

Proof of Theorem 2.4.3 We have already seen that a dimension group is a countable directed Riesz group. Conversely, let G be a countable unperforated directed Riesz group. By Lemma 2.4.5 the condition of Shen's Lemma (Lemma 2.4.6) is satisfied. Thus, G is a dimension group. ∎

Let us illustrate the proof on the example of the group $G = \mathbb{Z}[1/2]$ of dyadic rationals. The submonoid G^+ is generated by $S = \{1, 1/2, 1/4, \dots\}$. We start with $A_0 = B_0 = \{a_0\}$ and $\alpha_0(a_0) = 1$. Next, we find $A_1 = \{a_0, a_1\}$ with $\alpha_1(a_1) = 1/2$. Three iterations of the proof of Lemma 2.4.5 give $B_1 = \{b_0\}$

with $\eta(a_0) = 2b_0$ and $\eta(a_1) = b_0$. Continuing in this way (Exercise 2.22), we obtain G as the direct limit of the sequence

$$\mathbb{Z} \xrightarrow{2} \mathbb{Z} \xrightarrow{2} \mathbb{Z} \xrightarrow{2} : ,$$

which was expected (see Example 2.3.1).

2.5 Stationary Systems

Let us consider in more detail the dimension groups obtained as the direct limit of a sequence

$$\mathbb{Z}^d \xrightarrow{M} \mathbb{Z}^d \xrightarrow{M} \mathbb{Z}^d \ldots, \tag{2.10}$$

where at each step the matrix M is a fixed nonnegative integer matrix. Such a sequence is called a *stationary system*.

Recall that the direct limit of a stationary system has been characterized in Proposition 2.3.5 as a triple $(\Delta_M, \Delta_M^+, \mathbf{1}_M)$ called the ordered group of M. This unital ordered group is a dimension group, which has properties closely related to algebraic number theory. Indeed, the largest eigenvalue λ of M is an algebraic integer since it is a root of the polynomial $\det(xI - M)$. Consequently, all components of the corresponding eigenvector are in the algebraic field $\mathbb{Q}[\lambda]$.

We already know by Proposition 2.3.5 that, when M is primitive, the group $\mathcal{G} = (\Delta_M, \Delta_M^+, \mathbf{1}_M)$ is simple. We prove the following additional property. A subset C of $\mathbb{Q}^n$ is a *cone* if $\alpha x \in C$ for every $x \in C$ and $\alpha \in \mathbb{Q}_+$.

Proposition 2.5.1 (Elliott) *A dimension group* $\mathcal{G} = (\Delta_M, \Delta_M^+, \mathbf{1}_M)$ *with* M *primitive has a unique state.*

Proof We have seen in Proposition 2.3.6 that when M is primitive, we have $\Delta_M^+ = \{x \in \Delta_M \mid v \cdot x > 0\} \cup \{0\}$, where v is a row eigenvector of M relative to the maximal eigenvalue. We may further assume that $v \cdot \mathbf{1}_M = 1$. This implies, by Proposition 2.2.9 that $p \colon x \to v \cdot x$ is the unique state of $\mathcal{G}$. Indeed, let q be another state of $\mathcal{G}$. Since every vector of $\mathcal{R}_M$ is colinear to a vector of Δ_M, p and q can be uniquely extended to linear forms on $\mathcal{R}_M$. The set C of $x \in \mathcal{R}_M$ such that $p(x), q(x)$ have opposite signs is a nonempty cone. Every nonempty cone in $\mathbb{Q}^n$ contains integer points. Thus, C contains integer points, which are consequently in Δ_M, a contradiction. ∎

Example 2.5.2 Let M be the primitive matrix

$$M = \begin{bmatrix} 0 & 1 & 1 & 0 \\ 0 & 1 & 0 & 1 \\ 1 & 0 & 1 & 0 \\ 0 & 1 & 1 & 0 \end{bmatrix}.$$

Its eigenvalues are $-1, 0, 1, 2$ and the corresponding eigenvectors are

$$x = \begin{bmatrix} 2 \\ -1 \\ -1 \\ 2 \end{bmatrix}, y = \begin{bmatrix} 1 \\ 1 \\ -1 \\ -1 \end{bmatrix}, z = \begin{bmatrix} 0 \\ 1 \\ -1 \\ 0 \end{bmatrix}, t = \begin{bmatrix} 1 \\ 1 \\ 1 \\ 1 \end{bmatrix}.$$

The eventual range $\mathcal{R}_M$ is the space generated by x, z, t. Thus, we have

$$\Delta_M = \{\alpha x + \beta t + \gamma z \mid \alpha, \beta, \gamma \in \mathbb{Q}, M^k(\alpha x + \beta t + \gamma z) \in \mathbb{Z}^4 \text{ for } k \text{ large enough}\}.$$

The vector $v = \begin{bmatrix} 1/6 & 1/3 & 1/3 & 1/6 \end{bmatrix}$ is a left eigenvector corresponding to the eigenvalue 2 such that $v \cdot x = v \cdot z = 0$ and $v \cdot t = 1$. Thus, the unique state is $\alpha x + \beta t + \gamma z \mapsto \beta$.

We will now suppose that M is *unimodular*. This means that M has determinant ± 1 or, equivalently that M is an element of the group $GL(n, \mathbb{Z})$ of integer matrices with integer inverse. The dominating eigenvalue λ of M is then a unit of the ring $\mathbb{Z}[\lambda]$. Indeed, set $\det(xI - M) = x^d + a_{d-1}x^{d-1} + \cdots + a_1 x + a_0$. Then $\lambda(\lambda^{d-1} + a_{d-1}\lambda^{d-2} + \cdots + a_1) = -a_0$. Since $a_0 = \det(M) = \pm 1$, we conclude that λ is invertible.

Proposition 2.5.3 *Let $M \in GL(n, \mathbb{Z})$ be a primitive unimodular matrix with dominating eigenvalue λ. The group $\mathcal{G}_M / \mathrm{Inf}(\mathcal{G}_M)$ is isomorphic to $\mathbb{Z}[\lambda]$ with order unit the projection of $u = \begin{bmatrix} 1 & 1 & \ldots & 1 \end{bmatrix}^t$ in the quotient.*

Proof Let $p(x)$ be the minimal polynomial of λ. Set $\det(xI - M) = p(x)q(x)$ and let $E = \ker p(M)$, $F = \ker q(M)$. The subgroups E and F are invariant by M. Since p is irreducible and λ has multiplicity 1, p and q are relatively prime. This implies that $\mathbb{Z}^n = E \oplus F$. Indeed, let $a(x), b(x)$ be such that $a(x)p(x) + b(x)q(x) = 1$. For every $w \in \mathbb{Z}^n$, we have $w = a(M)p(M)w + b(M)q(M)w$. Since $p(M)q(M) = 0$, the first term is in F and the second one in E. Let v be a row eigenvector of M corresponding to λ. We have for every $w \in F$,

$$v \cdot w = v \cdot a(M)p(M)w = a(\lambda)p(\lambda)v \cdot w = 0.$$

On the other hand, for every nonzero vector $w \in E$, we have $v \cdot w \neq 0$ since otherwise E would contain an invariant subgroup, in contradiction with the fact that p is irreducible. Thus, F is the infinitesimal subgroup and E is isomorphic with $\mathbb{Z}[\lambda]$ via the map $w \mapsto v \cdot w$. ∎

Example 2.5.4 Let M be the primitive matrix

$$M = \begin{bmatrix} 1 & 1 & 0 \\ 2 & 1 & 1 \\ 0 & 1 & 0 \end{bmatrix}.$$

The eigenvalues of M are $\lambda = (3 + \sqrt{5})/2$, $(3 - \sqrt{5})/2$ and -1, and thus M is unimodular with dominating eigenvalue λ. We have $p(x) = x^2 - 3x + 1$ and $q(x) = x + 1$, and thus $\mathcal{G}_M / \mathrm{Inf}(\mathcal{G}_M)$ is isomorphic to $\mathbb{Z}[\lambda]$. The infinitesimal group $\mathrm{Inf}(\mathcal{G}_M)$ is generated by $[1, -2, 2]^t$, which is an eigenvector for the value -1. A left eigenvector of M for λ is $v = [2\lambda - 2, \lambda, 1]$.

Since $v \cdot u = 3\lambda - 1 = \lambda^2$ is a unit of $\mathbb{Z}[\lambda]$, the map $w \mapsto (1/\lambda^2)v \cdot w$ induces an isomorphism from $\mathcal{G}_M / \mathrm{Inf}(\mathcal{G}_M)$ onto $\mathbb{Z}[\lambda]$ with unit 1.

It is a natural question to ask when the group $\mathcal{G}_M / \mathrm{Inf}(\mathcal{G}_M)$ is isomorphic, as a unital group, to $\mathbb{Z}[\lambda]$ with unit 1. Actually, this happens if and only if $v \cdot u$ is a unit, as in the above example.

2.6 Exercises

Section 2.1

2.1 Let S be a submonoid of an abelian group G. Show that the subgroup generated by S is the set $S - S = \{s - t \mid s, t \in S\}$.

2.2 Show that an ordered group (G, G^+) is simple if and only if every nonzero element of G^+ is an order unit.

2.3 Show that a subgroup of $\mathbb{Z}^n$ can be generated by at most n elements. Hint: use induction on n.

Section 2.2

2.4 Show that the definition of an infinitesimal element in an unperforated ordered group does not depend on the choice of the order unit.

2.5 Let $G = (G, G^+, u)$ be an unperforated unital ordered group. Show that g is infinitesimal if and only if $-\varepsilon u \leq g \leq \varepsilon u$ for every $\varepsilon \in \mathbb{Q}_+$.

2.6 Let $\mathcal{G} = (G, G^+, u)$ be a simple unital group. Denote by $\dot{g}$ the image of g in $G/\operatorname{Inf}(G)$. Show that $G/\operatorname{Inf}(G)$ has a natural ordering defined by $\dot{g} \geq 0$ if $g + h \geq 0$ for some infinitesimal h, that $G/\operatorname{Inf}(G)$ is simple and that the infinitesimal subgroup of $G/\operatorname{Inf}(G)$ is trivial.

2.7 Let $\mathcal{G} = (G, G^+, u)$ be a unital ordered group. For $g \in G^+$, set

$$f_*(g) = \sup\{\varepsilon \in \mathbb{Q}_+ \mid \varepsilon u \leq g\},$$
$$f^*(g) = \inf\{\varepsilon \in \mathbb{Q}_+ \mid g \leq \varepsilon u\}.$$

Show that $\alpha(g) = f_*(g)$ and $\beta(g) = f^*(g)$ where α, β are as in Lemma 2.2.6 with $H = \mathbb{Z}u$.

2.8 Let $\mathcal{G} = (G, G^+, u)$ be a directed unital ordered group. Show that for every $g \in G^+$, one has the following minimax principle:

$$\inf\{p(g) \mid p \in S(\mathcal{G})\} = \sup\{\varepsilon \in \mathbb{Q}_+ \mid \varepsilon u \leq g\}.$$

Hint: use Exercise 2.7.

2.9 Let $\mathcal{G} = (G, G^+, u)$ be a directed unital ordered group. Show that for every $g \in G^+$ there is a state p on $\mathcal{G}$ such that $p(g) = \inf\{\varepsilon \in \mathbb{Q}_+ \mid g \leq \varepsilon u\}$.

2.10 Let $\mathcal{G} = (G, G^+, u)$ be an unperforated directed unital group. Show that for $z \in G$, if one has $p(z) \geq 0$ for every state p of $\mathcal{G}$, then $z + u \geq 0$.

Section 2.3

2.11 Let $(G_n)_{n\geq 0}$ be a sequence of abelian groups with connecting morphisms $i_{n+1,n} \colon G_n \to G_{n+1}$. For $m \leq n$, set $i_{n,m} = i_{n,n-1} \circ \cdots \circ i_{m+1,m}$, with the convention that $i_{n,n}$ is the identity. Let G be the quotient of the disjoint union of the G_n by the equivalence generated by the pairs $(g, i_{n+1,n}(g))$ for all $g \in G_n$ and all $n \geq 0$. Denote by $[g]$ the class of $g \in \cup G_n$.

Show that G is a group for the operation

$$[g + h] = [g + i_{n,m}(h)],$$

where $g \in G_n$ and $h \in G_m$ with $m < n$.

Show that G is isomorphic with the direct limit of the sequence $(G_n)_{n\geq 0}$.

2.12 Prove the universal property of direct limits (Eq. (2.5)).

2.13 Let $(\mathcal{G}_n)_{n\geq 0}$ be a sequence of unital ordered groups $\mathcal{G}_n = (G_n, G_n^+, 1_n)$ with connecting morphisms $i_{n+1,n}$. Let G be the quotient of the disjoint union of the G_n by the equivalence generated by the pairs $(g, i_{n+1,n}(g))$ for all $g \in G_n$ and all $n \geq 0$, as in Exercise 2.11. Let G^+ be the set of classes of the elements of $\cup G_n^+$ and let **1** be the class of 1_0. Show that $(G, G^+, 1)$ is isomorphic to the direct limit of the sequence $\mathcal{G}_n$.

2.14 Let

$$M = \begin{bmatrix} 1 & 1 \\ 1 & 1 \end{bmatrix}.$$

Show that $\Delta_M \sim \mathbb{Z}[1/2]$ and $\Delta_M^+ \sim \mathbb{Z}_+[1/2]$.

2.15 Two integral square matrices M, N are *shift equivalent over* $\mathbb{Z}$ *with lag* ℓ, written $M \sim_{\mathbb{Z}} N$, if there are rectangular integral matrices R, S such that

$$MR = RN, \qquad SM = NS, \tag{2.11}$$

and

$$M^\ell = RS, \qquad N^\ell = SR. \tag{2.12}$$

We denote this situation by $(R, S) : M \sim_{\mathbb{Z}} N$ (lag ℓ). When M, N are nonnegative, the matrices are *shift equivalent*, written $M \sim N$, if R, S can be chosen to be nonnegative integral matrices. We then denote $(R, S) : M \sim N$ (lag ℓ). Show that shift equivalence over $\mathbb{Z}$ (as well as shift equivalence) is an equivalence relation.

2.16 Show that two matrices that are shift equivalent over $\mathbb{Z}$ have the same nonzero eigenvalues.

2.17 For a square integral matrix M, denote by δ_M the restriction of M to Δ_M. Show that $M \sim_{\mathbb{Z}} N$ if and only if $(\Delta_M, \delta_M) \simeq (\Delta_N, \delta_N)$ (the latter means that there is a linear isomorphism $\theta : \Delta_M \to \Delta_N$ such that $\delta_N \circ \theta = \theta \circ \delta_M$).

2.18 Show that two nonnegative integral matrices are shift equivalent if and only if $(\Delta_M, \Delta_M^+, \delta_M) \simeq (\Delta_N, \Delta_N^+, \delta_N)$ (in the sense that there is a linear isomorphism $\theta : \Delta_M \to \Delta_N$ such that $\theta(\Delta_M^+) = \Delta_N^+$ and $\theta : \Delta_M \to \delta_N$ is such that $\delta_N \circ \theta = \theta \circ \delta_M$).

Section 2.4

2.19 A submonoid of an abelian group is *simplicial* if its positive cone is generated, as a monoid, by a finite independent set. Show that an ordered group is isomorphic to $\mathbb{Z}^n$ with the natural order if and only if its positive cone is simplicial.

2.20 Show that an ordered abelian group satisfies the Riesz interpolation property if and only if it satisfies the Riesz decomposition property.

2.21 Let (G, G^+) be an ordered group that satisfies the Riesz interpolation property. Show that for all $x, y_1, \ldots, y_k \in G^+$ such that $x \leq y_1 + y_2 + \cdots + y_k$, there exist $x_1, x_2, \ldots, x_k \in G^+$ such that $x = x_1 + x_2 + \cdots + x_k$ and $x_i \leq y_i$ for $1 \leq i \leq k$.

2.22 Let $G = \mathbb{Z}[1/2]$ be the group of dyadic rationals. Show that the proof of Lemma 2.4.6 gives successively the following values for A_n, B_n and the morphisms $\alpha_n, \beta_n, \eta_n$:

$$A_0 = \{a_0\}. \; A_i = \{b_{i-1}, a_i\} \text{ with } i > 0, \; B_i = \{b_i\},$$

$$\alpha_i(b_{i-1}) = \frac{1}{2^{i-1}}, \alpha_i(a_i) = \frac{1}{2^i},$$

$$\eta_i(b_{i-1}) = 2b_i, \eta_i(a_i) = b_i, \quad \beta_i(b_i) = \frac{1}{2^i}.$$

2.7 Notes

Ordered algebraic structures are a classical subject of which ordered groups (and a fortiori ordered abelian groups) are a particular case. See Fuchs (1963) for a general introduction to ordered groups and Goodearl (1986) for a more detailed presentation of the ordered groups described in this chapter. Many authors assume ordered groups to be directed (see, for example, Putnam (2018)). This simplifies the presentation but has the drawback of complicating the definition of subgroups of ordered groups. We follow the choice of Goodearl and Handelman (1976). This does not make any difference in the sequel since dimension groups are directed.

Note that, by a result of Glass and Madden (1984), the isomorphism of finitely presented lattice ordered groups is undecidable (see also the notes of Chapter 5).

2.7.1 States

The notion of state is closely related with the notion of trace in a C^*-algebra (see Chapter 9). Theorem 2.2.5 is a Hahn–Banach type existence theorem due to Goodearl and Handelman (1976).

Proposition 2.2.9 is Corollary 4.2 in Effros (1981) (it is stated there for a simple dimension group but actually holds in this slightly more general case). The result appears already in Effros et al. (1980), and variants of it appear in Goodearl (1986).

The definition of the infinitesimal group is from Giordano et al. (1995).

2.7.2 Direct Limits

The notion of direct limit is classical and can be formulated for other categories than groups, in particular for algebras, as we shall see in Chapter 9.

The group Δ_M defined by Eq. (2.7) is called in Lind and Marcus (1995) the dimension group of M. It is actually a dimension group, in the sense of the definition given in Section 2.4. Proposition 2.3.5 is essentially Theorem 7.5.13 in Lind and Marcus (1995).

2.7.3 Dimension Groups

The definition of dimension groups (Section 2.3) was introduced by G. Elliott (1976). Simplicial semigroups (Exercise 2.19) are taken from Elliott (1979).

A group with the Riesz (interpolation or decomposition) property was called a *Riesz group* in Fuchs (1965). Some authors add the requirement that the group is unperforated (see Davidson (1996), for example).

The Effros–Handelman–Shen Theorem (Theorem 2.4.3) is from Effros et al. (1980), (see also the expositions in Effros (1981, Theorem 3.1), Davidson (1996, Section IV.7) or Putnam (2018, Chapter 8)). Part of the proof is already in Shen (1979). In particular, Lemma 2.4.5 is Shen (1979, Theorem 3.1) and the argument of Lemma 2.4.6 is already in Elliott (1979).

The *unimodular conjecture* proposed by Effros and Shen (1979) asks whether any finitely generated dimension group G can be obtained as a direct limit

$$\mathbb{Z}^k \xrightarrow{M_1} \mathbb{Z}^k \xrightarrow{M_2} \mathbb{Z}^k \dots,$$

where $k \geq 1$ and all M_n are unimodular matrices. It was proved to hold when G is simple and has one state (Riedel, 1981a) but disproved in the general case (Riedel, 1981b).

2.7.4 Exercises

The minimax principle on ordered groups (Exercise 2.8) is due to Goodearl and Handelman (1976).

Our treatment of shift equivalence (Exercises 2.15, 2.17 and 2.18) follows Lind and Marcus (1995) where the pair (Δ_M, δ_M) is called the *dimension pair* of M and the triple $(\Delta_M, \Delta_M^+, \delta)$ is called the *dimension triple* of M. The statement of Exercise 2.18 can be expressed by the property that the dimension pair is a complete invariant of shift equivalence over $\mathbb{Z}$ and the dimension triple a complete invariant of shift equivalence, a result of Krieger (1980a) (see also Krieger (1977)). Since the dimension triple $(\Delta_M, \Delta_M^+, \delta)$ is invariant by shift equivalence, it is well defined on a shift of finite type X by choosing an edge shift X_M on a graph with adjacency matrix M conjugate to X (see Proposition 1.3.1). The group (Δ_M, Δ_M^+) can actually by defined on more general shift spaces (see Putnam (2014) for the definition on *Smale spaces*).

Note that shift equivalence is decidable; that is, using the notation of Exercise 2.15, one can effectively decide, given two matrices M, N, whether there are U, V and ℓ such that $(U, V) : M \equiv N$ (lag ℓ) (Kim and Roush (1988); see also Kim and Roush (1979)). Strong shift equivalence obviously implies shift equivalence. After remaining many years as a conjecture, known as the Williams Conjecture, the converse was disproved by Kim and Roush (1992), even for primitive matrices (Kim and Roush, 1997). The proof uses notions introduced in Boyle and Krieger (1987). Strong shift equivalence is not known to be decidable or not (see Boyle (2008)).

3

Ordered Cohomology Groups

In this chapter, we define coboundaries and cohomologous functions in the space of continuous integer valued functions on a topological dynamical system. These terms are used in homological algebra and are the dual notion of boundaries and homologous elements. We briefly explain these terms in the elementary setting of graphs.

Assume that $G = (V, E)$ is a graph on a set V of vertices with a set E of edges. Each edge e has its origin $\alpha(e)$ and its end $\omega(e)$. One defines the *boundary* operator $\partial \colon \mathbb{Z}(E) \to \mathbb{Z}(V)$ from the free abelian group on E to the free abelian group on V by $\partial(e) = \omega(e) - \alpha(e)$. The elements of $\mathrm{Im}(\partial)$ are called *boundaries* and the elements of $\ker(\partial)$ are the *cycles*. Elements of $\mathbb{Z}(E)$ equivalent modulo $\ker(\partial)$ are said to be *homologous*.

Identifying $\mathbb{Z}^V$ to $\mathrm{Hom}\,(\mathbb{Z}(V), \mathbb{Z})$, by duality, we have a *coboundary operator*

$$\partial^t \colon \mathbb{Z}^V \to \mathbb{Z}^E$$

(note that it operates in the reverse way). It is such that for $\phi \in \mathbb{Z}^V$,

$$\partial^t \phi(e) = \phi(\omega(e)) - \phi(\alpha(e)).$$

Indeed, by definition of the dual operator, we have

$$\partial^t \phi(e) = \phi(\partial(e)) = \phi(\omega(e) - \alpha(e)) = \phi(\omega(e)) - \phi(\alpha(e)).$$

The elements of $\mathrm{Im}(\partial^t)$ are called *coboundaries* and the elements of $\ker(\partial^t)$ *cocycles*. In what follows, we will choose for V a topological space X, use the graph with edges (x, Tx) for a transformation T on X and use the topological dual $C(X, \mathbb{Z})$ instead of $\mathbb{Z}^V$.

The chapter is organized as follows. In Section 3.1, we define the coboundary operator on continuous functions from a space X to $\mathbb{R}$. We prove in the next section an important result, due to Gottschalk and Hedlund, characterizing coboundaries in a minimal system (Theorem 3.2.5). It is used in the next section (Section 3.3) to define the ordered cohomology group $K^0(X, T)$ of a system (X, T). In Sections 3.6 and 3.7 we relate the group $K^0(X, T)$ to the operations of conjugacy and induction.

In Section 3.8, we introduce invariant Borel probability measures on topological dynamical systems. We recall the basic notions of invariant measure and of ergodic measure. We prove the unique ergodicity of primitive substitution shifts (Theorem 3.8.13). We relate in Section 3.9 coboundaries with invariant Borel probability measures and the notion of state. We prove the important result stating that, for a minimal Cantor system, the states of the dimension group are in one-to-one correspondence with invariant Borel probability measures (Theorem 3.9.3). We finally use this result to give a description of the dimension groups of Sturmian shifts (Theorem 3.9.6).

3.1 Coboundaries

Let (X, T) be a topological dynamical system. We denote by $C(X, \mathbb{R})$ the group of real-valued continuous functions on X and by $C(X, \mathbb{Z})$ the group of integer valued continuous functions. We denote by $C(X, \mathbb{R}_+)$ and $C(X, \mathbb{Z}_+)$ the corresponding sets of nonnegative functions.

As any function with values in a discrete space, an integer valued function f on X is continuous if and only if it is *locally constant*, that is, for every $x \in X$, there is a neighborhood of x on which f is constant. When X is a Cantor space, this neighborhood can be chosen clopen. When X is a space without nontrivial clopen sets, like the torus or any nontrivial closed interval of $\mathbb{R}$, then $C(X, \mathbb{Z})$ consists of constant functions.

Since X is compact, a function $f \in C(X, \mathbb{Z})$ takes only a finite number of values. Indeed, the family $(f^{-1}(\{n\}))_{n \in \mathbb{Z}}$ is a covering of X by open sets, which has a finite subcover. For every $f \in C(X, \mathbb{R})$, we define the *coboundary* of f as the function

$$\partial_T f = f \circ T - f.$$

Clearly, the map $f \mapsto \partial_T f$ is an endomorphism of both groups $C(X, \mathbb{Z})$ and $C(X, \mathbb{R})$. Note that the operator ∂_T is the coboundary operator (as introduced above) related to the graph with X as the set of vertices and edges from each $x \in X$ to Tx.

A function $f \in C(X, \mathbb{R})$ is a *coboundary* if there is a function $g \in C(X, \mathbb{R})$ such that $f = \partial_T g$. Two functions f, f' are *cohomologous* if $f - f'$ is a coboundary.

Note that if $f \in C(X, \mathbb{R})$ is a coboundary, then $f \circ T$ is also a coboundary. Indeed, if $f = g \circ T - g$, then $f \circ T = g \circ T^2 - g \circ T = \partial_T(g \circ T)$.

Example 3.1.1 Let $A = \{a, b\}$, $\delta \in \mathbb{R}$ and $(A^{\mathbb{Z}}, S)$ be the full shift on A. The continuous function f defined, for $x \in A^{\mathbb{Z}}$ by

$$f(x) = \begin{cases} \delta & \text{if } x \in [ab] \\ -\delta & \text{if } x \in [ba] \\ 0 & \text{otherwise} \end{cases}$$

is a coboundary. Indeed, it is the coboundary of any function g defined for $x \in A^{\mathbb{Z}}$ by

$$g(x) = \begin{cases} \alpha & \text{if } x \in [a] \\ \beta & \text{if } x \in [b] \end{cases}$$

for $\beta - \alpha = \delta$.

We now give a natural example of a real-valued continuous function on a Cantor space.

Example 3.1.2 Let $X = \{0, 1\}^{\mathbb{N}}$ be the one-sided full shift on $\{0, 1\}$. To every $x \in X$, we associate the real number

$$f(x) = \sum_{n \geq 0} x_n 2^{-n-1},$$

which is the value of x considered as an expansion $0.x_0 x_1 \ldots$ in base 2. The map f is continuous and we have

$$\partial_S f(x) = \begin{cases} f(x) & \text{if } x \leq 1/2 \\ f(x) - 1 & \text{otherwise.} \end{cases}$$

Proposition 3.1.3 *Let (X, T) be a minimal dynamical system and $f \in C(X, \mathbb{R})$. Then*

1. One has $\partial_T f = 0$ if and only if f is constant.
2. If $f \in C(X, \mathbb{Z})$ is a coboundary, it is the coboundary of some $h \in C(X, \mathbb{Z})$.

Proof 1. Suppose $\partial_T f = 0$. For $c \in \mathbb{R}$, the set $Y = f^{-1}(\{c\})$ is closed. Assume that Y is nonempty. Since $\partial_T f = 0$, the set Y is invariant by T. Hence, (X, T) being minimal, this forces $Y = X$.

2. Assume that $f = \partial_T g$ with $g \in C(X, \mathbb{R})$. Let $\tau \colon \mathbb{R} \to \mathbb{R}/\mathbb{Z}$ be the natural projection onto the torus $\mathbb{T} = \mathbb{R}/\mathbb{Z}$. Since $g \circ T - g$ belongs to $C(X, \mathbb{Z})$, we

have $\tau \circ (g \circ T - g) = 0$ and thus $\partial_T(\tau \circ g) = \tau \circ g \circ T - \tau \circ g = 0$. Since τ is continuous, the same argument as above implies that $\tau \circ g$ is constant. Thus, there exists $c \in \mathbb{R}$ such that $h(x) = g(x) - c$ is an integer for all $x \in X$. Since $\partial_T h = \partial_T g$, we obtain the conclusion. ∎

For $n > 0$, we set

$$f^{(n)} = f + f \circ T + \cdots + f \circ T^{n-1}$$

with $f^{(0)} = 0$. The family $f^{(n)}$ for $n \geq 0$ is called the *cocycle* associated with f. One has for all $n, m \geq 0$, the relation

$$f^{(m+n)} = f^{(m)} + f^{(n)} \circ T^m \tag{3.1}$$

called the *cocycle relation*. Indeed, we have

$$f^{(m+n)} = (f + f \circ T + \cdots + f \circ T^{m-1}) + (f \circ T^m + \cdots + f \circ T^{m+n-1})$$
$$= f^{(m)} + (f + f \circ T + \cdots + f \circ T^{n-1}) \circ T^m = f^{(m)} + f^{(n)} \circ T^m.$$

The following formula will be used often.

Proposition 3.1.4 *Let (X, T) be a dynamical system and $g \in C(X, \mathbb{R})$. If $f = \partial_T g$, then we have for all $n \geq 0$,*

$$f^{(n)} = g \circ T^n - g. \tag{3.2}$$

Proof We have $f^{(n)} = g \circ T - g + g \circ T^2 - g \circ T + \cdots + g \circ T^n - g \circ T^{n-1} = g \circ T^n - g$. ∎

3.2 Gottschalk and Hedlund Theorem

We first prove the following simple property.

Proposition 3.2.1 *Let (X, T) be a topological dynamical system. If $f \in C(X, \mathbb{R})$ is a coboundary, then the sequence $(f^{(n)})$ is bounded uniformly.*

Proof Let $f = \partial_T g$ for some $g \in C(X, \mathbb{R})$, then, using Proposition 3.1.4, one gets $f^{(n)} = g \circ T^n - g$ and thus all $|f^{(n)}|$ are bounded by $2 \sup |g|$. ∎

Proposition 3.2.1 is useful to prove that a continuous function is not a coboundary.

Example 3.2.2 Let $X = \{0, 1\}^{\mathbb{N}}$ and let f be the function defined by $f(x) = x_0$. Then $f^{(n)}(x)$ is the number of symbols 1 in $x_{[0,n-1]}$. Since f is not bounded on X, it is not a coboundary.

Figure 3.1 The shift X.

The following consequence will be used in the next section.

Corollary 3.2.3 *Let (X, T) be a recurrent topological dynamical system. If $f \in C(X, \mathbb{R})$ is a nonnegative coboundary, it is identically zero.*

Proof Let $x_0 \in X$ be a recurrent point. By Proposition 3.2.1, the sequence $f^{(n)}(x_0)$ is bounded. Since x_0 is recurrent, one has $f(T^n x_0) \geq 1/2 \sup f$ for infinitely many values of n. Thus, $f^{(n)}(x_0) \to \infty$ as $n \to \infty$ unless $\sup f = 0$. ∎

The statement is not true if the system is not recurrent, as shown by the following example.

Example 3.2.4 Let X be the shift space such that $\mathcal{L}(X) = a^* b^*$. Thus X is the set of labels of two-sided infinite paths in the graph of Figure 3.1. Let f be the characteristic function of the set of sequences $x \in X$ such that $x_0 = a$. Then $f \circ S$ is the characteristic function of the set of sequences $x \in X$ such that $x_1 = a$. Thus, $-\partial f$ is the characteristic function of the set reduced to the sequence $x \in X$ such that $x_0 = a$ and $x_1 = b$. This function is therefore a nonnegative coboundary that is not zero.

In the minimal case there is a converse of Proposition 3.2.1 by the following result. We will use it several times.

Theorem 3.2.5 (Gottschalk, Hedlund) *Let (X, T) be a minimal topological dynamical system and f be in $C(X, \mathbb{R})$. The following are equivalent.*

1. *f is a coboundary.*
2. *The sequence $(f^{(n)})$ is bounded uniformly.*
3. *There exists $x_0 \in X$ such that the sequence $(f^{(n)}(x_0))_{n \geq 0}$ is bounded.*

Proof We have already seen that (1) implies (2). Assertion (2) clearly implies (3).

The proof that (3) implies (1) is in three steps.

Step 1 For each clopen neighborhood U of x_0, we set

$$\Lambda(U) = \overline{\{f^{(n)}(x_0) \mid n \geq 0, T^n x_0 \in U\}} \text{ and } \Lambda = \bigcap_U \Lambda(U)$$

with U running over all the clopen neighborhoods of x_0. These sets are bounded and contain 0. We claim that $\Lambda = \{0\}$. To prove it, we show that for any $a \in \Lambda$, we have $2a \in \Lambda$, which will imply $a = 0$ since Λ is bounded. Suppose indeed that a belongs to Λ. Let U be a clopen neighborhood of x_0 and $\varepsilon > 0$. Since a belongs to $\Lambda(U)$, there exists $n \geq 0$ such that $T^n x_0 \in U$ and $|f^{(n)}(x_0) - a| < \varepsilon$. As the maps T^n and $f^{(n)}$ are continuous, there is a clopen neighborhood $V \subset U$ of x_0 such that $T^n y$ is in U and $|f^{(n)}(y) - a| < 2\varepsilon$ for all $y \in V$. Since a belongs to $\Lambda(V)$, there exists $m \geq 0$ with $T^m x_0 \in V$ and $|f^{(m)}(x_0) - a| < \varepsilon$. As $T^m x_0$ belongs to V, we have $T^{n+m} x_0 \in U$, $f^{(n+m)}(x_0) \in \Lambda(U)$ and $|f^{(n)} T^m(x_0) - a| < 2\varepsilon$. By the cocycle relation (3.1), we obtain

$$|f^{(n+m)}(x_0) - 2a| \leq |f^{(n)}(T^m x_0) - a| + |f^{(m)}(x_0) - a| < 3\varepsilon.$$

Since ε is arbitrary, this implies that $2a$ belongs to $\Lambda(U)$ and U being an arbitrary neighborhood of x_0, it implies that $2a$ also belongs to Λ, which proves the claim.

Step 2 Because each $\Lambda(U)$ is compact and $\Lambda(U \cap U') \subset \Lambda(U) \cap \Lambda(U')$, for every $\varepsilon > 0$, there exists a neighborhood U_ε of x_0 such that $\Lambda(U_\varepsilon) \subset [-\varepsilon, \varepsilon]$.

We claim that there exists a function $g \in C(X, \mathbb{R})$ such that $g(T^n(x_0)) = f^{(n)}(x_0)$ for all $n \geq 0$. For this, it is enough to prove that for every $x \in X$ and every sequence $n_i \to \infty$ of integers such that $T^{n_i} x_0 \to x$, the sequence $f^{(n_i)}(x_0)$ converges. We then define $g(x)$ as this limit. Fix $\varepsilon > 0$. By minimality, there exists $n \geq 0$ such that $T^n x_0$ is in U_ε.

Let W be a neighborhood of x such that $T^n y \in U_\varepsilon$ and $|f^{(n)}(y) - f^{(n)}(x)| < \varepsilon$ for all $y \in W$. For i large enough, $y = T^{n_i} x_0$ is in W. Then $T^n y = T^{n_i + n} x_0$ belongs to U_ε and consequently $|f^{(n_i + n)}(x_0)| \leq \varepsilon$. Moreover, we have $|f^{(n)}(y) - f^{(n)}(x)| = |f^{(n)}(T^{n_i} x_0) - f^{(n)}(x)| < \varepsilon$. Thus,

$$|f^{(n_i)}(x_0) + f^{(n)}(x)| \leq \varepsilon + |f^{(n_i)}(x_0) + f^{(n)}(T^{n_i} x_0)| = \varepsilon + |f^{(n_i + n)}(x_0)| \leq 2\varepsilon.$$

For large enough n_i, n_j, we obtain

$$|f^{(n_i)}(x_0) - f^{(n_j)}(x_0)| \leq |f^{(n_i)}(x_0) - f^{(n)}(x)| + |f^{(n)}(x) - f^{(n_j)}(x_0)| \leq 4\varepsilon.$$

It follows that $(f^{(n_i)}(x_0))$ is a Cauchy sequence and converges.

Figure 3.2 The golden mean shift.

Step 3 Since (X, T) is minimal, the orbit of x_0 is dense. For any $x = T^n x_0$ in this set, we have by construction

$$\partial_T g(x) = g \circ T^{n+1}(x_0) - g \circ T^n(x_0) = f^{(n+1)}(x_0) - f^{(n)}(x_0)$$
$$= f \circ T^n(x_0) = f(x).$$

By continuity, this extends to any $x \in X$ and this proves that f is a coboundary.   ∎

It can be shown that the function g such that $f = \partial g$ is determined uniquely up to a constant (Exercise 3.8). The following example illustrates the fact that Theorem 3.2.5 is false without the hypothesis of minimality.

Example 3.2.6 Let X be the golden mean shift, which is the set of labels of infinite paths in the graph of Figure 3.2. It is recurrent but not minimal since it contains $a^{\mathbb{Z}}$.

Let f be the characteristic function of the set of points $x \in X$ such that $x_0 = b$ and let $y = a^{\infty}$. Then $f^{(n)}(y) = 0$ for all $n \geq 0$, although f is not a coboundary.

We prove the following additional result. The proof uses the Baire Category Theorem (see Appendix B.4).

Proposition 3.2.7 *Let (X, T) be a minimal topological dynamical system. The following conditions are equivalent for $f \in C(X, \mathbb{Z})$.*

(i) *There exists $g \in C(X, \mathbb{Z})$ such that $f + \partial_T g \geq 0$.*
(ii) *There exists $g \in C(X, \mathbb{R})$ such that $f + \partial_T g \geq 0$.*
(iii) *The family of functions $(f^{(n)})_{n \geq 0}$ is uniformly bounded from below.*
(iv) *For every $x \in X$, the family of numbers $(f^{(n)}(x))_{n \geq 0}$ is bounded from below.*

Proof (i) $\Rightarrow$ (ii) is obvious.

(ii) $\Rightarrow$ (iii) If $f + \partial_T g \geq 0$, then $(f + \partial_T g)^{(n)} = f^{(n)} + (\partial_T g)^{(n)} \geq 0$. Hence, by Eq. (3.2), $f^{(n)} + g \circ T^n - g \geq 0$ and thus $f^{(n)} \geq g - g \circ T^n$ is bounded from below.

(iii) $\Rightarrow$ (iv) is trivial.

(iv) $\Rightarrow$ (i). For each $n \geq 0$, let g_n and g be defined by $g_n(x) = \inf\{f^{(k)}(x) \mid 0 \leq k \leq n\}$ and $g(x) = \inf\{f^{(n)}(x) \mid n \geq 0\} = \inf\{g_n(x) \mid n \geq 0\}$. For each $n \geq 0$ and $k \geq 1$, we obtain $g_{n+k}(x) = \inf\{g_{k-1}(x), f^{(k)}(x) + g_n \circ T^k(x)\}$ and $g(x) = \inf\{g_{k-1}(x), f^{(k)}(x) + g \circ T^k(x)\}$. In particular, $g(x) = \inf\{0, f(x) + g \circ T(x)\} \leq f(x) + g \circ T(x)$, which implies $f + \partial_T g \geq 0$. Because each g_n belongs to $C(X, \mathbb{Z})$, it is sufficient to prove that $g = g_n$ for some $n \geq 0$.

For each $n \geq 0$, let $K_n = \{x \in X \mid g(x) = g_n(x)\}$. For all $x \in X$, as each g_n takes only integer values, there exists an $n \geq 0$ such that $g(x) = g_n(x)$ and thus $x \in K_n$. Consequently, $X = \cup_{n \geq 0} K_n$. Since each K_n is closed, by the Baire Category Theorem, there exists $m \geq 0$ such that K_m has a non-empty interior. By minimality and compactness, there exists $p \geq 1$ such that $\cup_{1 \leq k \leq p} T^{-k} K_m = X$. Let $x \in X$ and $1 \leq k \leq p$ be such that $T^k x \in K_m$. We obtain

$$g(x) = \inf\{g_{k-1}(x), f^{(k)} + g \circ T^k(x)\} = \inf\{g_{k-1}, f^{(k)}(x) + g_m \circ T^k(x)\}$$
$$= g_{k+m}(x)$$

and thus $x \in K_{k+m} \subset K_{p+m}$. We conclude that $K_{p+m} = X$ and $g = g_{p+m}$. ∎

3.3 Ordered Group of a Dynamical System

Let (X, T) be a topological dynamical system. Since the coboundary operator on $C(X, \mathbb{Z})$ is a group morphism, its image $\partial_T C(X, \mathbb{Z})$ is a subgroup. We denote by $H(X, T, \mathbb{Z})$ the quotient group

$$H(X, T, \mathbb{Z}) = C(X, \mathbb{Z})/\partial_T C(X, \mathbb{Z}),$$

and by $H^+(X, T, \mathbb{Z})$ the image of $C(X, \mathbb{Z}_+)$ in this quotient.

We denote by $K^0(X, T)$ the triple

$$K^0(X, T) = (H(X, T, \mathbb{Z}), H^+(X, T, \mathbb{Z}), \mathbf{1}_X),$$

where $\mathbf{1}_X$ is the image in $H(X, T, \mathbb{Z})$ of the constant function with value 1 on X.

The proof of the following result uses Gottschalk and Hedlund Theorem (more precisely Corollary 3.2.3).

Proposition 3.3.1 *For every recurrent topological dynamical system (X, T), the triple $K^0(X, T)$ is a unital ordered group.*

Proof Assume that $f, f' \in C(X, \mathbb{Z}_+)$ are such that f is cohomologous to $-f'$. Then $f + f'$ is a nonnegative coboundary. By Corollary 3.2.3, we

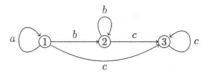

Figure 3.3 The shift X.

have $f + f' = 0$, which implies $f = f' = 0$. Thus, $H^+(X, T, \mathbb{Z}) \cap -H^+ (X, T, \mathbb{Z}) = \{0\}$. Consequently, $K^0(X, T)$ is an ordered group and 1_X is clearly a unit.   ∎

Following the term introduced by Boyle and Handelman, the group $K^0(X, T)$ is called the *ordered cohomology group* of the topological dynamical system (X, T).

Proposition 3.3.1 is not true without the hypothesis that X is recurrent, as illustrated by the following example.

Example 3.3.2 Let X be the shift space such that $\mathcal{L}(X) = a^*b^*c^*$. It is the set of labels of infinite paths in the graph of Figure 3.3.

We have, denoting by χ_U the characteristic function of the set U and by $[u]$ the cylinder $[u] = \{x \in X \mid u = x_{[0,|u|-1]}\}$,

$$\partial \chi_{[a]} = \chi_{[aa]} - \chi_{[a]} = -\chi_{[ab]} - \chi_{[ac]}.$$

Thus, $\chi_{[ab]} + \chi_{[ac]}$ is a coboundary. It can be verified that $\chi_{[ab]}$ is not a coboundary (Exercise 3.12). Thus, the image in $G = H(X, S, \mathbb{Z})$ of $\chi_{[ab]}$ is a nonzero element of $G^+ \cap -G^+$ and thus G is not an ordered group.

We give now a very simple example of computation of $K^0(X, T)$.

Example 3.3.3 Let (X, S) be the shift space formed of the two infinite sequences $x = (\cdots abab.abab \cdots)$ and $y = (\cdots baba.baba \cdots)$. Obviously, $Sx = y$ and $Sy = x$. The characteristic functions of x and y are also exchanged by S. Thus, $H(X, S, \mathbb{Z}) = \mathbb{Z}$ and $K^0(X, S) = (\mathbb{Z}, \mathbb{Z}_+, 1)$.

We can generalize the last example by considering the case of a finite set X and a permutation T on X. Then $H(X, T, \mathbb{Z}) = \mathbb{Z}^d$ where d is the number of orbits of the permutation T. Actually, a recurrent dynamical system (X, T) is periodic if and only if $H(X, T, \mathbb{Z}) = \mathbb{Z}$ (Exercise 3.14).

We now give an elementary argument to show that $H(X, S, \mathbb{Z})$ is isomorphic to $\mathbb{Z}^2$ when (X, S) is a Sturmian shift. We will see later (using return words, in Section 4.4) how this can be done by more general methods.

Example 3.3.4 Let X be a Sturmian shift on $A = \{a, b\}$. We denote by $\chi_{[w]}$ the characteristic function of the cylinder set $[w]$.

Set $A = \{a, b\}$. We show by induction on $|w|$ that $\chi_{[w]}$ is cohomologous to an element of the subgroup G generated by $\chi_{[a]}$ and $\chi_{[b]}$. This is true if $|w| = 1$. Assume that it holds for words of length n and consider a word w of length $n + 1$. Then $w = ux$ for some nonempty word u of length n and some letter $x \in \{a, b\}$. If u is not right-special, we have $uA^{\mathbb{N}} = uxA^{\mathbb{N}}$ and thus $\chi_{[ux]}$ is cohomologous to an element of G by the induction hypothesis. Otherwise, we have

$$\chi_{[u]} = \chi_{[ua]} + \chi_{[ub]}. \tag{3.3}$$

As w is either ua or ub, it suffices to show that $\chi_{[ua]}$ and $\chi_{[ub]}$ are cohomologuous to some elements in G. Set $u = yv$ with y a letter and v a word. Then va and vb cannot be both left-special. Assume that va is not left-special. Then $\chi_{[ua]} = \chi_{[yva]} = \chi_{[va]} \circ T$. By the induction hypothesis, $\chi_{[va]}$ is cohomologous to an element of G and thus also the map $\chi_{[ua]}$. By Eq. (3.3), this implies that $\chi_{[ub]}$ is also cohomologous to an element of G.

Since $C(X, \mathbb{Z})$ is generated by the functions $\chi_{[w]}$ (see Proposition 3.4.1), this shows that $H(X, S, \mathbb{Z})$ is generated by the projections of $\chi_{[a]}$ and $\chi_{[b]}$. Thus, the morphism sending (α, β) to the class of $\alpha\chi_{[a]} + \beta\chi_{[b]}$ is surjective from $\mathbb{Z}^2$ to $H(X, S, \mathbb{Z})$. It is injective because if $f = \alpha\chi_{[a]} + \beta\chi_{[b]}$ with α or β nonzero, then $f^{(n)}(x)$ is not bounded because otherwise the slope of x would be a rational number (by Corollary 1.5.13). Thus, f is not a coboundary by Theorem 3.2.5.

We may also consider the group $H(X, T, \mathbb{R}) = C(X, \mathbb{R})/\partial_T C(X, \mathbb{R})$ and define $H^+(X, T, \mathbb{R})$ as the image of $C(X, \mathbb{R})$ in $H(X, T, \mathbb{R})$. By Corollary 3.2.3, the pair $(H, X, T, \mathbb{R}), H^+(X, T, \mathbb{R}))$ is an ordered group. We observe the following relation between $H^+(X, T, \mathbb{Z})$ and $H^+(X, T, \mathbb{R})$.

Corollary 3.3.5 $H(X, T, \mathbb{Z})$ *is a subgroup of* $H(X, T, \mathbb{R})$ *and*

$$H^+(X, T, \mathbb{Z}) = H(X, T, \mathbb{Z}) \cap H^+(X, T, \mathbb{R}). \tag{3.4}$$

Proof The group $C(X, \mathbb{Z})$ is included in $C(X, \mathbb{R})$ and if $f \in C(X, \mathbb{Z})$ is in $\partial_T C(X, \mathbb{R})$, then by Proposition 3.1.3, it is in $\partial_T C(X, \mathbb{Z})$. Thus, the inclusion of $C(X, \mathbb{Z})$ in $C(X, \mathbb{R})$ defines an injection of $H(X, T, \mathbb{Z})$ in $H(X, T, \mathbb{R})$.

Next the left side of (3.4) is clearly included the right side. Suppose conversely that $f = \partial_T g + h$ with $f \in C(X, \mathbb{Z})$, $g \in C(X, \mathbb{R})$ and $h \in C(X, \mathbb{R}_+)$. For every $n \geq 0$, we have by Eq. (3.2)

$$f^{(n)}(x) = g(T^n x) - g(x) + h^{(n)}(x)$$

and the family $(f^{(n)})_{n \geq 0}$ is uniformly bounded from below. By Proposition 3.2.7, f is cohomologous in $C(X, \mathbb{Z})$ to some $f' \in C(X, \mathbb{Z}_+)$, which proves (3.4). ∎

The next proposition characterizes functions that are cohomologous to integer valued functions. We will not need it here, but it clarifies the relations between real and integer coboundaries.

Proposition 3.3.6 *Let (X, T) be a minimal Cantor system. Let U be a non-empty clopen set in X and let f be in $C(X, \mathbb{R})$.*

1. *If $f^{(n)}(x)$ belongs to $\mathbb{Z}$ for every $x \in U$ such that $T^n x \in U$, then f is cohomologous to some $g \in C(X, \mathbb{Z})$.*
2. *If $f^{(n)}(x)$ belongs to $\mathbb{Z}_+$ for every $x \in U$ such that $T^n x \in U$, then f is cohomologous to some $g \in C(X, \mathbb{Z}_+)$.*

We need before a classical lemma. We provide a proof for sake of completeness. Let $\tau \colon \mathbb{R} \to \mathbb{R}/\mathbb{Z}$ be the projection from $\mathbb{R}$ onto the torus $\mathbb{T} = \mathbb{R}/\mathbb{Z}$.

Lemma 3.3.7 *Let X be a Cantor space and $f \colon X \to \mathbb{T}$ be a continuous map. For every $\epsilon > 0$, there exists a continuous function $h_\epsilon \colon X \to [-\epsilon, 1]$ such that $f = \tau \circ h_\epsilon$.*

Proof Let $\epsilon > 0$. For each $x \in X$, there exists a neighborhood U_x and a continuous map $E_x \colon U_x \to \mathbb{R}$ such that $E_x(U_x)$ is included in $[E_x(x) - \epsilon, E_x(x) + \epsilon]$ and $f(y) = \tau \circ E_x(y)$, for all $y \in U_x$. Since X is a Cantor space, we can suppose U_x is a clopen set. Compactness of X yields points $x_1, \ldots, x_n$ such that $U_{x_1}, \ldots, U_{x_n}$ is a finite clopen covering of X. Then, for each $i \in [1, n]$ there exists a (possibly empty) clopen set V_i included in U_{x_i} such that $V_1, \ldots, V_n$ is a partition of X. Let $I = \{i \mid E_{x_i}(x_i) + \epsilon > 1\}$. Then the map $F_\epsilon = \sum_{i \notin I} \mathbf{1}_{V_i} E_{x(i)} + \sum_{i \in I} \mathbf{1}_{V_i} (E_{x(i)} - 1)$ fulfills the requirement. ∎

Proof of Proposition 3.3.6 1. Using the same method as in Step 2 of the proof of Theorem 3.2.5, we get that the sequence $(\tau(f^{(n_i)}(x_0))$ converges in $\mathbb{T}$ whenever $n_i \to \infty$ and $T^{n_i} x_0$ converges. Thus, there exists a continuous function $u \colon X \to \mathbb{T}$ such that $u(T^n x_0) = \tau(f^{(n)}(x_0))$ for every $n \geq 0$. If $x = T^n x_0$ for some $n \geq 0$, then

$$u(Tx) - u(x) = u(T^{n+1} x_0) - u(T^n x_0)$$
$$= \tau(f^{(n+1)}(x_0) - f^{(n)}(x_0)) = \tau(f(T^n x_0)) = \tau(f(x)).$$

By density, the same is true for every $x \in X$. By Lemma 3.3.7 there exists $h \in C(X, \mathbb{R})$ such that $\tau \circ h = u$. The function $g = f - \partial_T h$ belongs to $C(X, \mathbb{Z})$ and is cohomologous to f.

2. The family $(f^{(n)})$ is uniformly bounded from below, and the result follows from Proposition 3.2.7. ∎

3.4 Cylinder Functions

We consider in this section the case of shift spaces, in which the cylinder functions play an important role.

Let $X \subset A^{\mathbb{Z}}$ be a shift space. Recall that we denote by $\mathcal{L}_n(X)$ the set of words of length n in $\mathcal{L}(X)$. We denote by $R_n(X)$ the group of maps from $\mathcal{L}_n(X)$ into $\mathbb{R}$, by $Z_n(X)$ the group of maps from $\mathcal{L}_n(X)$ into $\mathbb{Z}$ and by $Z_n^+(X)$ the corresponding subset of nonnegative maps.

For $\phi \in R_n(X)$, the function $\underline{\phi} \colon X \to \mathbb{R}$ given by

$$\underline{\phi}(x) = \phi(x_{[0,n-1]})$$

is called the *cylinder function* associated with ϕ. It belongs to $C(X, \mathbb{Z})$ when ϕ belongs to $Z_n(X)$ and to $C(X, \mathbb{Z}_+)$ when ϕ belongs to $Z_n^+(X)$.

A function $f \in C(X, \mathbb{Z})$ is a cylinder function if and only if there exists $n \geq 1$ such that $f(x)$ depends only on $x_{[0,n-1]}$.

Proposition 3.4.1 *Let X be a shift space. Every function in $C(X, \mathbb{Z})$ (resp. $C(X, \mathbb{Z}_+)$) is cohomologous to some cylinder function (resp. nonnegative cylinder function).*

Proof Let $f \in C(X, \mathbb{Z})$. Since f is locally constant, there exists n such that $f(x)$ depends only on $x_{[-n,n]}$. Then $f(S^n x)$ depends only on $x_{[0,2n]}$ and $f \circ S^n$ is a cylinder function. Since $f \circ S^n - f = (\partial_S f)^{(n)}$ by Eq. (3.2), f is cohomologous to $f \circ S^n$ and the conclusion follows. Finally, if f belongs to $C(X, \mathbb{Z}_+)$, then $f \circ S^n$ is nonnegative. ∎

Proposition 3.4.2 *Let X be a shift space. If a cylinder function with integer values is a coboundary, it is the coboundary of some cylinder function.*

Proof Let $g \in C(X, \mathbb{Z})$ and suppose that $f = \partial_S g$ is a cylinder function. We may choose n large enough so that simultaneously $g(x)$ depends only on $x_{[-n,n]}$ and $f(x)$ depends only on $x_{[0,n]}$. Assume that $y_{[0,2n]} = z_{[0,2n]}$. Since $f(x)$ depends only on $x_{[0,n]}$, the value $f^{(n)}(x)$ depends only on $x_{[0,2n]}$ and thus $f^{(n)}(y) = f^{(n)}(z)$. Similarly, since $(S^n y)_{[-n,n]} = (S^n z)_{[-n,n]}$, we have $g(S^n y) = g(S^n z)$. Thus, by Eq. (3.2), we have

$$g(y) = g(S^n y) - f^{(n)}(y)$$
$$= g(S^n z) - f^{(n)}(z) = g(z)$$

and thus g is a cylinder function. ∎

3.5 Ordered Group of a Recurrent Shift Space

In this section, we show how to compute the ordered cohomology group of a recurrent shift space using the cylinder functions introduced in Section 3.4. We will first define an ordered group $\mathcal{G}_n(X)$ associated with cylinder functions corresponding to words of length n (Proposition 3.5.1). In a second part we show that the direct limit of these groups is the cohomology group $K^0(X, S)$ of the shift space X (Proposition 3.5.3).

3.5.1 Groups Associated with Cylinder Functions

For a word $w = a_1 a_2 \cdots a_n$ of length $n \geq 1$ with $a_i \in A$, we set $p_n(w) = a_1 \cdots a_{n-1}$ and $s_n(w) = a_2 \cdots a_n$. The map p_n (resp. s_n) is called the *prefix map* (resp. *suffix map*).

Next, given a shift space X, recall that $R_n(X)$ denotes the group of maps from the set $\mathcal{L}_n(X)$ of words of length n in $\mathcal{L}(X)$ into $\mathbb{R}$. We define, for $n \geq 1$, three group morphisms

$$p_{n-1}^*, s_{n-1}^*, \partial_{n-1} : R_{n-1}(X) \to R_n(X)$$

by

$$p_{n-1}^*(\phi) = \phi \circ p_n, \quad s_{n-1}^*(\phi) = \phi \circ s_n,$$

and

$$\partial_{n-1}(\phi) = s_{n-1}^*(\phi) - p_{n-1}^*(\phi) = \phi \circ s_n - \phi \circ p_n$$

for every $\phi \in R_{n-1}(X)$. These morphisms map $Z_{n-1}(X)$ into $Z_n(X)$ (recall that $Z_n(X)$ denotes the group of functions from $\mathcal{L}_n(X)$ into $\mathbb{Z}$). Moreover, p_{n-1}^* and s_{n-1}^* are injective and positive.

Note that one has the equality

$$s_n \circ p_{n+1} = p_n \circ s_{n+1} \tag{3.5}$$

since both sides send $a_1 \ldots a_{n+1}$ to $a_2 \cdots a_n$. It follows from (3.5) that

$$\partial_n \circ p_{n-1}^* = p_n^* \circ \partial_{n-1}, \tag{3.6}$$

as one may verify.

Note also that for every $\phi \in R_{n-1}(X)$, the cylinder functions associated with $p_{n-1}^*(\phi)$ and ϕ are the same, that is,

$$\underline{p_{n-1}^*(\phi)} = \underline{\phi}. \tag{3.7}$$

Note also that for $\phi \in R_{n-1}(X)$ and $\psi \in R_n(X)$ one has

$$\psi = \partial_{n-1}(\phi) \Leftrightarrow \underline{\psi} = \partial_S\underline{\phi} \tag{3.8}$$

and thus the cylinder function $\underline{\psi}$ associated with $\psi \in \partial_{n-1}R_{n-1}(X)$ is a coboundary. We denote by

$$G_n(X) = Z_n(X)/\partial_{n-1}Z_{n-1}(X) \tag{3.9}$$

the quotient of the group $Z_n(X)$ by its subgroup $\partial_{n-1}Z_{n-1}(X)$, we denote by $G_n^+(X)$ the image in $G_n(X)$ of $Z_n^+(X)$ and by $\mathbf{1}_n(X)$ the image in $G_n(X)$ of the constant function $\mathbf{1} \in Z_n(X)$.

Proposition 3.5.1 *For every recurrent shift space X, the triple*

$$\mathcal{G}_n(X) = (G_n(X), G_n^+(X), \mathbf{1}_n(X))$$

is a directed unital ordered group.

Proof The set $G_n^+(X)$ is a submonoid of $G_n(X)$ because $Z_n^+(X)$ is a submonoid of $Z_n(X)$, and it generates $G_n(X)$ because $Z_n^+(X)$ generates $Z_n(X)$. Let $\alpha, \beta \in Z_n^+(X)$ be such that $\alpha + \beta = \partial_{n-1}\phi$ for some $\phi \in Z_{n-1}(X)$. Then $\partial\underline{\phi}$ is a positive coboundary. By Corollary 3.2.3, this implies $\partial_S\underline{\phi} = 0$ and thus $\partial_{n-1}\phi = 0$. Thus, $\alpha + \beta = 0$, which implies $\alpha = \beta = 0$. This shows that $G_n^+(X) \cap (-G_n^+(X)) = \{0\}$ and thus $\mathcal{G}_n(X)$ is a directed ordered group. It is unital because $\mathbf{1}$ is a unit of $Z_n(X)$. $\blacksquare$

A direct proof of Proposition 3.5.1, that is, without using Corollary 3.2.3, is given in Exercise 3.13.

Note that, since $\partial_n \circ p_{n-1}^* = p_n^* \circ \partial_{n-1}$ by (3.6), the morphism p_n^* induces a morphism of the quotient groups

$$i_{n+1,n} \colon G_n(X) \to G_{n+1}(X),$$

which is a morphism of unital ordered groups.

Example 3.5.2 Let X be the Fibonacci shift. We have $\mathcal{L}_1(X) = \{a, b\}$ and $\mathcal{L}_2(X) = \{aa, ab, ba\}$. For $u \in \mathcal{L}_n(X)$, denote by ϕ_u the element of $Z_n(X)$, which takes the value 1 on u and 0 elsewhere, that is, such that $\chi_{[u]} = \underline{\phi_u}$. Since $\partial_1(\phi_a) = \phi_{ba} + \phi_{aa} - (\phi_{aa} + \phi_{ab}) = \phi_{ba} - \phi_{ab}$, the projections of ϕ_{ba}

and ϕ_{ab} in $G_2(X)$ are equal and $G_2(X) \simeq \mathbb{Z}^2$. The matrix of the projection of $Z_2(X)$ on $G_2(X)$ is

$$P = \begin{bmatrix} 1 & 0 & 0 \\ 0 & 1 & 1 \end{bmatrix}.$$

The matrix of the morphism $i_{2,1}$ is

$$\begin{bmatrix} 1 & 0 \\ 1 & 1 \end{bmatrix}$$

since, for example,

$$i_{2,1}\begin{bmatrix} 1 \\ 0 \end{bmatrix} = P(p_1^* \chi_{[a]}) = P \begin{bmatrix} 1 \\ 1 \\ 0 \end{bmatrix} = \begin{bmatrix} 1 \\ 1 \end{bmatrix}.$$

Note that Example 3.3.2 shows that Proposition 3.5.1 does not hold without the hypothesis that X is recurrent.

3.5.2 Ordered Cohomology Group

We now prove the following result, which describes the ordered cohomology group $K^0(X, S)$ of a recurrent shift space X.

Proposition 3.5.3 *For every recurrent shift space X, the unital ordered group $K^0(X, S)$ is the inductive limit of the family $G_n(X)$ with the morphisms $i_{n+1,n}$.*

Proof To every $\phi \in Z_n(X)$, we can associate the corresponding cylinder function $\underline{\phi}$ and its projection $\pi(\underline{\phi})$ in $H(X, S, \mathbb{Z})$. When ϕ is in $\partial_{n-1}(Z_{n-1}(X))$, the cylinder function $\underline{\phi}$ is a coboundary and $\pi(\underline{\phi}) = 0$. If ϕ belongs to $Z_n^+(X)$, then $\underline{\phi}$ is in $C(X, \mathbb{Z}_+)$, and the cylinder function associated with the constant function equal to 1 on $\mathcal{L}_n(X)$ is the constant function equal to 1 on X.

Thus, we have defined a morphism of unital ordered groups

$$j_n : G_n(X) \to K^0(X, S)$$

and clearly $j_{n+1} \circ i_{n+1,n} = j_n$. By the universal property of direct limits (Proposition 2.3.2) the sequence $(j_n)_{n \geq 1}$ induces a morphism j from the inductive limit of the $G_n(X)$ to the group $K^0(X, S)$. Let us show that j is an isomorphism.

Let $\phi \in Z_n(X)$, let α be its projection in $G_n(X)$ and suppose that $j_n(\alpha) = 0$. Then $\underline{\phi}$ is a coboundary and, by Proposition 3.4.2, it is the coboundary of

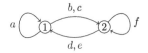

Figure 3.4 A shift of finite type.

some cylinder function ψ. Let $m \geq n$ be such that ψ belongs to $Z_m(X)$. Since $p_m^* \circ \cdots \circ p_n^*(\phi)$ belongs to $Z_{m+1}(X)$, we have by Eq. (3.7)

$$\phi = \underline{p_m^* \circ \cdots \circ p_n^*(\phi)}.$$

Now, since $\phi = \partial_T(\psi)$, we have by (3.8)

$$\underline{p_m^* \circ \cdots \circ p_n^*(\phi)} = \partial_T(\psi) \quad \Rightarrow \quad p_m^* \circ \cdots \circ p_n^*(\phi) = \partial_m(\psi),$$

which implies $p_m^* \circ \cdots \circ p_n^*(\phi) \in \partial_m(Z_m(X))$ and $i_{m,m-1} \circ \cdots \circ i_{n+1,n}(\alpha) = 0$. Thus, the image of α in the inductive limit is 0. This shows that j is injective.

Moreover, every $f \in C(X, \mathbb{Z})$ (resp. $C(X, \mathbb{Z}_+)$) is cohomologous to a cylinder function (resp. to a nonnegative cylinder function). It follows that j is onto and maps the positive cone of the inductive limit onto $H^+(X, S, \mathbb{Z})$. ∎

We now present an example that shows that the ordered cohomology group of a shift of finite type may fail to be a dimension group. This contrasts with the fact that, as we will see in Chapter 4, the ordered cohomology group of a minimal Cantor system is a simple dimension group (Theorem 4.3.4).

Example 3.5.4 Let X be the set of two-sided infinite paths in the graph represented in Figure 3.4. Since a can be preceded by a, d or e, we have

$$\chi_{[a]} \circ T = \chi_{[aa]} + \chi_{[da]} + \chi_{[ea]}.$$

Similarly, since a can be followed by a, b or c, we have $\chi_{[a]} = \chi_{[aa]} + \chi_{[ab]} + \chi_{[ac]}$.

Thus, $\partial \chi_{[a]} = \chi_{[da]} + \chi_{[ea]} - \chi_{[ab]} - \chi_{[ac]}$. We have then in $H(X, S, \mathbb{Z})$ the equality

$$\chi_{[d]} + \chi_{[e]} = \chi_{[da]} + \chi_{[ea]} + \sum_{x \in \{d,e\}, y \in \{b,c\}} \chi_{[xy]}$$

$$\simeq \chi_{[ab]} + \chi_{[ac]} + \sum_{x \in \{e,d\}, y \in \{b,c\}} \chi_{[xy]}$$

$$\simeq \chi_{[b]} + \chi_{[c]}.$$

But $\chi_{[b]} + \chi_{[c]} \simeq \chi_{[d]} + \chi_{[e]}$ implies

$$0 \leq \chi_{[b]} \leq \chi_{[d]} + \chi_{[e]}.$$

It can be shown, using an invariant Markov measure on X, that there is no decomposition $\chi_{[b]} = \phi_1 + \phi_2$ such that $\phi_1 \leq \chi_{[d]}$ and $\phi_2 \leq \chi_{[e]}$ (Exercise 3.32). Thus, $H(X, S, \mathbb{Z})$ is not a Riesz group.

3.6 Factor Maps and Conjugacy

Let (X, T) and (X', T') be two topological dynamical systems and let ϕ be a factor map from (X, T) to (X', T'). If $f \in C(X', \mathbb{Z})$ is the coboundary of g, then $f \circ \phi$ is the coboundary of $g \circ \phi$. Therefore, the group homomorphism $f \mapsto f \circ \phi$ from $C(X', \mathbb{Z})$ to $C(X, \mathbb{Z})$ induces a group homomorphism

$$\phi^* : H(X', T', \mathbb{Z}) \to H(X, T, \mathbb{Z}).$$

Clearly, the map ϕ^* is a morphism of ordered groups with order units since $\phi^*(H^+(X', T', \mathbb{Z}))$ is a subset of $H^+(X, T, \mathbb{Z})$ and $\phi^*(1_{X'}) = 1_X$.

Proposition 3.6.1 *If (X, T) is minimal, the morphism ϕ^* is injective and*

$$\phi^*(H^+(X', T', \mathbb{Z})) = \phi^*(H(X', T', \mathbb{Z})) \cap H^+(X, T, \mathbb{Z}).$$

Proof If $f \in C(X', \mathbb{Z})$ is such that $f \circ \phi$ is a coboundary, then by Theorem 3.2.5 the sequence of functions $((f \circ \phi)^{(n)})_{n \geq 0}$ is uniformly bounded. But, for all $n \geq 0$, $(f \circ \phi)^{(n)} = f^{(n)} \circ \phi$ and since, by minimality, the map ϕ is onto, $\|f^{(n)} \circ \phi\|_\infty = \|f^{(n)}\|_\infty$. Thus, the sequence $(f^{(n)})_{n \geq 0}$ is uniformly bounded and, by Theorem 3.2.5 again, f is a coboundary in $C(X', \mathbb{Z})$. This shows that ϕ^* is injective.

To prove the second assertion, we first note that the inclusion from left to right is clear since $\phi^*(H^+(X', T', \mathbb{Z}))$ is included in $H^+(X, T, \mathbb{Z})$. Conversely, consider $f \in C(X', \mathbb{Z})$ and $g \in C(X, \mathbb{Z})$ such that $f \circ \phi + \partial_T(g) \geq 0$. By Proposition 3.2.7, the sequence of functions $((f \circ \phi)^{(n)})_{n \geq 0}$ is bounded from below. By the same argument as in the proof of the first assertion, the sequence $(f^{(n)})_{n \geq 0}$ is bounded from below. Thus, by Proposition 3.2.7 again, there is a function $g' \in C(X', \mathbb{Z})$ such that $f + \delta_{T'}(g') \geq 0$. Thus, the class of f is in $H^+(X', T', \mathbb{Z})$ and its image by ϕ^* is in $\phi^*(H^+(X', T', \mathbb{Z}))$. ∎

We deduce from Proposition 3.6.1 that the ordered cohomology group is invariant under conjugacy.

Corollary 3.6.2 *If ϕ is a conjugacy from (X, T) onto (X', T'), then the map ϕ^* from $H(X', T', \mathbb{Z})$ to $H(X, T, \mathbb{Z})$ is an isomorphism.*

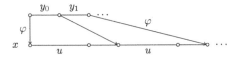

Figure 3.5 The unique y such that $\varphi(y) = x$.

Note that Proposition 3.4.2 can be deduced from Proposition 3.6.1. Indeed, let (Y, S) be the one-sided shift space associated with the minimal shift space (X, S) and $\theta : (X, S) \to (Y, S)$ be the natural morphism. It is a factor map. Assume that $f \in C(X, \mathbb{Z})$ is a cylinder function with integer values. Then $f = k \circ \theta$ for some $k \in C(Y, \mathbb{Z})$. If f is a coboundary, by Proposition 3.6.1, k is a coboundary. Thus, $k = \partial_S(h)$ for some $h \in C(Y, S)$. Then $f = \partial_S(h) \circ \theta = h \circ S \circ \theta - h \circ \theta = h \circ \theta \circ S - h \circ \theta$. Thus, f is the coboundary of $h \circ \theta$, which is a cylinder function.

Note also that this argument proves the following statement.

Corollary 3.6.3 *The map*

$$\theta^* : (H(Y, S, \mathbb{Z}), H_+(Y, S, \mathbb{Z}), 1_Y) \to (H(X, S, \mathbb{Z}), H_+(X, S, \mathbb{Z}), 1_X)$$

is an isomorphism of unital ordered groups.

3.7 Ordered Groups of Induced Systems

Let (X, T) be a minimal topological dynamical system and let U be a non-empty clopen subset of X. We have already seen the notion of induced system (U, T_U) in Section 1.1.3. Recall that $T_U(x) = T^{n(x)}(x)$ where $n(x) = \inf\{n > 0 \mid T^n x \in U\}$. The system (U, T_U) is again minimal.

As an example, which will be considered often in the next chapters, let us consider a minimal shift space X, a word $u \in \mathcal{L}(X)$ and the clopen set $U = [u]$. Let $\mathcal{R}'_X(u)$ be the set of left return words to u and let $\varphi : B^* \to \mathcal{R}'_X(u)$ be a coding morphism for $\mathcal{R}'_X(u)^*$. Every $x \in U$ can be written in a unique way as $x = \varphi(y)$ for some $y \in B^{\mathbb{Z}}$ (see Figure 3.5 or Proposition 1.4.29 if you are not entirely convinced).

Then $Y = \varphi^{-1}(U)$ is a shift space on B and we have the following important property.

Proposition 3.7.1 *The induced system (U, S_U) is isomorphic to (Y, T) where T is the shift on $B^{\mathbb{Z}}$.*

Proof Indeed, by the definition of return words, for every $x \in [u]$, the integer $n(x) = |\varphi(y_0)|$ is the length of the unique prefix of x^+ in $\mathcal{R}'_X(u)$. Thus,

$$S_U \circ \varphi(y) = S^{n(x)}(\varphi(y)) = \varphi \circ T(y)$$

showing that φ is a conjugacy from (Y, T) onto (U, S_U). ∎

The shift (Y, T) is called the *derivative shift* of X with respect to $[u]$. The morphism $\varphi \colon Y \to X$ is the *morphism associated* with the induction on $[u]$.

Example 3.7.2 Let X be the Fibonacci shift, which is the substitution shift generated by $\varphi \colon a \to ab, b \to a$. The system induced on $U = [a]$ is isomorphic to X. Indeed, the set of left return words to a is $\mathcal{R}'(a) = \{ab, a\}$ and thus φ is an isomorphism from X onto (U, T_U).

Let $I_U \colon C(X, \mathbb{Z}) \to C(U, \mathbb{Z})$ and $R_U \colon C(U, \mathbb{Z}) \to C(X, \mathbb{Z})$ be the morphisms of ordered groups defined by

$$I_U(f)(x) = f^{(n(x))}(x) \text{ for all } x \in U,$$

and let $R_U(f)$ be the map equal to f on U and equal to 0 elsewhere.

We will now show that an induction does not modify the cohomology group as an ordered group (although not as a unitary ordered group since the unit is not preserved; see the remark after the proof).

Proposition 3.7.3 *The maps I_U, R_U have the following properties.*

1. *For all f in $\partial_T C(X, \mathbb{Z})$, the map $I_U(f)$ belongs to $\partial_{T_U} C(U, \mathbb{Z})$.*
2. *For all g in $\partial_{T_U} C(U, \mathbb{Z})$, the map $R_U(g)$ belongs to $\partial_T C(X, \mathbb{Z})$.*
3. *$I_U \circ R_U = \mathrm{id}_{C(U,\mathbb{Z})}$ and the image of $R_U \circ I_U - \mathrm{id}_{C(X,\mathbb{Z})}$ is contained in $\partial_T C(X, \mathbb{Z}) \cap \ker(I_U)$.*

They induce reciprocal isomorphisms of ordered groups from $K^0(X, T)$ onto $K^0(U, T_U)$.

Proof 1. Let $f \in C(X, \mathbb{Z})$ be the coboundary of $g \in C(X, \mathbb{Z})$. Then for every $x \in U$, by Eq. (3.2), we have $f^{(n(x))} = g \circ T^{n(x)} - g$ and thus $I_U(f)$ is the coboundary of the restriction of g to U.

2. Suppose that $f \in C(U, \mathbb{Z})$ is the coboundary in (U, T_U) of some $g \in C(U, \mathbb{Z})$. For every $x \in X$, let $m(x) = \inf\{m \geq 0 \mid T^m(x) \in U\}$. Then the function $x \mapsto m(x)$ is continuous, coincides with $n(x)$ outside U and is zero on U. Let $h(x) = g(T^{m(x)}(x))$. We have

$$h(Tx) - h(x) = \begin{cases} f(x) & \text{if } x \in U \\ 0 & \text{if } x \notin U. \end{cases}$$

Indeed, if $x \in U$, then $m(x) = 0$, $m(Tx) = n(x) - 1$ and

$$h(T(x)) - h(x) = g(T^{n(x)}(x)) - g(x) = g(T_U(x)) - g(x) = f(x)$$

and if $x \notin U$, then $m(T(x)) = m(x) - 1$ and $h(T(x)) = h(x)$. This shows that $\partial_T(h) = R_U(f)$ and thus that R_U maps coboundaries of (U, T_U) to coboundaries of (X, T).

3. For any $f \in C(U, \mathbb{Z})$ and $x \in U$, we have

$$I_U \circ R_U(f)(x) = (R_U(f))^{(n(x))}(x)$$
$$= f(x) + R_U(f)(T(x)) \ldots + R_U(f)(T^{n(x)-1}(x)) = f(x).$$

Thus, $I_U \circ R_U(f) = f$ and $I_U \circ R_U$ is the identity. Next for every $f \in C(X, \mathbb{Z})$, we have

$$R_U \circ I_U(f) - f = \partial_T(k) \tag{3.10}$$

where $k(x) = f^{(m(x))}(x)$. Indeed, for every $x \in X$, recalling that $I_U(f)(x) = f^{(n(x))}(x)$, the above equation can be rewritten

$$(R_U \circ I_U(f))(x) - f(x) = f^{(m(T(x)))}(T(x)) - f^{(m(x))}(x).$$

Indeed, if x belongs to U, the value of the right side is $f^{(m(Tx))}(T(x)) - f^{(m(x))}(x) = f^{(n(x)-1)}(T(x)) - f^{(0)}(x) = f^{(n(x)-1)}(T(x)) = f^{(n(x))}(x) - f(x)$. Next, if x is not in U, the value of the left side is $-f(x)$ and the value of the right side is $f^{(n(x)-1)}(T(x)) - f^{(n(x))}(x) = -f(x)$. Equation (3.10) shows that $R_U \circ I_U(f) - f$ is the coboundary of k, which shows that the image of $R_U \circ I_U - \text{id}_{C(X,\mathbb{Z})}$ is contained in $\partial_T C(X, \mathbb{Z})$. Finally, since $I_U \circ R_U = \text{id}_{C(U,\mathbb{Z})}$, we have $I_U \circ R_U \circ I_U = I_U$. Thus, if $g = R_U \circ I_U(f) - f$, then $I_U(g) = I_U \circ R_U \circ I_U(f) - I_U(f) = I_U(f) - I_U(f) = 0$, showing that $g \in \ker(I_U)$.

By Assertion 1, I_U induces a morphism

$$i_U : H(X, T, \mathbb{Z}) \to H(U, T_U, \mathbb{Z}).$$

Since I_U maps $C(X, \mathbb{Z}_+)$ to $C(U, \mathbb{Z}_+)$, it is a morphism of ordered groups from $K^0(X, T)$ to $K^0(U, T_U)$.

By Assertion 2, R_U induces a morphism $r_U : H(U, T_U, \mathbb{Z}) \to H(X, T, \mathbb{Z})$. By Assertion 3, these morphisms are mutually inverse. ∎

Note that the isomorphism from $K^0(X, T)$ onto $K^0(U, T_U)$ is not an isomorphism of unital ordered groups, because the units are not the same. Indeed, the image by R_U of the constant function equal to 1 on U is the characteristic function of U in X and not the constant function equal to 1 on X. This phenomenon will play an important role when we compute the dimension group of an induced system (see Proposition 5.4.2).

3.8 Invariant Borel Probability Measures

Let (X, T) be a topological dynamical system. There is an important connection between the ordered cohomology group $H(X, T, \mathbb{Z})$ of (X, T) and the invariant Borel probability measures on (X, T).

3.8.1 Invariant Measures

Recall (see Appendix B.5) that a *Borel probability measure* on a topological space X is a Borel measure μ such that $\mu(X) = 1$.

It is useful to see how such a measure is defined in a shift space. Consider a shift space X on the alphabet A. A Borel probability measure μ on X determines a map $\pi : \mathcal{L}(X) \to [0, 1]$ by

$$\pi(u) = \mu([u]_X), \tag{3.11}$$

where $[u]_X = \{x \in X \mid x_{[0,|u|-1]} = u\}$ is the cylinder defined by u. This map satisfies, as a consequence of the equality $[u] = \cup_{a \in A}[ua]$, the *compatibility conditions* $\pi(\varepsilon) = 1$ and

$$\sum_{a \in A, ua \in \mathcal{L}(X)} \pi(ua) = \pi(u) \tag{3.12}$$

for every $u \in \mathcal{L}(X)$.

Conversely, any map satisfying these compatibility conditions defines a unique Borel probability measure satisfying (3.11) by the Carathéodory extension theorem (see Appendix B.5).

The following example gives the simplest possible measure on the full shift. It corresponds to a sequence of successive independent and identically distributed choices of the letters forming a sequence.

Example 3.8.1 Let $X = A^{\mathbb{Z}}$ be the full shift on A. Let $\pi : A \to [0, 1]$ be a map such that $\sum_{a \in A} \pi(a) = 1$. We extend π to multiplicative morphism from A^* into $[0, 1]$. Since the compatibility conditions (3.12) are satisfied, there is a unique Borel probability measure μ on X satisfying (3.11). It is called a *Bernoulli measure*. For $\mu(a) = 1/\operatorname{Card}(A)$, it is called the *uniform* Bernoulli measure.

As a more general example, where the successive choices are not independent, but depend on the previous one, we have the following important class of measures.

Example 3.8.2 Let P be an $A \times A$-stochastic matrix and let v be a stochastic row A-vector, that is, such that $\sum_{a \in A} v_a = 1$. Let $\pi : A^* \to [0, 1]$ be defined by $\pi(\varepsilon) = 1$ and

$$\pi(a_1 a_2 \cdots a_n) = v_{a_1} P_{a_1, a_2} \cdots P_{a_{n-1}, a_n} \tag{3.13}$$

for $n \geq 1$ and $a_i \in A$. Since v and P are stochastic, the map π satisfies the compatibility conditions (3.12). Indeed $\pi(\varepsilon) = \sum_{a \in A} \pi(a) = \sum_{a \in A} v_a = 1$ and next

$$\sum_{a \in A} \pi(a_1 \ldots a_n a) = \sum_{a \in A} \pi(a_1 \ldots a_n) P_{a_n, a} = \pi(a_1 \ldots a_n).$$

The probability measure on $A^{\mathbb{Z}}$ defined by π is called a *Markov measure* and the pair (v, P) a *Markov chain*.

Now let (X, T) be a topological dynamical system. A Borel measure μ on X is said to be *invariant* if one has $\mu(T^{-1}(U)) = \mu(U)$ for every Borel subset U of X. In this case, the triple (X, T, μ) is a measure-theoretic system (Exercise 3.26).

Let μ be an invariant Borel probability measure on a shift space X. The associated map $\pi : \mathcal{L}(X) \to [0, 1]$ such that $\pi(u) = \mu([u])$ satisfies, in addition to the compatibility conditions (3.12), the symmetric conditions

$$\pi(u) = \sum_{a \in A, au \in \mathcal{L}(X)} \pi(au). \tag{3.14}$$

Indeed, $\sum_{a \in A, au \in \mathcal{L}(X)} \pi(au) = \sum_{a \in A} \mu([au]) = \mu(T^{-1}([u])) = \mu([u]) = \pi(u)$. Conversely, for every map $\pi : \mathcal{L}(X) \to [0, 1]$ satisfying the compatibility conditions (3.12) and (3.14), there is by Carathéodory extension theorem (see Appendix B.5) a unique invariant Borel probability measure μ such that $\mu([u]) = \pi(u)$.

It may be puzzling that the compatibility conditions (3.12) and (3.14) are left-right symmetric of each other, although the notion of invariant measure is by no means the symmetric of the notion of measure. This apparent contradiction comes from the asymmetry of the definition of π, which is "future oriented" since $x \in [u]$ if x^+ begins by u (and not x^-).

For example, a Bernoulli measure is an invariant Borel probability measure, since (3.14) is obviously satisfied.

Example 3.8.3 A Markov chain (v, P) is called *stationary* if v is a left eigenvector of P for the eigenvalue 1. A Markov measure on $A^{\mathbb{Z}}$ defined by a Markov chain (v, P) is invariant if and only if (v, P) is stationary. Indeed, if $vP = v$, the map π defined by (3.13) satisfies for every $a \in A$ and $w = a_1 \ldots a_n$

$$\sum_{a \in A} \pi(aa_1 \ldots a_n) = \sum_{a \in A} (v_a P_{a,a_1}) P_{a_1,a_2} \cdots P_{a_{n-1},a_n}$$
$$= \pi(a_1 \ldots a_n).$$

Conversely, if μ is invariant, then the associated map π satisfies $v_b = \pi(b) = \sum_{a \in A} \pi(ab) = \sum_{a \in A} (v_a P)_b = (vP)_b$ and thus $vP = v$.

The set of invariant Borel probability measures on X is convex since, for invariant Borel probability measures μ, v and $\alpha, \beta \geq 0$ with $\alpha + \beta = 1$, $\alpha\mu + \beta v$ is an invariant Borel probability measure.

Theorem 3.8.4 (Krylov, Bogolyubov) *Every topological dynamical system has at least one invariant Borel probability measure.*

Proof Let (X, T) be a topological dynamical system and let $x \in X$. For every $n \geq 0$ we consider the probability measure $\delta_{T^n x}$, where δ_y denotes the Dirac measure at the point y. Consequently

$$\mu_N = \frac{1}{N} \sum_{n < N} \delta_{T^n(x)}$$

is also a Borel probability measure.

By Theorem B.5.5, the sequence (μ_N) has a cluster point μ for the weak-star topology. Let (μ_{N_i}) be a subsequence converging to μ. For every Borel subset U of X,

$$\mu(U) = \lim \frac{1}{N_i} \mathrm{Card}\{n < N_i \mid T^n(x) \in U\}$$
$$= \lim \frac{1}{N_i} \mathrm{Card}\{n < N_i \mid T^{n+1}(x) \in U\} = \mu(T^{-1}(U)).$$

Thus, μ is an invariant Borel probability measure. ∎

A different proof uses the *Markov–Kakutani fixed point theorem* (see Exercise 3.15).

Theorem 3.8.4 shows that every topological dynamical system may be considered as a measure-theoretic one.

3.8.2 Ergodic Measures

Recall that, for a topological dynamical system (X, T), a subset U of X is said to be *invariant* if $T^{-1}(U) = U$.

An invariant Borel probability measure on (X, T) is *ergodic* whenever $\mu(U)$ equals 0 or 1 for every invariant Borel subset U of X. One also says that the transformation T is ergodic with respect to μ or that the triple (X, T, μ) is ergodic.

As an example, a Bernoulli measure is ergodic (Exercise 3.17) and a Markov measure defined by (v, P) with $v > 0$ and $vP = v$ is ergodic if and only if P is irreducible (Exercise 3.19).

A basic result, that we state without proof, is the Birkhoff *Ergodic Theorem*.

Theorem 3.8.5 (Birkhoff) *Let μ be an ergodic measure on (X, T). For every integrable function f on X, the sequence $f_n = \frac{1}{n} f^{(n)}$ converges μ-almost everywhere to $\int f d\mu$.*

The functions f_n are sometimes called the *Birkhoff averages*. A real-valued measurable function f on X is *invariant* if $f = f \circ T$. Thus, a set U is invariant if and only if its characteristic function is invariant.

For two sets U, V we write $U = V$ mod μ if U, V differ by sets of measure 0. The following statement gives a useful variant of the definition of an ergodic measure.

Proposition 3.8.6 *The following conditions are equivalent for an invariant Borel probability measure μ on (X, T).*

(i) *μ is ergodic.*
(ii) *Every Borel set U such that $T^{-1}(U) = U$ mod μ is such that $\mu(U) = 0$ or $\mu(U) = 1$.*

The proof is left as Exercise 3.16.

Ergodicity is the measure-theoretic counterpart of minimality. Both notions are different. For example, a Bernoulli measure on the full shift is ergodic, although the full shift is not minimal. Conversely, an invariant Borel probability measure on a minimal shift need not be ergodic, as we shall see. We note the following relation between ergodicity and recurrence.

Proposition 3.8.7 *Let (X, T) be a topological dynamical system and let μ be an invariant Borel probability measure on (X, T). Assume that $\mu(U) > 0$ for every nonempty open set U. If μ is ergodic, then (X, T) is recurrent.*

The proof is left as an exercise (Exercise 3.22).

Recall from Appendix B.5 that a Borel probability measure ν is *absolutely continuous* with respect to μ, denoted $\nu \ll \mu$, if for every Borel set $U \subset X$ such that $\mu(U) = 0$ one has $\nu(U) = 0$.

The following result shows that the ergodic measures are the minimal invariant measures with respect to the preorder $\ll$.

Proposition 3.8.8 *Let μ, ν be invariant Borel probability measures on (X, T). If μ is ergodic and $\nu \ll \mu$, then $\mu = \nu$.*

Proof Since $\nu \ll \mu$, by the Radon–Nikodym Theorem (see Appendix B.5), there is a nonnegative μ-integrable function f such that $\nu(U) = \int_U f d\mu$ for every Borel set $U \subset X$.

Consider the Borel set $B = \{x \in X \mid f(x) > 1\}$. We will prove that B is invariant modulo a set of μ-measure zero. Note first that $\mu(T^{-1}(B) \setminus B) = \mu(B \setminus T^{-1}(B))$ and $\nu(T^{-1}(B) \setminus B) = \nu(B \setminus T^{-1}(B))$ since μ, ν are invariant measures. Indeed, one has

$$\mu(T^{-1}(B)) = \mu(T^{-1}(B) \setminus B) + \mu(B \cap T^{-1}(B)),$$

$$\mu(B) = \mu(B \setminus T^{-1}(B)) + \mu(B \cap T^{-1}(B))$$

and thus the claim. Assume that $\mu(B \setminus T^{-1}(B)) > 0$. Then we have

$$\mu(B \setminus T^{-1}(B)) < \int_{B \setminus T^{-1}(B)} f d\mu = \nu(B \setminus T^{-1}(B)) = \nu(T^{-1}(B) \setminus B)$$

$$= \int_{T^{-1}(B) \setminus B} f d\mu \leq \mu(T^{-1}(B) \setminus B) = \mu(B \setminus T^{-1}(B)),$$

which is absurd and thus $\mu(T^{-1}(B) \setminus B) = \mu(B \setminus T^{-1}(B)) = 0$. This, in turn, implies that $B = T^{-1}(B)$ modulo a set of μ-measure zero. By Proposition 3.8.6, since μ is ergodic, this implies that $\mu(B) = 0$ or 1. If $\mu(B) = 1$, then $\nu(X) > 1$, which is absurd. Thus, $\mu(B) = 0$. Since $\nu(X) = \int_X f d\mu = 1$, this implies that $f = 1$ almost everywhere or equivalently $\mu = \nu$. ∎

An *extreme point* of a convex set K is a point that does not belong to any open line segment in K. By the Krein–Milman Theorem (see Appendix B.5), the set $\mathcal{M}(X, T)$ of invariant Borel probability measures on (X, T) is the convex hull of its extreme points.

Proposition 3.8.9 *The ergodic measures are the extreme points of the set of invariant Borel probability measures.*

Proof Let μ be an extreme point of $\mathcal{M}(X, T)$. If μ is not ergodic, there is an invariant Borel set U such that $0 < \mu(U) < 1$. The complement V of U is also invariant. This defines two invariant Borel probability measures μ_1 and μ_2 by

$$\mu_1(W) = \frac{1}{\mu(U)}\mu(W \cap U), \quad \mu_2(W) = \frac{1}{\mu(U)}\mu(W \cap V).$$

But then $\mu = \mu(U)\mu_1 + (1 - \mu(U))\mu_2$ shows that μ is not an extreme point.

Now let μ be ergodic. Then $\mu = \alpha\mu_1 + (1 - \alpha)\mu_2$ for $0 \leq \alpha \leq 1$ and some extreme points $\mu_1, \mu_2 \in \mathcal{M}(X, T)$. Assume $\alpha > 0$. For any Borel set $U \subset X$, one has that $\mu(U) = 0$ implies $\mu_1(U) = 0$, that is, μ_1 is absolutely continuous with respect to μ. By Proposition 3.8.8, this implies that $\mu = \mu_1$. Thus, μ is an extreme point. ∎

In particular, we have the following important case.

Corollary 3.8.10 *If there is a unique invariant Borel probability measure, it is ergodic.*

The following statement gives additional properties of the set of ergodic measures.

Proposition 3.8.11 *Let (X, T) be a topological dynamical system.*

1. *Two distinct ergodic measures are mutually singular.*
2. *Every set of n distinct ergodic measures forms a linearly independent set.*

Proof Let μ, ν be two distinct ergodic measures on (X, T). Since μ, ν are distinct, there exists an integrable function f such that $\int f d\mu \neq \int f d\nu$. Let U be the set of points x such that $f_n(x)$ converges to $\int f d\nu$. Then, by the Birkhoff Theorem, $\mu(U) = \nu(X \setminus U) = 0$. Thus, μ, ν are mutually singular.

Let $\mu_1, \ldots, \mu_n$ be n distinct ergodic measures. Assume that $\sum_{k=1}^{n} \alpha_k \mu_k = 0$. Let $1 \leq i \leq n$ be an integer. For every $j \neq i$, since μ_j, μ_i are mutually singular, there is a Borel set G_J such that $\mu_j(G_j) = \mu_i(X \setminus G_j) = 0$ and consequently $\mu_i(G_j) = 1$. Then we have $\sum_{k=1}^{n} \alpha_k \mu_k(\cap_{j \neq i} G_j) = \alpha_i$ and thus $\alpha_i = 0$. This shows that $\mu_1, \ldots, \mu_n$ are linearly independent. ∎

3.8.3 Unique Ergodicity

In view of Corollary 3.8.10, a system with a unique invariant Borel probability measure is called *uniquely ergodic*. It is called *strictly ergodic* if it is minimal and uniquely ergodic.

It is easy to give examples of a system that is not uniquely ergodic when it is not minimal. For example, $\{0^\infty\} \cup \{1^\infty\}$ has clearly two ergodic measures. An example of a minimal system that is not uniquely ergodic is given in Exercise 3.25.

Theorem 3.8.12 (Oxtoby) *Let (X, T) be a topological dynamical system and μ be an invariant Borel probability measure on (X, T). The following conditions are equivalent.*

(i) (X, T) *is uniquely ergodic.*
(ii) $f_n(x) = \frac{1}{n} f^{(n)}(x)$ *converges uniformly on X to $\int f d\mu$ for every $f \in C(X, \mathbb{R})$.*
(iii) $f_n(x) = \frac{1}{n} f^{(n)}(x)$ *converges pointwise to $\int f d\mu$ for every $f \in C(X, \mathbb{R})$.*

Proof (i)$\Rightarrow$(ii). Suppose that (ii) does not hold. We use a diagonal argument to find a contradiction. We can find $\varepsilon > 0$, a map $g \in C(X, \mathbb{R})$ and a sequence (x_n) of points of X such that

$$\left| g_n(x_n) - \int g d\mu \right| > \varepsilon.$$

For every $n \geq 0$, the sum $\delta_{x,n} = \frac{1}{n} \sum_{i=0}^{n-1} \delta_{T^i x}$, where δ_x is the Dirac measure, is a Borel probability measure for which

$$\int f d\delta_{x,n} = f_n(x)$$

for every $f \in C(X, \mathbb{R})$ and $x \in X$. Let ν be a cluster point of the sequence $\delta_{x,n}$ for the weak-star topology. Refining (x_n) if necessary, we have

$$\lim f_n(x_n) = \int f d\nu$$

for every $f \in C(X, \mathbb{R})$. Hence,

$$\left| \int g d\mu - \int g d\nu \right| > \varepsilon$$

showing that μ, ν are distinct invariant probability measures. Thus, (i) does not hold.

(ii)$\Rightarrow$ (iii) is obvious.

(iii)$\Rightarrow$ (i). For any invariant Borel probability measure ν, we have

$$\int f_n d\nu = \frac{1}{n} \sum_{i < n} \int f \circ T^i d\nu = \int f d\nu. \tag{3.15}$$

On the other hand, by Eq. (3.15) and the Dominated Convergence Theorem, we have

$$\int f \, dv = \lim \int f_n \, dv = \int \lim f_n \, dv = \int \left(\int f \, d\mu \right) dv = \int f \, d\mu.$$
(3.16)

Thus, by Eq. (3.15) and (3.16), we conclude that $\mu = v$. ∎

Theorem 3.8.12 is a refinement for uniquely ergodic measures of Birkhoff's ergodic Theorem.

3.8.4 Unique Ergodicity of Primitive Substitution Shifts

In the case of substitution shifts, we have the following important result.

Theorem 3.8.13 *Primitive substitution shifts are uniquely ergodic.*

For two words u, v, we denote $|u|_v$ the number of factors of u equal to v. By convention, $|u|_\varepsilon = |u|$. We say that an infinite sequence $x \in A^{\mathbb{Z}}$ has *uniform frequencies* if for every factor v of x, all sequences

$$(f_{k,v}(x))_n = \frac{|x_k \cdots x_{k+n-1}|_v}{n}$$

converge and tend to the same limit $f_v(x)$ when $n \to \infty$, uniformly in k.

Note that x has uniform frequencies if and only if for every $v \in \mathcal{L}(x)$ there is an f_v such that for every ε there is an $N \geq 1$ such that for every $u \in \mathcal{L}(x)$ of length at least N, we have $||u|_v/|u| - f_v| \leq \varepsilon$. We express this property by saying that the frequency of v in u tends to f_v when $|u| \to \infty$ uniformly in u.

Proposition 3.8.14 *Let X be a minimal shift space. The following conditions are equivalent.*

(i) *Every $x \in X$ has uniform frequencies.*
(ii) *Some $x \in X$ has uniform frequencies.*
(iii) *X is uniquely ergodic.*

Moreover, in this case the unique invariant measure is given by $\mu([v]) = f_v(x)$ for every $x \in X$.

Proof (i) implies (ii) is clear. Assume (ii). For $v \in \mathcal{L}(X)$, set $\pi(v) = f_v(x)$. This map satisfies clearly the compatibility conditions (3.12) and (3.14). Thus, there is a unique invariant Borel probability measure μ such that $\mu([v]) = \pi(v)$. Set $g_{k,v} = \chi_{[v]} \circ S^k$. Since $|x_k \ldots x_{k+n}|_v = \sum_{j<n-|v|} \chi_{[v]}(S^{j+k}x) = (g_{k,v})^{(n-|v|)}(x)$, we have

$$\frac{1}{n}(g_{k,v})^{(n)}(x) \to_{n \to +\infty} \mu([v]) = \int \chi_{[v]} d\mu$$

uniformly in k. But the family of linear combinations of maps $g_{k,v}$ is dense in $C(X, \mathbb{R})$. Hence,

$$\frac{1}{n}g^{(n)}(x) \to_{n \to +\infty} \int g \, d\mu$$

for every continuous function g. By Oxtoby's Theorem (Theorem 3.8.12), this implies that (X, S) is uniquely ergodic.

Finally, by Oxtoby's Theorem again, (iii) implies (i). ■

Let $\varphi \colon A^* \to A^*$ be a primitive substitution and X its shift space. Let M be the incidence matrix of φ. Since M is a primitive matrix, by the Perron–Frobenius Theorem (Theorem B.3.1), the matrix M has a dominant eigenvalue λ_M and positive left and right eigenvectors $x = (x_a)$, $y = (y_b)$ relative to λ_M. We may assume that $\sum_{a \in A} x_a = 1$ and that $xy = \sum_{a \in A} x_a y_a = 1$. With this notation, we recall that in Lemma 1.4.23 we proved that there are some $\tau < \lambda_M$ and C such that for all $n \in \mathbb{N}$ and $a, b \in A$, one has

$$\left| |\varphi^n(a)|_b - |\varphi^n(a)| x_b \right| \leq C\tau^n, \tag{3.17}$$

$$\left| |\varphi^n(a)| - \lambda_M^n y_a \right| \leq C\tau^n. \tag{3.18}$$

From these inequalities we will first show, in order to prove Theorem 3.8.13, that the quantity $|u|_b/|u|$ converges to x_b uniformly in u when $|u|$ goes to infinity. But before we do that, we need the following decomposition lemma. Recall from Chapter 1 that, for a morphism $\varphi \colon A^* \to A^*$, we denote $|\varphi| = \max_{a \in A} |\varphi(a)|$.

Lemma 3.8.15 *For every nonempty word $u \in \mathcal{L}(X)$, there is some $m \geq 0$ and words $v_i, w_i \in \mathcal{L}(X)$ for $0 \leq i \leq m$, with v_m nonempty such that*

$$u = v_0 \varphi(v_1) \cdots \varphi^{m-1}(v_{m-1}) \varphi^m(v_m) \varphi^{m-1}(w_{m-1}) \cdots \varphi(w_1) w_0, \tag{3.19}$$

with $|v_i|, |w_i| \leq |\varphi|$.

Proof We use induction on $|u|$. The result is true if $|u| < |\varphi|$ choosing $m = 0$ and $v_0 = u$. Otherwise, by definition of $\mathcal{L}(X)$, there exists a nonempty word $u' \in \mathcal{L}(X)$ such that $u = v_0 \varphi(u') w_0$. Choosing u' of maximal length, we have, moreover, $|v_0|, |w_0| \leq |\varphi|$. By the induction hypothesis we have a decomposition (3.19) for u', that is,

$$u' = v_0' \varphi(v_1') \cdots \varphi^{m-1}(v_{m-1}') \varphi^m(v_m') \varphi^{m-1}(w_{m-1}') \cdots \varphi(w_1') w_0'.$$

In this way, we obtain

$$u = v_0\varphi(u')w_0 = v_0\varphi(v'_0)\cdots\varphi^m(v'_{m-1})\varphi^{m+1}(v'_m)\varphi^m(w'_{m-1})\cdots\varphi(w'_0)w_0,$$

which is of the form (3.19). ∎

Lemma 3.8.16 *Let $b \in A$. The quantity $|u|_b/|u|$ converges to x_b uniformly in $u \in \mathcal{L}(X)$ when $|u|$ goes to infinity.*

Proof Let $b \in A$. Take a decomposition of u given by (3.19). We obtain

$$|u|_b = \sum_{i \leq m} |\varphi^i(v_i)|_b + \sum_{i < m} |\varphi^i(w_i)|_b.$$

By (3.17) one has

$$\big|\, |\varphi^i(v_i)|_b - x_b|\varphi^i(v_i)|\,\big| \leq C\tau^i|v_i| \tag{3.20}$$

for $0 \leq i \leq m$ (and the analogue inequality for the w_i). Since

$$|u| = \sum_{i \leq m} |\varphi^i(v_i)| + \sum_{i < m} |\varphi^i(w_i)|,$$

we obtain

$$
\begin{aligned}
\big|\,|u|_b - x_b|u|\,\big| &= \Big|\,|u|_b - x_b\Big(\textstyle\sum_{i \leq m}|\varphi^i(v_i)| + \sum_{i < m}|\varphi^i(w_i)|\Big)\Big| \\
&= \sum_{i \leq m}\big|\,|\varphi^i(v_i)|_b - x_b|\varphi^i(v_i)|\,\big| + \sum_{i < m}\big|\,|\varphi^i(w_i)|_b - x_b|\varphi^i(w_i)|\,\big| \\
&\leq C\sum_{i \leq m}\tau^i(|v_i| + |w_i|) \leq 2C|\varphi|\sum_{i \leq m}\tau^i.
\end{aligned}
$$

Now observe that by (3.18) one gets $|u| \geq |\varphi^m(v_m)| \geq \lambda_M^m - C\tau^m$. This shows that $|u|_b/|u|$ converges to x_b uniformly in $u \in \mathcal{L}(X)$ when $|u|$ tends to infinity. ∎

To prove Theorem 3.8.13 there remains to extend Lemma 3.8.16 to arbitrary words $v \in \mathcal{L}(X)$ instead of b. For this, fix some $v \in \mathcal{L}(X)$ and set $\ell = |v|$. We consider the alphabet A_ℓ in one-to-one correspondence with $\mathcal{L}_\ell(X)$ via $f : \mathcal{L}_\ell(X) \to A_\ell$ and the ℓ-block presentation φ_ℓ of φ (see Section 1.4). Recall that, by Proposition 1.4.42, φ_ℓ is primitive. Let M_ℓ be the incidence matrix of φ_ℓ. Recall that M_ℓ has dominant eigenvalue λ_M (Proposition 1.4.43). Let $x^{(\ell)}$ be the left eigenvector of M_ℓ relative to λ_M with coefficients of sum 1.

Lemma 3.8.17 *The quantity $|u|_v/|u|$ converges to $x^{(\ell)}_{f(v)}$ uniformly in $u \in \mathcal{L}(X)$ when $|u|$ goes to infinity, where $\ell = |v|$.*

Proof Let $u \in \mathcal{L}_{\geq \ell}(X)$. Observe that $|u| = |f(u)| - \ell + 1$ and $|u|_v = |f(u)|_{f(v)} + \theta(u)$ with $0 \leq \theta(u) \leq \ell$. Lemma 3.8.16 applied to φ_ℓ allows us to conclude. ∎

Proof of Theorem 3.8.13 It is a consequence of Proposition 3.8.14 and Lemma 3.8.17. ∎

3.8.5 Computation of the Unique Invariant Borel Probability Measure

The computation of the unique invariant Borel probability measure can be done using the matrices M_ℓ of ℓ-block presentations of the substitution, as we shall see below.

Let $\varphi \colon A^* \to A^*$ be a primitive morphism and let X be the associated shift space. Let μ be the unique invariant Borel probability measure on X. For $\ell \geq 1$, let A_ℓ be an alphabet in bijection with $\mathcal{L}_\ell(X)$ via $f \colon \mathcal{L}_\ell(X) \to A_\ell$ and let $x^{(\ell)}$ be the unique positive row eigenvector of the matrix M_ℓ corresponding to the maximal eigenvalue λ and such that the sum of its components is 1.

Proposition 3.8.18 *For every $v \in \mathcal{L}_\ell(X)$,*

$$\mu([v]) = x_c^{(\ell)} \tag{3.21}$$

with $c = f(v)$.

Proof By Theorem 3.8.14, $\mu([v])$ is equal to the frequency $f_v(x)$ of the word v in any $x \in X$. By Lemma 3.8.17, the frequency of v in $\varphi^n(a)$ tends to $x_c^{(\ell)}$ when $n \to \infty$ and thus we have $f_v(x) = x_c^{(\ell)}$. ∎

We develop below the cases of the Fibonacci and of the Thue–Morse morphisms. For the Fibonacci shift, we already know that it is uniquely ergodic since this is true of every Sturmian shift (Exercise 3.23).

Example 3.8.19 Let X be the Fibonacci shift. Since the Fibonacci substitution is primitive, there is a unique invariant probability measure on X. Its values on the cylinder $[w]$ defined by words w of length at most 4 are shown on Figure 3.6 with $\rho = (\sqrt{5} - 1)/2$.

This is consistent with the value of the eigenvector of M_2

$$v^{(2)} = \begin{bmatrix} 2\rho - 1 & 1 - \rho & 1 - \rho \end{bmatrix}.$$

Note that the Fibonacci shift is Sturmian of slope $\alpha = (3 - \sqrt{5})/2$ (see Example 1.5.15). The unique invariant Borel probability measure μ on the Fibonacci

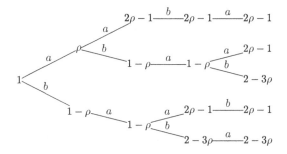

Figure 3.6 The invariant Borel probability measure on the Fibonacci shift.

shift can also be computed using the natural coding γ_α (Exercise 3.23), giving $\mu([a]) = 1 - \alpha$. Consistently with the above, $\rho = 1 - \alpha$.

Example 3.8.20 Consider the Thue–Morse morphism. The matrix M_2 is

$$M_2 = \begin{bmatrix} 0 & 1 & 1 & 0 \\ 0 & 1 & 0 & 1 \\ 1 & 0 & 1 & 0 \\ 0 & 1 & 1 & 0 \end{bmatrix}.$$

The unique invariant Borel probability measure on the Thue–Morse shift is shown in Figure 3.7. The values on the words of length 2 are consistent with the value of the eigenvector of M_2, which is $v^{(2)} = \begin{bmatrix} 1/6 & 1/3 & 1/3 & 1/6 \end{bmatrix}$.

3.9 Invariant Measures and States

We now come to an essential point and relate invariant measures and coboundaries.

Proposition 3.9.1 *Let μ be an invariant measure on the dynamical system (X, T). For every coboundary $f \in \partial_T C(X, \mathbb{R})$, one has $\int f d\mu = 0$.*

Proof When μ is a T-invariant Borel probability measure we have $\mu \circ T^{-1} = \mu$. Thus, for $f = \partial_T g$, by the change of variable formula (see Appendix B.5), $\int \partial_T g d\mu = \int g \circ T d\mu - \int g d\mu = 0$.   ∎

Proposition 3.9.1 implies that, for all T-invariant Borel probability measures μ on (X, T), the map $f \mapsto \int f d\mu$ defines a group homomorphism

$$\alpha_\mu \colon H(X, T, \mathbb{Z}) \to \mathbb{R}.$$

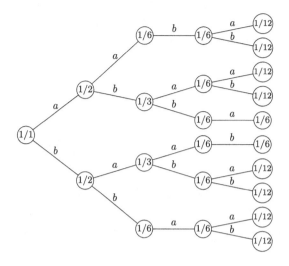

Figure 3.7 The invariant Borel probability measure on the Thue–Morse shift.

Proposition 3.9.2 *Let (X, T) be a recurrent dynamical system. The map α_μ is a morphism of unital ordered groups from $K^0(X, T)$ to $(\mathbb{R}, \mathbb{R}_+, 1)$.*

Proof By definition of $\int f d\mu$, we have $\alpha_\mu(H(X, T, \mathbb{Z}_+)) \subset \mathbb{R}_+$. Moreover, $\alpha_\mu(\mathbf{1}_X) = \mu(X) = 1$, since μ is a probability measure. ∎

The proof of the following statement uses the Carathéodory extension theorem (see Appendix B.5).

Theorem 3.9.3 *Let (X, T) be a recurrent dynamical system. The map $\mu \mapsto \alpha_\mu$ is a bijection from the space of T-invariant Borel probability measures on (X, T) onto the set of states of $K^0(X, T)$.*

Proof Let α be a state on $K^0(X, T)$. For a clopen set $U \subset X$, we set $\phi(U) = \alpha(\chi_U)$, where χ_U is the image in $H(X, T, \mathbb{Z})$ of the characteristic function of U. Thus, $\phi(U) \geq 0$ for every clopen set U, and $\phi(U \cup V) = \phi(U) + \phi(V)$ if U, V are disjoint clopen sets. Since X is a Cantor space, its topology is generated by the clopen sets. Thus, there is, by the Carathéodory Theorem, a unique probability measure μ on X such that $\mu(U) = \phi(U)$ for every clopen set U. This already shows that $\mu \mapsto \alpha_\mu$ is one-to-one. For every clopen set U, the difference between the characteristic functions of U and $T^{-1}U$ is a coboundary. Thus, these functions have the same image in $H(X, T, \mathbb{Z})$ and $\phi(U) = \phi(T^{-1}U)$. It follows that μ is T-invariant. Moreover, by construction

$\alpha = \alpha_\mu$ since $\alpha_\mu(\chi_U) = \int \chi_U d\mu = \mu(U) = \alpha(\chi_U)$ for every clopen set U. This shows that the map $\mu \mapsto \alpha_\mu$ is onto.   ∎

We derive directly from Theorem 3.9.3 the following interesting result.

Corollary 3.9.4 *Let (X, T) be a recurrent dynamical system. If $K^0(X, T)$ has a unique state, then (X, T) is uniquely ergodic.*

Observe also that Theorem 3.9.3 implies that there is a bijection between the invariant Borel probability measures on a minimal system (X, T) and an induced system (U, T_U) for $U \subset X$ clopen. Indeed, by Proposition 3.7.3, the ordered groups $K^0(X, T)$ and $K^0(U, T_U)$ are isomorphic. A direct proof of the bijection between invariant Borel probability measures is given in Exercise 3.33.

3.9.1 The Number of Ergodic Measures

We note another important consequence of Theorem 3.9.3. The $\mathbb{Q}$-rank of an abelian group G is the dimension over $\mathbb{Q}$ of $\mathbb{Q} \otimes_{\mathbb{Z}} G$ (see Appendix B.2).

Proposition 3.9.5 *Let (X, T) be a recurrent dynamical system. If the group $K^0(X, T)$ has $\mathbb{Q}$-rank n, there are at most n ergodic measures on (X, T). If, additionally, $K^0(X, T)$ is a simple finitely generated dimension group and $n \geq 2$, this number is at most $n - 1$.*

Proof Set $G = K^0(X, T)$ and let m be the number of ergodic measures on (X, T). Let us first show that $m \leq n$. Indeed, let V be the set of $\mathbb{Q}$-linear combinations of ergodic measures on (X, T) and W the set of $\mathbb{Q}$-linear combinations of states on G. By Proposition 3.8.11, the m ergodic measures form a basis of V and thus V has dimension m. On the other hand, an element $\sum \alpha_i p_i$ of W can be identified with a $\mathbb{Q}$-linear form on $\mathbb{Q} \otimes_{\mathbb{Z}} G$ by $(\sum \alpha_i p_i)(\alpha \otimes g) = \sum \alpha_i \alpha p_i(g)$. Thus, W has dimension at most n. Since the map $\mu \mapsto \alpha_\mu$ defines a linear injective map from V to W, the conclusion follows.

Suppose now that G is simple and finitely generated and that $n \geq 2$. Assume that $m = n$. Let $f_1, f_2, \ldots, f_n$ be n states generating a space W of dimension n. Let $v_1, v_2, \ldots, v_n \in \mathbb{Q} \otimes_{\mathbb{Z}} G$ be a dual basis. Fix $g \in G^+$ and set $g = \sum_{i=1} \alpha_i v_i$. Let $D = \{h \in G \mid 0 \leq h \leq g\}$. By Proposition 2.2.9, we have

$$D = \{h \in \mathbb{Z}^n \mid 0 \leq f_i(h) \leq \alpha_i \text{ for } 1 \leq i \leq n\}.$$

Since the last set is contained in the compact set

$$K = \left\{ \sum_{i=1}^n \beta_i v_i \mid 0 \leq \beta_i \leq \alpha_i \text{ for } 1 \leq i \leq n \right\},$$

it is finite. This implies that D contains minimal nonzero elements. Such an element generates a nonzero ideal J. Since G is simple, we have $J = G$ and thus $n = 1$, a contradiction. ∎

We will see later many examples where the above hypotheses are satisfied. Note that we already met the particular case where G is isomorphic to a subgroup of $\mathbb{Q}$, and thus has $\mathbb{Q}$-rank 1, in Example 2.2.3.

3.9.2 Dimension Groups of Sturmian Shifts

As an illustration of Theorem 3.9.3, let us prove the following statement, which gives the form of the cohomology group of Sturmian shifts. In the next statement the group $\mathbb{Z} + \alpha\mathbb{Z}$ of real numbers of the form $x + \alpha y$ is considered as a unital ordered group for the order induced by $\mathbb{R}$ and the order unit 1.

Theorem 3.9.6 *Let X be a Sturmian shift of slope α. Then $K^0(X, S) = \mathbb{Z} + \alpha\mathbb{Z}$.*

Proof We have already seen (Example 3.3.4) that $H(X, S, \mathbb{Z}) = \mathbb{Z}^2$ for every Sturmian shift. Let $\{0, 1\}$ be the alphabet of X. We can identify the group $H(X, S, \mathbb{Z})$ with the pairs $(x, y) \in \mathbb{Z}^2$ via the map $(x, y) \mapsto x[0] + y[1]$. We have seen (Exercise 3.23) that (X, S) is uniquely ergodic and that the unique invariant Borel probability measure μ is such that $\mu([0]) = 1 - \alpha$. By Theorem 3.9.3, the ordered group $K^0(X, S)$ has a unique state, which is α_μ. Thus, by Proposition 2.2.9, $H^+(X, S, \mathbb{Z}) = \{(x, y) \in \mathbb{Z}^2 \mid (1-\alpha)x + \alpha y > 0\} \cup \{0\}$, the order unit being $(1, 1)$. Finally, the map $(x, y) \to (x, y - x)$ identifies $K^0(X, T)$ and $\mathbb{Z} + \alpha\mathbb{Z}$. ∎

Example 3.9.7 Let X be the Fibonacci shift. It is a Sturmian shift of slope $\alpha = (3 - \sqrt{5})/2$ (Example 1.5.15). Thus, $K^0(X, S) = \mathbb{Z}[\alpha]$.

As an application of Theorem 3.9.6, we prove the following result.

Theorem 3.9.8 *Let X and Y be Sturmian shifts of slopes α and β respectively. The following conditions are equivalent.*

 (i) *X, Y are conjugate.*
 (ii) *The cohomology groups of X, Y are isomorphic.*
 (iii) *$\alpha = \beta$ or $\alpha = 1 - \beta$.*
 (iv) *X, Y differ by a permutation of the two letters.*

Proof (i)$\Rightarrow$ (ii). If X, Y are conjugate, their cohomology groups are isomorphic.

(ii)$\Rightarrow$(iii). By Theorem 3.9.6, the groups $\mathbb{Z}+\alpha\mathbb{Z}$ and $\mathbb{Z}+\beta\mathbb{Z}$ are isomorphic. Thus, there are $a, b, c, d \in \mathbb{Z}$ such that $\beta = a + b\alpha$ and $\alpha = c + d\beta$. Then $\alpha = c + ad + bd\alpha$ implies $bd = 1$. Thus, either $\alpha = \beta$ or $\alpha = 1 - \beta$.

(iii)$\Rightarrow$ (iv). If $\alpha = \beta$, then $X = Y$. If $\alpha = 1 - \beta$, let $\alpha = [0; 1 + d_1, d_2, \ldots]$ be the continued fraction expansion of α. We may assume $\alpha > 1/2$, that is $d_1 = 0$. Then $1 - \alpha = [0, 1 + d_2, d_3, \ldots]$. Thus, the directive sequences of X and Y are respectively $1^{d_2}0^{d_3} \cdots$ and $0^{d_2}1^{d_3} \cdots$. Thus, X, Y differ by a permutation of the letters.

(iv)$\Rightarrow$(i) is obvious. ∎

This statement is a striking illustration of the power of the cohomology group. It would probably be very difficult to prove the equivalence of (i) and (iv) without using the cohomology group.

3.9.3 Infinitesimal and Image Subgroups

We can also use Theorem 3.9.3 to give the form of the image subgroup and the infinitesimal subgroup for a minimal Cantor system (X, T).

The image subgroup of $K^0(X, T)$, denoted $I(X, T)$, is given by

$$I(X, T) = \cap_{\mu \in \mathcal{M}(X,T)} \left\{ \int f d\mu \mid f \in C(X, \mathbb{Z}) \right\}, \tag{3.22}$$

where $\mathcal{M}(X, T)$ denotes the set of invariant Borel probability measures on (X, T). This follows directly from the definition of the image subgroup given in Eq. (2.3) by Theorem 3.9.3.

The infinitesimal subgroup of $K^0(X, T)$, denoted $\text{Inf}(X, T)$ is given by

$$\text{Inf}(X, T) = \left\{ [f] \in H(X, T, \mathbb{Z}) \mid \int f d\mu = 0 \text{ for all } \mu \in \mathcal{M}(X, T) \right\}. \tag{3.23}$$

If (X, T) is uniquely ergodic, the quotient $K^0(X, T)/\text{Inf}(K^0(X, T))$ is isomorphic to $I(X, T)$ as a unital ordered group.

Example 3.9.9 Let X be the Fibonacci shift as in the previous example. Then $\text{Inf}(X, S) = \{0\}$ and $I(X, S) = \mathbb{Z} + \alpha\mathbb{Z}$.

3.10 Exercises

Section 3.1

3.1 Let (X, T) be an invertible topological dynamical system. For $f \in C(X, \mathbb{R})$ extend the definition of $f^{(n)}(x)$ to negative indexes by defining for $n \geq 1$ and $x \in X$

$$f^{(-n)}(x) = -f^{(n)}(T^{-n}x). \tag{3.24}$$

Show that, with this definition, the *cohomological equation*

$$f^{(n+m)}(x) = f^{(n)}(x) + f^{(m)}(T^n x) \tag{3.25}$$

holds for every $n \in \mathbb{Z}$ and $x \in X$.

3.2 A *homotopy* between topological spaces X, Y is a family $f_t \colon X \to Y$ ($t \in [0, 1]$) of maps such that the associated map $F \colon X \times [0, 1]$ given by $F(x, t) = f_t(x)$ is continuous. One says in this case that $f_0, f_1 \colon X \to Y$ are *homotopic*. Show that homotopy is an equivalence relation compatible with composition.

3.3 A continous function is *nullhomotopic* if it is homotopy equivalent to a constant function. Show that any function $f \colon X \to [a, b]$ from X to a closed interval $[a, b]$ of $\mathbb{R}$ is nullhomotopic.

3.4 A *continuous flow* is a pair $(X, (T_t)_{t \in \mathbb{R}})$ of a compact metric space X and a family $(T_t)_{t \in T}$ of homeomorphisms $T_t \colon X \to X$ such that

(i) the map $(x, t) \mapsto T_t(x)$ is continuous from $X \times \mathbb{R}$ to X,
(ii) $T_{t+s} = T_t \circ T_s$ for all $s, t \in \mathbb{R}$.

For every topological dynamical system (X, T), one can build a continuous flow called the *suspension flow* over (X, T) as follows. Consider the quotient $\tilde{X}$ of $X \times \mathbb{R}$ by the equivalence relation, which identifies $(x, s+1)$ and (Tx, s) for all $x \in X$ and $s \in \mathbb{R}$. Denote by $[(x, s)]$ the equivalence class of (x, s). Then define T_t on the quotient by

$$T_t([(x, s)]) = [(x, s+t)].$$

Show that we obtain in this way a continuous flow.

3.5 Show that the suspension flow over any periodic system can be identified with the torus $\mathbb{T} = \mathbb{R}/\mathbb{Z}$.

3.6 An *equivalence* between two continuous flows $(X, (T_t)_{t \in \mathbb{R}})$ and $(X', (T'_t)_{t \in \mathbb{R}})$ is a homeomorphism $\pi \colon X \to X'$, which maps orbits of T_t to orbits of T'_t in an orientation-preserving way, that is, for all $x \in X$, $\pi(T_t(x)) = T'_{f_x(t)}(\pi(x))$ for some monotonically increasing $f_x \colon \mathbb{R} \to \mathbb{R}$. Two flows are *equivalent* if there is an equivalence from one to the other.

Two (ordinary) topological dynamical systems are *flow equivalent* if their suspension flows are flow equivalent. Show that two conjugate dynamical systems are flow equivalent but that the converse is false.

3.7 Let (X, T) be a topological dynamical system and let $(\tilde{X}, (T_t)_{t \in \mathbb{R}})$ be its suspension flow (as defined in Exercise 3.4). The first *Čech cohomology group* of $\tilde{X}$, denoted $H^1(\tilde{X}, \mathbb{Z})$, is the group of continuous maps from $\tilde{X}$ to the torus $\mathbb{T} = \mathbb{R}/\mathbb{Z}$ modulo the group of nullhomotopic maps.

Show that $H^1(\tilde{X}, \mathbb{Z})$ is isomorphic to $H(X, T, \mathbb{Z})$. Hint: Consider the map $\pi : C(X, \mathbb{Z}) \to C(\tilde{X}, \mathbb{T})$, which associates to $f \in C(X, \mathbb{Z})$, the map $\pi(f) \in C(\tilde{X}, \mathbb{T})$ defined by $\pi(f)(x, s) = \tau(f(x)s)$ where $\tau : \mathbb{R} \to \mathbb{T}$ is the natural projection. Show that π induces an isomorphism from $H(X, T, \mathbb{Z})$ onto $H^1(\tilde{X}, \mathbb{Z})$.

Section 3.2

3.8 Show that if (X, T) is a topologically transitive system, then, for every $f \in C(X, \mathbb{R})$, two solutions of the equation $\partial g = f$ differ by a constant.

3.9 Let (X, T) be an invertible system. For $f \in C(X, \mathbb{R})$, assume that the sequence $(f^{(n)}(x))_{n \geq 0}$ is uniformly bounded. Set

$$g(x) = \sup_{n \in \mathbb{Z}} f^{(n)}(x),$$

where $f^{(n)}$ is defined for all $n \in \mathbb{Z}$ as in Exercise 3.1. Show that $\partial g = f$.

3.10 Let (X, T) be a minimal system. Let $f \in C(X, \mathbb{R})$ and $x_0 \in X$ be such that $(f^{(n)}(x_0))_{n \geq 0}$ is bounded. Show that $f^{(n)}(x)$ is bounded for all $x \in X$ and that $g(x) = \sup_{n \in \mathbb{Z}} f^{(n)}(x)$ is a continuous function such that $\partial g = f$. Hint: use Exercise 3.9.

3.11 Let (X, T) be a minimal invertible system and $f \in C(X, \mathbb{R})$ be such that $(f^{(n)}(x_0))_{n \geq 0}$ is bounded for some $x_0 \in X$.

Let (Y, S) be the dynamical system formed of $Y = X \times \mathbb{R}$ with $S(x, t) = (Tx, t + f(x))$. In this way, $S^n(x, t) = (T^n x, t + f^{(n)}(x))$ for all $n \geq 0$.

1. Show that the closure E of the set $\{S^n(x_0, 0) \mid n \geq 0\}$ is compact and S-invariant.
2. Let K be a minimal closed S-invariant nonempty subset of E. Show that $K = \{(x, g(x)) \mid x \in X\}$ for some $g \in C(X, \mathbb{R})$.
3. Show that $\partial g = f$.

Section 3.5

3.12 Show that for the shift space X of Example 3.3.2, $\chi_{[ab]}$ is not a coboundary.

3.13 Prove directly (that is, without using Corollary 3.2.3) that if X is recurrent, $\mathcal{G}_n(X)$ is an ordered group.

3.14 Prove that a recurrent system (X, T) is periodic of period n if and only if $K^0(X, T) = (\mathbb{Z}, \mathbb{Z}_+, n)$.

Section 3.8

3.15 The *Markov–Kakutani fixed point theorem* states that if V is a topological vector space and K is a convex and compact subset of V, then every continuous linear map T mapping K into itself has a fixed point in K.

Use this theorem to give a proof of Theorem 3.8.4. Hint: Consider the map $T^* \colon \mathcal{M}_X \to \mathcal{M}_X$ defined by $T^*\mu = \mu \circ T^{-1}$.

3.16 Prove Proposition 3.8.6. Hint: set $V = \cap_{n \geq 1} \cup_{i \geq n} T^{-i}U$. Show that $T^{-1}V = V$ and $U = V \bmod \mu$.

3.17 Show that if an invariant measure μ on (X, T) is ergodic, then

$$\lim_{n \to \infty} \frac{1}{n} \sum_{i=0}^{n-1} \mu(U \cap T^{-i}V) = \mu(U)\mu(V) \tag{3.26}$$

for all Borel sets U, V and that the converse holds provided (3.26) is true for all clopen sets U, V.

3.18 An invariant probability probability measure μ on (X, T) is *mixing* if

$$\lim_{n \to \infty} \mu(U \cap T^{-n}V) = \mu(U)\mu(V) \tag{3.27}$$

for all Borel sets U, V. Show that if μ is mixing, it is ergodic and that if (3.27) holds for all clopen sets U, V, then μ is mixing. Use this to prove that Bernoulli measures are mixing and thus ergodic.

3.19 Let μ be the invariant probability measure defined by a Markov chain (v, P) with $vP = v$ and $v > 0$. Show that μ is ergodic if and only if the matrix P is irreducible, and mixing if and only if P is primitive.

3.20 An invariant probability measure μ on (X, T) is *weakly mixing* if

$$\lim_{n \to \infty} \frac{1}{n} \sum_{i=0}^{n-1} |\mu(U \cap T^{-i}V) - \mu(U)\mu(V)| = 0 \qquad (3.28)$$

(by contrast, a mixing probability measure is also called *strongly mixing*). Show that mixing $\Rightarrow$ weakly mixing $\Rightarrow$ ergodic.

3.21 A dynamical system (X, T) is *topologically mixing* if for every non-empty open sets $U, V \subset X$, there is an N such that for all $n \geq N$, one has $U \cap T^{-n}V \neq \emptyset$. Let μ be an invariant probability measure on (X, T) such that $\mu(U) > 0$ for every nonempty open set U. Show that if μ is mixing, then (X, T) is topologically mixing.

3.22 Prove Proposition 3.8.7. Hint: prove that the set of recurrent points has measure 1.

3.23 Use Oxtoby's Theorem (Theorem 3.8.12) to prove that irrational rotations are uniquely ergodic. Conclude that Sturmian shifts are uniquely ergodic and that the unique invariant Borel probability measure μ on a Sturmian shift of slope α satisfies $\mu([0]) = 1 - \alpha$. Hint: use the fact that trigonometric polynomials are dense in the space $C(\mathbb{S}^1, \mathbb{C})$ of complex valued continuous functions defined on the unit circle.

3.24 Show that a rotation of the interval $[0, 1]$ is never mixing.

3.25 Let $(k_i)_{i \geq 0}$ be a sequence of positive integers such that k_i divides k_{i+1}. Let $x \in \{0, 1\}^{\mathbb{Z}}$ be the Toeplitz sequence with periodic structure $(k_i)_{i \geq 0}$ defined as follows. For $i \geq 1$, let

$$E_i = \cup_{m \in \mathbb{Z}} \{n \in \mathbb{Z} \mid |n - mk_i| \leq k_{i-1}\}.$$

For $n \in \mathbb{Z}$, let $p(n)$ be the least integer p such that $n \in E_p$. Set $x_n \equiv p(n)$ mod. 2. Show that if

$$\sum_{i \geq 1} \frac{k_{i-1}}{k_i} \leq \frac{1}{12}, \qquad (3.29)$$

then x is not uniquely ergodic.

3.26 Show that if (X, T) is a topological dynamical system and μ is an invariant Borel probability measure, then (X, T, μ) is a measure-theoretic dynamical system.

3.27 Let (X, T, μ) be a measure-theoretic dynamical system with T invertible. Let $H = L^2(X)$ be the Hilbert space of real-valued square integrable functions on X (modulo a.e. vanishing functions).

Show that the operator U defined by $Uf = f \circ T$ is a unitary operator from H to itself.

3.28 Let (X, T, μ) be a measure-theoretic dynamical system. Show that an element f of the Hilbert space $H = L^2(X)$ is a coboundary of the form $f = Ug - g$ for some $g \in H$, with U as in Exercise 3.27, if and only if $\| f^{(n)} \|$ is bounded. Hint: Assume $\| f^{(n)} \| \le k$ and consider the closure of the set of convex linear combinations of the $U^n f$. Apply the Schauder–Tychonov fixed point theorem to the map $h \mapsto f + Uh$.

3.29 Let $\varphi \colon A^* \to A^*$ be a primitive morphism and let X be the associated shift space. We indicate here a method to compute the frequency of the factors of length k in $\mathcal{L}(X)$ by a faster method than by using Formula (3.21).

Let M_k be the incidence matrix of φ_k. Let p be an integer such that $|\varphi^p(a)| > k - 2$ for all $a \in A$. Let U be the $\mathcal{L}_2(X) \times \mathcal{L}_k(X)$–matrix defined as follows. For $a, b \in A$ such that $ab \in \mathcal{L}_2(X)$ and $y \in \mathcal{L}_k(X)$, $U_{ab,y}$ is the number of occurrences of y in $\varphi^p(ab)$ that begin inside the prefix $\varphi^p(a)$. Show that

$$UM_k = M_2 U,$$

and that if v_2 is a row eigenvector of M_2 corresponding to the common dominant eigenvalue ρ of M_2 and M_k, then $v_k = v_2 U$ is an eigenvector of M_k corresponding to ρ.

3.30 Let $\mu \colon a \to ab, b \to ba$ be the Thue–Morse morphism. Show that for $k = 5$, $p = 3$, the matrix U of the previous problem (with the 12 factors of length 5 of the Thue–Morse sequence listed in alphabetical order) is

$$U = \begin{bmatrix} 1 & 0 & 1 & 1 & 0 & 1 & 1 & 0 & 1 & 1 & 0 & 1 \\ 0 & 1 & 1 & 0 & 1 & 1 & 1 & 1 & 0 & 1 & 1 & 0 \\ 1 & 1 & 0 & 1 & 1 & 0 & 0 & 1 & 1 & 0 & 1 & 1 \\ 1 & 0 & 1 & 1 & 0 & 1 & 1 & 0 & 1 & 1 & 0 & 1 \end{bmatrix}$$

and that the vector $v_2 U$ with $v_2 = \begin{bmatrix} 1 & 2 & 2 & 1 \end{bmatrix}$ is the vector with all components equal to 4. Conclude that the 12 factors of length 5 of the Thue–Morse sequence have the same frequency (see Example 3.8.20).

3.31 Let k, p be as in the previous exercise. Let V be the matrix of the map $\pi: \mathcal{L}_k(X) \to \mathcal{L}_2(X)$ that sends $a_0 a_1 \cdots a_{k-1}$ to $a_0 a_1$. Show that M_2 is shift equivalent over $\mathbb{Z}$ to M_k (see Exercise 2.15) and more precisely that $(U, V): M_2 \sim M_k$ (lag k). Conclude that M_2 and M_k have the same nonzero eigenvalues. Hint: use Exercise 2.16.

Section 3.9

3.32 Let X be the shift space of Example 3.5.4. Show that there is no decomposition $\chi_{[b]} = \phi_1 + \phi_2$ such that $\phi_1 \le \chi_{[d]}$ and $\phi_2 \le \chi_{[e]}$. Hint: To prove that $\chi_{[b]} \le \phi_1 \le \chi_{[d]}$, find an invariant Markov measure such that $\mu([a^{n-1}b]) > 0$ for all $n \ge 1$ while $\mu([d]) = 0$ and use Theorem 3.9.3.

3.33 Let (X, T) be a minimal dynamical system and let $U \subset X$ be a clopen set. Let $n(x) = \inf\{n > 0 \mid T^n x \in U\}$ be the entrance time in U and $T_U(x) = T^{n(x)}(x)$ be the transformation induced by T on U (see Section 3.7). Let also $X_n = \{x \in X \mid n(x) = n\}$. Show that for every invariant Borel probability measure ν on (U, T_U), the measure defined on X by

$$\hat{\nu}(V) = \frac{1}{\lambda} \sum_{n \ge 1} \nu(T^n(V \cap X_n)), \tag{3.30}$$

with $\lambda = \sum_{n \ge 1} n\nu(U \cap X_n)$ is an invariant Borel probability measure on (X, T). Show that the map $\nu \mapsto \hat{\nu}$ is a bijection from the set of invariant Borel probability measures on (U, T_U) onto the set of invariant Borel probability measures on (X, T).

3.34 Let $\varphi: A^* \to B^*$ be an injective nonerasing morphism. Let X be a shift space on A and let Y be the closure under the shift of $\varphi(X)$. For a nonempty word $w \in B^*$, a *context* of w is a pair (u, v) of words in B^* such that $uwv = \varphi(a_1 \cdots a_n)$ with $n \ge 0$, $a_i \in A$ and u (resp. v) a proper prefix of $\varphi(a_1)$ (resp. a proper suffix of $\varphi(a_n)$). Denote by $C(w)$ the set of contexts of w.

Let ν be a shift-invariant Borel probability measure on X. Set $\lambda = \sum_{a \in A} |\varphi(a)| \nu([a])$. For $w \in B^*$, let

$$\mu([w]) = \frac{1}{\lambda} \sum_{(u,v) \in C(w)} \nu(\varphi^{-1}[uwv]). \tag{3.31}$$

Show that μ defines an invariant Borel probability measure on Y.

3.11 Notes

For a general introduction to cohomology, see, for example, Hatcher (2001).

3.11.1 Gottschalk and Hedlund Theorem

The Gottschalk and Hedlund Theorem (Theorem 3.2.5) is from Gottschalk and
Hedlund (1955). A somewhat simpler proof appears in Katok and Hasselblatt
(1995, Theorem 2.9.4). It is reproduced as Exercise 3.10 using an addition due
to Petite (2019). Yet another proof (perhaps the most elegant one) is given in
Exercise 3.11. According to Petite (2019), it should be credited to Michael
Herman (through Sylvain Crovisier). Given $f \in C(X, T)$, one may consider
$\partial g = f$ as a functional equation with g as unknown. More generally, an
equation of the form

$$f(x) = \lambda g(Tx) - g(x), \tag{3.32}$$

where $T : X \to X$ is a given map, f is a given scalar function on X, λ is a given
constant and g is an unknown scalar function, is called in Katok and Hasselblatt
(1995) a *cohomological equation*. In this context, given a dynamical system
(X, T), a map $\alpha : \mathbb{Z} \times X \to \mathbb{R}$ is called a *one-cocycle* if it satisfies the identity
$\alpha(n + m, x) = \alpha(n, T^m x) + \alpha(m, x)$. Thus, for every map $f \in C(X, \mathbb{R})$, the
map $(n, x) \mapsto f^{(n)}(x)$ is a one-cocycle (see Exercise 3.1).

The analogue of the ordered cohomology group for real-valued functions
has been studied by Ormes (2000).

3.11.2 Ordered Cohomology Group

Boyle and Handelman (1996) have introduced the term *ordered cohomology
group* of a topological dynamical system (X, T) for the group $K^0(X, T)$,
which had before been defined only for minimal systems. They showed that
this group is a complete invariant for flow equivalence of irreducible shifts of
finite type. The group is actually defined for a class of systems that is larger
than recurrent dynamical systems, namely that of *chain recurrent* dynamical
systems. This direction had been explored previously by Poon (1989).

A word is in order here on the notation $K^0(X, T)$. The letter K is used in
reference to K-theory, and more precisely algebraic K-theory. In this theory,
an abelian group $K(\mathfrak{A})$ is associated with a ring $\mathfrak{A}$, based on the structure of
idempotents in the algebra of matrices over this ring. These algebras, actually
C^*-algebras, will be introduced in Chapter 9, and the group $K^0(X, T)$ will be
shown to be the group $K_0(\mathfrak{A})$ for an algebra $\mathfrak{A}$ associated with (X, T). In the

present case, the group is also the K_0 group of the cross product C^*-algebra arising from a dynamical system (see Boyle and Handelman (1996)). We warn readers familiar with topological K-theory that the group $K^0(X, T)$ is not the group $K^0(X)$ associated with X as in Atiyah (1967) (we thank Claude Schochet for this observation).

The proof of Lemma 3.3.7 follows the lines of the proof given in Giordano et al. (2018).

Example 3.5.4 showing that the ordered cohomology group of a shift of finite type may fail to be a dimension group is from Kim et al. (2001, Example 3.3).

3.11.3 Invariant Measures and States

The original reference for the Krylov–Bogolyubov Theorem (Theorem 3.8.4) is Krylov and Bogolioubov (1937).

The original reference for Oxtoby's Theorem (Theorem 3.8.12) is Oxtoby (1952). The example of a minimal non uniquely ergodic shift (Exercise 3.25) is from Oxtoby (1952). It is the first ever constructed sequence with this property.

The ergodic theorem, due to Birkhoff, is also called the *pointwise ergodic theorem* or *strong ergodic theorem*, in contrast with the *mean ergodic theorem*, which is due to von Neumann and states the weaker convergence in mean in an L^2-space. The pointwise ergodic theorem is sometimes also called the Birkhoff–Khinchin Theorem.

The fact that the shift associated with a primitive substitution is uniquely ergodic (Theorem 3.8.13) is due to Michel (1974). We follow the proof of Queffélec (2010). A one-sided version of Lemma 3.8.15 appears in Dumont and Thomas (1989). The computation of the invariant measure on the shift associated with a primitive substitution is developed in Queffélec (2010). Formula (3.21) is from Queffélec (2010, Corollary 5.14),

Theorem 3.9.3 is Theorem 5.5 in Herman et al. (1992), where it is credited to Kerov. The computation of the cohomology group of Sturmian shifts is classical. It appears in particular in Dartnell et al. (2000).

3.11.4 Exercises

The equivalent definition of $H(X, T, \mathbb{Z})$ in terms of Čech cohomology (Exercise 3.7) is taken from Parry and Tuncel (1982). The definition of the Čech cohomology is not the classical one but is equivalent. According to Parry and Tuncel (1982), it defines the *Brushlinski group*.

The variant of the Gottschalk and Hedlund Theorem for L^2 presented in Exercise 3.28 is from Parry and Tuncel (1982; Proposition II.2.11).

The Schauder–Tychonoff fixed point theorem can be found in Dunford and Schwartz (1988, Theorem V.10.5), while the Markov–Kakutani fixed point theorem appears as Theorem V.10.6 in Dunford and Schwartz (1988). The unique ergodicity of irrational rotations (Exercise 3.23) is known as the Kronecker–Weyl Theorem.

The notion of strongly or weakly mixing transformations (Exercises 3.18 and 3.20 are classical in Ergodic Theory (see Petersen (1983) for example). Rotations of the interval are never mixing (Exercise 3.24) nor even weakly mixing.

The relation of topological mixing (Exercise 3.21) with (measure-theoretic) mixing is described in Petersen (1983) together with a notion of topological ergodicity. The following result was proved in Kenyon et al. (2005). Let $\sigma : A^* \to A^*$ be a primitive morphism, let λ be its dominant eigenvalue and let θ be the maximal modulus of the other eigenvalues. If $\theta > 1$, the shift $X(\sigma)$ is topologically mixing if and only if for every $n \geq 1$, the gcd of the $|\sigma^n(a)|$ for $a \in A$ is 1.

The invariant Borel probability measure μ defined by Formula (3.31) is called the *contextual probability* measure on Y in Hansel and Perrin (1983) (see also Blanchard and Perrin (1980)). The following result is proved in Hansel and Perrin (1983). Assume that $X = A^{\mathbb{Z}}$, $Y = B^{\mathbb{Z}}$ and that π is a Bernoulli measure on $B^{\mathbb{Z}}$ such that $\nu([w]) = \pi([\varphi(w)])$ for every $w \in A^*$. Then the contextual probability measure μ is equal to π.

4

Partitions in Towers

In this chapter, we first present the important notion of partition in towers. It is the basis of many of the constructions that will follow. We prove the Theorem of Herman, Putnam and Skau asserting that every invertible minimal Cantor system can be represented by a sequence of partitions in towers (Theorem 4.1.6).

We next show how partitions in towers allow us to compute the ordered cohomology group of invertible minimal Cantor systems. We first define in Section 4.2 the ordered group associated with a partition. The definition uses the notions on induction introduced in Section 3.7. In Section 4.3, we prove that the ordered cohomology group $K^0(X, T)$ is the direct limit of the sequence of ordered groups associated with a sequence of partitions in towers. This allows us to prove, as a main result of this chapter, the Theorem of Herman, Putnam and Skau (a second one) asserting that the ordered cohomology group of a minimal invertible Cantor system is a simple dimension group (Theorem 4.3.4).

We will use these results to determine the dimension groups of some minimal shift spaces and relate them to several notions such as return words or Rauzy graphs. We will in particular consider episturmian shifts (Proposition 4.3.4), which will illustrate the use of return words. Next, we will use Rauzy graphs to give a description of dimension groups of substitution shifts (Proposition 4.3.4). These methods will appear again in a new light in the next chapter when we consider Bratteli diagrams.

4.1 Partitions in Towers

Let (X, T) be an invertible Cantor system. Let $B_1, \ldots, B_m$ be a family of disjoint nonempty open sets and let $h_1, \ldots, h_m$ be positive integers. Assume that the family

156

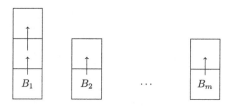

Figure 4.1 A partition in towers.

$$\mathfrak{P} = \{T^j B_i \mid 1 \le i \le m,\ 0 \le j < h_i\}$$

is a partition of X. It implies that each element of $\mathfrak{P}$ is a clopen set. We say that it is the *clopen partition in towers* (or *Kakutani–Rokhlin partition* or *KR-partition*) of (X, T) built on $B_1, \ldots, B_m$ with *heights* $h_1, \ldots, h_m$.

The number of towers is m, $\{T^j B_i \mid 0 \le j < h_i\}$ is the ith *tower*, h_i is its *height* and B_i its *basis*. The union $B(\mathfrak{P}) = \cup_i B_i$ is the *basis* of the partition $\mathfrak{P}$.

Since T is bijective, it sends the elements at the top of the towers back to the bottom, that is $T^{h_i} B_i$ is included in $B(\mathfrak{P})$ for $1 \le i \le m$. As a consequence, when such a partition exists, the integers h_i are unique.

Thus, informally speaking, a partition in towers gives an approximate description of the action of T on X. Each tower can be seen as a stack of clopen sets. The transformation consists in climbing one step up the stack except at the top level, where it goes back to the basis of some tower (see Figure 4.1).

A fundamental observation is the following link between partitions in towers and induction. Let $\mathfrak{P}$ be a partition in towers of a minimal system (X, T). Then the return time to the basis $B = B(\mathfrak{P})$ is bounded and thus the system (B, T_B) induced by (X, T) on the clopen set B is well defined. It is easy to verify that X is conjugate to the system obtained from (B, T_B) by the tower construction (Section 1.1.3) using the function $f(x) = h_i$ if x belongs to B_i.

Note also the important fact that if μ is an invariant Borel probability measure on (X, T) and $\mathfrak{P} = \{T^j B_i \mid 1 \le i \le m, 0 \le j < h_i\}$, then

$$\sum_{i=1}^{m} h_i \mu(B_i) = 1, \tag{4.1}$$

since $X = \cup_i \cup_j T^j B_i$ and $\mu(T^j B_i) = \mu(B_i)$.

4.1.1 Partitions and Return Words

We give two examples of partitions in towers of a shift space. The first example is related to return words. The partition relies on the location of two (possibly

Figure 4.2 Two occurrences of w in x.

overlapping) occurrences of a word w, with the second one occurring at a strictly positive index.

Proposition 4.1.1 *Let X be a minimal two-sided shift space. For every $w \in \mathcal{L}(X)$, the family*

$$\mathfrak{P} = \{S^j[vw] \mid v \in \mathcal{R}'_X(w), 0 \le j < |v|\}$$

is a partition in towers with basis the union of the cylinders $[vw]$ for $v \in \mathcal{R}'_X(w)$.

Proof Let $x \in X$. Since X is minimal, there is a smallest integer $i > 0$ such that $x_{[i,i+|w|-1]} = w$. By the definition of a left return word, there is a unique integer $j \ge 0$ such that $x_{[-j,i-1]}$ belongs to $\mathcal{R}'_X(w)$ (see Figure 4.2).

We set $v = x_{[-j,i-1]} \in \mathcal{R}'_X(w)$ and observe that we have $0 \le j < |v|$. Thus, x belongs to $S^j[vw]$. Since v and j are unique, this shows that $\mathfrak{P}$ is a partition. ∎

Example 4.1.2 Let X be the two-sided Fibonacci shift on the alphabet $A = \{a, b\}$. We have $ab \in \mathcal{L}(X)$ and $\mathcal{R}'_X(ab) = \{ab, aba\}$. The corresponding partition in towers is represented in Figure 4.3.

4.1.2 Partitions of Substitution Shifts

The second example is a partition of a substitution shift.

Proposition 4.1.3 *Let $\varphi: A^* \to A^*$ be a primitive morphism and let X be the associated shift space. If X is infinite, the family*

$$\mathfrak{P}(n) = \{S^j \varphi^n([a]) \mid a \in A, \ 0 \le j < |\varphi^n(a)|\}$$

is, for every $n \ge 0$, a partition in towers with basis $\varphi^n(X)$.

Proof We may assume $n \ge 1$. Set $\psi = \varphi^n$. Since φ is primitive and X is infinite, the morphism φ is recognizable on X (Theorem 1.4.35), and thus ψ is recognizable on X. This implies clearly that $\mathfrak{P}(n)$ is a partition. ∎

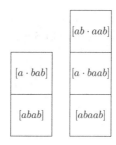

Figure 4.3 A partition in towers of the Fibonacci shift.

The partition $\mathfrak{P}(n)$ is called the *partition associated with the morphism* φ^n. We illustrate Proposition 4.1.3 with the following example.

Example 4.1.4 Let φ be the Fibonacci morphism and let X be the Fibonacci shift. The partition $\mathfrak{P}(n)$ corresponding to Proposition 4.1.3 has two towers with basis $\varphi^n([a])$ and $\varphi^n([b])$ respectively. Note that $\varphi^n([a]) \subset [\varphi^n(a)]$ but that the inclusion is strict for $n \geq 2$. Indeeed, we have $\varphi^2([a]) = [abaab]$. The partition in towers for $n = 2$ is identical to the partition represented in Figure 4.3.

4.1.3 Sequences of Partitions

A partition α is a *refinement* of a partition ρ if every element of α is a subset of an element of ρ or, equivalently, if every element of ρ is a union of elements of α. We also say that ρ is *coarser* than α. We denote $\alpha \geq \rho$.

Let $\mathcal{A}$ and $\mathcal{B}$ be two families of subsets of some set X. We set

$$\mathcal{A} \vee \mathcal{B} = \{A \cap B \mid A \in \mathcal{A}, \ B \in \mathcal{B}\}.$$

For finitely many families $\mathcal{A}_1, \mathcal{A}_2, \ldots, \mathcal{A}_n$ of subsets of X, we set

$$\bigvee_{i=1}^{n} \mathcal{A}_i = \mathcal{A}_1 \vee \mathcal{A}_2 \vee \cdots \vee \mathcal{A}_n$$

while $\mathcal{A} \cup \mathcal{B}$ is the union of the families $\mathcal{A}$ and $\mathcal{B}$.

We also say that a partition in towers $\mathfrak{P}$ with basis B is *nested* in a partition in towers $\mathfrak{P}'$ with basis B' if $B \subset B'$ and $\mathfrak{P}$ is a refinement, as a partition, of $\mathfrak{P}'$.

A sequence $(\mathfrak{P}(n))_{n \geq 0}$ of partitions in towers is *nested* if $\mathfrak{P}(n+1)$ is nested in $\mathfrak{P}(n)$ for all $n \geq 0$.

The following statement shows that the sequence of partitions of a primitive substitution shift defined in Proposition 4.1.3 is nested.

Proposition 4.1.5 *Let $\varphi: A^* \to A^*$ be a primitive morphism and let X be the associated shift space. If X is infinite, the sequence of partitions*

$$\mathfrak{P}(n) = \{S^j \varphi^n([a]) \mid a \in A, 0 \le j < |\varphi^n(a)|\}$$

is nested.

Proof If $\varphi(a)$ begins with b, we have $\varphi^{n+1}([a]) \subset \varphi^n([b])$. Thus, the base of $\mathfrak{P}(n+1)$ is contained in the base of $\mathfrak{P}(n)$. Next, if $0 \le j < |\varphi^{(n+1)}(a)|$, there is a factorization $\varphi(a) = xcy$ with $c \in A$ such that $|\varphi^n(x)| \le j < |\varphi^n(xc)|$. Set $k = j - |\varphi^n(x)|$. Then

$$S^j \varphi^{n+1}([a]) \subset S^k \varphi^n([c]),$$

with $0 \le k < |\varphi^n(c)|$. This shows that $\mathfrak{P}(n+1)$ refines $\mathfrak{P}(n)$.   ∎

We say that a sequence $(\mathfrak{P}(n))_{n \ge 0}$ with

$$\mathfrak{P}(n) = \{T^j B_i(n) \mid 0 \le j < h_i(n), 1 \le i \le t(n)\}$$

of KR-partitions of X with bases $B(n)$ is a *refining sequence* if it satisfies the three following conditions.

(KR1) $\cap_n B(n) = \{x\}$ for some $x \in X$, that is, the intersection of the bases consists in one point $x \in X$,
(KR2) the sequence $(\mathfrak{P}(n))$ is nested,
(KR3) $\cup_{n \ge 0} \mathfrak{P}(n)$ generates the topology of X (that is, every open set is a union of elements of the partitions $\mathfrak{P}(n)$).

Condition (KR3) can be expressed equivalently by the condition that the sequence of partitions $(\mathfrak{P}(n))$ tends to the point partition (that is, for every $\varepsilon > 0$, there is an n such that all elements of $\mathfrak{P}(n)$ are contained in a ball of radius ε).

None of the conditions (KR1) or (KR3) implies the other one (see Exercises 4.2, 4.3). It may be puzzling that (KR1) does not imply (KR3). After all, the element $T^j B_i(n)$ of $\mathfrak{P}(n)$ at height j is the image by T^j of the element $B_i(n)$ of $B(n)$, which tends to a point. The reason is that T, although continuous, is not an isometry and that the ratio of the diameters of $T^j B_i(n)$ and $B_i(n)$ can be unbounded. This is the phenomenon taking place in Exercise 4.2.

We will prove the following important statement.

Theorem 4.1.6 *Let (X, T) be a minimal invertible Cantor system. There exists a refining sequence of KR-partitions of X.*

The proof uses the following technical result.

Proposition 4.1.7 *Let (X, T) be a minimal invertible Cantor system. Let $\mathfrak{Q}$ be a clopen partition of X and B be a clopen set. Then there exists a clopen partition $B_1, \ldots, B_t$ of B and integers $(h_i)_{1 \le i \le t}$ such that*

$$\mathfrak{P} = \{T^j B_i \mid 0 \le j < h_i, \ 1 \le i \le t\}$$

is finer than $\mathfrak{Q}$.

Proof Let $r_B \colon X \to \mathbb{N}$ be the first return map to B defined by $r_B(x) = \inf\{n > 0 \mid T^n x \in B\}$. Since (X, T) is minimal, r_B is well defined, continuous and takes finitely many values. Let $r_1, r_2, \ldots, r_{t'}$ be the set of values taken by r_B. For every i with $1 \le i \le t'$, we define $B_i' = \{x \in B \mid r_B(x) = r_i\}$. Then

$$\mathfrak{P}' = \{T^j B_i' \mid 0 \le j < r_i, \ 1 \le i \le t'\}$$

is a clopen partition of X. Indeed, since $(B_i')_{1 \le i \le t'}$ is a partition of B, the family $\mathfrak{P}'$ is formed of disjoint sets. Their union is a nonempty closed invariant subset of X, and since (X, T) is minimal, it is equal to X. It is, however, not necessarily finer than $\mathfrak{Q}$. Let $\mathfrak{Q}' = \{P' \cap Q \mid P' \in \mathfrak{P}', Q \in \mathfrak{Q}\}$. It suffices to find $\mathfrak{P}$ finer than $\mathfrak{Q}'$ and which is a partition in towers. Let Q' be an atom of $\mathfrak{Q}'$. There exists a unique pair (i_0, j_0) with $1 \le i_0 \le t'$ and $0 \le j_0 < r_{i_0}$ such that $Q' \subset T^{j_0} B_{i_0}'$. Set $Q'' = T^{j_0} B_{i_0}' \setminus Q'$. We divide the tower i_0 into two new towers and obtain a new KR-partition $\mathfrak{P}''$ with $t' + 1$ towers

$$\mathfrak{P}'' = \{T^j B_i' \mid 0 \le j < r_i, 1 \le i \le t', i \ne i_0\}$$
$$\cup \{T^j Q' \mid -j_0 \le j < r_{i_0} - j_0\}$$
$$\cup \{T^j Q'' \mid -j_0 \le j < r_{i_0} - j_0\},$$

with a split of the i_0-tower propagating up and down the split of $T^{j_0} B_{i_0}'$ in two parts, namely Q' and Q''. We repeat this procedure for every atom of $\mathfrak{Q}'$. The result is the desired KR-partition. ∎

Proof of Theorem 4.1.6 Let $x \in X$. We start choosing a decreasing sequence of clopen sets $(B_n)_{n \ge 0}$ whose intersection is $\{x\}$ and an increasing sequence of partitions $(\mathfrak{P}'(n))_{n \ge 0}$ generating the topology. We apply Proposition 4.1.7 to $\mathfrak{Q} = \mathfrak{P}'(0)$ and $B = B_0$ to obtain $\mathfrak{P}(0)$.

Applying Proposition 4.1.7 iteratively for $n \ge 1$ to $B = B_n$ and by setting now

$$\mathfrak{Q} = \mathfrak{P}'(n) \vee \mathfrak{P}(n-1),$$

we obtain a partition $\mathfrak{P}(n)$ with basis B_n, which is finer than $\mathfrak{P}'(n)$ and $\mathfrak{P}(n-1)$.

Condition (KR1) holds because for each n, the basis of $\mathfrak{P}(n)$ is B_n and $\cap_n B_n = \{x\}$ by hypothesis.

Condition (KR2) holds because, by construction, the partition $\mathfrak{P}(n)$ is nested in $\mathfrak{Q}$, which is nested in $\mathfrak{P}(n-1)$.

Finally, condition (KR3) holds because $\cup_{n\geq 0}\mathfrak{P}(n)$ is finer than $\cup_{n\geq 0}\mathfrak{P}'(n)$, which generates the topology. ∎

Note that, by definition, in a nested sequence $(\mathfrak{P}(n))$ of partitions in towers, the sequence $B(\mathfrak{P}(n))$ is decreasing.

We give two simple examples illustrating Theorem 4.1.6. The first one is the group of p-adic integers (see Section 1.1.4).

Example 4.1.8 We show that the odometer on the group of p-adic integers can be represented by a sequence of partitions in towers with one tower. For $n \geq 0$, let $B(n) = p^n \mathbb{Z}_p$, that is, the ball of $\mathbb{Z}_p$ centered in 0 of radius p^n. Then the family $\mathfrak{P}(n) = \{T^j B(n) \mid 0 \leq j < p^n\}$ is, for each $n \geq 0$, a partition formed of one tower. It is easy to verify that the sequence $(\mathfrak{P}(n))$ satisfies the conditions of Theorem 4.1.6.

In the second example, we show how a nested sequence of partitions can be modified to become a refining sequence.

Example 4.1.9 Let X be the two-sided Fibonacci shift. For $n \geq 0$, let $\mathfrak{P}(n)$ be the partition $\mathfrak{P}(n) = \{S^j \varphi^n([c]) \mid c \in \{a, b\}, 0 \leq j < |\varphi^n(c)|\}$ (see Proposition 4.1.3). Properties (KR1) and (KR3) do not hold for this sequence of partitions. Indeed, $\varphi^2 \colon a \mapsto aba, b \mapsto ab$ has one right infinite fixed point x (the *Fibonacci sequence*) and two left infinite fixed points y, z. The two points (actually fixed points of φ) $y \cdot x$ and $z \cdot x$ belong to all $\varphi^n([a])$. The set $\varphi^n([a])$ being an atom of each $\mathfrak{P}(n)$ and being included in its basis, the sequence of partitions $(\mathfrak{P}(n))_n$ does not satisfy either (KR1) or (KR3).

Using instead the conjugate morphism $\psi \colon a \mapsto baa, b \mapsto ba$, the shift defined remains the same. The sequence of partitions $(\mathfrak{P}(n))$ associated with the substitutions ψ^n satisfies the three conditions because there is a unique fixed point (we shall see in Chapter 6 that this example describes a general situation).

4.2 Ordered Group Associated with a Partition

Let (X, T) be a minimal Cantor system and let

$$\mathfrak{P} = \{T^j B_i \mid 1 \leq i \leq m, 0 \leq j < h_i\}$$

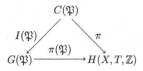

Figure 4.4 The morphism $\pi(\mathfrak{P})$.

be a KR-partition of (X, T) built on $B_1, \ldots, B_m$ and with heights $h_1, \ldots, h_m$. Denote by $C(\mathfrak{P})$ the subgroup of $C(X, \mathbb{Z})$ formed of the functions that are constant on every element of the partition $\mathfrak{P}$ and $C^+(\mathfrak{P}) = C(\mathfrak{P}) \cap C(X, \mathbb{Z}_+)$. The triple $(C(\mathfrak{P}), C^+(\mathfrak{P}), \chi_X)$ is a unital ordered group.

Next, denote by $G(\mathfrak{P})$ the subgroup of $C(B(\mathfrak{P}), \mathbb{Z})$ formed of the functions constant on the basis B_i of each tower, denote by $G^+(\mathfrak{P}) = G(\mathfrak{P}) \cap C(B(\mathfrak{P}), \mathbb{Z}_+)$ and $\mathbf{1}_{\mathfrak{P}}$ the function with value h_i on B_i. We define the *unital ordered group associated with* $\mathfrak{P}$ as the triple $(G(\mathfrak{P}), G^+(\mathfrak{P}), \mathbf{1}_{\mathfrak{P}})$.

Let $I(\mathfrak{P}) \colon C(\mathfrak{P}) \to G(\mathfrak{P})$ be the group morphism defined by

$$(I(\mathfrak{P})(f))(x) = f^{(h_i)}(x)$$

for every $f \in C(\mathfrak{P})$ and $x \in B_i$. It is the restriction to $C(\mathfrak{P})$ of the morphism $I_{B(\mathfrak{P})}$ introduced in Section 3.7. Thus, by Proposition 3.7.3, we have for all $f \in C(\mathfrak{P})$,

$$R(\mathfrak{P}) \circ I(\mathfrak{P})(f) - f \in \partial_T C(\mathfrak{P}) \cap \ker I(\mathfrak{P}), \quad I(\mathfrak{P}) \circ R(\mathfrak{P}) = \mathrm{id}_{G(\mathfrak{P})}, \quad (4.2)$$

where $R(\mathfrak{P}) \colon G(\mathfrak{P}) \to C(X, \mathbb{Z})$ is the restriction to $G(\mathfrak{P})$ of the map $R(\mathfrak{P})$ from $C(B(\mathfrak{P}))$ to $C(X, \mathbb{Z})$. Thus, $R(\mathfrak{P})(f)$ is equal to f on $B(\mathfrak{P})$ and equal to 0 elsewhere.

By Proposition 3.7.3, the kernel of the map $I(\mathfrak{P})$ consists in coboundaries for (X, T). Hence, there exists a morphism $\pi(\mathfrak{P}) \colon G(\mathfrak{P}) \to H(X, T, \mathbb{Z})$, which makes the diagram of Figure 4.4 commutative (we denote by π the canonical morphism from $C(X, \mathbb{Z})$ onto $H(X, T, \mathbb{Z}) = C(X, \mathbb{Z})/\partial_T C(X, \mathbb{Z})$).

Proposition 4.2.1 *The morphism $\pi(\mathfrak{P})$ defines a morphism of unital ordered groups from $(G(\mathfrak{P}), G^+(\mathfrak{P}), \mathbf{1}_{\mathfrak{P}})$ to $K^0(X, T)$.*

Proof It is clear that $\pi(\mathfrak{P})(G^+(\mathfrak{P}))$ is included in $H^+(X, T, \mathbb{Z})$. Next, for any $x \in B_i$,

$$(I(\mathfrak{P})(\chi_X))(x) = h_i.$$

Thus, $\pi(\mathfrak{P})(\mathbf{1}_{\mathfrak{P}}) = \pi(\chi_X) = \mathbf{1}_X$. ∎

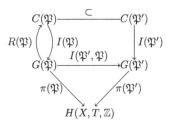

Figure 4.5 The morphism $I(\mathfrak{P}', \mathfrak{P})$.

Assume now that

$$\mathfrak{P}' = \{T^\ell B'_k \mid 1 \le k \le m', \, 0 \le \ell < h'_k\}$$

is another KR-partition in towers nested in $\mathfrak{P}$ with bases $B'_1, \ldots, B'_{m'}$ and heights $h'_1, \ldots, h'_{m'}$. We have then $C(\mathfrak{P}) \subset C(\mathfrak{P}')$ and $\ker I(\mathfrak{P}) \subset I(\mathfrak{P}')$.

The morphism $R(\mathfrak{P})$ maps $G(\mathfrak{P})$ into $C(\mathfrak{P})$, and thus into $C(\mathfrak{P}')$. The morphism

$$I(\mathfrak{P}', \mathfrak{P}) = I(\mathfrak{P}') \circ R(\mathfrak{P})$$

maps $G(\mathfrak{P})$ to $G(\mathfrak{P}')$ (see the commutative diagram in Figure 4.5). We have

$$I(\mathfrak{P}', \mathfrak{P}) \circ I(\mathfrak{P}) = I(\mathfrak{P}') \tag{4.3}$$

where the right side is actually the restriction of $I(\mathfrak{P}')$ to $C(\mathfrak{P})$. Indeed, for $f \in C(\mathfrak{P})$, we have by (4.2), $R(\mathfrak{P}) \circ I(\mathfrak{P})(f) - f \in \ker(I(\mathfrak{P})) \subset \ker(I(\mathfrak{P}'))$ and thus $I(\mathfrak{P}', \mathfrak{P}) \circ I(\mathfrak{P})(f) = I(\mathfrak{P}') \circ R(\mathfrak{P}) \circ I(\mathfrak{P})(f) = I(\mathfrak{P}')(f)$. The morphism $I(\mathfrak{P}', \mathfrak{P})$ is clearly a morphism of ordered groups. It is also a morphism of unital ordered groups because $I(\mathfrak{P}', \mathfrak{P})(1_\mathfrak{P}) = I(\mathfrak{P}', \mathfrak{P})(I(\mathfrak{P})(\chi_X))) = I(\mathfrak{P}')(\chi_X) = 1_{\mathfrak{P}'}$. Moreover,

$$\pi(\mathfrak{P}') \circ I(\mathfrak{P}', \mathfrak{P}) = \pi(\mathfrak{P}) \tag{4.4}$$

by commutativity of the diagram of Figure 4.5.

It can be useful to write the morphism $I(\mathfrak{P}', \mathfrak{P})$ in matrix form. We can make the following identifications

$$G(\mathfrak{P}) = \mathbb{Z}^m, \; G^+(\mathfrak{P}) = \mathbb{Z}^m_+, \quad G(\mathfrak{P}') = \mathbb{Z}^{m'}, \; G^+(\mathfrak{P}') = \mathbb{Z}^{m'}_+.$$

The units $1_\mathfrak{P}$ and $1_{\mathfrak{P}'}$ are identified with the vectors of heights.

Proposition 4.2.2 *The $m' \times m$-matrix $M(\mathfrak{P}', \mathfrak{P})$ of the morphism $I(\mathfrak{P}', \mathfrak{P})$ is given by*

$$M(\mathfrak{P}', \mathfrak{P})_{k,i} = \mathrm{Card}\{\ell \mid 0 \le \ell < h'_k, \, T^\ell B'_k \subset B_i\} \tag{4.5}$$

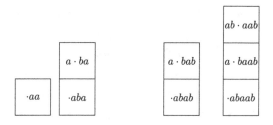

Figure 4.6 The partitions $\mathfrak{P}$ and $\mathfrak{P}'$.

for $1 \le k \le m'$ and $1 \le i \le m$.

Proof Let $f \in G(\mathfrak{P})$ be the function with value α_i on B_i for $1 \le i \le m$. Then, for $x \in B'_k$, we have by definition

$$I(\mathfrak{P}', \mathfrak{P})(f)(x) = I(\mathfrak{P}') \circ R_{B(\mathfrak{P})}(f)(x).$$

Set $g = R_{B(\mathfrak{P})}(f)$. By definition of $R_{B(\mathfrak{P})}$, we have

$$g(x) = \begin{cases} \alpha_i & \text{if } x \in B_i \\ 0 & \text{if } x \notin B(\mathfrak{P}). \end{cases}$$

Thus, for $x \in B'_k$,

$$I(\mathfrak{P}', \mathfrak{P})(f)(x) = I(\mathfrak{P}')(g)(x) = g^{(h'_k)}(x) = \sum_{i=1}^{m} M(\mathfrak{P}', \mathfrak{P})_{k,i}\alpha_i.$$

∎

Example 4.2.3 Let X be the two-sided Fibonacci shift and $\mathfrak{P}$ be the partition corresponding to the return words on a. We have $\mathcal{R}'_X(a) = \{a, ab\}$, and the partition $\mathfrak{P}$ is represented in Figure 4.6 on the left. Let $\mathfrak{P}'$ be the partition corresponding to the return words on ab (see Example 4.1.2) represented in Figure 4.6 on the right.

Since a is a prefix of ab, the partition $\mathfrak{P}'$ is nested in $\mathfrak{P}$. The matrix $M(\mathfrak{P}', \mathfrak{P})$ is

$$M(\mathfrak{P}', \mathfrak{P}) = \begin{bmatrix} 0 & 1 \\ 1 & 1 \end{bmatrix}.$$

For example, the second row is [1 1] because, in the second tower of $\mathfrak{P}'$, the lower element is contained in the basis of the second tower of $\mathfrak{P}$ and the upper element in the basis of the first one.

Note also that since $I(\mathfrak{P}', \mathfrak{P})$ is a morphism of unital ordered group, we have

$$M(\mathfrak{P}', \mathfrak{P})h = h', \tag{4.6}$$

where $h = (h_i)_{1 \leq i \leq m}$ and $h' = (h'_k)_{1 \leq k \leq m'}$, considered as column vectors.

The matrices $M(\mathfrak{P}', \mathfrak{P})$ are also useful in connexion with an invariant probability measure μ on (X, T). We have already seen in (4.1) a formula relating the measure of the bases with the height of the towers. In complement, when $\mathfrak{P}'$ is nested in $\mathfrak{P}$, we have the useful formula

$$\mu_B = \mu_{B'} M(\mathfrak{P}', \mathfrak{P}), \tag{4.7}$$

where $\mu_B = (\mu(B_i))_{1 \leq i \leq m}$ is considered as a row vector. Indeed, we have for $1 \leq i \leq m$,

$$B_i = \bigcup_{1 \leq k \leq m'} (\bigcup_{0 \leq \ell < h'_k, T^\ell B'_k \subset B_i} T^\ell B'_k),$$

with disjoint unions and thus

$$\mu(B_i) = \sum_{k=1}^{m} (\sum_{0 \leq \ell < h'_k, T^\ell B'_k \subset B_i} \mu(T^\ell B'_k))$$

$$= \sum_{k=1}^{m} \mu(B'_k) M(\mathfrak{P}', \mathfrak{P})_{k,i}.$$

Note that, using (4.6), Eq. (4.7) implies that $\langle \mu_B, h \rangle = \langle \mu_{B'}, h' \rangle$ in agreement with (4.1).

4.3 Ordered Groups of Sequences of Partitions

We now give a description of the ordered group of a minimal Cantor system (X, T) in terms of a refining sequence $(\mathfrak{P}(n))$, with

$$\mathfrak{P}(n) = \{T^j B_i(n) \mid 0 \leq j < h_i(n), 1 \leq i \leq t(n)\},$$

of KR-partitions. Denote for simplicity by $B(n)$ the base of the partition $\mathfrak{P}(n)$ and (see Section 4.2 for the definitions)

$$(G(n), G^+(n), \mathbf{1}_n) = (G(\mathfrak{P}(n)), \mathcal{G}^+(\mathfrak{P}(n)), \mathbf{1}_{\mathfrak{P}(n)}),$$

$$I(n+1, n) = I(\mathfrak{P}(n+1), \mathfrak{P}(n)),$$

$$C(n) = C(\mathfrak{P}(n)), \quad I(n) = I(\mathfrak{P}(n)), \quad \pi(n) = \pi(\mathfrak{P}(n)).$$

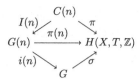

Figure 4.7 The morphism σ.

Proposition 4.3.1 *Let (X, T) be an invertible minimal Cantor system. Let $(\mathfrak{P}(n))$ be a refining sequence of KR-partitions. The ordered group $K^0(X, T)$ is the inductive limit of the unital ordered groups $(G(n), G^+(n), \mathbf{1}_n)$ with the connecting morphisms $I(n + 1, n)$.*

The proof is given below, but before that we need some lemmas, in which, for later use, we do not assume that the intersection of the bases has one element.

Let $(G, G^+, \mathbf{1})$ be the inductive limit of the sequence of ordered groups $(G(n), G^+(n)), \mathbf{1}_n)$ with the connecting morphisms $I(n + 1, n) = I(\mathfrak{P}(n + 1), \mathfrak{P}(n))$. Let $i(n)\colon G(n) \to G$ be the natural morphism. Note that, for $m > n$, we have $i(n) = i(m) \circ I(m, n)$ where

$$I(m, n) = I(m, m - 1) \circ \cdots \circ I(n + 1, n)$$

(see (2.4)).

Lemma 4.3.2 *Let (X, T) be an invertible minimal Cantor system and $(\mathfrak{P}(n))$ be a nested sequence of KR-partitions. Then, there is a unique morphism $\sigma\colon (G, G^+, \mathbf{1}) \to K^0(X, T)$ such that $\sigma \circ i(n) = \pi(n)$ for every $n \geq 0$.*

Proof By Proposition 4.2.1, the morphism $\pi(n)\colon G(n) \to H(X, T, \mathbb{Z})$ is for every $n \geq 0$ a morphism of unital ordered groups. By Proposition 2.3.9 there is a unique morphism $\sigma\colon (G, G^+, \mathbf{1}) \to K^0(X, T)$ such that $\sigma \circ i(n) = \pi(n)$. ∎

We illustrate the proof in Figure 4.7, where the upper part reproduces Figure 4.4 for $\mathfrak{P} = \mathfrak{P}(n)$.

The proof uses the following second lemma. A sequence

$$G_0 \overset{\varphi_1}{\to} G_1 \overset{\varphi_2}{\to} G_2 \to \cdots \overset{\varphi_n}{\to} G_n$$

of group morphisms is *exact* if the image of each morphism is the kernel of the next one.

Lemma 4.3.3 *Let (X, T) be an invertible minimal Cantor system. Let $(\mathfrak{P}(n))$ be a nested sequence of partitions such that $\mathfrak{P}(n)$ converges to the point* *partition. Let* $Y = \cap_{n \geq 0} B(n)$. *There is a group morphism* $r : C(Y, \mathbb{Z}) \rightarrow$ $\ker(\sigma)$ *such that the sequence of group morphisms*

$$0 \rightarrow \mathbb{Z} \rightarrow C(Y, \mathbb{Z}) \xrightarrow{r} G \xrightarrow{\sigma} H(X, T, \mathbb{Z}) \rightarrow 0 \qquad (4.8)$$

is exact, with $\mathbb{Z}$ *identified with the group of constant functions on* Y *where* G *is the inductive limit of the groups* $G(n)$ *and* σ *is defined in Lemma 4.3.2.*

Proof Recall that $C(n)$ is the set of functions in $C(X, \mathbb{Z})$ that are constant on every element of $\mathfrak{P}(n)$. Since the sequence $(\mathfrak{P}(n))$ tends to a partition in points, and since a continuous function is locally constant, we have $C(X, \mathbb{Z}) = \cup_n C(n)$.

This implies, as a first step, that σ is surjective by Proposition 2.3.9.

As a second step, let us define the morphism r. For this, let $u \in C(Y, \mathbb{Z})$, let $h \in C(X, \mathbb{Z})$ having u as restriction to Y and let $g = \partial_T h$. For n large enough, since $C(X, \mathbb{Z}) = \cup_n C(n)$, we have $g \in C(n)$. Let $f = I(n)(g)$ and let $\alpha = i(n)(f)$.

Since α is the image in G of a coboundary, it belongs to the kernel of σ. Indeed, we have (see Figure 4.7)

$$\sigma(\alpha) = \sigma(i(n)(f)) = \pi(n)(f)$$
$$= \pi(n) \circ I(n)(g) = \pi(g) = 0.$$

Thus, the map $r : u \mapsto \alpha$ is a group morphism from $C(Y, \mathbb{Z})$ onto $\ker(\sigma)$.

Let us finally prove that $\alpha = 0$ if and only if u is constant.

If u is constant, h is constant on Y and thus on some neighborhood of Y. For $m \geq n$ large enough, since the sequence $B(n)$ is decreasing, h is constant on $B(m)$. We may assume that m is large enough to have also $g = \partial_T h \in C(m)$. Then, by Eq. (3.2), for any $x \in B_i(m)$, since $T^{h_i(m)} x$ belongs to $B(m)$, we have

$$I(m)(g)(x) = g^{(h_i(m))}(x) = h(T^{h_i(m)} x) - h(x) = 0. \qquad (4.9)$$

Since $I(m)(g) = I(m, n) \circ I(n)(g) = I(m, n)(f)$, we have also $I(m, n)(f) = 0$ and thus $\alpha = i(n)(f) = i(m) \circ I(m, n)(f) = 0$.

Conversely, if $\alpha = 0$, there exists $m \geq n$ such that $0 = I(m, n)(f) = I(m)(g)$. Thus, by Eq. (4.9), h is constant on a set which is dense in $B(m)$, which implies that it is constant on $B(m)$ and therefore that u is constant.

It follows that the kernel of r is the group of constant functions on Y.

This completes the proof that the sequence of Eq. (4.8) is an exact sequence. ∎

Proof of Proposition 4.3.1 Since the intersection of the bases $B(n)$ consists in one point, we have $\ker(r) = C(Y, \mathbb{Z})$ and $\mathrm{Im}(r) = 0$. Thus, σ is an isomorphism, and the exact sequence displayed in (4.8) reduces to the isomorphism of G with $H(X, T, \mathbb{Z})$. ∎

It is often convenient to write the direct limit

$$G(0) \xrightarrow{I(1,0)} G(1) \xrightarrow{I(2,1)} G(2) \xrightarrow{I(3,2)} \cdots \qquad (4.10)$$

defining $K^0(X, T)$ in terms of matrices. Each $G(n)$ is isomorphic to $\mathbb{Z}^{t(n)}$ with the natural order, and the morphism $I(n+1, n)$ is represented by the matrix $M(n+1, n) = M(\mathfrak{P}(n+1), \mathfrak{P}(n))$ defined by (4.5). Thus, the direct limit (4.10) can be written

$$\mathbb{Z}^{t(0)} \xrightarrow{M(1,0)} \mathbb{Z}^{t(1)} \xrightarrow{M(2,1)} \mathbb{Z}^{t(2)} \xrightarrow{M(3,2)} \cdots \qquad (4.11)$$

The matrices $M(n+1, n)$, called the *connecting matrices*, will play an important role later when we introduce Bratteli diagrams.

We deduce from Proposition 4.3.1 the following important result.

Theorem 4.3.4 (Herman, Putnam, Skau) *For any invertible minimal Cantor system (X, T) the ordered group $K^0(X, T)$ is a simple dimension group.*

Proof By Theorem 4.1.6, there exists a sequence $(\mathfrak{P}(n))$ of partitions in towers satisfying the hypotheses of Proposition 4.3.1. Thus, $K^0(X, T)$ is the direct limit G of the ordered groups $(G(n))$. Since each $G(n)$ is isomorphic to some $\mathbb{Z}^m$ with the natural order, it follows that $K^0(X, T)$ is a dimension group.

For $n \geq 1$, let $[j, n] = i(n)(\chi_{B_j(n)})$, that is, the class in G of the characteristic function $\chi_{B_j(n)}$ of an element $B_j(n)$ of the basis of $\mathfrak{P}(n)$, and let $I(j, n)$ be the order ideal of G formed of the classes of the functions $f \in G$ such that $-k[j, n] \leq f \leq k[j, n]$ for some positive integer k. Next, suppose that I is any nonzero order ideal in G. Let g be a nonzero positive element of I. It must be represented by some strictly positive element in some $G(n)$ and, if $B_j(n)$ is an element of $B(n)$ on which g is positive, we have $0 \leq [j, n] \leq g$ and hence $[j, n]$ is also in I. It follows that $I(j, n)$ is included in I.

We have shown that if G contains a nonzero order ideal, then it contains one of the form $I(j, n)$. On the other hand, since (X, T) is minimal, one has $I(j, n) = G$ for every element $B_i(n)$ of $\mathfrak{P}(n)$. Indeed, consider the connecting matrix $M(m, n) = M(\mathfrak{P}(m), \mathfrak{P}(n))$ of the connecting morphism $I(\mathfrak{P}(m), \mathfrak{P}(n))$. Since (X, T) is minimal, there is an $m > n$ such that the matrix $M(m, n)$ has all its elements positive. Then, the map $I(\mathfrak{P}(m), \mathfrak{P}(n))(\chi_{B_j(n)}) \in G(m)$ is positive on each atom of $\mathfrak{P}(m)$. Thus, for convenient k, the map $kI(\mathfrak{P}(m), \mathfrak{P}(n))(\chi_{B_j(n)})$ can exceed any positive element g of $G(m)$. Hence,

$$k[j, n] = ki(n)(\chi_{B_j(n)}) = i(m) \circ kI(\mathfrak{P}(m), \mathfrak{P}(n))(\chi_{B_j(n)}) \geq i(m)(g).$$

This implies that g belongs to $I(j, n)$ and that G is simple.   ■

In agreement with Theorem 4.3.4, the group $K^0(X, T)$ for a minimal Cantor system (X, T), with T a homeomorphism, is called the *dimension group* of (X, T).

We give two examples illustrating Theorem 4.3.4. Many more examples will appear later.

Example 4.3.5 The dimension group of the odometer on the ring of 2-adic integers is the group of dyadic rationals. Indeed, we have seen in Example 4.1.8 that it can be represented by a sequence $(\mathfrak{P}(n))$ of partitions with one tower, and the map $i_{n+1,n}$ is easily seen to be the multiplication by 2. Thus, the dimension group $K^0(X, T)$ is the group $\mathbb{Z}[1/2]$ (see Example 2.3.1).

Example 4.3.6 We have seen before (Proposition 3.9.6) that the dimension group of a Sturmian shift of slope α is $\mathbb{Z} + \mathbb{Z}\alpha$, which is a simple dimension group.

Note that, although Theorem 4.3.4 applies only to invertible systems, we know by Corollary 3.6.3 that the ordered cohomology group of a minimal one-sided shift is isomorphic to the ordered cohomology group of its associated two-sided system. Thus, the ordered cohomology group of a one-sided minimal shift is also a simple dimension group.

4.4 Dimension Groups and Return Words

In this section, we show how to use return words to compute the dimension group of a minimal shift space. This method has the advantage of using, in general, abelian groups of smaller dimension than with the cylinder functions (as seen in Section 3.5.1). For example, in a Sturmian shift space, the number of return words is constant, while the word complexity is linear.

In the first part, we show that the group $K^0(X, T)$ is, for every minimal shift space, the direct limit of a sequence of groups associated with return words. In the second part we illustrate the use of return words to compute the dimension groups of episturmian shifts.

4.4.1 Sequences of Return Words

Let X be a minimal shift space. We have already introduced, in Section 1.2.4, the sets $\mathcal{R}_X(w)$ and $\mathcal{R}'_X(w)$ of right and left return words to $w \in \mathcal{L}(X)$.

We fix a point $x \in X$ and we denote $W_n(x) = \mathcal{R}'_X(x_{[0,n-1]})$. Let $G_n(x)$ be the group of maps from $W_n(x)$ to $\mathbb{Z}$, $G_n^+(x)$ the subset of nonnegative ones and $\mathbf{1}_n(x)$ the map that associates to $v \in W_n(x)$ its length.

Since $x_{[0,n-1]}$ is a prefix of $x_{[0,n]}$, the set $W_{n+1}(x)$ is contained in the submonoid generated by the set $W_n(x)$. This means that for every $v \in W_{n+1}(x)$ there is a decomposition $v = w_1(v)w_2(v) \cdots w_{k(v)}(v)$ as a concatenation of elements of $W_n(x)$. The decomposition is, moreover, easily seen to be unique. For every $\phi \in G_n(x)$, let $i_{n+1,n}(\phi) \in G_{n+1}(x)$ be defined by

$$(i_{n+1,n}(\phi))(v) = \sum_{i=1}^{k(v)} \phi(w_i(v)).$$

It is clear that $i_{n+1,n} : (G_n(x), G_n^+(x), \mathbf{1}_n(x)) \to (G_{n+1}(x), G_{n+1}^+(x), \mathbf{1}_{n+1}(x))$ is a morphism of unital ordered groups. Indeed, in particular, for $v \in W_{n+1}(x)$,

$$i_{n+1,n}(\mathbf{1}_n(x))(v) = \sum_{i=1}^{k(v)} \mathbf{1}_n(x)(w_i(v)) = \sum_{i=1}^{k(v)} |w_i(v)| = |v| = \mathbf{1}_{n+1}(x)(v).$$

We will prove the following result using the partition in towers associated with return words (see Proposition 4.1.1). However, we will not be able to use Proposition 4.3.1 because the sequence of partitions in towers associated with the sets $W_n(x)$ do not satisfy the hypotheses of Proposition 4.3.1 (the intersection of the bases does not consist in one point).

Proposition 4.4.1 *Let X be a minimal shift space. The group $K^0(X, S)$ is the inductive limit of the family $(G_n(x), G_n^+(x), \mathbf{1}_n(x))$ with the morphisms $i_{n+1,n}$.*

Proof By Proposition 4.1.1, for every $n > 0$, the family $\mathfrak{P}_n = \{S^j[vx_{[0,n-1]}] \mid v \in W_n(x), 0 \le j < |v|\}$ is a partition in towers with basis $[x_{[0,n-1]}]$. We can identify $G(n) = G(\mathfrak{P}_n)$ with $G_n(x)$ (see Section 4.2) so that

$$(G_n(x), G_n^+(x), \mathbf{1}_n(x)) = (G(n), G^+(n), \mathbf{1}_n).$$

The morphisms $i_{n+1,n}$ are then identified to the morphisms $I(n + 1, n)$. Let $(G, G^+, \mathbf{1})$ be the direct limit of the sequence of ordered groups $(G(n), G^+(n), \mathbf{1}_n)$ with the connecting morphisms $I(n + 1, n)$. By Lemma 4.3.2, there is a unique morphism $\sigma : G \to H(X, T, \mathbb{Z})$ such that $\sigma \circ I(n) = \pi(n)$ for every $n \ge 1$. We show that σ is an isomorphism.

Let $y, z \in S^j[vx_{[0,n-1]}]$ for some $n \geq 1$, some $v \in W_n(x)$ and some j with $0 \leq j < |v|$. Then y, z have the same prefix of length $|v| - j + n$ and thus the same prefix of length n. Thus, $y_{[0,n-1]} = z_{[0,n-1]}$. It follows that the partition $\mathfrak{P}_n$ is finer than the partition defined by cylinders of the words of length n, and that $C(\mathfrak{P}_n)$ contains every cylinder function corresponding to some $\phi \in Z_n(X)$ (recall that $Z_n(X)$ denotes the group of functions from $\mathcal{L}_n(X)$ into $\mathbb{Z}$). Consequently $\cup_{n \geq 1} C(\mathfrak{P}_n)$ contains every cylinder function. Since every $f \in C(X, \mathbb{Z})$ is cohomologous to some cylinder function (Proposition 3.4.1), the morphism σ is onto and maps G^+ onto $H^+(X, S, \mathbb{Z})$.

Assume now that $\alpha \in G$ is in the kernel of σ. For some n, α is the image in G of some $f \in C(\mathfrak{P}_n)$: $\alpha = i(n) \circ I(n)(f)$. As we have $0 = \sigma(\alpha) = \sigma \circ i(n) \circ I(n)(f) = \pi(f)$, the map f is a coboundary (see Figure 4.7). The function $g = R_{B(\mathfrak{P}_n)} \circ I(n)(f)$ is, by Proposition 3.7.3, cohomologous to f and thus is also a coboundary. It is a cylinder function because it is constant on $[vx_{[0,n-1]}]$ for each $v \in W_n(x)$ and null outside $[x_{[0,n-1]}]$. Thus, by Lemma 3.4.2, it is the coboundary of some cylinder function h. For m large enough, h is constant on the basis $x_{[0,m-1]}$ of $\mathfrak{P}_m$. This implies that $I(m)(f) = I(m)(g) = 0$ because $I(m)(g)(x) = h(T^{n(x)}(x)) - h(x) = 0$ for all $x \in \mathfrak{P}_m$. Hence, the image α of f in G is 0.   ∎

Example 4.4.2 Let X be the two-sided Fibonacci shift. Let $x \in X$ be such that $x_0 x_1 \cdots$ is the Fibonacci sequence. We have

$$W_1(x) = \{a, ab\}, \ W_2(x) = \{ab, aba\}, \ W_3(x) = \{ab, aba\},$$
$$W_4(x) = \{aba, abaab\}.$$

In general, one has $W_{n+1}(x) = W_n(x)$ if $x_{[0,n-1]}$ is not right special and otherwise $W_{n+1}(x) = \{v, vu\}$ if $W_n(x) = \{u, v\}$ with $|u| < |v|$. Indeed, the right-special prefixes of x are its palindrome prefixes u_n and $\mathcal{R}'_X(u_n) = \{\varphi^n(a), \varphi^n(b)\}$ (this results from Eq. (1.34)). The set $W_n(x)$ has two elements for every $n \geq 1$ and thus $G_n(x) = \mathbb{Z}^2$ for every $n \geq 1$.

The fact that $W_n(x)$ has always two elements is true for every Sturmian shift, and thus also that $G_n(x) = \mathbb{Z}^2$ for all $n \geq 1$, as we have already seen in Example 3.3.4.

4.4.2 Dimension Groups of Arnoux–Rauzy Shifts

We now use return words to describe the dimension group of an arbitrary Arnoux–Rauzy shift. Recall from Section 1.5 that if s is a standard episturmian sequence, there is a sequence $x = a_0 a_1 \cdots$ called its directive sequence such that $s = \text{Pal}(x)$. Moreover, the words $u_n = \text{Pal}(a_0 \cdots a_{n-1})$ are the palindrome prefixes of s, and the set of left return words to u_n is, by Eq. (1.34),

$$\mathcal{R}'_X(u_n) = \{L_{a_0\cdots a_{n-1}}(a) \mid a \in A\}.$$

Denote by M_a the incidence matrix of the Rauzy automorphism L_a. It is a unimodular matrix since it is triangular with coefficients 1 on the diagonal. Thus, for example, if $A = \{a, b\}$, we have

$$M_a = \begin{bmatrix} 1 & 0 \\ 1 & 1 \end{bmatrix}, \quad M_b = \begin{bmatrix} 1 & 1 \\ 0 & 1 \end{bmatrix}.$$

The following result allows us to compute the dimension group of a strict episturmian shift, also called Arnoux–Rauzy shift.

Proposition 4.4.3 *Let s be a strict and standard episturmian sequence on the alphabet A, let $x = a_0 a_1 \cdots$ be its directive sequence and let X be the shift generated by s. The dimension group of X is the direct limit of the sequence*

$$\mathbb{Z}^A \xrightarrow{M_{a_0}} \mathbb{Z}^A \xrightarrow{M_{a_1}} \mathbb{Z}^A \xrightarrow{M_{a_2}} \cdots. \tag{4.12}$$

Proof Set $u_n = \mathrm{Pal}(a_0 \cdots a_{n-1})$ and $\alpha_n = |u_n|$. We set $W_n(s) = \mathcal{R}'_X(s_{[0,n-1]})$ and we use the notation of the previous section with s in place of x. We identify $G_n(s)$ with $\mathbb{Z}^A$ via the bijection $a \mapsto L_{a_0\cdots a_{n-1}}(a)$ from A onto $W_n(s)$. It is enough to prove that the matrix of the map $i_{\alpha_{n+1},\alpha_n}$ is the matrix M_{a_n}. Note that for every $a \in A$, since $L_{a_0\cdots a_n}(a) = L_{a_0\cdots a_{n-1}}(L_{a_n}(a))$, we have, denoting $L_{[0,n]} = L_{a_0\cdots a_n}$,

$$L_{[0,n]}(a) = \begin{cases} L_{[0,n-1]}(a_n)L_{[0,n-1]}(a) & \text{if } a \neq a_n \\ L_{[0,n-1]}(a) & \text{otherwise.} \end{cases} \tag{4.13}$$

This gives the decomposition of an element of $W_{\alpha_{n+1}}(s)$ as a product of elements of $W_{\alpha_n}(s)$. Consider now $\phi \in G_{\alpha_n}(s)$. We consider ϕ as a column vector with components ϕ_a for $a \in A$ (via the identification above). Then, by (4.13), we have

$$(i_{\alpha_{n+1},\alpha_n}\phi)(b) = \begin{cases} \phi_{a_n} + \phi_b & \text{if } b \neq a_n \\ \phi_b & \text{otherwise.} \end{cases}$$

This shows that $i_{\alpha_{n+1},\alpha_n}\phi = M_{a_n}\phi$ and completes the proof. ∎

The description of the direct limit G of a sequence given in (4.12) with unimodular matrices M_{a_i} is similar to the case where all matrices M_{a_i} are equal (see Proposition 2.3.5).

Proposition 4.4.4 *The direct limit of a sequence*

$$\mathbb{Z}^A \xrightarrow{M_0} \mathbb{Z}^A \xrightarrow{M_1} \mathbb{Z}^A \xrightarrow{M_2} \cdots,$$

with all matrices M_i unimodular, is isomorphic to $(G, G^+, \mathbf{1})$ where $G = \mathbb{Z}^A$,

$$G^+ = \{x \in G \mid M_n \cdots M_0 x \in \mathbb{Z}_+^A \text{ for } n \text{ large enough}\}, \qquad (4.14)$$

and $\mathbf{1}$ is the vector with all components equal to 1.

Proof Indeed, let

$$\Delta = \{(x_n)_{n \geq 0} \mid x_n \in \mathbb{Z}^A \text{ for all } n, \ x_{n+1} = M_n x_n \text{ for } n \text{ large enough}\},$$

$$\Delta^0 = \{(x_n)_{n \geq 0} \in \Delta \mid x_n = 0 \text{ for } n \text{ large enough}\},$$

and

$$\Delta^+ = \{(x_n)_{n \geq 0} \in \Delta \mid x_n \in \mathbb{Z}_+^A \text{ for } n \text{ large enough}\}.$$

Let $(x_n)_{n \geq 0} \in \Delta$ be such that $x_{n+1} = M_n x_n$ for all $n \geq m$. Since $M_{m-1} \cdots M_0$ is unimodular, there is a unique $x \in \mathbb{Z}^A$ such that $M_{m-1} \cdots M_0 x = x_m$. This defines a morphism from Δ onto G with kernel Δ^0. Since this morphism sends Δ^+ onto G^+ and the vector $\mathbf{1}$ to itself, this proves the claim. ∎

We give two examples with the Fibonacci shift (which is Sturmian and thus we already know its dimension group by Theorem 3.9.6) and the Tribonacci shift.

Example 4.4.5 Let s be the Fibonacci sequence, which generates the Fibonacci shift X (see Example 1.4.2). The directive sequence of s is $x = (ab)^\omega$. Indeed, one has by Justin's Formula $x = L_{ab}(x)$, whence the result since $L_{ab} = \varphi^2$ where φ is the Fibonacci morphism. It follows from Proposition 4.4.3 that the dimension group of X is the ordered group Δ_M of the matrix $M = M_b M_a = M(\varphi)^2$ where φ is the Fibonacci morphism. Thus, we prove again that the dimension group of X is the group of algebraic integers $\mathbb{Z} + \frac{1+\sqrt{5}}{2}\mathbb{Z}$, in agreement with Theorem 3.9.6.

Example 4.4.6 Let s be the Tribonacci, sequence which is the fixed point of the morphism $\varphi: a \mapsto ab, b \mapsto ac, c \mapsto a$ (see Example 1.5.6). Its directive sequence is, as we have seen, $x = (abc)^\omega$. Thus, the dimension group of the Tribonacci shift X generated by s is the group Δ_M of the incidence matrix M of the morphism φ. We have

$$M = \begin{bmatrix} 1 & 1 & 0 \\ 1 & 0 & 1 \\ 1 & 0 & 0 \end{bmatrix}.$$

The dominant eigenvalue is the positive real number λ such that $\lambda^3 = \lambda^2 + \lambda + 1$. A corresponding row eigenvector is $[\lambda^2, \lambda, 1]$. Thus, $K^0(X, S)$ is isomorphic to $\mathbb{Z}[\lambda]$.

In these two examples, there is only one state for the group $K^0(X, S)$. This is actually the case in general for Arnoux–Rauzy shifts, as we shall see now.

4.4.3 Unique Ergodicity of Arnoux–Rauzy Shifts

We have already seen that Sturmian shifts are uniquely ergodic (Exercise 3.23). In fact, the following holds more generally.

Theorem 4.4.7 *Every Arnoux–Rauzy shift is uniquely ergodic.*

Let s be a strict and standard episturmian sequence on the alphabet A and let $x = a_0 a_1 \cdots$ be its directive sequence. Set $M_n = M_{a_n}^t$ (thus, M_n is the composition matrix of L_{a_n}) and $M_{[0,n)} = M_0 \cdots M_{n-1}$. We use the notation $\|M\|_\infty$ for the matrix norm induced by the L^∞-norm (see Appendix B.4).

Lemma 4.4.8 *There exists a sequence of matrices $\tilde{M}_n$ such that, for all n, $\|\tilde{M}_{[0,n)}\|_\infty \le 1$ and*

$$\tilde{M}_{[0,n)} x = M_{[0,n)} x \tag{4.15}$$

for every $x \in (M_{[0,n)})^{-1} \mathbf{1}^\perp$ where $\mathbf{1}$ is the vector with all coefficients equal to 1.

Proof Set

$$v^{(n)} = \frac{{}^t\mathbf{1} M_{[0,n)}}{\|{}^t\mathbf{1} M_{[0,n)}\|_1}.$$

We prove by induction on $n \ge 0$ that for every $n \ge 1$,

$$\|v^{(n)}\|_\infty \le \frac{1}{d-1} \tag{4.16}$$

with $d = \mathrm{Card}(A)$. It is true for $n = 0$ since one has $v^{(0)} = \begin{bmatrix} 1/d & \cdots & 1/d \end{bmatrix}$. Next, assume that it holds for n. We can write $v^{(n)} = (1/(d-1))(\alpha_a)_{a \in A}$ with $0 \le \alpha_a \le 1$ and $\sum \alpha_a = (d-1)$. Since $v^{(n+1)} = v^{(n)} M_n / \|v^{(n)} M_n\|_1$ one obtains

$$\|v^{(n+1)}\|_\infty = \frac{\max_a(\alpha_{a_n} + \alpha_a)}{a_n + \sum_{a \ne a_n}(\alpha_{a_n} + \alpha_a)} = \frac{\max_a(\alpha_{a_n} + \alpha_a)}{(d-1)(\alpha_{a_n} + 1)} \le \frac{1}{d-1}.$$

This proves (4.16).

Define $\tilde{M}_n = M_n - V_n$, where V_n is the matrix with all rows equal to 0 except the row of index a_n, which is equal to $w^{(n)} = v^{(n+1)}/\|v^{(n+1)}\|_\infty$.

We first prove that $\|\tilde{M}_n\|_\infty \le 1$. Let x be such that $\|x\|_\infty = 1$. If $a \ne a_n$ then $((M_n - V_n)x)_a = (M_n x)_a = x_a$ and thus $|((M_n - V_n)x)_a| \le 1$. For $a = a_n$, we have,

$$
|((M_n - V_n)x)_a| = \left| \sum_b \left(1 - w_b^{(n)}\right) x_b \right| \le \sum_b \left|1 - w_b^{(n)}\right| = d - \sum_b w_b^{(n)}
$$
$$
= d - \frac{\|v^{(n+1)}\|_1}{\|v^{(n+1)}\|_\infty} = d - \frac{1}{\|v^{(n+1)}\|_\infty}.
$$

This proves that $\|\tilde{M}_n\|_\infty \le 1$, and it implies that $\|\tilde{M}_{[0,n)}\|_\infty \le 1$ since $\| \ \|_\infty$ is an induced norm.

Finally, we prove (4.15) by induction on n. It holds trivially for $n = 0$. For $x \in (M_{[0,n+1)})^{-1} \mathbf{1}^\perp$, we have $v^{(n+1)}x = 0$ and thus $\tilde{M}_n x = M_n x - V_n x = M_n x$. We have then

$$
\tilde{M}_{[0,n+1)}x = \tilde{M}_{[0,n)}\tilde{M}_n x = \tilde{M}_{[0,n]}M_n x.
$$

Since $M_n x$ belongs to $(M_{[0,n)})^{-1}\mathbf{1}^\perp$, we obtain, using the induction hypothesis,

$$
\tilde{M}_{[0,n+1)}x = \tilde{M}_{[0,n]}M_n x = M_{[0,n)}M_n x = M_{[0,n+1)}x. \qquad \blacksquare
$$

Example 4.4.9 Assume that $A = \{a, b, c\}$ and that $a_0 = a$. Then

$$
M_0 = \begin{bmatrix} 1 & 1 & 1 \\ 0 & 1 & 0 \\ 0 & 0 & 1 \end{bmatrix}, \quad v^{(1)} = \begin{bmatrix} \frac{1}{5} & \frac{2}{5} & \frac{2}{5} \end{bmatrix} \text{ and } w^{(1)} = \begin{bmatrix} \frac{1}{2} & 1 & 1 \end{bmatrix}.
$$

Thus, we find

$$
\tilde{M}_0 = \begin{bmatrix} \frac{1}{2} & 0 & 0 \\ 0 & 1 & 0 \\ 0 & 0 & 1 \end{bmatrix}.
$$

The simplex

$$
S = \left\{ x \in \mathbb{R}_+^A \mid \|x\|_1 = 1, \|x\|_\infty \le \frac{1}{d-1} \right\}, \tag{4.17}
$$

to which belong all vectors $v^{(n)}$, plays an important role for Arnoux–Rauzy shifts (see Exercise 4.4).

Proof of Theorem 4.4.7 Let $v \in \cap_{n \geq 0} M_{[0,n)} \mathbb{R}_+^A$ with $\|v\|_1 = 1$. We will prove that v is unique, which will imply by Eq. (4.14) that the dimension group of x has a unique state, whence the conclusion that X is uniquely ergodic by Theorem 3.9.3.

Let π_n be the projection on $(M_{[0,n)})^{-1} \mathbf{1}^{\perp}$ along $(M_{[0,n)})^{-1} v$. Note that $\pi_0 M_{[0,n)} = M_{[0,n)} \pi_n$. For a word w, denote by $\ell(w) \in \mathbb{Z}^A$ the vector $(|w|_a)_{a \in A}$. By Lemma 4.4.8, we have

$$\|\pi_0 M_{[0,n)} \ell(a_n)\|_\infty = \|M_{[0,n)} \pi_n(\ell(a_n))\|_\infty = \|\tilde{M}_{[0,n)} \pi_n(\ell(a_n))\|_\infty$$
$$\leq \|\pi_n(\ell(a_n))\|_\infty. \tag{4.18}$$

Set $y = \ell(a_n) - \pi_n(\ell(a_n))$ and $M_{[0,n)} y = \lambda v$. We have

$$\lambda = \lambda \langle v, \mathbf{1} \rangle = \langle M_{[0,n)} \ell(a_n), \mathbf{1} \rangle.$$

Thus, λ is a positive integer. We claim that

$$\|\pi_n(\ell(a_n))\|_\infty \leq 1. \tag{4.19}$$

Indeed, since $\pi_n(\ell(a_n)) = \ell(a_n) - y$, we have

$$\pi_n(\ell(a_n))_a = \begin{cases} 1 - y_{a_n} & \text{if } a = a_n \\ -y_a & \text{otherwise.} \end{cases}$$

Since v is in $M_{[0,n+1)} \mathbb{R}_+^A$, we have that $M_{[0,n)}^{-1} v$ belongs to $M_n \mathbb{R}_+^A$ and thus y_{a_n} is larger than or equal to (the sum of) the other coordinates. If $y_{a_n} > 1$, then $\pi_n(\ell(a_n))$ has all its components negative or zero, a contradiction with the fact that $M_{[0,n)} \pi_n(\ell(a_n)) \mathbf{1} = 0$. This proves (4.19).

Set $L_n = L_{a_n}$ and $L_{[0,m)} = L_{a_0 \cdots a_{m-1}}$. We can write for every $k \geq 1$,

$$s_{[0,k)} = L_{[0,m-1)}(p_{m-1}) L_{[0,m-2)}(p_{m-2}) \cdots L_{[0,1)}(p_1) p_0,$$

where each p_n is either empty or equal to a_n for some $m = m(k)$ (this is a one-sided version of Lemma 3.8.15). Indeed, this is true for $k = 1$, choosing $m = 0$ and $p_0 = a_0$. Set next $u_{[0,k)} = L_0(t_{[0,\ell)}) p_0$, where $t = \text{Pal}(a_1 a_2 \cdots)$ is the episturmian sequence with directive sequence $a_1 a_2 \cdots$ and $p_0 = a_0$ or ε. Then, since $\ell < k$, using the induction hypothesis we can write

$$t_{[0,\ell)} = L_{[1,m-1)}(p_{m-1}) \cdots L_{[1,2)}(p_2) p_1,$$

with $p_n = a_n$ or ε. But then $u_{[0,k)} = L_0(t_{[0,\ell)}) p_0$ has the required expression. Then, using (1.12),

$$\ell(s_{[0,k)}) = \sum_{n=0}^{m-1} \ell(L_{[0,n)}(p_n)) = \sum_{n=0}^{m-1} M_{[0,n)} \ell(p_n). \tag{4.20}$$

Therefore, using (4.20), (4.19) and (4.18),

$$\|\pi_0 \ell(s_{[0,k]})\|_\infty = \left\|\left\|\sum_{n=0}^{m-1} \pi_0 M_{[0,n)} \ell(p_n)\right\|\right\|_\infty \leq \sum_{n=0}^{m-1} \|\pi_0 M_{[0,n)} \ell(p_n)\|_\infty$$

$$\leq \sum_{n=0}^{m-1} \|\pi_n(\ell(a_n))\|_\infty \leq m.$$

Since $m(k)/\|\ell(s_{[0,k]})\|_\infty \to 0$ when $k \to \infty$ (see Exercise 1.61), the direction of $\ell(s_{[0,k)})$ converges to that of v and thus v is determined by s. ∎

Example 4.4.10 Consider again the Tribonacci shift (see Example 4.4.6). The vector $v \in \cap_{n \geq 0} M_{[0,n)} \mathbb{R}_+^A$ with $\|v\|_1 = 1$ is $v = \begin{bmatrix} 1/\lambda & 1/\lambda^2 & 1/\lambda^3 \end{bmatrix}^t$. It is the eigenvector corresponding to the maximal eigenvalue λ of the matrix M.

4.5 Dimension Groups and Rauzy Graphs

We now show how to use Rauzy graphs to compute the dimension group of a minimal shift space. We begin with considerations valid for all graphs. We use the fundamental group $G(\Gamma)$ of a connected graph Γ to define the ordered cohomology group of a connected graph. We then define the notion of Rauzy graphs associated with a recurrent shift space X. We show that the direct limit sequence of ordered cohomology groups of the Rauzy graphs is the group $K^0(X, S)$ (Proposition 4.5.6).

4.5.1 Group of a Graph

Let $\Gamma = (V, E)$ be a finite oriented strongly connected graph with V as the set of vertices and E as the set of edges. We have already met basic notions concerning graphs such as paths or cycles, but we recall them now for convenience (see also Appendix B.2 where more details are given). For an edge $e \in E$, we denote by $s(e)$ its *source* (also called its *origin*) and by $r(e)$ its *range* (also called its *end*). There may be several edges with the same source and range (thus, G is actually a *multigraph*).

Two edges e, f are *consecutive* if the range of e is the source of f. A *path* in Γ is a sequence of consecutive edges. A *cycle* is a path $(e_1, \ldots, e_n)$ such that the source of e_1 is the range of e_n. To every path $p = (e_1, \ldots, e_n)$ in Γ, we associate its *composition* $\kappa(p) = e_1 + \cdots + e_n$, which is an element of the free abelian group $\mathbb{Z}(E)$ on the set E. The *group of cycles* of Γ, denoted $\Sigma(\Gamma)$, is the subgroup of $\mathbb{Z}(E)$ spanned by the compositions of the cycles of Γ.

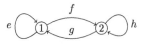

Figure 4.8 A connected graph.

The elements of $\mathbb{Z}(E)$ can be represented by row vectors indexed by E. Thus, the elements of $\Sigma(\Gamma)$ are also represented by row vectors indexed by E.

We consider, for $v \in V$, the fundamental group $G(\Gamma, v)$ of Γ formed of the generalized cycles from v to itself (see Appendix B.2). Since Γ is strongly connected, it is a free group, and a basis of $G(\Gamma, v)$ is obtained as follows. Choose a spanning tree T of G rooted at v, that is, a set of edges of G such that for every $w \in V$ there is a unique path p_w from v to w using edges in T. Then the set of $p_{s(e)}ep_{r(e)}^{-1}$ is a basis of $G(\Gamma, v)$.

The group $\Sigma(\Gamma)$ is the abelianization of any of the groups $G(\Gamma, v)$. Indeed, any cycle $p = p_{s(e)}ep_{r(e)}^{-1}$ can be written $(p_{s(e)}eq)(p_{r(e)}q)^{-1}$, where q is a path from $r(e)$ to v and then $\kappa(p) = \kappa(p_{s(e)}eq) - \kappa(p_{r(e)}q)$. Moreover, if p is a cycle from v' to v', let q be a path from v to v'. Then $r = qpq^{-1}$ is a cycle from v to itself and $\kappa(r) = \kappa(p)$.

Let $C(\Gamma) = \text{Hom}(\Sigma(\Gamma), \mathbb{Z})$ and let $C_+(\Gamma)$ be the submonoid of $C(\Gamma)$ formed by the morphisms giving a nonnegative value to every cycle of Γ.

It will be convenient, given a basis B of $\Sigma(\Gamma)$, to represent an element of $C(\Gamma)$ as a column vector indexed by B.

The coboundary homomorphism $\partial \colon \mathbb{Z}^V \to \mathbb{Z}^E$ is defined by

$$(\partial \phi)(e) = \phi(r(e)) - \phi(s(e))$$

for every $\phi \in \mathbb{Z}^V$ and $e \in E$. We denote $H(\Gamma) = \mathbb{Z}^E / \partial \mathbb{Z}^V$ and $H_+(\Gamma) = \mathbb{Z}_+^E / \partial \mathbb{Z}^V$.

We identify $\mathbb{Z}^E$ with $\text{Hom}(\mathbb{Z}(E), \mathbb{Z})$ by duality. In this identification, an element $u = (u_e) \in \mathbb{Z}^E$ is identified with the morphism

$$\sum_{e \in E} \alpha_e e \mapsto \sum_{e \in E} \alpha_e u_e,$$

which is the scalar product of $\alpha = \sum \alpha_e e$ and u. Thus, it will be convenient to consider the elements of $\mathbb{Z}^E$ as column vectors indexed by E.

Example 4.5.1 Consider the graph Γ of Figure 4.8.

We use the basis $\{e, fg, fhf^{-1}\}$ of $G(\Gamma, 1)$ and the corresponding basis $B = \{e, f + g, h\}$ of $\Sigma(\Gamma)$. Thus, $\Sigma(\Gamma)$ is formed of integer row vectors of size 3 and is isomorphic to $\mathbb{Z}^3$, while $C(\Gamma)$ is formed of integer column vectors of the same size. The matrix of the coboundary map is the *incidence matrix* of the graph Γ, which is the $V \times E$ matrix defined by

$$D_{v,e} = \begin{cases} 1 & \text{if } v = r(e) \text{ and } v \neq s(e) \\ -1 & \text{if } v = s(e) \text{ and } v \neq r(e) \\ 0 & \text{otherwise.} \end{cases}$$

In our example, we find

$$D = \begin{array}{c} {} \\ 1 \\ 2 \end{array} \begin{array}{cccc} e & f & g & h \\ \left[\begin{array}{cccc} 0 & -1 & 1 & 0 \\ 0 & 1 & -1 & 0 \end{array} \right] \end{array}.$$

The group $\partial \mathbb{Z}^V$ is the group generated by the rows of the matrix D, and thus it is, in this example, isomorphic to $\mathbb{Z}$. Accordingly, the group $H(\Gamma)$ is isomorphic to $\mathbb{Z}^3$.

Let $\Gamma = (V, E)$ be a strongly connected graph and let $\rho \colon \mathbb{Z}^E \to C(\Gamma)$ be the morphism assigning to an element of $\text{Hom}(\mathbb{Z}(E), \mathbb{Z})$ its restriction to $\Sigma(\Gamma)$. Thus, for $u \in \mathbb{Z}^E$ and for a cycle $(e_1, e_2, \ldots, e_n)$ in Γ, we have

$$\rho(u)(e_1 + \cdots + e_n) = u_{e_1} + \cdots + u_{e_n}.$$

Then ρ is a positive morphism such that $\rho(\mathbb{Z}^E_+) = C_+(\Gamma)$. Given a basis B of $\Sigma(\Gamma)$, the matrix of the morphism ρ is the matrix having as rows the elements of B (considered as row vectors indexed by E).

The following statement is just the dual of the classical statement that the sequence

$$0 \to \Sigma(\Gamma) \xrightarrow{\kappa} \mathbb{Z}(E) \xrightarrow{\beta} \mathbb{Z}(V) \xrightarrow{\gamma} \mathbb{Z} \to 0, \tag{4.21}$$

where κ is the composition map, $\beta(e) = r(e) - s(e)$ and $\gamma(v) = 1$ identically, is exact (Exercise 4.8). We give, however, a direct proof for the sake of clarity.

Proposition 4.5.2 *For every strongly connected graph Γ, the sequence*

$$0 \to \mathbb{Z} \xrightarrow{\gamma} \mathbb{Z}^V \xrightarrow{\partial} \mathbb{Z}^E \xrightarrow{\rho} C(\Gamma) \to 0,$$

where $\gamma(i)$ is the constant map equal to i, is exact and there is an isomorphism from $C(\Gamma)$ onto $H(\Gamma)$, sending $C_+(\Gamma)$ onto $H_+(\Gamma)$.

Proof The equality $\text{Im}(\gamma) = \ker(\partial)$ results from the hypothesis that Γ is strongly connected.

Let us now show that $\text{Im}(\partial) = \ker(\rho)$. If θ belongs to $\mathbb{Z}^V$ and $\psi = \partial\theta$, then $\psi(\kappa(p)) = \theta(v') - \theta(v)$ for every path p from v to v' and thus $\rho(\psi) = 0$. Conversely, suppose that $\rho(\psi) = 0$. Let T be a spanning tree of Γ rooted at v, and let p_w be the path in T from v to w, for every $w \in V$. Let $\theta \in \mathbb{Z}^V$ be defined by $\theta(w) = \psi(\kappa(p_w))$. Then it is easy to verify that $\psi = \partial\theta$ and thus that ψ belongs to $\text{Im}(\partial)$.

We now verify that ρ is surjective. Let T be a spanning tree of Γ rooted at $v \in V$, and let B' be the corresponding basis of $G(\Gamma, v)$ given by (B.2.2). Then the set $B = \{\kappa(p) \mid p \in B'\}$ is a basis of $C(\Gamma)$. It is enough to show that for each $\pi = \kappa(p) \in B$ there is some $\psi \in \mathbb{Z}^E$ such that $\rho(\psi)$ is the unit vector u_π (recall that the elements of $C(\Gamma)$ are column vectors indexed by B). Let $e \in E \setminus T$ be the unique edge such that $p = p_{s(e)} e p_{r(e)}^{-1}$. Let $\psi \in \mathbb{Z}^E$ be the characteristic function of e. We claim that $\rho(\psi) = u_\pi$. Indeed, we have $\rho(\psi)(\pi) = \psi(\kappa(p)) = 1$ because e appears only once in the path p. Next, for every other element $\pi' = \kappa(p')$ of B, we have $\rho(\psi)(\pi') = \psi(\kappa(p')) = 0$ because e does not appear in p'. This proves the claim. Since ψ belongs to $\mathbb{Z}_+^E$, this shows also that $\rho(\mathbb{Z}_+^E) = C_+(\Gamma)$. ∎

Let $\mathbf{1}_\Gamma$ be the function that associates to each cycle its length. The triple $(C(\Gamma), C_+(\Gamma), \mathbf{1}_\Gamma)$ is an ordered group called the *ordered cohomology group* associated with the graph $\Gamma = (V, E)$.

Example 4.5.3 Consider again the graph Γ of Figure 4.8. As in Example 4.5.1, we use the basis $\{e, fg, fhf^{-1}\}$ of $G(\Gamma, 1)$ and the corresponding basis $B = \{e, f + g, h\}$ of $\Sigma(\Gamma)$. Thus, $\Sigma(\Gamma)$ is isomorphic to $\mathbb{Z}^3$. The corresponding matrix of the morphism ρ is

$$P = \begin{bmatrix} 1 & 0 & 0 & 0 \\ 0 & 1 & 1 & 0 \\ 0 & 0 & 0 & 1 \end{bmatrix}.$$

The rows of P are the coefficients of the elements of B in $\mathbb{Z}(E)$. The group $H(\Gamma)$ is the group of linear maps from $\Sigma(\Gamma)$ to $\mathbb{Z}$, and thus it is isomorphic to $\mathbb{Z}^3$. The submonoid $H_+(\Gamma)$ is formed by the nonnegative linear maps on the space generated by B. Thus, it is isomorphic to $\mathbb{N}^3$, and the ordered cohomology group of Γ is $\mathbb{Z}^3$ with the natural ordering. The order unit is the vector

$$P \begin{bmatrix} 1 \\ 1 \\ 1 \end{bmatrix} = \begin{bmatrix} 1 \\ 2 \\ 1 \end{bmatrix}.$$

Figure 4.9 The graph Γ.

We give below an example of a graph with an ordered cohomology group, which is not isomorphic to $\mathbb{Z}^n$ with the natural ordering.

Example 4.5.4 Let Γ be the graph represented in Figure 4.9.

We take this time $\{e + g, f + g, f + h\}$ as the basis of $\Sigma(\Gamma)$. Thus, the matrix P is

$$P = \begin{bmatrix} 1 & 0 & 1 & 0 \\ 0 & 1 & 1 & 0 \\ 0 & 1 & 0 & 1 \end{bmatrix}.$$

The ordered cohomology group is again $\mathbb{Z}^3$, but this time $H_+(\Gamma) = \{(\alpha, \beta, \gamma) \in \mathbb{Z}^3_+ \mid \alpha + \gamma \geq \beta\}$. Indeed, let $\theta = \begin{bmatrix} \alpha & \beta & \gamma \end{bmatrix}^t$ be an element of $H(\Gamma)$. Then θ belongs to $H_+(\Gamma)$ if and only if its value on any cycle of Γ is nonnegative, that is, if and only if $\alpha, \beta, \gamma \geq 0$ and

$$\theta(e + h) = \theta(e + g) - \theta(f + g) + \theta(f + h) = \alpha - \beta + \gamma$$

is nonnegative, that is, if and only if, $\alpha + \gamma \geq \beta$. The order unit is $\begin{bmatrix} 2 & 2 & 2 \end{bmatrix}^t$. The ordered group is not a Riesz group because, in $H(\Gamma)$, the sum of the first two colums of P is equal to the sum of the two last ones, although none of these vectors can be written as a sum of positive vectors.

4.5.2 Rauzy Graphs

Let X be a shift space on the alphabet A. Recall from Section 1.2.5 that the *Rauzy graph* of X of order n is the graph $\Gamma_n(X)$ with $\mathcal{L}_{n-1}(X)$ as the set of vertices and $\mathcal{L}_n(X)$ as the set of edges. The edge w goes from u to v if $w = ua = bv$ with $a, b \in A$.

Example 4.5.5 Let X be the Fibonacci shift. The Rauzy graphs of order $n = 1, 2, 3$ are represented in Figure 4.10 (with the edge from $w = ua$ labeled a).

There is a positive morphism from the group $C(\Gamma_n(X))$ to the group $C(\Gamma_{n+1}(X))$. Indeed, the prefix map defines a projection from the graph $\Gamma_{n+1}(X)$ onto the graph $\Gamma_n(X)$. The set of edges of the first one is mapped onto the set of edges of the second one, and the set of cycles onto the set of cycles.

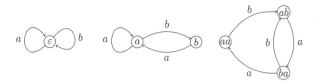

Figure 4.10 The Rauzy graphs of order $n = 1, 2, 3$ of the Fibonacci shift.

It follows that this projection defines a positive morphism from $\Sigma(\Gamma_{n+1}(X))$ to $\Sigma(\Gamma_n(X))$ and by duality from $C(\Gamma_n(X))$ to $C(\Gamma_{n+1}(X))$.

Proposition 4.5.6 *For any recurrent shift space X, the unital ordered group $K^0(X, S)$ is the direct limit of the sequence of ordered groups associated with the sequence of its Rauzy graphs.*

Proof Since the set of vertices of $\Gamma_n(X)$ is $\mathcal{L}_{n-1}(X)$ and its set of edges is $\mathcal{L}_n(X)$, we can identify $H(\Gamma_n(X))$ to $G_n(X)$ (defined by (3.9)) and $H_+(\Gamma_n(X))$ to $G_n^+(X)$.

The unital ordered group $(C(\Gamma_n(X)), C_+(\Gamma_n(X)), 1_{\Gamma_n(X)})$ can be identified, by Proposition 4.5.2, with $(G_n(X), G_n^+(X), 1_n(X))$. Since the morphism from $C(\Gamma_n(X))$ to $C(\Gamma_{n+1}(X))$ induced by taking the prefixes is the same as the morphism $i_{n+1,n}$ from $G_n(X)$ to $G_{n+1}(X)$, the result follows from Proposition 4.4.1. ∎

Example 4.5.7 We consider again the Fibonacci shift X. We already know that its dimension group is isomorphic to $\mathbb{Z}[\alpha]$ with $\alpha = (1 + \sqrt{5})/2$ (see Examples 4.5.7 and 4.4.5). We will not prove it again, but our point is to put in evidence the isomorphisms used in the proof of Proposition 4.5.6.

The prefix p_n of length n of the Fibonacci word x is the vertex labeled 1 in $\Gamma_{n+1}(X)$ in Figures 1.7 and 4.11. Thus, the simple cycles starting at 1 in $\Gamma_{n+1}(X)$ are labeled by the right return words to p_n (conjugate to the left return words forming the set denoted $W_n(x)$ in Example 4.4.2). For example, the vertex 1 of $\Gamma_5(X)$ is the word $p_4 = abaa$. The simple cycles starting at 1 are labeled by baa and $babaa$, which are conjugate to the elements of $W_4(x) = \{aba, abaab\}$ by the conjugacy $u \to (abaa)u(abaa)^{-1}$.

The prefix map from $\Gamma_4(X)$ to $\Gamma_3(X)$ sends the vertices 2 and 4 to the vertex ba. Since $W_2(x) = W_3(x)$, the morphism from $\Sigma(\Gamma_3(X))$ to $\Sigma(\Gamma_4(X))$ induced by the prefix map is the identity. By contrast, the morphism from $\Sigma(\Gamma_4(X))$ to $\Sigma(\Gamma_5(X))$ is given by the matrix

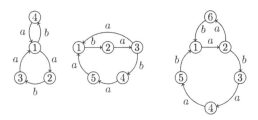

Figure 4.11 The Rauzy graphs of order 4, 5, 6 of the Fibonacci shift.

$$\begin{array}{c} \\ \begin{matrix} ab & aba \end{matrix} \\ \begin{matrix} aba \\ abaab \end{matrix} \begin{bmatrix} 0 & 1 \\ 1 & 1 \end{bmatrix} \end{array},$$

where the rows are indexed by the words of $W_4(X)$, in bijection with the (composition of the) cycles starting at 1 in $\Gamma_5(X)$. The columns are indexed by the words of $W_3(X)$. Finally, the morphism from $C(\Gamma_4(X))$ to $C(\Gamma_5(X))$ is obtained by transposition of this matrix.

We will need several times (in the next section, and later in Chapter 7) the following result relating return words and Rauzy graphs.

Proposition 4.5.8 *Let X be a minimal shift space and let $u \in \mathcal{L}(X)$. There exists an $n \geq 1$ with the following property. Let $x \in \mathcal{L}_n(X)$ be a word ending with u, and let S be the set of labels of paths from x to itself in $\Gamma_{n+1}(X)$. Then S is contained in the submonoid $\mathcal{R}_X(u)^*$ generated by $\mathcal{R}_X(u)$.*

Proof Let n be the maximal length of the words in $u\mathcal{R}_X(u)$. Consider $y \in S$. Since y is the label of a path from x to x in $\Gamma_{n+1}(X)$, the word xy ends with x. Thus, there is a unique factorization $y = y_1 y_2 \cdots y_k$ in nonempty words y_i where for each i with $1 \leq i \leq k$, the word uy_i ends with u and has no other occurrence of u except as a prefix or as a suffix. But, by the choice of n, the prefix of length n of uy_i has a factor u other than as a prefix and thus $|uy_i| \leq n$. Now since uy_i is the label of a path of length at most n in $\Gamma_n(X)$, it is in $\mathcal{L}(X)$. This implies that y_i is in $\mathcal{R}_X(u)$ and proves the claim. ∎

4.6 Dimension Groups of Substitution Shifts

We show in this section how to compute the dimension group of a substitutive shift.

We begin by establishing a connection between the 2-block presentation of the shift space and its Rauzy graph of order two.

4.6.1 Two-Block Presentation and Rauzy Graph

Let $\varphi \colon A^* \to A^*$ be a substitution, and let X be the shift space associated with φ.

We have seen that there is a matrix M associated with φ, called its composition matrix. Unfortunately, the composition (or incidence) matrix of the substitution does not determine the ordered group of the substitution shift associated with φ, as shown by the following example. We show below that it is, however, determined by the incidence matrix M_2 of the 2-block presentation φ_2 of φ (see Section 1.4 for the definition of the matrix M_2).

Example 4.6.1 Let X be the Thue–Morse shift. The matrix M associated with the Thue–Morse morphism is

$$M = \begin{bmatrix} 1 & 1 \\ 1 & 1 \end{bmatrix},$$

and it is the same as for the morphism $a \mapsto ab, b \mapsto ab$. The shift space corresponding to the second morphism has two elements, and its dimension group is $\mathbb{Z}$. But the dimension group of the Thue–Morse shift is not isomorphic to $\mathbb{Z}$. Indeed (see Example 3.8.20), one has $\mu([aa]) \neq \mu([ab])$ for the unique invariant measure on X and thus, by Proposition 3.9.1, the difference $\chi_{[aa]} - \chi_{[ab]}$ of the characteristic functions of the cylinders $[ab]$ and $[ba]$ is not a coboundary (we will see shortly that actually the group $H(X, S, \mathbb{Z})$ is isomorphic to $\mathbb{Z}[1/2] \times \mathbb{Z}$). Thus, the dimension groups are not the same for the two shift spaces.

We prove a statement connecting the endomorphism of $R_2(X)$ defined by the matrix M_2 with the fundamental group $\Sigma(\Gamma_2(X))$ of the Rauzy graph $\Gamma_2(X)$. Recall that $R_n(X)$ denotes the group of maps from $\mathcal{L}_n(X)$ to $\mathbb{R}$ and that $M_2(X)$ operates on the left on the elements of $R_2(X)$ considered as column vectors.

Recall from Section 3.5 that the map $\partial_1 \colon Z_1(X) \to Z_2(X)$ is the morphism defined by $\partial_1 \phi(ab) = \phi(b) - \phi(a)$. Since $\mathcal{L}_1(X)$ (resp. $\mathcal{L}_2(X)$) is the set of vertices (resp. of edges) of the graph $\Gamma_2(X)$, it coincides with the map ∂ of Section 4.5.1 and we shall denote it simply ∂.

Let $\tau \colon A^* \to A^*$ be the endomorphism that sends each letter $a \in A$ to the first letter of $\varphi(a)$. We define an endomorphism I of $R_1(X)$ by

$$I(\phi)(a) = \phi(\tau(a))$$

for every $\phi \in R_1(X)$ and $a \in A$.

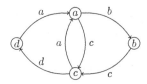

Figure 4.12 The graph $\Gamma_2(X)$.

Let $a \in A$ be such that that $\varphi(a)$ begins with a (we may always find such a, up to replacing φ by some power). Let P be the matrix whose rows are the compositions of return words to a.

The rows of P generate the group $\Sigma(\Gamma_2(X))$. Indeed, by Proposition 4.5.8 applied with $u = a$, there is an integer n such that the set labels of paths from x to itself in $\Gamma_{n+1}(X)$, where $x \in \mathcal{L}_n(X)$ ends with a, are generated by $\mathcal{R}_X(a)$. This implies, by projection from $\Gamma_{n+1}(X)$ to $\Gamma_2(X)$, that every cycle from a to a in $\Gamma_2(X)$ is a product of return words to a or their inverses. Thus, the compositions of return words to a generate the group $\Sigma(\Gamma_2(X))$.

When the compositions of return words to a form a basis of $\Sigma(\Gamma)$, the matrix P is the matrix of the morphism ρ in Proposition 4.5.2.

We illustrate this on an example.

Example 4.6.2 Let $A = \{a, b, c, d\}$ and let $\varphi \colon a \mapsto ab, b \mapsto cda, c \mapsto cd, d \mapsto abc$ (we shall consider again this morphism below and later in Example 7.1.4). We have $\mathcal{L}_2(X) = \{ab, ac, bc, ca, cd, da\}$, and the graph $\Gamma_2(X)$ is represented in Figure 4.12.

We have $\mathcal{R}_X(a) = \{bca, bcda, cda\}$. The choice for the matrix P is then

$$
\begin{array}{c c}
 & \begin{array}{c c c c c c} ab & ac & bc & ca & cd & da \end{array} \\
P = \begin{array}{c} bca \\ bcda \\ cda \end{array} & \left[\begin{array}{c c c c c c}
1 & 0 & 1 & 1 & 0 & 0 \\
1 & 0 & 1 & 0 & 1 & 1 \\
0 & 1 & 0 & 0 & 1 & 1
\end{array} \right].
\end{array}
$$

For example, the first row corresponds to the return word bca since it gives the sequence $(ab)(bc)(ca)$ of 2-blocks.

Proposition 4.6.3 *We have $M_2 \circ \partial = \partial \circ I$, so that $\partial(R_1(X))$ and $\partial(Z_1(X))$ are stable by M_2. Moreover, assuming that the rows of P are the compositions of return words to a letter $a \in A$ such that $\varphi(a)$ begins with a, there is a unique nonnegative matrix N_2 such that*

$$
P M_2 = N_2 P.
$$

If φ is a primitive morphism, the matrix N_2 is primitive.

Proof We first prove that $M_2 \circ \partial = \partial \circ I$. Let $\phi \in R_1(X)$. Then

$$\partial \circ I(\phi)(ab) = (\partial \circ \phi \circ \tau)(ab)$$
$$= \phi \circ \tau(b) - \phi \circ \tau(a) = \phi(\tau(b)) - \phi(\tau(a)).$$

To evaluate $M_2 \circ \partial(\phi)(ab)$, set $\varphi(a) = a_1 a_2 \cdots a_k$ and $\varphi(b) = b_1 b_2 \cdots b_\ell$. Then, identifying A_2 and $\mathcal{L}_2(X)$, we have

$$\varphi_2(ab) = (a_1 a_2)(a_2 a_3) \cdots (a_k b_1) \tag{4.22}$$

where φ_2 is the 2-block presentation of φ. Next, considering accordingly M_2 as an endomorphism of $R_2(X)$, we have for any $\psi \in R_2(X)$, using Eq. (4.22),

$$(M_2 \psi)(ab) = \psi(a_1 a_2) + \psi(a_2 a_3) + : + \psi(a_k b_1).$$

We obtain

$$(M_2 \circ \partial \phi)(ab) = \phi(a_2) - \phi(a_1) + \phi(a_3) - \phi(a_2) + : + \phi(b_1) - \phi(a_k)$$
$$= \phi(b_1) - \phi(a_1) = \phi(\tau(b)) - \phi(\tau(a))$$

and thus the conclusion.

Since $M_2 \circ \partial = \partial \circ I$, the subgroups $\partial(R_1(X))$ and $\partial(Z_1(X))$ are stable by M_2.

Let $p = (a_1 a_2)(a_2 a_3) \cdots (a_{n-1} a_n)(a_n a_1)$ be a cycle in $\Gamma_2(X)$, which is a row of P and thus with $a_1 = a$. We can consider p as a word on the alphabet $\mathcal{L}_2(X)$ and compute its image by the morphism φ_2. Since the rows of P correspond to return words to a, the first letter of $\varphi(a_1)$ is a. Then the first element of $\varphi_2(a_1 a_2)$ is equal to ab for some $b \in A$, and the last element of $\varphi_2(a_n a_1)$ is equal to ca for some $c \in A$ (as we have seen above computing $\varphi_2(ab)$). It follows that $\varphi_2(p)$ is a cycle starting at a in $\Gamma_2(X)$ and thus a composition of return words to a. By the choice of P, it is a nonnegative combination of rows of P.

This implies that $P M_2 = N_2 P$, where N_2 is the nonnegative matrix of the map sending the composition of a cycle p on the composition of $\varphi_2(p)$.

Let us prove the last assertion. If φ is primitive, then φ_2 is primitive (Proposition 1.4.42) and thus M_2 is primitive. Thus, there is some $k > 0$ such that $\varphi_2^k(p)$ is a cycle starting at a in $\Gamma_2(X)$ that is a composition of at least one of each return word to a. Consequently $N_2^k > 0$ and therefore N_2 is primitive. ∎

Note that the first two assertions of Proposition 4.6.3 are related. Indeed, the matrix of the map ∂ is the transpose of the incidence matrix D of the graph $\Gamma_2(X)$. The rows of D generate the orthogonal of the space generated by the rows of P (because (4.21) is an exact sequence). It is thus equivalent to say that the space generated by the rows of P is stable by M_2 and that the space

$\partial(R_1(X))$ is invariant by M_2. This duality is described in more detail in the following example.

Example 4.6.4 Consider again the substitution shift of Example 4.6.2. Using the bijection $\{ab, ac, bc, ca, cd, da\} \rightarrow \{x, y, z, t, u, v\}$, we find $\varphi_2 \colon x \mapsto xz, y \mapsto xz, z \mapsto uvy, t \mapsto uv, u \mapsto uvx, v \mapsto xzt$. The matrices M, M_2 are

$$M = \begin{bmatrix} 1 & 1 & 0 & 0 \\ 1 & 0 & 1 & 1 \\ 0 & 0 & 1 & 1 \\ 1 & 1 & 1 & 0 \end{bmatrix}, \quad M_2 = \begin{bmatrix} 1 & 0 & 1 & 0 & 0 & 0 \\ 1 & 0 & 1 & 0 & 0 & 0 \\ 0 & 1 & 0 & 0 & 1 & 1 \\ 0 & 0 & 0 & 0 & 1 & 1 \\ 0 & 0 & 0 & 0 & 1 & 1 \\ 1 & 0 & 1 & 1 & 0 & 0 \end{bmatrix}.$$

The matrices P and N_2 are then

$$P = \begin{bmatrix} 1 & 0 & 1 & 1 & 0 & 0 \\ 1 & 0 & 1 & 0 & 1 & 1 \\ 0 & 1 & 0 & 0 & 1 & 1 \end{bmatrix}, \quad N_2 = \begin{bmatrix} 0 & 1 & 1 \\ 1 & 1 & 1 \\ 1 & 1 & 0 \end{bmatrix}$$

so that N_2 has dimension less than M. The matrix D is

$$D = \begin{array}{c} \\ a \\ b \\ c \\ d \end{array} \begin{array}{cccccc} ab & ac & bc & ca & cd & da \\ \begin{bmatrix} -1 & -1 & 0 & 1 & 0 & 1 \\ 1 & 0 & -1 & 0 & 0 & 0 \\ 0 & 1 & 1 & -1 & -1 & 0 \\ 0 & 0 & 0 & 0 & 1 & -1 \end{bmatrix} \end{array}.$$

One may easily verify that the rows of D generate a vector space of dimension 3 orthogonal to the space generated by the rows of P.

Example 4.6.5 Let $\varphi \colon a \mapsto ab, b \mapsto a$ be the Fibonacci morphism and let X be the Fibonacci shift. Let $f \colon x \mapsto aa, y \mapsto ab, z \mapsto ba$ be a bijection from $A_2 = \{x, y, z\}$ onto $L_2(X) = \{aa, ab, ba\}$. The morphism φ_2 is $x \mapsto yz, y \mapsto yz, z \mapsto x$. The associated matrices M and M_2 and the vector v spanning the group $\partial_1(R_1(X))$ are (see Example 4.4.5)

$$M = \begin{bmatrix} 1 & 1 \\ 1 & 0 \end{bmatrix}, \quad M_2 = \begin{bmatrix} 0 & 1 & 1 \\ 0 & 1 & 1 \\ 1 & 0 & 0 \end{bmatrix}, \quad v = \begin{bmatrix} 0 \\ 1 \\ -1 \end{bmatrix}.$$

The Rauzy graph $\Gamma_2(X)$ is represented in Figure 4.13 with the edges labeled by the corresponding element of A_2.

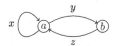

Figure 4.13 The Rauzy graph $\Gamma_2(X)$.

The matrices P corresponding to the basis $\{x, yz\}$ and the matrix N_2 associated with the action of φ_2 on the rows of P are

$$P = \begin{bmatrix} 0 & 1 & 1 \\ 1 & 0 & 0 \end{bmatrix}, \qquad N_2 = \begin{bmatrix} 1 & 1 \\ 1 & 0 \end{bmatrix}.$$

Thus, we conclude that $N_2 = M$ in this case.

4.6.2 Dimension Group of a Substitution Shift

Recall first from Section 2.3 that, for a $d \times d$-matrix M, $\mathcal{R}_M$ denotes the eventual range of M and $\mathcal{K}_M$ its eventual kernel.

For M nonnegative, we have defined a unital ordered group $(\Delta_M, \Delta_M^+, \mathbf{1}_M)$. By Eq. (2.7), the group Δ_M is given by

$$\Delta_M = \{v \in \mathcal{R}_M \mid \text{for some } k \geq 1, M^k v \in \mathbb{Z}^d\},$$

with positive cone

$$\Delta_M^+ = \{v \in \mathcal{R}_M \mid \text{for some } k \geq 1, M^k v \in \mathbb{Z}_+^d\}$$

and order unit $\mathbf{1}_M$ equal to the projection on $\mathcal{R}_M$ along $\mathcal{K}_M$ of the vector with all its components equal to 1.

The following statement gives a surprisingly simple way to compute the dimension group of a substitution shift. Indeed, it shows that the dimension group of a primitive substitution shift is defined by the matrix N_2 of Proposition 4.6.3 and the nonnegative matrix P such that $PM_2 = N_2 P$.

We shall see later another method to compute this group (Theorem 6.2.14).

Proposition 4.6.6 *Assume that φ is a primitive morphism and that the shift X associated with φ is infinite. The dimension group $K^0(X, S)$ is isomorphic to $(\Delta_{N_2}, \Delta_{N_2}^+, P\,\mathbf{1}_{M_2})$.*

Note that the order unit is not the vector $\mathbf{1}_{N_2}$ but the projection on $G_2(X)$ of the vector $\mathbf{1}_{M_2}$ (see Example 4.6.11). The proof relies on several lemmas.

We first introduce a sequence of partitions in towers associated with the primitive morphism φ.

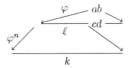

Figure 4.14 Representing $S^k(\varphi^{n+1}([ab]) \subset \varphi^n([cd])$.

Lemma 4.6.7 *The sequence* $(\mathfrak{P}(n))_{n \geq 0}$ *with*

$$\mathfrak{P}(n) = \{S^j \varphi^n([ab]) \mid ab \in \mathcal{L}_2(X), 0 \leq j < |\varphi^n(a)|\} \tag{4.23}$$

is a nested sequence of partition in towers of X.

Proof Let $f \colon \mathcal{L}_2(X) \to A_2$ be a bijection and let $X^{(2)}$ be the second higher block presentation of X. We consider the sequence $(\mathfrak{Q}(n))$ of partitions in towers of the second higher block presentation $X^{(2)}$ of X associated with the 2-block presentation φ_2^n of φ^n as in Proposition 4.1.3. Thus,

$$\mathfrak{Q}(n) = \{S^j \varphi_2^n([u]) \mid u \in A_2, 0 \leq j < |\varphi_2^n(u)|\}.$$

Let $\pi \colon A_2^* \to A^*$ be the morphism assigning to $u \in A_2$ the first letter of $f(u)$. The extension of π to $A_2^{\mathbb{Z}}$ defines an isomorphism from $X^{(2)}$ onto X. Let $(\mathfrak{P}(n))$ be the image of the sequence $(\mathfrak{Q}(n))$ by the isomorphism π. Since $\pi([u]) = [ab]$ when $f(u) = ab$, the partition $\mathfrak{P}(n)$ is given by Eq. (4.23). By Proposition 4.1.5, the sequence $\mathfrak{P}(n)$ is nested. ∎

Denote $G(n) = G(\mathfrak{P}(n))$, $G^+(n) = G^+(\mathfrak{P}(n))$ and $1_n = 1_{\mathfrak{P}(n)}$. Let $(G, G^+, 1)$ be the inductive limit of the sequence $(G(n), G^+(n), 1_n))$ with the morphisms $I(n+1, n) = I(\mathfrak{P}(n+1), \mathfrak{P}(n))$.

Lemma 4.6.8 *The map from* $Z_2(X)$ *to* $C(X, \mathbb{Z})$ *sending* $\phi \in Z_2(X)$ *to the map equal to* $\phi(ab)$ *on the cylinder* $[ab]$ *defines an isomorphism of unital ordered groups from* $(\Delta_{M_2}, \Delta_{M_2}^+, 1_{M_2})$ *onto* $(G, G^+, 1)$.

Proof By associating to each $\phi \in Z_2(X)$ the function equal to $\phi(ab)$ on $\varphi^n([ab])$, we can identify $G(n)$ to $Z_2(X)$, $G^+(n)$ to $Z_2^+(X)$ and 1_n to the map $ab \mapsto |\varphi^n(a)|$. Given k with $0 \leq k < |\varphi^n(a)|$, we have (see Figure 4.14)

$$S^k(\varphi^{n+1}([ab]) \subset \varphi^n([cd]) \Leftrightarrow \left\{ \begin{array}{l} \text{there exists } \ell \text{ with } 0 \leq \ell < |\varphi(a)| \text{ such that} \\ S^\ell([ab]) \subset [cd] \text{ and } k = |\varphi^n(\varphi([ab])_{[0, \ell-1]})|. \end{array} \right.$$

Therefore, for every $u, v \in A_2$ with $f(u) = ab$ and $f(v) = cd$, the number

$$\mathrm{Card}\{k \mid 0 \leq k < |\varphi^n(a)|, S^k \varphi^{n+1}([ab]) \subset \varphi^n([cd])\}$$

Figure 4.15 The morphism $\pi(n)$.

is the number of occurrences of v in $\varphi_2(u) = (M_2)_{uv}$. Consequently, we may identify the morphism $I(n + 1, n))$ to the morphism $M_2 \colon Z_2(X) \to Z_2(X)$.

Thus, the inductive limit $(G, G^+, 1)$ associated with the sequence $(\mathfrak{P}(n))$ of partitions in towers can be identified to $(\Delta_{M_2}, \Delta_{M_2}^+, \mathbf{1}_{M_2})$. ∎

Recall from Proposition 4.2.1 that there is a morphism $\pi(n) \colon G(n) \to H(X, S, \mathbb{Z})$ making the diagram of Figure 4.15 commutative and from Lemma 4.3.2 that there is a unique morphism $\sigma \colon (G, G^+, 1) \to K^0(X, S)$ such that $\sigma \circ I(n) = \pi(n)$ for every $n \geq 0$.

Recall also from Section 3.5.1 that $\partial_1 \colon R_1(X) \to R_2(X)$ is the morphism defined by $(\partial_1 \phi)(ab) = \phi(b) - \phi(a)$.

Lemma 4.6.9 *The morphism* $\sigma \colon G \to H(X, S, \mathbb{Z})$ *is onto and its kernel is* $\Delta_{M_2} \cap \partial_1(R_1(X))$.

Proof We first show that the morphism σ is surjective. Since every function in $C(X, \mathbb{Z})$ is cohomologous to a cylinder function by Proposition 3.4.1, it is enough to consider a cylinder function ϕ associated with $\phi \in Z_m(X)$ for some $m \geq 1$. Choose n so large that $|\varphi^n(a)| > m$ for every letter $a \in A$. Since all elements of the atoms of the partition $\mathfrak{P}(n)$ have the same prefix of length m, the cylinder function ϕ is constant on every element of the partition $\mathfrak{P}(n)$ and thus belongs to $C(n)$. Its image by $I(n)$ is sent by $\pi(n)$ to the class of ϕ modulo the coboundaries (see Figure 4.15).

We now prove that the kernel of σ is $\Delta_{M_2} \cap \partial_1(Z_1(X))$. Let ϕ be in G, or equivalently in Δ_{M_2}. In particular, ϕ belongs to $R_2(X)$ and there exists $k \geq 0$ such that $M_2^k \phi \in Z_2(X)$. Assume that $\sigma(\phi) = 0$. By definition this means that $\pi(\tilde{\phi}) = 0$, where $\pi \colon C(n) \to H(X, S, \mathbb{Z})$ is the natural projection and $\tilde{\phi}$ is defined by

$$\tilde{\phi}(x) = \begin{cases} (M_2^k \phi)(ab) & \text{if } x \in \varphi^k([ab]) \text{ for some } ab \in \mathcal{L}_2(X) \\ 0 & \text{otherwise.} \end{cases}$$

Note that, as in the proof of Proposition 4.6.3, we identify A_2 with $\mathcal{L}_2(X)$ and consequently consider M_2 as an endomorphism of $R_2(X)$.

Let $g \in C(X, \mathbb{Z})$ be such that $\tilde{\phi} = g \circ S - g$. We claim that there exists $n \geq 0$ such that g is constant on the set $\varphi^n([ab])$ for every $ab \in \mathcal{L}_2(X)$.

Since g is continuous, it is locally constant and there is an $m \geq 1$ such that $g(x)$ depends only on $x_{[-m,m]}$. Choose $n > k$ so large that $|\varphi^n(a)| > m$ for every letter a. Let $ab \in \mathcal{L}_2(X)$ and $y, z \in \varphi^n([ab])$. Since g depends only on $x_{[-m,m]}$, $g(S^m x)$ depends only on $x_{[0,2m]}$. Since y, z belong to $\varphi^n([ab])$ and $|\varphi^n(a)|, |\varphi^n(b)| > m$, y and z share the same $2m$ first coordinates and therefore $g(S^m z) = g(S^m y)$. On the other hand, for all $0 \leq j < |\varphi^n(a)|$, $S^j y$ and $S^j z$ are in the same atom of the partition $\mathfrak{P}(n)$. Since $m < |\varphi^n(a)|$ and since $\tilde{\phi}$ is constant on the atoms of $\mathfrak{P}(n)$, we obtain $\tilde{\phi}^{(m)}(y) = \tilde{\phi}^{(m)}(z)$. Since finally $g = g \circ S^m - \tilde{\phi}^{(m)}$ by Eq. (3.2), we conclude that $g(y) = g(z)$.

Let $\psi \in Z_2(X)$ be such that $\psi(ab) = g(x)$ for $x \in \varphi^n([ab])$. Then if x is in $\varphi^n([ab])$ and $S^{|\varphi^n(a)|}x$ in $\varphi^n([bc])$, we have with $\ell = |\varphi^n(a)|$,

$$\psi(bc) - \psi(ab) = g(S^\ell x) - g(x)$$
$$= \tilde{\phi}^{(\ell)}(x).$$

Recall that

$$\tilde{\phi}^{(\ell)}(x) = \tilde{\phi}(x) + \tilde{\phi} \circ S(x) + \cdots + \tilde{\phi}(S^{\ell-1}(x))$$

and note that the term $\tilde{\phi} \circ S^j(x)$ of this sum is equal to $(M^k \phi)(cd)$ if there is $cd \in \mathcal{L}_2(X)$ such that $S^j \varphi^n([ab]) \subset \varphi^k([cd])$ and equal to 0 otherwise. Thus,

$$\tilde{\phi}^{(\ell)}(x) = \sum_{cd \in \mathcal{L}_2(X)} \text{Card}\{0 \leq j < \ell \mid S^j \varphi^n([ab]) \subset \varphi^k([cd])\}(M_2^k \phi)(cd)$$

$$= \sum_{cd \in \mathcal{L}_2(X)} M_2^{n-k}(ab, cd)(M_2^k \phi)(cd) = (M_2^n \phi)(ab).$$

Consequently, we have for every $abc \in \mathcal{L}_3(X)$,

$$\psi(bc) - \psi(ab) = (M_2^n \phi)(ab).$$

Choose m so large that $|\varphi^m(a)| \geq 2$ for every $a \in A$ and define $\theta \in Z_1(X)$ by

$$\theta(a) = \psi(a_1 a_2)$$

for $a \in A$ if $\varphi^m(a) = a_1 \cdots a_\ell$.

If ab belongs to $\mathcal{L}_2(X)$ with $\varphi^m(a) = a_1 \cdots a_r$ and $\varphi^m(b) = b_1 \cdots b_s$, we obtain

$$(M_2^{n+m}\phi)(ab) = (M_2^n \phi)(a_1 a_2) + \cdots + (M_2^n \phi)(a_r b_1)$$
$$= \psi(b_1 b_2) - \psi(a_1 a_2) = \theta(b) - \theta(a)$$
$$= (\partial_1 \theta)(ab).$$

If follows that $M_2^{n+m}\phi$ belongs to $\partial_1(Z_1(X))$. Choosing m large enough, we may assume that $M_2^{n+m}\phi$ is in $\mathcal{R}_{M_2}$. Since M_2 defines an automorphism of $\mathcal{R}_{M_2}$, and since, by Proposition 4.6.3, the subspace $\partial_1(R_1(X))$ is invariant by

M_2, we conclude that ϕ is an element of $\Delta_{M_2} \cap \partial_1(R_1(X))$. Thus, the kernel of σ is included in $\Delta_{M_2} \cap \partial_1(R_1(X))$. Since the converse inclusion is obvious, the conclusion follows. ∎

Proof of Proposition 4.6.6 By Lemma 4.6.8, the ordered groups $(\Delta_{M_2}, \Delta_{M_2}^+, 1_{M_2})$ and $(G, G^+, 1)$ can be identified. By Lemma 4.6.9 the morphism σ defines an isomorphism from $\Delta_{M_2}/(\Delta_{M_2} \cap \partial_1(R_1(X)))$ onto $H(X, S, \mathbb{Z})$. But since $PM_2 = N_2 P$, we have also $PM_2^k = N_2^k P$ for every $k \geq 1$. Thus, the projection $v \mapsto Pv$ maps Δ_{M_2} onto Δ_{N_2} and we obtain

$$H(X, S, \mathbb{Z}) \simeq \frac{\Delta_{M_2}}{\Delta_{M_2} \cap \partial_1(R_1(X))} \simeq \Delta_{N_2}.$$

Similarly, we have $H^+(X, S, \mathbb{Z}) \simeq \Delta_{N_2}^+$. Finally, the map σ sends 1 to 1_X, and we conclude that $K^0(X, S)$ is isomorphic to $(\Delta_{N_2}, \Delta_{N_2}^+, P1_{M_2})$. ∎

Observe that Proposition 4.6.6 allows us to give a very simple proof of Theorem 3.8.13. Indeed, if X is a primitive substitution shift, then $K^0(X, S)$ is the limit of a stationary system defined by a primitive matrix and thus has a unique state by Proposition 2.5.1. Thus, X is uniquely ergodic by Corollary 3.9.3.

We give two examples of computation of the dimension group of a substitution shift. Other examples are treated in the exercises.

Example 4.6.10 Let $\varphi: a \mapsto ab, b \mapsto a$ be the Fibonacci morphism, and let X be the Fibonacci shift. As seen in Example 4.6.5, we have $N_2 = M$ and

$$M = \begin{bmatrix} 1 & 1 \\ 1 & 0 \end{bmatrix}, \quad P = \begin{bmatrix} 0 & 1 & 1 \\ 1 & 0 & 0 \end{bmatrix}.$$

The maximal eigenvalue of M is $\lambda = (1 + \sqrt{5})/2$ and the vector $[\lambda \ 1]$ is a left eigenvector of M. Thus, $H(X, T, \mathbb{Z}) = \mathbb{Z}^2$ and $H^+(X, T, \mathbb{Z}) = \{(\alpha, \beta) \mid \alpha\lambda + \beta \geq 0\}$. The order unit is

$$P \begin{bmatrix} 1 \\ 1 \\ 1 \end{bmatrix} = \begin{bmatrix} 2 \\ 1 \end{bmatrix}.$$

Thus, using the map $(\alpha, \beta) \mapsto \alpha\lambda + \beta$, we see that the dimension group of X is isomorphic to the group of algebraic integers $\mathbb{Z} + \mathbb{Z}\lambda$, with the order induced by the reals and the order unit $2\lambda + 1$. (see Example 2.3.7 and Example 4.4.2).

To obtain a normalized subgroup with 1 as unit, we consider the automorphism of $\mathbb{Z}^2$ such that $(\alpha, \beta) \mapsto (\alpha - \beta, 2\beta - \alpha)$. This is actually the automorphism defined by the matrix M^{-2} since

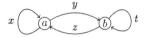

Figure 4.16 The Rauzy graph $\Gamma_2(X)$.

$$M^{-2}\begin{bmatrix} \alpha \\ \beta \end{bmatrix} = \begin{bmatrix} 1 & -1 \\ -1 & 2 \end{bmatrix}\begin{bmatrix} \alpha \\ \beta \end{bmatrix} = \begin{bmatrix} \alpha - \beta \\ 2\beta - \alpha \end{bmatrix}.$$

Then $(2, 1)$ maps to $(1, 0)$. Since

$$\alpha\lambda + \beta = \lambda^2(\lambda(\alpha - \beta) + (2\beta - \alpha)),$$

we do not change the order. Thus, we conclude that the dimension group of X is isomorphic with $\mathbb{Z} + \lambda\mathbb{Z}$, with the order induced by the reals and 1 as order unit.

Example 4.6.11 Let $\varphi\colon a \mapsto ab, b \mapsto ba$ be the Thue–Morse morphism on the alphabet $A = \{a, b\}$, and let X be the Thue–Morse shift. We have $\mathcal{L}_2(X) = A^2$. Set $A_2 = \{x, y, z, t\}$ and let $f\colon A_2 \to A^2$ be the bijection $x \mapsto aa, y \mapsto ab, z \mapsto ba, t \mapsto bb$. Then φ_2 is the substitution $x \mapsto yz, y \mapsto yt, z \mapsto zx, t \mapsto zy$. The Rauzy graph $\Gamma_2(X)$ is represented in Figure 4.16.

We take for P the matrix whose rows are the compositions of the words in $a\mathcal{R}_X(a)$. The matrices M, M_2, P and N_2 are

$$M = \begin{bmatrix} 1 & 1 \\ 1 & 1 \end{bmatrix}, \quad M_2 = \begin{bmatrix} 0 & 1 & 1 & 0 \\ 0 & 1 & 0 & 1 \\ 1 & 0 & 1 & 0 \\ 0 & 1 & 1 & 0 \end{bmatrix},$$

$$P = \begin{bmatrix} 1 & 0 & 0 & 0 \\ 0 & 1 & 1 & 0 \\ 0 & 1 & 1 & 1 \end{bmatrix}, \quad N_2 = \begin{bmatrix} 0 & 1 & 0 \\ 1 & 0 & 1 \\ 1 & 1 & 1 \end{bmatrix}.$$

Note that we already saw M_2 in Example 2.5.2 where we have computed Δ_{M_2}. The eventual range $\mathcal{R}_{N_2}$ of N_2 is generated by the vectors

$$v = \begin{bmatrix} 1 \\ 2 \\ 3 \end{bmatrix}, \quad w = \begin{bmatrix} 1 \\ -1 \\ 2 \end{bmatrix},$$

the first one being an eigenvector for the maximal eigenvalue 2 and the second one for the eigenvalue -1. Now, for $\alpha v + \beta w \in \mathcal{R}_{N_2}$, since

$$N_2^k(\alpha v + \beta w) = 2^k \alpha v + (-1)^k \beta w = \begin{bmatrix} 2^k \alpha + (-1)^k \beta \\ 2^{k+1} \alpha + (-1)^{k+1} \beta \\ 3 \cdot 2^k \alpha + 2(-1)^k \beta \end{bmatrix},$$

we have $N_2^k(\alpha v + \beta w) \in \mathbb{Z}^3$ if and only if $\alpha = \frac{m}{3 \cdot 2^k}$ and $\beta = \frac{n}{3}$ with $m, n \in \mathbb{Z}$ and $m + (-1)^k n \equiv 0 \bmod 3$. The order unit is

$$P \begin{bmatrix} 1 \\ 1 \\ 1 \\ 1 \end{bmatrix} = \begin{bmatrix} 1 \\ 2 \\ 3 \end{bmatrix} = v.$$

Thus, since 2 is invertible modulo 3, the conguence modulo 3 is well defined in $\mathbb{Z}[1/2]$. If α, β are as above, we have $3\alpha + 3\beta \equiv m2^{-k} + n \equiv m(-1)^k + n \equiv 0 \bmod 3$. Thus,

$$\Delta_{N_2} \simeq \{(\alpha, \beta) \mid 3\alpha \in \mathbb{Z}[1/2], 3\beta \in \mathbb{Z}, 3\alpha + 3\beta \equiv 0 \bmod 3\},$$

with

$$\Delta_{N_2}^+ \simeq \{(\alpha, \beta) \in \Delta_{N_2} \mid \alpha > 0\} \cup \{(0, 0)\},$$

with order unit $(1, 0)$.

Actually, using the map $(\alpha, \beta) \to (\alpha + \beta, 3\beta)$, we find that the group $K^0(X, S)$ is isomorphic to $\mathbb{Z}[1/2] \times \mathbb{Z}$, with the positive cone $\{(\alpha, \beta) \mid 3\alpha - \beta > 0\} \cup \{(0, 0)\}$ and the unit $(1, 0)$. Using a change of sign on the second component, we can also express the result with the inequality $3\alpha + \beta > 0$.

To close the loop, let us express the unique trace on $K^0(X, S)$, which, by Proposition 3.9.3, has the form α_μ where μ is the unique invariant probability measure on X (see Example 3.8.20). We have

$$\alpha_\mu(\alpha v + \beta w) = \alpha$$

since this map is a positive unital morphism from $K^0(X, S)$ to $(\mathbb{R}, \mathbb{R}_+, 1)$. We find, for example (in agreement with the value given in Example 3.8.20), $\mu([aa]) = 1/6$ since the characteristic function of the cylinder $[aa]$ can be identified with the vector

$$P \begin{bmatrix} 1 \\ 0 \\ 0 \\ 0 \end{bmatrix} = \begin{bmatrix} 1 \\ 0 \\ 0 \end{bmatrix} = \begin{bmatrix} 1/2 \\ 0 \\ 1/2 \end{bmatrix} + \begin{bmatrix} 1/2 \\ 0 \\ -1/2 \end{bmatrix} = 1/6(v + 2w) + \begin{bmatrix} 1/2 \\ 0 \\ -1/2 \end{bmatrix},$$

where the last expression is the decomposition in $\mathcal{R}_{N_2} \oplus \mathcal{K}_{N_2}$.

4.7 Exercises

Section 4.1

4.1 Let $\mathfrak{P} = \{T^j B_i \mid 1 \le i \le m, 0 \le j < h_i\}$ be a partition in towers nested in a partition $\mathfrak{P}' = \{T^h B'_k \mid 1 \le k \le m', 0 \le h \le h'_k\}$. Show that if

$$T^j B_i \subset T^h B'_k,$$

then

$$0 \le j - h \le h_i - h'_k.$$

4.2 Let σ be the morphism defined by $\sigma: a \mapsto abb, b \mapsto aab$. Let $\mathfrak{P}(n)$ be the partition with basis $\sigma^n(X)$ associated with σ^n. Let $\mathfrak{P}'(n)$ be the partition obtained by merging the two towers of $\mathfrak{P}(n)$, that is,

$$\mathfrak{P}'(n) = \{T^j \sigma^n(X) \mid 0 \le j < 3^n\}.$$

Show that the sequence $(\mathfrak{P}'(n))$ satisfies (KR1) and (KR2) but not (KR3).

4.3 Let σ be the morphism $a \mapsto acb, b \mapsto bcb, c \mapsto abb$. Let $\mathfrak{P}(n)$ be the partition associated with σ^n. Show that the sequence $(\mathfrak{P}(n))$ satisfies (KR2) and (KR3) but not (KR1).

Section 4.4

4.4 Let $S = \{x \in \mathbb{R}^A_+ \mid \|x\|_1 = 1, \|x\|_\infty \le \frac{1}{d-1}\}$ (see Figure 4.17 for $d = 3$). Show that for every $a \in A$ and $x \in S$, the vector $x M_a$ is colinear to an element of S.

4.5 For a positive vector $x \in \mathbb{R}^n$, denote $\|x\| = \frac{\max x_i}{\min x_i}$. For positive vectors x, y denote $xy = (x_i y_i)_{1 \le i \le n}$ and $x^{-1} = (x_i^{-1})_{1 \le i \le n}$. Set $d(x, y) = \|x y^{-1}\|$ and

$$\delta(x, y) = \log d(x, y). \tag{4.24}$$

Show that δ is a distance on the set S_n of positive vectors x such that $\sum_{i=1}^n x_i = 1$. It is called the *projective distance*.

4.6 Show that if M is a positive m $\times$ n matrix, then

$$\delta(Mx, My) \le \delta(x, y)$$

for every pair of positive *vectors* $x, y \in \mathbb{R}^n$, where δ is the projective distance.

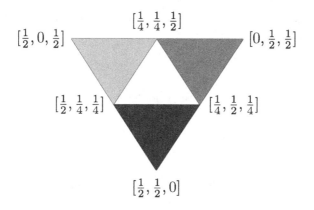

Figure 4.17 The simplex S.

4.7 For an m $\times$ n matrix M, set $d(M) = \max_{i,j} d(m_i, m_j)$ where m_i is the row of index i of M. Set also $\tau(r) = (\sqrt{r} - 1)/(\sqrt{r} + 1)$ for $r \geq 1$.

A real number $c > 0$ is called a *contraction coefficient* for a positive m $\times$ n matrix M if

$$d(Mx, My) < d(x, y)^c \qquad (4.25)$$

whenever $d(M) > 1$ and $d(x, y) > 1$.

Prove, using the steps indicated below, that for every positive m $\times$ n matrix M, the real number $\tau(d(M))$, called the *Birkhoff contraction coefficient*, is the smallest contraction coefficient of M.

1. Show that it is enough to prove (4.25) for $m = 2$.
2. Set

$$F(r, s, y) = \frac{\langle rs, y \rangle \langle \mathbf{1}, y \rangle}{\langle r, y \rangle \langle s, y \rangle}$$

where $\langle x, y \rangle = \sum x_i y_i$ denotes the usual scalar product of x, y. Show that (4.25) holds if and only if

$$F(r, s, y) < \|s\|^{\tau(\|r\|)} \qquad (4.26)$$

for every positive $r, s, y \in \mathbb{R}^n$ with r, s nonconstant and where $\mathbf{1}$ is the vector with all components equal to 1.

3. Show that it is enough to prove (4.25) for $n = m = 2$. Hint: use the fundamental Theorem of Linear Programming (see Appendix B.3), which implies that among the set of nonnegative vectors y maximizing $\langle rs, y \rangle$ under the

linear constraints $\langle r, y \rangle = 1$ and $\langle s - 1, y \rangle = 0$, there is one with at most two nonzero coordinates.

4. Prove (4.26) for $n = 2$.

Section 4.5

4.8 Let G be a strongly connected graph. Show that the sequence

$$0 \to \mathbb{Z}(\Gamma) \xrightarrow{\kappa} \mathbb{Z}(E) \xrightarrow{\beta} \mathbb{Z}(V) \xrightarrow{\gamma} \mathbb{Z} \to 0,$$

where κ is the composition map, $\beta(e) = r(e) - s(e)$ and $\gamma(v) = 1$ identically, is exact.

Section 4.6

4.9 Consider the Chacon binary morphism defined by $\sigma : 0 \mapsto 0010, 1 \mapsto 1$ (see Exercise 1.33). Show that the dimension group of the associated shift space X is isomorphic to $\mathbb{Z}[1/3] \times \mathbb{Z}$ with positive cone $\mathbb{Z}_+[1/3]$ and unit $(1, 1)$.

4.10 Let $\varphi : a \mapsto ab, b \mapsto ac, c \mapsto a$ be the Tribonacci morphism and let X be the corresponding substitution shift. Show, using Proposition 4.6.6, that the dimension group of X is the group $\mathbb{Z}[\lambda]$ where λ is the positive real solution of $\lambda^3 = \lambda^2 + \lambda + 1$ (see also Example 4.4.6 where we found the same result using return words).

4.11 Show that the dimension group of the Rudin–Shapiro shift is $\mathbb{Z}[1/2]^3 \times \mathbb{Z}$ with positive cone $\mathbb{Z}_+[1/2] \times \mathbb{Z}[1/2]^2 \times \mathbb{Z}$.

4.8 Notes

Kakutani–Rokhlin partitions owe their name to a result in ergodic theory called *Rokhlin's Lemma* or *Kakutani–Rokhlin Lemma*, which states that every aperiodic measure-theoretic dynamical system can be represented by an arbitrary high tower of measurable sets (Rokhlin, 1948; Kakutani, 1943).

4.8.1 Partitions in Towers

Proposition 4.1.7 is (Putnam, 1989, Lemma 3.1), where the credit of the construction is given to Vershik. We follow the presentation of (Durand, 2010, Proposition 6.4.2). Theorem 4.1.6 is due to Herman et al. (1992).

Theorem 4.3.4 is also from Herman et al. (1992). The proof of the fact that the dimension group is simple is from Putnam (2010, Theorem 2.14).

4.8.2 Dimension Groups and Return Words

The results of Section 4.4.1, in particular Proposition 4.4.1, are from Host (1995) (see also Host (2000)). The unique ergodicity of Arnoux–Rauzy shifts is a result of Delecroix et al. (2013). Lemma 4.4.8 is from Avila and Delecroix (2013).

The ergodic properties of products of sequences of matrices have been investigated, especially in the context of nonhomogeneous Markov chains defined by sequences of stochastic matrices (instead of powers of a single stochastic matrix, as in an ordinary finite Markov chain). Classical references are Hartfiel (2002) and Seneta (2006). The projective distance (Exercise 4.5), also called *Hilbert projective distance*, can be found in Hartfiel (2002) (including the link with Hilbert's original definition) and Seneta (2006). Birkhoff contraction coefficient (Exercise 4.7) was introduced in Birkhoff (1957) (see also Birkhoff (1967), pp. 383–386). The original proof by Birkhoff uses projective geometry. The elementary proof presented here is from Carroll (2004).

4.8.3 Dimension Groups and Rauzy Graphs

The notion of fundamental group of a graph used in Section 4.5 is classical in algebraic topology (see Lyndon and Schupp (2001) for a more detailed introduction to its direct definition on a graph and its connection with spanning trees of the graph). The ordered group $C(\Gamma)$ associated with a graph Γ is called a *graph group* in Boyle and Handelman (1996), while a direct limit of graph groups as in Proposition 4.5.6 is called a *graphical group*.

4.8.4 Dimension Group of a Substitution Shift

The results of this section, in particular Proposition 4.6.6, are from Host (1995) (see also Host (2000)).

5

Bratteli Diagrams

We now introduce Bratteli diagrams. We will see that, adding an order on the diagram and provided this order is what we will call proper, an ordered Bratteli diagram defines in a natural way a topological dynamical system. We prove the Bratteli–Vershik representation theorem: Any minimal topological dynamical system defined on a Cantor set can be obtained in this way (Theorem 5.3.3). Thus, every minimal Cantor system (X, T) can be represented by an ordered Bratteli diagram, called a BV-representation of (X, T).

We will next introduce an equivalence on dynamical systems called Kakutani equivalence and prove that it can be characterized by a transformation on BV-representations.

We then prove one of the major results presented in this book, namely the Strong Orbit Equivalence Theorem (Theorem 5.5.1). This result shows that the dimension group is a complete invariant for the so-called strong orbit equivalence. As a complement, the Orbit Equivalence Theorem (Theorem 5.5.3) shows that the quotient of the dimension group by the infinitesimal subgroup is a complete invariant for orbit equivalence.

In Section 5.6, we develop a systematic study of equivalences on Cantor spaces. We introduce the notion of étale equivalence relation and prove that both the relations of orbit equivalence and of cofinality in Bratteli diagrams are étale equivalences. In the last section (Section 5.7), we discuss the link with the notion of entropy (which had not been considered before in this book).

5.1 Bratteli Diagrams

A *Bratteli diagram* is an infinite directed graph $\mathcal{B} = (V, E)$ where the *vertex* set V and the *edge* set E can be partitioned into nonempty finite sets

$$V = V(0) \cup V(1) \cup V(2) \cup \cdots \quad \text{and} \quad E = E(1) \cup E(2) \cup \cdots$$

with the following properties:

1. $V(0) = \{v(0)\}$ is a one-point set,
2. $r(E(n)) \subseteq V(n)$, $s(E(n)) \subseteq V(n-1)$, $n = 1, 2, \ldots$,

where $r : E \to V$ is called the *range map* and $s : E \to V$ the *source map*. They satisfy $s^{-1}(\{v\}) \neq \emptyset$ for all $v \in V$ and $r^{-1}(\{v\}) \neq \emptyset$ for all $v \in V \setminus V(0)$.

We use the terminology of graphs to handle Bratteli diagrams (see also Appendix B.2). Note that Bratteli diagrams are actually multigraphs (there can be several edges between two vertices). Since edges can only go from $V(n)$ to $V(n+1)$, they are *acyclic*, that is, there is no nontrivial cycle. Some particular terms are of common use for acyclic graphs. In particular, the vertex $v(0)$ is called the *root*. A *successor* of a vertex $v \in V$ is a vertex w such that $v = s(e)$ and $w = r(e)$ for some $e \in E$. Two edges $e, f \in E$ are *consecutive* if $r(e) = s(f)$. A *path* is a sequence $(e_1, e_2, \ldots, e_n)$ of consecutive edges. The *source* of the path is $s(e_1)$ and its *range* is $r(e_n)$. A *descendant* of a vertex v is a vertex w such that there is a path from v to w, that is, a path with source v and range w.

It is convenient to represent the Bratteli diagram by a picture with the set $V(n)$ of vertices at (horizontal) level n, and the set $E(n)$ of edges (downward directed) connecting the vertices at level $n-1$ with those at level n (see Figure 5.1). Although a Bratteli diagram is an oriented graph, we do not draw the edges with arrows, since the orientation is implicit.

The *adjacency matrix* $M(n)$ is the $V(n) \times V(n-1)$-matrix defined for every $n \geq 1$ by

$$M(n)_{r,s} = \mathrm{Card}\{e \in E(n) \mid r(e) = r, s(e) = s\}. \tag{5.1}$$

Note that $M(n)$ is not exactly the adjacency matrix of the graph (V, E) because the edges are taken from range to source (or, equivalently, the matrix is

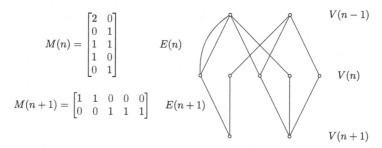

Figure 5.1 Representation of a diagram between the levels $n-1$ and $n+1$.

transposed) and moreover because the adjacency matrix of the graph (V, E) is
a $V \times V$-matrix instead of a $V(n) \times V(n-1)$-matrix.

We say that two Bratteli diagrams (V, E) and (V', E') are *isomorphic* whenever there exists a pair of bijections $f : V \to V'$, preserving the degrees, and
$g : E \to E'$, intertwining the respective source and range maps:

$$s' \circ g = f \circ s \text{ and } r' \circ g = f \circ r.$$

Let $k, l \in \mathbb{N}$ with $1 \leq k < l$ and let $E_{k,l}$ denote the set of paths from
$V(k-1)$ to $V(l)$. Specifically,

$$E_{k,l} = \{(e_k, \ldots, e_l) \mid e_i \in E(i), k \leq i \leq l, r(e_i) = s(e_{i+1}), k \leq i \leq l-1\}.$$

Remark that the adjacency matrix of $E_{k,l}$ is $M(l) \cdots M(k)$. As a
consequence, the number of paths from $s \in V(k-1)$ to $r \in V(l)$ is
$(M(l) \cdots M(k))_{r,s}$. We define $r(e_k, \ldots, e_l) = r(e_l)$ and $s(e_k, \ldots, e_l) = s(e_k)$.

A Bratteli diagram (V, E) is *stationary* if there exists k such that $k =$
Card$(V(n))$ for all n, and if (by an appropriate labeling of the vertices) the
adjacency matrices between level n and $n+1$ are the same $k \times k$ matrix M for
all $n = 1, 2, \ldots$. In other words, beyond level 1 the diagram repeats itself (see
Figure 5.4). Clearly we may label the vertices in $V(n)$ as $v(n, a_1), \ldots, v(n, a_k)$,
where $A = \{a_1, \ldots, a_k\}$ is a set of k distinct symbols. The matrix M is called
the *matrix of the stationary diagram*.

5.1.1 Telescoping and Simple Diagrams

Given a Bratteli diagram (V, E) and a sequence $m_0 = 0 < m_1 < m_2 < \ldots$ of
integers, we define the *telescoping* of (V, E) with respect to (m_n) as the new
Bratteli diagram (V', E'), where $V'(n) = V(m_n)$ and $E'(n) = E_{m_{n-1}+1, m_n}$
and the range and source maps are as above (see Figure 5.2).

We say that $\mathcal{B} = (V, E)$ is a *simple Bratteli diagram* if there exists a telescoping $\mathcal{B}' = (V', E')$ of $\mathcal{B}$ such that the adjacency matrices of $\mathcal{B}'$ have only
nonzero entries at each level.

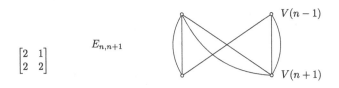

$$\begin{bmatrix} 2 & 1 \\ 2 & 2 \end{bmatrix} \qquad E_{n,n+1}$$

Figure 5.2 Telescoping between the levels $n-1$ and $n+1$ in Figure 5.1.

We will use the following characterization of simple diagrams. Let (V, E) be a Bratteli diagram. A set $W \subset V$ is *directed* if every edge having its source in W has also its range in W. In symbols, for every $e \in E$,

$$s(e) \in W \Rightarrow r(e) \in W.$$

It is *hereditary* if it satisfies for every $v \in V$ the following condition. If every edge with source v has its range in W, then v itself is in W. In symbols, for every $v \in V$,

$$r(e) \in W \text{ for every edge } e \text{ such that } v = s(e) \Rightarrow v \in W.$$

Proposition 5.1.1 *A Bratteli diagram is simple if and only if there is no nonempty set $W \subset V$ both directed and hereditary other than V.*

Proof Set $\mathcal{B} = (V, E)$. Assume first that $\mathcal{B}$ is simple. Let $W \subset V$ be a nonempty directed and hereditary set. Since W is nonempty, there is at least one w in W. Let n be such that $w \in V(n)$. Since $\mathcal{B}$ is simple, there is an $m > n$ such that there is a path from w to every vertex in $V(m)$. Since W is directed, this implies $V(m) \subset W$. Since W is hereditary, this implies that all vertices of $V(n)$ for $n \leq m$ are in W. Thus, $v(0)$ belongs to W, which implies $V = W$.

Conversely, assume that $\mathcal{B}$ is not simple. Let $v \in V(n)$ be such that for every $m > n$ there is some $w \in V(m)$ that cannot be reached from v. Consider the set W of vertices $w \in V$ for which there is an integer $p = p(w)$ such that all descendants of w in $V(p)$ are descendants of v. It is a directed set by definition. Suppose that some vertex $w \in V$ is such that all its successors belong to W. Let p be the supremum of the integers $p(u)$ for u successor of w. Then all descendants of w in $V(p)$ are descendants of v and thus w is in W. This shows that W is hereditary. Finally there is at least one vertex in $V(n)$ that is not in W since otherwise taking the supremum p of the integers $p(u)$ for $u \in V(n)$, we find that all vertices in $V(p)$ are descendants of v. Thus, W is a nonempty directed and hereditary set strictly contained in V. ∎

Example 5.1.2 Consider the Bratteli diagram represented in Figure 5.3.

This diagram is not simple because the vertices of the lower level can never reach any of those at top level. Accordingly, the vertices at lower level (excluding the root) form a directed and hereditary set.

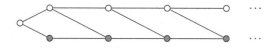

Figure 5.3 A non-simple Bratteli diagram.

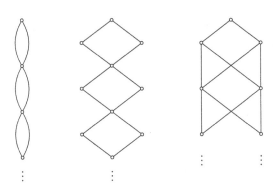

Figure 5.4 Three equivalent stationary Bratteli diagrams.

We denote by $\sim$ the *telescoping equivalence* on Bratteli diagrams as the equivalence relation generated by isomorphism and telescoping. It is not hard to show that $\mathcal{B}^1 = (V^1, E^1) \sim \mathcal{B}^2 = (V^2, E^2)$ if and only B^1, B^2 have a common *intertwining*, that is, if there exists a Bratteli diagram $\mathcal{B} = (V, E)$ such that telescoping $\mathcal{B}$ to odd levels $0 < 1 < 3 <:$ yields a telescoping of either $\mathcal{B}^1$ or $\mathcal{B}^2$, and telescoping $\mathcal{B}$ to even levels $0 < 2 < 4 <:$ yields a telescoping of the other (Exercise 5.2).

Example 5.1.3 Consider the stationary Bratteli diagram in the middle of Figure 5.4. Telescoping at even levels gives the diagram on the left, and telescoping at odd levels gives the diagram on the right. Thus, the left and right diagrams are equivalent.

5.1.2 Dimension Group of a Bratteli Diagram

Let (V, E) be a Bratteli diagram. Let $G(n) = \mathbb{Z}^{V(n)}$. Let also $h_v(n)$ be the number of paths from $V(0)$ to $v \in V(n)$. Let $h(n) = (h_v(n))_{v \in V(n)}$. Then, by definition of $M(n)$, we have for $n \geq 2$,

$$h(n) = M(n)h(n-1), \qquad (5.2)$$

with $h(n)$ considered as a column vector.

We consider $G(n)$ as a unital ordered group with the usual order and the order unit $h(n)$. The *dimension group* of $\mathcal{B} = (V, E)$, denoted $\mathrm{DG}(\mathcal{B})$, is the direct limit of the sequence

$$G(0) \xrightarrow{M(1)} G(1) \xrightarrow{M(2)} G(2) \ldots$$

defined by the adjacency matrices $M(n)$ (acting on the left on the elements of $G(n-1)$ considered as column vectors). Equation (5.2) shows that $M(n)$ defines a morphism of unital ordered groups from $G(n-1)$ to $G(n)$.

Example 5.1.4 The dimension group of the diagram represented in Figure 5.4 on the left is the group $\mathbb{Z}[1/2]$ (see Example 2.3.1).

The following important result shows that the group $DG(V, E)$ is a complete invariant for the telescoping equivalence.

Theorem 5.1.5 (Elliott) *Two Bratteli diagrams $\mathcal{B} = (V, E)$ and $\mathcal{B}' = (V', E')$ are telescoping equivalent if and only if the unital ordered groups $DG(\mathcal{B})$ and $DG(\mathcal{B}')$ are isomorphic.*

Proof Consider a sequence $m_0 = 0 < m_1 < m_2 <:$ and the corresponding telescoping $\mathcal{B}' = (V', E')$ of $\mathcal{B} = (V, E)$. The dimension group of $\mathcal{B}'$ is the direct limit of the sequence

$$G'(0) \xrightarrow{M'(1)} G'(1) \xrightarrow{M'(2)} :,$$

where $G'(n) = G(m_n)$ and $M'(n) = M(m_n) \cdots M(m_{n-1} + 1)$. This does not change the direct limit, and thus telescoping does not change the dimension group.

Conversely, set $G = DG(\mathcal{B})$ and $G' = DG(\mathcal{B}')$. We shall construct a Bratteli diagram $\mathcal{C} = (W, F)$ that contracts to a contraction of $\mathcal{B}$ on odd levels and to a contraction of $\mathcal{B}'$ on even levels. It suffices to give the sets of vertices $W(n)$ and the adjacency matrices $N(n)$ between consecutive levels.

We set $W(1) = V(1)$ and $N(1) = M(1)$. Looking at the canonical generators of $\mathbb{Z}^{V(1)}$ as elements of G', we can consider that they are elements of some $\mathbb{Z}^{V'(n_2)}$. We set $W(2) = V'(n_2)$ and denote $N(2)$ the matrix of the map it defines from $\mathbb{Z}^{V(1)}$ to $\mathbb{Z}^{V'(n_2)}$. Again, the elements of $\mathbb{Z}^{V'(n_2)}$ can be considered as elements of G and thus belong to some $\mathbb{Z}^{V(n_3)}$. We set $W(3) = V(n_3)$ and we call $N(3)$ the map that it defines from $\mathbb{Z}^{V'(n_2)}$ to $\mathbb{Z}^{V(n_3)}$. Proceeding like this, we obtain the sequence

$$\mathbb{Z} \xrightarrow{N(1)} \mathbb{Z}^{V(1)} \xrightarrow{N(2)} \mathbb{Z}^{V'(n_2)} \xrightarrow{N(3)} \mathbb{Z}^{V(n_3)} \cdots,$$

which is sufficient to define the Bratteli diagram $\mathcal{C}$ we are looking for. ∎

We illustrate this result with the following example.

Example 5.1.6 Consider the two diagrams of Figure 5.4 on the left and on the right. We have already seen in Example 5.1.4 that the dimension group of the first one is $\mathbb{Z}[1/2]$ obtained as the direct limit of the sequence $\mathbb{Z} \xrightarrow{2} \mathbb{Z} \xrightarrow{2} \cdots$. The dimension group of the second one is the direct limit of the sequence $\mathbb{Z}^2 \xrightarrow{M} \mathbb{Z}^2 \xrightarrow{M} \cdots$ where M is the matrix

$$M = \begin{bmatrix} 1 & 1 \\ 1 & 1 \end{bmatrix}.$$

Since $\mathcal{R}_M = \left\{ \begin{bmatrix} x & x \end{bmatrix}^t \mid x \in \mathbb{R} \right\}$, the isomorphism of the dimension groups is a consequence of the commutative diagram below.

$$
\begin{array}{ccc}
x & \longrightarrow & \begin{bmatrix} x & x \end{bmatrix}^t \\
\downarrow{\scriptstyle 2} & & \downarrow{\scriptstyle M} \\
2x & \longrightarrow & \begin{bmatrix} 2x & 2x \end{bmatrix}^t
\end{array}
$$

Theorem 5.1.5 means that the properties of a Bratteli diagram, or at least of its equivalence class for telescoping, should be read on its dimension group. A first step in the direction is the following statement.

Proposition 5.1.7 *A Bratteli diagram is simple if and only if its dimension group is simple.*

Proof Let $\mathcal{B} = (V, E)$ be a Bratteli diagram and $G = \mathrm{DG}(\mathcal{B})$. Let us first suppose that $\mathcal{B}$ is simple. We have to show that every nonzero element $g \in G^+$ is an order unit. Let indeed $g \in G^+$ be nonzero and let $h \in G^+$. We can choose $n \geq 1$ such that $g = i_n(x)$ and $h = i_n(y)$ with $x, y \in G(n)^+$. Let $v \in V(n)$ be such that $x_v > 0$. Since $\mathcal{B}$ is simple, we have $M(m) \cdots M(n+1)x > 0$ for all m large enough. Then $M(m) \cdots M(n+1)y \leq N M(m) \cdots M(n+1)x$ for N large enough (where the symbol $\leq$ is to be taken componentwise). This implies that $h < Ng$ and thus G is simple.

Conversely, assume that $\mathcal{B}$ is not simple. Let W be a nonempty directed and hereditary set strictly contained in V. Let H be the set of $h \in \mathrm{DG}(\mathcal{B})$ defined by a sequence $x = (x_n)$ with $x_n \in \mathbb{Z}^{V(n)}$ having the property that for some $n \geq 1$ we have $x_{n,v} = 0$ for every $v \notin W$ and $x_{m+1} = M(m+1)x(m)$ for all $m \geq n$. Since W is directed, every matrix $M(m)$ has the form

$$M(m) = {}^W \begin{bmatrix} \begin{array}{c|c} \overset{W}{} & \\ \hline 0 & \end{array} \end{bmatrix}.$$

Thus, if x satisfies this property for n, it holds for every $m \geq n$. Thus, H is a subgroup of G, which is clearly an ideal. Since $W \neq \emptyset$, we can choose $w \in W \cap V(n)$ and x such that $x_{n,w} > 0$. Then $x_{m,v} > 0$ for every descendant of w, which implies that the class of x is not 0. Therefore, we have $H \neq \{0\}$. Since W is strictly contained in V and since it is hereditary, we have $H \neq G$. Thus, G is not simple. ∎

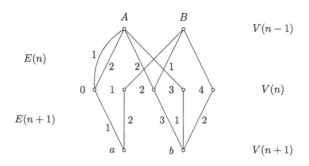

Figure 5.5 Order on the diagram of Figure 5.1.

Example 5.1.8 Consider again the non-simple Bratteli diagram of Example 5.1.2. All matrices $M(n)$ for $n \geq 1$ are equal to

$$M = \begin{bmatrix} 1 & 1 \\ 0 & 1 \end{bmatrix}.$$

The dimension group G is $\mathbb{Z}^2$ with the lexicographic order, that is $(\mathbb{Z}^2, \mathbb{Z}^+ \times \mathbb{Z} \cup \{0\} \times \mathbb{Z}_+, 0)$ (see Example 2.2.10). The set $\{0\} \times \mathbb{Z}$ is an order ideal.

5.1.3 Ordered Bratteli Diagrams

An *ordered Bratteli diagram* $(V, E, \leq)$ is a Bratteli diagram (V, E) together with a partial order $\leq$ on E such that edges e, e' in E are comparable if and only if $r(e) = r(e')$, in other words, we have a linear order on each set $r^{-1}(\{v\})$, where v belongs to $V \setminus V(0)$ (see Figure 5.5).

If $(V, E, \leq)$ is an ordered Bratteli diagram and $k < l$ are in $\mathbb{Z}_+$, then the set $E_{k+1,l}$ of paths from $V(k)$ to $V(l)$ may be given an induced order, called the reverse *lexicographic order*, as follows:

$$(e_{k+1}, e_{k+2}, \ldots, e_l) < (f_{k+1}, f_{k+2}, \ldots, f_l)$$

if and only if for some i with $k + 1 \leq i \leq l$, $e_j = f_j$ for $i < j \leq l$ and $e_i < f_i$ (this is the right-to-left version of the usual lexicographic order). It is a simple observation that if $(V, E, \leq)$ is an ordered Bratteli diagram and (V', E') is a telescoping of (V, E) as defined above, then with the induced order $\leq'$, $(V', E', \leq')$ is again an ordered Bratteli diagram. We say that $(V', E', \leq')$ is a *telescoping* of $(V, E, \leq)$ (see Figure 5.6).

Again there is an obvious notion of isomorphism between ordered Bratteli diagrams. Let $\approx$ denote the equivalence relation on ordered Bratteli diagrams generated by isomorphism and by telescoping. One can show that $\mathcal{B}^1 \approx \mathcal{B}^2$, where $\mathcal{B}^1 = (V^1, E^1, \leq^1)$, $\mathcal{B}^2 = (V^2, E^2, \leq^2)$, if and only if there exists an

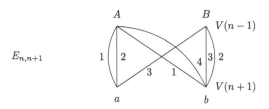

$E_{n,n+1}$

Figure 5.6 Telescoping of the diagram of Figure 5.5.

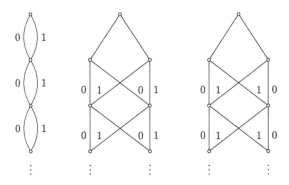

Figure 5.7 Three ordered Bratteli diagrams.

ordered Bratteli diagram $\mathcal{B} = (V, E, \leq)$ such that telescoping $\mathcal{B}$ to odd levels $0 < 1 < 3 <:$ yields a telescoping of either $\mathcal{B}^1$ or $\mathcal{B}^2$, and telescoping $\mathcal{B}$ to even levels $0 < 2 < 4 <:$ yields a telescoping of the other (Exercise 5.2). This is analogous to the situation for the equivalence relation $\sim$ on Bratteli diagrams as we discussed above.

Example 5.1.9 The two first ordered diagrams of Figure 5.7 are equivalent. The third one is not (although all three are equivalent as unordered Bratteli diagrams).

The following notion will be important when we deal with Bratteli diagrams and shift spaces. Fix $n \geq 1$ and let us consider $V(n-1)$ and $V(n)$ as alphabets. For every letter $a \in V(n)$, consider the ordered list $(e_1, \ldots, e_k)$ of edges of $E(n)$, which range at a, and let $(a_1, \ldots, a_k)$ be the ordered list of the labels of the sources of these edges. This defines a morphism $\tau(n): a \mapsto a_1 \cdots a_k$ from $V(n)^*$ to $V(n-1)^*$ that we call *the morphism read on $E(n)$*. For example, in Figure 5.5, the morphism we read on:

- $E(n)$ is $\tau(n): 0 \mapsto AA, \ 1 \mapsto B, \ 2 \mapsto BA, \ 3 \mapsto A, \ 4 \mapsto B$,
- $E(n+1)$ is $\tau(n+1): a \mapsto 01, \ b \mapsto 342$,

and on Figure 5.6 the morphism we read on $E_{n,n+1}$ is $\sigma : a \mapsto AAB, b \mapsto ABBA$. We can check, of course, that we have $\sigma = \tau_n \circ \tau_{n+1}$. Note that the matrix $M(n)$ is the incidence matrix of the morphism $\tau(n)$. In our example, we find

$$M(n) = \begin{bmatrix} 2 & 0 \\ 0 & 1 \\ 1 & 1 \\ 0 & 1 \end{bmatrix}, \quad M(n+1) = \begin{bmatrix} 1 & 1 & 0 & 0 & 0 \\ 0 & 0 & 1 & 1 & 1 \end{bmatrix}.$$

Note that the elements of the matrices $M(n)$ are integers (since $M(n)_{r,s}$ is the number of edges from s to r), whereas the label of an edge in the picture of a diagram (which is also an integer) does not represent a multiplicity of the edge but its order with respect the other edges with the same range.

The ordered Bratteli diagram $(V, E, \leq)$ is *stationary* if (V, E) is stationary, and the ordering on the edges with range $v(n, a_i)$ is the same as the ordering on the edges with range $v(m, a_i)$ for $m, n = 2, 3, \ldots$ and $i = 1, \ldots, k$. In other words, beyond level 1 the diagram with the ordering repeats itself (see Figure 5.7). In terms of morphisms read on a diagram, an ordered Bratteli diagram $\mathcal{B}$ is stationary if and only if all morphisms read on $\mathcal{B}$ at different levels are the same. Therefore, a stationary oredered Bratteli diagram is essentially the same thing as a morphism.

5.2 Dynamics for Ordered Bratteli Diagrams

We shall see now how one can define the dynamics on the set of paths in a Bratteli diagram.

5.2.1 The Bratteli Compactum

Let $(V, E, \leq)$ be an ordered Bratteli diagram. Let X_E denote the associated *infinite path space*, that is,

$$X_E = \{(e_1, e_2, \ldots) \mid e_i \in E(i), r(e_i) = s(e_{i+1}), i = 1, 2, \ldots\}.$$

Two paths in X_E are said to be *cofinal* if they have the same tails, that is, the edges agree from a certain level on. We denote by R_E this equivalence, called the equivalence of *cofinality* on the set X_E. We will see that cofinality is an important notion for Bratteli diagrams when we will define the dynamics on ordered Bratteli diagrams.

The set X_E is a closed subset of $\prod_{i \geq 1} E(i)$. Since every $E(i)$ is finite, the product is compact and thus X_E is compact. A basis for the topology is the family of *cylinder sets*

$$[e_1, e_2, \ldots, e_k]_E = \{(f_1, f_2, \ldots) \in X_E \mid f_i = e_i, 1 \le i \le k\}.$$

Each $[e_1, \ldots, e_k]_E$ is also closed, as is easily seen. When it will be clear from the context we will write $[e_1, \ldots, e_k]$ instead of $[e_1, \ldots, e_k]_E$. Endowed with this topology, we call X_E the *Bratteli compactum* associated with $(V, E, \le)$. Let d_E be the distance on X_E defined by $d_E((e_n)_n, (f_n)_n) = \frac{1}{2^k}$, where $k = \inf\{i \mid e_i \ne f_i\}$. It clearly defines the topology of the cylinder sets.

If (V, E) is a simple Bratteli diagram, then either X_E is finite or X_E has no isolated points, and so is a Cantor space (see Exercise 5.5). Moreover, each class of the equivalence R_E (corresponding to cofinality) is dense in X_E (Exercise 5.6).

Let $x = (e_1, e_2, \ldots)$ be an element of X_E. We will say that e_n is the nth label of x and denote it by $x(n)$. We denote by $X_E^{\max}$ the set of elements x of X_E such that $x(n)$ is a maximal edge for all n and by $X_E^{\min}$ the analogous set for the minimal edges.

It is not difficult to show that $X_E^{\max}$ and $X_E^{\min}$ are nonempty (see Exercise 5.7). Moreover, for every $v \in V$, the set of minimal edges forms a *spanning tree* of the graph (V, E). This means that for every $v \in V$, there is a unique path formed of minimal edges from v to $v(0)$ (Exercise 5.7). The same holds for maximal edges.

The ordered Bratteli diagram $(V, E, \le)$ is *properly ordered* if it is simple and if $X_E^{\max}$ and $X_E^{\min}$ both are a one-point set: $X_E^{\max} = \{x_{\max}\}$ and $X_E^{\min} = \{x_{\min}\}$.

Example 5.2.1 The Bratteli diagrams of Figure 5.7 on the left and center are properly ordered while the diagram on the right is not. Indeed, there are two paths labeled with $0, 0, 0, \ldots$ and two paths labeled $1, 1, 1, \ldots$.

Note that every simple Bratteli diagram can be properly ordered (Exercise 5.8).

5.2.2 The Vershik Map

We can now define, for a properly ordered Bratteli diagram $(V, E, \le)$, a map $T_E \colon X_E \to X_E$, called the *Vershik map* (or the *lexicographic map*), associated with $(V, E, \le)$.

We set $T_E(x_{\max}) = x_{\min}$. If $x = (e_1, e_2, \ldots) \ne x_{\max}$, let k be the least integer such that e_k is not a maximal edge. Let f_k be the successor of e_k (relative to the order $\le$ so that $r(e_k) = r(f_k)$). Define

$$
\begin{array}{ccccccc}
0 & 0 & 0 & 0 & 0 & \cdots \\
1 & 0 & 0 & 0 & 0 & \cdots \\
0 & 1 & 0 & 0 & 0 & \cdots \\
1 & 1 & 0 & 0 & 0 & \cdots \\
0 & 0 & 1 & 0 & 0 & \cdots \\
1 & 0 & 1 & 0 & 0 & \cdots \\
0 & 1 & 1 & 0 & 0 & \cdots \\
1 & 1 & 1 & 0 & 0 & \cdots
\end{array}
$$

Figure 5.8 The addition of 1 in base 2.

$$T_E(x) = (f_1, \ldots, f_{k-1}, f_k, e_{k+1}, e_{k+2}, \ldots), \tag{5.3}$$

where $(f_1, \ldots, f_{k-1})$ is the minimal edge in $E_{1,k-1}$ with range equal to $s(f_k)$.

Thus, the image by T_E of a point $x \neq x_{\max}$ is its successor in the reverse lexicographic order. Moreover, x and $T_E(x)$ are clearly cofinal.

The map T_E is clearly continuous. It is, moreover, invertible (Exercise 5.9). We call the resulting pair (X_E, T_E) a *Bratteli–Vershik dynamical system*.

Proposition 5.2.2 *Let* $(V, E, \leq)$ *be a properly ordered Bratteli diagram. The system* (X_E, T_E) *is a minimal invertible dynamical system. When* X_E *is infinite, it is a Cantor system.*

The proof is left as an exercise (Exercise 5.10). We will see below with Theorem 5.3.3 that the converse is also true.

In the sequel, BV will refer to *Bratteli–Vershik*.

Example 5.2.3 Consider the Bratteli diagram (V, E) of Figure 5.7 on the left. The system (X_E, T_E) is isomorphic to the odometer $(\mathbb{Z}_2, T)$, where $T(x) = x + 1$. Indeed, as is well known, the addition of 1 in base 2 consists, on the representation in base 2 of numbers with a fixed number of digits, in taking the next sequence in the lexicographic order. Since the representation in base 2 of 2-adic numbers is written with the least significant digit on the left, addition of 1 corresponds to the next element in the reverse lexicographic order for right infinite sequences (see Figure 5.8 below).

5.3 The Bratteli–Vershik Model Theorem

Let (X, T) be an invertible minimal Cantor dynamical system. The properly ordered Bratteli diagram $(V, E, \leq)$ is a *BV-representation* of (X, T) if (X_E, T_E) is isomorphic to (X, T). We will show, as a main result of this chapter, that every invertible minimal Cantor system has a BV-representation.

5.3.1 From Partitions in Towers to Bratteli Diagrams

We will first show how to associate to any nested sequence of partitions of an invertible system (X, T) an ordered Bratteli diagram.

Let

$$\mathfrak{P}(n) = \{T^j B_i(n) \mid 0 \leq j < h_i(n), 1 \leq i \leq t(n)\}$$

be a nested sequence of KR-partitions of (X, T). We may suppose that $\mathfrak{P}(0) = \{X\}$. Hence $t(0) = 1$, $h_1(0) = 1$ and $B_1(0) = X$.

Let $V(n) = \{(n, 1), \ldots, (n, t(n))\}$, for $n \geq 0$. Thus, the set of vertices is, at each level n, the set of towers of the partition $\mathfrak{P}(n)$.

The set of edges records the inclusions of the elements of $\mathfrak{P}(n)$ in the elements of $\mathfrak{P}(n-1)$. Specifically, let $E(n)$ be the set of quadruples (n, t', t, j) satisfying

$$T^j B_t(n) \subseteq B_{t'}(n-1), \tag{5.4}$$

for $1 \leq t' \leq t(n-1)$, $1 \leq t \leq t(n)$, $0 \leq j \leq h_t(n) - 1$ and $n \geq 1$. Note that, in particular,

1. the index j is such that $T^j B_t(n)$ is contained in the basis $B(n-1)$ of $\mathfrak{P}(n-1)$,
2. not all indices j with $0 \leq j \leq h_t(n) - 1$ appear in these quadruples.

The range and source maps are given by

$$s(n, t', t, j) = (n-1, t') \text{ and } r(n, t', t, j) = (n, t). \tag{5.5}$$

Two edges $e_1 = (n_1, t'_1, t_1, j_1)$ and $e_2 = (n_2, t'_2, t_2, j_2)$ are comparable whenever $n_1 = n_2$ and $t_1 = t_2$. In this case we define $e_1 \geq e_2$ if $j_1 \geq j_2$. It is straightforward to verify that $(V, E, \leq)$ is an ordered Bratteli diagram.

It is useful to remark, from (5.5), that $((n, t'_n, t_n, j_n))_n$ is an infinite path of $(V, E, \leq)$ if and only if $t_{n-1} = t'_n$ for all $n \geq 1$. Hence the paths of the Bratteli diagram have the form $((n, t_{n-1}, t_n, j_n))_n$ with $1 \leq t_{n-1} \leq t(n-1)$, $1 \leq t_n \leq t(n)$, $0 \leq j_n \leq h_{t_n}(n)$ and

$$T^{j_n} B_{t_n}(n) \subset B_{t_{n-1}}(n-1). \tag{5.6}$$

Note that (5.6) implies (by Exercise 4.1) that

$$0 \leq j_n \leq h_{t_n}(n) - h_{t_{n-1}}(n-1).$$

Note also that (n, t_{n-1}, t_n, j_n) is a minimal edge if and only if $j_n = 0$ and is maximal if and only if

$$j_n = h_{t_n}(n) - h_{t_{n-1}}(n-1). \tag{5.7}$$

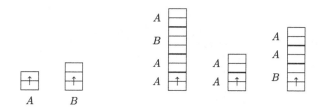

Figure 5.9 The partition $\mathfrak{P}(1)$ consists of two towers called A and B. The map T acts vertically except for the last levels, where it goes back to the base. The elements of the partition $\mathfrak{P}(2)$ can be seen as the piling up of vertical pieces of the towers A and B.

Additionally, if (n, t_{n-1}, t_n, j_n) is not a maximal edge, its successor is an edge (n, t'_{n-1}, t_n, j'_n) with

$$j'_n = j_n + h_{t_{n-1}}(n-1) \tag{5.8}$$

since j'_n is the least integer such that $T^{j'_n - j_n} B_{t_n}(n) \subset B(n-1)$.

For example, suppose that $\mathfrak{P}(n)$ is a refining sequence of KR-partitions such that (see Figure 5.9)

1. $\mathfrak{P}(1) = \{A(1), TA(1), B(1), TB(1), T^2 B(1)\}$ and
2. $\mathfrak{P}(2) = \{T^j B_i(2) \mid 0 \leq j < h_i(2), 1 \leq i \leq t(2)\}$ with
   (a) $t(2) = 3, h_1(2) = 9, h_2(2) = 4, h_3(2) = 7$,
   (b) $B_1(2) \subseteq A(1), T^2 B_1(2) \subseteq A(1), T^4 B_1(2) \subseteq B(1), T^7 B_1(2) \subseteq A(1)$,
   (c) $B_2(2) \subseteq A(1), T^2 B_2(2) \subseteq A(1)$,
   (d) $B_3(2) \subseteq B(1), T^3 B_3(2) \subseteq A(1), T^5 B_3(2) \subseteq A(1)$.

Note that every element of $\mathfrak{P}(1)$ is a union of elements of $\mathfrak{P}(2)$. In particular, we have

$$B(1) = T^4 B_1(2) \cup B_3(2).$$

The corresponding Bratteli diagram is represented in Figure 5.10 with the value of j indicated on the edge (n, t', t, j). Note that only the order of these indices matters for the Bratteli diagram, and not their particular value.

The next two propositions give properties of the ordered Bratteli diagram $(V, E, \leq)$ associated as above with the nested sequence of partitions $(\mathfrak{P}(n))$. The first one relates the paths in the diagram and the nested sequence of partitions.

Proposition 5.3.1 *For every infinite path $(e_1, e_2, \ldots)$ with $e_n = (n, t_{n-1}, t_n, j_n)$ in $(V, E, \leq)$, the following assertions hold.*

1. *We have $0 \leq \sum_{i=1}^{n} j_i \leq h_{t_n}(n) - 1$ for every $n \geq 1$.*

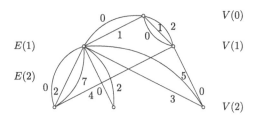

Figure 5.10 The first levels of the Bratteli diagram associated with the sequence $\mathfrak{P}(0)$, $\mathfrak{P}(1)$ and $\mathfrak{P}(2)$.

2. *The sequence*

$$C_n = T^{\sum_{i=1}^{n} j_i} B_{t_n}(n) \tag{5.9}$$

is decreasing, that is, $C_1 \supset C_2 \supset: \supset C_n \supset:$ and the map $\phi: (e_n) \to \bigcap_{n \geq 1} C_n$ sends distinct paths to disjoint sets.

3. *If the sequence $\mathfrak{P}(n)$ generates the topology of X, then $\text{Card}(\bigcap_{n \geq 1} C_n) = 1$.*

Proof 1. Let us prove the first assertion by induction on n. It is true for $n = 1$ since $j_1 \leq h_{t_1}(1) - 1$. Next, for $n \geq 2$, since $T^{j_n} B_{t_n}(n) \subset B_{t_{n-1}}(n-1)$, we have $j_n \leq h_{t_n}(n) - h_{t_{n-1}}(n-1)$ (see Exercise 4.1). Since, by induction hypothesis, we have $j_{n-1} + : + j_1 \leq h_{t_{n-1}}(n-1) - 1$ we conclude that $j_n + \cdots + j_1 \leq h_{t_n}(n) - 1$.

2. By (5.6), we have for every $n \geq 1$,

$$C_{n+1} = T^{\sum_{i=1}^{n+1} j_i} B_{t_{n+1}}(n+1) = T^{\sum_{i=1}^{n} j_i} T^{j_{n+1}} B_{t_{n+1}}(n+1)$$
$$\subset T^{\sum_{i=1}^{n} j_i} B_{t_n}(n) = C_n.$$

Since $(C_n)_{n \geq 1}$ is a decreasing sequence of closed sets and since X is compact, its intersection is nonempty. Since $0 \leq \sum_{i=1}^{n} j_i \leq h_{t_n}(n) - 1$ for every $n \geq 1$, C_n is an element of the partition $\mathfrak{P}(n)$. Let (e_n) and (e'_n) be distinct paths and let $(C_n), (C'_n)$ be the associated sequences. We have $e_m \neq e'_m$ for some $m \geq 1$. Then C_m and C'_m are distinct elements of the partition $\mathfrak{P}(m)$. Indeed, set $e_m = (m, t_{m-1}, t_m, j_m)$ and $e'_m = (m, t'_{m-1}, t'_m, j'_m)$. If $t_{m-1} \neq t'_{m-1}$, then C_m, C'_m are contained in distinct towers of $\mathfrak{P}(m-1)$. Otherwise, if $t_m \neq t'_m$, the sets C_m and C'_m belong to distinct towers. Finally if $t'_{m-1} = t_{m-1}$ and $t'_m = t_m$, since $j_m \neq j'_m$, they are at distinct heights of the tower $B_{t_m}(m)$. Thus, C_m, C'_m are disjoint, which implies $\phi((e_n)) \cap \phi((e'_n)) = \emptyset$.

3. If the sequence of partitions generates the topology, the intersection of all C_n is reduced to one point. ∎

In the second statement, we assume an hypothesis on $(\mathfrak{P}(n))_{n \geq 1}$ mildly weaker than that of being a refining sequence (recall that there exist nested sequences of partitions satisfying (KR1) but not (KR3); see Exercise 4.2).

Proposition 5.3.2 *If $(\mathfrak{P}(n))_{n \geq 1}$ is a nested sequence satisfying (KR1), the Bratteli diagram $(V, E, \leq)$ is properly ordered.*

Proof First, the diagram (V, E) is simple. Indeed, fix $m \geq 1$. Since the intersection of the bases is reduced to one point, there is, by minimality, an $n \geq 1$ such that every $x \in B(n)$ visits every clopen set $B_t(m)$ for $1 \leq t \leq t(m)$ before returning to $B(n)$. This implies that there is a path from every vertex in $V(m)$ to every vertex in $V(n)$ and thus that (V, E) is simple.

Let us show now that $X_E^{\min}$ consists of a single path. Let (e_n) with $e_n = (n, t_{n-1}, t_n, j_n)$ be an infinite path of $X_E^{\min}$. The edges comparable to e_n are the edges of the form (n, t, t_n, j) for some t, and exactly one of them is of the form $(n, t, t_n, 0)$. It is clearly a minimal edge. Hence $j_n = 0$ for all n. But, by Proposition 5.3.1,

$$\phi(e_n) = \cap_n T^0 B_{t_n}(n) = \cap_n B(n),$$

which consists of a single point by hypothesis. Since ϕ sends distinct paths to disjoint sets, the path (e_n) is the unique path of $X_E^{\min}$.

Similarly, if $(e_n) \in X_E^{\max}$, then $j_n = h_{t_n}(n) - h_{t_{n-1}(n-1)}$ by (5.7). Then,

$$\phi(e_n) = \cap_n T^{h_{t_n}(n)-1} B_{t_n}(n),$$

which is reduced to one point because $T(\phi(e_n))$ belongs to the intersection of the bases and T is invertible, whence the conclusion again. ∎

5.3.2 The BV-Representation Theorem

We can now state and prove the BV-representation theorem.

Theorem 5.3.3 (Herman, Putnam, Skau) *For every minimal and invertible Cantor system (X, T), there exists a properly ordered Bratteli diagram $(V, E, \leq)$ such that (X, T) is isomorphic to (X_E, T_E).*

More precisely, for every refining sequence $(\mathfrak{P}(n))$ of KR-partitions of (X, T), the Bratteli diagram $(V, E, \leq)$ associated with $\mathfrak{P}(n)$ is such that (X, T) is isomorphic to (X_E, T_E).

Proof By Theorem 4.1.6, there exists a refining sequence of partitions $(\mathfrak{P}(n))$ of (X, T). By Proposition 5.3.2, the ordered Bratteli diagram $(V, E, \leq)$

associated with $(\mathfrak{P}(n))$ is properly ordered. This allows us to consider the Cantor system (X_E, T_E). It is minimal by Proposition 5.2.2.

Consider the map $\phi \colon X_E \to X$ defined by

$$\phi((n, t_{n-1}, t_n, j_n)_n) = x \text{ where } \{x\} = \cap_{n \geq 1} C_n, \tag{5.10}$$

with $C_n = T^{\sum_{i=1}^{n} j_i} B_{t_n}(n)$. It is well defined (Proposition 5.3.1) and is a homeomorphism (see Exercise 5.14). Note that $(\mathfrak{P}(n))_n$ being a decreasing sequence of partitions, we also have $\{x\} = \cap_{n \geq N} C_n$ for all N.

There remains to show that it commutes with the dynamics. Let $e = (e_n)_n$ be an infinite path of X_E with $e_n = (n, t_{n-1}, t_n, j_n)$. Suppose first that e is not the maximal path. Then there exists n_0 such that $T_E(e) = e_1' \cdots e_{n_0-1}' e_{n_0}' e_{n_0+1} e_{n_0+2} \cdots$ where $e_n' = (n, t_{n-1}', t_n', j_n')$, with $j_n' = 0$, $1 \leq n \leq n_0 - 1$ and $e_{n_0}' = (n_0, t_{n_0-1}', t_{n_0}, j_{n_0}')$ is the successor of e_{n_0}. Note that, for $1 \leq n \leq n_0 - 1$, the edges e_n being maximal, we have $j_n = h_{t_n}(n) - h_{t_{n-1}}(n-1)$.

Since (e_{n_0}') is the successor of e_{n_0}, we have by (5.8), $j_{n_0}' = j_{n_0} + h_{t_{n_0-1}}(n_0 - 1)$. Hence,

$$\sum_{1 \leq n \leq n_0} j_n = j_{n_0} + \sum_{1 \leq n \leq n_0} h_{t_n}(n) - h_{t_{n-1}}(n-1) = j_{n_0} + h_{t_{n_0-1}}(n_0 - 1) - 1,$$

while

$$\sum_{1 \leq n \leq n_0} j_n' = j_{n_0} + h_{t_{n_0-1}}(n_0 - 1) = \sum_{1 \leq n \leq n_0} j_n + 1.$$

Set $C_n' = \cap_n T^{\sum_{i=1}^{n} j_i'} B_{t_n'}(n)$. Then $\phi \circ T_E(e) = \phi(e') = \cap_n C_n'$. Since $\sum_{i=1}^{n_0} j_i' = \sum_{i=1}^{n_0} j_i + 1$ and $j_n' = j_n$ for $n > n_0$, we have $C_n' = T C_n$ for $n \geq n_0$. This shows that $\phi \circ T_E(e) = T \circ \phi(e)$ and thus the conclusion.

Suppose now that e is the maximal path. Let $x_{\min}$ be the minimal path of $(V, E, \leq)$. Then we have to prove that $\phi(x_{\min}) = T(\phi(e))$. But since $\mathfrak{P}(0) = \{X\}$, we have $h_{t_0}(0) = 1$ and consequently

$$T(\phi(e)) = T \left(\cap_{n \geq 1} T^{\sum_{i=1}^{n} h_{t_i}(i) - h_{t_{i-1}}(i-1)} B_{t_n}(n) \right)$$
$$= \cap_{n \geq 1} T^{h_{t_n}(n)} B_{t_n}(n) \subseteq \cap_{n \geq 1} \cup_{1 \leq i \leq t(n)} B_i(n) = \{\phi(x_{\min})\} .$$

We conclude using the invertibility of T. ∎

Note that a periodic minimal system, although not a Cantor system, has also a BV-representation. Indeed, the system $X = \{x, Tx, \ldots, T^{n-1}x\}$ is isomorphic to (X_E, T_E), where (V, E) has one vertex at each level, n edges from

$V(0)$ to $V(1)$ and one edge at all other levels. We shall have more to say about the case of periodic systems in Chapter 6.

5.3.3 From Bratteli Diagrams to Partitions in Towers

We now show that the construction used in Theorem 5.3.3 can be done backwards, using the fact that the system (X_E, T_E) has a natural sequence of partitions.

Proposition 5.3.4 *Let* $(V, E, \leq)$ *be a properly ordered Bratteli diagram. For every* $n \geq 1$*, the family*

$$\mathfrak{P}(n) = \{[e_1, \ldots, e_n] \mid (e_1, \ldots, e_n) \in E_{1,n}\} \tag{5.11}$$

is a partition in towers of (X_E, T_E) *with basis the cylinders* $[e_1, \ldots, e_n]$ *formed of minimal edges and the sequence* $(\mathfrak{P}(n))$ *is a refining sequence.*

Proof The basis of the partition $\mathfrak{P}(n)$ is the union of the cylinders corresponding to minimal paths and can be identified with $V(n)$. This sequence of partitions is a refining sequence. Indeed, condition (KR1) is satisfied because $(V, E, \leq)$ is properly ordered, and thus that there is a unique minimal element in X_E. It is clear that the sequence $(\mathfrak{P}(n))$ is nested and thus (KR2) is satisfied. Finally, (KR3) is also satisfied, by definition of the topology on X_E. ∎

The sequence of partitions $(\mathfrak{P}(n))$ defined by (5.11) is called the *natural sequence of partitions* associated with $(V, E, \leq)$.

Proposition 5.3.5 *Each* $\mathfrak{P}(n)$ *of the natural sequence of partitions is a partition in* $t(n) = \mathrm{Card}(V(n))$ *towers. For each* $v \in V(n)$*, the height of tower* $B_v(n)$ *is*

$$h_v(n) = (M(n) \cdots M(1))_{v,1}. \tag{5.12}$$

Proof Since, by (5.2), the number of paths from $V(0) = \{1\}$ to $v \in V(n)$ is equal to $(M(n) \cdots M(1))_{v,1}$, the height of the tower $B_v(n)$ with $v \in V(n)$ is given by (5.12). ∎

Note that $M(n) \cdots M(1)$ is actually a column vector so the right-hand side could also be denoted $(M(n) \cdots M(1))_v$. From (5.12) one deduces the following relation between the heights of different partitions $\mathfrak{P}(n)$ and $\mathfrak{P}(m)$, $m < n$,

$$h(n) = M(n) \cdots M(m+1)h(m), \tag{5.13}$$

where $h(n) = (h_v(n))_{v \in V(n)}$ is considered as a column vector. This useful formula is related to another one concerning the row vector $\mu(n) = (\mu(B_v(n))_{v \in V(n)}$, where μ is an invariant Borel probability measure on (X_E, T_E),

$$\mu(m) = \mu(n) M(n) \cdots M(m+1), \tag{5.14}$$

which results directly from (4.7).

For every minimal invertible Cantor system (X, T), the Bratteli diagram $G = (V, E)$ built from a refining sequence $(\mathfrak{P}(n))$ of KR-partitions is such that (X_E, T_E) is, by Theorem 5.3.3, topologically conjugate to (X, T). It is easy to verify that the natural sequence of partitions and the original partition coincide up to the conjugacy (Exercise 5.15).

The above correspondence between refining sequences of partitions and properly ordered Bratteli diagrams invites to a translation of the properties of each other. Thus, telescoping a diagram corresponds to taking a subsequence of the refining sequence. We will see next another translation with the adjacency matrices of a diagram.

5.3.4 Dimension Groups and BV-Representation

We have shown in Proposition 4.3.1 that, for any minimal invertible Cantor system (X, T), the dimension group $K^0(X, T)$ is a direct limit of the groups $G(n)$ associated with a sequence $(\mathfrak{P}(n))$ of KR-partitions. We will now see how this group is defined directly in terms of a BV-representation of (X, T).

Theorem 5.3.6 *Let $(V, E, \leq)$ be a properly ordered Bratteli diagram. The group $K^0(X_E, T_E)$ is the dimension group $\mathrm{DG}(V, E)$ of the diagram (V, E).*

This relies on the important fact that, as we shall see now, the adjacency matrices $M(n)$ are precisely the connecting matrices $M(n+1, n)$ introduced in Section 4.3.

Lemma 5.3.7 *Let $(V, E, \leq)$ be a properly ordered Bratteli diagram with adjacency matrices $(M(n))$. Let $(\mathfrak{P}(n))$ be the natural sequence of partitions associated with $(V, E, \leq)$. Let $G(n) = G(\mathfrak{P}(n))$ be the ordered group associated with the partition $\mathfrak{P}(n)$ and let $M(n+1, n)$ be the matrix of the connecting morphism $I(n+1, n)$. Then*

$$M(n, n-1) = M(n). \tag{5.15}$$

The group $K^0(X_E, T_E)$ is isomorphic to the direct limit of the groups $G(n)$ with the connecting matrices $M(n)$.

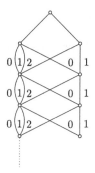

Figure 5.11 A properly ordered Bratteli diagram.

Proof One has by Eq. (4.5),

$$M(n, n - 1)_{k,i} = \mathrm{Card}\{j \mid 0 \le j \le h_k(n), T_E^j B_k(n) \subset B_i(n - 1)\}$$

for $1 \le i \le t(n)$ and $0 \le j < h_i(n)$. Thus, by (5.6), we have also

$$M(n, n - 1)_{k,i} = \mathrm{Card}\{e \in E(n) \mid r(e) = k, s(e) = i\} = M(n)_{k,i}.$$

By Proposition 4.3.1, the group $K^0(X_E, T_E)$ is the direct limit of the groups $(G(n), G^+(n), \mathbf{1}_n)$ where $G(n)$ can be identified with $\mathbb{Z}^{t(n)}$ and with the morphisms $I(n, n - 1)$ defined (by Eq. (4.5)) by the connecting matrices $M(n - 1, n)$. Since $M(n - 1, n) = M(n)$, this proves the second assertion. ∎

Proof of Theorem 5.3.6 Consider the natural sequence $(\mathfrak{P}(n))$ of partitions associated with $(V, E, \le)$. The basis of the partition $\mathfrak{P}(n)$ is, as we have seen, the union of the cylinders corresponding to minimal paths and can be identified with $V(n)$. Set $t_n = \mathrm{Card}(V(n))$. By Lemma 5.3.7, the group $K^0(X_E, T_E)$ is the direct limit $\mathbb{Z}^{t(0)} \overset{M(1)}{\to} \mathbb{Z}^{t(1)} \overset{M(2)}{\to} \cdots$, which is precisely $\mathrm{DG}(V, E)$. ∎

Example 5.3.8 Let (V, E) be the properly ordered Bratteli diagram represented in Figure 5.11. The adjacency matrix at each level is

$$M = \begin{bmatrix} 2 & 1 \\ 1 & 1 \end{bmatrix} = \begin{bmatrix} 1 & 1 \\ 1 & 0 \end{bmatrix}^2.$$

Thus, the dimension group $K^0(X_E, T_E)$ is $\mathbb{Z} + \lambda\mathbb{Z}$ with $\lambda = (1 + \sqrt{5})/2$ (see Example 2.3.7).

5.4 Kakutani Equivalence

Two minimal Cantor dynamical systems (X, T) and (Y, S) are *Kakutani equivalent* if they have (up to isomorphism) a common induced system, that is, there exist nonempty clopen sets $U \subseteq X$ and $V \subseteq Y$ such that the induced systems (X_U, T_U) and (Y_V, S_V) are isomorphic (see Exercise 5.17 for a proof that it is really an equivalence relation).

Two systems that are conjugate are Kakutani equivalent, but the converse is false. For example, the system with two points is Kakutani equivalent with the system with one point, but they are not conjugate.

Let us relate Kakutani equivalence to Bratteli diagrams. If $(V, E, \leq)$ is a properly ordered Bratteli diagram, we may change it into a new properly ordered Bratteli diagram $(V', E', \leq')$ by making a finite change, that is, by adding and/or removing any finite number of edges (vertices) and then making arbitrary choices of linear orderings of the edges meeting at the same vertex (for a finite number of vertices). So $(V, E, \leq)$ and $(V', E', \leq')$ are cofinally identical, that is, they only differ on finite initial portions. (Observe that this defines an equivalence relation on the family of properly ordered Bratteli diagrams.) We have the following nice characterization of the Kakutani equivalence.

Theorem 5.4.1 (Giordano, Putnam, Skau) *Let (X_E, T_E) be the dynamical system associated with the properly ordered Bratteli diagram $(V, E, \leq)$. Then the minimal invertible Cantor dynamical system (X, T) is Kakutani equivalent to (X_E, T_E) if and only if (X, T) is isomorphic to $(X_{E'}, T_{E'})$, where $(V', E', \leq')$ is obtained from $(V, E, \leq)$ by a finite change as described above.*

An interesting particular case of this result is the following. Let U be a clopen set of (X_E, T_E). It is a finite union of cylinder sets. We can suppose that they all have the same length, that is, for some n, $U = \cup_{p \in P} [p]$ where P is a set of paths from level 0 to level n. To obtain a BV-representation of the induced system on U it suffices to take the properly ordered Bratteli diagram $(V', E', \leq')$, which consists of all the paths starting with an element of P endowed with the induced ordering. It is not too much work to prove that the induced system on U is isomorphic to $(X_{E'}, T_{E'})$.

Conversely, we will have several occasions to use the following statement, which describes the construction of a BV-representation for a system from one of an induced system (see, for example, Proposition 6.2.14).

Proposition 5.4.2 *Let $(V, E, \leq)$ be a properly ordered Bratteli diagram and let $(V', E', \leq')$ be obtained by suppressing for all $\upsilon \in V(1)$ all edges from*

Figure 5.12 The BV-representation of a system and of its derivative.

$V(0)$ *to v except the minimal one. Then $(X_{E'}, T_{E'})$ is the system induced by (X_E, T_E) on the clopen set U formed by the $x = (e_1, e_2, \ldots)$ such that e_1 is minimal.*

Proof Let $x = (e_1, e_2, \ldots)$ be such that e_1 is minimal. Then the return time $n(x)$ of x to U is the number of edges from $V(0)$ to $v = r(e_1)$. Thus, $T_{E'}(x) = T_E^{n(x)}(x)$. ∎

We illustrate Proposition 5.4.2 with the following simple example.

Example 5.4.3 Let $X = \mathbb{Z}/3\mathbb{Z} \times \mathbb{Z}_2$ where $\mathbb{Z}_2$ is the group of 2-adic integers. Consider the transformation T on X defined by

$$T(x, y) = \begin{cases} (x + 1, y) & \text{if } x + 1 \neq 0 \\ (0, y + 1) & \text{otherwise.} \end{cases}$$

The transformation T defines a dynamical system on X (called the odometer in base $(3 \cdot 2^{n-1})$, as we shall see in Chapter 6). The subset U of X formed of the pairs $(0, y)$ is a clopen set, and the transformation induced by T on U is that addition of 1 in the group $\mathbb{Z}_2$. The systems (X, T) and the system induced on U have the BV-representations shown in Figure 5.12. The diagram on the right is (up to the first level) the usual BV-representation of $\mathbb{Z}_2$ (see Example 5.2.3). The diagram on the left is the same except for the first level made of three edges in agreement with Proposition 5.4.2.

5.5 Orbit Equivalence

5.5.1 Strong Orbit Equivalence

We say that two invertible dynamical systems (X, T) and (Y, S) are *orbit equivalent* whenever there exists a homeomorphism $\phi \colon X \to Y$ sending orbits to orbits

$$\phi\left(\{T^n(x) \mid n \in \mathbb{Z}\}\right) = \{S^n(\phi(x)) \mid n \in \mathbb{Z}\},$$

for all $x \in X$. This induces the existence of maps $\alpha \colon X \to \mathbb{Z}$ and $\beta \colon X \to \mathbb{Z}$ satisfying for all $x \in X$,

$$\phi \circ T(x) = S^{\alpha(x)} \circ \phi(x) \text{ and } \phi \circ T^{\beta(x)}(x) = S \circ \phi(x).$$

These maps are called the *orbit cocycles* associated with ϕ.

When α and β have at most one point of discontinuity, we say that (X, T) and (Y, S) are *strongly orbit equivalent* (SOE). It is natural to consider such a definition because one can show that, if α is continuous, then (X, T) is conjugate to (Y, S) or to (Y, S^{-1}) (Exercise 5.20) and thus, the corresponding equivalence is not really a new one.

The following Strong Orbit Equivalence Theorem characterizes strong orbit equivalence by means of Bratteli diagrams and dimension groups.

Recall from Section 5.1.1 that two Bratteli diagrams (V, E) and (V', E') are telescoping equivalent if they have a common *intertwining*, that is, a Bratteli diagram (W, F) such that telescoping to odd levels gives a telescoping of (V, E) and telescoping to even levels gives a telescoping of (V', E').

Theorem 5.5.1 (Giordano, Putnam, Skau) *Let (X, T) and (X', T') be two invertible minimal Cantor dynamical systems. The following are equivalent.*

(i) *There exist two BV-representations, $(V, E, \le)$ of (X, T) and $(V', E', \le')$ of (X', T'), such that (V, E) and (V', E') have a common intertwining.*

(ii) *There exist two BV-representations, $(V, E, \le)$ of (X, T) and $(V', E', \le')$ of (X', T'), and a homeomorphism $\psi \colon X_E \to X_{E'}$ such that $\psi(x)(n)$ depends only on $x(1) \ldots x(n)$ and $\psi(x_u) = x'_u$, $u \in \{\min, \max\}$, and having the property that if x and y are cofinal from level n on, then $\psi(x)$ and $\psi(y)$ are cofinal from level $n + 1$.*

(iii) *(X, T) and (X', T') are strong orbit equivalent.*

(iv) *The dimension groups $K^0(X, T)$ and $K^0(X', T')$ are isomorphic as unital ordered groups.*

Proof Let us show that (i) implies (ii). Let (W, F) be a common intertwining of (V, E) and (V', E'). Since a BV-representation is assumed to be simple, we can also suppose that all adjacency matrices have entries at least equal to two. This means that every pair of vertices in consecutive levels has at least two connecting edges.

Let $x_{\min}$ and $x_{\max}$ be the minimal and maximal paths of $(V, E, \le)$, and let $x'_{\min}$ and $x'_{\max}$ be those for $(V', E', \le')$. There are unique paths $\tilde{x}_{\min}$ and $\tilde{x}'_{\min}$ in (W, F) that contract respectively to $x_{\min}$ and $x'_{\min}$. Choose a path $z_{\min}$ in (W, F) passing through the same vertices as $\tilde{x}_{\min}$ does at odd levels and through the same vertices as $\tilde{x}'_{\min}$ at even levels. We similarly construct a path

$z_{\max}$ by taking care that it does not share any common edge with $z_{\min}$. This is possible because the adjacency matrices have entries larger than two.

Let us define two homeomorphisms $\phi\colon X_F \to X_E$ and $\phi'\colon X_F \to X_{E'}$. In constructing $z_{\min}$, for each even n, we matched a pair of edges in $F(n) \circ F(n+1)$ with an edge in $E(n/2)$, namely

$$(z_{\min}(n), z_{\min}(n+1)) \to x_{\min}(n/2)\,,$$

and we matched a pair in $F(n+1) \circ F(n+2)$ with an edge in $E'((n+2)/2)$, namely

$$(z_{\min}(n+1), z_{\min}(n+2)) \to x'_{\min}((n+2)/2).$$

In the same way, $(z_{\max}(n), z_{\max}(n+1))$ is matched with $x_{\max}(n/2)$ and $(z_{\max}(n+1), z_{\max}(n+2))$ with $x'_{\max}((n+2)/2)$. Now, for all even n, we extend these matchings in an arbitrary way to bijections respecting the range and source maps from $F(n) \circ F(n+1)$ to $E(n/2)$ and from $F(n+1) \circ F(n+2)$ to $E'((n+2)/2)$. This defines two homeomorphisms $\phi\colon X_F \to X_E$ and $\phi'\colon X_F \to X_{E'}$. The homeomorphism $\psi = \phi' \circ \phi^{-1}$ has the desired properties.

Let us show that (ii) implies (iii). We will show that (X_E, T_E) and $(X_{E'}, T_{E'})$ are SOE. In a minimal BV-representation, two points belong to the same orbit if and only if they are cofinal, except when it is the orbit of the minimal path. This implies that ψ maps orbits to orbits with the possible exception of the orbit of the minimal paths. But since $\psi(x_u) = x'_u$, $u \in \{\min, \max\}$, this is also true for the orbit of the minimal paths. Consequently, there are maps $\alpha\colon X_E \to \mathbb{Z}$ and $\beta\colon X_E \to \mathbb{Z}$ uniquely defined by the relations

$$\psi \circ T_E(x) = T_{E'}^{\alpha(x)} \circ \psi(x) \text{ and } \psi \circ T_E^{\beta(x)}(x) = T_{E'} \circ \psi(x)$$

for all $x \in X_E$. It remains to prove that α and β are continuous with the possible exception of $x_{\max}$ and $x'_{\max}$. We do it for α. It is similar for β.

Let $x = (x_n)_n \in X_E \setminus \{x_{\max}\}$ and $k = \alpha(x)$. Let n_0 be such that $(x_1, \ldots, x_{n_0})$ has a non-maximal edge and the minimum number of paths from any vertex in V_{n_0-1} to V_0 is greater than k.

Let y belong to the cylinder $[x_1, \ldots, x_{n_0+1}]$. It suffices to show that $\alpha(y) = k$. The paths $T_E(x)$ and $T_E(y)$ start with the same $n_0 + 1$ first edges. Thus, from the property of ψ, the points $\psi \circ T_E(x)$ and $\psi \circ T_E(y)$ start with the same $n_0 + 1$ first edges $f_1, f_2, \ldots, f_{n_0+1}$:

$$\psi \circ T_E(x) = (f_1, f_2, \ldots, f_{n_0+1}, x'_{n_0+2}, \ldots) \text{ and}$$
$$\psi \circ T_E(y) = (f_1, f_2, \ldots, f_{n_0+1}, y'_{n_0+2}, \ldots).$$

For the same reason, and because x and $T_E(x)$, and, y and $T_E(y)$ are cofinal from $n_0 + 1$, $\psi(x)$ and $\psi(y)$ start with the same edges $g_1, g_2, \ldots, g_{n_0+1}$ and

$$\psi(x) = (g_1, g_2, \ldots, g_{n_0+1}, x'_{n_0+2}, \ldots) \text{ and}$$

$$\psi(y) = (g_1, g_2, \ldots, g_{n_0+1}, y'_{n_0+2}, \ldots).$$

But since there are at least k paths from any vertex in V_{n_0-1} to V_0, we deduce that $T_E^k([g_1, g_2, \ldots, g_{n_0+1}]) = [f_1, f_2, \ldots, f_{n_0+1}]$ because $\psi \circ T_E(x) = T_{E'}^k \circ \psi(x)$. Therefore, $\psi \circ T_E(y) = T_{E'}^k \circ \psi(y)$ and $\alpha(y) = k$.

Let us show that (iii) implies (iv). Let $\psi : (X, T) \to (X', T')$ be a SOE map. Remark that (X', T') is isomorphic to $(X, \psi^{-1} \circ T' \circ \psi)$. Hence we can suppose $X' = X$ and set $S = \psi^{-1} \circ T' \circ \psi$, Then we have

$$T(x) = S^{\alpha(x)}(x) \text{ and } S(x) = T^{\beta(x)}(x)$$

where α and β are continuous everywhere, except possibly on some y.

Let A be a clopen set not containing y. Since α is continuous on A, the set $\alpha(A)$ is compact and consequently finite: there exist $n_1, \ldots, n_k$ such that $A = \cup_{1 \leq i \leq k} A \cap \alpha^{-1}(\{n_i\})$. Recall that the characteristic function of the set A is denoted by χ_A. Hence,

$$TA = \bigcup_{1 \leq i \leq k} S^{n_i}(A \cap \alpha^{-1}(\{n_i\}))$$

and thus, taking the characteristic function of each side,

$$\chi_A \circ T^{-1} = \sum_{1 \leq i \leq k} \chi_{A \cap \alpha^{-1}(\{n_i\})} \circ S^{-n_i} .$$

But since $f - f \circ S^{-n} = (\sum_{1 \leq i \leq n} f \circ S^{-i}) \circ S - (\sum_{1 \leq i \leq n} f \circ S^{-i})$, we deduce that $\chi_A \circ T^{-1} - \chi_A$ belongs to $\partial_S(C(X, \mathbb{Z}))$.

Now suppose that A contains y. Note that $\chi_A \circ T^{-1} - \chi_A = \chi_{X \setminus A} - \chi_{X \setminus A} \circ T^{-1}$. But since y is not contained in $X \setminus A$ we deduce from the previous case that $\chi_A \circ T^{-1} - \chi_A$ belongs to $\partial_S(C(X, \mathbb{Z}))$. Since T is invertible we proved that for every clopen set U, $\chi_U \circ T - \chi_U$ belongs to $\partial_S(C(X, \mathbb{Z}))$ and consequently that $\partial_T(C(X, \mathbb{Z})) \subseteq \partial_S(C(X, \mathbb{Z}))$. Proceeding similarly with the equality $S(x) = T^{\beta(x)}(x)$ we obtain $\partial_T(C(X, \mathbb{Z})) = \partial_S(C(X, \mathbb{Z}))$.

This shows that $H(X, T, \mathbb{Z}) = H(X, S, \mathbb{Z})$ and thus $K^0(X, T) = K^0(X, S)$.

Finally the implication (iv) implies (i) results of Theorem 5.1.5.   ∎

We give below an example of strongly orbit equivalent invertible minimal Cantor systems which are not conjugate.

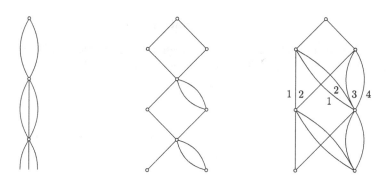

Figure 5.13 Three Bratteli diagrams.

Example 5.5.2 Consider the primitive substitution $\sigma : a \mapsto ab, b \mapsto a^2b^2$. As we shall see in Chapter 6, the shift $X(\sigma)$ is conjugate to (X_E, T_E), where (V, E) is the Bratteli diagram represented in Figure 5.13 on the right (the morphism σ is read on (V, E)). The adjacency matrix of the diagram is

$$M = \begin{bmatrix} 1 & 1 \\ 2 & 2 \end{bmatrix},$$

which has eigenvalues 0, 3. We claim that the dimension group is isomorphic to $\mathbb{Z}[1/3]$ with the usual ordering but with order unit 2, that is, the group $\frac{1}{2}\mathbb{Z}[1/3]$. Indeed, in the basis

$$u = \begin{bmatrix} 1 \\ 2 \end{bmatrix}, v = \begin{bmatrix} 1 \\ -1 \end{bmatrix}$$

we have

$$\begin{bmatrix} 1 \\ 1 \end{bmatrix} = \frac{2}{3}u + \frac{1}{3}v.$$

Since multiplication by 3 is an automorphism, this proves the claim. Thus, $X(\sigma)$ has the same dimension group as the odometer in base $p_n = 2.3^{n-1}$. Indeed, the dimension group of this odometer is the direct limit of the sequence

$$\mathbb{Z} \xrightarrow{2} \mathbb{Z} \xrightarrow{3} \mathbb{Z} \xrightarrow{3} \dots,$$

which is the subgroup of $\mathbb{Q}$ formed of the p/q with q dividing some 2.3^n, which is $\frac{1}{2}\mathbb{Z}[1/3]$ (we shall see odometers in more detail in Chapter 6 and in particular their dimension groups in Proposition 6.1.2).

The shift $X(\sigma)$ and the odometer in base 2.3^{n-1} are thus strong orbit equivalent by Theorem 5.5.1. The BV-representation of this odometer is represented in Figure 5.13 on the left. An intertwining of the Bratteli diagrams is represented in the middle.

5.5.2 Orbit Equivalence

Concerning orbit equivalence, we have the following additional Orbit Equivalence Theorem that we state without a complete proof. Recall from Chapter 2 that we denote by $\mathrm{Inf}(G)$ the infinitesimal subgroup of a unital ordered group G.

Theorem 5.5.3 *Let (X, T) and (X', T') be two invertible minimal Cantor systems and let $G = K^0(X, T)$, $G' = K^0(X', T')$ be their dimension groups. The following conditions are equivalent.*

(i) *The systems (X, T) and (X', T') are orbit equivalent.*
(ii) *The groups $G / \mathrm{Inf}(G)$ and $G' / \mathrm{Inf}(G')$ are isomorphic.*
(iii) *There exists a homeomorphism from X onto X' exchanging the invariant Borel probability measures.*

We give below the proof of some of the implications. We omit the proof of (ii)$\Rightarrow$ (i), which is far more difficult (see Section 5.9 for a reference).

Proof (i)$\Rightarrow$ (iii). We may assume that $X = X'$ since, otherwise, we may replace X by $\phi(X)$, where $\phi \colon X \to X'$ is a homeomorphism sending orbits to orbits. Set for typographical convenience $T_1 = T$ and $T_2 = T'$, Let $n_1, n_2 \colon X \to \mathbb{Z}$ be the associated orbit cocycles, that is, such that $T_2(x) = T_1^{n_1(x)}(x)$ and $T_1(x) = T_2^{n_2(x)}(x)$. Let μ_1 be a T_1-invariant Borel probability measure on X. Then for every Borel subset E of X, we have

$$\mu_1(T_2(E)) = \mu_1(\cup_k T_2(A_k \cap E)) = \sum_k \mu_1(T_2(A_k \cap E))$$

$$= \sum_k \mu_1(T_1^k(A_k \cap E)) = \sum_k \mu_1(A_k \cap E) = \mu_1(E),$$

where A_k is the closed set $A_k = \{x \in X \mid n_1(x) = k\}$. Thus, μ_1 is T_2-invariant.
    (iii)$\Rightarrow$ (ii) For $f \in C(X, \mathbb{Z})$, the class of f modulo coboundaries is in $\mathrm{Inf}(K^0(X, T))$ if and only if $\int f d\mu = 0$ for every invariant Borel probability measure μ on (X, T). Thus, the implication follows from Theorem 3.9.3. ∎

We give below an example of orbit equivalent shifts, which are not strongly orbit equivalent.

Example 5.5.4 Let σ be the primitive and proper substitution $a \to aab, b \to abb$. The corresponding substitution shift is a Toeplitz shift (this can be shown

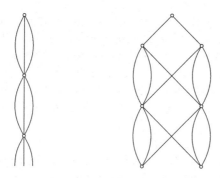

Figure 5.14 Bratteli diagrams of orbit equivalent systems.

in the same way as in Example 1.6.1). A BV-representation of $X(\sigma)$ is given in Figure 5.14 on the right. The adjacency matrix of this Bratteli diagram is

$$M = \begin{bmatrix} 2 & 1 \\ 1 & 2 \end{bmatrix}.$$

It has eigenvalues 1, 3. The dimension group is $G = \mathbb{Z}[1/3] \times \mathbb{Z}$ with $G^+ = (\mathbb{Z}_+[1/3] \setminus \{0\}) \times \mathbb{Z} \cup \{(0, 0)\}$ and $1_G = (1, 1)$. The infinitesimal subgroup is generated by $(0, 1)$. Thus, (see Example 2.2.11), the quotient $G/\mathrm{Inf}(G)$ is isomorphic to $\mathbb{Z}[1/3]$. By Theorem 5.5.3, the shift $X(\sigma)$ is orbit equivalent to the odometer defined on $\mathbb{Z}_3$ whose BV-representation is shown on the left.

Actually, Example 5.5.4 is a particular case of a more general property. Indeed, every constant-length primitive morphism is orbit equivalent to an odometer (Exercise 6.22).

5.5.3 Kakutani Orbit Equivalence

Two dynamical systems (X, T) and (X', T') are *Kakutani orbit equivalent* (resp. *Kakutani strong orbit equivalent*) if they are Kakutani equivalent to orbit equivalent (resp. strong orbit equivalent) systems.

The fact that these are really equivalence relations will come as a consequence of the results below. In the following result (Kakutani Strong Orbit Equivalence Theorem), we use dimension groups as ordered groups with no prescribed unit because we use induction, which preserves the dimension group as ordered groups, but may change the unit (see Proposition 3.7.3).

Theorem 5.5.5 *Two minimal invertible Cantor systems are Kakutani strong orbit equivalent if and only if their dimension groups are isomorphic as ordered groups.*

Proof Let (X, T) and (X', T') be Kakutani strong orbit equivalent. Let (Y, S) and (Y', S') be strong orbit equivalent systems that are Kakutani equivalent to (X, T) and (X', T') respectively. By Proposition 3.7.3, $K^0(Y, S)$ is isomorphic as ordered group to $K^0(X, T)$. Similarly $K^0(Y', S') \simeq K^0(X', T')$. By Theorem 5.5.1, the dimension groups $K^0(Y, S)$ and $K^0(Y', S')$ are isomorphic (this time as unital ordered groups). Thus, $K^0(X, T) \simeq K^0(X', T')$ as ordered groups.

Conversely, assume that $\varphi \colon K^0(X, T) \to K^0(X', T')$ is an isomorphism of ordered groups. Let $u = \varphi(\mathbf{1}_X)$ and let $f \in C(X', \mathbb{Z})$ be such that u is the class of f modulo coboundaries. We may assume that $X' = X_{E'}$ and $T' = T_{E'}$ for some properly ordered Bratteli diagram $(V', E', \leq)$. Let n be such that f is constant on every cylinder $[p]$ defined by a path from 0 to level n in (V', E'). Let $(W, F, \leq)$ be the Bratteli diagram obtained from $(V', E', \leq)$ by replacing every path $p \in E'_{[0,n)}$ by $f(p)$ edges. The dimension group of (W, F) is now isomorphic to $K^0(X, T)$ with order unit $\mathbf{1}_X$. By Theorem 5.5.1, the systems (X, T) and (X_F, T_F) are strong orbit equivalent. By Theorem 5.4.1, the systems (X_F, T_F) and $(X_{E'}, T_{E'})$ are Kakutani equivalent. Therefore (X, T) and (X', T') are Kakutani strong orbit equivalent.   ∎

Example 5.5.6 The shift $X(\sigma)$ of Example 5.5.2 is Kakutani strong orbit equivalent to the odometer in base (3^n). Indeed, the dimension group is $\mathbb{Z}[1/3]$ in both cases (with different units).

Concerning Kakutani orbit equivalence, we have the following analogous result.

Theorem 5.5.7 *Two minimal invertible Cantor systems (X, T) and (X', T') are Kakutani orbit equivalent if and only if the groups $K^0(X, T)/\operatorname{Inf}(K^0(X, T))$ and $K^0(X', T')/\operatorname{Inf}(K^0(X', T'))$ are isomorphic as ordered groups.*

The proof is similar to the proof of Theorem 5.5.1, using this time the orbit equivalence theorem (Theorem 5.5.3) instead of the strong orbit equivalence.

If σ, σ' are two primitive morphisms with constant lengths n, n', the shifts $X(\sigma), X(\sigma')$ are Kakutani orbit equivalent if and only if n, n' have the same prime factors (Exercise 6.23).

5.6 Equivalences on Cantor Spaces

We systematically consider in this section pairs (X, R) of a topological space X and an equivalence relation R on X. After all, the orbit equivalence suggests the idea of studying this notion for its own sake. We will show that this can be successfully realized. We have in mind two basic examples.

The first one is the cofinality equivalence R_E on the space X_E of paths in Bratteli diagram (V, E). Recall that

$$R_E = \{(x, y) \in X_E \times X_E \mid \text{ for some } N \geq 1, x_n = y_n \text{ for all } n \geq N\}.$$

Note that R_E is related to the transformation $S(e_n, e_{n+1}, \ldots) = (e_{n+1}, e_{n+2}, \ldots)$, which is the one-sided shift on the set X formed by the sequences $(e_n, e_{n+1}, \ldots)$ for some $n \geq 1$ with $e_m \in E(m)$ for $m \geq n$. In fact, R_E is formed of the pairs (x, y) of sequences in X_E such that $S^N x$ is equal to $S^N y$ for some $N \geq 1$.

The second one is the orbit equivalence

$$R_T = \{(x, y) \in X \times X \mid T^n x = y \text{ for some } n \in \mathbb{Z}\}$$

when (X, T) is a topological dynamical system.

Let us give a first example of how one may study pairs (X, R) of an equivalence R on a space X on their own.

A relation R on a space X is *minimal* if for every $x \in X$, the equivalence class of x is dense in X. A set $Y \subset X$ is *invariant*, or *R-invariant*, if it is saturated by R, that is, if Y is a union of classes of R or, still equivalently, if for every pair $x, x' \in X$ of elements in the same class of R, if $x \in Y$, then $x' \in Y$.

We have then the following statement.

Proposition 5.6.1 *If R is minimal, the only closed R-invariant sets in X are X and $\emptyset$.*

Proof Assume that Y is a closed nonempty R-invariant subset of X. Then Y is dense in X and thus $Y = X$. ∎

The converse of Proposition 5.6.1 is not true (Exercise 5.21). We will see below that the converse is also true under an additional condition.

The orbit equivalence R_T of a dynamical system (X, T) is obviously minimal if and only if the system is minimal.

More interestingly, the cofinality equivalence R_E on a Bratteli diagram is minimal if and only if the diagram is simple.

Indeed, let (V, E) be a simple Bratteli diagram and let x, y be elements in X_E. For every $n \geq 1$ there is an $m \geq n$ such that there is a path $(z_{n+1}, \ldots, z_{m-1})$ from $r(y_n)$ to $s(x_m)$. Let $z^{(n)} = (y_0, \ldots, y_n, z_{n+1} \ldots, z_{m-1}, x_m, x_{m+1}, \ldots)$. Then x and $z^{(n)}$ are cofinal and $\lim z^{(n)} = y$. This shows that the class of x is dense in X_E.

Conversely, assume that (V, E) is not simple. Then, by Proposition 5.1.1, we can find a proper subset W of V that is directed and hereditary. Let Y be the set of infinite paths y in X_E that pass by a vertex w in W. By definition, the set Y contains all paths that coincide with y until w. Thus, Y is open. Its complement is a closed set Z. The set Z is also invariant because if $z \in Z$ is cofinal to t, then t cannot pass by a vertex in W and thus t is in Z. Finally, the definition of W implies that the set Z is nonempty and strictly included in X_E. This shows that R_E is not minimal.

5.6.1 Local Actions and Étale Equivalences

Let X, Y be topological spaces. If $U \subset X$ and $V \subset Y$ are clopen sets, a homeomorphism $\gamma : U \to V$ is called a *partial homeomorphism*.

We denote by $s(\gamma) = U$ the *source* of γ and by $r(\gamma) = V$ the *range* of γ.

If $\gamma_1 : U_1 \to V_1$ and $\gamma_2 : U_2 \to V_2$ are such partial homeomorphisms, we denote $\gamma_1 \cap \gamma_2$ the map equal to γ_1 on the set where the two functions agree. Thus, for the intersection to be a partial homeomorphism, we require this set to be open.

A function $f : X \to Y$ is a *local homeomorphism* if for every $x \in X$, there is a clopen set $U \subset X$ containing x such that $f(U) \subset Y$ is open and that $f|_U$ is a partial homeomorphism.

For a relation ρ on X, we may consider its *inverse*, which is the set of pairs (y, x) for $(x, y) \in \rho$. The *composition* $\sigma \circ \rho$ of two relations σ, ρ on X is defined as usual by

$$\sigma \circ \rho = \{(x, y) \in X \times X \mid (x, z) \in \sigma \text{ and } (z, y) \in \rho \text{ for some } z \in X\}.$$

Finally the *intersection* of two relations is well defined since a relation on X is just a subset of $X \times X$.

We will find it convenient to consider a map $\gamma : X \to X$ as a relation on X, via the identification of γ and its graph $\{(x, \gamma(x)) \mid x \in X\}$. When $\gamma : U \to V$ is a partial homeomorphism, its inverse (considered as a function or as a relation) $\gamma^{-1} : V \to U$ is again a partial homeomorphism. When $\gamma' : U' \to V'$ is another partial homeomorphism, the composition $\gamma \circ \gamma'$ always exists, even without the requirement that $U' = V$ as usual for maps (note that the notation of composition for relations reverses the order of the factors). If $U' \cap V$ is empty, then $\gamma \circ \gamma'$ is empty.

We denote by s, r the two canonical projections from $X \times X$ onto X defined by $s(x, y) = x$ and $r(x, y) = y$. This is consistent with the notation $s(\gamma) = U$ and $r(\gamma) = V$ for a partial homeomorphism $\gamma : U \to V$. For an open set U, we denote id_U the identity map on U, that is, $\mathrm{id}_U = \{(x, x) \mid x \in U\}$.

A collection Γ of partial homeomorphisms of X is a *local action* if:

(i) The collection of sets $U \subset X$ such that id_U belongs to Γ forms a base of the topology.

(ii) The family Γ is closed under taking inverses, composition and intersection.

Note that for $\gamma_1, \gamma_2 \in \Gamma$, the intersection $\gamma_1 \cap \gamma_2$ is the map that is equal to γ_1 (and γ_2) on the set of points x where $\gamma_1(x) = \gamma_2(x)$. Thus, the condition $\gamma_1 \cap \gamma_2 \in \Gamma$ implies that if γ_1 and γ_2 agree on some point x, they agree on a neighborhood of x. Regarding a local action Γ as a set of binary relations on X, we can consider the union $\cup\Gamma$ of all its elements. Thus, $(x, y) \in \cup\Gamma$ if and only if $y = \gamma(x)$ for some $\gamma \in \Gamma$.

Proposition 5.6.2 *Let Γ be a local action and let $R = \cup\Gamma$. Then*

1. *R is an equivalence relation.*
2. *Γ is a basis for a topology on R.*
3. *With this topology, the source and range maps $s, r \colon \cup\Gamma \to X$ are local homeomorphisms.*

Proof 1. For every $x \in X$, there is by condition (i) a set $U \subset X$ containing x such that $\mathrm{id}_U \in \Gamma$. Thus, R is reflexive. Since Γ is closed by inverse, R is symmetric. Finally, since Γ is closed by composition, the relation R is transitive (note that we did not use the closure by intersection).

2. This results from the fact that the elements of Γ cover R and that Γ is closed under intersection.

3. For $\gamma \in \Gamma$, denote $s_\gamma = s|_\gamma$. Then $s_\gamma \colon (x, \gamma(x)) \mapsto x$ is a bijection from γ to $s(\gamma)$. We claim that s_γ is a homeomorphism from γ (as a subset of R with the topology from Γ) to $s(\gamma)$ (as a subset of X with usual topology).

To show that s_γ is continuous, consider a clopen set $U \subset X$ such that $\mathrm{id}|_U \in \Gamma$. The set

$$s_\gamma^{-1}(U \cap s(\gamma)) = \mathrm{id}_U \circ \gamma$$

is in Γ since Γ is closed by composition. Thus, it is an open set for the topology of R. Since such sets U generate the topology of X by definition of a local action, we conclude that s_γ is continuous.

To show that s_γ^{-1} is continuous, consider $\gamma' \in \Gamma$. We have $s_\gamma(\gamma \cap \gamma') = s(\gamma \cap \gamma')$. Since $\gamma \cap \gamma' \in \Gamma$, the set $s(\gamma \cap \gamma')$ is clopen. Thus, s_γ^{-1} is also continuous. This proves the claim.

The proof concerning r is symmetric. ∎

Let X be a topological space. An equivalence relation R on X with its topology is *étale* if its topology arises from a local action Γ on X as in Proposition 5.6.2.

Note that the topology on R defined by a local action will in general not be the topology induced by X on the product $X \times X$.

The following result shows that the new notion can be used to obtain a result that is more precise than Proposition 5.6.1.

Proposition 5.6.3 *An étale equivalence relation R on a Cantor set X is minimal if and only if the only closed R-invariant subsets of X are X and $\emptyset$.*

Proof We have already seen that the condition is necessary. To prove the converse, consider $x \in X$. To show that its class $[x]_R$ is dense, consider its closure $\overline{[x]_R}$. We will show that it is R-invariant. For this, let $(y, z) \in R$ with $y \in \overline{[x]_R}$. Choose a sequence (y_n) in $[x]_R$ with limit y. Since R is étale, there is a partial homeomorphism $\gamma \in R$ such that $(y, z) \in \gamma$. By definition of the topology on R, all (y_n) are in $s(\gamma)$ for n large enough. For such n, we set $z_n = \gamma(y_n)$. Since γ is continuous, the sequence (z_n) converges to z. But each y_n is in $[x]_R$ and (y_n, z_n) is also in R, so that z_n is in $[x]_R$. This shows that z is in $\overline{[x]_R}$ and thus that $\overline{[x]_R}$ is R-invariant. By hypothesis, this forces $\overline{[x]_R} = X$ and thus the class of x is dense, whence the conclusion that R is minimal.   ∎

We are now going to show that each of the two examples of relations on a Cantor space given above are étale relations.

5.6.2 The Étale Relation R_E

Consider the cofinality relation R_E on a Bratteli diagram (V, E). We first prove the following statement, which associates to it a local action. Let (V, E) be a Bratteli diagram. For $n \geq 1$ and $p, q \in E_{0,n}$ with $r(p) = r(q)$, let

$$\gamma(p, q) = \{(x, y) \in X_E \times X_E \mid x \in [p], y \in [q], x_k = y_k \ (k > n)\}.$$

Then each $\gamma(p, q)$ is a partial homeomorphism from $[p]$ to $[q]$.

Proposition 5.6.4 *The set Γ_E of all partial homeomorphisms $\gamma(p, q)$ is a local action.*

Proof Since $\gamma(p, p) = \mathrm{id}_{[p]}$, the family Γ_E contains all $\mathrm{id}_{[p]}$ and thus the family of $U \subset X_E$ such that id_U is in Γ_E is a basis of the topology of X. We have $\gamma(p, q)^{-1} = \gamma(q, p)$ and thus Γ_E is closed under taking inverses. Next

$\gamma(p, q) \circ \gamma(q, r) = \gamma(p, r)$ and thus Γ_E is closed under composition. Finally, $\gamma(p, q) \cap \gamma(p', q')$ is empty or equal to $\gamma(p, q)$ if $p = p'$ and $q = q'$. Thus, Γ_E is closed under intersection. ∎

Theorem 5.6.5 *Let (V, E) be a Bratteli diagram. The cofinality equivalence R_E is an étale equivalence relation with respect to the topology defined by Γ_E.*

Proof Let us show that $R_E = \cup \Gamma_E$. If x, y are cofinal, there is an $N \geq 1$ such that $x_n = y_n$ for all $n \geq N$. Set $p = x_0 x_1 \cdots x_{N-1}$ and $q = y_0 y_1 \cdots y_{N-1}$. Then $(x, y) \in \gamma(p, q)$ and thus $(x, y) \in \Gamma_E$. The converse is obvious. ∎

5.6.3 The Étale Relation R_T

We will prove that the orbit equivalence R_T on a minimal Cantor system is étale. We first prove that one may associate to R_T a local action.

Proposition 5.6.6 *Let (X, T) be a minimal Cantor system. Let Γ_T be the set of all partial homeomorphisms of the form $T^n|_U$ for $n \in \mathbb{Z}$ and $U \subset X$ clopen. Then Γ_T is a local action.*

Proof Since X is a Cantor space, the clopen sets form a basis of the topology. And for every clopen set U, we have $\mathrm{id}_U = T^0|_U$ and thus id_U is in Γ_T. Next, if $\gamma = T^n|_U$, then γ^{-1} is the restriction of T^{-n} to $T^n U$. Thus, Γ_T is closed by taking inverses. It is also closed under composition since for $\gamma = T^n|_U$ and $\gamma' = T^{n'}|_{U'}$ with $U' = T^n U$, we have $\gamma' \circ \gamma = T^{n+n'}|_U$. Finally, since (X, T) is minimal $T^n x = T^{n'} x$ implies $n = n'$. Thus, $\gamma \cap \gamma'$ is either empty or equal to γ. ∎

Theorem 5.6.7 *Let (X, T) be a minimal Cantor dynamical system. The equivalence R_T on (X, T) defined by*

$$R_T = \{(x, T^n x) \mid x \in X, n \in \mathbb{Z}\}$$

is étale with respect to the topology defined by the local action Γ_T.

Proof This follows from the fact that $R_T = \cup \Gamma_T$. ∎

5.7 Entropy and Bratteli Diagrams

The *topological entropy* of a dynamical system (X, T) is defined as

$$h(T) = \sup_{\alpha} h(T, \alpha) \tag{5.16}$$

where α runs over the open covers of X (see Appendix B.6). The topological entropy of a shift space (X, S), denoted by $h(X, S)$, or simply $h(S)$, is actually (see Theorem B.6.2) the growth rate of the number $p_n(X)$ of finite words of length n occurring in elements of X, that is,

$$h(S) = \lim_{n \to \infty} \frac{1}{n} \log p_n(X). \tag{5.17}$$

Thus, $0 \le h(S) \le \log \mathrm{Card}(A)$ for a shift on the alphabet A. One can show easily that the limit exists and is equal to $\inf \frac{1}{n} \log p_n(X)$, using the fact that $p_{n+m}(X) \le p_n(X) p_m(X)$ (see Exercise 5.22).

As seen in Appendix B.6, entropy is invariant under conjugacy (see Exercise 5.23 for the case of shift spaces). It is a measure of the "size" of a shift space. For example, an edge shift on a graph with adjacency matrix M has entropy $\log \lambda$ where λ is the dominant eigenvalue of M (Exercise 5.24). One could expect that minimal shifts have entropy zero. However, one can show that there are minimal shifts of arbitrary entropy. Indeed, one has the following result.

Theorem 5.7.1 (Grillenberger) *For every $k \ge 2$ and $0 < h < \log k$, there exists a minimal shift space (X, S) on k symbols such that $h(X, S) = h$.*

See Exercise 5.26 for the existence of minimal shifts with arbitrary large entropy.

We will show that entropy is far from being invariant by strong orbit equivalence and that actually every minimal Cantor dynamical system is SOE to a minimal Cantor dynamical system of entropy zero.

Let $(V, E, \le)$ be a properly ordered Bratteli diagram. For $n \ge 1$, let A_n be the set of paths $(e_1, \ldots, e_n)$ of length n from $V(0)$ to $V(n)$. For $x = (e_1, e_2, \ldots) \in X_E$, denote $\pi_n(x) = (e_1, \ldots, e_n)$. Let (X_n, S_n) be the shift space on the alphabet A_n defined by

$$X_n = \{(\pi_n(T_E^k x))_{k \in \mathbb{Z}} \mid x \in X\}. \tag{5.18}$$

The topological entropy of T_E is given by

$$h(T_E) = \lim_{n \to \infty} h(S_n). \tag{5.19}$$

Indeed, let α_n be the partition of X_E defined by the map π_n. Then $h(S_n) = h(T_E, \alpha_n)$ and, by Theorem B.6.1, we have $h(T_E) = \lim_{n \to \infty} h(T_E, \alpha_n)$.

Hence, we need first to compute $h(S_n)$. To this end, we will need the following shift spaces. When W is a set of finite words, we denote by W^∞ the set of all bi-infinite sequences formed by concatenation of words belonging to W. Let S_W denote the shift map on W^∞. It is clear (W^∞, S_W) is a shift space.

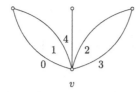

Figure 5.15 An example of a consecutive order viewed from a vertex $v \in V(n+1)$ to $V(n)$.

Lemma 5.7.2 *Let W be a set of m finite words of length at least l. Then*

$$h(S_W) \leq \frac{\log m}{l}.$$

Proof Let k be the maximal length of the words in W. Let w be a word of length n occurring in some sequence of W^∞. Then there exist r words $u_1, \ldots,$ u_r of W, a prefix s and suffix p of some words in W such that $w = s u_1 \cdots u_r p$. Since $r \leq \frac{n}{l}$, we deduce that there are at most $k^2 m^{2+\frac{n}{l}}$ words in $\mathcal{L}_n(W^\infty)$, which ends the proof. ∎

An order on a Bratteli diagram is a *consecutive order* if, whenever edges e, f and g have the same range, e and g have the same source and $e \leq f \leq g$, then e and f have the same source (see Figure 5.15).

Proposition 5.7.3 *Let $(V, E, \leq)$ be a properly ordered Bratteli diagram where $\leq$ is a consecutive order. Suppose that*

$$\lim_{n \to +\infty} \frac{\log(\eta(n + 1)\,\mathrm{Card}(V(n)))}{\eta(n + 1)} = 0,$$

where $\eta(n)$, $n \geq 1$, is the minimum number of edges from a vertex at level $n - 1$ to a vertex at level n. Then, $h(T_E) = 0$.

Proof Let A_n be the set of paths of length n in (V, E) starting at $V(0)$ and X_n be the shift space on A_n defined by Eq. (5.18). Let u be a vertex at level n, and let $p_1, \ldots, p_s$ be the paths from level 0 to u, listed in increasing order. We set $W(u) = p_1 \cdots p_s$. We can consider that it belongs to A_n^s. Now, assume that v is a vertex at level $n + 1$, that $a_1, \ldots, a_t$ are the edges from u to v, listed in increasing order, and that y is an infinite path in the Bratteli diagram such that $y_1 \cdots y_{n+1} = p_1 a_1$. Then,

$$\left(\pi_n\left(T_E^k(y)\right)\right)_{0 \leq k \leq st - 1} = W(u)^t.$$

Note that t is larger than $\eta(n+1)$. Therefore, X_n is included in $\mathcal{W}^\infty$ where

$$\mathcal{W} = \left\{ W(u)^t \mid u \in V(n),\ \eta(n+1) \leq t < 2\eta(n+1) \right\},$$

and consequently, $h(S_n) \leq h(S_\mathcal{W})$. Since $\mathcal{W}$ consists of at most $\mathrm{Card}(V(n))$ $\eta(n+1)$ finite words of length at least $\eta(n+1)$, from Lemma 5.7.2 we obtain

$$h(S_\mathcal{W}) \leq \frac{\log\left(\eta(n+1)\,\mathrm{Card}(V(n))\right)}{\eta(n+1)}.$$

This completes the proof.   ∎

Theorem 5.7.4 *Any minimal Cantor dynamical system is strongly orbit equivalent to a minimal Cantor dynamical system of entropy zero.*

Proof From Theorem 5.3.3, it suffices to consider a minimal Bratteli–Vershik dynamical system (X_E, T_E). Let $\eta(n)$, $n \geq 1$, be the minimum number of edges from a vertex at level $n - 1$ to a vertex at level n.

From Theorem 5.3.3, we know, by contracting if needed, that we can assume the adjacency matrices of $B = (V, E, \leq)$ to have strictly positive entries. Hence, contracting again if needed, we can suppose that

$$\lim_{n \to +\infty} \frac{\log(\eta(n+1)\,\mathrm{Card}(V(n)))}{\eta(n+1)} = 0\,.$$

Consider $(V', E', \leq')$ where $V' = V$, $E' = E$ and $\leq'$ is a consecutive order. Then, from Proposition 5.7.3, $(X_{E'}, T_{E'})$ has zero entropy and, from Theorem 5.5.1, is strongly orbit equivalent to (X_E, T_E).   ∎

Note that this proof actually implies that all minimal BV-dynamical systems with a consecutive order have entropy zero.

We finally quote without proof the following, more general, statement.

Theorem 5.7.5 (Sugisaki) *Let $\alpha \in [1, +\infty[$ and let (X, T) be a minimal Cantor dynamical system. There exists a minimal shift space of entropy $\log \alpha$ which is strongly orbit equivalent to (X, T).*

5.8 Exercises

Section 5.1

5.1 Show that the following conditions are equivalent for a Bratteli diagram (V, E).

(i) For each $m \geq 0$ and every vertex v in $V(m)$, there exists $n > m$ such that, for every $w \in V(n)$, there is a path from v to w.

(ii) For each $m \geq 0$, there exists $n > m$ such that, for every vertex v in $V(m)$ and for every $w \in V(n)$, there is a path from v to w.

(iii) The Bratteli diagram (V, E) is simple.

5.2 Show that the equivalence on Bratteli diagrams (ordered or not) generated by telescoping is given by $(V, E) \equiv (V', E')$ if they have a common intertwining.

5.3 Two nonnegative integer square matrices M, N are said to be C^*-*equivalent* if there are sequences R_n, S_n of nonnegative matrices and k_n, ℓ_n of integers such that

$$M^{k_n} = R_n S_n, \quad N^{\ell_n} = S_n R_{n+1} \tag{5.20}$$

for all $n \geq 1$. Show that C^*-equivalence is an equivalence relation containing shift equivalence and that M, N are C^*-equivalent if and only if the stationary Bratteli diagrams with matrices M and N are equivalent modulo intertwining.

5.4 Let M, N be the matrices

$$M = \begin{bmatrix} 1 & 1 & 0 & 0 & 0 \\ 0 & 1 & 1 & 0 & 0 \\ 0 & 0 & 1 & 1 & 0 \\ 0 & 0 & 0 & 1 & 1 \\ 1 & 0 & 0 & 0 & 1 \end{bmatrix} = I + P, \quad N = \begin{bmatrix} 0 & 1 & 1 & 0 & 0 \\ 0 & 0 & 1 & 1 & 0 \\ 1 & 0 & 0 & 0 & 1 \\ 1 & 1 & 0 & 0 & 0 \\ 0 & 0 & 0 & 1 & 1 \end{bmatrix} = QM, \tag{5.21}$$

where P and Q denote the matrices of the permutations (12345) and (123541) respectively. Show that M, N are C^*-equivalent (it can be shown that they are not shift equivalent; see Section 5.9 for a reference).

Section 5.2

5.5 Show that if $(V, E, \leq)$ is a simple Bratteli diagram, the Bratteli compactum X_E is either finite or a Cantor space.

5.6 Show that a Bratteli diagram (V, E) is simple if and only if each class of the equivalence R_E of cofinality is dense in R_E.

5.7 Show that $X_E^{\max}$ and $X_E^{\min}$ are nonempty sets for every ordered Bratteli diagram $(V, E, \leq)$ and more precisely that the set of maximal (resp. minimal) edges forms a spanning tree of (V, E).

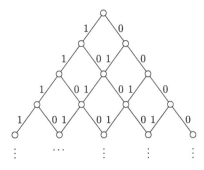

Figure 5.16 The Pascal triangle diagram.

5.8 Show that every simple Bratteli diagram can be properly ordered.

5.9 Let $(V, E, \leq)$ be a properly ordered Bratteli diagram. Show that the Vershik map $T_E: X_E \to X_E$ is one-to-one.

5.10 Show that for every properly ordered Bratteli diagram $(V, E, \leq)$, the system (X_E, T_E) is minimal and invertible.

5.11 Show that a telescoping of Bratteli diagrams from $(V, E, \leq)$ to $(V', E', \leq')$ induces a conjugacy from (X_E, T_E) to $(X_{E'}, T_{E'})$.

5.12 Let $(V, E, \leq)$ be a properly ordered stationary diagram with a $t \times t$-matrix M. Show that the system (X_E, T_E) has a unique invariant probability measure μ given by

$$\mu([u]) = \frac{\alpha_{r(u)}}{\lambda^{n-1}}, \tag{5.22}$$

where u is a path of length n from $V(0)$ to $r(u)$, λ is the maximal eigenvalue of M and $\alpha = (\alpha_i)_{1 \leq i \leq t}$ is the eigenvector of M relative to λ with coefficients of sum 1.

5.13 Let (V, E) be the ordered Bratteli diagram represented in Figure 5.16. We identify X_E with $\{0, 1\}^{\mathbb{N}}$. We consider on $X_E = \{0, 1\}^{\mathbb{N}}$ the Bernoulli measure defined by $\mu([0]) = p$, $\mu([1]) = 1 - p$. Show that

$$X_E^{\min} = 1^*0^{\omega} \cup 0^{\omega} \text{ and } X_E^{\max} = 0^*1^{\omega} \cup 0^{\omega}.$$

The system (X_E, T_E) where T_E defined on $X_E \setminus X_E^{\max}$ by (5.3), by $T_E(0^n1^{\omega}) = 1^n0^{\omega}$ for every $n \geq 1$ $T(0^{\omega}) = 0^{\omega}$ and $T(1^{\omega}) = 1^{\omega}$ is a measure-theoretic dynamical system called the *Pascal adic system*.

Section 5.3

5.14 Prove that the map ϕ in the proof of Theorem 5.3.3 is a homeomorphism.

5.15 Let (X, T) be a minimal invertible Cantor system. Let $\mathfrak{P}(n)$ be a refining sequence of partitions in towers with bases $B(n) = \cup_{1 \leq t \leq t(n)} B_t(n)$ and let (V, E) be the corresponding Bratteli diagram. Let $\phi \colon X_E \to X$ be the conjugacy from (X_E, T_E) onto (X, T) defined by (5.10). Show that

$$B_{t_n}(n) = \phi([e_1, \ldots, e_n]), \qquad (5.23)$$

where $(e_1, \ldots, e_n)$ is the minimal path from $V(0)$ to $(n, t_n) \in V(v)$. Conclude that ϕ sends the partition $\{[e_1, \ldots, e_n] \mid (e_1, \ldots, e_n) \in E_{1,n}\}$ to the partition $\mathfrak{P}(n)$.

Section 5.4

5.16 Let (X_1, T_1) and (X_2, T_2) be minimal topological dynamical systems. Recall that (X_1, T_1) is a *derivative* of (X_2, T_2) if (X_1, T_1) is isomorphic to an induced transformation of (X_2, T_2). We also say in this case that (X_2, T_2) is a *primitive* of (X_1, T_1). Show that two minimal transformations S, T have a common derivative if and only if they have a common primitive.

5.17 Show that Kakutani equivalence is an equivalence relation.

5.18 Prove that the family of systems isomorphic to a stationary BV-dynamical system is stable under Kakutani equivalence.

5.19 Show that the number of right-asymptotic (resp. left-asymptotic) components is invariant by Kakutani equivalence.

Section 5.5

5.20 Let (X, T) and (Y, S) be two invertible Cantor minimal systems such that there is homeomorphism $\phi \colon X \to Y$, which sends orbits to orbits, that is such that there are two maps α, β such that $\phi \circ T(x) = S^{\alpha(x)} \circ \phi(x)$ and $\phi \circ T^{\beta(x)}(x) = S \circ \phi(x)$. Our aim is to show that if α is continuous, then (X, T) is conjugate to (Y, S) or to (Y, S^{-1}).

1. Show that, replacing S by $\phi^{-1} \circ S \circ \phi$, we may assume that $X = Y$. Set $T^k(x) = S^{f_k(x)}(x)$. Show that every f_k is continuous and that $k \mapsto f_k(x)$ is a family of bijections from $\mathbb{Z}$ to $\mathbb{Z}$ that satisfies the cohomological equation

$$f_k(T^j x)) = f_{k+j}(x) - f_j(x). \qquad (5.24)$$

2. Show that for every integer $M > 0$ there is an integer $\overline{M}$ such that $[-M, M] \subset \{f_k(x) \mid -\overline{M} \le k \le \overline{M}\}$ for all $x \in X$.

3. For $m > 0$, set

$$A_m = \{x \mid \forall n \ge m, \ f_n(x) > 0 \text{ and } f_{-n}(x) < 0\},$$
$$B_m = \{x \mid \forall n \ge m, \ f_n(x) < 0 \text{ and } f_{-n}(x) > 0\}.$$

Show that there is $K > 0$ such that $X = A_K$ or $X = B_K$. Assume that $X = A_K$ (the other case is symmetric).

4. Set

$$P_m(x) = \text{Card}\{f_i(x) \mid f_i(x) > 0, |i| \le m\},$$
$$N_m(x) = \text{Card}\{f_i(x) \mid f_i(x) < 0, |i| \le m\}.$$

Show that $a(x) = (N_M(x) - P_M(x))/2$ is an integer (which is independent of M for $M \ge K$). Show that $a(Sx) = a(x) - j + 1$, where j is such that $f_j(x) = 1$.

5. Show that $g(x) = T^{a(x)}x$ is a conjugacy from (X, T) to (X, S).

Conclude that (X, T) is conjugate to (Y, S) or to (Y, S^{-1}).

Section 5.6

5.21 Show that the converse of Proposition 5.6.1 is not true. Hint: Let $x \in \mathbb{Z}_2$ and consider, on the odometer $(\mathbb{Z}_2, T)$, the equivalence R whose classes are the orbits of T distinct of the orbit of x plus the two sets $\{x + 2n \mid n \in \mathbb{Z}\}$ and $\{x + 2n + 1 \mid n \in \mathbb{Z}\}$.

Section 5.7

5.22 Let (u_n) be a sequence of real numbers such that $u_{n+m} \le u_n + u_m$ for all $n, m \ge 0$ (such a sequence is called *subadditive*). Show that the limit $\lim u_n/n$ exists and is equal to $\inf u_n/n$ (*Fekete's Lemma*). Conclude that the limit in the definition of topological entropy (5.17) exists.

5.23 Show that entropy is invariant under conjugacy.

5.24 Let (X, S) be the edge shift on a graph G. Let M be the adjacency matrix of G and let λ be its dominant eigenvalue. Show that $h(X, S) = \log \lambda$.

5.25 Let $m_j(k)$ be defined for $j, k \ge 1$ by $m_1(k) = 1$ and $m_j(k) = \prod_{i=0}^{j-2} k!^{(i)}$ for $k \ge 2$, where $k!^{(0)} = k$ and $k!^{(j+1)} = (k!^{(j)})!$.

Let $\lambda_j(k)$ be defined for $j \geq 1$ by $m_{j+1} = m_j e^{m_j \lambda_j}$ or, equivalently, by $e^{m_{j+1}\lambda_{j+1}} = (e^{m_j\lambda_j})!$. Show that $\lambda_j(k)$ is nonincreasing with j and that $\lambda(k) = \lim_{j\to\infty} \lambda_j(k)$ satisfies $\lim_{k\to\infty}(\log k - \lambda(k)) = 1$. Hint: use the Stirling Formula,

$$1 \leq n!(2\pi n)^{-1/2} n^{-n} e^n \leq e^{\frac{1}{12n}}. \tag{5.25}$$

5.26 Let $U_1 = \{0, 1, \ldots, k-1\}$ and for $j \geq 2$,

$$U_{j+1} = \{u_{\sigma(1)} \cdots u_{\sigma(\mathrm{Card}(U_j))} \mid u_i \in U_j, \sigma \in \mathfrak{S}_{\mathrm{Card}(U_j)}\},$$

where $\mathfrak{S}_n$ is the *symmetric group* on $\{1, 2, \ldots, n\}$. Show that the words of U_j have length m_j and that $\mathrm{Card}(U_j) = e^{m_j\lambda_j}$, where m_j, λ_j are as in Exercise 5.25. For each $j \geq 1$, let $\ell_j, r_j \in U_j$ be such that ℓ_j is a suffix of ℓ_{j+1} and r_j is a prefix of r_{j+1}. Then define x by $x_{[-m_j, m_j)} = \ell_j r_j$.

Show that x is uniformly recurrent and such that $h(x) = \lambda(k)$. Show that for $k = 2$, with $\ell_1 = r_1 = 0$, the sequence x is the Thue–Morse sequence.

5.9 Notes

Let us begin by a historical overview of the contents of this chapter. In 1972 Ola Bratteli (1972) introduced special infinite graphs subsequently called *Bratteli diagrams*, which conveniently encoded the successive embeddings of an ascending sequence $(\mathfrak{A}_n)_{n\geq 0}$ of finite dimensional semi-simple algebras over $\mathbb{C}$ ("multi-matrix algebras"). The sequence $(\mathfrak{A}_n)_{n\geq 0}$ determines a so-called approximately finite dimensional (AF) C^*-algebra (see Chapter 9). Bratteli proved that the equivalence relation on Bratteli diagrams generated by the operation of telescoping is a complete isomorphism invariant for AF-algebras (see Theorem 9.3.21).

From a different direction came the extremely fruitful idea of A. M. Vershik (1985) to associate a dynamics (called an *adic transformation*) with Bratteli diagrams (*Markov compacta*) by introducing a lexicographic order on the infinite paths of the diagram. By a careful refining of Vershik's construction, Herman, Putnam and Skau (1992) succeeded in showing that every minimal Cantor dynamical system is isomorphic to a Bratteli–Vershik dynamical system.

5.9.1 Bratteli Diagrams

The BV-representation theorem (Theorem 5.3.3) saying that (X, T) can be topologically realized as a BV-dynamical system is the main result of

Herman et al. (1992). Recall that Vershik (1985) obtained such a result in a measure-theoretic context.

Theorem 5.1.5 is due to Elliott (1976) (see also Krieger (1980b)).

5.9.2　Kakutani Equivalence

Kakutani equivalence was introduced in the context of measure-theoretic systems (Kakutani, 1943). Theorem 5.4.1 is from Giordano et al. (1995). Exercises 5.16 and 5.18 are from Ornstein et al. (1982) (the context is for measure-theoretic systems, but the arguments transpose easily).

5.9.3　Strong Orbit Equivalence

The Strong Orbit Equivalence Theorem (Theorem 5.5.1) is due to Giordano et al. (1995). We follow the proof proposed in Glasner and Weiss (1995), giving more details.

Exercise 5.20 is due to Boyle (1983) (see Giordano et al. (1995), Theorem 2.4). See also Boyle and Tomiyama (1998). Two systems (X, T) and (Y, S) such that (X, T) is conjugate to (Y, S) or to (Y, S^{-1}) are called *flip equivalent*. Theorems 5.5.5 and 5.5.7 are from Giordano et al. (1995).

5.9.4　Equivalences on Cantor Spaces

The consideration of étale equivalence relations to formulate results on orbit equivalence in topological dynamical systems is due to Putnam (2010). Étale equivalence relations are a particular case of the notion of *étale groupoid* introduced by Renault (2014). We follow here the beautiful book of Putnam (2018).

5.9.5　Entropy

We suggest Walters (1982) or Lind and Marcus (1995) for an introduction to topological entropy.

The fact that there exist minimal systems of arbitrary entropy (Theorem 5.7.1) is due to Grillenberger (1972/73) (we give in Exercises 5.25 and 5.26 the proof that the entropy of a minimal system can be arbitrarily large). The fact that every minimal Cantor dynamical system is SOE to a minimal Cantor dynamical system of entropy zero (Theorem 5.7.4) is from Boyle and Handelman (1994).

In Boyle and Handelman (1994) the authors use a different lemma instead of Lemma 5.7.2. They show that $h(S_W) = \frac{\log m}{l}$, whenever W is a set of m distinct finite words of length l.

The original reference to Fekete's Lemma (Exercise 5.22) is Fekete (1923).

Theorem 5.7.4 shows that there can be dynamical systems with different entropies in a strong orbit equivalence class. Hence it is natural to ask whether all entropies can be realized inside a given class. M. Boyle and D. Handelman (1994) showed that it is true in the class of the odometer ($\mathbb{Z}_2, x \mapsto x + 1$). Later, F. Sugisaki (2003) proved that it is true in any strong orbit equivalence class. Theorem 5.7.5, showing that the realizations can be chosen to be subshifts, is from Sugisaki (2007).

5.9.6 Exercises

The solution of Exercise 5.2 has been kindly provided to us by Ian Putnam. Exercise 5.3 is from Bratteli et al. (2000), where a proof that the matrices of Eq. (5.21) are not shift equivalent is given (we have not included it here, as it uses technical tools from algebraic number theory, which are out of our scope).

The decidability of C^*-equivalence of matrices was proved in Bratteli et al. (2001). It proves the decidability of the telescoping equivalence of stationary diagrams. Actually, the equivalence of Bratteli diagrams is undecidable in general, as shown by Mundici and Panti (1993) (see also Mundici (2004)). The undecidability is proved by a reduction from the problem of isomorphism of lattice ordered groups, proved undecidable by Glass and Madden (1984). The latter, being itself obtained by a reduction from the problem of piecewise linear homeomorphism for compact polyhedra, was proved undecidable by Markov (1958).

The Pascal adic system (Exercise 5.13) is from Méla and Petersen (2003).

6

Substitution Shifts and Generalizations

In this chapter, we describe the BV-representations of substitution shifts and their generalizations, namely linearly recurrent shifts and S-adic shifts. We treat before the case of odometers, which will be needed after and show that they can be represented by Bratteli diagrams with one vertex at each level.

We describe next the construction of BV-representations for substitution shifts. The main result is Theorem 6.2.1, which states that the family of Bratteli–Vershik systems associated with infinite, stationary and properly ordered Bratteli diagrams is (up to isomorphism) the disjoint union of the family of infinite minimal substitution shifts and the family of stationary odometers.

In the following sections, we treat the cases of linearly recurrent shifts (Section 6.3) and of S-adic shifts (Section 6.4). In Section 6.5, we give a description of the dimension group of unimodular S-adic shifts (Theorem 6.5.4). This will be applied, in Chapter 7, to the case of dendric shifts. Finally, in Section 6.6, we prove a result characterizing substitutive sequences by a finiteness property of the set of their derivatives (Theorem 6.6.1).

6.1 Odometers

We have already met odometers in Chapter 1 as an example of Cantor systems and several times in other chapters. We study here these systems in more detail.

6.1.1 The Ring of (p_n)-adic Integers

Let $(p_n)_{n \geq 0}$ be a strictly increasing sequence of natural nonzero integers such that $p_0 = 1$ and p_n divides p_{n+1} for all $n \geq 0$. We endow the set $X = \prod_{n \geq 1} \mathbb{Z}/p_n\mathbb{Z}$ with the product topology of the discrete topologies. This topology, called the (p_n)-adic topology, is also the topology defined by the (p_n)-adic distance. This distance is defined for $x = (x_n)$ and $y = (y_n)$

by $d(x, y) = 1/p_n$, where n is minimal such that $x_n \neq y_n$. We also set $d(x, x) = 0$. The set

$$\mathbb{Z}_{(p_n)} = \{(x_n)_{n \geq 1} \in X \mid x_n \equiv x_{n+1} \mod p_n\}$$

is a ring for componentwise addition and multiplication, called the ring of (p_n)-*adic integers*. It is a Cantor space and a compact topological ring (see Exercise 6.1). We may identify $\mathbb{Z}$ to a dense subset of $\mathbb{Z}_{(p_n)}$ via the map $x \mapsto (x \mod p_n)$. Actually, $\mathbb{Z}_{(p_n)}$ is the completion of $\mathbb{Z}$ for the (p_n)-adic topology. A basis of the topology is given by the sets

$$[s_1, s_2, \ldots, s_m] = \{(x_n) \in \mathbb{Z}_{(p_n)} \mid s_i \text{ representative of } x_i, 1 \leq i \leq m\},$$

with $0 \leq s_i < p_i$ for $1 \leq i \leq m$. The neutral element is $0 = (0, 0, \ldots)$, and the unit for multiplication is $(1, 1, \ldots)$. More generally, the elements of $\mathbb{Z}_{(p_n)}$ that are in $\mathbb{Z}$, are the eventually periodic sequences (x_n).

When $p_n = p^n$ for all $n \geq 1$, where $p \geq 2$ is a prime number, the ring $\mathbb{Z}_{(p_n)}$ is the classical ring of p-*adic integers* $\mathbb{Z}_p$ already met several times (see Example 4.1.8). A p-adic integer is usually introduced as an infinite expansion $x = a_0 + a_1 p + a_2 p^2 + \cdots$ with $0 \leq a_i < p^{i+1}$, which defines a convergent infinite sum in the p-adic topology. These expansions also exist for (p_n)-adic integers, as we shall see now.

To every element $x = (x_n)_{n \geq 0}$ of $\mathbb{Z}_{(p_n)}$, we associate its (p_n)-*adic expansion* defined as follows. Set $q_n = p_n/p_{n-1}$ for $n \geq 1$. Set $s_0 = 0$ and for each $n \geq 1$, let s_n be the representative of x_n such that $0 \leq s_n < p_n$. Set $a_0 = s_1$ and inductively $a_n = (s_{n+1} - s_n)/p_n$ in such a way that

$$s_n = a_0 + a_1 p_1 + \cdots + a_{n-1} p_{n-1}, \tag{6.1}$$

with $0 \leq a_n < q_{n+1}$. Then x is the limit in $\mathbb{Z}_{(p_n)}$ of the sequence (s_n) and thus

$$x = a_0 + a_1 p_1 + a_2 p_2 + \cdots \tag{6.2}$$

is a converging infinite sum called the (p_n)-*adic expansion* of x. The expansion (6.2) with $0 \leq a_n < q_{n+1}$ is unique.

The sum of (p_n)-adic expansions is computed using a carry from left to right, as for expansions of integers in a fixed basis written with the less significant digit first. More precisely, if $x = a_0 + a_1 p_1 + \cdots$ and $x' = a_0' + a_1' p_1 + \cdots$, then the expansion of $y = x + x'$ is $y = b_0 + b_1 p + \cdots$ where the coefficients b_i are defined by

$$a_0 + a_0' = b_0 + q_1 t_1$$
$$a_1 + a_1' + t_1 = b_1 + q_2 t_2$$

$$\cdots$$

and so on, with $0 \leq b_n < q_{n+1}$ (Exercise 6.3).

For example, the 2-adic expansion of -1 is $-1 = 1 + 2 + 2^2 + \cdots$ since

$$1 + (1 + 2 + 2^2 + \cdots) = 2 + (2 + 2^2 + \cdots) = 4 + (2^2 + \cdots) = \cdots = 0.$$

Two particular cases are of special interest.

The already mentioned case of p-adic integers defines the ring $\mathbb{Z}_p$, which is an integral domain, that is, $xy = 0$ implies $x = 0$ or $y = 0$. Its field of fractions is the *field of p-adic numbers*, denoted $\mathbb{Q}_p$. As an example, 2 is not invertible in $\mathbb{Z}_2$ since $x = 2y$ implies $x_0 = 0$. Thus, $1/2$ is an element of $\mathbb{Q}_2$ that is not in $\mathbb{Z}_2$. On the contrary, $1/3$ belongs to $\mathbb{Z}_2$ and has the 2-adic expansion

$$\frac{1}{3} = 1 - \frac{2}{3} = 1 + 2\frac{1}{1 - 2^2} = 1 + 2(1 + 2^2 + 2^4 + \cdots) = 1 + 2 + 2^3 + 2^5 + \cdots.$$

At the exact opposite of the case of p-adic integers, set $p_n = (n + 1)!$ for all $n \geq 1$. The ring $\mathbb{Z}_{(n!)}$ is called the ring of *profinite integers* (the term profinite refers to the general notion of profinite group or ring, see Exercise 6.9). The corresponding expansions

$$x = c_0 + c_1 2! + c_2 3! + \cdots$$

with $0 \leq c_i \leq i + 1$ give in particular a representation of integers called the *factorial number system*. (see Exercise 6.4).

Let $T: \mathbb{Z}_{(p_n)} \to \mathbb{Z}_{(p_n)}$ be the map $x \mapsto x + 1$ where $1 = (1, 1, \ldots)$. The pair $(\mathbb{Z}_{(p_n)}, T)$ is called the *odometer in base* (p_n). It is a minimal Cantor dynamical system. Indeed, the orbit of 0 is dense, and 0 is in the closure of the orbit of every point.

The odometer in base $(n!)$ is called the *universal odometer*.

Note that the odometer in base (p_n) can be expressed as follows in terms of the odometer in base $p'_n = p_{n+1}/p_1$. We may identify $\mathbb{Z}_{(p_n)}$ with $\mathbb{Z}/p_1\mathbb{Z} \times \mathbb{Z}_{(p'_n)}$ and for $(x, y) \in \mathbb{Z}/p_1\mathbb{Z} \times \mathbb{Z}_{(p'_n)}$,

$$T(x, y) = \begin{cases} (x + 1, y) & \text{if } x + 1 \neq 0 \\ (0, y + 1) & \text{otherwise.} \end{cases}$$

This shows that $(\mathbb{Z}_{(p_n)}, T)$ is a skew product of the finite system $(\mathbb{Z}/p_1\mathbb{Z}, +1)$ and $\mathbb{Z}_{(p'_n)}$ or equivalently, that the system induced by $\mathbb{Z}_{(p_n)}$ on the clopen set $[0]$ is the odometer $\mathbb{Z}_{(p'_n)}$ (see Exercise 6.10).

6.1.2 BV-Representation of Odometers

Let us compute a BV-representation of the odometer in base $(p_n)_{n \geq 1}$. Set $\mathfrak{P}(0) = \mathbb{Z}_{(p_n)}$ and, for $n \geq 1$, set $B(n) = [0^n]$, $t(n) = 1$, $h(n) = p_n$ and

$$\mathfrak{P}(n) = \{T^j B(n) \mid 0 \leq j \leq h(n) - 1\}. \tag{6.3}$$

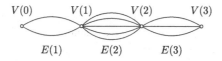

Figure 6.1 The three first levels of the BV-representation of $(\mathbb{Z}_{(p_n)}, R)$.

Then $(\mathfrak{P}(n))_n$ is a refining sequence of KR-partitions (see Exercise 6.7). Remark that $T^j B(n) = [j_1 j_2 \cdots j_n]$ where $j_i \equiv j \mod p_i$. The edges of the BV-representation of $(\mathbb{Z}_{(p_n)}, T)$ given in Section 5.3 are of the form $(n, 1, 1, l)$, with $0 \leq l \leq q_n - 1$ where $q_1 = p_1$ and $q_n = p_n/p_{n-1}$ for $n \geq 2$.

Thus, an odometer has a BV-representation with one vertex at each level. The converse is also clearly true. We can thus state the following nice simple result.

Theorem 6.1.1 *A Cantor dynamical system is an odometer if and only if it has a BV-representation with one vertex at each level.*

For example, if $p_1 = 2$, $p_2 = 10$ and $p_3 = 30$, the first three levels are given in Figure 6.1.

An odometer is *stationary* if it has a stationary BV-representation. It is easy to see that the odometer in base (p_n) is stationary if and only if the set of prime divisors of the p_n is finite or, equivalently, if it has a BV-representation with pq^n edges at level n (see Exercise 6.8). Or if it is isomorphic to an odometer in base $(pq^n)_n$.

For a sequence (p_n) such that $p_0 = 1$ and p_n divides p_{n+1} for $n \geq 0$, let us denote by $\mathbb{Z}[(p_n)]$ the subgroup of $\mathbb{Q}$ with order unit 1 and order induced by $\mathbb{Q}$, formed of the p/q with q dividing some p_n. Thus, for example, $\mathbb{Z}[1/2] = \mathbb{Z}[(2^n)]$.

Proposition 6.1.2 *The dimension group of the odometer in base (p_n) is isomorphic to $\mathbb{Z}[(p_n)]$.*

Proof The dimension group of $(\mathbb{Z}_{(p_n)}, T)$ is the direct limit of the sequence $\mathbb{Z} \xrightarrow{p_1} \mathbb{Z} \xrightarrow{p_2} \mathbb{Z} \cdots$ whence the statement. Note that the order unit is 1. ∎

For example, the dimension group of the odometer in base (2^n) is the group of dyadic rationals.

Corollary 6.1.3 *Every odometer is uniquely ergodic.*

Proof Since the dimension group is isomorphic to a subgroup of the ordered group $\mathbb{Q}$, and since subgroups of $\mathbb{Q}$ have a unique state (Example 2.2.3), this follows from Corollary 3.9.4. ∎

The unique ergodic probability measure of the odometer in base (p_n) is known as the *Haar probability measure* on the group $\mathbb{Z}_{(p_n)}$.

Corollary 6.1.4 *An invertible minimal Cantor system (X, T) is orbit equivalent to an odometer if and only if the quotient of the group $K^0(X, T)$ by its infinitesimal subgroup is isomorphic to a subgroup of $\mathbb{Q}$.*

Proof The condition is necessary since, by Proposition 6.1.2, the dimension group of an odometer is a subgroup of $\mathbb{Q}$ and, as such, has a unique state that is the identity (see Example 2.2.3). Thus, it has a trivial infinitesimal subgroup. Conversely, let (X, T) be an invertible minimal Cantor system and let $G = K^0(X, T)$. Assume that $G/\operatorname{Inf}(G)$ is isomorphic to a subgroup of $\mathbb{Q}$. Let P be the set of denominators of the elements of $G/\operatorname{Inf}(G)$ written in reduced form. For $n \geq 1$, let p_n be the product of all $p^n \in P$ with p prime and $p \leq n$. Then $G/\operatorname{Inf}(G)$ is the dimension group of the odometer in base (p_n), which is, by Theorem 5.5.3, orbit equivalent to (X, T). ∎

We will use in the next section the following statement, which identifies systems isomorphic to an odometer in a nonobvious way.

Proposition 6.1.5 *Let $\mathcal{B} = (V, E, \leq)$ be an ordered Bratteli diagram such that*

(i) *The incidence matrices $M(n)$ are all of rank 1 and distinct of the scalar 1.*
(ii) *The order is such that $e \leq e'$ if $r(e) = r(e')$ and $s(e) \prec s(e')$ where $\prec$ is an order fixed on each $V(n)$.*

Then (X_E, T_E) is topologically conjugate to an odometer. If $\mathcal{B}$ is additionally stationary, it is conjugate to a stationary odometer.

Proof Since $M(n)$ has rank 1, we have $M(n) = x_n y_n$ with x_n a nonnegative integer column vector and y_n a nonnegative integer row vector. Consider the ordered Bratteli diagram $(V', E', \leq)$ with incidence matrices $M'(1) = M(1)$ and $M'(2n) = y_{n+1}$, $M'(2n + 1) = x_{n+1}$ for $n \geq 1$. By condition (ii), we can choose the order on $(V', E', \leq')$ such that the telescoping with respect to $0, 1, 3, 5, \ldots$ gives $(V, E, \leq)$. Now the telescoping of G' with respect to $0, 2, 4, \ldots$ is the odometer in base $(p_n)_{n \geq 1}$ with $p_n = y_n x_n \ldots y_2 x_2 y_1 x_1$ (and (p_n) is strictly increasing since $y_n x_n \neq 1$). If $\mathcal{B}$ is stationary, we have $x_n = x$

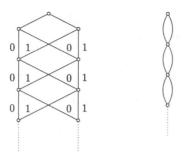

Figure 6.2 The Bratteli diagram with matrix M.

and $y_n = y$ for all $n \geq 2$ and thus $p_n = q^{n-1}p$ with $q = yx$ and $p = y_1 x_1$, which gives a stationary odometer. ∎

For example, the diagram represented in Figure 6.2 on the left has incidence matrices

$$M(1) = \begin{bmatrix} 1 \\ 1 \end{bmatrix}, \quad M(n) = \begin{bmatrix} 1 & 1 \\ 1 & 1 \end{bmatrix} = \begin{bmatrix} 1 \\ 1 \end{bmatrix} \begin{bmatrix} 1 & 1 \end{bmatrix}$$

for $n \geq 2$, which have rank 1. The corresponding odometer is shown on the right.

Note finally the following simple but important observation.

Proposition 6.1.6 *The family of odometers is disjoint from the family of minimal shift spaces.*

The proof uses the notion of expansive system. A dynamical system (X, T) is *expansive* if the distance d on X satisfies the following condition. There exists ϵ such that for all pairs of distinct points (x, y), there exists n with $d(T^n x, T^n y) \geq \epsilon$. We say that ϵ is a *constant of expansivity* of (X, T). Expansivity is a property invariant by conjugacy (but the constant may not be the same).

A shift space is clearly expansive. But an odometer is not expansive. Indeed, for every $x, y \in \mathbb{Z}_{(p_n)}$, one has $d(T^n x, T^n y) = d(x + n, y + n) = d(x, y)$. This proves Proposition 6.1.6.

Actually, a topological dynamical system is a shift space if and only if it is expansive (Exercise 6.12).

At the opposite of expansive systems, a topological dynamical system (X, T), endowed with the distance d, is said to be *equicontinuous* whenever for every $\epsilon > 0$, there exists $\delta > 0$ such that

$$d(x, y) < \delta \Rightarrow \sup_{n \in \mathbb{Z}} d(T^n x, T^n y) < \epsilon.$$

For example, an odometer is equicontinuous (with $\delta = \epsilon$). We shall come back to expansive systems in Section 6.3.

6.1.3 Equal Path Number Property

A Bratteli diagram has the *equal path number property* if for all $n \geq 1$ and $u, v \in V(n)$ we have $\mathrm{Card}(r^{-1}(u)) = \mathrm{Card}(r^{-1}(v))$. Note that this implies that the number of paths from u or v to $v(0)$ are the same.

Note that the equal path number property is preserved by telescoping.

Note also that the representation of odometers given in Section 6.1.2 shares this property. We prove the following result, which connects all these notions together.

Theorem 6.1.7 (Gjerde Johansen) *A minimal shift is Toeplitz if and only if it has a BV-representation* (X_E, T_E) *where* $(V, E, \leq)$ *has the equal path number property.*

Proof Let (X_E, T_E) be a BV-representation of the minimal shift (X, S), associated with the sequence of refining KR-partition $(\mathfrak{P}(n))$, which has the equal path number property. Let $\phi \colon X_E \to X$ be a conjugacy. Taking $(\mathfrak{P}(n))_{n \geq n_0}$ instead of $(\mathfrak{P}(n))$ preserves the equal path number property. Consequently, we may assume that the partitions $\mathfrak{P}(n)$ are such that all points x, y in $\phi(B(n))$ satisfy $x_{[-n,n]} = y_{[-n,n]}$. Then, the sequence $x = \phi(x_{min})$ is a Toeplitz sequence. Indeed, one has $x_{n+kp} = x_n$ for all $k \in \mathbb{Z}$ with $k = \mathrm{Card}(r^{-1}(\{u\}))$ for all $u \in V(n)$.

Conversely, let (X, S) be a Toeplitz shift and let $x \in X$ be a Toeplitz sequence. For a clopen set $U \subset X$ and $y \in X$, we set

$$\mathrm{Per}_p(y, U) = \{n \in \mathbb{Z} | S^{n+kp}(y) \in U \text{ for all } k \in \mathbb{Z}\}.$$

We first claim that for all $n \in \mathbb{N}$ there exist $p > 0$ and a clopen partition $\{C, S^{-1}C, \ldots, S^{-p+1}C\}$ such that C is included in the cylinder set $U = [x_{[-n,-1]} . x_{[0,n]}]$.

The sequence x being Toeplitz, there exists $p > 0$ such that $\mathrm{Per}_p(x, U)$ is nonempty. We can suppose that p is minimal, that is, if q is such that $0 \leq q < p$ then $\mathrm{Per}_q(x, U)$ is empty.

The set $\{y \in X | \mathrm{Per}_p(y, U) \neq \emptyset\}$ being nonempty, closed and S-invariant, it is equal to X by minimality. Thus, the set $\mathrm{Per}_p(y, U)$ is nonempty for all $y \in X$.

Let C be the closed subset $\{y \in X | \mathrm{Per}_p(y, U) = \mathrm{Per}_p(x, U)\}$. Observe that $\{C, S^{-1}C, \ldots, S^{-p+1}C\}$ is a partition as p is minimal. Since $S^{-p}C = C$, the nonempty closed set $\cup_{i=0}^{p-1} S^{-i} C$ is S-invariant set. Thus, it is equal to X. The set C being closed, it is a clopen partition of X. Moreover, because 0 belongs to $\text{Per}_p(x, U)$, it also belongs to $\text{Per}_p(y, U)$ for all $y \in C$. Hence y belongs to U and C is included in U. This proves the claim.

Let us now proceed as in Theorem 4.1.6 with the exception that the sequence of clopen sets C_n will not be chosen at the initial step but defined as adequate cylinder sets $[u.v]$ including x.

We start choosing an increasing sequence of partitions $(\mathfrak{P}'(n))_n$ generating the topology. Let $U_1 = [x_{[-n_1, n_1]}]$, $n_1 = 1$ and let C_1 be as in the claim (for the period $p_1 = p$).

We apply Proposition 4.1.7 to $\mathfrak{Q} = \mathfrak{P}'(1)$ and $C = C_1$ to obtain the KR-partition $\mathfrak{P}(1)$. Observe that the height of each tower is p_1. The base is $C_1 = \cup B_i \subset U_1$ where the B_i's are atoms of $\mathfrak{P}(1)$. One can suppose that x belongs to B_1 and thus there exists some $n_2 > n_1$ such that the cylinder set $U_2 = [x_{[-n_2, -1]}.x_{[0, n_2]}]$ is included in B_1.

The sequence x being Toeplitz, there exists $p_2 > 0$, taken minimal, such that 0 belongs to $\text{Per}_{p_2}(x, U_2)$. We apply the claim to obtain the clopen set C_2 and the period p_2.

Applying Proposition 4.1.7 iteratively for $n \geq 2$ to $C = C_n$ and by setting now

$$\mathfrak{Q} = \mathfrak{P}'(n) \vee \mathfrak{P}(n-1),$$

we obtain KR-partition $\mathfrak{P}(n)$ with basis C_n, which is finer than $\mathfrak{P}'(n)$ and $\mathfrak{P}(n-1)$ and whose heights are all equal to p_n for some p_n. Moreover, the sequence $(\mathfrak{P}(n))_n$ of KR-partition is nested, generates the topology and the intersection of the bases is included in $\cap_n U_n = \{x\}$.

Thus, the BV-representation corresponding to this partition has the equal path number property. ∎

The class of BV-systems having the equal path number property is larger than that of Toeplitz shifts. Indeed, one can show that there exist BV-systems having the equal path number property that are neither expansive nor equicontinuous (see Section 6.8 for a reference).

Example 6.1.8 Let $\sigma: 0 \to 01, 1 \to 00$ be the morphism generating the period-doubling sequence (see Example 1.6.1), which is a Toeplitz sequence. A BV-representation of the corresponding shift is shown in Figure 6.3 (see Exercise 6.14 for the derivation of this representation). It can be verified that it has the equal path number property.

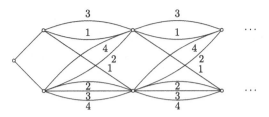

Figure 6.3   A BV-representation of the period doubling shift.

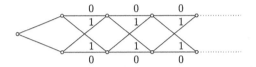

Figure 6.4   The Thue–Morse morphism read on a Bratteli diagram.

6.2 Substitution Shifts

We will now consider the BV-representation of substitution shifts. Let us note that we work in all this section with bi-infinite sequences and two-sided shifts. We first need a new definition.

Let $(V, E, \leq)$ be a stationary properly ordered Bratteli diagram. The morphism read on $E(n)$ is constant from $n \geq 2$. We call it the *morphism read on* $(V, E, \leq)$.

Given a morphism σ, it is not possible in general to use the stationary Bratteli diagram $\mathcal{B} = (V, E, \leq)$, with σ read on $\mathcal{B}$, to represent the shift space corresponding to σ. For example, in the case of the Thue–Morse morphism $a \mapsto ab, b \mapsto ba$ the Bratteli diagram is given in Figure 6.4.

It is clear that it has two maximal and two minimal paths. Hence this representation does not give a properly ordered Bratteli diagram.

6.2.1 Main Result

We will show that one has, however, the following result. It implies in particular that every infinite minimal substitution shift has a stationary BV-representation.

Theorem 6.2.1 *The family of infinite Bratteli–Vershik systems associated with stationary, properly ordered Bratteli diagrams is (up to isomorphism) the disjoint union of the family of infinite substitution minimal shifts and the family of stationary odometers.*

Furthermore, we will see that the correspondence in question is given by an explicit and algorithmically effective construction.

A morphism σ on the alphabet A is *left proper* (resp. *right proper*) if there is a letter $b \in A$ such that, for every $a \in A$, b is the first letter (resp. the last letter) of $\sigma(a)$. It is called *proper* if it is left and right proper. A morphism is called *eventually proper* if there is an integer $p \geq 1$ such that σ^p is proper. An eventually proper morphism σ has exactly one fixed point, which is $\sigma^\omega(r \cdot \ell)$.

Proposition 6.2.2 *The morphism read on a stationary, properly ordered, Bratteli diagram is primitive and eventually proper.*

Proof Let $\sigma : A^* \to A^*$ be the morphism read on $(V, E, \leq)$. Since $(V, E, \leq)$ is properly ordered, it is simple. For every $a, b \in A$, since $(V, E, \leq)$ is simple, there is a path from $(1, b)$ to some (n, a). Then b occurs in $\sigma^n(a)$, showing that σ is primitive.

Let $i(a)$ be the first letter of $\sigma(a)$. For every $a \in A$ and $n \geq 1$, the source of the minimum edge with range (n, a) is $(n-1, i(a))$. Thus, if the minimal value of the integers $\mathrm{Card}(i^n(A))$ were larger than 1, there would exist more than one minimal infinite path, a contradiction with the hypothesis that $(V, E, \leq)$ is properly ordered. This shows that there exists n such that $i^n(a)$ is the same for all $a \in A$. A symmetric argument holds for the last letter. Thus, σ is eventually proper. ∎

Recall that a shift space X is said to be *periodic* if there exist $x \in X$ and an integer k such that $X = \{x, Sx, \ldots, S^{k-1}x\}$. Thus, a shift space X is periodic if and only if the dynamical system (X, S) is periodic. Likewise, it is said to be *aperiodic* if the system is aperiodic, that is, it does not contain any periodic point. Thus, a periodic shift is the same as a minimal finite shift space. Observe that the property of being periodic is decidable for a substitution shift (see Exercise 1.39).

6.2.2 Diagrams with Simple Hat

We say that a Bratteli diagram (V, E) has a *simple hat* whenever it has only simple edges between the top vertex and any vertex of the first level. Note that the Bratteli diagram associated with a nested sequence of partitions as in Section 5.3 has a simple hat.

The following result gives a proof of one direction of Theorem 6.2.1 in the particular case of diagrams with a simple hat.

Proposition 6.2.3 *Let $(V, E, \leq)$ be a stationary, properly ordered Bratteli diagram with a simple hat and such that X_E is not reduced to one point.*

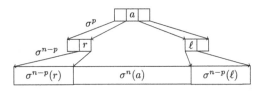

Figure 6.5 The sequence $\mathfrak{P}(n)$ generates the topology.

Let σ be the morphism read on $(V, E, \le)$, and let X be the substitution shift associated with σ.

1. *If X is not periodic, then it is isomorphic to (X_E, T_E).*
2. *If X is periodic, then (X_E, T_E) is isomorphic to a stationary odometer.*

If X_E (and thus also X) is reduced to one point, the statement is false, since (X_E, T_E) is not an odometer. We will use the following lemma.

Lemma 6.2.4 *Let $\sigma : A^* \to A^*$ be a primitive and eventually proper morphism. If $X(\sigma)$ is infinite, the family*

$$\mathfrak{P}(n) = \{S^j \sigma^n([a]) \mid a \in A,\ 0 \le j < |\sigma^n(a)|\}$$

is a refining sequence of partitions in towers.

Proof By Proposition 4.1.3 the partition $\mathfrak{P}(n)$ is for every $n \ge 1$ a KR-partition of $X(\sigma)$ with basis $\sigma^n(X)$. The sequence $\mathfrak{P}(n)$ is clearly nested. Since σ is eventually proper, the intersection of the bases is reduced to one point, namely the unique fixed point of σ. Finally, $\mathfrak{P}(n)$ tends to the partition in points. Indeed, let $p \ge 1$ be such that all $\sigma^p(a)$ begin with ℓ and end with r. Then (see Figure 6.5)

$$\sigma^n([a]) \subset [\sigma^{n-p}(r) \cdot \sigma^n(a)\sigma^{n-p}(\ell)].$$

Thus, all words in $S^j \sigma^n([a])$ coincide on $[-m, m]$ for $m = \min_{b \in B} |\sigma^{n-p}(b)|$. This shows that $\mathfrak{P}(n)$ is a refining sequence of KR-partitions. ∎

Proof of Proposition 6.2.3 Assume first that X is aperiodic. By Proposition 6.2.2, the morphism σ is primitive and eventually proper. By Lemma 6.2.4, the family $(\mathfrak{P}(n))$ is a refining sequence of partitions in towers.

Since $(V, E, \le)$ has a simple hat, the Bratteli diagram associated with the sequence of partitions $(\mathfrak{P}(n))$ is clearly equal to $(V, E, \le)$. Thus, by Theorem 5.3.3, (X, S) is isomorphic to (X_E, T_E).

Assume now that X is periodic. Replacing σ by some power does not modify X (and replaces $(V, E, \le)$ by a periodic telescoping) so that we may assume

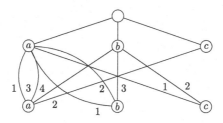

Figure 6.6 A BV-representation of the Tribonacci shift.

that all words $\sigma(a)$ for $a \in A$ begin with ℓ and end with r. The unique fixed point $x = \sigma^\omega(r \cdot \ell)$ of σ is then periodic. Set $x = \cdots ww \cdot www \cdots$ with w as short as possible. Then w is a primitive word, that is, it is not a power of a shorter word. Let n be large enough so that $|\sigma^n(a)| \geq |w|$ for every $a \in A$. Then each $\sigma^{2n}(a)$ is a word of period $|w|$ which begins with w since it begins with $\sigma^n(\ell)$, and ends with w since it ends with $\sigma^n(r)$.

Since X_E is infinite, we have $|w| > 1$. Since w is primitive, this forces each $\sigma^{2n}(a)$ to be a power of w. Thus, replacing again σ by one of its powers, we may assume that every $\sigma(a)$ is a power of w: $\sigma(a) = w^{k_a}$. In this case, the incidence matrix of the diagram $(V, E, \leq)$ is such that for every $a, b \in A$,

$$M_{a,b} = k_a |w|_b.$$

Thus, the hypotheses of Proposition 6.1.5 are satisfied and (X_E, T_E) is conjugate to a stationary odometer. ∎

Example 6.2.5 The stationary diagram associated with the primitive and proper morphism $a \mapsto acab, b \mapsto aab, c \mapsto ab$ gives a BV-representation of the Tribonacci shift (see Figure 6.6). Indeed, this morphism is conjugate to $\sigma^2 : a \mapsto abac, b \mapsto aba, c \mapsto ab$ where $\sigma : a \mapsto ab, b \mapsto ac, c \mapsto a$ is the Tribonacci morphism. Thus, the statement results from Proposition 6.2.3.

We illustrate the periodic case in the following example.

Example 6.2.6 Let $(V, E, \leq)$ be the stationary Bratteli diagram represented in Figure 6.7 on the left. The morphism read on $(V, E, \leq)$ is $\sigma : a \to ab, b \to c, c \to abc$. The substitution shift defined by σ is periodic since $\sigma(abc) = (abc)^2$. The corresponding odometer is represented in Figure 6.7 on the right.

A *partition matrix* is a $t \times \ell$-matrix P with coefficients $0, 1$ such that every column of P has exactly one coefficient equal to 1. Such a matrix defines a

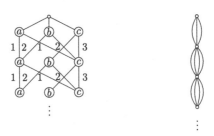

Figure 6.7 The periodic case.

partition θ of the set $\{1, 2, \ldots, \ell\}$ of indices of the columns grouping together x, y if $P_{i,x} = P_{i,y} = 1$. It also defines a map π from $\{1, 2, \ldots, \ell\}$ onto $\{1, 2, \ldots, k\}$ by $\pi(x) = i$ if i is the index such that $P_{i,x} = 1$.

Example 6.2.7 The matrix

$$P = \begin{bmatrix} 1 & 0 & 0 \\ 0 & 1 & 1 \end{bmatrix}$$

is a partition matrix. The corresponding partition θ is $\{1\}$, $\{2, 3\}$ and the map π is $1 \to 1, 2 \to 2, 3 \to 2$.

The following result will allow us to reduce to the case of a Bratteli diagram with a simple hat.

Proposition 6.2.8 *For every stationary and properly ordered Bratteli diagram* $(V, E, \leq)$*, there is a stationary properly ordered Bratteli diagram* $(V', E', \leq')$ *with a simple hat such that* (X_E, T_E) *and* $(X_{E'}, T_{E'})$ *are isomorphic.*

More precisely, for every ordered Bratteli diagram $\mathcal{B}$*, there is an ordered Bratteli diagram* $\mathcal{B}'$ *such that the following holds. There exist two nonnegative matrices* P, Q*, with* P *a partition matrix, such that the matrices* M, M' *of* $\mathcal{B}$ *and* $\mathcal{B}'$ *satisfy*

$$M = PQ, \quad M' = QP$$

and $M(1) = P[1 \ 1 \ldots 1]^t$*. Moreover, if* $\mathcal{B}$ *is properly ordered, then* $\mathcal{B}'$ *is properly ordered.*

Proof Let M be the $t \times t$-matrix equal to the incidence matrices $M(n)$ of the diagram $\mathcal{B}$, for all $n \geq 2$ and let $v = M(1)$. Set $\ell = \sum_{i=1}^{t} v_i$.

Let P be the $t \times \ell$ partition matrix defined by

$$P_{i,j} = \begin{cases} 1 & \text{if } \sum_{k<i} v_k < j \leq \sum_{k \leq i} v_k \\ 0 & \text{otherwise.} \end{cases}$$

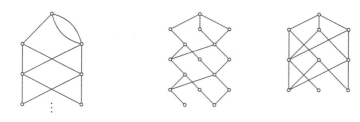

Figure 6.8 The transformation of $(V, E, \leq)$ by splitting vertices.

Thus, the rows of P are the characteristic vectors of elements of the partition of $\{1, 2, \ldots, \ell\}$ into the t sets $V_1 = \{1, 2, \ldots, v_1\}, \ldots, V_t = \{\ell - v_t + 1, \ldots, \ell\}$. We may assume, replacing if necessary M by some power, that, for every i, the set of edges with range i has at least v_i elements. We choose an $\ell \times t$ matrix Q such that $M = PQ$. This is equivalent to splitting the set of edges entering the vertex i in v_i nonempty subsets (the sum of the rows of Q with index in V_i is then the row of index i of M). The edges in each subset keep the order induced by the order on B. Let B' be the Bratteli diagram with incidence matrices $M'(1) = w = \begin{bmatrix} 1 \ 1 \ldots 1 \end{bmatrix}^t$ and $M'(n) = QP$ for $n \geq 2$. We order the diagram B' by the order induced by that of B. Since $v = Pw$, B and B' can both be obtained by telescoping from the Bratteli diagram C with incidence matrices $(w, P, Q, P, Q, \ldots)$. If B is properly ordered, C is properly ordered and consequently B' also. ∎

We illustrate the construction with the following example.

Example 6.2.9 Let $(V, E, \leq)$ be the Bratteli diagram represented in Figure 6.8 on the left. The matrices M, P, Q, M' are

$$M = \begin{bmatrix} 1 & 1 \\ 1 & 1 \end{bmatrix}, \quad P = \begin{bmatrix} 1 & 0 & 0 \\ 0 & 1 & 1 \end{bmatrix}, \quad Q = \begin{bmatrix} 1 & 1 \\ 1 & 0 \\ 0 & 1 \end{bmatrix}, \quad M' = \begin{bmatrix} 1 & 1 & 1 \\ 1 & 0 & 0 \\ 0 & 1 & 1 \end{bmatrix}.$$

The diagram with matrices $([1\ 1\ 1]^t, P, Q, P, \ldots)$ is represented in the middle of Figure 6.8 and the diagram with simple hat $(V', E', \leq')$ with matrices $([1\ 1\ 1]^t, M', M', \ldots)$ on the right.

6.2.3 A Useful Result

The following result, which we call the *Rauzy Lemma*, will be the key to building a BV-representation of an aperiodic minimal substitution shift. It allows one to replace a pair (τ, ϕ) of morphisms with τ primitive by a pair (ζ, θ) with ζ primitive and θ letter-to-letter (see Figure 6.9).

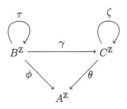

Figure 6.9 The eventually proper substitution ζ.

Proposition 6.2.10 *Let* $y \in B^{\mathbb{Z}}$ *be an admissible fixed point of a primitive morphism* τ *on the alphabet* B *and let* $\phi: B^* \to A^*$ *be a nonerasing morphism. Set* $x = \phi(y)$ *and let* X *be the subshift spanned by* x.

There exist a primitive morphism ζ *on an alphabet* C, *an admissible fixed point* z *of* ζ, *a morphism* $\gamma: B^* \to C^*$ *and a letter-to-letter morphism* $\theta: C^* \to A^*$ *such that:*

1. $\theta(z) = x$ *and* $\phi = \theta \circ \gamma$. *In particular, X is primitive substitutive.*
2. *If ϕ is a circular morphism, then θ is a conjugacy from $X(\zeta)$ onto X. In particular, X is conjugate to a primitive substitution shift.*
3. *If τ is eventually proper, then ζ is eventually proper.*

Proof The proof below is very simple, but the notation is, in an unavoidable way, a bit heavy. By substituting τ by a power of itself if needed, we can assume that $|\tau(b)| \geq |\phi(b)|$ for all $b \in B$. We define:

- an alphabet C by $C = \{b_p \mid b \in B, \; 1 \leq p \leq |\phi(b)|\}$,
- a letter-to-letter morphism $\theta: C^* \to A^*$ by $\theta(b_p) = (\phi(b))_p$,
- a morphism $\gamma: B^* \to C^*$ by $\gamma(b) = b_1 b_2 \cdots b_{|\phi(b)|}$.

Clearly $\theta \circ \gamma = \phi$. We define a morphism ζ on C as follows. For b in B and $1 \leq p \leq |\phi(b)|$, we define a morphism $\rho: C^* \to B^*$ by

$$\rho(b_p) = \begin{cases} (\tau(b))_p & \text{if } 1 \leq p < |\phi(b)| \\ (\tau(b))_{[|\phi(b)|,|\tau(b)|]} & \text{if } p = |\phi(b)| \end{cases}$$

and we set $\zeta = \gamma \circ \rho$. Hence, for every $b \in B$, $\rho(\gamma(b)) = \rho(b_1) \cdots \rho(b_{|\phi(b)|}) = \tau(b)$, that is, $\rho \circ \gamma = \tau$ or equivalently

$$\zeta \circ \gamma = \gamma \circ \tau. \tag{6.4}$$

It follows that $\zeta^n \circ \gamma = \gamma \circ \tau^n$ for all $n \geq 0$. We claim that ζ is primitive. Let n be an integer such that b occurs in $\tau^n(a)$ for all $a, b \in B$. Let b_p and c_q be in C. By construction, $\zeta(b_p)$ contains $\gamma(\tau(b)_p)$ as a factor, thus $\zeta^{n+1}(b_p)$ contains $\zeta^n(\gamma(\tau(b)_p)) = \gamma(\tau^n(\tau(b)_p))$ as a factor. By the choice of n, the letter c occurs in $\tau^n\left(\tau(b)_p\right)$, thus $\gamma(c)$ is a factor of $\gamma\left(\tau^n\left(\tau(b)_p\right)\right)$, and also of $\zeta^{n+1}(b_p)$. Since c_q is a letter of $\gamma(c)$, c_q occurs in $\zeta^{n+1}(b_p)$ and our claim is proved.

Proof of 1. Let $z = \gamma(y)$. By (6.4) we get $\zeta(z) = \gamma(\tau(y)) = \gamma(y) = z$, and z is a fixed point of ζ. By construction, the sequence z is uniformly recurrent, thus it is an admissible fixed point of ζ. Moreover, $\theta(z) = \theta(\gamma(y)) = \phi(y) = x$, and 1 is proved.

Proof of 2. Since θ commutes with the shift and maps z to x, and by minimality of the subshifts, it maps $X(\zeta)$ onto X. There remains to prove that $\theta: X(\zeta) \to X$ is injective. Let $X(\tau)^\phi = \{(t, i) \mid t \in X(\tau), 0 \le i < |\phi(t_0)|\}$ and let $\hat{\phi}: X(\tau)^\phi \to X$ be defined by $\hat{\phi}(t, i) = S^i\phi(t)$. Since ϕ is a circular morphism, ϕ is recognizable on $X(\tau)$ by Proposition 1.4.32. Consequently, the map $\hat{\phi}$ is a homeomorphism from $X(\tau)^\phi$ onto X. Let $\hat{\gamma}: X(\tau)^\phi \to X(\zeta)$ be similarly defined by $\hat{\gamma}(t, i) = S^i\gamma(t)$. Then $\hat{\phi} = \theta \circ \hat{\gamma}$. Since $X(\zeta)$ is minimal, $\hat{\gamma}$ is surjective and thus, $\hat{\phi}$ being injective, θ is also injective.

Proof of 3. Since $\zeta^n \circ \gamma = \gamma \circ \tau^n$, we may replace τ by τ^n and thus assume that τ is proper. Let $l \in B$ be the first letter of $\tau(b)$ for every $b \in B$. Let $b_p \in C$, and $c = \tau(b)_p$. By definition of ζ, the first letter of $\zeta(b_p)$ is c_1, and the first letter of $\zeta^2(b_p)$ is the first letter of $\zeta(c_1)$, that is, l_1. Thus, ζ^2 is left proper. By the same method, if r is the last letter of $\tau(b)$ for every $b \in B$, then the last letter of $\zeta^2(b_p)$ is $r_{|\phi(r)|}$ for every $b_p \in C$ and thus ζ^2 is also right proper. We conclude that ζ is eventually proper. ∎

Note that the proof of Proposition 6.2.10 is very close to that of Proposition 6.2.8 (see Exercise 6.18).

Consider a morphism $\sigma: A^* \to A^*$ generating an aperiodic minimal shift X. By Proposition 1.4.13, replacing σ by one of its powers, we can choose an admissible fixed point $x = \sigma^\omega(r \cdot \ell)$ of σ where r, ℓ are nonempty words and $r\ell \in \mathcal{L}(\sigma)$. Note that since x is a fixed point of σ, the word $\sigma(r)$ ends with r and the word $\sigma(\ell)$ begins with ℓ.

Let $\mathcal{R}_X(r\ell)$ be the set of right return words to $r\ell$. Every word in $\mathcal{R}_X(r\ell)$ ends with ℓ. Set

$$\mathcal{R}_X(r \cdot \ell) = \ell\mathcal{R}_X(r\ell)\ell^{-1}.$$

This is the set of words of the form ℓu for $u\ell \in \mathcal{R}_X(r\ell)$. Thus, w is in $\mathcal{R}_X(r \cdot \ell)$ if and only if $rw\ell$ is in $\mathcal{L}(X)$ and contains exactly two occurrences of $r\ell$, one as a prefix and one as a suffix. In particular, all words in $\mathcal{R}_X(r \cdot \ell)$ begin with ℓ and end with r. Moreover, the set $\mathcal{R}_X(r \cdot \ell)$ satisfies the following properties.

(i) Every word u such that $ru\ell$ begins and ends with $r\ell$ and is in $\mathcal{L}(X)$, is a concatenation of words of $\mathcal{R}_X(r \cdot \ell)$.

(ii) The set $\mathcal{R}_X(r \cdot \ell)$ is a circular code. More precisely, no word in $r\mathcal{R}_X(r \cdot \ell)\ell$ overlaps nontrivially a product of words of $\mathcal{R}_X(r \cdot \ell)$, that is, if $ru\ell \in r\mathcal{R}_X(r \cdot \ell)\ell$ is a factor of $ru_1u_2 \cdots u_n\ell$ with $u_i \in \mathcal{R}_X(r \cdot \ell)$, then $u = u_i$ for some i.

Since X is minimal, the set $\mathcal{R}_X(r \cdot \ell)$ is finite. Let $\phi \colon B^* \to A^*$ be a coding morphism for $\mathcal{R}_X(r \cdot \ell)$ (see Section 1.4.3 for the definition).

Since x is uniformly recurrent, it has an infinite number of occurrences of $r\ell$ at positive and at negative indices. By the above properties of $\mathcal{R}_X(r \cdot \ell)$, there is a unique element y of $B^{\mathbb{Z}}$ such that $\phi(y) = x$.

We now define as follows a morphism τ on the alphabet B. For every $b \in B$, $r(\sigma \circ \phi(b))\ell$ begins and ends with $r\ell$ and is in $\mathcal{L}(X)$. This implies that $\sigma \circ \phi(b)$ belongs to $\mathcal{R}_X(r \cdot \ell)^*$ and thus that $\sigma \circ \phi(b) = \phi(w)$ for some unique $w \in B^*$. We set $\tau(b) = w$.

This defines a morphism τ on the alphabet B, characterized by

$$\phi \circ \tau = \sigma \circ \phi. \tag{6.5}$$

It follows that $\phi \circ \tau^n = \sigma^n \circ \phi$ for each $n \geq 0$. Note that, since X is infinite, we have $\mathrm{Card}(B) \geq 2$.

Example 6.2.11 Consider the Fibonacci morphism $\varphi \colon a \to ab, b \to a$ generating the Fibonacci shift $X(\varphi)$. The sequence $x = \varphi^{2\omega}(a \cdot a)$ is an admissible fixed point of $\varphi^2 \colon a \to aba, b \to ab$. We have $\mathcal{R}_X(a \cdot a) = \{aba, ababa\}$. Set $B = \{a, b\}$ with $\phi(a) = aba$ and $\phi(b) = ababa$. Then $\phi \circ \tau(a) = \varphi^2 \circ \phi(a) = \varphi^2(aba) = abaababa = \phi(ab)$ so that $\tau(a) = ab$. Similarly, we find $\tau(b) = abb$.

The following statement shows that τ, being a primitive morphism on more than one letter, is in particular a substitution (Proposition 1.4.15).

Proposition 6.2.12 *The morphism τ defined by (6.5) is primitive, eventually proper and aperiodic. Moreover,*

1. the shift $X(\tau)$ is isomorphic to the shift induced by X on the cylinder $[r \cdot \ell]$;
2. the spectral radiuses of the incidence matrices of σ, τ are equal.

Proof Let us show that τ is eventually proper. For this, let n be such that $|\sigma^n(\ell)| > |\phi(y_0)|$. Then for every $b \in B$, the first letter of $\tau^n(b)$ is y_0. Indeed, $\phi \circ \tau^n(b) = \sigma^n \circ \phi(b)$ begins with $\sigma^n(\ell)$ (see Figure 6.10). But $\sigma^n(\ell)$ is a prefix of $\phi(y^+) = x^+ = \sigma^n(x^+)$. Thus, by the choice of n, the word $\phi(y_0)\ell$ is a prefix of $\sigma^n(\ell)$. This implies that $\tau^n(b)$ begins with y_0. A similar argument

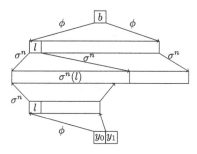

Figure 6.10 Proof that τ is eventually proper.

shows that if n is such that $|\sigma^n(r)| > |\phi(y_{-1})|$, then every $\tau^n(b)$ end with y_{-1}. Thus, τ is eventually proper.

Let $k > 0$ be an occurrence of $r.\ell$ large enough so that every return word $w \in \mathcal{R}(r.\ell)$ appears in the decomposition of $x_{[0,k)}$, that is, every $b \in B$ occurs in the finite word $u \in B^+$ defined by $\phi(u) = x_{[0,k)}$. Let n be so large that $|\sigma^n(\ell)| > k$. Let $b, c \in B$. As above, $x_{[0,k)}\ell$ is a prefix of $\sigma^n(\ell)$, which is a prefix of $\sigma^n(\phi(b)) = \phi(\tau^n(b))$. Thus, u is a prefix of $\tau^n(b)$, and c occurs in $\tau^n(b)$, hence τ is primitive.

Moreover, $\phi(\tau(y)) = \sigma(\phi(y)) = \sigma(x) = x = \phi(y)$, whence $\tau(y) = y$ since $\mathcal{R}_X(r \cdot \ell)$ is circular, and y is the unique fixed point of τ. Since $\phi(y) = x$ is not periodic, y is not periodic. Thus, τ is aperiodic.

Assertion 1 follows from the fact that the shift $X(\tau)$ is the derivative shift of X with respect to $[r \cdot \ell]$ (Proposition 3.7.1). Finally, let $M(\sigma), M(\tau)$ be the incidence matrices of σ, τ and let $\lambda(\sigma), \lambda(\tau)$ be their respective spectral radiuses. By (6.5), we have $M(\phi)M(\sigma) = M(\tau)M(\phi)$. Let v be a positive row eigenvector of $M(\tau)$ relative to $\lambda(\tau)$. Then $vM(\phi)M(\sigma) = vM(\tau)M(\phi) = \lambda(\tau)M(\phi)$ and thus $vM(\phi)$ is a nonnegative eigenvector of $M(\sigma)$ relative to $\lambda(\tau)$. By the Perron–Frobenius Theorem (Theorem B.3.1), this implies $\lambda(\sigma) = \lambda(\tau)$. ∎

We are now ready to prove Theorem 6.2.1.

Proof of Theorem 6.2.1 Let first $(V, E, \leq)$ be a stationary properly ordered Bratteli diagram. By Proposition 6.2.8, we may assume that $(V, E, \leq)$ has a simple hat. Then, using Proposition 6.2.3, we conclude that (X_E, T_E) is either an aperiodic minimal substitution shift or an odometer. This proves the theorem in one direction.

Let us now prove the converse implication. If (X, T) is an odometer, we have seen (Section 6.1) that (X, T) has a BV-representation with a stationary, properly ordered Bratteli diagram. Moreover, (X, T) cannot be at the same time a minimal substitution shift by Proposition 6.1.6.

Let finally $\sigma: A^* \rightarrow A^*$ be a morphism generating an aperiodic minimal shift space X. Let $\tau: B^* \rightarrow B^*$ be the primitive, eventually proper and aperiodic morphism defined by Proposition 6.2.12.

Let $\zeta: C^* \rightarrow C^*$ be the morphism given by Proposition 6.2.10 together with an admissible fixed point $z \in C^{\mathbb{Z}}$ and a letter-to-letter morphism $\theta: C^* \rightarrow A^*$. Since ϕ is a coding morphism for $\mathcal{R}_X(r \cdot \ell)$ and $\mathcal{R}_X(r \cdot \ell) = \phi(B)$ is a circular code, by Assertion 2 of Proposition 6.2.10, θ is an isomorphism from $(X(\zeta), S)$ onto (X, S). Moreover, by Assertion 3, since τ is eventually proper, the morphism ζ is eventually proper. Let $(V, E, \leq)$ be the stationary Bratteli diagram such that ζ is the morphism read on $(V, E, \leq)$. By Proposition 6.2.2, it is properly ordered. By Proposition 6.2.3, the systems $(X(\zeta), S)$ and (X_E, T_E) are isomorphic. This concludes the proof. ∎

Note that the preceding results imply the following property of substitution shifts.

Corollary 6.2.13 *Every minimal substitution shift is isomorphic to a primitive eventually proper substitution shift.*

Proof If X is finite, it is generated by the morphism sending all letters to the same word w, which is primitive and proper. Otherwise, by Theorem 6.2.1, it is isomorphic to (X_E, T_E), with $(V, E, \leq)$ a stationary, properly ordered Bratteli diagram. By Proposition 6.2.3, the system (X_E, T_E) is isomorphic to the substitution shift $(X\sigma, S)$, where σ is the morphism read on $(V, E, \leq)$. But by Proposition 6.2.2, the morphism σ is primitive and eventually proper, which proves the statement. ∎

Let us illustrate this result with the case of the Chacon binary morphism $\sigma: 0 \rightarrow 0010, 1 \rightarrow 1$. The morphism is not primitive, but the corresponding shift space is minimal (see Exercise 1.33).

Let $x = \sigma^\omega(0.0)$. This is the *Chacon binary sequence*. Using the return words to 0.0, we see that $x = \phi(y)$ where $\phi: B^* \rightarrow \{0, 1\}^*$ with $B = \{a, b, c\}$ is defined by $\phi(a) = 0$, $\phi(b) = 010$ and $\phi(c) = 01010$, and $y = \tau^\omega(b.a)$ where τ is defined by $\tau(a) = ab$, $\tau(b) = acb$ and $\tau(c) = accb$. According to the proof of Proposition 6.2.10 we need to use τ^2 instead of τ and we define

1. $C = \{a_1, b_1, b_2, b_3, c_1, c_2, c_3, c_4, c_5\}$,

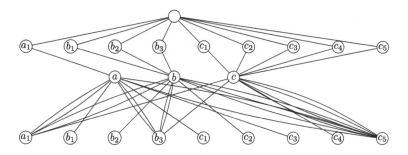

Figure 6.11 The BV-representation of the Chacon binary shift.

2. $\theta : C^* \to \{0, 1\}^*$ given by the following table

α	a_1	b_1	b_2	b_3	c_1	c_2	c_3	c_4	c_5
$\theta(\alpha)$	0	0	1	0	0	1	0	1	0

3. $\gamma : B^* \to C^*$ defined by $\gamma(a) = a_1$, $\gamma(b) = b_1 b_2 b_3$, $\gamma(c) = c_1 c_2 c_3 c_4 c_5$,
4. the substitution $\zeta = \gamma \circ \rho$ with $\rho : C^* \to B^*$ defined by the following table

α	a_1	b_1	b_2	b_3	c_1	c_2	c_3	c_4	c_5
$\rho(\alpha)$	$abacb$	a	b	$accbacb$	a	b	a	c	$cbaccbacb$

A BV-representation of the Chacon shift (that is, the subshift generated by x)
is isomorphic to (X_E, T_E), where $(V, E, \leq)$ is a stationary, properly ordered
Bratteli diagram with a simple hat such that ζ is the morphism read on it. This
diagram is obtained by telescoping at odd levels the diagram of Figure 6.11.

As already mentioned (and developed in Exercise 6.18), an alternative route
to obtain a Bratteli diagram with a simple hat is to use Proposition 6.2.8. In the
present example, we would start with $M = M(\tau)$ and write $M = PQ$, where
P is the partition matrix corresponding to ϕ:

$$
M = \begin{bmatrix} 2 & 2 & 1 \\ 3 & 3 & 3 \\ 4 & 4 & 5 \end{bmatrix}, \quad
P = \begin{bmatrix} 1 & 0 & 0 & 0 & 0 & 0 & 0 & 0 & 0 \\ 0 & 1 & 1 & 1 & 0 & 0 & 0 & 0 & 0 \\ 0 & 0 & 0 & 0 & 1 & 1 & 1 & 1 & 1 \end{bmatrix}, \quad
Q = \begin{bmatrix} 2 & 2 & 1 \\ 1 & 0 & 0 \\ 0 & 1 & 0 \\ 2 & 2 & 3 \\ 1 & 0 & 0 \\ 0 & 1 & 0 \\ 1 & 0 & 0 \\ 0 & 0 & 1 \\ 2 & 3 & 4 \end{bmatrix}.
$$

The matrix

$$M' = QP = \begin{bmatrix} 2 & 2 & 2 & 2 & 1 & 1 & 1 & 1 & 1 \\ 1 & 0 & 0 & 0 & 0 & 0 & 0 & 0 & 0 \\ 0 & 1 & 1 & 1 & 0 & 0 & 0 & 0 & 0 \\ 2 & 2 & 2 & 2 & 3 & 3 & 3 & 3 & 3 \\ 1 & 0 & 0 & 0 & 0 & 0 & 0 & 0 & 0 \\ 0 & 1 & 1 & 1 & 0 & 0 & 0 & 0 & 0 \\ 1 & 0 & 0 & 0 & 0 & 0 & 0 & 0 & 0 \\ 0 & 0 & 0 & 0 & 1 & 1 & 1 & 1 & 1 \\ 2 & 3 & 3 & 3 & 4 & 4 & 4 & 4 & 4 \end{bmatrix}$$

defines the stationary diagram looked for.

6.2.4 Dimension Groups and BV-Representation of Substitution Shifts

We now derive from the previous results a description of the dimension group of substitution shifts.

Let $\sigma : A^* \to A^*$ be a substitution generating an aperiodic minimal shift X. As in the proof of Theorem 6.2.1, changing, if necessary, σ for some power of σ, let $x = \sigma^\omega(r \cdot \ell)$ be an admissible fixed point of σ. Let $\mathcal{R}_X(r \cdot \ell) = \ell \mathcal{R}_X(r\ell)\ell^{-1}$ and let $\phi : B^* \to A^*$ be a coding morphism for $\mathcal{R}_X(r \cdot \ell)$. Let $\tau : B^* \to B^*$ be the morphism such that $\phi \circ \tau = \sigma \circ \phi$.

Theorem 6.2.14 *Let σ be a substitution generating an aperiodic minimal shift X and let $\tau : B^* \to B^*$ be as above. Let M be the composition matrix of τ. Then $K^0(X, S) = (\Delta_M, \Delta_M^+, v)$ where v is the image in Δ_M of the vector with components $|\phi(b)|$ for $b \in B$.*

Proof By Proposition 6.2.12, the substitution τ is primitive, eventually proper and aperiodic. Thus, by Proposition 6.2.3, the system (X_τ, S) is isomorphic to (X_E, T_E), where $B = (V, E, \leq)$ is the Bratteli diagram with τ read on B. By Theorem 5.3.6, the dimension group of $(X(\tau), S)$ is isomorphic to $(\Delta_M, \Delta_M^+, \mathbf{1}_M)$. Since $(X(\tau), S)$ is the system induced by X on the clopen set $[r \cdot \ell]$, the result follows from Proposition 3.7.3. ∎

The result can, of course, also be deduced from Proposition 5.4.2.

We will use the following simple result (note that it is a particular case of Theorem 5.4.1 on Kakutani equivalence).

Let $\tau : B^* \to B^*$ be a primitive eventually proper aperiodic morphism and let $X(\tau)$ be the associated substitution shift. Let $(V, E, \leq)$ be a properly ordered simple Bratteli diagram with simple hat such that τ is the morphism read on $(V, E, \leq)$. By Proposition 6.2.3, the shift $(X(\tau), S)$ is conjugate to (X_E, T_E). We identify the set $V \setminus \{0\}$ to $B \times \mathbb{N}$. Let $\phi: B^* \to A^*$ be a morphism recognizable on $X(\tau)$ and let $Y = X(\tau, \phi)$ be the corresponding substitutive shift.

Proposition 6.2.15 *The shift Y is conjugate to $(X_{E'}, T_{E'})$, where $(V', E', \leq)$ is the Bratteli diagram obtained from $(X, E, \leq)$ by replacing each edge from 0 to $(b, 1)$ by $|\phi(b)|$ edges $(0, b, i)$ with $0 \leq i < |\phi(b)|$.*

Proof Let U be the clopen subset of $X_{E'}$ formed of the paths with a first edge of the form $(0, b, 0)$ for some $b \in B$. The system induced by $(X_{E'}, T_{E'})$ on U is clearly X_E by Theorem 5.4.1. Thus, $X_{E'}$ is the primitive of X_E relative to the function $f(x) = |\phi(x_0)|$. On the other hand, since ϕ is recognizable on $X(\tau)$, by Proposition 1.4.30, Y is the primitive of X relative to the function $f(x) = |\phi(x_0)|$. Thus, $X_{E'}$ and Y are conjugate. ∎

We will see below several examples of application of this result.

6.2.5 Some Examples

We will illustrate the preceding results on some classical examples.

Example 6.2.16 Let X be the Fibonacci shift generated by $\sigma: a \to ab, b \to a$. Consider the admissible fixed point $\sigma^{2\omega}(a \cdot a)$. Let $\phi: \{\alpha, \beta\}^* \to \{a, b\}^*$ be a coding morphism for $\mathcal{R}_X(a \cdot a) = \{ababa, aba\}$. The morphism $\tau: B^* \to B^*$ such that $\phi \circ \tau = \sigma^2 \circ \phi$ is $\tau: \alpha \to \beta\alpha\alpha, \beta \to \beta\alpha$ (see Example 6.2.11). A Bratteli diagram with simple hat corresponding to τ is shown in Figure 6.12 on the left. A BV-representation of X (according to Proposition 6.2.15) is shown on the right.

The matrix of both diagrams is M^2 where

$$M = \begin{bmatrix} 1 & 1 \\ 1 & 0 \end{bmatrix}, \quad M^2 = \begin{bmatrix} 2 & 1 \\ 1 & 1 \end{bmatrix}, \quad M^3 = \begin{bmatrix} 3 & 2 \\ 2 & 1 \end{bmatrix}$$

and the hat of the diagram on the right is the image of a simple hat by the matrix M^3. Thus, both diagrams are telescoping equivalent, and the first one is also a BV-representation of the Fibonacci shift. Note that we could have obtained this result directly, observing that the morphism $\rho: a \mapsto baa, b \mapsto ba$ is a primitive and proper morphism, which is conjugate of σ^2. However, the method used to obtain it through the morphism τ is a general method working

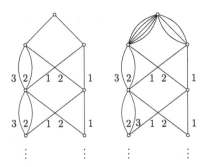

Figure 6.12 The BV-representation of the Fibonacci shift.

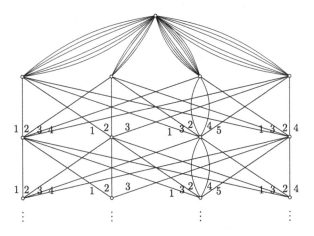

Figure 6.13 The BV-representation of the Thue–Morse shift.

in all cases (we shall see yet another way to obtain a BV-representation of the Fibonacci shift in Example 7.2.6).

All this is consistent with the fact that $K^0(X, S) = \mathbb{Z}[\lambda]$ with $\lambda = (1 + \sqrt{5})/2$ (see Examples 3.9.7 and 4.6.10).

Example 6.2.17 Now let X be the Thue–Morse shift generated by $\sigma: 0 \to 01, 1 \to 10$. Let $x = \sigma^{2\omega}(0 \cdot 0)$. Set $B = \{1, 2, 3, 4\}$ and let $\phi: B^* \to \{0, 1\}^*$ be the coding morphism for $\mathcal{R}_X(0 \cdot 0) = \{011010, 0110, 01011010, 010110\}$. Let $\tau: B^* \to B^*$ be the morphism such that $\phi \circ \tau = \sigma^2 \circ \phi$. We have $\tau: 1 \to 1234, 2 \to 124, 3 \to 13234, 4 \to 1324$. Using again Proposition 6.2.15, we obtain the BV-representation of Figure 6.13.

The dimension group $K^0(X, S)$ is thus $(\Delta_M, \Delta_M^+, v)$ with

$$
M = \begin{bmatrix} 1 & 1 & 1 & 1 \\ 1 & 1 & 0 & 1 \\ 1 & 1 & 2 & 1 \\ 1 & 1 & 1 & 1 \end{bmatrix}, \text{ and } v = \begin{bmatrix} 6 \\ 4 \\ 8 \\ 6 \end{bmatrix}.
$$

This gives (fortunately) the same result as in Example 4.6.11, where we had computed $K^0(X, S)$ as $(\Delta_{N_2}, \Delta_{N_2}^+, w)$ with

$$
N_2 = \begin{bmatrix} 0 & 1 & 0 \\ 1 & 0 & 1 \\ 1 & 1 & 1 \end{bmatrix} \text{ and } w = \begin{bmatrix} 1 \\ 2 \\ 3 \end{bmatrix}.
$$

Actually, we have

$$
N_2 = \begin{bmatrix} 0 & 1 \\ 1 & 0 \\ 1 & 1 \end{bmatrix} \begin{bmatrix} 1 & 0 & 1 \\ 0 & 1 & 0 \end{bmatrix} \text{ with } \begin{bmatrix} 1 & 0 & 1 \\ 0 & 1 & 0 \end{bmatrix} \begin{bmatrix} 0 & 1 \\ 1 & 0 \\ 1 & 1 \end{bmatrix} = \begin{bmatrix} 1 & 2 \\ 1 & 0 \end{bmatrix},
$$

while

$$
M = \begin{bmatrix} 1 & 1 \\ 1 & 0 \\ 1 & 2 \\ 1 & 1 \end{bmatrix} \begin{bmatrix} 1 & 1 & 0 & 1 \\ 0 & 0 & 1 & 0 \end{bmatrix}
$$

$$
\text{with } \begin{bmatrix} 1 & 1 & 0 & 1 \\ 0 & 0 & 1 & 0 \end{bmatrix} \begin{bmatrix} 1 & 1 \\ 1 & 0 \\ 1 & 2 \\ 1 & 1 \end{bmatrix} = \begin{bmatrix} 3 & 2 \\ 1 & 2 \end{bmatrix} = \begin{bmatrix} 1 & 2 \\ 1 & 0 \end{bmatrix}^2.
$$

This shows that M and N_2 are C^*-equivalent and thus have the same ordered group. The order units also coincide because

$$
\begin{bmatrix} 1 & 1 & 0 & 1 \\ 0 & 0 & 1 & 0 \end{bmatrix} v = \begin{bmatrix} 16 \\ 8 \end{bmatrix} = \begin{bmatrix} 3 & 2 \\ 1 & 2 \end{bmatrix} \begin{bmatrix} 4 \\ 2 \end{bmatrix} \text{ while } \begin{bmatrix} 1 & 0 & 1 \\ 0 & 1 & 0 \end{bmatrix} w = \begin{bmatrix} 4 \\ 2 \end{bmatrix}.
$$

In Example 4.6.11 we obtained the dimension group as the group of a matrix of dimension 3 instead of 4 as here. Actually the method used in Example 4.6.11 can be used also to build a BV-representation of the Thue–Morse shift with 3 vertices at each level $n \geq 1$ (Exercise 6.19).

6.3 Linearly Recurrent Shifts

In this section, we study linearly recurrent shifts, introduced in Chapter 1. This family contains the primitive substitution shifts (Proposition 1.4.24). The main result is a characterization of the BV-representation of aperiodic linearly recurrent shifts (Theorem 6.3.5).

6.3.1 Properties of Linearly Recurrent Shifts

The class of linearly recurrent shift spaces is clearly closed under conjugacy (Exercise 6.27). Minimal shifts need not be linearly recurrent since a minimal shift is, in general, not K-power free for any $K \geq 1$. A simple example is given below.

Example 6.3.1 Let u_n be the sequence of words defined by $u_{-1} = b$, $u_0 = a$ and $u_{n+1} = u_n^n u_{n-1}$ for $n \geq 1$. The subshift generated by the infinite word u having all u_n as prefixes is not LR. This results from the property of LR shift spaces with constant K to be $(K + 1)$th power free (Proposition 6.3.3).

We will use Theorem 6.2.1 to prove the following result, which gives an easily verifiable condition for a substitution shift to be LR (the condition is, for example, visibly satisfied by the Chacon binary morphism). It proves that the property of being LR is decidable for a substitution shift. Indeed, it reduces the problem to that of $\mathcal{L}(X) \cap (A \setminus \{e\})^*$ being finite (see Exercise 6.26).

Proposition 6.3.2 *Let $\sigma \colon A^* \to A^*$ be a substitution and let X be the corresponding substitution subshift. Let $e \in A$ be a letter such that $\lim_{n \to \infty} |\sigma^n(e)| = \infty$. Then the following are equivalent.*

(i) *The letter e occurs with bounded gaps in $\mathcal{L}(X)$ and for every $a \in A$, one has $|\sigma^n(e)|_a \geq 1$ for some $n \geq 1$.*
(ii) *X is minimal.*
(iii) *X is linearly recurrent.*

Proof (i)$\Rightarrow$ (ii) We have to show that for every $n \geq 1$, the word $\sigma^n(e)$ occurs in every long enough word of $\mathcal{L}(X)$. Since e occurs with bounded gaps, there is k such that every word of $\mathcal{L}_k(X)$ contains e. Choose w of length $|w| \geq (3 + k) \max_{a \in A} |\sigma^n(a)|$. Then w contains a factor $\sigma^n(u)$ for some word $u \in \mathcal{L}_k(X)$ and thus contains $\sigma^n(e)$.

(ii)$\Rightarrow$ (iii) By Corollary 6.2.13 X is isomorphic to a primitive substitution shift. Thus, the assertion results from Proposition 1.4.24.

(iii)$\Rightarrow$ (i) is obvious. ∎

A fourth equivalent statement could be added, namely that (X, S) is uniquely ergodic (see Section 6.8 for a reference).

We will prove now some important properties of LR shifts.

Proposition 6.3.3 *Let X be an aperiodic shift space and suppose that it is linearly recurrent with constant K. Then:*

1. *Every word of $\mathcal{L}_n(X)$ appears in every word of $\mathcal{L}_{(K+1)n-1}(X)$.*
2. *The number of distinct factors of length n of X is at most Kn.*
3. *$\mathcal{L}(X)$ is $(K + 1)$-power free, that is, for every nonempty word u, if u^n belongs to $\mathcal{L}(X)$, then $n \leq K$.*
4. *For all $u \in \mathcal{L}(X)$ and for all $w \in \mathcal{R}_X(u)$, we have $\frac{1}{K}|u| < |w| \leq K|u|$.*
5. *If $u \in \mathcal{L}(X)$, then $\mathrm{Card}(\mathcal{R}_X(u)) \leq K(K + 1)^2$.*

Proof The first three assertions are already in Proposition 1.2.9.

4. Suppose that for $u \in \mathcal{L}(X)$ and $w \in \mathcal{R}_X(u)$, we have $|u| \geq K|w|$. Since u is a suffix of uw, by Proposition 1.2.1, the length of w is a period of uw. Set $n = \lfloor K \rfloor$. Then $|w^{n+1}| = (n + 1)|w| \leq (K + 1)|w| \leq |uw|$ and thus w^{n+1} is a suffix of uw, a contradiction with assertion 3.

5. Let $u, v \in \mathcal{L}(X)$ with $|v| = (K + 1)^2|u|$. By assertion 1, every element of $\mathcal{L}(X)$ of length $(K + 1)|u|$ appears in v and thus all words of $u\mathcal{R}_X(u)$ are factors of v. By Assertion 4, we have

$$\mathrm{Card}(\mathcal{R}_X(u))|u|/K \leq \sum_{w \in \mathcal{R}_X(u)} |w|. \tag{6.6}$$

Now, every word in $u\mathcal{R}_X(u)$ occurs in v. Considering all successive occurrences of u in v, we have a factor of v of the form ux with $x = x_1 x_2 \cdots x_k$ where all x_i are in $\mathcal{R}_X(u)$ and every occurrence of u in v is either as a prefix of ux or as a suffix of some ux_i. Since the x_i are all the factors of v that belong to $\mathcal{R}_X(u)$, every word in $\mathcal{R}_X(u)$ appears as some x_i. This implies that

$$\sum_{w \in \mathcal{R}_X(u)} |w| \leq |v| = (K + 1)^2|u|, \tag{6.7}$$

whence the conclusion $\mathrm{Card}(\mathcal{R}_X(u)) \leq K(K + 1)^2$. ∎

Note the following corollary of Proposition 6.3.3.

Corollary 6.3.4 *The factor complexity of a minimal substitution shift is at most linear.*

Indeed, by Proposition 6.3.2, a minimal substitution shift is linearly recurrent. This statement is not true for a substitution shift that is not minimal (Exercise 6.28).

We now come back to return words. Let X be a minimal shift space. For u, v such that uv belongs to $\mathcal{L}(X)$, we consider, as in the proof of Theorem 6.2.1, the set $\mathcal{R}_X(uv)$ of right return words to uv and the set

$$\mathcal{R}_X(u \cdot v) = v\mathcal{R}_X(uv)v^{-1},$$

called the set of *return words* to $u \cdot v$. As we will manipulate return words, it is important to observe that a finite word $w \in A^+$ is a return word to $u.v$ in X if and only if

1. uwv is in $\mathcal{L}(X)$, and
2. v is a prefix of wv and u is a suffix of uw, and
3. the finite word uwv contains exactly two occurrences of the finite word uv.

If v is the empty word, we have $\mathcal{R}_X(u \cdot v) = \mathcal{R}_X(u)$, and if u is the empty word, then $\mathcal{R}(u \cdot v) = \mathcal{R}'_X(v)$, the set of left return words to v. When it will be clear from the context, we will use to $\mathcal{R}(u \cdot v)$ in place of $\mathcal{R}_X(u \cdot v)$.

Let $B_{u \cdot v}$ be a finite alphabet in bijection with $\mathcal{R}(u \cdot v)$ by $\theta_{u \cdot v} \colon B_{u \cdot v} \to \mathcal{R}(u \cdot v)$. The map $\theta_{u \cdot v}$ extends to a morphism from $B^*_{u \cdot v}$ to A^*, and the set $\theta_{u \cdot v}(B^*_{u \cdot v})$ consists of all concatenations of return words to $u \cdot v$.

The set $\mathcal{R}(u \cdot v)$ is a circular code and, more precisely, no word of $\mathcal{R}(u \cdot v)$ overlaps nontrivially a product of words of $\mathcal{R}(u \cdot v)$. In particular, the map $\theta_{u \cdot v} \colon B^*_{u \cdot v} \to A^*$ is one-to-one.

6.3.2 BV-Representation of Linearly Recurrent Shifts

The following result characterizes linearly recurrent shift spaces in terms of BV-representations. Recall that a dynamical system (X, T), endowed with the distance d, is *expansive* if there exists $\epsilon > 0$ such that for all pairs of points (x, y), $x \neq y$, there exists n with $d(T^n x, T^n y) \geq \epsilon$. We say that ϵ is a *constant of expansivity* of (X, T). It is easy to see that the shift spaces are expansive but that odometers are not.

Theorem 6.3.5 *An aperiodic shift space is linearly recurrent if and only if it has a BV-representation satisfying the following conditions.*

(i) *Its incidence matrices have positive entries and belong to a finite set of matrices.*

(ii) *For every $n \geq 1$, the morphism read on $E(n)$ is proper.*

Proof Let X be an aperiodic LR shift space. It suffices to construct a sequence of KR-partitions having the desired properties.

From Proposition 6.3.3 there exists an integer K such that for all words u occurring in some $x \in X$ and all $w \in \mathcal{R}(u)$, we have

$$\frac{|u|}{K} \leq |w| \leq K|u|.$$

Moreover, every word of $\mathcal{L}_n(X)$ occurs in every word of $\mathcal{L}_{(K+1)n-1}(X)$.

We set $\alpha = (K+1)^2$. Let $x = (x_n)_n$ be an element of X. For each non-negative integer n, we set $u_n = x_{-\alpha^n} \cdots x_{-2}x_{-1}$, $v_n = x_0x_1 \cdots x_{\alpha^n-1}$ and $\mathcal{R}_n = \mathcal{R}(u_n.v_n)$.

By the choice of K, every word of $\mathcal{R}_X(u_n \cdot v_n)$ appears in $u_{n+1}v_{n+1}$. Indeed, $\alpha = (K+1)^2$ implies $2\alpha^{n+1} = (K+1)^2 2\alpha^n$ and thus $|u_{n+1}v_{n+1}| = (K+1)^2|u_nv_n|$.

Now define for all $n \geq 0$, $\mathcal{P}(n) = \{S^j[u_n.wv_n] \mid w \in \mathcal{R}(u_n \cdot v_n), \ 0 \leq j < |w|\}$.

The verification that $(\mathcal{P}(n))_n$ is a sequence of KR-partitions having the desired properties is left to the reader.

Now let $(V, E, \leq)$ be a BV-representation of an aperiodic shift space satisfying (i) and (ii). Let $(M(n))$ be the associated sequence of incidence matrices satisfying (i) and let $h(n) = (h_v(n))_{v \in V(n)}$ be the vector of heights of the BV-representation. Since (X_E, T_E) is conjugate to a shift space, it is expansive. Let $\epsilon > 0$ be a constant of expansivity of (X_E, T_E). Let n_0 be a level of $(V, E, \leq)$ such that all cylinders $[e_1, \ldots, e_{n_0}]$ are included in a ball of radius $\epsilon/2$. For any vertex $v \in V(n_0)$ let h_v denote the number of paths from $V(0)$ to v. Now consider the alphabet

$$A = \{(v, j) \mid v \in V(n_0), \ 0 \leq j < h_v\},$$

the map $C \colon X_E \to A$ defined, for $e = (e_n)_n \in X_E$, by

$$C(e) = (r(e_{n_0}), j)$$

if $(e_n)_{1 \leq n \leq n_0}$ is the jth finite path in $(V, E, \leq)$ from $V(0)$ to $r(e_{n_0})$ with respect to the order on $(V, E, \leq)$, and finally define $\varphi \colon X_E \to A^{\mathbb{Z}}$ by

$$\varphi(e) = \big(C \circ T_E^n(e)\big)_{n \in \mathbb{Z}}.$$

By the choice of ϵ, the system (X_E, T_E) is isomorphic to (X, S) where $X = \varphi(X_E)$ and S is the shift on A. There remains to show that (X, S) is LR.

Let $K = \sup_n \max_{v \in V(n)} \sum_{v' \in V(n-1)} M(n)_{v,v'}$. Condition (i) implies that K is finite and, by (5.2),

$$\frac{h_v(n)}{h_{v'}(n)} \leq \frac{K \max_w h_w(n-1)}{\max_w h_w(n-1)} \leq K$$

for all n and $v, v' \in V(n)$. For all n, let $\tau_n \colon V(n)^* \to V(n-1)^*$ be the morphism read on $E(n)$.

Let $v \in V(n_0)$ and let w be a return word to $u = (v, 0)(v, 1) \cdots, (v, h_v - 1)$. It is a word of $\mathcal{L}(X)$ and $|u| = h_v$. By Proposition 5.3.5 one has $h_v = (M(n_0) \cdots M(2)M(1))_v = |\tau_1 \circ \tau_2 \circ \cdots \circ \tau_{n_0}(v)|$. The incidence matrices having positive entries, the vertex v occurs at least once in $\tau_{n_0+1}(v')$ for all $v' \in V(n_0 + 1)$. Hence one has

$$|w| \leq 2|\tau_1 \circ \cdots \circ \tau_{n_0} \circ \tau_{n_0+1}| \leq 2|\tau_1 \circ \cdots \circ \tau_{n_0}||\tau_{n_0+1}|$$
$$\leq 2K \max_{v' \in V(n_0)} h_{v'} \leq 2K^2 |u|. \tag{6.8}$$

We set $W = \{(v, 0)(v, 1) \cdots, (v, h_v - 1) \mid v \in V(n_0)\}$ and we define the morphism $\sigma \colon V(n_0) \to A^*$ by $\sigma(v) = (v, 0)(v, 1) \cdots, (v, h_v - 1)$.

The elements of X are concatenations of finite words belonging to W. Indeed, each time $e = (e_n) \in X_E$ is such that $r(T_E^k(e)_{n_0}) = v$, for some k with $(e_1, \ldots e_{n_0})$ a minimal path, then $\varphi(e)_{[k,k+h_v]} = (v, 0)(v, 1) \cdots, (v, h_v - 1)$ and, by (ii), the point $T_E^{k+h_v}(e)$ passes through a minimal path from $V(n_0)$ to $V(0)$.

The elements of X are also concatenations of finite words belonging to $\sigma \circ \tau_{n_0+1}(V(n_0 + 1))$, and more generally, of finite words belonging to $\sigma \circ \tau_{n_0+1} \circ \cdots \circ \tau_n(V(n))$ for all $n \geq n_0 + 1$. As for (6.8), we can prove that all return words to some elements of $\sigma \circ \tau_{n_0+1} \circ \cdots \circ \tau_n(V(n))$ satisfy the same inequality as K does not depend on n.

Now let u be any nonempty finite word appearing in some word of X and w be a return word to u. There exists n such that

$$\max_{v \in V(n)} |\sigma \circ \tau_{n_0+1} \circ \cdots \circ \tau_n(v)| \leq |u| < \max_{v \in V(n+1)} |\sigma \circ \tau_{n_0+1} \cdots \circ \tau_{n+1}(v)|.$$

Then u is a factor of some $\sigma \circ \tau_{n_0+1} \circ \cdots \circ \tau_{n+1}(vv')$, v and v' belonging to $V(n + 1)$. From Condition (i) and Condition (ii), we deduce that vv' is a factor of some $\tau_{n+2} \circ \tau_{n+3}(v'')$, $v'' \in V(n + 3)$. Then, the word u is a factor of $\sigma \circ \tau_{n_0+1} \circ \cdots \circ \tau_{n+3}(v'')$. Consequently,

$$|w| \leq 2K^2 |\sigma \circ \tau_{n_0+1} \circ \cdots \circ \tau_{n+3}(v'')|$$
$$\leq 2K^2 |\sigma \circ \tau_{n_0+1} \circ \cdots \circ \tau_n||\tau_{n+1}||\tau_{n+2}||\tau_{n+3}| \leq 2K^5 |u|$$

and (X, S) is LR.   ∎

The *rank* of a Bratteli diagram (V, E) is the maximal cardinality of the sets $V(n)$. It can be finite or infinite.

We say that a minimal Cantor system (X, T) has *topological rank* k if k is the smallest integer such that (X, T) has a BV-representation of rank k. We denote by rank(X) the topological rank of X.

Let us call *linearly recurrent* any Cantor dynamical system having a BV-representation satisfying conditions (i) and (ii) of Theorem 6.3.5. By Proposition 5.2.2, linearly recurrent systems are minimal. By definition, they have finite rank.

Actually, more can be said about these dynamical systems. Indeed, we have the following theorem that we quote without proof. It can be seen as an extension of Proposition 6.2.3. It uses the notion of equicontinuous system introduced in Section 6.1.

Theorem 6.3.6 (Downarowicz, Maass) *Let (X, T) be a minimal Cantor system with topological rank $k \geq 1$. Then, (X, T) is expansive if and only if $k \geq 2$. Otherwise it is equicontinuous.*

Another result concerning systems of finite topological rank is the following. It concerns the numbers of asymptotic components (defined in Exercise 1.19).

Theorem 6.3.7 *A minimal Cantor system of topological rank 2 has at most one nontrivial right-asymptotic (resp. left-asymptotic) component.*

The proof, which is not given here, uses Theorem 6.4.18 to be proved below in Section 6.4.4.

Example 6.3.8 The topological rank of the Thue–Morse shift is 3. Indeed, we have seen a BV-representation with three vertices at each level $n \geq 1$ (Exercise 6.19). By Theorem 6.3.7, the rank cannot be 2 since there are two nontrivial asymptotic components (Exercise 1.30).

The topological rank of a minimal shift space X satisfies the inequality

$$\text{rank}_{\mathbb{Q}}(K^0(X, S)) \leq \text{rank}(X). \tag{6.9}$$

Indeed, if X has a BV-representation with at most k vertices at each level, the group $K^0(X, S)$ is a subgroup of $\mathbb{Q}^k$ and thus $\text{rank}_{\mathbb{Q}}(K^0(X, S)) \leq k$.

For a primitive substitution shift, it also satisfies the inequality

$$\text{rank}(X) \leq r, \tag{6.10}$$

where r is the minimal value of $\text{Card}(\mathcal{R}_X(w))$ for some $w \in \mathcal{R}_X(w)$. Indeed, let σ be a primitive substitution and let $X = X(\sigma)$. For $w \in \mathcal{L}(X)$, let $\phi \colon B^* \to A^*$ be a coding morphism for $\mathcal{R}_X(w)$. Let $\tau \colon B^* \to B^*$ be the morphism such that $\phi \circ \tau = \sigma \circ \phi$ (as in Eq. 6.5)). Then τ is primitive eventually proper and aperiodic by Proposition 6.2.12. Some power of τ has a conjugate that is proper and thus X has a BV-representation of rank Card(B).

6.3.3 Unique Ergodicity of Linearly Recurrent Systems

We will prove the following property of linearly recurrent Cantor systems. It generalizes the property of unique ergodicity for primitive substitution shifts (Theorem 3.8.13).

Theorem 6.3.9 *Every linearly recurrent Cantor system is uniquely ergodic.*

Let (X, T) be a linearly recurrent system. By definition, there is a BV-representation $(V, E, \leq)$ satisfying the conditions of Theorem 6.3.5, and thus such that the incidence matrices have positive entries and belong to a finite set of matrices.

Let $\mathfrak{P}(n) = \{T^j B_k(n); 1 \leq k \leq t(n), 0 \leq j < h_k(n)\}, n \geq 0$ be the refining sequence of partitions associated with (X_E, V_E) as in Section 5.3.3. Let $M(n) = (m_{l,k}(n), 1 \leq l \leq t(n), 1 \leq k \leq t(n-1)), n \geq 1$, be the associated sequence of matrices.

The following lemma expresses a property of linearly recurrent systems, which could actually be taken for a definition (Exercise 6.29).

Lemma 6.3.10 *There is an integer $L \geq 1$ such that*

$$h_i(n) \leq L h_j(n-1) \tag{6.11}$$

for every $n \geq 1$, $1 \leq i \leq t(n)$ and $1 \leq j \leq t(n-1)$.

Proof Let $h(n)$ be the vector with components $h_i(n)$ for $1 \leq i \leq t(n)$. By (5.12), we have $h(n) = M(n)h(n-1)$.

Since there is a finite number of distinct matrices $M(n)$, we can define the integer $K = \max_{n,l} \sum_k m_{l,k}(n)$. Let i_0 be such that $h_{i_0}(n-1) = \max_i h_i(n-1)$. Then

$$h_{i_0}(n-1) \leq h_j(n) \leq K h_{i_0}(n-1),$$

where the first inequality holds because $M(n)$ is a positive integer matrix. This implies that for every j, k and $n \geq 1$, we have

$$\frac{1}{K} \leq \frac{h_j(n)}{h_k(n)} \leq K$$

and thus

$$h_j(n) \leq K h_{i_0}(n-1) \leq K^2 h_i(n-1).$$

Consequently (6.11) will hold with $L = K^2$. ∎

Note that for each T–invariant probability measure μ and for every $n \geq 1$ and $1 \leq k \leq t(n-1)$, we have, as already observed in (4.1),

$$\sum_{k=1}^{t(n)} h_k(n)\mu(B_k(n)) = 1, \tag{6.12}$$

and also, by (5.14),

$$\mu(n-1) = \mu(n)M(n) \tag{6.13}$$

where $\mu(n) = (\mu(B_k(n)))_{1 \leq k \leq t(n)}$.

Let us observe that the real numbers $\mu(B_k(n-1))$ should be positive. Indeed, from Eq. (6.12) there is at least one index k for which $\mu(B_k(n))$ is not zero. Hence, the coefficients $m_{l,k}(n)$ being positive, it is also the case for $\mu(B_k(n-1))$.

To prove that linearly recurrent systems are uniquely ergodic we need the following lemma, in which L is the integer given by Lemma 6.3.10.

Lemma 6.3.11 *Let μ be an invariant measure of (X, T). Then, for all $n \geq 0$ and $1 \leq k \leq t(n)$, we have*

$$\frac{1}{L} \leq h_k(n)\mu(B_k(n)) \leq 1.$$

Proof The inequality on the right is obvious from Eq. (6.12).

Fix k with $1 \leq k \leq t(n)$. By Eq. (6.13), since all the entries of $M(n+1)$ are positive, we get

$$\mu(B_k(n)) \geq \sum_{\ell=1}^{t(n+1)} \mu(B_l(n+1)).$$

By Lemma 6.3.10, for every l we have

$$h_k(n) \geq h_l(n+1)/L, \tag{6.14}$$

thus,

$$h_k(n)\mu(B_k(n)) \geq \sum_{\ell=1}^{t(n+1)} \frac{h_l(n+1)}{L}\mu(B_l(n+1)) = \frac{1}{L}.$$

∎

Proof of Theorem 6.3.9 Let (X, T) be a linearly recurrent system. Given a T–invariant probability measure μ, the vectors $\mu(n) = (\mu(B_k(n)))_{1 \leq k \leq t(n)}$ satisfy the relations

$$\mu(n-1) = \mu(n)M(n). \tag{6.15}$$

Conversely, let the nonnegative vectors $\mu(n)$ satisfy these conditions. Since the partitions $\mathfrak{P}(n)$ are clopen and span the topology of X, it is immediate to check that there exists a unique invariant probability measure μ on X such that $\mu(B_k(n)) = \mu(n)_k$ for every $n \geq 0$ and $k \in \{1, \ldots, t(n)\}$.

By Lemma 6.3.11 and Eq. (6.14), for the constant constant $\delta = 1/L^2$ one has

$$\mu(n)_i \geq \frac{1}{Lh_i(n)} \geq \frac{1}{L^2 h_k(n-1)} \geq \frac{1}{L^2}\mu(n-1)_k = \delta\mu(n-1)_k$$

for every $n \geq 1$ and $(i, k) \in \{1, \ldots, t(n)\} \times \{1, \ldots, t(n-1)\}$, and every invariant measure μ. Without loss of generality we can assume $\delta < 1/2$. Let μ, μ' be two invariant measures, and $\mu(n)_k, \mu'(n)_k$ be defined as above. We define

$$S_n = \max_k \frac{\mu'(n)_k}{\mu(n)_k} = \frac{\mu'(n)_i}{\mu(n)_i}, \quad s_n = \min_k \frac{\mu'(n)_k}{\mu(n)_k} = \frac{\mu'(n)_j}{\mu(n)_j} \quad \text{and } r_n = \frac{S_n}{s_n}$$

for some i, j. For every $k \in \{1, \ldots, t(n-1)\}$, we have

$$\mu'(n-1)_j = \sum_{l \neq j} \mu'(n)_l m_{l,k}(n) + \mu'(n)_j m_{j,k}(n)$$

$$\leq S_n \sum_{l \neq j} \mu(n)_l m_{l,k}(n) + s_n \mu(n)_j m_{j,k}(n)$$

$$= S_n \mu(n-1)_k - (S_n - s_n)\mu(n)_j m_{j,k}(n) \leq S_n \mu(n-1)_k$$

$$- (S_n - s_n)\mu(n)_j$$

$$\leq \mu(n-1)_k s_n \big(r_n(1 - \delta) + \delta\big).$$

And, in a similar way, for every $k \in \{1, \ldots, t(n-1)\}$, we have

$$\mu'(n-1)_k \geq \mu(n-1)_k s_n \big(\delta r_n + (1 - \delta)\big).$$

We deduce that

$$r_{n-1} \leq \phi(r_n) \text{ where } \phi(x) = \frac{(1-\delta)x + \delta}{\delta x + (1-\delta)}.$$

The function ϕ is increasing on $[0, +\infty)$ and tends to $(1-\delta)/\delta$ at $+\infty$. Writing $\phi^m = \phi \circ \cdots \circ \phi$ (m times), for every $n, m \geq 0$, we have $1 \leq r_n \leq \phi^m(r_{n+m}) \leq \phi^{m-1}((1-\delta)/\delta)$. Taking the limit with $m \to +\infty$, we get $r_n = 1$ and thus $\mu = \mu'$. ∎

Another proof of Theorem 6.3.9 is proposed in Exercise 6.30.

6.4 $\mathcal{S}$-adic Representations

We introduce now the notion of an $\mathcal{S}$-adic representation of a shift, which generalizes the representation of shifts as substitution shifts. We will replace the iteration of a morphism by the application of an arbitrary sequence of morphisms. The concept is, of course, close to that of a BV-representation but more flexible. For example, every stationary Bratteli diagram defines a substitution shift (with the substitution read on the diagram), but not conversely since, as we have seen, one has to use a conjugacy to obtain a stationary Bratteli diagram for a substitution shift (when the substitution is not proper). The same situation occurs for $\mathcal{S}$-adic representations.

6.4.1 Directive Sequence of Morphisms

For a morphism $\sigma : A^* \to B^*$, we denote as usual $|\sigma| = \max_{a \in A} |\sigma(a)|$ and $\langle \sigma \rangle = \min_{a \in A} |\sigma(a)|$.

Let $\mathcal{S}$ be a family of nonerasing morphisms. Let $(A_n)_{n \geq 0}$ be a sequence of finite alphabets and let $\tau = (\tau_n)_{n \geq 0}$ be a sequence of morphisms with $\tau_n : A_{n+1}^* \to A_n^*$ and $\tau_n \in \mathcal{S}$. We thus have an infinite sequence of morphisms,

$$\cdots \xrightarrow{\tau_2} A_2^* \xrightarrow{\tau_1} A_1^* \xrightarrow{\tau_0} A_0^*.$$

For $0 \leq n < N$, we define $\tau_{[n,N)} = \tau_n \circ \tau_{n+1} \circ \cdots \circ \tau_{N-1}$ and $\tau_{[n,N]} = \tau_n \circ \tau_{n+1} \circ \cdots \circ \tau_N$. For $n \geq 0$, the *language* $\mathcal{L}^{(n)}(\tau)$ *of level n* associated with τ is defined by

$$\mathcal{L}^{(n)}(\tau) = \{w \in A_n^* \mid w \text{ occurs in } \tau_{[n,N)}(a) \text{ for some } a \in A_N \text{ and } N > n\}.$$

Thus, $\mathcal{L}^{(n)}(\tau)$ is the set of factors of $\tau_n(\mathcal{L}^{(n+1)}(\tau))$ for $n \geq 0$ and

$$\cdots \xrightarrow{\tau_1} \mathcal{L}^{(1)}(\tau) \xrightarrow{\tau_0} \mathcal{L}^{(0)}(\tau).$$

More generally, $\mathcal{L}^{(n)}(\tau)$ is the set of factors of $\tau_{[n,n+m)}(\mathcal{L}^{(n+m)}(\tau))$.

The language $\mathcal{L}^{(n)}(\tau)$ defines a shift space $X^{(n)}(\tau)$ called the *shift generated by* $\mathcal{L}^{(n)}(\tau)$. More precisely, $X^{(n)}(\tau)$ is the set of points $x \in A_n^{\mathbb{Z}}$ such that $\mathcal{L}(x) \subseteq \mathcal{L}^{(n)}(\tau)$. We have then, since the τ_n are nonerasing, the sequence

$$\cdots \xrightarrow{\tau_1} X^{(1)}(\tau) \xrightarrow{\tau_0} X^{(0)}(\tau).$$

Note that it may happen that $\mathcal{L}(X^{(0)}(\tau))$ is strictly contained in $\mathcal{L}^{(0)}(\tau)$ or even empty (if $|\tau_{[0,n)}|$ is bounded).

We say that τ is a *directive sequence of morphisms* if

$$\mathcal{L}(X^{(0)}(\tau)) = \mathcal{L}^{(0)}(\tau).$$

Note first that τ is a directive sequence if and only if the language $\mathcal{L}^{(0)}(\tau)$ is extendable. Indeed, the condition is necessary. Conversely, if $\mathcal{L}^{(0)}(\tau)$ is extendable, the shift space Y such that $\mathcal{L}(Y) = \mathcal{L}^{(0)}(\tau)$ is equal to $X^{(0)}(\tau)$ and thus τ is directive sequence. Next, when τ is a directive sequence, then $|\tau_{[0,n)}|$ tends to infinity with n.

Thus, a constant sequence $\tau = (\sigma, \sigma, \ldots)$ is a directive sequence if and only if σ is a substitution.

When τ is a directive sequence of morphisms, we set $X(\tau) = X^{(0)}(\tau)$ and $\mathcal{L}(\tau) = \mathcal{L}^{(0)}(\tau)$. We call $(X(\tau), S)$ the *S-adic shift* with *directive sequence* τ. We also say that τ is an *S*-adic *representation* of $X = X(\tau)$.

As a first example, let us see that *S*-adic representations capture the notion of substitutive shift.

Example 6.4.1 Every substitutive shift $X(\sigma, \phi)$, where $\sigma \colon B^* \to B^*$ is a substitution and $\phi \colon B^* \to A^*$ is nonerasing, has an *S*-adic representation with $S = \{\sigma, \phi\}$. Indeed, $X(\sigma, \phi)$ has the *S*-adic representation $(\phi, \sigma, \sigma, \ldots)$.

As a second example, we recover the Arnoux–Rauzy shifts.

Example 6.4.2 Let s be a strict standard episturmian sequence with directive sequence $x = a_0 a_1 \cdots$ (see Section 1.5). The shift generated by s has the *S*-adic representation $\tau = (\tau_n)_{n \geq 0}$ where $\tau_n = L_{a_n}$ and where each $L_a \colon A^* \to A^*$ is the Rauzy automorphism defined for $b \in A$ by $L_a(b) = ab$ for $b \neq a$ and $L_a(a) = a$. Indeed, every word in $\mathcal{L}(s)$ is a factor of some $L_{a_0 \cdots a_{n-1}}(a_n)$ that is in $\mathcal{L}^0(\tau)$. Conversely, since s is strict, every letter appears infinitely often in x. This implies that all letters appear in $\mathrm{Pal}(a_n a_{n+1} \cdots)$ and thus that $L_{a_0 \cdots a_{n-1}}(a)$ appears in $s = L_{a_0 \cdots a_{n-1}}(\mathrm{Pal}(a_n a_{n+1} \cdots))$ for every $a \in A$. This shows that $\mathcal{L}^0(\tau) = \mathcal{L}(s)$.

As for BV-representations, we have the notion of *telescoping* of a directive sequence. Given a sequence of morphisms $\tau = (\tau_n)$ and a sequence (n_m) of integers with $n_0 = 0 < n_1 < n_2 < \cdots$, the telescoping of τ with respect to (n_m) is the sequence of morphisms $\tau' = (\tau_{[n_m, n_{m+1})})_{m \geq 0}$. If τ is a directive sequence, then τ' is a directive sequence. Moreover τ and τ' define the same shift $X(\tau) = X(\tau')$.

As for substitution shifts, in which fixed points play an important role, we have for directive sequences the notion of *limit point*. We only develop the analog of one-sided fixed points, although the two-sided version raises no additional difficulties. Let $\tau = (\tau_n)_{n \geq 0}$ be a directive sequence of morphisms with $\tau_n \colon A_{n+1}^* \to A_n^*$. For every $n \geq 0$, let $Y^{(n)}(\tau)$ denote the set of one-sided sequences $y \in A_n^{\mathbb{N}}$ such that $\mathcal{L}(y) \subset \mathcal{L}^{(n)}(\tau)$. A sequence $x \in Y^{(0)}(\tau)$ is called a limit point of τ if there is a sequence $(w^{(n)})$ of sequences $w^{(n)} \in Y^{(n)}(\tau)$ such that $w^{(n)} = \tau_n(w^{(n+1)})$ with $x = w^{(0)}$. Such a sequence is also called an *S-adic sequence*.

Example 6.4.3 Let σ be a substitution with fixed point $x = \sigma^\omega(a)$. Then x is a limit point of the directive sequence $(\sigma, \sigma, \ldots)$. Indeed, the constant sequence $w^{(n)} = \sigma^\omega(a)$ is such that $w^{(n+1)} = \sigma(w^{(n)})$ and $x = w^{(0)}$.

Observe that, although not every sequence is a substitutive sequence, every sequence has an S-adic representation, that is, is a limit point of a directive sequence of morphisms (Exercise 6.32).

6.4.2 Primitive Directive Sequences

We say that the sequence $\tau = (\tau_n)_{n \geq 0}$ of morphisms is *primitive* if, for every $n \geq 0$, there exists $N > n$ such that for all $a \in A_N$, the word $\tau_{[n,N)}(a)$ contains occurrences of all letters of A_n. Thus, the constant sequence $(\sigma, \sigma, \ldots)$ is primitive if and only if σ is a primitive morphism.

Set $M_{[n,N)} = M_{\tau_n} M_{\tau_{n+1}} \cdots M_{\tau_{N-1}}$ where M_{τ_n} is the composition matrix of τ_n. Then τ is primitive if for any $n \geq 1$, there exists $N > n$ such that $M_{[n,N)} > 0$.

A sequence of morphisms $\tau = (\tau_n)_{n \geq 0}$, with $\tau_n : A_{n+1}^* \to A_n^*$, has a *bottleneck* if $\mathrm{Card}(A_n) = 1$ for some $n \geq 0$. Observe that if τ is primitive without bottleneck, then $\langle \tau_{[n,N)} \rangle$ goes to infinity when N increases. Note that if τ has a bottleneck, then X_0 is periodic.

When τ is primitive, we can use alternative definitions of a limit point x.

(i) There is a sequence (a_n), $a_n \in A_n$, of letters such that x is the limit of $\tau_{[0,n)}(a_n^\omega)$.

(ii) There is a sequence (a_n), $a_n \in A_n$, of letters such that $\{x\} = \bigcap_{n \geq 0}[\tau_{[0,n)}(a_n)]$ where $[w]$ is the cylinder defined by w.

(See Exercise 6.31.)

The following result generalizes the fact that every growing morphism has a power with an admissible fixed point (Proposition 1.4.6).

Proposition 6.4.4 *Every primitive sequence of morphisms without bottleneck has a limit point.*

Proof Let $\tau = (\tau_n)$ be a primitive directive sequence. For $a \in A_{n+1}$, denote by $f_{[0,n)}(a)$ the first letter of $\tau_{[0,n)}(a)$. Since A_0 is finite, there is a letter a in the intersection of the decreasing sequence of sets $\cdots \subset f_{[0,n)}(A_n) \subset \cdots \subset f_{[0,1)}(A_1)$. By construction, there is a sequence (a_n) such that $a_0 = a$ and $a_n = f_n(a_{n+1})$. Set $u_n = \tau_{[0,n)}(a_n)$. By construction, every u_n is a prefix of u_{n+1}, and since τ is primitive without bottleneck, their lengths tend to infinity. Let $u^{(0)}$ be the unique one-sided sequence having all u_n as prefixes. By shifting the sequences (τ_n) and (a_n) (at the first step, replace $(\tau_n)_{n \geq 0}$ by $(\tau_n)_{n \geq 1}$ and so on), we define a sequence $u^{(k)}$ of sequences such that $u^{(k)} = \tau_k(u^{(k+1)})$. Thus, $u^{(0)}$ is a limit point of τ. ∎

The following proposition generalizes the fact that a primitive substitution shift is minimal (Proposition 1.4.16).

Proposition 6.4.5 *If τ is a primitive sequence of morphisms, the shift $X(\tau)$ is minimal. Moreover, if τ has no bottleneck, it is a directive sequence.*

Proof Let us show that $\mathcal{L}(\tau)$ is uniformly recurrent. For this, let $u \in \mathcal{L}(\tau)$. By definition of $\mathcal{L}(\tau)$, there is an $n > 1$ and a letter $a \in A_n$ such that u appears in $\tau_{[0,n)}(a)$. Since τ is primitive, there is an $N > n$ such that for all $b \in A_N$, the letter a appears in $\tau_{[n,N)}(b)$. Then u is a factor of every $\tau_{[0,N)}(b)$ for $b \in A_N$ and thus in every word of $\mathcal{L}(\tau)$ of length $2|\tau_{[0,N)}|$. This shows that $X(\tau)$ is uniformly recurrent.

Assume now that there is no bottleneck. Since $\langle \tau_{[0,n)} \rangle \to \infty$ with n, the language $\mathcal{L}^{(0)}(\tau)$ is extendable and thus τ is a directive sequence. ∎

Recall that a morphism $\sigma \colon A^* \to B^*$ is said to be left proper (resp. right proper) when there exists a letter $b \in B$ such that for all $a \in A$, the word $\sigma(a)$ starts with b (resp. ends with b). Thus, it is proper if it is both left and right proper.

We say that a sequence $\tau = (\tau_n)$ of morphisms is *left proper* (resp. *right proper*, resp. *proper*) whenever each morphism τ_n is left proper (resp. right proper, resp. proper).

We also say that a shift is a left proper (resp. right proper, resp. primitive) $\mathcal{S}$-adic shift if there exists a left proper (resp. right proper, resp. primitive) sequence of morphisms τ such that $X = X(\tau)$.

Let us give another way to define $X(\tau)$ when τ is primitive and proper. For a non-erasing morphism $\sigma \colon A^* \to B^*$, let $\Omega(\sigma)$ be the closure of $\cup_{k \in \mathbb{Z}} S^k \sigma(A^{\mathbb{Z}})$.

Lemma 6.4.6 *Let $\tau = (\tau_n \colon A_{n+1}^* \to A_n^*)_{n \geq 0}$ be a primitive and proper sequence of morphisms without bottleneck. Then,*

$$X(\tau) = \cap_{n \geq 0} \Omega(\tau_{[0,n]}).$$

Proof It is equivalent to prove that

$$\mathcal{L}(\tau) = \cap_{n \geq 1} \text{Fac}(\tau_{[0,n)}(A_n^*)),$$

where Fac(L) denotes the set of factors of the words in L. Consider first $w \in \mathcal{L}(\tau)$. By definition of $\mathcal{L}(\tau)$, there is an $n \geq 2$ and a letter $a \in A_n$ such that w appears in $\tau_{[0,n)}(a)$. Since τ is primitive, for all $N > n$ large enough and every letter $b \in A_N$, the letter a is a factor of $\tau_{[n,N)}(b)$. This proves that $w \in \text{Fac}(\tau_{[0,N)}(A_N))$ for arbitrary large N and thus that $w \in \cap_{n \geq 1} \text{Fac}(\tau_{[0,n)}(A_n^*))$.

Conversely, let $\ell_n \in A_n$ (resp. $r_n \in A_n$) be the first letter (resp. last letter) of all $\tau_n(a)$ for $a \in A_{n+1}$. Let $w \in \cap_{n \geq 1} \text{Fac}(\tau_{[0,n)}(A_n^*))$. Since τ is primitive we may assume, telescoping if necessary the sequence τ_n, that $\langle \tau_n \rangle \geq 2$. For each $n \geq 1$, there is some $w_n \in A_n^*$ such that w is a factor of $\tau_{[0,n)}(w_n)$ and we can choose w_n of minimal length. Then $|w| \geq 2^n(|w_n| - 2)$. Thus, there is an $n \geq 1$ such that w_n has length at most 2. If $|w_n| = 1$, we obtain the conclusion $w \in \mathcal{L}(\tau)$. Assume that w_n has length 2. If $w_n \neq r_n \ell_n$, then w_n is a factor of some $\tau_{n+1}(a)$ for $a \in A_{n+1}$ and we are done. Otherwise, consider a letter $c \in A_{n+2}$ and two consecutive letters d, e of $\tau_{n+1}(c)$. Then $\tau_n \circ \tau_{n+1}(c)$ has a factor $r_n \ell_n$ and thus, we can choose $w_{n+2} = c$. ∎

Observe that the hypotheses that τ is both primitive and proper cannot be dropped in the previous statement. Without these hypotheses, the inclusion $X(\tau) \subset \cap_{n \in \mathbb{N}} \Omega(\tau_{[0,n]})$ still holds (under the mild assumption that $A_n \subset \text{Fac}(\tau_n(A_{n+1}))$), but not the reverse inclusion, as shown by the following examples.

Example 6.4.7 Take for the directive sequence τ the constant sequence equal to τ, defined by $\tau(0) = 0010$ and $\tau(1) = 1$ (this is the Chacon binary morphism). The directive sequence τ is not primitive, and $\cdots 111 \cdots$ belongs to $\cap_{n \in \mathbb{N}} \Omega(\tau_{[0,n]})$ but not to $X(\tau)$.

Example 6.4.8 In the case of the non-proper constant sequence given by τ with $\tau(0) = 0100$ and $\tau(1) = 101$, the fixed point $\tau^\omega(1 \cdot 1)$ belongs to $\cap_{n \in \mathbb{N}} \Omega(\tau_{[0,n]})$, but not to $X(\tau)$, since 11 appears in no element of $X(\tau)$.

With a left proper morphism $\sigma : A^* \to B^*$ such that $b \in B$ is the first letter of all images $\sigma(a)$, $a \in A$, we associate the right proper morphism $\overline{\sigma} : A^* \to B^*$ defined by $b\overline{\sigma}(a) = \sigma(a)b$ for all $a \in A$.

The following statement shows how to replace a left proper primitive directive sequence by one which is proper (a dual result holds, of course, for a right proper sequence).

Lemma 6.4.9 *Let X be an S-adic shift generated by the primitive and left proper directive sequence $\tau = (\tau_n)_{n \geq 0}$ without bottleneck. Then X is also generated by the primitive and proper directive sequence $\tilde{\tau} = (\tilde{\tau}_n)_{n \geq 1}$, where for all n, $\tilde{\tau}_n = \tau_{2n-1}\overline{\tau}_{2n}$.*

In particular, if τ is unimodular, then so is $\tilde{\tau}$.

Proof Each morphism $\tilde{\tau}_n$ is trivially proper. For all $x \in A_{n+1}^{\mathbb{Z}}$, one has $\overline{\tau}_n(x) = S\tau_n(x)$. By Lemma 6.4.6 this proves that $\tilde{\tau}$ is an S-adic representation of X. Finally, it is clear that the unimodularity of τ is preserved in this process. ∎

The next proposition provides a general construction to get a primitive proper S-adic representation of any aperiodic minimal shift space X.

Proposition 6.4.10 *An aperiodic shift X is minimal if and only if it has a primitive proper S-adic representation for some (possibly infinite) set S of morphisms.*

We first prove the following lemma.

Lemma 6.4.11 *Let X be an aperiodic minimal shift on A. Let $(u_n)_{n \geq 0}$ be a sequence of words in $\mathcal{L}(X)$, with $u_0 = \varepsilon$, such that u_n is, for every $n \geq 0$, a proper suffix of u_{n+1} and let $\alpha_n \colon A_n^* \to A^*$ be, for $n \geq 1$, a coding morphism for $\mathcal{R}_X(u_n)$. The sequence $\tau = (\tau_n)$ of morphisms such that $\alpha_n \circ \tau_n = \alpha_{n+1}$ for every $n \geq 0$ is a primitive right proper S-adic representation of X without bottleneck.*

Proof Since u_n is a proper suffix of u_{n+1}, and since X is aperiodic, we may assume that each word in $\mathcal{R}_X(u_n)$ ends with u_n. Since $\mathcal{R}_X(u_{n+1})$ is contained in $\mathcal{R}_X(u_n)^*$, we have $\alpha_{n+1}(A_{n+1}^*) \subset \alpha_n(A_n^*)$ and thus there exists a morphism $\tau_n \colon A_{n+1}^* \to A_n^*$ such that $\alpha_{n+1} = \alpha_n \circ \tau_n$ for $n \geq 0$. Since $u_0 = \varepsilon$, we have $\alpha_0 = \mathrm{id}$ and thus $\alpha_n = \tau_{[0,n)}$. The sequence (τ_n) is an S-adic representation of X. Indeed, every $w \in \mathcal{L}(X)$ is a factor of some u_n, whence of some $\alpha_n(A_n)$ and conversely. We cannot have $\mathrm{Card}(A_n) = 1$ since otherwise X would be periodic. Thus, there is no bottleneck. For every $n > 1$, since X is minimal, there is an $N > n$ such that every word $\tau_{[0,n)}(b)$ for $b \in A_n$ appears in every $\alpha_N(a)$ for $a \in A_N$. This implies that every $b \in A_n$ appears in every $\tau_{[n,N)}(a)$ for $a \in A_N$. Thus, τ is primitive. Finally, every τ_n is clearly right proper. ∎

There is a dual version of this lemma obtained by considering a sequence of words (u_n) such that u_n is a proper prefix of u_{n+1} and replacing $\mathcal{R}_X(u_n)$ by $\mathcal{R}'_X(u_n)$.

Proof of Proposition 6.4.10 The condition is sufficient by Proposition 6.4.5.

Let us prove the converse. Let (u_n) be a sequence of words as in Lemma 6.4.11. In this way, τ is a primitive right proper $\mathcal{S}$-adic representation of X. By Lemma 6.4.9, we can modify the morphisms τ_n to make them proper. ∎

Even for minimal shifts with at at most linear factor complexity, the set of morphisms $\mathcal{S} = \{\tau_n \mid n \geq 0\}$ considered in Proposition 6.4.10 is usually infinite, and the sequence of alphabets $(A_n)_{n\geq 0}$ is usually of unbounded cardinality (see Section 6.8 for a reference).

6.4.3 Unimodular $\mathcal{S}$-adic Shifts

We also say that the sequence $\tau = (\tau_n)$ of morphisms is *unimodular* whenever, for all $n \geq 1$, $A_n = A$ and the matrix $M(\tau_n)$ is unimodular, that is, has determinant of absolute value 1. We say that a shift is a *unimodular* (resp. *proper*, resp. *primitive*) $\mathcal{S}$-adic shift if there exists a unimodular (resp. proper, resp. primitive) directive sequence of morphisms τ such that $X = X(\tau)$.

Lemma 6.4.12 *All primitive unimodular proper $\mathcal{S}$-adic shifts on an alphabet A with more than one letter are aperiodic.*

Proof Let τ be a primitive unimodular proper directive sequence on the alphabet A of cardinality $d \geq 2$. Suppose that the shift $X(\tau)$ is periodic, that is, $X(\tau) = \{x, Sx, \ldots, S^{p-1}x\}$ with p minimal.

Since τ is primitive, there is some $n \geq 1$ such that $|\tau_{[0,n)}(a)| \geq p$ for all $a \in A$.

Let $b \in A$ and set $\tau_n(b) = b_0b_1 \cdots b_k$. Since the directive sequence τ is proper, $b_0b_1 \cdots b_kb_0$ is also a word in $\mathcal{L}(\tau_n)$. Then $w = \tau_{[0,n)}(b_0b_1 \cdots b_kb_0)$ is a word that has both period p (as a factor of x) and period $|\tau_{[0,n)}(b)|$. By Fine–Wilf Theorem (Exercise 1.14), we have $|\tau_{[0,n)}(b)| \equiv 0 \bmod p$. But then every column of the matrix $M(\tau_{[0,n)})$ has a sum divisible by p, and thus its determinant is a multiple of p, which contradicts the unimodularity of τ. ∎

6.4.4 Recognizability and Unimodular $\mathcal{S}$-adic Shifts

We have seen in Section 1.4 the definition of recognizability of morphisms. Let us recall this definition here. Let $\varphi \colon A^* \to B^*$ be a nonerasing morphism. Let X with $X \subset A^{\mathbb{Z}}$ be a shift space and let Y be the closure of $\varphi(X)$ under the shift. Every $y \in Y$ has a representation as $y = S^k\varphi(x)$ with $x \in X$ and $0 \leq k < |\varphi(x_0)|$. We say that φ is *recognizable* on the shift X for the point y if y has only one such representation.

We say that φ is *recognizable* on X if it is recognizable on X for every point $y \in Y$. We also say that φ is *recognizable* on X for aperiodic points if φ is recognizable on X for every aperiodic point $y \in Y$. Finally, we say that φ is *recognizable for aperiodic points* if it is recognizable on the full shift for aperiodic points.

Recall from Section 1.4.4 that recognizability can expressed using the system $X^\varphi = \{(x, i) \mid x \in X, 0 \leq i < |\varphi(x_0)|\}$ and the morphism $\widehat{\varphi} \colon X^\varphi \to Y$ defined by $\widehat{\varphi}(x, i) = S^i \varphi(x)$. Then φ is recognizable on X for y if $\widehat{\varphi}^{-1}(y)$ has only one element.

Example 6.4.13 The Thue–Morse morphism $\varphi \colon a \mapsto ab, b \mapsto ba$ is recognizable for aperiodic points. Indeed, any sequence containing aa or bb has at most one factorization in $\{ab, ba\}$.

A nonerasing morphism $\varphi \colon A^* \to B^*$ is *left marked* if every word $\varphi(a)$ for $a \in A$ begins with a distinct letter. In particular, φ is injective on A and $\varphi(A)$ is a prefix code.

Proposition 6.4.14 *If* $\varphi \colon A^* \to B^*$ *is left marked, then it is recognizable for aperiodic points.*

The proof is left as Exercise 6.35.

Example 6.4.15 The Thue–Morse morphism $\sigma \colon a \to ab, b \to ba$ is left marked. Thus, it is recognizable for aperiodic points (as we have already seen in Example 6.4.13).

Proposition 6.4.16 *Let* $\sigma \colon A^* \to B^*$ *and* $\tau \colon B^* \to C^*$ *be morphisms. Let* $X \subset A^{\mathbb{Z}}$ *be a shift space and let* Y *be the subshift generated by* $\sigma(X)$. *Then*

1. *$\tau \circ \sigma$ is recognizable on X if and only if σ is recognizable on X and τ is recognizable on Y.*
2. *If σ is recognizable on X for aperiodic points and τ is recognizable on Y for aperiodic points, then $\tau \circ \sigma$ is recognizable on X for aperiodic points.*
3. *If $\tau \circ \sigma$ is recognizable on X for aperiodic points, then τ is recognizable on Y for aperiodic points.*

Proof Let Z be the subshift of $C^{\mathbb{Z}}$ generated by $\tau(Y)$. Set $\rho = \tau \circ \sigma$. We have $\hat{\rho} = \hat{\tau} \circ \alpha$ (see Figure 6.14) where $\alpha \colon X^\rho \to Y^\tau$ is the following map. For each $(x, k) \in X^\rho$, there is a unique pair (i, j) such that $k = |\tau(b_0 \cdots b_{i-1})| + j$

Figure 6.14 The map α.

with $j < |\tau(b_i)|$ and $\sigma(x_0) = b_0 \cdots b_{|\sigma(x_0)|}$. Set $\alpha(x, k) = (\hat{\sigma}(x, i), j)$. Then

$$\hat{\tau} \circ \alpha(x, k) = \hat{\tau}(\hat{\sigma}(x, i), j)$$
$$= S^j \tau(S^i \sigma(x)) = S^{j + |\tau(b_0 \cdots b_{i-1})|} \tau \circ \sigma(x) = S^k \tau \circ \sigma(x)$$
$$= \hat{\rho}(x, k).$$

1. It is clear that α is injective if and only if $\hat{\sigma}$ is injective. Since α is surjective, it follows that $\hat{\rho}$ is injective if and only if $\hat{\sigma}$ and $\hat{\tau}$ are injective. This proves the first assertion.

Assertions 2 and 3 follow easily since $\hat{\tau}$ sends periodic points to periodic points. ∎

Recall (see Exercise 1.37) that a morphism $\sigma \colon A^* \to C^*$ is called *indecomposable* if, for every decomposition $\sigma = \alpha \circ \beta$ with $\alpha \colon B^* \to C^*$ and $\beta \colon A^* \to B^*$, one has $\mathrm{Card}(B) \geq \mathrm{Card}(A)$. In particular, an indecomposable morphism is nonerasing and one has also $\mathrm{Card}(C) \geq \mathrm{Card}(A)$.

A sufficient condition for a morphism to be indecomposable can be formulated in terms of its incidence matrix $M(\sigma)$. If the rank of $M(\sigma)$ is equal to $\mathrm{Card}(A)$, then σ is indecomposable. Indeed, if $\sigma = \alpha \circ \beta$ with $\beta \colon A^* \to B^*$ and $\alpha \colon B^* \to C^*$, then

$$M(\sigma) = M(\beta)M(\alpha).$$

If $\mathrm{rank}(M(\sigma)) = \mathrm{Card}(A)$, then $\mathrm{Card}(A) = \mathrm{rank}(M(\sigma)) \leq \mathrm{rank}(M(\alpha)) \leq \mathrm{Card}(B)$. Thus, σ is indecomposable.

Proposition 6.4.17 *An indecomposable morphism $\sigma \colon A^* \to C^*$ is injective on $A^{\mathbb{N}}$. More precisely, if $\sigma \colon A^* \to C^*$ is not injective on $A^{\mathbb{N}}$, there is a decomposition $\sigma = \alpha \circ \beta$ with $\alpha \colon B^* \to C^*$ and $\beta \colon A^* \to B^*$ such that α is injective on $B^{\mathbb{N}}$, $\mathrm{Card}(B) < \mathrm{Card}(A)$ and every $b \in B$ appears as the first letter of $\beta(a)$ for some $a \in A$.*

The proof is Exercise 1.37.

We will now prove the following result. It holds in particular when $M(\sigma)$ is unimodular.

Theorem 6.4.18 *An indecomposable morphism is recognizable for aperiodic points.*

Proof Let $\sigma : A^* \to B^*$ be an indecomposable morphism. We use an induction on $\ell(\sigma) = \sum_{a \in A}(|\sigma(a)| - 1)$. Since σ is indecomposable, it is nonerasing and the minimal possible value of $\ell(\sigma)$ is 0. In this case, σ is a bijection from A to B and thus it is recognizable.

Assume now that σ is not recognizable for aperiodic points. Thus, there exist $x, x' \in A^{\mathbb{N}}$ and w with $0 < |w| < |\sigma(x_0)|$ such that $\sigma(x) = w\sigma(x')$ for some proper suffix w of $\sigma(a')$. Set $\sigma(a') = vw$. We can then write $\sigma = \sigma_1 \circ \tau_1$ with $\tau_1 : A^* \to A_1^*$ and $\sigma_1 : A_1^* \to B^*$ and $A_1 = A \cup \{a''\}$ where a'' is a new letter. We have $\tau_1(a') = a'a''$ and $\tau_1(a) = a$ otherwise. Next, $\sigma_1(a') = v$, $\sigma_1(a'') = w$ and $\sigma_1(a) = \sigma(a)$ otherwise. Since $\ell(\tau_1) > 0$, we have $\ell(\sigma_1) < \ell(\sigma)$ by Proposition 1.4.1.

Since τ_1 is injective on $A^{\mathbb{N}}$, σ_1 is not injective on $A_1^{\mathbb{N}}$. By Proposition 6.4.17, we can write $\sigma_1 = \sigma_2 \circ \tau_2$ with $\sigma_2 : A_2^* \to B^*$ and $\tau_2 : A_1^* \to A_2^*$ for some alphabet A_2 such that $\mathrm{Card}(A_2) < \mathrm{Card}(A_1)$ and that every letter $c \in A_2$ appears as the first letter of some $\tau_2(a)$ for $a \in A_1$. Then, by Proposition 1.4.1, we have

$$\ell(\sigma_1) \geq \ell(\sigma_2) + \ell(\tau_2). \tag{6.16}$$

If $\mathrm{Card}(A_2) < \mathrm{Card}(A)$, then σ is decomposable and we are done. Otherwise, we have $\mathrm{Card}(A_2) = \mathrm{Card}(A)$. We may also assume that σ_2 and $\tau_2 \circ \tau_1$ are indecomposable since otherwise σ is decomposable.

Since σ_2 is indecomposable and since $\ell(\sigma_2) \leq \ell(\sigma_1) < \ell(\sigma)$, by the induction hypothesis, σ_2 is recognizable for aperiodic points.

Since $\sigma = \sigma_2 \circ \tau_2 \circ \tau_1$, we have also

$$\ell(\sigma) \geq \ell(\sigma_2) + \ell(\tau_2 \circ \tau_1). \tag{6.17}$$

Thus, if $\ell(\sigma_2) > 0$, the inequality $\ell(\tau_2 \circ \tau_1) < \ell(\sigma)$ holds. Since $\tau_2 \circ \tau_1$ is indecomposable, we obtain that $\tau_2 \circ \tau_1$ is recognizable for aperiodic points by induction hypothesis. Thus, by Proposition 6.4.16, σ is recognizable for aperiodic points.

Let us finally assume that $\ell(\sigma_2) = 0$. Then σ is left marked because σ_2 is a bijection from A_2 onto B and every letter of A_2 appears as initial letter of $\tau_2(a)$ for some $a \in A_1$. We obtain the conclusion by Proposition 6.4.14. ∎

We will use the following corollary.

Corollary 6.4.19 *Let $\sigma : A^* \to B^*$ be a morphism such that $M(\sigma)$ has rank $\mathrm{Card}(A)$. Then σ is recognizable for aperiodic points.*

This follows directly from Theorem 6.4.18. Indeed, a morphism $\sigma : A^* \to B^*$ such that $M(\sigma)$ has rank $\text{Card}(A)$ is indecomposable.

Example 6.4.20 The matrix of the morphism $\sigma : a \to ab, b \to aa$ is

$$M(\sigma) = \begin{bmatrix} 1 & 2 \\ 1 & 0 \end{bmatrix},$$

which has rank 2. The morphism is not recognizable for a^∞ but for every other point y, the occurrence of a letter b is enough to determine the representation of y as $y = S^k \sigma(x)$ with $0 \le k < 2$. For example, $y_0 = b$ forces $k = 1$ and $x_0 = a$. All the other values of x_n are then determined.

We use Corollary 6.4.19 to prove the following.

Proposition 6.4.21 *Let X be an S-adic shift with unimodular and proper directive sequence of morphisms τ. Then X is aperiodic and minimal if and only if τ is primitive.*

Proof Recall that any S-adic shift with a primitive directive sequence is minimal (Proposition 6.4.5) and that aperiodicity is proved in Lemma 6.4.12.

We only have to show that the condition is necessary. We assume that X is aperiodic and minimal. For all $n \ge 1$, $X^{(n)}(\tau)$ is trivially aperiodic. Let us show that it is minimal.

Assume by contradiction that for some $n \ge 1$, the shift $X^{(n)}(\tau)$ is minimal, but not $X^{(n+1)}(\tau)$. There exist $u \in \mathcal{L}(X^{(n+1)}(\tau))$ and $x \in X^{(n+1)}(\tau)$ such that u does not occur in x. Since τ is unimodular, by Corollary 6.4.19, the morphism τ_n is recognizable for aperiodic points and thus $\{\tau_n([v]) \mid v \in \mathcal{L}(X^{(n+1)}(\tau)), |v| = |u|\}$ is a finite clopen partition of $\tau_n(X^{(n+1)}(\tau))$. Thus, considering $y = \tau_n(x)$, by minimality of $X^{(n)}(\tau)$, there exists $k \ge 0$ such that $S^k y$ is in $\tau_n([u])$. Take $z \in [u]$ such that $S^k y = \tau_n(z)$. Since y is aperiodic and since we also have $S^k y = S^{k'} \tau_n(S^\ell x)$ for some $\ell \in \mathbb{N}$ and $0 \le k' < |\tau_n(x_\ell)|$, we obtain that $\tau_n(z) = S^{k'} \tau_n(S^\ell x)$ with $z \in [u]$, $S^\ell x \notin [u]$ and $0 \le k' < |\tau_n(x_\ell)|$. This contradicts the fact that τ_n is recognizable for aperiodic points.

We now show that $\lim_{n \to +\infty} \langle \tau_{[0,n)} \rangle = +\infty$. We again proceed by contradiction, assuming that $(\langle \tau_{[0,n)} \rangle)_{n \ge 1}$ is bounded. Then there exists $N > 0$ and a sequence $(a_n)_{n \ge N}$ of letters in A such that for all $n \ge N$, $\tau_n(a_{n+1}) = a_n$. We claim that there are arbitrary long words of the form a_N^k in $\mathcal{L}(X^{(N)}(\tau))$, which contradicts the fact that $X^{(N)}(\tau)$ is minimal and aperiodic. Since τ is proper, for all $n \ge N$ and all $b \in A$, $\tau_n(b)$ starts and ends with a_n. Since $|\tau_{[0,n)}|$ goes to infinity, there exists a sequence $(b_n)_{n \ge N}$ of letters in A such that $|\tau_{[N,n)}(b_n)|$

goes to infinity and for all $n \geq N$, b_n occurs in $\tau_n(b_{n+1})$. This implies that there exists $M \geq N$ such that for all $n \geq M$, $b_n \neq a_n$ and, consequently, that $\tau_n(b_{n+1}) = a_n u_n$ for some word u_n containing b_n. It is then easily seen that, for all $k \geq 1$, a_M^k is a prefix of $\tau_{[M,M+k)}(b_{M+k})$, which proves the claim.

We finally show that τ is primitive. If not, there exist $N \geq 1$ and a sequence $(a_n)_{n \geq N}$ of letters in A such that for all $n > N$, a_N does not occur in $\tau_{[N,n)}(a_n)$. Since $(|\tau_{[N,n)}(a_n)|)_n$ goes to infinity, this shows that there are arbitrarily long words in $\mathcal{L}(X^{(N)}(\tau))$ in which a_N does not occur. We conclude that $X^{(N)}(\tau)$ is not minimal, a contradiction.   ∎

We are now ready to describe a BV-representation for unimodular proper $\mathcal{S}$-adic shifts. The Bratteli diagram *associated* with a sequence $\tau = (\tau_n)_{n \geq 0}$ of morphisms is the diagram with simple hat $(V, E, \leq)$ such that τ_n is the morphism read on $E(n+2)$ for $n \geq 0$. The following statement generalizes Proposition 6.2.15.

Theorem 6.4.22 *Let X be an $\mathcal{S}$-adic shift with a primitive unimodular and proper directive sequence of morphisms τ. Then the Bratteli diagram associated with τ is a BV-representation of X.*

Proof Since τ is primitive, the shift X is aperiodic and minimal. Thus, by Corollary 6.4.19, every morphism τ_n is recognizable on $X^{(n+1)}(\tau)$. Therefore the sequence $(\mathfrak{P}(n))$ with

$$\mathfrak{P}(n) = \{S^k \tau_{[0,n]}([a]) \mid 0 \leq k < |\tau_{[0,n]}(a)|, a \in A\} \qquad (6.18)$$

is a nested sequence of partition in towers. It satisfies (KR1) because τ is proper. Indeed, since τ is proper, there are letters (ℓ_n, r_n) such that $\tau_n(a)$ belongs to $\ell_n A^* r_n$ for all $n \geq 0$. By telescoping the sequence τ, we may assume that $\ell_n = \ell$ and $r_n = r$ for all $n \geq 1$. The intersection of the bases of the partitions $\mathfrak{P}(n)$ is the unique element of $\cap_{n \geq 1} \tau_{[0,n)}([r \cdot \ell])$. Finally, condition (KR3) is also satisfied for the same reason as in the case of a proper substitution (see Lemma 6.2.4).

The Bratteli diagram associated with the sequence $(\mathfrak{P}(n))$ is clearly the diagram associated with τ.   ∎

Example 6.4.23 We have obtained in Example 6.2.16 the Fibonacci shift as the substitutive shift $X = X(\tau, \phi)$ where τ, ϕ are the proper and unimodular morphisms $\tau: u \mapsto uv, v \mapsto uvv$ and $\phi: u \mapsto aba, v \mapsto ababa$. Thus, $(\phi, \tau, \tau, \ldots)$ is a unimodular proper $\mathcal{S}$-adic representation of the Fibonacci shift. The coreponding BV-representation is represented in Figure 6.15

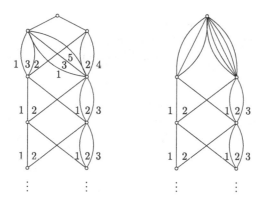

Figure 6.15 A BV-representation of the Fibonacci shift as a unimodular proper $\mathcal{S}$-adic shift.

on the left. The BV-representation obtained in Example 6.2.16 and shown in Figure 6.15 on the right is a telescoping of the one on the left.

6.5 Dimension Groups of Unimodular Shifts

In this section we first prove a key result of this chapter, namely Theorem 6.5.1, which states that, for primitive unimodular proper $\mathcal{S}$-adic shifts, the group $K^0(X, S)$ is generated, as an additive group, by the classes of the characteristic functions of letter cylinders. We then deduce a simple expression for the dimension group of primitive unimodular proper $\mathcal{S}$-adic subshifts.

6.5.1 From Letters to Factors

We recall that χ_U stands for the characteristic function of the set U.

Theorem 6.5.1 *Let X be a primitive unimodular proper $\mathcal{S}$-adic shift. Any function $f \in C(X, \mathbb{Z})$ is cohomologous to some integer linear combination of the form $\sum_{a \in A} \alpha_a \chi_{[a]} \in C(X, \mathbb{Z})$. Moreover, the classes $[\chi_{[a]}]$, $a \in A$, are $\mathbb{Q}$-independent.*

Proof By Theorem 6.4.22, X has a BV-representation $(V, E, \leq)$ such that τ_n is the morphism read on $E(n+2)$ for every $n \geq 0$. By Theorem 5.3.6, this implies that $K^0(X, T)$ is the dimension group of $(V, E, \leq)$, that is, the direct limit

$$\mathbb{Z}^A \overset{M(0)}{\to} \mathbb{Z}^A \overset{M(1)}{\to} \mathbb{Z}^A \ldots, \tag{6.19}$$

where $M(n)$ is the incidence matrix of τ_n. Since the matrices $M(n)$ are unimodular, by Proposition 4.4.4, the dimension group of $(V, E, \leq)$ is $\mathbb{Z}^A$. The isomorphism from $K^0(X, S)$ onto $\mathrm{DG}(V, E)$ sends the class of $\chi_{[a]}$ to the unit vector $e_a \in \mathbb{Z}^A$. This proves the statement. ∎

We now derive two corollaries from Theorem 6.5.1 dealing respectively with invariant measures and with the image subgroup. We denote by $\mathcal{M}(X, S)$ the set of invariant probability measures on (X, S).

Corollary 6.5.2 *Let X be a primitive unimodular proper S-adic shift over the alphabet A with $\mathrm{Card}(A) \geq 2$. There are at most $\mathrm{Card}(A) - 1$ ergodic measures on X. If $\mu, \mu' \in \mathcal{M}(X, S)$ coincide on the letters, then they are equal, that is, if $\mu([a]) = \mu'([a])$ for all a in A, then $\mu(U) = \mu'(U)$, for any clopen subset U of X.*

The first assertion results from Proposition 3.9.5.

Corollary 6.5.3 *Let X be a primitive unimodular proper S-adic shift over the alphabet A. The image subgroup of X satisfies*

$$I(X, S) = \bigcap_{\mu \in \mathcal{M}(X,S)} \left\{ \sum_{a \in A} \mathbb{Z}\mu([a]) \right\}.$$

6.5.2 An Explicit Description of the Dimension Group

Theorem 6.5.1 allows us to give a precise description of the dimension group of primitive unimodular proper S-adic shifts.

Theorem 6.5.4 *Let X be a primitive unimodular proper S-adic shift over the alphabet A. The linear map $\Phi \colon H(X, S) \to \mathbb{Z}^A$ defined by $\Phi([\chi_{[a]}]) = e_a$, where $\{e_a \mid a \in A\}$ is the canonical base of $\mathbb{Z}^A$, defines an isomorphism of dimension groups from $K^0(X, S)$ onto*

$$\left(\mathbb{Z}^A, \{x \in \mathbb{Z}^A \mid \langle x, \mu \rangle > 0 \text{ for all } \mu \in \mathcal{M}(X, S)\} \cup \{\mathbf{0}\}, \mathbf{1} \right), \qquad (6.20)$$

where the entries of $\mathbf{1}$ are equal to 1 and $\boldsymbol{\mu}$ is the vector $(\mu([a]))_{a \in A}$.

Proof By Theorem 6.5.1, the map Φ is an isomorphism of ordered groups from $K^0(X, S)$ onto the direct limit (6.19). The positive cone $H^+(X, S, \mathbb{Z})$ is the set of classes $[f]$ of $f \in C(X, \mathbb{Z})$ such that $p([f]) > 0$ for every state p of $K^0(X, S)$. By Theorem 3.9.3, the states of $K^0(X, S)$ are of the form

$$[f] \mapsto \int f d\mu$$

for some invariant probability measure μ. This gives the conclusion. ∎

Remark 6.5.5 We cannot remove the hypothesis of being left or right proper in Theorem 6.5.4. Consider indeed the subshift X defined by the primitive unimodular non-proper substitution τ defined over $\{a, b\}^*$ as $\tau: a \mapsto aab, b \mapsto ba$. The dimension group of X is isomorphic to

$$\left(\mathbb{Z}^3, \left\{ \mathbf{x} \in \mathbb{Z}^3 : \langle \mathbf{x}, \mathbf{v} \rangle > 0 \right\}, (1, 2, 1) \right)$$

where $\mathbf{v} = [2\lambda, \lambda + 1, 1]$ with $\lambda = (1 + \sqrt{5})/2$ (Exercise 6.33). Thus, although the matrix of τ is unimodular, the dimension group is $\mathbb{Z}^3$ and not $\mathbb{Z}^2$.

6.6 Derivatives of Substitutive Sequences

In this section, we will prove a finiteness result characterizing substitutive shifts. It can be viewed as a counterpart of Boshernitzan–Carroll Theorem concerning interval exchange transformations (Theorem 8.3.1).

Let X be a minimal shift space and let u be a word of $\mathcal{L}(X)$. Let $\varphi_u : A_u^* \to A^*$ be a coding morphism for the set $\mathcal{R}'_X(u)$ of left return words to u. We normalize the alphabet A_u by setting $A_u = \{0, 1, \ldots, n - 1\}$ with $n = \mathrm{Card}(\mathcal{R}'_X(u))$. Recall from Section 3.7 that the *derivative shift* of X with respect to the cylinder $[u]$ is isomorphic to the shift $Y = \{y \in A_u^{\mathbb{Z}} \mid \varphi_u(y) \in X\}$. For $x \in X$ and $u \in \mathcal{L}(x)$, the *derivative sequence* of x with respect to u or u-derivative of x, denoted $\mathcal{D}_u(x)$ is defined as follows. Let $i \geq 0$ be the least integer such that $(S^i x)^+$ begins with u, that is, such that $S^i x$ is in $[u]$. Then $\mathcal{D}_u(x)$ is the unique $y \in Y$ such that $S^i x = \varphi_u(y)$. In particular, when u is a prefix of x^+, we have $\mathcal{D}_u(x) = y$ where y is the unique word of Y such that $x = \varphi_u(y)$. Thus, by definition, we have in this case

$$x = \varphi_u(\mathcal{D}_u(x)). \tag{6.21}$$

6.6.1 A Finiteness Condition

We will prove the following characterization of primitive substitutive sequences by a finiteness condition.

Theorem 6.6.1 *Let x be a uniformly recurrent two-sided sequence. The following conditions are equivalent:*

(i) *The sequence x is primitive substitutive.*

(ii) *The set of u-derivative sequences of x is finite, u being a word of $\mathcal{L}(x)$.*

(iii) *The set of u-derivative sequences of x is finite, u being a prefix of x^+.*

Note that, as a corollary of Theorem 6.6.1, we obtain that every shift of a substitutive sequence is substitutive. Indeed, if $y = S^n x$ is a shift of a substitutive sequence x, we have $\mathcal{L}(x) = \mathcal{L}(y)$ and the set of u-derivatives of x and y are the same. Thus, y is substitutive. A direct proof is given in Exercise 6.37.

Note also that a uniformly recurrent substitutive two-sided sequence is actually primitive substitutive (see Corollary 6.2.13). Thus, we can replace condition (i) by

(i') *The sequence x is substitutive.*

Observe finally that in the definition of a substitutive sequence x, one may relax all conditions on the pair (σ, ϕ) of morphisms defining x, provided x is a two-sided infinite sequence (see Exercise 6.38).

We begin by considering the easy case where x is periodic.

Proposition 6.6.2 *Every periodic two-sided sequence is purely substitutive and has a finite number of derivatives.*

Proof Let $x = v^\infty$ (recall that $v^\infty = \cdots vv \cdot vv \cdots$). Let φ be the morphism sending each letter a to v. Then x is clearly a fixed point of φ.

Since X has only finitely many different cylinders $[u]$, the number of its derivatives is also finite.   ∎

6.6.2 Sufficiency of the Condition

The implication (iii)$\Rightarrow$ (i) will result from Proposition 6.6.3.

Proposition 6.6.3 *Let x be an aperiodic uniformly recurrent two-sided sequence such that the set of derivatives $\mathcal{D}_u(x)$, for u prefix of x^+, is finite. Then x is primitive substitutive.*

Proof There exists a sequence of prefixes $(u_n)_{n \geq 1}$ of x^+ such that $|u_n| < |u_{n+1}|$ and $\mathcal{D}_{u_n}(x) = \mathcal{D}_{u_{n+1}}(x)$, for all $n \geq 1$. Clearly this implies, for all $n \geq 0$, that u_n is a prefix of u_{n+1}. Take $u = u_1$.

We denote by $\mathcal{R}'(u)$ the set $\mathcal{R}'_X(u)$ where X is the shift generated by x. The sequence x being uniformly recurrent we can choose N so large that every factor of length N of x has an occurrence of each ru for $r \in \mathcal{R}'(u)$. Since x is not periodic, the minimal length of a word in $\mathcal{R}'(u_n)$ cannot be bounded independently of n. Otherwise, there exists a word v such that vu_n begins with u_n for all n large enough. As a consequence, $|v|$ is a period of all u_n and thus of x^+. Since x is uniformly recurrent, this forces the period of all words in $\mathcal{L}(x)$ to be bounded and thus x is periodic, a contradiction. Thus, there exists some $w = u_l$ such that $|rw| > N$ for all $r \in \mathcal{R}'(w)$.

Let $\varphi_u : A_u^* \to A^*$ be a coding morphism for $\mathcal{R}'(u)$. Since $\mathcal{R}'(u)$ is a suffix code, the map φ_u is injective. Since $\mathcal{D}_u(x) = \mathcal{D}_w(x)$ we have $A_u = A_w$. Set $B = A_u = A_w$. There is a morphism $\tau : B^* \to B^*$ such that $\varphi_u \circ \tau = \varphi_w$. Indeed, since $\mathcal{R}'(w) \subset \mathcal{R}'(u)^*$ and since φ_u is injective, there is for every $x \in B^*$ a unique $y \in B^*$ such that $\varphi_u(y) = \varphi_w(x)$ and we set $y = \tau(x)$. By the choice of u and w, for every i, j in B the word $\varphi_u(j)w$ appears in $\varphi_w(i)$. By definition of return words, this implies that j appears in $\tau(i)$. This means that τ is a primitive substitution. We have

$$\varphi_u \circ \tau(\mathcal{D}_u(x)) = \varphi_w(\mathcal{D}_w(x)) = x = \varphi_u(\mathcal{D}_u(x)).$$

Since φ_u is injective by Proposition 1.4.29, we obtain $\tau(\mathcal{D}_u(x)) = \mathcal{D}_u(x)$. Hence $\mathcal{D}_u(x)$ is a fixed point of τ. Since $x = \varphi_u(\mathcal{D}_u(x))$, by Rauzy's Lemma (Proposition 6.2.10), x is primitive substitutive. ∎

6.6.3 Necessity of the Condition

We now consider a primitive substitutive sequence x. By definition, there is a primitive substitution $\sigma : B^* \to B^*$, a fixed point y of σ and a letter-to-letter morphism $\phi : B^* \to A^*$ such that $x = \phi(y)$. We first prove the following preliminary result.

Proposition 6.6.4 *Let y be a two-sided fixed point of a primitive substitution $\sigma : B^* \to B^*$. Let u be a nonempty prefix of y^+. The derivative sequence $\mathcal{D}_u(y)$ is the fixed point of a primitive substitution $\sigma_u : B_u^* \to B_u^*$, which satisfies*

$$\varphi_u \circ \sigma_u = \sigma \circ \varphi_u. \tag{6.22}$$

Proof For every $i \in B_u$, the word u is a prefix of $\sigma(u)$ and $\varphi_u(i)u$. Next,

$$\varphi_u(i)u \in \mathcal{L}(X) \Rightarrow \sigma(\varphi_u(i))u \in \mathcal{L}(X).$$

Hence $\sigma(\varphi_u(i))$ belongs to $\mathcal{R}'_X(u)^+$. We can therefore define a morphism $\sigma_u : B_u^* \to B_u^*$ by (6.22).

For all $n \geq 1$ we have $hi_u \circ \sigma_u^n = \sigma^n \circ \varphi_u$. Let $i, j \in B_u$. For n large enough, since y is uniformly recurrent, the word $\varphi_u(j)u$ appears in $\sigma^n(\varphi_u(i))$.

By definition of return words, this implies that j is a factor of $\sigma_u^n(i)$. Thus, σ_u is primitive. Moreover

$$\varphi_u \circ \sigma_u(\mathcal{D}_u(y)) = \sigma \circ \varphi_u(\mathcal{D}_u(y)) = \sigma(y) = y = \varphi_u(\mathcal{D}_u(y)). \qquad (6.23)$$

Since φ_u is injective by Proposition 1.4.29, it follows that $\sigma_u(\mathcal{D}_u(y)) = \mathcal{D}_u(y)$. Thus, $\mathcal{D}_u(y)$ is a fixed point of σ_u. $\blacksquare$

The substitution $\sigma_u : B_u^* \to B_u^*$ will be called a *return substitution*.

Example 6.6.5 If $\sigma : B^* \to B^*$ is the substitution defined by $\sigma(a) = aba$ and $\sigma(b) = aa$. Then $\mathcal{R}'(a) = \{a, ab\}$. Set $B_a = \{1, 2\}$ with $\varphi_a(1) = a$ and $\varphi_a(2) = ab$. Then the return substitution σ_a is given by $\sigma_a(1) = 21$ and $\sigma_a(2) = 2111$.

Proposition 6.6.6 *Let $\phi : B^* \to A^*$ be a letter-to-letter morphism. Let $y \in B^{\mathbb{Z}}$ be uniformly recurrent, let $x = \phi(y)$, and let u be a prefix of x^+. Then, there exists a prefix v of y^+ and a morphism $\lambda_u : B_v^* \to A_u^*$ such that $\varphi_u \circ \lambda_u = \phi \circ \varphi_v$ and $\lambda_u(\mathcal{D}_v(y)) = \mathcal{D}_u(x)$.*

Proof Let v be the unique prefix of y^+ such that $\phi(v) = u$. If w is a return word to v, then $\phi(w)$ is a concatenation of return words to u. The morphism φ_u being one-to-one, we can define a morphism $\lambda_u : B_v^* \to A_u^*$ by $\varphi_u \circ \lambda_u = \phi \circ \varphi_v$. We have, as in Eq. (6.23),

$$\varphi_u \circ \lambda_u(\mathcal{D}_v(y)) = \phi \circ \varphi_v(\mathcal{D}_v(y)) = \phi(y) = x = \varphi_u(\mathcal{D}_u(x)).$$

This implies that $\lambda_u(\mathcal{D}_v(y)) = \mathcal{D}_u(x)$. $\blacksquare$

Consequently, to prove that (i) implies (iii), it suffices to prove that the sets $\{\sigma_v \mid v \text{ prefix of } y^+\}$ and $\{\lambda_u \mid u \text{ prefix of } x^+\}$ are finite.

Proposition 6.6.7 *Let $y \in B^{\mathbb{Z}}$ be a fixed point of a primitive substitution σ, let $\phi : B^* \to A^*$ be a letter-to-letter substitution and let $x = \phi(y)$. The sets $\{\sigma_v \mid v \text{ prefix of } y^+\}$ and $\{\lambda_u \mid u \text{ prefix of } x^+\}$ are finite.*

Proof The periodic case is easy to check hence we suppose that y is aperiodic. By Proposition 1.4.24, the sequence y is linearly recurrent, say with constant K. We start by proving that the set

$$\{\sigma_v : B_v^* \to B_v^* \mid v \text{ prefix of } y^+\}$$

is finite. For this, it suffices to prove that $\text{Card}(B_v)$ and $|\sigma_v(i)|$ are bounded independently of v and $i \in B_v$. Set as usual $|\sigma| = \sup\{|\sigma(a)| \mid a \in A\}$ and $\langle \sigma \rangle = \inf\{|\sigma(a)| \mid a \in A\}$.

Let v be a nonempty prefix of y^+, let i be an element of B_v and let $w = \varphi_v(i)$ be a left return word to v. Since y is not periodic, we have $|w| \leq K|v|$ and thus $|\sigma(w)| \leq |\sigma|K|v|$. The length of each element of $\mathcal{R}'(v)$ is larger than $|v|/K$ by Assertion 4 of Proposition 6.3.3. Now, since wv is in $\mathcal{L}(y)$ and since wv begins with v, we have that the word $\sigma(w)v$ is in $\mathcal{L}(y)$ and begins with v. Thus, we can write $\sigma(w) = x_1 x_2 \cdots x_k$ with $k \geq 1$ and $x_i \in \mathcal{R}'(v)$ for $1 \leq i \leq k$. Since $|x_i| \geq [v]/K$, we have

$$|v||\sigma|K \geq |\sigma(w)| \geq k|v|/K,$$

and we conclude that $k \leq |\sigma|K^2$. Therefore $\sigma_v(i) = i_1 i_2 \cdots i_k$ with $x_j = \varphi_v(i_j)$. Thus, we obtain finally that $|\sigma_v(i)| \leq |\sigma|K^2$. Moreover, we know from Assertion 5 of Proposition 6.3.3 that $\mathrm{Card}(B_v) \leq K(K+1)^2$. This ends the first part of the proof.

Now let u be a factor of x and v be a factor of y such that $\phi(v) = u$. The length of a return word to u in x is bounded by the length of a return word to v in y and thus x is linearly recurrent with constant K. Consequently, since x is nonperiodic, $\langle \varphi_u \rangle \geq |u|/K$ by Proposition 6.3.3 again. We have then for every prefix u of x^+,

$$|\lambda_u(i)|(1/K)|u| \leq |\lambda_u(i)|\langle \varphi_u \rangle \leq |\varphi_u(\lambda_u(i))|$$
$$= |\phi(\varphi_v(i))| = |\varphi_v(i)| \leq |\varphi_v| \leq K|v| = K|u|.$$

Hence $|\lambda_u(i)| \leq K^2$. This completes the proof. ∎

6.6.4 End of the Proof

We now conclude the proof of the main result.

Proof of Theorem 6.6.1 We have already proved that (iii) implies (i) (Proposition 6.6.3). We have also proved that (i) implies (iii) (Proposition 6.6.7).

Since (ii) clearly implies (iii), there only remains to prove that (i) implies (ii). We start with some notation. Let t be a word with prefix s. By $s^{-1}t$ we mean the word r such that $t = sr$. In this way we have $ss^{-1}t = t$.

Since the sequence x is primitive substitutive, there is a primitive substitution σ, an admissible fixed point y of σ and a letter-to-letter morphism ϕ such that $x = \phi(y)$.

Since y is a fixed point of a primitive substitution, by Proposition 1.4.24 it is linearly recurrent, say with constant K. It follows that x is linearly recurrent with the same constant K.

Let u be a word of $\mathcal{L}(x)$ and v be such that vu is a prefix of x^+ and u has exactly one occurrence in vu. Since v is a suffix of a word in $\mathcal{R}'_X(u)$, we have $|v| \leq K|u|$ and thus $|vu| \leq (K+1)|u|$.

If w is a left return word to vu then u is a prefix of $v^{-1}wvu$ and $v^{-1}wvu = (v^{-1}wv)u$ is a word of $\mathcal{L}(x)$. Hence $v^{-1}wv$ is a concatenation of return words to u. Thus, we can define $\phi_{v,u} \colon B_{vu}^* \to B_u^*$ by

$$\varphi_u \circ \phi_{v,u}(i) = v^{-1}\varphi_{vu}(i)v, \qquad (6.24)$$

for all $i \in B_{vu}$. We have

$$\phi_{v,u}(\mathcal{D}_{vu}(x)) = \mathcal{D}_u(x). \qquad (6.25)$$

Indeed, by definition of v, the integer $i = |v|$ is the least integer $i \geq 0$ such that $S^i(x)$ begins with u. Thus, by definition of $\mathcal{D}_u(x)$, we have

$$S^{-|v|}(x) = \varphi_u(\mathcal{D}_u(x)). \qquad (6.26)$$

Now, by definition of the morphism $\phi_{u,v}$, if $\varphi_{vu}(y) = x$, we have, using iteratively (6.24) on the prefixes of y,

$$\varphi_u \circ \phi_{v,u}(y) = S^{-|v|}\varphi_{vu}(y) = S^{-|v|}(x). \qquad (6.27)$$

We conclude, using Eq. (6.26) and (6.27) and the equality $y = \mathcal{D}_{vu}(x)$, that

$$\varphi_u(\phi_{v,u}(y)) = \varphi_u(\mathcal{D}_u(x))$$

whence (6.25), since φ_u is injective.

The set $\{\mathcal{D}_u(x); u = x_{[0,n]}, n \geq 0\}$ being finite, it suffices to prove that the set

$$H = \{\phi_{v,u} \colon B_{vu} \to B_u; vu = x_{[0,n]}, |vu| \leq (K+1)|u|, n \geq 0\}$$

is finite to conclude. By Proposition 6.3.3 the length of every word in $\mathcal{R}_X(u)$ is at least $|u|/K$. Thus, for every $i \in B_{vu}$, we have $|\varphi_u(\phi_{v,u}(i))| \geq |\phi_{v,u}(i)||u|/K$. Therefore, for all $i \in B_{vu}$ we have, using (6.24),

$$|\phi_{v,u}(i)| \leq \frac{|\varphi_u(\phi_{v,u}(i))|}{|u|/K} = \frac{|\varphi_u(i)|}{|u|/K} \leq K^2.$$

Moreover $\mathrm{Card}(B_s) \leq K(K+1)^2$ for all words $s \in \mathcal{L}(x)$, hence H is finite. ∎

Example 6.6.8 Let $\varphi \colon a \to ab, b \to a$ be the Fibonacci morphism and let $x = \varphi^{\omega}(a)$ be the Fibonacci word. Let $u_1, u_2, \ldots$ be the palindrome prefixes of x. Since the directive word of x is $(ab)^{\omega}$, we have by Eq. (1.34) $\varphi_{u_{2n}} = (L_{ab})^n$ and $\varphi_{u_{2n+1}} = (L_{ab})^n L_a$. Let $\overline{x}$ be the result of exchanging a, b in x. We have $x = L_a(\overline{x})$ because $x = \mathrm{Pal}((ab)^{\omega}) = L_a(\mathrm{Pal}(ba)^{\omega})) = L_a(\overline{x})$. Next $x = L_{ab}(x)$ by a similar computation (or also because $L_{ab} = \varphi^2$). Thus, $\mathcal{D}_{u_{2n}}(x) = x$ and $\mathcal{D}_{u_{2n+1}} = \overline{x}$.

If u is any factor of x, we have $\mathcal{R}'_x(u) = v^{-1}\mathcal{R}'_X(u_n)v$, where v is the shortest word such that vu is a prefix of some palindrome prefix u_n. Consequently, $\mathcal{D}_u(x) = \mathcal{D}_{u_n}(x)$. Thus, the derivatives of x with respect to a factor of x are x and $\bar{x}$.

6.6.5 A Variant of the Main Result

We have considered in the first part of this section substitutive sequences. We now consider substitutive shifts.

The following variant of Theorem 6.6.1 characterizes substitutive shifts. In the following statement, we consider two shifts as different if they cannot be identified by renaming the alphabet.

Theorem 6.6.9 *The following conditions are equivalent for a minimal shift space* X:

(i) X *is primitive substitutive.*

(ii) *There is a finite number of different derivative systems on cylinder sets* $[u]$ *for* $u \in \mathcal{L}(X)$.

(iii) X *is conjugate to a primitive substitution shift.*

Proof (i)$\Rightarrow$(ii). Consider $X = X(\sigma, \phi)$ with $\sigma : B^* \to B^*$ primitive and $\phi : B^* \to A^*$ letter-to-letter. Let y be an admissible fixed point of some power of σ and let $x = \phi(y)$. Let $u \in \mathcal{L}(X)$, let $U = [u]$ and let (X_U, S_U) be the shift induced by X on U. Since X is minimal, we have $u \in \mathcal{L}(x)$ and $\mathcal{D}_u(x) \in X_U$. Since X_U is also minimal, it is generated by $\mathcal{D}_u(x)$. By Theorem 6.6.1, since x is uniformly recurrent and primitive substitutive, there is a finite number of derivative sequences $\mathcal{D}_u(x)$. Thus, there is only a finite number of derivative shifts (X_U, T_U).

(ii)$\Rightarrow$(iii). Suppose that there are a finite number of systems induced by X on cylinder sets $U = [u]$ for $u \in \mathcal{L}(X)$. If X is finite, then it is a substitution shift by Proposition 6.6.2. Otherwise, there exist $uv, v \in \mathcal{L}(X)$ such that the derivatives of X with respect to $[uv]$ and $[v]$ are equal. Arguing as in the proof of Proposition 6.6.3, we may also assume that every word of $\mathcal{R}'_X(v)$ appears in the decomposition of every word $\mathcal{R}'_X(uv)$. Let $\varphi, \psi : \{0, 1\}^* \to \{0, 1\}^*$ be coding morphisms for $\mathcal{R}'_X(v)$ and $\mathcal{R}'_X(uv)$. Since $\mathcal{R}'_X(uv) \subset \mathcal{R}'_X(v)^*$, there is a morphism $\sigma : \{0, 1\}^* \to \{0, 1\}^*$ such that $\psi = \phi \circ \alpha$. Since every word of $\mathcal{R}'_X(v)$ appears in the decomposition of every word of $\mathcal{R}'_X(uv)$, the morphism σ is primitive. Then $X = X(\sigma, \phi)$. By Rauzy's Lemma (Proposition 6.2.10), X is conjugate to a primitive substitution shift.

(iii)$\Rightarrow$(i) Let $\gamma\colon Y \to X$ be a conjugacy from a primitive substitution shift $Y = X(\sigma)$ with σ primitive onto X. Every conjugacy is a composition of a k-block code $\gamma_k\colon Y \to Y^{(k)}$ and a letter-to-letter morphism $\phi\colon Y^{(k)} \to X$. We have $Y^{(k)} = X(\sigma_k)$ where σ_k is the k-block presentation of σ, which is primitive by Proposition 1.4.42. Thus, $Y^{(k)}$ is a primitive substitution shift and consequently X is primitive substitutive.   ∎

Note that this shows that the class of primitive substitutive shifts is closed under conjugacy.

Example 6.6.10 Consider again the Fibonacci shift $X = X(\varphi)$ as in Example 6.6.8. There are only two different shifts induced by X on clopen sets $[u]$ for $u \in \mathcal{L}(X)$, namely X and its image $\overline{X}$ by exchange of a, b.

6.7 Exercises

Section 6.1

6.1 Prove that the set $\mathbb{Z}_{(p_n)}$ of (p_n)-adic integers is a Cantor space and a compact topological ring, which is the completion of $\mathbb{Z}$ for the (p_n)-adic topology.

6.2 Show that two odometers (X, T) and (X', T') are topologically conjugate if and only if the groups X and X' are isomorphic.

6.3 Let $(q_n)_{n\geq 1}$ be an increasing sequence of natural integers and let $Y = \{(y_n)_{n\geq 0} \mid 0 \leq y_n < q_{n+1}\}$ be the group of (q_n)-adic expansions. Set $p_1 = q_1$ and $p_{n+1} = p_n q_{n+1}$ for $n \geq 1$. Show that the map $\varphi\colon y \mapsto x$ defined by $x_1 = y_0$ and $x_{n+1} = x_n + y_n p_n$ for $n \geq 1$ is homeomorphism from Y onto $\mathbb{Z}_{(p_n)}$.

6.4 Show that every integer $x \in \mathbb{Z}$ has a unique factorial expansion

$$x = c_0 + c_1 2! + \cdots$$

with $0 \leq c_i \leq i + 1$.

6.5 Show that the factorial expansion of -1 is $-1 = 1 + 2 \cdot 2! + 3 \cdot 3! + \cdots$.

6.6 Let $(G_n)_{n\geq 1}$ be a sequence of groups (or rings) and let $\varphi_n\colon G_{n+1} \to G_n$ be a sequence of morphisms. Set $X = \prod_{n\geq 1} G_n$. The *inverse limit* of the sequence (G_n, φ_n) is the set

$$X = \{(x_n)_{n \geq 1} \in X \mid \varphi_n(x_{n+1}) = x_n\}.$$

Show that the ring $\mathbb{Z}_{(p_n)}$ is the inverse limit of the ring $\mathbb{Z}/p_n\mathbb{Z}$ with $\varphi_n(x_{n+1}) = x_{n+1} \bmod p_n$.

6.7 Show that the sequence of partitions $\mathfrak{P}(n)$ defined by Eq. (6.3) is a refining sequence.

6.8 A *supernatural number* is a formal product $\prod_p p^{n_p}$ where for each prime number p we have $n_p \in \mathbb{N} \cup \infty$. If $(p_n)_{n \geq 1}$ is a sequence of integers with $p_n \mid p_{n+1}$ for all $n \geq 1$ and (X, T) is the associated odometer, we associate with X the supernatural integer $\sigma(X) = \prod_p p^{n_p}$ where p^{n_p} is the maximal power of p that divides some p_n (and thus all p_m for $m \geq n$). Show that two odometers (X, T) and (X', T') are topologically conjugate if and only if $\sigma(X) = \sigma(X')$.

6.9 A *profinite* group (or ring) is an inverse limit of finite groups (or rings). Show that for every odometer (X, T), the space X is a profinite ring.

6.10 Show that the odometer in base (p_n) is a skew product of the finite system $(\mathbb{Z}/p_1\mathbb{Z}, +1)$ and the odometer in base (p'_n) with $p'_n = p_{n+1}/p_1$.

6.11 Let (X, T) be an invertible topological dynamical system that is expansive. Let ε be the expansivity constant. Let γ be a finite cover of X by sets $C_1, C_2, \ldots, C_r$ of diameter at most ε.

Show that the diameter of the elements of the cover $\vee_{j=-n}^{n} T^{-j}\gamma$ converges to 0 when $j \to \infty$ (if α, β are covers of X, then $\alpha \vee \beta$ is formed of the $A \cap B$ for $A \in \alpha$ and $B \in \beta$ and $T^{-j}\alpha$ by the $T^{-j}A$ for $A \in \alpha$).

6.12 Let (X, T) be an invertible topological dynamical system that is expansive with constant ε. We aim at proving that (X, T) is conjugate to a shift space.

For this, let $\{B_0, B_1, \ldots, B_{k-1}\}$ be a cover by open balls with radius ε. Let $C_0 = \overline{B_0}$ and for $n \geq 1$, let $C_n = \overline{B_n} \setminus (B_0 \cup \cdots \cup B_{n-1})$.

1. Show that $\gamma = \{C_0, \ldots, C_{k-1}\}$ is a closed cover of X with $\mathrm{diam}(C_i) < \varepsilon$ for each i, $C_i \cap C_j = \partial C_i \cap \partial C_j$ if $i \neq j$ and ∂C_i having no interior (we denote $\partial C = C \setminus \mathrm{int}(C)$ where $\mathrm{int}(C)$ is the interior of C).

2. Set $D = \cup_{i=0}^{k-1} \partial C_i$ and $D_\infty = \cup_{n \in \mathbb{Z}} T^n D$. For $x \in X \setminus D_\infty$, let $y = \psi(x)$ be the sequence in $A^{\mathbb{Z}}$ with $A = \{0, \ldots, k-1\}$ defined by $y_n = i$ if $T^n(x) \in C_i$. Show that ψ extends to a conjugacy from X to a subshift of $A^{\mathbb{Z}}$.

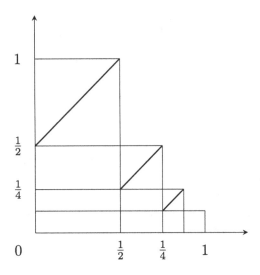

Figure 6.16 The map R.

6.13 Let R be the map (see Figure 6.16) from $[0, 1]$ to itself defined by

$$R(x) = x + \frac{1}{2^{n+1}} \quad \text{if} \quad 1 - \frac{1}{2^n} \le x < 1 - \frac{1}{2^{n+1}}.$$

Let $\varphi \colon \mathbb{Z}_2 \to [0, 1]$ be the map defined by $\varphi(x) = \sum x_n 2^{-n}$ if $x = \sum x_n 2^n$. Show that φ is a measure-theoretic isomorphism from the dyadic odometer onto $([0, 1], R)$.

Section 6.2

6.14 Show that Figure 6.3 is a BV-representation of the period-doubling shift and that its dimension group is $\mathbb{Z}[1/2] \times \mathbb{Z}$ with positive cone $\mathbb{Z}_+[1/2]$ and unit $(3, 0)$.

6.15 Show that the image of a minimal substitution shift by a nonerasing morphism is isomorphic to a primitive substitution shift. Hint: use Rauzy's Rauzy Lemma (Proposition 6.2.10).

6.16 The primitive morphism $\tau \colon 0 \to 0012, 1 \to 12, 2 \to 012$ is the *Chacon ternary morphism*. Show that $w_n = \tau^n(0)$ satisfies the recurrence relation $w_{n+1} = w_n w_n 1 w_n'$ where w_n' is obtained from w_n by changing the initial letter 0 into a 2. Deduce from this that the 1-block map $\theta \colon 0, 2 \to 0, 1 \to 1$ defines a conjugacy from the substitution shift $X(\tau)$ defined by τ to the substitution shift $X(\sigma)$ defined by the Chacon binary morphism $\sigma \colon 0 \to 0010, 1 \to 1$.

6.17 Show that the factor complexity of the Chacon ternary shift is $p_n(X) = 2n + 1$ (Hint: show that the bispecial words are 0 and the words $\alpha^n(012), \alpha^n(120)$ for $n \geq 0$ where $\alpha(w) = 012\tau(w)$ and where τ is the Chacon ternary morphism).

6.18 Use Proposition 6.2.8 to prove Proposition 6.2.10. Develop the construction on the example of the Thue–Morse morphism $\tau: 0 \rightarrow 01, 1 \rightarrow 10$ and $\phi: 0 \rightarrow ab, 1 \rightarrow a$.

6.19 Let X be the Thue–Morse shift generated by $\varphi: a \rightarrow ab, b \rightarrow ba$. Let $f: \{aa, ab, ba, bb\} \rightarrow A_2 = \{x, y, z, t\}$ and let $\varphi_2: x \rightarrow yz, y \rightarrow yt, z \rightarrow zx, t \rightarrow zy$ be the 2-block presentation of φ. Let $B = \{u, v, w\}$ and let $\phi: B^* \rightarrow A_2^*$ be a coding morphism for $f(a\mathcal{R}_X(a)) = f(\{aa, aba, abba\}) = \{x, yz, ytz\}$. Show that there is a morphism τ such that $\varphi_2 \circ \phi = \phi \circ \tau$ and derive a BV-representation of X with three vertices at each level $n \geq 1$.

6.20 Use the set $\mathcal{R}_X(b \cdot a)$ to derive a stationary BV-representation of the Rudin–Shapiro shift with nine vertices at each level.

6.21 Let $\sigma: A^* \rightarrow A^*$ be a primitive eventually proper aperiodic morphism. Let $\alpha = (\alpha_a)_{a \in A}$ be the left eigenvector with sum 1 of the incidence matrix M of σ for the eigenvalue $\lambda = \lambda_M$. Use Exercise 5.12 to show that the image subgroup of $K^0(X(\sigma), S)$ is the subgroup of $\mathbb{Q}$ generated by the numbers α_a/λ^n for $a \in A$ and $n \geq 1$.

6.22 Show that if σ is a primitive substitution of constant length, then $X(\sigma)$ is orbit equivalent to a stationary odometer.

6.23 Let σ, σ' be two constant-length primitive substitutions of respective lengths n, n'. Show that the shifts $X(\sigma)$ and $X(\sigma')$ are Kakutani orbit equivalent if and only if the prime divisors of n and n' are the same.

6.24 Let $\sigma: a \mapsto ab, b \mapsto ac, c \mapsto db, d \mapsto dc$ be the Rudin–Shapiro morphism (see Exercise 1.31). Show that the image subgroup of $K^0(X(\sigma), S)$ is $\mathbb{Z}[1/2]$. Conclude that the Rudin–Shapiro shift and the Thue–Morse shift are Kakutani orbit equivalent but not orbit equivalent.

6.25 Let σ be the morphism $0 \mapsto 00000111, 1 \mapsto 01001$. Show that the shift generated by σ is strong shift equivalent to the Fibonacci shift but not conjugate to it. Hint: Show that the incidence matrix of φ is $M(\sigma)^4$ where σ is the Fibonacci shift. To prove that the shifts are not conjugate, show that $\mathcal{L}(\sigma)$ contains words with exponent 5.

Section 6.3

6.26 Let $\sigma : A^* \to A^*$ be a nonerasing morphism and let $B \subset A$. Show that $\mathcal{L}(X) \cap B^*$ is infinite if and only if there is a letter $a \in A$ and integers $n, m \geq 1$ such that $\sigma^n(a) = uav$ with uv nonempty and $\sigma^m(uav) \in B^*$.

6.27 Show that if X is LR with constant K, then its kth block presentation $(X^{(k)}, S)$ is LR with constant $K(k-1) + 1$. Conclude that the class of LR shifts is closed under conjugacy.

6.28 Consider the morphism $\sigma : a \mapsto abd, b \mapsto bb, c \mapsto c, d \mapsto dc$. Show that the factor complexity $p_n(X)$ of the shift $X = X(\sigma)$ is not linear.

6.29 Show that a Cantor system (X, T) is linearly recurrent if and only if there is a refining sequence $(\mathfrak{P}(n))$ of partitions in towers with positive matrices $M(n)$ and with heights $(h_i(n))_{1 \leq i \leq t(n)}$ satisfying (6.11) for some $L \geq 1$ and for which the morphisms read on $E(n)$ are eventually proper.

6.30 For a probability measure μ on a shift space X, denote

$$\varepsilon_n(\mu) = \min\{\mu(u) \mid u \in \mathcal{L}_n(X)\}.$$

Show that a shift space X is linearly recurrent if and only if there is an invariant probability measure μ on X such that $\inf n \varepsilon_n(\mu) > 0$. Derive from this an alternative proof that linearly recurrent shifts are uniquely ergodic.

Section 6.4

6.31 Show that the following conditions are equivalent for a one-sided sequence x.

(i) x is a limit point of a primitive $\mathcal{S}$-adic sequence τ of morphisms.
(ii) There is a primitive $\mathcal{S}$-adic sequence τ and a sequence (a_n) of letters $a_n \in A_n$ such that $x = \lim \tau_{[0,n)}(a_n^\omega)$.
(iii) There is a primitive $\mathcal{S}$-adic sequence τ and a sequence (a_n) of letters $a_n \in A_n$ such that $\{x\} = \cap_n[\tau_{[0,n)}(a_n)]$ where $[w]$ denotes the cylinder $\{y \in A_0^{\mathbb{N}} \mid y_{[0,|w|)} = w\}$.

6.32 Show that for every sequence $x \in A^{\mathbb{N}}$, there are morphisms $(\sigma_a)_{a \in A}$: $(A \cup \#)^* \to (A \cup \#)^*$ and $\phi : (A \cup \#)^* \to A^*$ where $\#$ is a letter not in A such that $x = \lim \phi \circ \sigma_0 \circ \sigma_1 \cdots \sigma_n(\#)$ and thus that x is a limit point of the sequence $(\varphi, \sigma_0, \sigma_1, \ldots)$.

6.33 Let X be the shift generated by the substitution $\tau: a \to aab, b \to ba$. Show that the dimension group of X is isomorphic to

$$\left(\mathbb{Z}^3, \left\{\mathbf{x} \in \mathbb{Z}^3: \langle \mathbf{x}, \mathbf{v} \rangle > 0\right\} \cup \{0\}, (1, 2, 1)\right)$$

where $\mathbf{v} = [2\lambda, \lambda + 1, 1]$ with $\lambda = (1 + \sqrt{5})/2$.

6.34 Let $\varphi: A^* \to B^*$ be an injective morphism and let $U = \varphi(A)$. If φ is not injective on $A^{\mathbb{N}}$, there exist words u, v, w such that u, uv, vw, wv are in U^* but v is not in U^*.

6.35 A nonerasing morphism $\varphi: A^* \to B^*$ is *left marked* if every word $\varphi(a)$ for $a \in A$ begins with a distinct letter. In particular, φ is injective on A and $\varphi(A)$ is a prefix code.

Prove that if $\varphi: A^* \to B^*$ is left marked, then it is recognizable for aperiodic points. Hint: use Exercise 6.34.

Section 6.6

6.36 Show that the sequence 001^ω is substitutive but not purely substitutive.

6.37 Show that every shift of a substitutive sequence is substitutive. Hint: consider a kth higher block presentation of the shift generated by the sequence.

6.38 Let $\tau: B^* \to B^*$ be a morphism prolongable on $a \in B$ and let $y = \tau^\omega(a)$. Let $\phi: B^* \to A^*$ be a morphism such that $x = \phi(y)$ is an infinite word. Show that the sequence x is substitutive. Hint: adapt the proof of Rauzy's Lemma (Proposition 6.2.10) to the case where τ is not primitive and ϕ is possibly erasing.

6.39 Let $\sigma: \{0, 1, 2\}^* \to \{0, 1, 2\}^*$ be the morphism $0 \to 01222, 1 \to 10222, 2 \to \varepsilon$ and let $x = \sigma^\omega(0)$. Show that x is not the fixed point of a nonerasing substitution. Hint: show that erasing the letter 2 in x gives the Thue–Morse sequence.

6.8 Notes

6.8.1 Odometers

Odometers, also called *solenoids* (Katok and Hasselblatt (1995)) or *adding machines* (Brown (1976)), are a classical object in dynamical systems theory. The ring $\mathbb{Z}_p$ of p-adic integers is contained in a field, called the field of p-adic numbers. Classical references to p-adic numbers and p-adic analysis are Koblitz (1984) and Robert (2000). The factorial representation of integers (Exercise 6.4) is described in Knuth (1998). Supernatural numbers (Exercise 6.8), also called *generalized natural numbers* or *Steinitz numbers*, are used to define orders of profinite groups. The fact that they give a complete invariant for odometers (Exercise 6.8) is a result proved by Glimm (1960) in the context of *uniformly hyperfinite algebras* of *UHF-algebras* (see Chapter 9).

A Haar probability measure can be defined for every compact metric group G as a Borel probability measure μ on G such that $\mu(E + g) = \mu(E)$ for every Borel subset E of G and every $g \in G$. Its existence and unicity can be proved in general, even for nonabelian groups (see Rudin (1991)).

The notion of expansive system is classical in topological dynamics (see Katok and Hasselblatt (1995), for example). Exercises 6.11 and 6.12 are from Walters (1982, Theorem 5.24) (see also Kurka (2003)).

The notion of equicontinuity is classical in analysis for a family of functions. The condition defining an equicontinuous dynamical system is equivalent to the equicontinuity of the family (T^n) of maps from X to itself.

The equal path number property and Theorem 6.1.7 are from Gjerde and Johansen (2000).

6.8.2 Substitution Shifts

In Vershik and Livshits (1992) the authors showed that when σ is a primitive morphism, then the subshift it generates can be represented (in a measure-theoretic sense) by an ordered Bratteli diagram $\mathcal{B}$ where σ is the morphism read on $\mathcal{B}$.

Theorem 6.2.1 was first proven in Forrest (1997). The proofs given in that paper are mostly of an existential nature and do not state a method to compute effectively the BV-representation associated with substitution shifts. Another proof, given in Durand et al. (1999), provides such an algorithm. This method is described in Rust and Balchin (2017) as part of a set of computer-assisted tools concerning substitutions. Proposition 6.2.3 is from Durand et al. (1999).

Proposition 6.2.8 is Forrest (1997, lemma 15). As mentioned in this paper, the proof uses an important technique called *state-splitting* or *symbol splitting*.

We actually use in the proof of Proposition 6.2.8 an *output split*. See Lind and Marcus (1995) for a systematic presentation of state splitting.

The Rauzy Lemma (Proposition 6.2.10) is also from Durand et al. (1999). It is a modification of an unpublished result of Rauzy.

Corollary 6.2.13 appears (with another proof) in Maloney and Rust (2018). A similar result, concerning the more general class of minimal substitutive shifts appears in Durand (2013b), namely that every uniformly recurrent substitutive sequence (resp. shift) is primitive substitutive.

For more details about the Chacon substitution, see Ferenczi (1995) or Fogg (2002).

6.8.3 Linearly Recurrent Shifts

Linearly recurrent shifts, also called *linearly repetitive* shifts, were introduced in Durand et al. (1999).

Proposition 6.3.2 is from Damanik and Lenz (2006), where a direct proof is given (see also Shimomura (2019)). The fact that the equivalent conditions of Proposition 6.3.2 are also equivalent to unique ergodicity is from Durand (2000).

Proposition 6.3.3 is proved in Durand et al. (1999). The corollary asserting that the factor complexity of a primitive substitution shift is at most linear can be found in Michel (1976) (see also Pansiot (1984)). Exercise 6.28 is from Allouche and Shallit (2003, Example 10.4.1). The factor complexity of substitutive shifts has been extensively studied. By a result of Ehrenfeucht et al. (1975), one has always $p_n(X) = O(n^2)$. See Allouche and Shallit (2003) for a survey of this question.

Theorem 6.3.5 is from Durand (2003). Theorem 6.3.6 is from Downarowicz and Maass (2008). It is shown additionally in Donoso et al. (2021) that an expansive system has finite rank if and only if it has a primitive and recognizable $\mathcal{S}$-adic representation. Theorem 6.3.7 is also from Donoso et al. (2021). A generalization to arbitrary finite rank appears in Espinoza and Maass (2020).

The unique ergodicity of linearly recurrent Cantor systems (Theorem 6.3.9) is from Cortez et al. (2003). This result has been generalized by Bezuglyi et al. (2013, Corollary 4.14) to the case of an infinite set of positive matrices $M(n)$ such that $\|M(n)\|_1 \leq Cn$ for some constant C. The proof uses the Birkhoff contraction coefficient (Exercise 4.7).

A related result is the following. If X is a minimal shift space such that, for some integer $K \geq 1$, one has $\liminf p_X(n)/n \leq K$, then X has at most K ergodic probability measures. This result, due to Boshernitzan (1984), is presented in Berthé and Rigo (2010, Chapter 7) with a proof based on the notion of deconnectability of Rauzy graphs. The proof gives an improved statement, due to Monteil (2009): if $\limsup p_X(n)/n < K$, with $K \geq 3$, the number of ergodic probability measures is at most $K - 2$.

6.8.4 S-adic Shifts

The notion of S-adic shift was introduced in Ferenczi (1996), using a terminology initiated by Vershik and coined out by Bernard Host. For more information, see Fogg (2002), Berthé et al. (2019b) or Berthé and Delecroix (2014). See also Thuswaldner (2020) for a recent survey on S-adic shifts.

Proposition 6.4.4 is from Arnoux et al. (2014). Proposition 6.4.5 is from Durand (2000, Lemma 7) and Lemma 6.4.9 is a weaker version of Durand and Leroy (2012, Corollary 2.3).

The fact that the cardinality of the sets of return words may be unbounded in a shift of linear complexity (see the remark after Proposition 6.4.10) is from Durand et al. (2013, Example 3.17). It is proved in Ferenczi (1996) that a minimal shift with at most linear complexity has a finite primitive S-adic representation. The converse is not true but the existence of an additional condition making it true is known as the S-*adic conjecture* (see Fogg (2002) or Durand et al. (2013)).

The recognizability of a primitive morphism on its associated shift (Mosse Theorem 1.4.35) was extended by Bezuglyi et al. (2009) to arbitrary substitutions generating an aperiodic shift. The notion of recognizability for aperiodic points is from Berthé et al. (2019b), who proved in particular Theorem 6.4.18 (Berthé et al. (2019b, Theorem 3.1)). Note that Theorem 6.4.18 can actually be deduced easily from the main result of Karhumäki et al. (2003).

The original reference for the Fine–Wilf Theorem (Exercise 1.14) is Fine and Wilf (1965).

Proposition 6.4.17 is Berstel et al. (2009, Exercise 5.1.5), where it is proved as a variant of a result called the *Defect Theorem* (see Lothaire (1997) and the references of Exercise 1.37).

Exercise 6.35 is from Berthé et al. (2019b, Lemma 3.3).

Theorem 6.5.1 is from Berthé et al. (2021). The proof given here (using Theorem 6.4.22) is shorter. Observe that in Theorem 6.5.1, we can relax the assumption of minimality. Indeed, one checks that the proof given in Berthé et al. (2021) works if we assume that X is aperiodic (recognizability then holds by Corollary 6.4.19) and that $\min_{a \in A} |\tau_{[0,n)}(a)|$ goes to infinity. Corollary 6.5.2 extends a statement initially proved for interval exchanges by Ferenczi and Zamboni (2008). In Corollaries 6.5.2 and 6.5.3, the assumption of being proper can be dropped. The proof then uses the measure-theoretical Bratteli–Vershik representation of primitive unimodular S-adic subshift given in Berthé et al. (2019b, Theorem 6.5).

Note that one recovers, with the description of dimension groups of primitive unimodular proper $\mathcal{S}$-adic shifts of Theorem 6.5.4, the results obtained in the case of interval exchanges, in Putnam (1989). See also Putnam (1992) and Gjerde and Johansen (2002).

The BV-representation of non-minimal Cantor systems can also be considered. In Medynets (2006) the author shows that for Cantor dynamical systems without periodic points (but not necessarily minimal) a BV-representation can also be given. It is applied in Bezuglyi et al. (2009) to subshifts generated by non-primitive substitutions. The authors show that they have stationary BV-representations as in the minimal case.

6.8.5 Derivatives of Substitutive Sequences

Substitutive sequences have been considered early by Cobham (1968), who has proved the statement that every infinite sequence $x = \phi(\tau^{\omega}(a))$ is substitutive, whatever be the morphisms $\phi\colon B^* \to A^*$ and $\tau\colon B^* \to B^*$ (Exercise 6.38). This result was proved independently by Pansiot (1983) (see the presentation in Allouche and Shallit (2003)). We follow the proof given in Cassaigne and Nicolas (2003). The effective computability of the representation as a substitutive sequence was proved by Honkala (2009) and Durand (2013a).

Theorem 6.6.1 is from Durand (1998). It is closely related with the results of Holton and Zamboni (1999), who proved independently that conditions (i) and (ii) are equivalent. Theorem 6.6.9 is from Holton and Zamboni (1999, Theorem 1.3).

6.9 Exercises

The notion of profinite group (Exercise 6.9) generalizes the profinite integers of Section 6.1. See Almeida et al. (2020) for a description of the link between profinite groups or semigroups and symbolic dynamics.

The map described in Exercise 6.13 is called the *van der Corput map* (see Fogg (2002)).

The Chacon ternary morphism (Exercises 6.16 and 6.17) is from Fogg (2002).

Exercises 6.21, 6.22 and 6.23 are from Yuasa (2002). The statement of Exercise 6.23 is formulated in Yuasa (2002) under the weaker hypothesis that the maximal eigenvalue of the incidence matrix is rational (instead of being of constant length).

The fact that dimension groups are not enough to characterise primitive substitution shifts up to conjugacy (Exercise 6.25) is proved in Bezuglyi and Karpel (2014) with a different argument based on the factor complexity instead of the exponent of words. The conjugacy of primitive substitution shifts is known to be decidable (Durand and Leroy, 2020).

Exercise 6.30 is from Delecroix (2015) (see also Besbes et al. (2013)) where the result is credited to Boshernitzan (private communication). The condition $\inf n\varepsilon_n(X) > 0$ is a reinforcement of the condition $\limsup n\varepsilon_n(X) > 0$, which is proved to be equivalent to unique ergodicity in Boshernitzan (1992).

7

Dendric Shifts

In this chapter, we define the important class of dendric shifts, which are defined by a condition on the possible extensions of a word in their language. We prove a striking property of the sets of return words, namely that for every minimal dendric shift, the set of return words forms a basis of the free group (Theorem 7.1.15). We show that they have a finite $\mathcal{S}$-adic representation (Theorem 7.1.40). We illustrate these results on the class of Sturmian shifts (Section 7.2), which are a particular case of interval exchange shifts, considered in the next chapter. We next present the class of specular shifts (Section 7.3), which is built to generalize the class of linear involutions, itself a natural generalization of interval exchange transformations, and also presented in the next chapter.

7.1 Dendric Shifts

Let X be a shift space on the alphabet A. For $w \in \mathcal{L}(X)$ and $n \geq 1$, we denote

$$
\begin{aligned}
L_X(w) &= \{a \in A \mid aw \in \mathcal{L}(X)\}, \\
R_X(w) &= \{b \in A \mid wb \in \mathcal{L}(X)\}, \\
E_X(w) &= \{(a, b) \in L_X(w) \times R_X(w) \mid awb \in \mathcal{L}(X)\}.
\end{aligned}
$$

The *extension graph* of w, denoted $\mathcal{E}_X(w)$, is the undirected bipartite graph whose set of vertices is the disjoint union of $L_X(w)$ and $R_X(w)$ and whose edges are the elements of $E_X(w)$.

When the context is clear, we denote $L(w)$, $R(w)$, $E(w)$ and $\mathcal{E}(w)$ instead of $L_X(w)$, $R_X(w)$, $E_X(w)$ and $\mathcal{E}_X(w)$.

When we need to distinguish the disjoint copies of $L(w)$ and $R(w)$ forming the vertices of the extension graph $\mathcal{E}(w)$, we denote them by $1 \otimes L(w)$ and $R(w) \otimes 1$.

In general, the extension graphs $\mathcal{E}(w)$ are rather arbitrary except that they are bipartite since the edges connect a vertex of $R(w) \otimes 1$ to one of $1 \otimes L(w)$ (or conversely). They need not be connected and may contain nontrivial cycles. In order to define precisely the nontrivial cycles, we introduce the notion of reduced path. A path in an undirected graph is *reduced* if it does not contain successive equal edges. Thus, in a reduced path $(e_1, e_2, \ldots, e_n)$, we have $e_i \neq e_{i+1}$ for $1 \leq i < n$. Recall that an undirected graph is a tree if it is connected and acyclic (see Appendix B.2).

For any $w \in \mathcal{L}(X)$, since any vertex of $L_X(w)$ is connected by an edge to at least one vertex of $R_X(w)$, the bipartite graph $\mathcal{E}_X(w)$ is a tree if and only if there is a unique reduced path between every pair of vertices of $L_X(w)$ (resp. $R_X(w)$).

The shift X is said to be *eventually dendric* with *threshold* $m \geq 0$ if $\mathcal{E}_X(w)$ is a tree for every word $w \in \mathcal{L}_{\geq m}(X)$. It is said to be *dendric* if we can choose $m = 0$. Thus, a shift X is dendric if and only if $\mathcal{E}_X(w)$ is a tree for every word $w \in \mathcal{L}(X)$.

When X is a dendric shift (resp. eventually dendric shift), we also say that $\mathcal{L}(X)$ is a *dendric set* (resp. eventually dendric set).

An important observation is that, in any shift space, for a word $w \in \mathcal{L}(X)$ that is not bispecial, the graph $\mathcal{E}_X(w)$ is always a tree. Indeed, if w is not left-special, all vertices of $R_X(w)$ are connected to the unique vertex of $L(w)$ and thus $\mathcal{E}_X(w)$ is a tree. Also, if w is bispecial and such that $\mathcal{E}_X(w)$ is a tree, then w is neutral. Indeed, recall from Section 1.2.3 that w is neutral if its multiplicity,

$$m_X(w) = e_X(w) - \ell_X(w) - r_X(w) + 1,$$

is zero. But in an acyclic graph with n vertices, e edges and c connected components, we have

$$e - n = c \tag{7.1}$$

and thus $m_X(w) = 0$. Conversely, if $\mathcal{E}_X(w)$ is acyclic, then $m_X(w) = c - 1$ where c is the number of connected connected components of the graph $\mathcal{E}_X(w)$. In particular, if $\mathcal{E}_X(w)$ is acyclic and $m_X(w) = 0$, then $\mathcal{E}_X(w)$ is a tree.

We begin with an example of a non-minimal dendric shift.

Example 7.1.1 Let X be the shift space such that $\mathcal{L}(X) = a^*ba^*$ (we denote $a^* = \{a^n \mid n \geq 0\}$). The bispecial words are the words in a^*. Their extension graph is the tree represented in Figure 7.1. Thus, X is a dendric shift.

An important example of minimal dendric shifts is formed by *strict episturmian shifts* (also called *Arnoux–Rauzy shifts*).

Figure 7.1 The graph $\mathcal{E}_X(a)$.

Figure 7.2 The graph $\mathcal{E}(a)$.

Proposition 7.1.2 *Every Arnoux–Rauzy shift is dendric. In particular, every Sturmian shift is dendric.*

Proof Let X be an Arnoux–Rauzy shift. For every bispecial word $w \in \mathcal{L}(X)$, there is exactly one letter ℓ such that ℓw is right-special and one letter r such that wr is left-special. Thus, the extension graph $\mathcal{E}(w)$ has exactly two vertices ℓ, r that have degree more than one, with $\ell \in L(w)$ and $r \in R(w)$. Each vertex distinct from ℓ, r is connected to either ℓ or r by an edge but not to both and ℓ, r are connected by an edge. Thus, $\mathcal{E}(w)$ is a tree. ∎

Example 7.1.3 Let X be the *Fibonacci shift*, which is generated by the morphism $a \mapsto ab, b \mapsto a$. It is a Sturmian shift (Example 1.5.2). The graph $\mathcal{E}(a)$ is shown in Figure 7.2.

A shift space X is said to be *eventually dendric of characteristic c* if

1. for any $w \in \mathcal{L}_{\geq 1}(X)$, the extension graph $\mathcal{E}(w)$ is a tree, and
2. the graph $\mathcal{E}(\varepsilon)$ is a disjoint union of c trees.

An eventually dendric shift of characteristic $c \geq 1$ is eventually dendric with threshold 1 since the extension graphs of all nonempty words are trees. It is a dendric shift if $c = 1$.

Example 7.1.4 Let X be the shift generated by the morphism $a \mapsto ab, b \mapsto cda, c \mapsto cd, d \mapsto abc$. It is dendric of characteristic 2 (Exercise 7.1) and called the *Cassaigne shift*. The extension graph $\mathcal{E}(\varepsilon)$ is shown in Figure 7.3.

Example 7.1.5 Let X be the *Tribonacci shift*, which is the substitution shift generated by the *Tribonacci substitution* $\sigma : a \mapsto ab, b \mapsto ac, c \mapsto a$. It is an Arnoux–Rauzy shift (see Example 1.5.3) and thus a dendric shift.

Figure 7.3 The extension graph $\mathcal{E}(\varepsilon)$ for the Cassaigne shift.

The following statement shows that eventually dendric shifts have at most linear complexity. Recall from Chapter 1 that we denote $p_n(X) = \text{Card}(\mathcal{L}_n(X))$ and $s_n(X) = p_{n+1}(X) - p_n(X)$.

Proposition 7.1.6 *Let X be an eventually dendric shift on the alphabet A. Then X has at most linear complexity, that is, $p_n(X) \le Kn$ for all n, for some constant K. If X is dendric, then*

$$p_n(X) = (\text{Card}(A) - 1)n + 1. \tag{7.2}$$

Proof Let $b_n(X) = s_{n+1}(X) - s_n(X)$. Since X is eventually dendric, there is $n \ge 1$ such that the extension graph of every word in $\mathcal{L}_{\ge n}(X)$ is a tree. Then $b_p(X) = 0$ for every $p \ge n$. Indeed, by Proposition 1.2.17, we have $b_p(X) = \sum_{w \in \mathcal{L}_p(X)} m(w)$. Since all words of length p in $\mathcal{L}(X)$ are neutral, the conclusion follows. Thus, $s_p(X) = s_{p+1}(X)$ for every $p \ge n$, whence our conclusion.

If X is dendric, then $b_n(X) = 0$ for all $n \ge 1$ and thus $s_n(X)$ is constant. Since $s_1(X) = \text{Card}(A) - 1$ by the assumption $A \subset \mathcal{L}(X)$, this implies $p_n(X) = (\text{Card}(A) - 1)n + 1$. ∎

A more precise formula can be given for the complexity of eventually dendric shifts (Exercise 7.2).

Corollary 7.1.7 *The minimal dendric shifts on two letters are the Sturmian shifts.*

Proof We have already seen that a Sturmian shift is dendric (Proposition 7.1.2). Conversely, if X is a minimal dendric shift on two letters, its factor complexity is $p_n(X) = n + 1$ by Proposition 7.1.6. Since a minimal shift of complexity $n + 1$ is Sturmian (see Section 1.5), we conclude that X is Sturmian. ∎

On more than two letters, the class of dendric shifts is larger than the class of Arnoux–Rauzy shifts since it contains, as we shall see in Chapter 8, the class of interval exchange shifts.

Figure 7.4 The graph $\mathcal{E}(U, w, V)$.

The converse of Proposition 7.1.6 is not true, as shown by the following example.

Example 7.1.8 The *Chacon ternary shift* is the substitution shift X on the alphabet $A = \{0, 1, 2\}$ generated by the Chacon ternary substitution $\tau: 0 \to 0012, 1 \to 12, 2 \to 012$. Its complexity is $p_n(X) = 2n+1$ (see Exercise 6.17). It is not eventually dendric (Exercise 7.3).

7.1.1 Generalized Extension Graphs

We will need to consider extension graphs that correspond to extensions by words instead of letters. Let X be a shift space. For $w \in \mathcal{L}(X)$, and $U, V \subset \mathcal{L}(X)$, let $L(U, w) = \{\ell \in U \mid \ell w \in \mathcal{L}(X)\}$, let $R(w, V) = \{r \in V \mid wr \in \mathcal{L}(X)\}$ and let $E(U, w, V) = \{(\ell, r) \in U \times V \mid \ell wr \in \mathcal{L}(X)\}$. The *generalized extension graph* of w relative to U, V is the following undirected graph $\mathcal{E}(U, w, V)$. The set of vertices is made of two disjoint copies of $L(U, w)$ and $R(w, V)$. The edges are the elements of $E(U, w, V)$. The extension graph $\mathcal{E}(w)$ defined previously corresponds to the case where $U, V = A$.

Example 7.1.9 Let X be the Fibonacci shift. Let $w = a$, $U = \{aa, ba, b\}$ and let $V = \{aa, ab, b\}$. The graph $\mathcal{E}(U, w, V)$ is represented in Figure 7.4.

The following property shows that in a dendric shift, not only the extension graphs but, under appropriate hypotheses, all generalized extension graphs are acyclic.

Proposition 7.1.10 *Let X be a shift space and $n \geq 1$ be such that for every $w \in \mathcal{L}_{\geq n}(X)$, the graph $\mathcal{E}(w)$ is acyclic. Then, for every $w \in \mathcal{L}_{\geq n}(X)$, every finite suffix code U and every finite prefix code V, the generalized extension graph $\mathcal{E}(U, w, V)$ is acyclic.*

The proof uses the following lemma. We consider a finite suffix code U, a finite prefix code V and a word $\ell \in \mathcal{L}(X)$ such that $a\ell \in U$ for every $a \in L_X(\ell)$. Set $U' = (U \setminus A\ell) \cup \ell$ (see Figure 7.5).

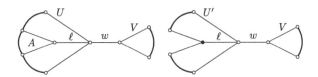

Figure 7.5 The sets U and U'. The set U is represented as the set of paths from the root of a tree located at the left end of w to its set of nodes, indicated by heavy lines, and similarly for U', V.

Lemma 7.1.11 *If the graphs $\mathcal{E}(U', w, V)$ and $\mathcal{E}(A, \ell w, V)$ are acyclic, then the graph $\mathcal{E}(U, w, V)$ is acyclic.*

Proof Assume that $\mathcal{E}(U, w, V)$ contains a cycle C. If the cycle does not use any of the vertices in U', it defines a cycle in the graph $\mathcal{E}(A, \ell w, V)$ obtained by replacing each vertex $a\ell$ of the graph $\mathcal{E}(U, w, V)$ for $a \in L_X(\ell)$ by a vertex a. Since $\mathcal{E}(A, \ell w, V)$ is acyclic, this is impossible. If it uses a vertex of U', it defines a cycle of the graph $\mathcal{E}(U', w, V)$ obtained by replacing each possible vertex $a\ell$ by ℓ (and suppressing the possible identical successive edges created by the identification). This is impossible since $\mathcal{E}(U', w, V)$ is acyclic. Thus, $\mathcal{E}(U, w, V)$ is acyclic.

■

Proof of Proposition 7.1.10 We show by induction on the sum of the lengths of the words in U, V that for any $w \in \mathcal{L}_{\geq n}(X)$, the graph $\mathcal{E}(U, w, V)$ is acyclic.

Let $w \in \mathcal{L}_{\geq n}(X)$. We may assume that $U = L(U, w)$ and $V = R(w, V)$ and also that $U, V \neq \emptyset$. If $U, V \subset A$, the property is true.

Otherwise, assume, for example, that U contains words of length at least 2. Let $u \in U$ be of maximal length. Set $u = a\ell$ with $a \in A$. Then $U' = (U \setminus A\ell) \cup \ell$ is a suffix code and $\ell w \in \mathcal{L}(X)$ since $U = L(U, w)$.

By induction hypothesis, the graphs $\mathcal{E}(U', w, V)$ and $\mathcal{E}(A, \ell w, V)$ are acyclic. By Lemma 7.1.11, the graph $\mathcal{E}(U, w, V)$ is acyclic.   ■

We prove now a similar statement concerning connectedness. For $w \in \mathcal{L}(X)$, we say that a suffix code $U \subset \mathcal{L}(X)$ is X-*maximal* for w if every word v such that vw belongs to $\mathcal{L}(X)$ is comparable for the suffix order with a word in U. For $w = \varepsilon$, we say X-maximal instead of X-maximal for ε. Thus, a suffix code is X-maximal if it is not strictly contained in any suffix code $U' \subset \mathcal{L}(X)$. The same definitions hold symmetrically for prefix codes. Thus, a prefix code $V \subset \mathcal{L}(X)$ is X-maximal for w if every word v such that $wv \in \mathcal{L}(X)$ is comparable for the prefix order with a word of V.

This definition is motivated by the following obvious statement.

Proposition 7.1.12 *Let X be a recurrent shift. For every $w \in \mathcal{L}(X)$, the set $\mathcal{R}_X(w)$ is an X-maximal prefix code for w and the set $\mathcal{R}'_X(w)$ is an X-maximal suffix code for w.*

We can now formulate the following complement to Proposition 7.1.10.

Proposition 7.1.13 *Let X be an eventually dendric shift with threshold n. For every $w \in \mathcal{L}_{\geq n}(X)$, every finite suffix code $U \subset \mathcal{L}(X)$ that is X-maximal for w and every finite prefix code $V \subset \mathcal{L}(X)$ that is X-maximal for w, the generalized extension graph $\mathcal{E}(U, w, V)$ is a tree.*

For a shift space X, two finite sets $U, V \subset \mathcal{L}(X)$ and $w \in \mathcal{L}(X)$, denote $\ell(U, w) = \mathrm{Card}(L(U, w))$, $r(w, V) = \mathrm{Card}(R(w, V))$ and $e(U, w, V) = \mathrm{Card}(E(U, w, V))$. Next, we define

$$m(U, w, V) = e(U, w, V) - \ell(U, w) - r(w, V) + 1.$$

Thus, for $U = V = A$, the integer $m(U, w, V)$ is the multiplicity $m(w)$ of w.

Lemma 7.1.14 *Let X be a shift space and $n \geq 0$ be such that $m(w) = 0$ for every $w \in \mathcal{L}_{\geq n+1}(X)$. Then for every $w \in \mathcal{L}_{\geq n}(X)$, every finite suffix code U that is X-maximal for w and every finite prefix code V that is X-maximal for w, we have $m(U, w, V) = m(w)$.*

Proof We use an induction on the sum of the lengths of the words in U and in V. We may assume that Uw, wV are included in $\mathcal{L}(X)$.

If U, V contain only words of length 1, since U (resp. V) is a suffix (resp. prefix) code that is X-maximal for w, we have $U = L(w)$ and $V = R(w)$ and there is nothing to prove. Assume next that one of them, say V, contains words of length at least 2. Let $p \in \mathcal{L}(X)$ be such that $pA \cap \mathcal{L}(X) \subset V$. Set $V' = (V \setminus pA) \cup \{p\}$. If wp does not belong to $\mathcal{L}(X)$, then $m(U, w, V) = m(U, w, V')$, and the conclusion follows by induction hypothesis. Thus, we may assume that wp is in $\mathcal{L}(X)$. Then

$$e(U, w, V) - e(U, w, V') = e(U, w, pA) - e(U, w, p)$$
$$= e(U, wp, A) - \ell(U, wp)$$

and

$$r(w, V) - r(w, V') = r(w, pA) - r(w, p) = r(wp, A) - 1.$$

This shows that

$$m(U, w, V) - m(U, w, V') = e(U, wp, A) - \ell(U, wp) - r(wp, A) + 1$$
$$= m(U, wp, A).$$

By induction hypothesis, we have $m(U, w, V') = m(w)$. But we have also $m(U, wp, A) = 0$ since $|wp| \geq n + 1$, whence the conclusion. ∎

Proof of Proposition 7.1.13 Let $w \in \mathcal{L}_{\geq n}(X)$. By Proposition 7.1.10, the graph $\mathcal{E}_{U,V}(w)$ is acyclic. Since, by Lemma 7.1.14, we have $m_{U,V}(w) = 0$, it follows from (7.1) that $\mathcal{E}_{U,V}(w)$ is a tree. ∎

7.1.2 Return Theorem

We will now prove the following result (called the *Return Theorem*).

Theorem 7.1.15 *Let X be a minimal dendric shift on the alphabet A with $A \subset \mathcal{L}(X)$. For every $u \in \mathcal{L}(X)$, the set $\mathcal{R}_X(u)$ is a basis of the free group on A.*

Note that, in the particular case of an Arnoux–Rauzy shift X, the property results directly from Eq. (1.34). Indeed, by Formula (1.34), the set of return words $\mathcal{R}'_X(u)$ is for every $u \in \mathcal{L}(X)$ conjugate to a set of the form $\alpha(A)$ where α is an automorphism of the free group on A (actually a product of Rauzy automorphisms).

A shift space X is *neutral* if every word u in $\mathcal{L}(X)$ is neutral. A dendric shift is, of course, neutral, but the converse is false (see Exercise 7.4). The first step of the proof of Theorem 7.1.15 is the following statement. It shows in particular that, under the hypotheses below, the cardinality of sets of return words is constant. We had already met this property in the case of strict episturmian shifts (see Eq. (1.34)).

Theorem 7.1.16 *If X is a recurrent neutral shift such that $A \subset \mathcal{L}(X)$, then for every $u \in \mathcal{L}(X)$, one has $\mathrm{Card}(\mathcal{R}_X(u)) = \mathrm{Card}(A)$.*

Note the following surprising consequence.

Corollary 7.1.17 *Every recurrent neutral shift is minimal.*

Indeed, if X is a recurrent neutral shift, all sets of return words are finite by Theorem 7.1.16. Thus, the shift is minimal. Thus, we could weaken the hypothesis in Theorem 7.1.15 to require X to be only recurrent.

The proof of Theorem 7.1.16 uses the following lemma. Recall from Section 1.2.3 that for $v \in \mathcal{L}(X)$, we denote $r_X(v) = \mathrm{Card}\{a \in A \mid va \in \mathcal{L}(X)\}$ or equivalently $r_X(v) = \mathrm{Card}(R_X(v))$.

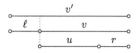

Figure 7.6 The set U is a suffix code.

Lemma 7.1.18 *Let X be a neutral shift. For every $v \in \mathcal{L}(X)$, set $\rho(v) = r_X(v) - 1$. Then one has*

$$\sum_{a \in L(v)} \rho(av) = \rho(v). \tag{7.3}$$

Proof Since v is neutral, we have $e_X(v) - \ell_X(v) - r_X(v) + 1 = 0$. Thus,

$$\sum_{a \in L(v)} \rho(av) = \sum_{a \in L(v)} (r_X(av) - 1) = e_X(v) - \ell_X(v)$$
$$= r_X(v) - 1 = \rho(v).$$

∎

Proof of Theorem 7.1.16 Let U be the set of proper prefixes of $u\mathcal{R}_X(u)$ that are not proper prefixes of u. We claim that U is an X-maximal suffix code. Indeed, assume first that v, v' belong to U, with v being a proper suffix of v'. Set $v' = \ell v$ with ℓ nonempty. Then u is a proper prefix of v and thus $v = ur$. This implies that $v' = \ell ur$, with the fact that v is a prefix of $u\mathcal{R}_X(u)$ (see Figure 7.6). This shows that U is a suffix code.

Now, since X is recurrent, every long enough word in $\mathcal{L}(X)$ has a factor equal to u. Thus, it has a suffix that begins with u and has no other factor equal to u. This suffix is in U. This shows that U is an X-maximal suffix code and proves the claim concerning U.

Set, as in Lemma 7.1.18, $\rho(v) = r_X(v) - 1$ for every $v \in \mathcal{L}(X)$.

Consider first the tree formed by the set P of prefixes of $u\mathcal{R}_X(u)$. The children of $p \in P$ are the $pa \in P$ for $a \in A$. Since $u\mathcal{R}_X(u)$ is a prefix code, the leaves are the elements of $u\mathcal{R}_X(u)$. The internal nodes are the elements of $Q = P \setminus u\mathcal{R}_X(u)$. As in any finite tree, the number of leaves minus 1 is equal to the sum over the internal nodes v of the integers $d(v) - 1$, where $d(v)$ is the number of children of v. For $v \in Q$, since $\mathcal{R}_X(u)$ is by Proposition 7.1.12 an X-maximal prefix code for w, we have

$$d(v) = \begin{cases} r_X(v) & \text{if } v \in U \\ 1 & \text{otherwise.} \end{cases}$$

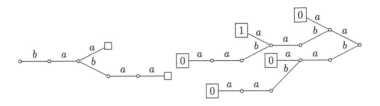

Figure 7.7 The trees $\mathcal{R}_X(u)$ and S.

Thus,

$$\text{Card}(\mathcal{R}_X(u)) - 1 = \text{Card}(u\mathcal{R}_X(u)) - 1 = \sum_{v \in Q}(d(v) - 1) = \sum_{v \in U}\rho(v). \quad (7.4)$$

Consider now the tree formed by the set S of suffixes of U. The root is ε and the children of a word $v \in S$ are the words $av \in S$ with $a \in A$. Since U is an X-maximal suffix code, the leaves of the tree S are the elements of U and the children of $v \in S \setminus U$ are all the av for $a \in L_X(v)$. Since for every internal node of S, the sum of the $\rho(av)$ taken over the children of v is equal to $\rho(v)$ (by Lemma 7.1.18), we have

$$\sum_{v \in U}\rho(v) = \rho(\varepsilon) = \text{Card}(A) - 1, \quad (7.5)$$

with the last equality resulting from the hypothesis $A \subset \mathcal{L}(X)$. Comparing (7.4) and (7.5), we obtain the desired equality. $\blacksquare$

We illustrate the proof with the following example.

Example 7.1.19 Let X be the Fibonacci shift and let $u = aa$. We have $\mathcal{R}_X(u) = \{baa, babaa\}$ and thus $U = \{aa, aab, aaba, aabab, aababa\}$. The tree formed by the set $\mathcal{R}_X(u)$ is represented in Figure 7.7 on the left. The tree S is represented on the right with the value of ρ indicated on the leaves. The unique leaf of S with a nonzero value of ρ is the unique right-special word that belongs to U, namely $aaba$.

Observe that the X-maximal suffix code U used in the proof is closely related with the partition in towers built from the set $\mathcal{R}_X(u)$ as in Proposition 4.1.1 (see Exercise 7.6).

Having proved Theorem 7.1.15, we now come to a second part of the proof of Theorem 7.1.15 in which we will prove that $\mathcal{R}_X(w)$ generates the free group on A.

In a graph $G = (V, E)$ labeled by an alphabet A, we consider for every edge e from v to w with label a, an *inverse edge* e^{-1}, which goes from w to v and

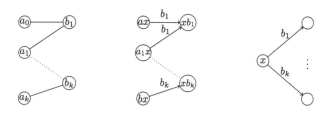

Figure 7.8 The path in $\mathcal{E}(w)$, the path in $\Gamma_{n+1}(X)$ and the folding.

is labeled a^{-1}. A *generalized path* in G is a sequence formed of consecutive edges or their inverses. The label of a generalized path is the reduced word, which is the reduction of the label of the path. Thus, it is an element of the free group on A.

Lemma 7.1.20 *Let X be a dendric shift such that $A \subset \mathcal{L}(X)$. For every $n \geq 1$, the group defined by the Rauzy graph $\Gamma_n(X)$ with respect to one of its vertices is the free group on A.*

Proof We will show that a sequence of Stallings foldings (see Section B.2.5 for the definition) reduces any Rauzy graph $\Gamma_{n+1}(X)$ to $\Gamma_n(X)$.

Consider two vertices ax, bx of $\Gamma_{n+1}(X)$ differing only by the first letter. The shift X being dendric, the extension graph $\mathcal{E}(x)$ of x is a tree, and there is a path $a_0, b_1, \ldots, a_{k-1}, b_k, a_k$ in $\mathcal{E}(x)$ from $a = a_0$ to $b = a_k$ (see Figure 7.8 on the left). The successive Stallings foldings at $xb_1, \ldots, xb_k$ identify the vertices $a_0 x, \ldots, a_k x$ (see Figure 7.8 in the middle). In this way, $\Gamma_{n+1}(X)$ is mapped onto $\Gamma_n(X)$.

Thus, the groups defined by the Rauzy graphs $\Gamma_n(X), \Gamma_{n-1}(X), \ldots, \Gamma_1(X)$ are all identical. Since $A \subset \mathcal{L}(X)$, the graph $\Gamma_1(X)$ defines the free group on A, and thus the same is true for $\Gamma_n(X)$. $\blacksquare$

We illustrate the proof of Lemma 7.1.20 with the following example.

Example 7.1.21 Consider the Fibonacci shift and the Rauzy graphs $\Gamma_n(X)$ for $n = 1, 2, 3$ represented in Figure 7.9. Since there are edges labeled b in $\Gamma_3(X)$ from aa and ba to ab, a Stallings folding merges aa and ba. The result is $\Gamma_2(X)$. Similarly, since there are edges labeled a from the vertices a and b to vertex a in $\Gamma_2(X)$, we merge the vertices a and b. The result is $\Gamma_1(X)$.

The following lemma shows in particular that the group defined by a strongly connected labeled graph is positively generated, that is, generated by a set of words on A (without inverses).

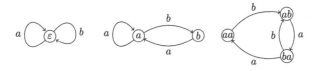

Figure 7.9 The Rauzy graphs of order $n = 1, 2, 3$ of the Fibonacci shift.

Figure 7.10 The decomposition of the path π.

Lemma 7.1.22 *Let G be a strongly connected labeled graph and x be a vertex. The group defined by G with respect to x is generated by the set S of labels of paths from x to x in G.*

Proof The subgroup $\langle S \rangle$ generated by the set S is clearly contained in the group defined by G with respect to x. Conversely, consider a generalized path π from x to x labeled y. We use an induction on the number r of inverse edges used in the path π to show that y belongs to the subgroup $\langle S \rangle$. If $r = 0$, then y is in S. Otherwise, we can write $y = ua^{-1}v$ where $x \xrightarrow{u} p \xrightarrow{a^{-1}} q \xrightarrow{v} x$ is a factorization of the path π. Since G is strongly connected, there are (ordinary) paths $p \xrightarrow{t} x$ and $x \xrightarrow{w} q$. Then

$$y = utt^{-1}a^{-1}w^{-1}wv = (ut)(wat)^{-1}wv.$$

By definition we have $wat \in S$ (see Figure 7.10) and by induction hypothesis, we have $ut, wv \in \langle S \rangle$. This shows that $y \in \langle S \rangle$ and concludes the proof. ∎

We are now ready for the proof of Theorem 7.1.15.

Proof of Theorem 7.1.15 Let $n \geq 1$ be such that the property of Proposition 4.5.8 holds for $x \in \mathcal{L}_n(X)$.

The inclusion $S \subset \mathcal{R}_X(u)^*$ implies the inclusion $\langle S \rangle \subset \langle \mathcal{R}_X(u) \rangle$. But, by Lemma 7.1.22, S generates the group defined by $\Gamma_{n+1}(X)$. By Lemma 7.1.20, this group is the whole free group on A. Thus, $\mathcal{R}_X(u)$ generates the free group on A. Since any generating set of $F(A)$ having $\mathrm{Card}(A)$ elements is a basis, and since $\mathcal{R}_X(u)$ has $\mathrm{Card}(A)$ elements by Theorem 7.1.16, this implies our conclusion. ∎

Example 7.1.23 Let $A = \{u, v, w\}$, and let $X = X(\sigma)$ be the shift generated by the substitution $\sigma : u \to vuwwv, v \to vuww, w \to vuwv$. The shift X is minimal since σ is primitive. It is also dendric. Consider indeed the morphism $\phi : u \to aa, v \to ab, w \to ba$. Then we have $\phi \circ \sigma = \varphi^3 \circ \phi$, where φ is the Fibonacci morphism. Indeed, we have

$$\phi \circ \sigma(u) = \phi(vuwwv) = abaababaab = \varphi^3(aa)$$

and similarly for v, w. This shows that X is obtained by reading the Fibonacci shift with nonoverlapping blocks of length 2 and thus that X is dendric (this is actually a particular case of Theorem 7.1.26). We have

$$\mathcal{R}_X(u) = \{wwvu, wwvvu, wvvu\},$$

which is a basis of the free group on $\{u, v, w\}$. Note that $\phi(\mathcal{R}_X(u))$ is a basis of a subgroup of index 2 of the free group on $\{a, b\}$.

7.1.3 Derivatives of Minimal Dendric Shifts

Let X be a minimal shift space and let $u \in \mathcal{L}(X)$. Let φ be a bijection from an alphabet B onto the set $\mathcal{R}_X(u)$ extended as usual to a morphism from B^* into A^*. The shift space $Y = \varphi^{-1}(X)$ is called the *derivative* of X with respect to u. Actually, the shift space Y is the induced system of X on the cylinder set $[u]$ (see Section 3.7). We will use (in the proof of Theorem 7.1.40) the following closure property of the family of minimal dendric shifts.

Theorem 7.1.24 *Any derivative of a minimal dendric shift is a minimal dendric shift on the same number of letters.*

Proof Let X be a minimal dendric shift on the alphabet A such that $\mathcal{L}(X)$ contains A let $u \in \mathcal{L}(X)$ and let φ be a bijection from an alphabet B onto $U = \mathcal{R}_X(u)$. By Theorem 7.1.16, the set $\mathcal{R}_X(u)$ has $\text{Card}(A)$ elements. Thus, we may choose $B = A$.

Set $Y = \varphi^{-1}(X)$. Since the shift space Y is an induced system of X, it is minimal.

Consider $y \in \mathcal{L}(Y)$ and set $x = \varphi(y)$. Let φ' be the bijection from A onto $U' = \mathcal{R}'_X(u)$ such that $u\varphi(b) = \varphi'(b)u$ for every $b \in B$. For $a, b \in B$, we have

$$(a, b) \in \mathcal{E}(y) \Leftrightarrow (\varphi'(a), \varphi(b)) \in \mathcal{E}(U', ux, U),$$

where $\mathcal{E}(U', ux, U)$ denotes the generalized extension graph of ux relative to U', U (see Section 7.1.1). Indeed,

$$ayb \in \mathcal{L}(Y) \Leftrightarrow u\varphi(a)x\varphi(b) \in \mathcal{L}(X) \Leftrightarrow \varphi'(a)ux\varphi(b) \in \mathcal{L}(X).$$

By Proposition 7.1.12, the set U' is an X-maximal suffix code for u and the set U is an X-maximal prefix code for u. By Proposition 7.1.13 the generalized extension graph $\mathcal{E}(U', ux, U)$ is a tree. Thus, $\mathcal{E}(y)$ is also a tree. This shows that Y is a dendric shift.   ∎

Example 7.1.25 Let X be the Tribonacci shift (see Example 7.1.5). It is the dendric shift generated by the morphism σ defined by $\sigma(a) = ab, \sigma(b) = ac$, $\sigma(c) = a$. Let $x = \sigma^\omega(a)$. We have $\mathcal{R}_X(a) = \{a, ba, ca\}$. Let $\varphi \colon A^* \to \mathcal{R}_X(a)^*$ be the morphism defined by $\varphi(a) = a$, $\varphi(b) = ba$, $\varphi(c) = ca$ and let $\varphi' \colon A^* \to \mathcal{R}'_X(a)^*$ be such that $a\varphi(a') = \varphi'(a')a$ for all $a' \in A$. We have $\sigma = \varphi' \circ \pi$ where π is the morphism realizing the circular permutation $\pi = (abc)$ on A. Set $z = \varphi'^{-1}(x)$. Since $\varphi' \circ \pi(x) = x$, we have $z = \pi(x)$. The shift Z generated by z is the derivative of X with respect to a. Since $Z = \pi(X)$, it is obtained by a permutation of the letters and is, of course, dendric.

7.1.4 Bifix Codes in Dendric Shifts

A *bifix code* on the alphabet A is a set U of words on A, which is both a prefix code and a suffix code. For example, for every $n \geq 1$, a set of words of length n is a bifix code.

Given a shift space X, a bifix code $U \subset \mathcal{L}(X)$ is called *X-complete* if it is an X-maximal prefix and suffix code. For example, $U = \mathcal{L}_n(X)$ is, for every $n \geq 1$, an X-complete bifix code.

Let X be a shift space and let $U \subset \mathcal{L}(X)$ be a finite X-complete bifix code. Let f be a coding morphism for U. Then $f^{-1}(\mathcal{L}(X))$ is factorial and extendable. The shift space Y such that $\mathcal{L}(Y) = f^{-1}(\mathcal{L}(X))$ is called the *decoding* of X by U. We denote $Y = f^{-1}(X)$.

As a particular case, when $U = \mathcal{L}_n(X)$ for some $n \geq 1$, the corresponding decoding $Y = f^{-1}(X)$ is called the decoding of X by *nonoverlapping blocks* of length n. The map f is then a conjugacy from Y onto the system (X, S^n) (see Example 7.1.27).

The following result expresses a closure property of the family of dendric shifts.

Theorem 7.1.26 *The decoding of a dendric shift X by a finite X-complete bifix code is a dendric shift.*

Proof Let $U \subset \mathcal{L}(X)$ be a finite X-maximal bifix code. Let $f: B^* \to A^*$ be a coding morphism for U and let $Y = f^{-1}(X)$. For $w \in \mathcal{L}(Y)$ and $a, b \in B$, we have

$$(a, b) \in E_Y(w) \Leftrightarrow (f(a), f(b)) \in E(U, f(w), U)$$

and thus the graph $\mathcal{E}_Y(w)$ is a tree by Proposition 7.1.13. This shows that Y is a dendric shift. ∎

It can be proved that the above closure property also holds for the family of minimal dendric shifts (Exercise 7.16).

Example 7.1.27 Let X be the Fibonacci shift on $\{a, b\}$ and consider the bifix code $U = \{aa, ab, ba\}$. The corresponding decoding of X is the shift of Example 7.1.23.

We will prove the following result.

Theorem 7.1.28 *Let X be a dendric shift on the alphabet A. A finite bifix code $U \subset \mathcal{L}(X)$, which is a basis of the free group on A, is equal to A.*

The following example shows that the hypothesis that X is dendric is necessary in Theorem 7.1.28.

Example 7.1.29 Let $A = \{a, b, c\}$ and $U = \{ab, acb, acc\}$. The set U is a bifix code. It is also a basis of the free group on A. Indeed, we have $accb = (acb)(ab)^{-1}(acb)$ and $b = (acc)^{-1}(accb)$. Next, $a = (ab)b^{-1}$ and $c = a^{-1}(acb)b^{-1}$. Thus, a, b, c belong to the group generated by U. Observe that we can verify directly that no dendric shift X can be such that $U \subset \mathcal{L}(X)$. Indeed, this would force ab, cb, cc, ac to belong to $\mathcal{L}(X)$ and thus the extension graph of ε to contain a cycle (see Figure 7.13).

To prove Theorem 7.1.28, we introduce the following notion. Let U be a bifix code and let P (resp. S) be the set of proper prefixes (resp. suffixes) of the words of U. The *incidence graph* of U is the following undirected graph. Its set of vertices is the disjoint union of P and S. The edges are $(\varepsilon, \varepsilon)$ and the pairs (p, s) is in $P \times S$ such that $ps \in U$.

Example 7.1.30 Let X be the Fibonacci shift and let $U = \mathcal{L}_3(X)$. The incidence graph of U is represented in Figure 7.11 (in each of the three parts, the vertices on the left are in P and those on the right in S).

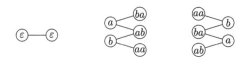

Figure 7.11 The incidence graph of $U = \mathcal{L}_3(X)$.

Proposition 7.1.31 *Let X be a dendric shift and let $U \subset \mathcal{L}(X)$ be a bifix code. Let P (resp. S) be the set of proper prefixes (resp. proper suffixes) of U and let G be the incidence graph of U. Then the following assertions hold.*

(i) *The graph G is acyclic.*
(ii) *The intersection of $P' = P \setminus \{\varepsilon\}$ (resp. $S' = S \setminus \{\varepsilon\}$) with each connected component of G is a suffix (resp. prefix) code.*
(iii) *For every reduced path $(v_1, u_1, \ldots, u_n, v_{n+1})$ in G with $u_1, \ldots, u_n \in P'$ and $v_1, \ldots, v_{n+1}$ in S', the longest common prefix of v_1 and v_{n+1} is a proper prefix of all $v_1, \ldots, v_n, v_{n+1}$.*
(iv) *Symmetrically, for every reduced path $(u_1, v_1, \ldots, v_n, u_{n+1})$ in G with $u_1, \ldots, u_{n+1} \in P'$ and $v_1, \ldots, v_n \in S'$, the longest common suffix of u_1, u_{n+1} is a proper suffix of $u_1, u_2, \ldots, u_{n+1}$.*

Proof Assertions (iii) and (iv) imply Assertions (i) and (ii). Indeed, assume that (iii) holds. Consider a reduced path $(v_1, u_1, \ldots, u_n, v_{n+1})$ in G with $u_1, \ldots, u_n \in P'$ and $v_1, \ldots, v_{n+1}$ in S'. If $v_1 = v_{n+1}$, then v_1 is a prefix of all v_i and in particular of v_2, a contradiction since U is a bifix code. Thus, G is acyclic and (i) holds. Next, if v_1, v_{n+1} are comparable for the prefix order, their longest common prefix is one of them, a contradiction with (iii) again. The assertion on P' is proved in an analogous way using Assertion (iv).

We prove simultaneously (iii) and (iv) by induction on $n \geq 1$.

The assertions hold for $n = 1$. Indeed, if $u_1 v_1, u_1 v_2 \in U$ and if $v_1 \in \mathcal{L}(X)$ is a prefix of $v_2 \in S'$, then $u_1 v_1$ is a prefix of $u_1 v_2$, a contradiction with the hypothesis that U is a prefix code. The same holds symmetrically for $u_1 v_1, u_2 v_1 \in U$ since U is a suffix code.

Let $n \geq 2$ and assume that the assertions hold for any path of length at most $2n - 2$. We treat the case of a path $(v_1, u_1, \ldots, u_n, v_{n+1})$ in G with $u_1, \ldots, u_n \in P'$ and $v_1, \ldots, v_{n+1}$ in S'. The other case is symmetric.

Let p be the longest common prefix of v_1 and v_{n+1}. We may assume that p is nonempty since otherwise the statement is obviously true. Any two elements of the set $L = \{u_1, \ldots, u_n\}$ are connected by a path of length at most $2n - 2$ (using elements of $\{v_2, \ldots, v_n\}$). Thus, by induction hypothesis, L is a suffix code. Similarly, any two elements of the set $R = \{v_1, \ldots, v_n\}$ are connected by

Figure 7.12 The coset graphs of $\{a, bab\}$ and of $\{a, bab, baab\}$.

a path of length at most $2n - 2$ (using elements of $\{u_1, \ldots, u_{n-1}\}$). Thus, R is a prefix code. We cannot have $v_1 = p$ since otherwise, using the fact that $u_n p$ is a prefix of $u_n v_{n+1}$ and thus in $\mathcal{L}(X)$, the generalized extension graph $\mathcal{E}_{L,R}(\varepsilon)$ would have the cycle $(p, u_1, v_2, \ldots, u_n, p)$, a contradiction since $\mathcal{E}_{L,R}(\varepsilon)$ is acyclic by Proposition 7.1.10. Similarly, we cannot have $v_{n+1} = p$.

Set $W = p^{-1} R$ and $R' = (R \setminus pW) \cup p$. Since R is a prefix code and since p is a proper prefix of R, the set R' is a prefix code. Suppose that p is not a proper prefix of all $v_2, \ldots, v_n$. Then there exist i, j with $1 \le i < j \le n + 1$ such that p is a proper prefix of v_i, v_j but not of any $v_{i+1}, \ldots, v_{j-1}$. Then $v_{i+1}, \ldots, v_{j-1} \in R'$ and there is the cycle $(p, u_i, v_{i+1}, u_{i+1}, \ldots, v_{j-1}, u_{j-1}, p)$ in the graph $\mathcal{E}_{U,V'}(\varepsilon)$. This is in contradiction with Proposition 7.1.10 because, V' being a prefix code, $\mathcal{E}_{L,R'}(\varepsilon)$ is acyclic. Thus, p is a proper prefix of all $v_2, \ldots, v_n$. ∎

Let X be a dendric shift and let $U \subset \mathcal{L}(X)$ be a bifix code. Let P be the set of proper prefixes of the words of U. Let θ_U be the equivalence on P defined by $p \equiv q \bmod \theta_U$ if p, q are in the same connected component of the incidence graph of U. Note that, since U is bifix, the class of ε is reduced to ε. The *coset graph* of U is the following labeled graph. The set of vertices is the set R of classes of θ_U. There is an edge labeled a from the class of p to the class of q in each of the following cases:

(i) $q = pa$,
(ii) $q = \varepsilon$ and $pa \in U$.

Example 7.1.32 Let X be the Fibonacci shift. The coset graph of $\{a, bab\}$ is shown in Figure 7.12 on the left and the coset graph of $\{a, bab, baab\}$ on the right.

A *simple path* from a vertex v to itself in a graph is a path that is not a concatenation of two nonempty paths from v to itself.

Proposition 7.1.33 *Let X be a dendric shift and let $U \subset \mathcal{L}(X)$ be a finite bifix code. Let P be the set of proper prefixes of U and let $H = \langle U \rangle$ be the subgroup generated by U. Let also C be the coset graph of U.*

1. *For every $p, q \in P$, $p \equiv q$ mod θ_U implies $Hp = Hq$.*
2. *If $p \equiv p'$ mod θ_U and if $p \xrightarrow{a} q$, $p' \xrightarrow{a} q'$ are edges in C, then $q \equiv q'$ mod θ_U.*
3. *Every $u \in U$ is the label of a simple path from ε to itself in C.*
4. *The graph C is the Stallings graph of the subgroup $\langle U \rangle$ generated by U.*

Proof Let $C = (R, E)$ be the coset graph of U and let G be its incidence graph.

1. The first assertion is clear since θ_U is the equivalence on P generated by the pairs p, q such that there is an s with $ps, qs \in U$ and thus $p, q \in Hs^{-1}$.

2. We assume that $q = pa$ and $q' = p'a$. The other cases are similar. Let s, s' be such that $qs, q's' \in U$. Let $p = u_0, v_1, u_1, \ldots, v_n, u_n = p'$ be a path from p to p' in the incidence graph G. Set $v_0 = as$ and $v_{n+1} = as'$. Then $(v_0, u_0, \ldots, u_n, v_{n+1})$ is a path in G. But since the letter a is a common prefix of v_0 and v_{n+1}, by Proposition 7.1.31 it is also a common prefix of all v_i. Set $v_i = av_i'$ for $0 \le i \le n + 1$. Then $(u_0 a, v_1', u_1 a, \ldots, v_n', u_n a)$ is a path from q to q' in the coset graph C and thus $q \equiv q'$ mod θ_U.

3. This follows from the fact that C can be obtained by Stallings foldings from the graph on P with edges $p \xrightarrow{a} q$ if either $pa = q$ or $q = \varepsilon$ and $pa \in U$.

4. Let K be the group defined by the coset graph C. Let us show that K is equal to H. By construction, we have $U \subset K$ and thus $H \subset K$. The converse follows easily from Assertion 1.

Let us finally show that C is Stallings reduced. Assume that $p, q \in P$ are such that there are edges with the same label a from the class $\bar{p}, \bar{q}$ of p, q to the same vertex $\bar{r}$. Let v be the label of a path from $\bar{r}$ to ε that does not pass by ε before. Then $pav, qav \in U$ and thus $p \equiv q$ mod θ_U, which implies that $\bar{p} = \bar{q}$. This shows that C is Stallings reduced. ∎

We are now ready to prove Theorem 7.1.28.

Proof of Theorem 7.1.28 Let X be a dendric shift on the alphabet A. Let $U \subset \mathcal{L}(X)$ be a bifix code that is a basis of the free group on A. By Proposition 7.1.33, the coset graph C of U has only one vertex ε and loops $\varepsilon \xrightarrow{a} \varepsilon$ for every $a \in A$. Since, by Proposition 7.1.33 again, every word of U is the label of a simple path from ε to itself in the coset graph of U, we have $U \subset A$ and thus $U = A$. ∎

7.1.5 Tame Automorphisms

An automorphism α of the free group on A is *positive* if $\alpha(a) \in A^+$ for every $a \in A$. We say that a positive automorphism of the free group on A is *tame* if it belongs to the submonoid generated by the permutations of A and the automorphisms $\alpha_{a,b}$, $\tilde{\alpha}_{a,b}$ defined for $a, b \in A$ with $a \neq b$ by

$$\alpha_{a,b}(c) = \begin{cases} ab & \text{if } c = a \\ c & \text{otherwise} \end{cases} \quad \text{and} \quad \tilde{\alpha}_{a,b}(c) = \begin{cases} ba & \text{if } c = a \\ c & \text{otherwise.} \end{cases}$$

Thus, $\alpha_{a,b}$ places a letter b after each a and $\tilde{\alpha}_{a,b}$ places a letter b before each a. The above automorphisms and the permutations of A are called the *elementary* positive automorphisms on A. The monoid of positive automorphisms is not finitely generated as soon as the alphabet has at least three generators (see Section 7.8).

A basis U of the free group is *positive* if $U \subset A^+$. A positive basis U of the free group is *tame* if there exists a tame automorphism α such that $U = \alpha(A)$.

Example 7.1.34 The set $U = \{ba, cba, cca\}$ is a tame basis of the free group on $\{a, b, c\}$. Indeed, one has the following sequence of elementary automorphisms:

$$(b, c, a) \xrightarrow{\alpha_{c,b}} (b, cb, a) \xrightarrow{\tilde{\alpha}_{a,c}^2} (b, cb, cca) \xrightarrow{\alpha_{b,a}} (ba, cba, cca).$$

The fact that U is a basis can be checked directly since $(cba)(ba)^{-1} = c$, $c^{-2}(cca) = a$ and finally $(ba)a^{-1} = b$.

The following result will play a key role in the proof of the main result of this section (Theorem 7.1.38).

Proposition 7.1.35 *A set $U \subset A^+$ is a tame basis of the free group on A if and only if $U = A$ or there is a tame basis V of the free group on A and $u, v \in V$ such that $U = (V \setminus v) \cup uv$ or $U = (V \setminus u) \cup uv$.*

Proof Assume first that U is a tame basis of the free group on A. Then $U = \alpha(A)$, where α is a tame automorphism of $\langle A \rangle$. Then $\alpha = \alpha_1 \alpha_2 \cdots \alpha_n$ where the α_i are elementary positive automorphisms. We use an induction on n. If $n = 0$, then $U = A$. If α_n is a permutation of A, then $U = \alpha_1 \alpha_2 \cdots \alpha_{n-1}(A)$ and the result holds by induction hypothesis. Otherwise, set $\beta = \alpha_1 \cdots \alpha_{n-1}$ and $V = \beta(A)$. By induction hypothesis, V is tame. If $\alpha_n = \alpha_{a,b}$, set $u = \beta(a)$ and $v = \beta(b) = \alpha(b)$. Then

$$U = \alpha(A \setminus a) \cup \alpha(a) = \beta(A \setminus a) \cup \beta(ab)$$
$$= (V \setminus u) \cup uv$$

and thus the condition is satisfied. The case where $\alpha_n = \tilde{\alpha}_{a,b}$ is symmetrical.

Figure 7.13 The graph $\mathcal{E}(\varepsilon)$.

Conversely, assume that V is a tame basis and that $u, v \in V$ are such that $U = (V \setminus u) \cup uv$. Then, there is a tame automorphism β of $F(A)$ such that $V = \beta(A)$. Set $a = \beta^{-1}(u)$ and $b = \beta^{-1}(v)$. Then $U = \beta \circ \alpha_{a,b}(A)$ and thus U is a tame basis.   ∎

We note the following corollary.

Corollary 7.1.36 *A tame basis of the free group that is a bifix code is the alphabet.*

Proof Assume that U is a tame basis that is not the alphabet. By Proposition 7.1.35 there is a tame basis V and $u, v \in V$ such that $U = (V \setminus v) \cup uv$ or $U = (V \setminus u) \cup uv$. In the first case, U is not prefix. In the second one, it is not suffix.   ∎

Example 7.1.37 The set $U = \{ab, acb, acc\}$ is a basis of the free group on $\{a, b, c\}$ (see Example 7.1.29). The set U is bifix and thus it is not a tame basis by Corollary 7.1.36.

The following result is a remarkable consequence of Theorem 7.1.28.

Theorem 7.1.38 *Any basis of the free group included in the language of a minimal dendric shift is tame.*

Proof Let X be a minimal dendric shift. Let $U \subset \mathcal{L}(X)$ be a basis of the free group on A. We use an induction on the sum $\lambda(U)$ of the lengths of the words of U. If U is bifix, by Theorem 7.1.28, we have $U = A$. Next assume, for example, that U is not prefix. Then there are nonempty words u, v such that $u, uv \in U$. Let $V = (U \setminus uv) \cup v$, so that $V \subset \mathcal{L}(X)$. Then V is a basis of the free group and $\lambda(V) < \lambda(U)$. By induction hypothesis, V is tame. Since $U = (V \setminus v) \cup uv$, U is tame by Proposition 7.1.35.   ∎

Example 7.1.39 The set $U = \{ab, acb, acc\}$ is a basis of the free group that is not tame (see Example 7.1.37). Accordingly, the extension graph $\mathcal{E}(\varepsilon)$ relative to the set of factors of U is not a tree (see Figure 7.13).

7.1.6 S-adic Representation of Dendric Shifts

Let S be a set of morphisms and $\tau = (\tau_n)_{n \geq 0}$ be a directive sequence of morphisms in S with $\tau_n \colon A_{n+1}^* \to A_n^*$ and $A_0 = A$. When τ is primitive and proper, we have by Lemma 6.4.6

$$\mathcal{L}(\tau) = \bigcap_{n \geq 1} \mathrm{Fac}(\tau_{(0,n)}(A_n^*)), \qquad (7.6)$$

where $\mathrm{Fac}(L)$ denotes the set of factors of the words in L.

For dendric shifts, the next theorem significantly improves the "only if" part of Proposition 6.4.10. Indeed, for such shifts, the set S can be replaced by the set S_e of elementary positive automorphisms. In particular, A_n is equal to A for all n.

Theorem 7.1.40 *Every minimal dendric shift on A, with* $\mathrm{Card}(A) \geq 2$, *has*

1. *a primitive proper unimodular S-adic representation* $\tau = (\tau_n)$ *where every* τ_n *is an automorphism of the free group on A and also*
2. *a primitive S_e-adic representation.*

Proof Since $\mathrm{Card}(A) \geq 2$, X is infinite by (7.2) and thus aperiodic. By Lemma 6.4.11, for any sequence $(u_n)_{n \geq 0}$ of words of $\mathcal{L}(X)$ such that $u_0 = \varepsilon$ and u_n is a proper suffix of u_{n+1}, the sequence of morphisms $(\tau_n)_{n \geq 0}$, defined by $\alpha_{n+1} = \alpha_n \circ \tau_n$ where α_n is a coding morphism for $\mathcal{R}_X(u_n)$, is a primitive proper S-adic representation of X with $S = \{\tau_n \mid n \geq 0\}$. By Theorem 7.1.15, the set $\mathcal{R}_X(u_1)$ is a basis of the free group on A. This implies that the matrix $M(\tau_n)$ is unimodular. This proves the first assertion.

To prove Assertion 2, all we need to do is to consider such a sequence $(u_n)_{n \geq 0}$ such that τ_n is tame for all n.

Let $u_1 = a^{(0)}$ be a letter in A. Since X is dendric, the set $\mathcal{R}_X(u_1)$ has $\mathrm{Card}(A)$ elements by Theorem 7.1.16. Let $\tau_0 \colon A \to \mathcal{R}_X(u_1)$ be a bijection. By Theorem 7.1.15 again, since the set $\mathcal{R}_X(u_1)$ is a basis of the free group on A, by Theorem 7.1.38, it is a tame basis. Thus, the morphism $\tau_0 \colon A^* \to A^*$ is a tame automorphism. Let $a^{(1)} \in A$ be a letter and set $u_2 = \tau_0(a^{(1)})$. Thus, $u_2 \in \mathcal{R}_X(u_1)$ and u_1 is a suffix of u_2. By Theorem 7.1.24, the derivative shift $X^{(1)} = \tau_0^{-1}(X)$ is a minimal dendric shift on the alphabet A. We thus reiterate the process with $a^{(1)}$ and we conclude by induction with $u_n = \tau_0 \cdots \tau_{n-2}(a^{(n-1)})$ for all $n \geq 2$. ∎

We illustrate Theorem 7.1.40 by the following example.

Example 7.1.41 Let $A = \{a, b, c\}$, let σ be the substitution defined by $\sigma(a) = ac$, $\sigma(b) = bac$, $\sigma(c) = cbac$ and let X be the substitution shift generated by σ. It can be shown that X is dendric (Exercise 7.17). We have $\sigma = \alpha_{a,c}\alpha_{b,a}\alpha_{c,b}$. Thus, S has the $\mathcal{S}_e$-adic representation $(\sigma_n)_{n \geq 0}$ given by the periodic sequence $\sigma_{3n} = \alpha_{a,c}$, $\sigma_{3n+1} = \alpha_{b,a}$, $\sigma_{3n+2} = \alpha_{c,b}$.

The converse of Theorem 7.1.40 is not true, as shown by Example 7.1.42 (see also Exercise 7.19).

Example 7.1.42 Let $A = \{a, b, c\}$ and let $\sigma : a \mapsto ac, b \mapsto bac, c \mapsto cb$. The substitution shift generated by σ (it is generated by the fixed point $\sigma^\omega(a)$) is not dendric since $bb, bc, cb, cc \in \mathcal{L}(X)$ and thus $\mathcal{E}(\varepsilon)$ has a cycle, although σ is a tame automorphism since $\sigma = \alpha_{a,c}\alpha_{c,b}\alpha_{b,a}$.

7.1.7 BV-Representation and Dimension Groups of Dendric Shifts

We use the $\mathcal{S}$-adic representation of minimal dendric shifts to describe a BV-representation.

Theorem 7.1.43 *Every minimal dendric shift on A has a BV-representation $(V, E, \leq)$ with simple hat such that the morphism read on $E(n)$ is for every $n \geq 2$ an automorphism of the free group on A.*

Proof Let X be a minimal dendric shift on the alphabet A. We may assume that Card$(A) \geq 2$. By Theorem 7.1.40, there is a primitive unimodular proper $\mathcal{S}$-adic representation $\tau = (\tau_n)$ of X such that every τ_n is an automorphism of $F(A)$. The conclusion then follows by Theorem 6.4.22. ∎

Note that Theorem 7.1.43 gives in particular a BV-representation of Arnoux–Rauzy shifts. Note that, as we have already seen, the converse of Theorem 7.1.43 is not true (see Exercise 7.19).

Recall that $\mathcal{M}(X, S)$ denotes the set of invariant measures on a shift space X.

Theorem 7.1.44 *The dimension group of a minimal dendric shift X on the alphabet A is $(G, G^+, 1_G)$ with $G = \mathbb{Z}^A$, $G^+ = \{x \in \mathbb{Z}^A \mid \langle x, \mu \rangle > 0, \mu \in \mathcal{M}(X, S)\} \cup 0$ and $1_G = 1$ where 1 is the vector with all components equal to 1 and μ is the vector $(\mu([a])_{a \in A}$.*

Proof By Theorem 7.1.40, X has a primitive proper and unimodular $\mathcal{S}$-adic representation. Thus, the form of the dimension group is given by Theorem 6.5.4. ∎

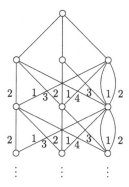

Figure 7.14 The BV-representation of the dendric shift X.

Example 7.1.45 Consider again the dendric shift $X = X(\sigma)$ generated by the automorphism $\sigma : a \rightarrow ac, b \rightarrow bac, c \rightarrow cbac$ of Example 7.1.41. Since every word in the image of σ ends with ac, the morphism $\tau : x \mapsto c\sigma(x)c^{-1}$ (where c^{-1} is the inverse of c in the free group) is proper and $X = X(\tau)$. Since τ is proper, the shift X has a stationary BV-representation with matrix $M(\sigma) = M(\tau)$ shown in Figure 7.14.

This implies by Theorem 5.3.6 that the dimension group of X is the group of the matrix $M(\sigma)$. The matrix $M(\sigma)$ is

$$M(\sigma) = \begin{bmatrix} 1 & 0 & 1 \\ 1 & 1 & 1 \\ 1 & 1 & 2 \end{bmatrix}.$$

The dominant eigenvalue is the largest root λ of $\lambda^3 - 4\lambda^2 + 5\lambda - 1 = 0$. The vector $w = \begin{bmatrix} \lambda & \lambda - 1 & (\lambda - 1)^2 \end{bmatrix}$ is a corresponding eigenvector. The map $x \mapsto \langle x, w \rangle$ sends $K^0(X, S)$ onto $\mathbb{Z}[\lambda]$. The image of the unit vector $\mathbf{1}_M$ is λ^2, which is a unit of $\mathbb{Z}[\lambda]$. Thus, the dimension group of X is isomorphic to $\mathbb{Z}[\lambda]$.

7.2 Sturmian Shifts

We illustrate the preceding results on the family of Sturmian shifts. We have seen that Sturmian shifts are dendric (Proposition 7.1.2). In the particular case of dendric shifts, all the general results concerning dendric shifts can be formulated more precisely. We have already seen that for the Return Theorem. We give below the complete description of the S-adic representations of Sturmian shifts.

7.2.1 BV-Representation of Sturmian Shifts

We define the morphisms ρ_n and γ_n, $n \geq 1$ from $\{0, 1\}$ to $\{0, 1\}^*$ by

$$\begin{array}{ll} \rho_n(0) = 01^n & \gamma_n(0) = 10^{n+1} \\ \rho_n(1) = 01^{n+1} & \text{and} \quad \gamma_n(1) = 10^n \, . \end{array}$$

The morphisms ρ_n, γ_n are related as follows to the elementary automorphisms L_0, L_1 introduced in Section 1.5. We have for every $n \geq 1$ and every $u \in \{0, 1\}^*$,

$$1^n \rho_n(u) = L_{1^n 0}(u) 1^n, \quad 0^n \gamma_n(u) = L_{0^n 1}(u) 0^n,$$

as one verifies easily for $u = 0, 1$, which implies the identities for all u.

The following result will be used to give a BV-representation of Sturmian shifts.

Proposition 7.2.1 *Let X be a Sturmian shift of directive sequence $\Delta = 0^{d_1} 1^{d_2} \cdots$ with $d_1 \geq 0$ and $d_n \geq 1$ for $n \geq 2$. There is a Sturmian shift Y and an $n \geq 1$ such that either $X = \rho_n(Y)$ or $X = \gamma_n(Y)$. More precisely, if $d_1 > 0$, then $n = d_1$, $X = \gamma_n(Y)$ and Y is the Sturmian shift of directive sequence $1^{d_2-1} 0^{d_3} \cdots$. If $d_1 = 0$, then $n = d_2$, $X = \rho_n(Y)$ and Y is the Sturmian shift of directive sequence $0^{d_3-1} 1^{d_4} \cdots$.*

Proof Let X^+ be the one-sided shift space associated with X and let x be the standard Sturmian sequence that belongs to X^+. By Theorem 1.5.4, we have $x = \text{Pal}(\Delta)$. Assume that $d_1 > 0$ and set $\Delta = 0^n 1 \Delta'$ with $n = d_1$. Then, by Justin's Formula (1.32), we have $x = L_{0^n 1}(\text{Pal}(\Delta'))$. We have seen that $L_{0^n 1}(u) 0^n = 0^n \gamma_n(u)$ for every $u \in \{0, 1\}^*$. Thus, $x = 0^n \gamma_n(y)$, where $y = \text{Pal}(\Delta')$ is Sturmian. This shows that $X^+ = \gamma_n(Y^+)$, where Y^+ is the one-sided shift generated by y. Finally, we obtain $X = \gamma_n(Y)$, where Y is the two-sided shift associated with Y^+. The case where Δ begins with 1 is analogous. ∎

Example 7.2.2 Let $X = X(\sigma)$ with $\sigma : 0 \to 01$, $1 \to 010$. Actually, X is the Sturmian shift of slope $\alpha = \sqrt{2} - 1$. Indeed, we have $\alpha = [0; 2, 2, \ldots]$ and thus the directive word of c_α is $011001100 \cdots = (0110)^\omega$. The standard word with slope α is thus the fixed point of the morphism $L_{0110} : 0 \to 01010$, $1 \to 0101001$, which is σ^2. We have $\sigma(0)0 = 0\gamma_1(1)$ and $\sigma(1)0 = 0\gamma_1(0)$. Thus, $X = \gamma_1(Y)$, where Y is obtained from X by exchanging 0 and 1.

Theorem 7.2.3 *Let X be a Sturmian shift on $\{0, 1\}$. There exists a sequence $(\zeta_n)_{n \in \mathbb{N}}$ taking values in $\{\rho_1, \gamma_1, \rho_2, \gamma_2, \ldots\}$ such that $(\mathcal{P}(n))_n$ is a refining sequence of KR-partitions with $\mathcal{P}(1) = \{[0], [1]\}$ and, for $n \geq 2$,*

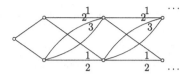

Figure 7.15 The BV-representation of the Sturmian shift X.

$$\mathcal{P}(n) = \left\{ S^k \zeta_1 \cdots \zeta_n ([a]) \mid 0 \leq k < |\zeta_1 \cdots \zeta_{n-1}(a)|, \ a \in \{0, 1\} \right\}.$$

Proof We apply iteratively Proposition 7.2.1 to obtain the sequence (ζ_n). To see that condition (KR1) (the intersection of the bases is reduced to a point) is satisfied, note that if, for example, $\zeta_n = \rho_i$, then

$$\zeta_1 \cdots \zeta_n ([a]) \subseteq \left[\zeta_1 \cdots \zeta_{n-1}(1).\zeta_1 \cdots \zeta_n (a) \zeta_1 \cdots \zeta_{n-1}(0) \right].$$

∎

Let X be a Sturmian shift and $(\mathcal{P}(n))_n$ be the sequence of partitions given by Corollary 7.2.3. With such a sequence is associated an ordered Bratteli–Vershik diagram $B = (V, E, \leq)$, which can be described as follows. For all $n \geq 1$, V_n consists of two vertices, and the substitution read on E_{n+1} is ζ_n, with $E(1)$ consisting of a simple hat. We have thus proved the following statement.

Corollary 7.2.4 *A shift space is Sturmian if and only if it has a BV-representation with simple hat, with two vertices at every level and such that the substitution read on E_{n+1} is some ρ_i or γ_i.*

Example 7.2.5 Let $X = X(\sigma)$ with σ as in Example 7.2.2. The sequence (ζ_n) can be chosen to be $\zeta_n = \tau$ where $\tau \colon 0 \to 10, 1 \to 100$ (the composition of γ_1 with the exchange of 0, 1). The corresponding Bratteli diagram is shown in Figure 7.15.

Example 7.2.6 Let X be the Fibonacci shift. The directive sequence is $\Delta = (01)^\omega$ (Example 1.5.5). According to Proposition 7.2.1, we have $X = \gamma_1(X)$. Thus, we obtain directly the BV-representation of Figure 6.12.

7.2.2 Linearly Recurrent Sturmian Shifts

We obtain as a corollary of Theorem 7.2.3, the following result.

Corollary 7.2.7 *A Sturmian shift X of slope $\alpha = [a_0; a_1, \ldots]$ is linearly recurrent if and only if the coefficients a_i are bounded.*

Proof Assume first that the coefficients are not bounded. By Exercise 1.65, the shift X is not linearly recurrent.

Conversely, if the coefficients a_i are bounded, the sequence (ζ_n) has a finite number of terms and X is linearly recurrent by Theorem 7.2.3.   ∎

We can add one more equivalent condition in Corollary 7.2.7, namely: $\mathcal{L}(X)$ is K-power free for some $K \geq 2$ (Exercise 7.20).

We note the following additional result.

Theorem 7.2.8 *Let X be a Sturmian shift with slope α. Then X is a substitutive shift if and only if α is quadratic.*

Proof Assume first that X is Sturmian with a slope α, which is quadratic. By Lagrange Theorem (see Appendix B.1), the continued fraction expansion of $\alpha = [0; 1 + d_1, d_2, \ldots]$ is eventually periodic. The directive word $x = 0^{d_1} 1^{d_2} \cdots$ of the standard word $s = c_\alpha$ is eventually periodic. Set $x = uy$ with $y = v^\omega$ and $t = \mathrm{Pal}(y)$. Then, by Justin's Formula, we have $s = L_u(t)$ and $t = L_v(t)$. Since v contains occurrences of 0 and 1, the morphism L_v is primitive. Thus, by Rauzy's Lemma (Proposition 6.2.10), X is substitutive.

Conversely, let X be a substitutive Sturmian shift. By Theorem 6.6.9, there exist $uv, v \in \mathcal{L}(X)$ such that the derivatives of X with respect to $[uv]$ and $[v]$ are equal. We have $\mathcal{R}'_X(uv) \subset \mathcal{R}'_X(v)^*$ and may also assume that every word of $\mathcal{R}'_X(v)$ appears in the decomposition of every word of $\mathcal{R}'_X(v)$. Since X is Sturmian, the sets $\mathcal{R}'_X(v), \mathcal{R}'_X(uv)$ have two elements. Let $\varphi, \psi : \{0, 1\}^* \to \{0, 1\}^*$ be coding morphisms for $\mathcal{R}'_X(v)$ and $\mathcal{R}'_X(uv)$. Since $\mathcal{R}'_X(uv) \subset \mathcal{R}'_X(v)^*$, there is a morphism $\sigma : \{0, 1\}^* \to \{0, 1\}^*$ such that $\psi = \phi \circ \alpha$. Since every word of $\mathcal{R}'_X(v)$ appears in the decomposition of every word of $\mathcal{R}'_X(v)$, the morphism σ is primitive. Then $X = X(\sigma, \phi)$. Since $M(\sigma)$ has dimension two, its spectral radius is a quadratic algebraic number. Thus, by Eq. (3.21), the frequencies of the letters are quadratic numbers. This implies by Corollary 1.5.13 that the slope of X is also quadratic.   ∎

7.2.3 Derivatives of Episturmian Shifts

We have seen that a derivative of a minimal dendric shift is a minimal dendric shift on the same alphabet (Theorem 7.1.24). This implies, since the Sturmian shifts are the minimal dendric shifts on two letters, that a derivative of a Sturmian shift is Sturmian. We prove the following more general statement.

Theorem 7.2.9 *Any derivative of an episturmian (resp. Arnoux–Rauzy) shift is episturmian (resp. Arnoux–Rauzy) on the same alphabet.*

Proof Let $s = \text{Pal}(x)$ be the standard episturmian sequence with directive sequence $x = a_0 a_1 \cdots$. Every derivative of the shift space X generated by s is conjugate to the derivative of X with respect to some palindrome prefix $u_n = \text{Pal}(a_0 a_1 \cdots a_{n-1})$ of s. By (1.34), the set of left return words to u_n is $\mathcal{R}'_X(u_n) = \{L_{a_0 \cdots a_{n-1}}(a) \mid a \in A\}$. This shows that the derivative X' of X with respect to u_n is generated by the standard episturmian word $s' = \text{Pal}(x')$ with $x' = a_n a_{n+1} \cdots$. Thus, X' is episturmian. If X is Arnoux–Rauzy, then every letter appears infinitely often in x and thus also in x', which implies that X' is also an Arnoux–Rauzy shift. ∎

Example 7.2.10 Let X be the Tribonacci shift (see Example 7.1.25). It is the shift generated by the standard episturmian sequence $s = \text{Pal}(abc)^\omega$. The derivative of X with respect to a is generated by $s' = \text{Pal}(bca)^\omega$.

7.3 Specular Shifts

We end this chapter with the description of a family of shifts that generalizes dendric shifts and is built as an abstract model for the transformations called linear involutions described in the next chapter.

7.3.1 Specular Groups

We begin with the definition of a class of groups that generalizes free groups. We consider an alphabet A with an involution $\theta \colon A \to A$, possibly with some fixed points. We also consider the group G_θ generated by A with the relations $a\theta(a) = 1$ for every $a \in A$. Thus, $\theta(a) = a^{-1}$ for $a \in A$. The set A is called a *natural* set of generators of G_θ.

When θ has no fixed point, we can set $A = B \cup B^{-1}$ by choosing a set of representatives of the orbits of θ for the set B. The group G_θ is then the free group on B, denoted F_B.

In general, the group G_θ is a free product of a free group and a finite number of copies of $\mathbb{Z}/2\mathbb{Z}$, that is $G_\theta = \mathbb{Z}^{*i} * (\mathbb{Z}/2\mathbb{Z})^{*j}$, where i is the number of orbits of θ with two elements and j the number of its fixed points. Such a group will be called a *specular group* of type (i, j). These groups are very close to free groups, as we will see. The integer $\text{Card}(A) = 2i + j$ is called the *symmetric rank* of the specular group $\mathbb{Z}^{*i} * (\mathbb{Z}/2\mathbb{Z})^{*j}$. Two specular groups are isomorphic if and only if they have the same type. Indeed, the commutative image of a group of type (i, j) is $\mathbb{Z}^i \times (\mathbb{Z}/2\mathbb{Z})^j$ and the uniqueness of i, j follows from the fundamental theorem of finitely generated abelian groups (see Appendix B.2).

Example 7.3.1 Let $A = \{a, b, c, d\}$ and let θ be the involution that exchanges b, d and fixes a, c. Then $G_\theta = \mathbb{Z} * (\mathbb{Z}/2\mathbb{Z})^2$ is a specular group of symmetric rank 4.

The Cayley graph of a specular group G_θ with respect to the set of natural generators A is a regular tree where each vertex has degree Card(A). The specular groups are actually characterized by this property.

By the Kurosh Subgroup Theorem, any subgroup of a free product $G_1 * G_2 * \cdots * G_n$ is itself a free product of a free group and of groups conjugate to subgroups of the G_i (see Appendix B.2). Thus, we have, replacing the Nielsen–Schreier Theorem of free groups, the following result.

Theorem 7.3.2 *Any subgroup of a specular group is specular.*

It also follows from the Kurosh Theorem that the elements of order 2 in a specular group G_θ are the conjugates of the j fixed points of θ, and this number is thus the number of conjugacy classes of elements of order 2. Indeed, an element of order 2 generates a subgroup conjugate to one of the subgroups generated by the letters.

A word on the alphabet A is θ-*reduced* (or simply reduced) if it has no factor of the form $a\theta(a)$ for $a \in A$. It is clear that any element of a specular group is represented by a unique reduced word.

A subset of a group G is called *symmetric* if it is closed under taking inverses. A set X in a specular group G is called a *monoidal basis* of G if it is symmetric, if the monoid that it generates is G and if any product $x_1 x_2 \cdots x_m$ of elements of X such that $x_k x_{k+1} \neq 1$ for $1 \leq k \leq m - 1$ is distinct of 1. The alphabet A is a monoidal basis of G_θ, and the symmetric rank of a specular group is the cardinality of any monoidal basis (two monoidal bases have the same cardinality since the type is invariant by isomorphism).

Let H be a subgroup of a specular group G. Let Q be a set of reduced words on A that is a prefix-closed set of representatives of the right cosets Hg of H. Such a set is traditionally called a *Schreier transversal* for H (the proof of its existence is classical in the free group and it is the same in any specular group). Let

$$U = \{paq^{-1} \mid a \in A, \ p, q \in Q, \ pa \notin Q, \ pa \in Hq\}. \qquad (7.7)$$

Each word x of U has a unique factorization paq^{-1} with $p, q \in Q$ and $a \in A$. The letter a is called the *central part* of x. The set U is a monoidal basis of H, called the *Schreier basis* relative to Q (the proof is the same as in the free group, see Appendix B.2).

Figure 7.16 The representation of G by permutations on the cosets of H.

One can deduce directly Theorem 7.3.2 from these properties of U. Indeed, let $\varphi\colon B \to U$ be a bijection from a set B onto U that extends to a morphism from B^* onto H. Let $\sigma\colon B \to B$ be the involution sending each b to c where $\varphi(c) = \varphi(b)^{-1}$. Since the central parts never cancel, if a nonempty word $w \in B^*$ is σ-reduced, then $\varphi(w) \neq 1$. This shows that H is isomorphic to the group G_σ. Thus, H is specular.

If H is a subgroup of index n of a specular group G of symmetric rank r, the symmetric rank s of H is

$$s = n(r - 2) + 2. \tag{7.8}$$

This formula replaces Schreier's Formula (which corresponds to the case $j = 0$). It can be proved as follows. Let Q be a Schreier transversal for H and let U be the corresponding Schreier basis. The number of elements of U is $nr - 2(n - 1)$. Indeed, this is the number of pairs $(p, a) \in Q \times A$ minus the $2(n - 1)$ pairs (p, a) such that $pa \in Q$ with pa reduced or $pa \in Q$ with pa not reduced. This gives Formula (7.8).

Example 7.3.3 Let G be the specular group of Example 7.3.1. Let H be the subgroup formed by the elements represented by a reduced word of even length. The set $Q = \{1, a\}$ is a prefix-closed set of representatives of the two cosets of H. The representation of G by permutations on the cosets of H is shown in Figure 7.16. The monoidal basis corresponding to Formula (7.7) is $U = \{ab, ac, ad, ba, ca, da\}$. The symmetric rank of H is 6, in agreement with Formula (7.8) and H is a free group of rank 3.

Example 7.3.4 Let again G be the specular group of Example 7.3.1. Consider now the subgroup K stabilizing 1 in the representation of G by permutations on the set $\{1, 2\}$ of Figure 7.17. We choose $Q = \{1, b\}$. The set U corresponding to Formula (7.7) is $U = \{a, bad, bb, bcd, c, dd\}$. The group K is isomorphic to $\mathbb{Z} * (\mathbb{Z}/2\mathbb{Z})^{*4}$.

Any specular group $G = G_\theta$ has a free subgroup of index 2. Indeed, let H be the subgroup formed of the reduced words of even length. It has clearly

Figure 7.17 The representation of G by permutations on the cosets of K.

index 2. It is free because it does not contain any element of order 2 (such an element is conjugate of a fixed point of θ and thus is of odd length).

We will need two more properties of specular groups. Both are well known to hold for free groups (see Appendix B.2).

A group G is called *residually finite* if for every element $g \neq 1$ of G, there is a morphism φ from G onto a finite group such that $\varphi(g) \neq 1$.

Proposition 7.3.5 *Any specular group is residually finite.*

Proof Let K be a free subgroup of index 2 in the specular group G. Let $g \neq 1$ be in G. If $g \notin K$, then the image of g in G/K is nontrivial. Assume $g \in K$. Since K is free, it is residually finite. Let N be a normal subgroup of finite index of K such that $g \notin N$. Consider the representation of G on the right cosets of N. Since $g \notin N$, the image of g in this finite group is nontrivial. ∎

A group G is said to be *Hopfian* if every surjective morphism from G onto G is also injective. By a result of Malcev, any finitely generated residually finite group is Hopfian. We thus deduce from Proposition 7.3.5 that any specular group is Hopfian. As a consequence, we have the following result, which will be used later.

Proposition 7.3.6 *Let G be a specular group of type (i, j) and let $U \subset G$ be a symmetric set with $2i + j$ elements. If U generates G, it is a monoidal basis of G.*

Proof Let A be a set of natural generators of G. Considering the commutative image of G, we obtain that U contains j elements of order 2. Thus, there is a bijection φ from A onto U such that $\varphi(a^{-1}) = \varphi(a)^{-1}$ for every $a \in A$. The map φ extends to a morphism from G to G, which is surjective since U generates G. Then φ being surjective, it is also injective since G is Hopfian, and thus U is a monoidal basis of G. ∎

7.3.2 Specular Shifts

We assume given an involution θ on the alphabet A generating the specular group G_θ.

A symmetric, factorial and extendable set S of reduced words on the alphabet A is called a *laminary set* on A relative to θ. Thus, the elements of a laminary set S are elements of the specular group G_θ and the set S is contained in G_θ.

A *specular shift* is a shift space X such that $\mathcal{L}(X)$ is a laminary set on A, which is also dendric of characteristic 2. Thus, in a specular shift, the extension graph of every nonempty word is a tree and the extension graph of the empty word is a union of two disjoint trees. If X is a specular shift, we also say that $\mathcal{L}(X)$ is a *specular set*.

Proposition 7.3.7 *The factor complexity of a specular shift on d letters is* $p_n(X) = (d-2)n + 2$ *for $n \geq 1$.*

Proof This follows easily for the fact that all words are neutral except the empty word, which has multiplicity -1. ∎

The following is a very simple example of a specular shift.

Example 7.3.8 Let $A = \{a, b\}$ and let θ be the identity on A. Then the periodic shift formed of the infinite repetitions of ab is a specular shift.

As a second example, we find every dendric shift giving rise to a specular shift (that we may consider as degenerate).

Example 7.3.9 Let $A = B \cup B^{-1}$ be a symmetric alphabet and let $\theta: b \to b^{-1}$. For every dendric shift Y on the alphabet B, the set $L = \mathcal{L}(Y) \cup \mathcal{L}(Y)^{-1}$ is a laminary set that is dendric of characteristic 2. Thus, the shift X such that $\mathcal{L}(X) = L$ is specular.

The next example is more interesting.

Example 7.3.10 The Cassaigne shift (Example 7.1.4) is the eventually dendric shift of characteristic 2 generated by the morphism

$$\sigma: a \mapsto ab, \quad b \mapsto cda, \quad c: \mapsto cd, \quad d \mapsto abc.$$

We will see later (Example 7.3.15) that X is specular relative to the involution $\theta = (bd)$.

The following result shows in particular that in a specular shift the two trees forming $\mathcal{E}(\varepsilon)$ are isomorphic since they are exchanged by the bijection $(a, b) \to (b^{-1}, a^{-1})$. To distinguish the disjoint copies of $L(w)$ and $R(w)$

forming the vertices of the extension graph $\mathcal{E}(w)$, we denote them by $1 \otimes L(w)$ and $R(w) \otimes 1$.

Proposition 7.3.11 *Let X be a specular shift. Let $\mathcal{T}_0, \mathcal{T}_1$ be the two trees such that $\mathcal{E}_X(\varepsilon) = \mathcal{T}_0 \cup \mathcal{T}_1$. For any $a, b \in A$ and $i = 0, 1$, one has $(1 \otimes a, b \otimes 1) \in \mathcal{T}_i$ if and only if $(1 \otimes b^{-1}, a^{-1} \otimes 1) \in \mathcal{T}_{1-i}$*

Proof Assume that $(1 \otimes a, b \otimes 1)$ and $(1 \otimes b^{-1}, a^{-1} \otimes 1)$ are both in $\mathcal{T}_0$. Since $\mathcal{T}_0$ is a tree, there is a path from $1 \otimes a$ to $a^{-1} \otimes 1$. We may assume that this path is reduced, that is, does not use consecutively twice the same edge. Since this path is of odd length, it has the form $(u_0, v_1, u_1, \ldots, u_p, v_p)$ with $u_0 = 1 \otimes a$ and $v_p = a^{-1} \otimes 1$. Since $\mathcal{L}(X)$ is symmetric, we also have a reduced path $(v_p^{-1}, u_p^{-1}, \cdots, u_1^{-1}, u_0^{-1})$, which is in $\mathcal{E}(\varepsilon)$ (for $u_i = 1 \otimes a_i$, we denote $u_i^{-1} = a_i^{-1} \otimes 1$ and similarly for v_i^{-1}) and thus in $\mathcal{T}_0$ since $\mathcal{T}_0, \mathcal{T}_1$ are disjoint. Since $v_p^{-1} = u_0$, these two paths have the same origin and end. But if a path of odd length is its own inverse, its central edge has the form (x, y) with $x = y^{-1}$, as one verifies easily by induction on the length of the path. This is a contradiction with the fact that the words of $\mathcal{L}(X)$ are reduced. Thus, the two paths are distinct. This implies that $\mathcal{E}(\varepsilon)$ has a cycle, a contradiction. ∎

We say that a laminary set S is *orientable* if there exist two factorial sets S_+, S_- such that $S = S_+ \cup S_-$ with $S_+ \cap S_- = \{\varepsilon\}$ and for any $x \in S$, one has $x \in S_-$ if and only if $x^{-1} \in S_+$ (where x^{-1} is the inverse of x in G_θ).

The following result shows in particular that for any dendric shift X on the alphabet B, the set $\mathcal{L}(X) \cup \mathcal{L}(X)^{-1}$ is a specular set on the alphabet $A = B \cup B^{-1}$.

Theorem 7.3.12 *Let X be a specular shift on the alphabet A. Then, $\mathcal{L}(X)$ is orientable if and only if there is a partition $A = A_+ \cup A_-$ of the alphabet A and a dendric shift Y on the alphabet $B = A_+$ such that $\mathcal{L}(X) = \mathcal{L}(Y) \cup \mathcal{L}(Y)^{-1}$.*

Proof Let X be a specular shift on the alphabet A, which is orientable. Let (S_+, S_-) be the corresponding pair of subsets of $S = \mathcal{L}(X)$. The sets S_+, S_- are biextendable, since S is. Set $A_+ = A \cap S_+$ and $A_- = A \cap S_-$. Then $A = A_+ \cup A_-$ is a partition of A and, since S_-, S_+ are factorial, we have $S_+ \subset A_+^*$ and $S_- \subset A_-^*$. Let $\mathcal{T}_0, \mathcal{T}_1$ be the two trees such that $\mathcal{E}(\varepsilon) = \mathcal{T}_0 \cup \mathcal{T}_1$. Assume that a vertex of $\mathcal{T}_0$ is in A_+. Then all vertices of $\mathcal{T}_0$ are in A_+ and all vertices of $\mathcal{T}_1$ are in A_-. Moreover, $\mathcal{E}_{S_+}(\varepsilon) = \mathcal{T}_0$ and $\mathcal{E}_{S_-}(\varepsilon) = \mathcal{T}_1$. Thus, $S_+ = \mathcal{L}(Y)$ with Y a dendric shift and $S_- = \mathcal{L}(Y)^{-1}$. ∎

7.3.3 Doubling Maps

We now introduce a construction that allows one to build specular shifts.

A *transducer* is a labeled graph with vertices in a set Q and edges labeled in $\Sigma \times A$. The set Q is called the set of *states*, the set Σ is called the *input alphabet* and A is called the *output alphabet*. The graph obtained by erasing the output letters is called the *input automaton*. Similarly, the *output automaton* is obtained by erasing the input letters.

Let $\mathcal{A}$ be a transducer with set of states $Q = \{0, 1\}$ on the input alphabet Σ and the output alphabet A. We assume the following:

1. Every letter of Σ acts on Q as a permutation.
2. The output labels of the edges are all distinct.

We define two maps $\delta_0, \delta_1 \colon \Sigma^* \to A^*$ corresponding to the choice of 0 and 1 respectively as initial vertices. Thus, $\delta_0(u) = v$ (resp. $\delta_1(u) = v$) if the path starting at state 0 (resp. 1) with input label u has output v. The pair $\delta = (\delta_0, \delta_1)$ is called a *doubling map* and the transducer $\mathcal{A}$ a *doubling transducer*. The *image* of a set T on the alphabet Σ by the doubling map δ is the set $S = \delta_0(T) \cup \delta_1(T)$.

If $\mathcal{A}$ is a doubling transducer, we define an involution $\theta_{\mathcal{A}}$ on A as follows. For any $a \in A$, let (i, α, a, j) be the edge with input label α and output label a. We define $\theta_{\mathcal{A}}(a)$ as the output label of the edge starting at $1 - j$ with input label α. Thus, $\theta_{\mathcal{A}}(a) = \delta_i(\alpha) = a$ if $i + j = 1$ and $\theta_{\mathcal{A}}(a) = \delta_{1-i}(\alpha) \neq a$ if $i = j$.

Recall that the *reversal* of a word $w = a_1 a_2 \cdots a_n$ is the word $\tilde{w} = a_n \cdots a_2 a_1$. A set S of words is closed under reversal if $w \in S$ implies $\tilde{w} \in S$ for every $w \in S$.

Theorem 7.3.13 *For any dendric shift X on the alphabet Σ, such that $\mathcal{L}(X)$ is closed under reversal and any doubling map δ, the image of $\mathcal{L}(X)$ by δ is a specular set relative to the involution $\theta_{\mathcal{A}}$.*

Proof Set $T = \mathcal{L}(X)$ and $S = \delta_0(T) \cup \delta_1(T)$. The set S is clearly biextendable. Assume that $x = \delta_i(y)$ for $i \in \{0, 1\}$ and $y \in T$. Let j be the end of the path starting at i and with input label y. Since each letter acts on the two elements of Q as the identity or as a transposition, there is a path labeled $\tilde{y}$ from j to i and also a path labeled $\tilde{y}$ from $1 - j$ to $1 - i$. Thus, $x^{-1} = \delta_{1-j}(\tilde{y})$. Since T is closed under reversal, $x^{-1} \in \delta_{1-j}(T)$. This shows that S is symmetric and that it is laminary.

Next, for any nonempty word $x = \delta_i(y)$, the graph $\mathcal{E}_S(x)$ is isomorphic to the graph $\mathcal{E}_T(y)$. Indeed, let j be the end of the path with origin i and input

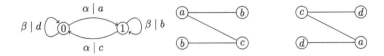

Figure 7.18 A doubling transducer and the extension graph $\mathcal{E}_S(\varepsilon)$.

label y. For $a, b \in A$, one has $axb \in S$ if and only if $cyd \in T$ where c (resp. d) is the input label of the edge with output label a (resp. b) ending in i (resp. with origin j).

Finally, the graph $\mathcal{E}_S(\varepsilon)$ is the union of two trees isomorphic to $\mathcal{E}_T(\varepsilon)$. Indeed, consider the map π from $S \cap A^2$ onto $\{0, 1\}$, which assigns to $ab \in S \cap A^2$ the state i, which is the end of the edge of $\mathcal{A}$ with output label a (and the origin of the edge with output label b). Set $S_i = \pi^{-1}(i)$. We have a partition $S \cap A^2 = S_0 \cup S_1$ such that each S_i is isomorphic to $\mathcal{E}_T(\varepsilon)$ and moreover $ab \in S_i$ if and only if $(ab)^{-1} \in S_{1-i}$. Thus, S is specular. ∎

We now give several examples of specular shifts obtained by a doubling map. The first one is obtained by doubling the Fibonacci shift.

Example 7.3.14 Let $\Sigma = \{\alpha, \beta\}$ and let X be the Fibonacci shift. Let δ be the doubling map given by the transducer of Figure 7.18 on the left.

Then θ_A is the involution θ of Example 7.3.1 and the image of $\mathcal{L}(X)$ by δ is a specular set S on the alphabet $A = \{a, b, c, d\}$. The graph $\mathcal{E}_S(\varepsilon)$ is represented in Figure 7.18 on the right.

Note that S is the set of factors of the fixed point $g^\omega(a)$ of the morphism

$$g: a \mapsto abcab, \quad b \mapsto cda, \quad c \mapsto cdacd, \quad d \mapsto abc.$$

The morphism g is obtained by applying the doubling map to the cube f^3 of the Fibonacci morphism f in such a way that $g^\omega(a) = \delta_0(f^\omega(\alpha))$.

In the next example (due to Julien Cassaigne), the specular set is obtained using a morphism of smaller size.

Example 7.3.15 Let $A = \{a, b, c, d\}$. Let X be the shift generated by the morphism $f : \alpha \mapsto \alpha\beta, \beta \mapsto \alpha\beta\alpha$. It is a Sturmian shift. Indeed, $x = f^\omega(\alpha)$ is the characteristic sequence of slope $-1 + \sqrt{2}$ (see Example 7.2.2). The sequence $s_n = f^n(\alpha)$ satisfies $s_n = s_{n-1}^2 s_{n-2}$ for $n \geq 2$. The image Y of X by the doubling automaton of Figure 7.18 is the set of factors of the fixed point $\sigma^\omega(a)$ of the morphism σ from A^* into itself defined by

$$\sigma(a) = ab, \quad \sigma(b) = cda, \quad \sigma(c) = cd, \quad \sigma(d) = abc.$$

Thus, the shift Y is the Cassaigne shift of Example 7.1.4.

7.3.4 Odd and Even Words

We introduce a notion that plays, as we shall see, an important role in the study of specular shifts. Let X be a specular shift on the alphabet A with $A \subset \mathcal{L}(X)$. Any letter $a \in A$ occurs exactly twice as a vertex of $\mathcal{E}(\varepsilon)$, one as an element of $L(\varepsilon)$ and one as an element of $R(\varepsilon)$. A letter $a \in A$ is said to be *even* if its two occurrences appear in the same tree. Otherwise, it is said to be *odd*. Observe that if a specular shift X is recurrent, there is at least one odd letter.

Example 7.3.16 Let X be the shift of period ab as in Example 7.3.8. Then a and b are odd.

A word $w \in S$ is said to be *even* if it has an even number of odd letters. Otherwise, it is said to be *odd*. The set of even words has the form $U^* \cap S$ where $U \subset S$ is a bifix code, called the *even code*. The set U is the set of even words without a nonempty even prefix (or suffix). A word u is an *internal factor* of a word w if $w = pus$ with p, s nonempty.

Proposition 7.3.17 *Let X be a minimal specular shift. The even code is a finite bifix code that is an X-maximal prefix and suffix code.*

Proof The even code U is bifix by definition. To prove that it is an X-maximal prefix code, let us verify that any $w \in S$ is comparable for the prefix order with an element of the even code X. If w is even, it is in U^*. Otherwise, since S is recurrent, there is a word u such that $wuw \in S$. If u is even, then wuw is even and thus $wuw \in U^*$. Otherwise, wu is even and thus $wu \in U^*$. This shows that U is X-maximal. A word of the form awb with a, b odd and w even cannot be an internal factor of a word of U. Indeed, if $pawbq$ is even, either p, q are even and then p, awb, q are in U^* or p, q are odd and pa, w, bq are in U^*. Since X is minimal, this implies that U is finite. ∎

Example 7.3.18 Let X be the Cassaigne shift of Example 7.1.4. The letters b, d are even and the letters a, c are odd. The even code is

$$U = \{abc, ac, b, ca, cda, d\}.$$

Denote by $\mathcal{T}_0, \mathcal{T}_1$ the two trees such that $\mathcal{E}(\varepsilon) = \mathcal{T}_0 \cup \mathcal{T}_1$. We consider the directed graph $\mathcal{G}$ with vertices $0, 1$ and edges all the triples (i, a, j) for $0 \le i, j \le 1$ and $a \in A$ such that $(1 \otimes b, a \otimes 1) \in \mathcal{T}_i$ and $(1 \otimes a, c \otimes 1) \in \mathcal{T}_j$ for some $b, c \in A$. The graph $\mathcal{G}$ is called the *parity graph* of S. Observe that for every letter $a \in A$ there is exactly one edge labeled a because a appears exactly once as a left (resp. right) vertex in $\mathcal{E}(\varepsilon)$.

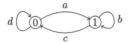

Figure 7.19 The parity graph.

Note that, when X is a specular shift obtained by a doubling map using a transducer $\mathcal{A}$, the parity graph of X is the output automaton of $\mathcal{A}$.

Example 7.3.19 Let X be the specular shift of Example 7.3.14. The parity graph of X is represented in Figure 7.19. It is the output automaton of the doubling transducer of Figure 7.18.

Proposition 7.3.20 *Let X be a specular shift and let $\mathcal{G}$ be its parity graph. Let $S_{i,j}$ be the set of words in $S = \mathcal{L}(X)$ that are the label of a path from i to j in the graph $\mathcal{G}$.*

(1) The family $(S_{i,j} \setminus \{\varepsilon\})_{0 \le i,j \le 1}$ is a partition of $S \setminus \{\varepsilon\}$.
(2) For $u \in S_{i,j} \setminus \{\varepsilon\}$ and $v \in S_{k,\ell} \setminus \{\varepsilon\}$, if $uv \in S$, then $j = k$.
(3) $S_{0,0} \cup S_{1,1}$ is the set of even words.
(4) $S_{i,j}^{-1} = S_{1-j,1-i}$.

Proof We first note that for $a, b \in A$ such that $ab \in S$, there is a path in $\mathcal{G}$ labeled ab. Since $(a, b) \in E(\varepsilon)$, there is a k such that $(1 \otimes a, b \otimes 1) \in \mathbb{T}_k$. Then we have $a \in S_{i,k}$ and $b \in S_{k,j}$ for some $i, j \in \{0, 1\}$. This shows that ab is the label of a path from i to j in $\mathcal{G}$.

Let us prove by induction on the length of a nonempty word $w \in S$ that there exists a unique pair i, j such that $w \in S_{i,j}$. The property is true for a letter, by definition of the extension graph $\mathcal{E}(\varepsilon)$ and for words of length 2 by the above argument. Let next $w = ax$ be in S with $a \in A$ and x nonempty. By induction hypothesis, there is a unique pair (k, j) such that $x \in S_{k,j}$. Let b be the first letter of x. Then the edge of $\mathcal{G}$ with label b starts in k. Since ab is the label of a path, we have $a \in S_{i,k}$ for some i and thus $ax \in S_{i,j}$. The other assertions follow easily (Assertion (4) follows from Proposition 7.3.11). ∎

Note that Assertion (4) implies that no nonempty even word is its own inverse. Indeed, $S_{0,0}^{-1} = S_{1,1}$ and $S_{1,1}^{-1} = S_{0,0}$.

Proposition 7.3.21 *Let X be a specular shift and let $S = \mathcal{L}(X)$. If $x, y \in S$ are nonempty words such that $xyx^{-1} \in S$, then y is odd.*

Proof Let i, j be such that $x \in S_{i,j}$. Then $x^{-1} \in S_{1-j,1-i}$ by Assertion (4) of Proposition 7.3.20 and thus $y \in S_{j,1-j}$ by Assertion (2). Thus, y is odd by Assertion (3). ∎

Recall that for a shift space X, a finite bifix code $U \subset \mathcal{L}(X)$ that is an X-maximal prefix and suffix code and a coding morphism f for U, the shift space Y such that $\mathcal{L}(Y) = f^{-1}(\mathcal{L}(X))$ is called a *decoding* of X by U. We denote $Y = f^{-1}(X)$.

The following result gives, in the particular case of the even code, a result on the decoding of a specular shift (compare with Theorem 7.1.26 for the case of a dendric shift).

Theorem 7.3.22 *The decoding of a minimal specular shift by the even code is a union of two minimal dendric shifts. More precisely, let X be a minimal specular shift and let f be a coding morphism for the even code. The shifts Y_0, Y_1 such that $\mathcal{L}(Y_0) = f^{-1}(S_{0,0})$ and $\mathcal{L}(Y_1) = f^{-1}(S_{1,1})$ are isomorphic minimal dendric shifts.*

Proof We show that the shift Y_0 such that $\mathcal{L}(Y_0) = f^{-1}(S_{0,0})$ is a minimal dendric shift. The proof for $f^{-1}(S_{1,1})$ is the same. Set $T_0 = \mathcal{L}(Y_0)$.

Set $S = \mathcal{L}(X)$. Since X is minimal, for every $u \in S$, there exists $n \geq 1$ such that u is a factor of any word w in S of length n. But if $u, w \in S_{0,0}$ are such that $w = \ell u r$, then $\ell, r \in S_{0,0}$. Thus, Y_0 is minimal.

We now show that Y_0 is dendric. Let U be the even code. Set $U_0 = U \cap S_{0,0}$, $U_1 = U \cap S_{1,1}$ and $\mathcal{E}_0(w) = \mathcal{E}_{U_0,U_0}(w)$. It is enough to show that the graph $\mathcal{E}_0(w)$ is a tree for any $w \in S_{0,0}$.

Assume first that w is nonempty. Note first that $\mathcal{E}_0(w) = \mathcal{E}_{U,U}(w)$. Indeed, if $x, y \in U$ are such that $xwy \in S$, one has $x, y \in U_0$ and $xwy \in S_{0,0}$. But the graph $\mathcal{E}_{U,U}(w)$ is a tree by Proposition 7.1.13.

Suppose now that $w = \varepsilon$. First, since $\mathcal{E}(\varepsilon)$ is a union of two trees, it is acyclic, and thus the graph $\mathcal{E}_0(\varepsilon)$ is acyclic by Proposition 7.1.10. Next, since every nonempty word in S is neutral, by Lemma 7.1.14, we have $m_{U,U}(\varepsilon) = m(\varepsilon) = -1$. This implies that $\mathcal{E}_{U,U}(\varepsilon)$ is a union of two trees. Since $\mathcal{E}_{U,U}(\varepsilon)$ is the disjoint union of $\mathcal{E}_0(\varepsilon)$ and $\mathcal{E}_{U_1,U_1}(\varepsilon)$, this implies that each one is a tree.

Clearly, Y_0 and Y_1 are isomorphic. Indeed, let $f \colon B^* \to A^*$ be a coding morphism for U. Set $B_0 = f^{-1}(U_0)$ and $B_1 = f^{-1}(U_1)$. Then $\alpha \colon B_0 \to B_1$ defined by $\alpha(b) = f^{-1}(b^{-1})$ defines an isomorphism from Y_0 onto Y_1. ∎

Example 7.3.23 Let X be the Cassaigne shift space of Example 7.1.4. We have seen that it generated by the morphism

$$\sigma : a \mapsto ab, \quad b \mapsto cda, \quad c \mapsto cd, \quad d \mapsto abc.$$

The even code U is given in Example 7.3.18. Let $\Sigma = \{a, b, c, d, e, f\}$ and let g be the coding morphism for X given by

$$a \mapsto abc, \quad b \mapsto ac, \quad c \mapsto b, \quad d \mapsto ca, \quad e \mapsto cda, \quad f \mapsto d.$$

The decoding of X by U is a union of two dendric shifts, which are generated by the two morphisms

$$a \mapsto afbf, \ b \mapsto af, \ f \mapsto a$$

and

$$c \mapsto e, \ d \mapsto ec, \ e \mapsto ecdc.$$

These two morphisms are actually the restrictions to $\{a, b, f\}$ and $\{c, d, e\}$ of the morphism $g^{-1}\sigma g$.

7.3.5 Complete Return Words

Let X be a shift space and let $U \subset \mathcal{L}(X)$ be a bifix code. A *complete return word* to U is a word of $\mathcal{L}(X)$ with a proper prefix in U, a proper suffix in U but no internal factor in U. We denote by $\mathcal{C}R_X(U)$ the set of complete return words to U.

The set $\mathcal{C}R_X(U)$ is a bifix code. If X is minimal, $\mathcal{C}R_X(U)$ is finite for any finite set U. For $x \in \mathcal{L}(X)$, we denote $\mathcal{C}R_X(x)$ instead of $\mathcal{C}R_X(\{x\})$.

Example 7.3.24 Let X be the specular shift of Example 7.3.14. One has

$$\mathcal{C}R_X(a) = \{abca, abcda, acda\}, \mathcal{C}R_X(b) = \{bcab, bcdacdab, bcdacdacdab\},$$
$$\mathcal{C}R_X(c) = \{cabc, cdabc, cdac\}, \mathcal{C}R_X(d) = \{dabcabcabcd, dabcabcd, dacd\}.$$

The *kernel* of a bifix code U is the set of words of U that are factors of another word in U. The following result is a generalization of Theorem 7.1.16. The proof is very similar (Exercise 7.22).

Theorem 7.3.25 *Let X be a minimal specular shift on the alphabet A such that $A \subset \mathcal{L}(X)$. For every finite nonempty bifix code $U \subset \mathcal{L}(X)$ with empty kernel, we have*

$$\mathrm{Card}(\mathcal{C}R_X(U)) = \mathrm{Card}(U) + \mathrm{Card}(A) - 2. \tag{7.9}$$

The following example illustrates Theorem 7.3.25.

Example 7.3.26 Let X be the Cassaigne shift (Example 7.1.4). We have

$$\mathcal{C}R_X(\{a, b\}) = \{ab, acda, bca, bcda\}.$$

It has four elements in agreement with Theorem 7.3.25.

7.3.6 Right Return Words

We now come to right return words in specular shifts. Note that when S is a laminary set, $\mathcal{R}_S(x)^{-1} = \mathcal{R}'_S(x^{-1})$.

Proposition 7.3.27 *Let X be a specular shift and let $u \in \mathcal{L}(X)$ be a nonempty word. All the words of $\mathcal{R}_X(x)$ are even.*

Proof If $w \in \mathcal{R}_X(u)$, we have $uw = vu$ for some $v \in \mathcal{L}(X)$. If u is odd, assume that $u \in S_{0,1}$. Then $w \in S_{1,1}$. Thus, w is even. If u is even, assume that $u \in S_{0,0}$. Then $w \in S_{0,0}$ and w is even again. ∎

We now establish the following result, which replaces, for specular shifts, Theorem 7.1.16 for dendric shifts.

Theorem 7.3.28 *Let X be a minimal specular shift on the alphabet A with $A \subset \mathcal{L}(X)$. For every $u \in \mathcal{L}(X)$, the set $\mathcal{R}_X(u)$ has $\mathrm{Card}(A) - 1$ elements.*

Proof This follows directly from Theorem 7.3.25 with $U = \{u\}$ since one has $\mathrm{Card}(\mathcal{R}_X(u)) = \mathrm{Card}(\mathcal{C}R_X(u))$. ∎

Example 7.3.29 Let X be the specular shift of Example 7.3.14. We have

$$\mathcal{R}_X(a) = \{bca, bcda, cda\},$$
$$\mathcal{R}_X(b) = \{cab, cdacdab, cdacdacdab\},$$
$$\mathcal{R}_X(c) = \{abc, dabc, dac\},$$
$$\mathcal{R}_X(d) = \{abcabcd, abcabcabcd, acd\}.$$

7.3.7 Mixed Return Words

Let S be a laminary set. For $w \in S$ such that $w \neq w^{-1}$, we consider complete return words to the set $X = \{w, w^{-1}\}$.

Example 7.3.30 Let X be the substitution shift generated by the substitution $\tau: a \rightarrow cb^{-1}, b \rightarrow c, c \rightarrow ab^{-1}$. We shall verify later that X is specular (Example 8.4.4). We have

$$CRS(\{a, a^{-1}\}) = \{ab^{-1}cba^{-1}, ab^{-1}cbc^{-1}a, a^{-1}cb^{-1}c^{-1}a,$$
$$ab^{-1}c^{-1}ba^{-1}, a^{-1}cbc^{-1}a, a^{-1}cb^{-1}c^{-1}ba^{-1}\},$$
$$CRS(\{b, b^{-1}\}) = \{ba^{-1}cb, ba^{-1}cb^{-1},$$
$$bc^{-1}ab^{-1}, b^{-1}cb, b^{-1}c^{-1}ab^{-1}, b^{-1}c^{-1}b\},$$
$$CRS(\{c, c^{-1}\}) = \{cba^{-1}c, cbc^{-1},$$
$$cb^{-1}c^{-1}, c^{-1}ab^{-1}c, c^{-1}ab^{-1}c^{-1}, c^{-1}ba^{-1}c\}.$$

The following result shows that, at the cost of taking return words to a set of two words, we recover a situation similar to that of dendric shifts.

Theorem 7.3.31 *Let X be a minimal specular shift on the alphabet A such that $A \subset \mathcal{L}(X)$. For any $w \in \mathcal{L}(X)$ such that $w \neq w^{-1}$, the set of complete return words to $\{w, w^{-1}\}$ has $\mathrm{Card}(A)$ elements.*

Proof The statement results directly from Theorem 7.3.25. ∎

Example 7.3.32 Let X be the specular shift of Example 7.3.14. In view of the values of $CRS(b)$ and $CRS(d)$ given in Example 7.3.24, we have

$$CRS(\{b, d\}) = \{bcab, bcd, dab, dacd\}.$$

Two words u, v are said to *overlap* if a nonempty suffix of one of them is a prefix of the other. In particular, a nonempty word overlaps with itself.

We now consider the return words to $\{w, w^{-1}\}$ with w such that w and w^{-1} do not overlap. This is true for every w in a laminary set S where the involution θ has no fixed point (in particular, when X is the natural coding of a linear involution, as we shall see). In this case, the group G_θ is free and for any $w \in S$, the words w and w^{-1} do not overlap.

With a complete return word u to $\{w, w^{-1}\}$, we associate a word $N(u)$ obtained as follows. If u has w as prefix, we erase it, and if u has a suffix w^{-1}, we also erase it. Note that these two operations can be made in any order since w and w^{-1} cannot overlap.

The *mixed return words to w* are the words $N(u)$ associated with complete return words u to $\{w, w^{-1}\}$. We denote by $\mathcal{M}R_X(w)$ the set of mixed return words to w in X.

Note that $\mathcal{M}R_X(w)$ is symmetric and that $w\mathcal{M}R_X(w)w^{-1} = \mathcal{M}R_S(w^{-1})$. Note also that if S is orientable, then

$$\mathcal{M}R_X(w) = \mathcal{R}_X(w) \cup \mathcal{R}_X(w)^{-1} = \mathcal{R}_X(w) \cup \mathcal{R}'_X(w^{-1}).$$

Example 7.3.33 Let X be the substitution shift generated by the morphism $f: a \rightarrow cb^{-1}, b \rightarrow c, c \rightarrow ab^{-1}$ extended to an automorphism of

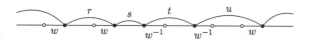

Figure 7.20 A uniformly recurrent infinite word factorized as an infinite product
$\cdots rstu \cdots$ of mixed return words to w.

the free group on $\{a, b, c\}$. We shall see later that X is actually specular
(Example 8.4.4). We have

$$\mathcal{M}R_S(a) = \{b^{-1}cb, b^{-1}cbc^{-1}a, a^{-1}cb^{-1}c^{-1}a,$$
$$b^{-1}c^{-1}b, a^{-1}cbc^{-1}a, a^{-1}cb^{-1}c^{-1}b\},$$
$$\mathcal{M}R_S(b) = \{a^{-1}cb, a^{-1}c, c^{-1}a, b^{-1}cb, b^{-1}c^{-1}a, b^{-1}c^{-1}b\},$$
$$\mathcal{M}R_S(c) = \{ba^{-1}c, b, b^{-1}, c^{-1}ab^{-1}c, c^{-1}ab^{-1}, c^{-1}ba^{-1}c\}.$$

Observe that any uniformly recurrent bi-infinite word x such that $F(x) = S$
can be uniquely written as a concatenation of mixed return words (see
Figure 7.20). Note that successive occurrences of w may overlap but that
successive occurrences of w and w^{-1} cannot.

We have the following cardinality result.

Theorem 7.3.34 *Let X be a minimal specular shift on the alphabet A such
that $A \subset \mathcal{L}(X)$. For every $w \in S$ such that w, w^{-1} do not overlap, the set
$\mathcal{M}R_X(w)$ has $\mathrm{Card}(A)$ elements.*

Proof This is a direct consequence of Theorem 7.3.31 since $\mathrm{Card}(\mathcal{M}R_X$
$(w)) = \mathrm{Card}(\mathcal{C}R_X(\{w, w^{-1}\}))$ when w and w^{-1} do not overlap. ∎

Note that the bijection between $\mathcal{C}R_X(\{w, w^{-1}\})$ and $\mathcal{M}R_X(w)$ is illustrated in
Figure 7.20.

Example 7.3.35 Let X be the specular shift of Example 7.3.14. The value of
$\mathcal{C}R_X(\{b, d\})$ is given in Example 7.3.32. Since b, d do not overlap, the set

$$\mathcal{M}R_X(b) = \{cab, c, dac, dab\}$$

has four elements in agreement with Theorem 7.3.34.

7.3.8 The Return Theorem for Specular Shifts

By Theorem 7.1.15, the set of right return words to a given word in a minimal
dendric shift on the alphabet A such that $A \subset \mathcal{L}(X)$ is a basis of the free group
on A. We will see a counterpart of this result for specular shifts.

Let S be a specular set. The *even subgroup* is the group formed by the even words. It is a subgroup of index 2 of G_θ with symmetric rank $2(\text{Card}(A) - 1)$ by (7.8) generated by the even code. Since no even word is its own inverse (by Proposition 7.3.20), it is a free group. Thus, its rank is $\text{Card}(A) - 1$.

The following result replaces, for specular shifts, the Return Theorem of dendric shifts (Theorem 7.1.15).

Theorem 7.3.36 *Let X be a minimal specular shift. For every $w \in \mathcal{L}(X)$, the set of right return words to w is a basis of the even subgroup.*

Proof Set $S = \mathcal{L}(X)$. We first consider the case where w is even. Let $f : B^* \to A^*$ be a coding morphism for the even code $U \subset S$. Consider the partition $(S_{i,j})$, as in Proposition 7.3.20, and set $U_0 = U \cap S_{0,0}, U_1 = U \cap S_{1,1}$. By Theorem 7.3.22, the shift $f^{-1}(X)$ is the union of two minimal dendric shifts, Y_0 and Y_1 on the alphabets $B_0 = f^{-1}(U_0)$ and $B_1 = f^{-1}(U_1)$ respectively. We may assume that $w \in S_{0,0}$. Then $\mathcal{R}_X(w)$ is the image by f of the set $R = \mathcal{R}_{Y_0}(f^{-1}(w))$. By Theorem 7.1.15, the set R is a basis of the free group on B_0. Thus, $\mathcal{R}_S(w)$ is a basis of the image of $F(B_0)$ by f, which is the even subgroup.

Suppose now that w is odd. Since the even code is an X-maximal bifix code, there exists an odd word u such that $uw \in S$. Then $\mathcal{R}_X(uw) \subset \mathcal{R}_X(w)^*$. By what precedes, the set $\mathcal{R}_X(uw)$ generates the even subgroup, and thus the group generated by $\mathcal{R}_X(w)$ contains the even subgroup. Since all words in $\mathcal{R}_X(w)$ are even, the group generated by $\mathcal{R}_X(w)$ is contained in the even subgroup, whence the equality. We conclude by Theorem 7.3.28.   ∎

Example 7.3.37 Let X be the specular shift of Example 7.3.14. The sets of right return words to a, b, c, d are given in Example 7.3.29. Each one is a basis of the even subgroup.

Concerning mixed return words, we have the following statement.

Theorem 7.3.38 *Let X be a minimal specular shift. For every $w \in \mathcal{L}(X)$ such that w, w^{-1} do not overlap, the set $\mathcal{M}\mathcal{R}_X(w)$ is a monoidal basis of the group G_θ.*

Proof Set $S = \mathcal{L}(X)$. Since w and w^{-1} do not overlap, we have $\mathcal{R}_X(w) \subset \mathcal{M}\mathcal{R}_X(w)^*$. Thus, by Theorem 7.3.36, the group $\langle \mathcal{M}\mathcal{R}_X(w) \rangle$ contains the even subgroup. But $\mathcal{M}\mathcal{R}_X(w)$ always contains odd words. Indeed, assume that $w \in S_{i,j}$. Then $w^{-1} \in S_{1-j,1-i}$ and thus any $u \in \mathcal{M}\mathcal{R}_X(w)$ such that $wuw^{-1} \in S$ is odd. Since the even group is a maximal subgroup of G_θ, this implies that

$\mathcal{M}R_X(w)$ generates the group G_θ. Finally, since $\mathcal{M}R_S(w)$ has $\mathrm{Card}(A)$ elements by Theorem 7.3.34, we obtain the conclusion by Proposition 7.3.6. ∎

Example 7.3.39 Let X be the specular shift of Example 7.3.14. We have seen in Example 7.3.35 that $\mathcal{M}R_X(b) = \{c, cab, dab, dac\}$. This set is a monoidal basis of G_θ in agreement with Theorem 7.3.38.

7.3.9 BV-Representation and Dimension Groups of Specular Shifts

We have the following description of dimension groups of minimal specular shifts. It shows that they are dimension groups of dendric shifts, except possibly for the order unit.

Theorem 7.3.40 *A minimal specular shift X on a k-letter alphabet A has a BV-representation with $k - 1$ vertices at each level. Its dimension group is, as an ordered group, isomorphic to the dimension group of a minimal dendric shift on $k - 1$ letters.*

Proof Let $w \in \mathcal{L}(X)$. By Theorem 7.3.36, the set $\mathcal{R}_X(w)$ is a basis of the even group. Let Y be the shift space induced by X on $[w]$. Assume that w is even and, for instance, that $w \in S_{0,0}$. Let f is a coding morphism for the even code U with $U_0 = U \cap S_{0,0}$. Let Y_0 be the shift such that $\mathcal{L}(Y_0) = f^{-1}(S_{0,0})$ Then $\mathcal{R}_X(w) \subset U_0^*$ and thus $Y \subset Y_0$. Since Y_0 is a minimal dendric shift by Proposition 7.3.22, we have $Y = Y_0$ and we obtain the conclusion that Y is a minimal dendric shift. Since $\mathrm{Card}(\mathcal{R}_X(w)) = k - 1$, this completes the proof by Proposition 5.4.2. ∎

Example 7.3.41 Let X be the Cassaigne shift, which is generated by the morphism $\varphi: a \to ab, b \to cda, c \to cd, d \to abc$ (see Example 7.3.10). The set of return words to a is $\mathcal{R}_X(a) = \{bca, bcda, cda\}$. It is a basis of the even group, itself generated by the even code $U = \{abc, ac, b, ca, cda, d\}$. Let $f: \{ab, ac, bc, ca, cd, da\} \to A_2 = \{u, v, w, x, y, z\}$ and let $\varphi_2: u \to uw, v \to uw, w \to yzv, x \to yz, y \to yz, z \to uwx$ be the 2-block presentation of φ. Let $B = \{r, s, t\}$ and let $\phi: B^* \to A_2^*$ be a coding morphism for $f(a\mathcal{R}_X(a)) = \{uwx, uwyz, vyz\}$. The morphism $\tau: r \to st, s \to str, t \to sr$ is such that $\varphi_2 \circ \phi = \phi \circ \tau$. The matrix $M(\tau)$ is

$$M(\tau) = \begin{bmatrix} 0 & 1 & 1 \\ 1 & 1 & 1 \\ 1 & 1 & 0 \end{bmatrix}.$$

The matrix $M(\tau)$ has eigenvalues $-1, 1 - \sqrt{2}$ and $\lambda = 1 + \sqrt{2}$, which is its dominant eigenvalue. A row eigenvector corresponding to λ is $w = \begin{bmatrix} 1 & \sqrt{2} & 1 \end{bmatrix}$. The dimension group is thus $G = \mathbb{Z}^3$ with $G^+ = \{(a_r, a_s, a_t) \in \mathbb{Z}^3 \mid a_r + a_s\sqrt{2} + a_t > 0\} \cup \{0\}$ and unit $u = \begin{bmatrix} 3 & 4 & 3 \end{bmatrix}^t$ (the unit is given by the lengths of the words of $\phi(B)$). The infinitesimal group is generated by the eigenvector $\begin{bmatrix} 1 & 0 & -1 \end{bmatrix}^t$ corresponding to the eigenvalue -1. The quotient is the image of G by the map $v \to w \cdot v$. It is isomorphic to $\frac{1}{2}\mathbb{Z}[\sqrt{2}]$ (the unit is sent by this map to $6 + 4\sqrt{2} = 2(1 + \sqrt{2})^2$).

It is interesting to make the following observation. We have seen before (Example 7.3.15) that the shift X is obtained by a doubling map from the Sturmian shift Y generated by the morphism $\sigma: a \to ab, b \to aba$. Since the map sending a, c to a and b, d to b is a morphism from X onto Y, we know from Proposition 3.6.1 that there is a natural embedding of $K^0(Y, S)$ in $K^0(X, T)$. Let us look in more detail at how this is related to the doubling map.

The morphism σ is eventually proper and thus the dimension group $K^0(Y, S)$ is the group of the matrix

$$M(\sigma) = \begin{bmatrix} 1 & 1 \\ 2 & 1 \end{bmatrix}.$$

The group is found to be $\mathbb{Z}[\sqrt{2}]$. Thus, up to the unit, $K^0(Y, S)$ is the same as the quotient of $K^0(X, T)$ by the infinitesimal group. This can be verified directly as follows. We have in X, first writing down all possible extensions of c and next using the form of $\mathcal{R}_X(a)$,

$$\chi_{[c]} = \chi_{[ab \cdot ca]} + \chi_{[a \cdot cda]} + \chi_{ab \cdot cda}$$
$$\sim \chi_{[abca]} + \chi_{[acda]} + \chi_{[abcda]}$$
$$= \chi_{[a]},$$

where $\sim$ denotes the cohomology equivalence. Since

$$M(\varphi) = \begin{bmatrix} 1 & 1 & 0 & 0 \\ 1 & 0 & 1 & 1 \\ 0 & 0 & 1 & 1 \\ 1 & 1 & 1 & 0 \end{bmatrix}$$

the values of the invariant probability measure μ of X on a, b, c, d are proportional to the left eigenvector of $M(\varphi)$ for $\lambda = 1 + \sqrt{2}$, which is $[\sqrt{2}, 1, \sqrt{2}, 1]$. Thus, using as basis the characteristic functions of $[a], [b], [d]$, we find the group $K^0(X, S)$ as $G = \{(\alpha, \beta, \delta) \in \mathbb{Z}^3 \mid \alpha\sqrt{2} + \beta + \delta > 0\} \cup \{0\}$. The natural embedding of $H(Y, S, \mathbb{Z})$ in $H(X, S, \mathbb{Z})$ is induced by the map $(\alpha, \beta) \mapsto (2\alpha, \beta, \beta)$. The image is, as expected, embedded in $\frac{1}{2}\mathbb{Z}[\sqrt{2}]$.

7.4 Exercises

Section 7.1

7.1 Let X be the substitution shift generated by the substitution $a \to ab, b \to cda, c \to cd, d \to abc$. Show that the graph of every nonempty word in $\mathcal{L}(X)$ is a tree.

7.2 Show that if X is eventually dendric with threshold m, its complexity is, for $n \geq m$, given by $p_n(X) = Kn + L$ with $K = s_m(X)$ and $L = p_m(X) - m s_m(X)$.

7.3 Show that the Chacon ternary shift is not eventually dendric.

7.4 Let $B = \{1, 2, 3\}$ and $A = \{a, b, c, d\}$. Let $\tau : A^* \to B^*$ be the morphism $a \to 12, b \to 2, c \to 3$ and $d \to 13$. Let X be the shift on A generated by the morphism $a \to ab, b \to cda, c \to cd, d \to abc$ of Example 7.1.4. Show that the shift $Y = \tau(X)$ is neutral.

7.5 Let X be a shift space. A bifix code $U \subset \mathcal{L}(X)$ is X-*maximal* if it is not strictly contained in another bifix code $V \subset \mathcal{L}(X)$. Show that, if X recurrent, every finite X-maximal bifix code is also X-maximal as a prefix code (and, symmetrically as a suffix code). Hint: show that if U is neither X-maximal as a prefix code and as a suffix code, it is not X-maximal as a bifix code.

7.6 Let X be a minimal shift space. For $u \in \mathcal{L}(X)$, let $\mathfrak{P} = \{S^i[vu] \mid v \in \mathcal{R}'_X(u), 0 \leq i < |v|\}$ be the partition in towers of Proposition 4.1.1. Show that the set

$$T = \{t \in \mathcal{L}(X) \mid [s \cdot t] \in \mathfrak{P} \text{ for some } s\}$$

is an X-maximal prefix code.

7.7 Let X be a minimal dendric shift on the alphabet A. Denote by $F(A)$ the free group on A. Let $U \subset \mathcal{L}(X)$ be a finite X-maximal bifix code. Let H be the subgroup of $F(A)$ generated by U. Show that $H \cap \mathcal{L}(X) = U^* \cap \mathcal{L}(X)$. Hint: consider the coset graph of U and the set V of labels of simple paths from ε to ε.

7.8 Let X be a recurrent shift space and let $U \subset \mathcal{L}(X)$ be a finite X-maximal bifix code. A *parse* of a word w is a triple (s, x, p) such that $w = sxp$ with s a proper suffix of a word in U, $x \in U^*$ and p a proper prefix of a word of U. Let $d_U(w)$ be the number of parses of w. Show that for every $u \in \mathcal{L}(X)$ and $a \in R_X(u)$,

$$d_U(ua) = \begin{cases} d_U(u) & \text{if } ua \text{ has a suffix in } U \\ d_U(u) + 1 & \text{otherwise.} \end{cases} \qquad (7.10)$$

Show that one has for $u, v, w \in \mathcal{L}(X)$, the inequality

$$d_U(v) \le d_U(uvw) \qquad (7.11)$$

with equality if v is not a factor of a word in U.

7.9 Let X be a recurrent shift space and let U be a finite X-maximal bifix code. The X-*degree* of U, denoted $d_U(X)$ is the maximum of the numbers $d_U(w)$ for $w \in \mathcal{L}(X)$. Show that every word in $\mathcal{L}(X)$ that is not a factor of a word in U has $d_U(X)$ parses.

7.10 Show that the X-degree of $\mathcal{L}_n(X)$ is equal to n.

7.11 Let X be a recurrent shift space. Let $U \subset \mathcal{L}(X)$ be a finite X-maximal bifix code. Set $d = d_U(X)$. Show that the set S of nonempty proper suffixes of U is a disjoint union of $d - 1$ X-maximal prefix codes. Hint: consider for $2 \le i \le d$ the set S_i of proper suffixes s of U such that $d_U(s) = i$.

7.12 Let X be a minimal dendric shift on the alphabet A such that $A \subset \mathcal{L}(X)$. Show that for every finite X-maximal bifix code U, one has

$$\text{Card}(U) = (\text{Card}(A) - 1)d_U(X) + 1. \qquad (7.12)$$

Hint: use Exercise 7.11.

7.13 Let X be a minimal dendric shift on the alphabet A such that $A \subset \mathcal{L}(X)$. Show that a finite bifix code $U \subset \mathcal{L}(X)$ is X-maximal with X-degree d if and only if it is a basis of a subgroup of index d of the free group $F(A)$.

Hint: Consider the set Q of suffixes of a word w having $d_U(X)$ parses. Show that the set V of words v such that $Qv \subset \langle U \rangle Q$ is a subgroup of the free group containing $\mathcal{R}_X(w)$. This implies that U generates a subgroup of index $d = d_U(X)$. Conclude using Exercise 7.12 and Schreier's Formula (see Appendix B.2).

7.14 Let $\varphi: A^* \to G$ be a morphism from A^* onto a finite group G and let H be a subgroup of G of index d. Let U be the bifix code which generates the submonoid $\varphi^{-1}(H)$.

Show that for every minimal dendric shift X, the set $U \cap \mathcal{L}(X)$ is a basis of a subgroup of index d of $F(A)$. Hint: use Exercise 7.13.

7.15 Let $\varphi \colon A^* \to G$ be a morphism from A^* onto a finite group G. Show, using Exercise 7.14, that for every minimal dendric shift on the alphabet A, the restriction of φ to $\mathcal{L}(X)$ is surjective.

7.16 Show that the decoding of a minimal dendric shift X by a finite X-complete bifix code is a minimal dendric shift. Hint: use Exercises 7.13 and 7.15 to prove that the decoding is recurrent and use Corollary 7.1.17.

7.17 Show that the substitution shift generated by $\sigma \colon a \to ac, b \to bac, c \to cbac$ of Example 7.1.41 is dendric.

7.18 Let X be a shift space. A bispecial word $w \in \mathcal{L}(X)$ is called *regular* if there are unique $a \in L(w)$ and $b \in R(w)$ such that aw is right special and wb is left special. In other words, the graph $\mathcal{E}(w)$ is a tree with paths of length at most 3. The shift X is said to satisfy the *regular bispecial condition* if it there is an $n \geq 1$ such that every bispecial word in $\mathcal{L}_{\geq n}(X)$ is regular. Show that X is eventually dendric if and only if it satisfies the regular bispecial condition.

7.19 A primitive $\mathcal{S}$-adic shift X is called a *Brun shift* if it is generated by a directive sequence $\tau = (\tau_n)$ such that every τ_n is an elementary automorphism $\beta_{ab} = \tilde{\alpha}_{ba}$ (which places a before b) and if for all n, we have that $\tau_n \circ \tau_{n+1}$ is either equal to β_{ab}^2 or to $\beta_{ab}\alpha_{ca}$ for some $a, b, c \in A$ with $a \neq b$ and $a \neq c$.

   1. Prove that the morphism $\sigma \colon a \to cbccba, b \to cbccb, c \to cbccbacbc$ generates a Brun shift that is not dendric.

   2. Show that a Brun shift is a proper unimodular $\mathcal{S}$-adic shift.

Section 7.2

7.20 Let X be a Sturmian shift of slope $\alpha = [a_0; a_1, \ldots]$. Show that the following conditions are equivalent.

  (i) X is linearly recurrent.
  (ii) The coefficients a_i are bounded.
 (iii) $\mathcal{L}(X)$ is K-power free for some $K \geq 1$.

Section 7.3

7.21 Let X be a recurrent specular shift. Show that the even code is an X-maximal bifix code of X-degree 2.

7.22 Prove Theorem 7.3.25. Hint: proceed as for the proof of Theorem 7.1.16.

7.5 Notes

7.5.1 Dendric Shifts

The languages of dendric shifts were introduced in Berthé et al. (2015a) under the name of *tree sets*. The language of the shift of Example 7.1.4 is a tree set of characteristic 2 (Berthé et al. (2017a, Example 4.2)) and it is actually a specular set. Tree sets of characteristic $c \geq 1$ were introduced in Berthé et al. (2017a) (see also Dolce and Perrin (2017a)). The shift spaces of Example 7.1.4 and of Exercise 7.4 (a neutral shift, which is not dendric) are due to Julien Cassaigne (2015).

It can be proved that the class of eventually dendric shifts is closed under conjugacy (see Dolce and Perrin (2020)).

Theorem 7.1.15 is from Berthé et al. (2015a). Theorem 7.1.16 was proved earlier in Balková et al. (2008) and was proved even earlier for episturmian shifts in Justin and Vuillon (2000).

Theorem 7.1.24 is from Berthé et al. (2015b, Theorem 6.5.19). It generalizes the fact that the derivative word of a Sturmian word is Sturmian (see Justin and Vuillon (2000)).

Theorem 7.1.28 is from Berthé et al. (2015b). It is obtained as a corollary of a result (called the *Finite Index Basis Theorem*) proved in Berthé et al. (2015c) (see Exercise 7.13).

The statement proved in Exercise 7.16 is the main result of Berthé et al. (2015b). The proof presented here (using Corollary 7.1.17) is shorter.

The fact that the group of positive automorphisms of a free group on three letters is not finitely generated is from Tan et al. (2004). The word "tame" used for tame automorphisms (as opposed to wild) is used here on analogy with its use in ring theory (see Cohn (1985)). The tame automorphisms as introduced here should, strictly speaking, be called positive tame automophisms since the group of all automorphisms, positive or not, is tame in the sense that it is generated by the elementary automorphisms.

Theorem 7.1.40 is from Berthé et al. (2015b).

In the case of a ternary alphabet, a characterization of tree sets by their $\mathcal{S}$-adic representation can be proved, using a *Büchi condition* on the alphabet $\mathcal{S}_e$ recognizing the set of $\mathcal{S}_e$-adic representations of uniformly recurrent tree sets (Leroy (2014a,b)). A Büchi condition uses a finite graph (V, E) with edges labeled by an alphabet S and two subsets U, W of V. A sequence $x \in S^{\mathbb{N}}$ satisfies the Büchi condition defined by (V, E) and U, W if it is the label of an infinite path in (V, E) starting in U and passing infinitely often through W (see Perrin and Pin (2004) for background on this type of condition).

The Brun shifts of Exercise 7.19 arise in the generalization of continued fractions expansion to triples of integers instead of pairs (Brun, 1958). The study of these algorithms as dynamical system was initiated in Berthé et al. (2019a). The relation of these shifts with dendric shifts was studied in Labbé and Leroy (2016). It is shown in Avila and Delecroix (2013), based on an analysis similar to that done for Arnoux–Rauzy shifts, that Brun shifts (on a ternary alphabet) are uniquely ergodic.

The regular bispecial condition (Exercise 7.18) was introduced by Damron and Fickenscher (2020). They proved that for an irreducible shift satisfying this condition, the number of ergodic measures is at most $(K + 1)/2$, where K is the limiting value of $p_{n+1}(X) - p_n(X)$. This generalizes a result proved independently by Katok (1973) and Veech (1978) concerning interval exchange transformations.

An interesting generalization of dendric shifts has been proposed in Goulet–Ouellet (2021). In this paper, the condition that the extension graph of every word in the language is a tree is replaced by the property for every word in the language of being *suffix-connected*. This condition still implies a number of properties such as the fact that the sets of return words generate the free group (without being always a basis).

7.5.2 Sturmian Shifts

Theorem 7.2.1 is already in Morse and Hedlund (1940). Assertion 1 follows from Theorem 7.1 in Morse and Hedlund (1940), and Assertion 2 is Theorem 7.1.

Corollary 7.2.7 is from Durand et al. (1999). It is, of course, related to the fact that a Sturmian word of slope α is k-power free if and only if the coefficients of the expansion of α as a continued fraction are bounded (Exercise 1.64).

Theorem 7.2.8 is from Dartnell et al. (2000). It is proved as a consequence of the fact that a Sturmian shift is a substitution shift if and only if the sequence $(\zeta_n)_n$ is ultimately periodic (see also Kurka (2003) and Araújo and Bruyère (2005)). For similar theorems characterizing purely substitutive Sturmian words, see the references in Lothaire (2002). In particular, it is proved in Allauzen (1998) that, for $0 < \alpha < 1$, the characteristic sequence c_α of slope α is purely substitutive if and only if α is quadratic and its conjugate is > 1.

It is not surprising that the condition on α for c_α to be purely substitutive is more restrictive than for the shift of slope α to be substitutive. Indeed, if α is quadratic, the Sturmian shift of slope α contains a purely substitutive sequence, but it need not be the characteristic sequence c_α.

7.5.3 Specular Shifts

The notion of specular shift was introduced in Berthé et al. (2017a). The idea of considering laminary sets is from Coulbois et al. (2008) (see also Lopez and Narbel (2013)).

For a proof of the Kurosh Subgroup Theorem, see Magnus et al. (2004). Specular groups are characterized by the property of their Cayley graphs to be regular (see Harpe (2000)). See also Harpe (2000) concerning the notion of virtually free group. Actually, specular groups can be studied as groups acting on trees as developed in the Bass–Serre theory (Serre, 2003).

A group having a free subgroup of finite index is called *virtually free*. On the other hand, a finitely generated group is said to be *context-free* if, for some presentation, the set of words equivalent to 1 is a context-free language. By Muller and Schupp's theorem (1983), a finitely generated group is virtually free if and only if it is context-free. Thus, a specular group is context-free. One may verify this directly as follows. A context-free grammar generating the words equivalent to 1 for the natural presentation of a specular group $G = G_\theta$ is the grammar with one nonterminal symbol σ and the rules

$$\sigma \to a\sigma a^{-1}\sigma \quad (a \in A), \quad \sigma \to 1. \tag{7.13}$$

The proof that the grammar given by Eq. (7.13) generates the set of words equivalent to 1 is similar to that used in Berstel (1979) for the so-called Dyck-like languages.

Theorem 7.3.22 is the counterpart for minimal specular shifts of the main result of Berthé et al. (2015b, Theorem 6.1) asserting that the family of uniformly recurrent tree sets of characteristic 1 is closed under maximal bifix decoding.

Theorem 7.3.25 is proved in Dolce and Perrin (2017a, Theorem 3).

The definition of mixed return words comes from the fact that, when S is the natural coding of a linear involution, we are interested in the transformation induced on $I_w \cup \sigma_2(I_w)$ (see Berthé et al. (2017b)). The natural coding of a point in I_w begins with w while the natural coding of a point z in $\sigma_2(I_w)$ "ends" with w^{-1} in the sense that the natural coding of $T^{-|w|}(z)$ begins with w^{-1}.

A geometric proof and interpretation of Theorem 7.3.38 is given in Berthé et al. (2017b). It is shown that the set of mixed return words is a symmetric basis of a fundamental group corresponding to a surface built above the linear involution.

8

Interval Exchange Transformations

In this chapter, we study a class of dynamical systems obtained by iterating simple geometric transformations on an interval. These transformations, called interval exchange, form a classical family of dynamical systems. They can be seen as a generalization of the rotations of the circle, already met several times. They define, by a natural coding, shift spaces that turn out to be dendric shifts. Thus, we obtain, after Arnoux–Rauzy shifts, a large class of dendric shifts for which we will describe more precisely return words and S-adic representations.

In the first part, we introduce interval exchange transformations. We define the natural coding of an interval exchange transformation, which is shown to be a dendric shift (Proposition 8.1.13). We introduce the notion of regular interval exchange transformation and prove that a regular transformation is minimal (Theorem 8.1.2). We develop the notion of Rauzy induction and characterize the subintervals, which are the domain of a transformation obtained by iterating the induction (Theorem 8.1.25). We generalize Rauzy induction to a two-sided version and characterize the intervals reached by this more general induction (Theorem 8.2.2). We link these transformations with automorphisms of the free group (Theorem 8.2.14). We also relate these results with the theorem of Boshernizan and Carroll, giving a finiteness condition on the systems induced by an interval exchange when the lengths of the intervals belong to a quadratic field (Theorem 8.3.2).

In the second part, we present linear involutions, which form a larger family by adding the possibility of symmetries in addition to translations. We introduce the notion of connection and relate it to the minimality of the transformation (Proposition 8.4.6). We show that the natural coding of a linear involution without connection is specular (Theorem 8.4.9).

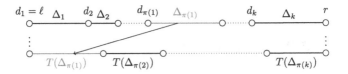

Figure 8.1 A k-interval exchange transformation.

8.1 Interval Exchange Transformations

A *semi-interval* is a nonempty subset of the real line of the form $[\ell, r) = \{z \in \mathbb{R} \mid \ell \leq z < r\}$. Thus, it is a left-closed and right-open interval. For two semi-intervals Δ, Γ, we denote $\Delta < \Gamma$ if $x < y$ for every $x \in \Delta$ and $y \in \Gamma$. A partition $\{\Delta_1, \ldots, \Delta_k\}$ of the semi-interval I indexed by $1, \cdots, k$ is *ordered* if $\Delta_1 < \cdots < \Delta_k$.

Let $\{\Delta_1, \ldots, \Delta_k\}$ be an ordered partition of the semi-interval $I = [\ell, r)$ into $k \geq 2$ disjoint semi-intervals. A *k-interval exchange transformation* on I is an onto map $T : [\ell, r) \to [\ell, r)$ where $T : \Delta_i \to [\ell, r)$ is a translation.

We sometimes denote (I, T) the interval exchange on analogy with the notation of dynamical systems (this is actually a measure-theoretic dynamical system, see below).

For every i with $1 \leq i \leq k$, there is a real number α_i such that

$$T(x) = x + \alpha_i$$

for every $x \in \Delta_i$. The numbers α_i are called the *translation values* of T.

We denote by π the permutation of the set $\{1, \ldots, k\}$ such that $T(\Delta_{\pi(1)}), \ldots, T(\Delta_{\pi(k)})$ is an ordered partition (see Figure 8.1). Thus, $\pi(i)$ is the index of the interval sent to the ith position. As always when dealing with permutations, there is an unavoidable and unfortunate tendency to confuse this permutation with its inverse π^{-1}, which gives the position of the image of the jth interval.

If λ_i denotes the length of the interval Δ_i and $\lambda = (\lambda_1, \ldots, \lambda_k)$, the transformation T is defined by λ and the permutation π. We denote $T = T_{\lambda, \pi}$.

One may imagine an interval exchange transformation as the "physical" operation presented in Figure 8.1. Break the semi-interval I into k pieces and rearrange them in a different order.

Such a transformation is not continuous and thus does not fit well in the framework of topological dynamical systems. Actually, since T is a piecewise isometry on the semi-interval I, it preserves the Lebesgue measure. Thus, an interval exchange transformation is a measure-theoretic dynamical system.

We shall see below how the interval can be modified to obtain a topological dynamical system.

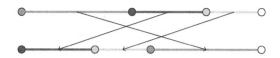

Figure 8.2 A 3-interval exchange transformation.

For $k = 2$, an interval exchange transformation on $I = [0, 1)$ is a rotation. We will often consider the semi-interval interval $[0, 1)$ to simplify the notation. Indeed, for $0 < \alpha < 1$, set $\Delta_1 = [0, 1 - \alpha)$ and $\Delta_2 = [1 - \alpha, 1)$. The corresponding interval exchange transformation T is $T(x) = x + \alpha \bmod 1$. In this particular case, the transformation is continuous provided one identifies the two endpoints of the interval (see Example 1.1.5).

An interval exchange transformation is invertible since a translation is one-to-one. Its inverse is again an interval exchange transformation, on the intervals $\Delta'_i = T(\Delta_i)$.

A power T^n of an interval exchange transformation is again an interval exchange transformation. Since we have seen that T^{-1} is also an interval exchange transformation, we may assume $n > 0$. Consider, for a word $w = i_0 i_1 \cdots i_{n-1}$ of length n on the alphabet $\{1, \ldots, k\}$, the nonempty sets of the form

$$I(w) = \Delta_{i_0} \cap T^{-1} \Delta_{i_1} \cap \cdots \cap T^{-n+1} \Delta_{i_{n-1}}. \tag{8.1}$$

These sets are semi-intervals (Exercise 8.8). Thus, T^n is an exchange of the nonempty semi-intervals $I(w)$.

Example 8.1.1 A 3-interval exchange transformation is represented in Figure 8.2. The associated permutation is the cycle $\pi = (123)$.

8.1.1 Minimal Interval Exchange Transformations

The *separation points* of an interval exchange are the left end points $d_1 = 0, d_2, \ldots, d_k$ of the semi-intervals Δ_i for $1 \le i \le k$. We denote by $\mathrm{Sep}(T)$ the set of separation points of T.

An interval exchange transformation T on $I = [0, 1)$ is called *regular* if the orbits of the nonzero separation points are infinite and disjoint. Note that the orbit of 0 cannot be disjoint of the others, since one has $T(d_i) = 0$ for some i with $1 \le i \le k$, but that the orbit of 0 is infinite when T is regular.

A *connection* is a triple (i, j, n) for $2 \le i, j \le k$, and $n > 0$ such that $T^n(d_i) = d_j$. An interval exchange transformation is regular if and only if there is no connection. Indeed, a connection (i, j, n) with $i = j$ corresponds to a finite orbit and with $i \neq j$ to intersecting orbits.

As an example, a rotation of angle α is regular if and only if α is irrational. Indeed, α is rational if and only if the orbit of α is finite (in which case all orbits are finite).

We say that an interval exchange is *minimal* if the orbit of every point is dense.

Theorem 8.1.2 (Keane) *A regular interval exchange transformation is minimal.*

The converse is not true. Indeed, consider the rotation of irrational angle α on $I = [0, 1)$ as a 3-interval exchange with the partition $[0, 1 - 2\alpha), [1 - 2\alpha, 1 - \alpha), [1 - \alpha, 0)$ (see Figure 8.3). The transformation is minimal, as any rotation of irrational angle but $T(1 - 2\alpha) = 1 - \alpha$ and thus there is a connection.

We first prove the following result. Let T be an interval exchange on $I = [0, 1)$. We shall consider the transformation T_1 induced by T on a semi-interval $\Delta_1 = [a, b) \subset [0, 1)$. It is defined by $T_1(z) = T^{n(x)}(z)$ where $n(z)$ is the least integer $n \geq 1$ such that $T^n(z)$ belongs to Δ_1. The return time $n(z)$ exists, even if T is not minimal, by the Poincaré Recurrence Theorem, because T preserves the Lebesgue measure. In a particular case, the induction on a semi-interval is called a Rauzy induction and we will have more to say on this later.

Theorem 8.1.3 (Rauzy) *Let T be a k-interval exchange transformation on $I = [0, 1)$ and let $X_1 = [a, b) \subset [0, 1)$ be a semi-interval. The transformation T_1 induced by T on X_1 is a k_1-interval exchange transformation with $k_1 \leq k + 2$ on semi-intervals $\Delta'_1, \ldots, \Delta'_{k_1}$. For each index i with $1 \leq i \leq k_1$, the return time to X_1 of all $x \in \Delta'_i$ is the same.*

Proof Consider the set $Y = \{a, b, d_2, \ldots, d_k\}$, where $d_2, \ldots, d_k$ are the nonzero separation points. For $y \in Y$, let $s(y)$ be the least integer $s \geq 0$ if it exists such that $T^{-s(y)}(y)$ belongs to (a, b). The points $T^{-s(y)}(y)$ divide the semi-interval $[a, b)$ in $k_1 \leq k + 2$ semi-intervals $\Delta'_1, \ldots, \Delta'_{k_1}$. For each semi-interval Δ'_i consider the number n_i which is the least $n \geq 1$ such that $T^n(\Delta'_i) \cap [a, b) \neq \emptyset$.

Then, for $1 \leq p \leq n_i$ the transformations T^p are continuous on Δ'_i and we have $T^{n_i}(\Delta'_i) \subset [a, b)$. Indeed, otherwise, for some p with $1 \leq p \leq n_i - 1$, the semi-interval $T^p(\Delta'_i)$ contains one of the points $y \in Y$ and then $s(y) = p$ so that the point $T^{-p}(y)$ would lie within Δ'_i, a contradiction with the definition of the semi-intervals Δ_i'. This shows that n_i is the return time of all $x \in \Delta_i'$ and thus that T_1 is the interval exchange of $[a, b)$ corresponding to the intervals Δ_i'. ∎

Proof of Theorem 8.1.2 Let us first show that T has no periodic point. Assume that T^n has a fixed point, that is, a point $x \in I$ such that $T^n x = x$. Since T^n is an interval exchange transformation on the intervals $\Delta_{i_0,\dots,i_{n-1}}^{(n)}$, one of these intervals is formed of fixed points. But the left end of this interval is in the orbit of some of the separation points d_i. This contradicts the hypothesis that the orbit of all nonzero separation points is infinite or disjoint unless $i = 1$ and $n = 1$. But in this case, the interval Δ_1 is fixed by T, which implies that $T(d_j) = d_2$ for some $j \geq 2$, a contradiction again with the hypothesis.

Suppose now that the orbit $O(z)$ of some $z \in [0, 1)$ is not dense. We can find a semi-interval $[a, b)$ disjoint from $O(z)$. Let T_1 be the transformation induced by T on $[a, b)$. By Theorem 8.1.3, T_1 is an interval exchange on semi-intervals Δ_j' and the return time to $[a, b)$ of every point $x \in \Delta_j'$ is constant and equal to an integer n_j. Set

$$F = \bigcup_j \bigcup_{n=0}^{n_j-1} T^n(\Delta_j').$$

The set F can be written as a union of a finite number of nonintersecting semi-intervals. Hence its connected components are also semi-intervals, say F_s and their number is finite. Let G be the set of left end points of all the semi-intervals F_s. By definition, the set F is invariant by T and for every $x \in G$, either $T(x)$ is in G or is equal to some of the d_i for $1 \leq i \leq k$ (this happens if $T(x)$ has on its left arbitrarily close elements of F). Since G is finite and since T has no periodic points, there is for every $x \in G$ some $n \geq 0$ such that $T^n(x) = d_i$ with $1 \leq i \leq k$. Similarly, there is for every $x \in G$ an integer $m > 0$ such that $T^{-m}(x) = d_j$ with $2 \leq i \leq k$. Then, we have

$$T^n(x) = T^{n+m}(d_j) = d_i.$$

Since the orbits of $d_2, \dots, d_k$ are disjoint, the only possibility is $i = 1$. Then $d_j = T^{-1}(d_1)$ and thus $m = 1$, $n = 0$ and $x = 0$. Thus, $G = \{0\}$, which implies $F = [0, 1)$, a contradiction with the fact that, by construction, one has $F \cap O(z) = \emptyset$. ∎

The following necessary condition for minimality of an interval exchange transformation is useful. A permutation π of an ordered set A is called *reducible* if there exists an element $b \in A$ such that the set B of elements strictly less than b is nonempty and such that $\pi(B) = B$. Otherwise it is called *irreducible*.

If an interval exchange transformation $T = T_{\lambda,\pi}$ is minimal, the permutation π is irreducible. Indeed, if B is a set as above, the set of orbits of the points in the set $S = \cup_{a \in B} \Delta_a$ is invariant and strictly included in $[\ell, r[$. The following example shows that the irreducibility of π is not sufficient for T to be minimal.

Example 8.1.4 Let $A = \{a, b, c\}$ and λ be such that $\lambda_a = \lambda_c$. Let π be the transposition (ac). Then π is irreducible but $T_{\lambda,\pi}$ is not minimal since it is the identity on Δ_b.

A k-interval exchange transformation $T = T_{\lambda,\pi}$ on $I = [0, 1)$ is *irrational* if π is irreducible and if the numbers $\lambda_1, \ldots, \lambda_k$ are linearly independent over $\mathbb{Q}$ (note that $\lambda_1, \ldots, \lambda_k, 1$ are not independent since $\sum_{i=1}^{k} \lambda_i = 1$). The following result gives a sufficient condition for minimality of interval exchange transformations.

Theorem 8.1.5 (Keane) *An irrational interval exchange transformation is minimal.*

The proof is left as Exercise 8.4. Note that the condition is not necessary (see Example 8.1.7).

The iteration of a k-interval exchange transformation is, in general, an interval exchange transformation operating on a larger number of semi-intervals.

Proposition 8.1.6 *Let T be a regular k-interval exchange transformation. Then, for every $n \geq 1$, T^n is a regular $n(k - 1) + 1$-interval exchange transformation.*

Proof Since T is regular, the set $\cup_{i=0}^{n-1} T^{-i}(d)$, where d runs over the set of $k - 1$ nonzero separation points of T, has $n(k - 1)$ elements. These points partition the interval $[\ell, r[$ in $n(k - 1) + 1$ semi-intervals on which T is a translation. ∎

8.1.2 Natural Coding

To every interval exchange transformation (I, T), we associate a shift space called its *natural coding*. As for the notion introduced for rotations in Section 1.5.3, we will see that it is injective (provided T is minimal) but not continuous. It is defined as follows.

Set $A = \{1, 2, \ldots, k\}$ and let $\Sigma_T : I \to A^{\mathbb{Z}}$ be the map defined by $\Sigma_T(z) = (x_n)_{n \in \mathbb{Z}}$ with

$$x_n = a \quad \text{if } T^n(z) \in \Delta_a.$$

Note that the right infinite sequence $\Sigma_T(z)^+$ satisfies

$$\Sigma_T(z)^+ = a_0 a_1 \cdots a_{n-1} \Sigma_T(T^n z)^+ \tag{8.2}$$

with $T^i(z) \in \Delta_{a_i}$ for $0 \leq i < n$.

The two-sided sequence $x = \Sigma_T(z)$ is called the *natural coding* of z. The natural coding of T, denoted $X(T)$, is the closure in $A^{\mathbb{Z}}$ of $\Sigma_T(I)$.

An *interval exchange shift* is the natural coding $X(T)$ of an interval exchange T. It is said to be *regular* if the interval exchange is regular. We denote $\mathcal{L}(T) = \mathcal{L}(X(T))$.

We now list the properties of the map Σ_T as we did for rotations in Proposition 1.5.12.

1. The map Σ_T satisfies $\Sigma_T \circ T = S \circ \Sigma_T$ where S denotes as usual the shift.
2. The map Σ_T is continuous except at a countable set of points, namely the set $\mathcal{O}(T) = \{T^n(d_a) \mid n \in \mathbb{Z}, a \in A\}$, which is the union of the orbits of the separation points. Indeed, let $z \in I$, which is not in $\mathcal{O}(T)$. For every $n \geq 1$, there is a neighborhood U of z such that the natural coding of all points of U coincides with $\Sigma_T(z)$ on the indices i with $-n \leq i \leq n$.
3. If T is minimal, the map Σ_T is injective. Indeed, let $z < z'$ be such that $\Sigma_T(z) = \Sigma_T(z')$. Then there is no n such that $T^{-n}(d_a) \in (z, z')$ for some $a = 2, \ldots, k$, a contradiction.
4. The map Σ_T is actually a measure-theoretic isomorphism from (I, T) with the Lebesgue measure onto the system formed by the shift space $X(T)$ with the measure induced by Σ_T. Indeed, Σ_T is measurable because, by (8.1), $\Sigma_T^{-1}([u])$ is a semi-interval for every $u \in \mathcal{L}(T)$, it interleaves T with the shift by Assertion 1, and preserves the measure by definition. As a consequence, the shift space $X(T)$ is minimal if and only if T is minimal. Indeed, the orbit of z is dense if and only if the orbit of $\Sigma_T(z)$ is dense.

Example 8.1.7 Consider the interval exchange transformation T represented in Figure 8.3 with $\alpha = (3 - \sqrt{5})/2$. It is the rotation $x \mapsto x + 2\alpha$ represented as a regular interval exchange on three semi-intervals: $\Delta_a = [0, 1 - 2\alpha)$, $\Delta_b = [1 - 2\alpha, 1 - \alpha)$, and $\Delta_c = [1 - \alpha, 1)$.

Since T is minimal, the set $\mathcal{L}(T)$ is uniformly recurrent. The words of length at most 6 of the set $\mathcal{L}(T)$ are represented in Figure 8.4.

The shift $X = X(T)$ is actually a primitive substitution shift, called the *squared Fibonacci shift*. Indeed, since T is the rotation of angle 2α, the shift $X(T)$ is the coding by nonoverlapping blocks of length 2 of the Fibonacci shift, which is the natural coding of the rotation of angle α. But the Fibonacci

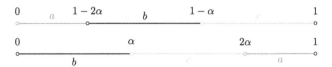

Figure 8.3 The rotation of angle 2α.

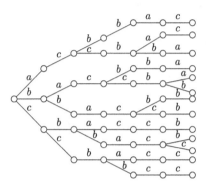

Figure 8.4 The words of length ≤ 6 of the set $\mathcal{L}(T)$.

morphism $\varphi: a \mapsto ab, b \mapsto a$ is such that $\varphi^3: a \mapsto abaab, b \mapsto aba$ sends words of even length to words of even length. Using the coding $a \mapsto aa, b \mapsto ab, c \mapsto ba$, the action of φ^3 on $\{a, b, c\}$ is $\psi: a \mapsto baccb, b \to bacc, c \mapsto bacb$. Thus, $X = X(\psi)$.

Exactly as we had in Theorem 1.5.16, we have the following statement concerning minimal interval exchanges. A map $f: X \to Y$ is *two-to-one* if $\mathrm{Card}(f^{-1}(y)) = 2$ for every $y \in Y$.

Theorem 8.1.8 *Let (I, T) be a minimal interval exchange and let $X = X(T)$. There exists a unique continuous map $f: X \to I$ such that its restriction to $\Sigma_T(I)$ is the inverse of Σ_T. Moreover, we have the following properties.*

1. $T \circ f = f \circ S$.
2. f is one-to-one on $X \setminus \Sigma_T(\mathcal{O}(T))$ where $\mathcal{O}(T) = \{T^n(d_a) \mid n \in \mathbb{Z}, a \in A\}$ is the union of the orbits of the separation points.
3. f is two-to-one on $\Sigma_T(\mathcal{O}(T))$.

Proof Assume that $I = [0, 1)$. Set $s = \Sigma_T(0)$. Since X is minimal, the orbit of s is dense and there is for every $t \in X$ a sequence (k_n) of integers such that $t = \lim S^{k_n}(s)$. Set $z_n = T^n(0)$. Then, since T is minimal, for every $\varepsilon > 0$, every interval of length ε contains some $T^n 0$ and thus there exists $\eta > 0$ such that for all $z, z' \in I$,

$$d(\Sigma_T(z), \Sigma_T(z')) < \eta \Rightarrow |z - z'| < \varepsilon. \tag{8.3}$$

This implies that (z_n) is a Cauchy sequence. Let z be its limit in $[0, 1]$. We define $f(t) = z$ if $z \neq 1$ and $f(z) = 0$ otherwise. If $t = \Sigma_T(z)$, the sequence (z_n) above converges to z and thus $f(t) = z$. This shows that f is onto and that its restriction to $\Sigma_T(I)$ is the inverse of Σ_T.

If $z \notin \mathcal{O}(T)$, then Σ_T is continuous at z and thus $f(t) = z$ implies $t = \Sigma_T(z)$. Otherwise, z belongs to the orbit of some separation point d_a. If $d_a \neq 0$, the set $f^{-1}(d_a)$ is formed of the two sequences $\lim_{y \to d_a^-} \Sigma_T(y)$ and $\Sigma_T(d_a)$. The map f will then be two-to-one on all points of the orbit of d_a. Finally, $f^{-1}(0)$ is made of the two sequences $\lim_{y \to 1^-} \sigma_T(y)$ and $s = \Sigma_T(0)$ and the conclusion follows also. The fact that $T \circ f = f \circ S$ is easy to verify by the definition of f. ∎

8.1.3 Cantor Version of Interval Exchange

The natural coding can be made continuous at the cost of modifying the space $I = [0, 1]$ as described below.

Suppose that T is minimal. Let $\mathrm{Sep}(T) = \{d_1, \ldots, d_k\}$ be the set of separation points and set $\mathcal{O}(T) = \{T^j d \mid j \in \mathbb{Z}, d \in \mathrm{Sep}(T)\}$. We define

$$\tilde{I} = (I \setminus \mathcal{O}(T)) \bigcup \{z^-, z^+ \mid z \in \mathcal{O}(T)\},$$

where $0^- = 1$. Defining $x < z^- < z^+ < y$ for all $z \in \mathcal{O}(T)$ and $x, y \in I \setminus \mathcal{O}(T)$ such that $x < z < y$ (with the exception of $0^- \geq x$ for all $x \in \tilde{I}$), this extends the natural order on $I = [0, 1)$ to $\tilde{I}$. Endowed with the order topology, the set $\tilde{I}$ is a Cantor space because $\mathcal{O}(T)$ is dense in I.

Let $\tilde{T} : \tilde{I} \to \tilde{I}$ defined by $\tilde{T}(x) = T(x)$ if $x \in I \setminus \mathcal{O}(T)$ and $\tilde{T}(z^\epsilon) = T(z)^\epsilon$ if z is in $\mathcal{O}(T)$ and ϵ in $\{+, -\}$. The pair $(\tilde{I}, \tilde{T})$ is a minimal Cantor system, we will refer to as the *Cantor version of the interval exchange T*.

Proposition 8.1.9 *The map $\phi : \tilde{I} \to I$ defined by $\phi(z^+) = \phi(z^-) = z$ and $\phi(z) = z$ when $z \notin \mathcal{O}(T)$ is an onto continuous map, which is one-to-one everywhere except on a countable set of points. Moreover, $\phi \circ \tilde{T} = T \circ \phi$.*

Proof The map ϕ is clearly, by definition, an onto continuous map and such that $\phi \circ \tilde{T} = T \circ \phi$. Let $f : X \to I$ be the map defined by Theorem 8.1.8. Let $\alpha : X \to \tilde{I}$ be the map defined as follows. For $t \in X$, set

$$\alpha(t) = \begin{cases} f(t) & \text{if } f(t) \notin \mathcal{O}(T) \\ f(t)^+ & \text{if } f(t) \in \mathcal{O}(T) \text{ and } \Sigma_T(f(t)) = t \\ f(t)^- & \text{otherwise.} \end{cases}$$

It is easy to verify that $f = \phi \circ \alpha$ and that α is a conjugacy from X onto $\tilde{I}$. Thus, $\phi \circ \tilde{T} = f \circ \alpha^{-1} \circ \tilde{T} = f \circ S \circ \alpha^{-1} = T \circ f \circ \alpha^{-1} = T \circ \phi$. Finally, ϕ is one-to-one except on the countable set $\phi^{-1}(\mathcal{O}(T))$.   ∎

8.1.4 The Semi-intervals $I(w)$ and $J(w)$

Let T be a k-interval exchange transformation on the semi-interval I corresponding to semi-intervals Δ_i. Let $A = \{1, 2, \ldots, k\}$ and let X be the natural coding of T. For $w \in A^*$, we already defined by Eq. (8.1) a semi-interval $I(w)$. As an equivalent definition, we have $I(\varepsilon) = I$ and

$$I(au) = \Delta_a \cap T^{-1}(I(u)). \tag{8.4}$$

Every nonempty $I(w)$ is a semi-interval (Exercise 8.8) and $I(a) = \Delta_a$.

Proposition 8.1.10 *One has $w \in \mathcal{L}(X)$ if and only if $I(w) \neq \emptyset$.*

Proof The statement follows from the fact that w belongs to $\mathcal{L}(X)$ if and only if $[w]$ is nonempty and that an easy induction shows that $[w] \neq \emptyset$ if and only if $I(w) \neq \emptyset$.   ∎

Besides the semi-intervals $I(w)$, we will also need symmetrically the sets $J(w)$ defined, for $w \in A^*$, by $J(\varepsilon) = I$ and by

$$J(ua) = T(J(u)) \cap T(\Delta_a) \tag{8.5}$$

for $a \in A$ and $u \in A^*$. As for the $I(w)$, the nonempty sets $J(w)$ are semi-intervals (Exercise 8.9).

For $x = \Sigma_T(z)$ and $w \in \mathcal{L}_m(T)$, one has

$$x_0 x_1 \cdots x_{m-1} = w \Leftrightarrow z \in I(w) \tag{8.6}$$

and symmetrically

$$x_{-m} x_{-m+1} \cdots x_{-1} = w \Leftrightarrow z \in J(w), \tag{8.7}$$

as one may verify using the definition of $I(w)$ and $J(w)$. Thus, $I(w)$ is the semi-interval formed of all points $z \in I$ such that x^+ begins with w and $J(w)$ is the semi-interval formed of all points $z \in I$ such that x^- ends with w.

We will use later the following remark in which $\Sigma_T : I \to X(T)$ denotes the natural coding of (I, T) and $\Sigma_T^+(z) = (\Sigma_T(z))^+$.

Figure 8.5 The transformation induced on $J(b)$.

Proposition 8.1.11 *Let T be a regular interval exchange transformation and let $X = X(T)$ be its natural coding. For $w \in \mathcal{L}(X)$, let T' be the transformation induced by T on $J(w)$. One has $u \in \mathcal{R}_X(w)$ if and only if*

$$\Sigma_T^+(z) = u\Sigma_T^+(T'(z))$$

for some $z \in J(w)$. The shift $X(T')$ is the derivative of $X(T)$ with respect to $[w]$.

Proof Assume first that u belongs to $\mathcal{R}_X(w)$. Then for every $z \in J(w) \cap I(u)$, it follows from (8.6) and (8.7) that the least integer $i \geq 1$ such that $T^i(z)$ is in $J(w)$ is $i = |u|$. Thus, we have $T'(z) = T^{|u|}(z)$ and

$$\Sigma_T^+(z) = u\Sigma_T^+(T^{|u|}(z)) = u\Sigma_T^+(T'(z)).$$

Conversely, assume that $\Sigma_T^+(z) = u\Sigma_T^+(T'(z))$ for some $z \in J(w)$. Then $T^{|u|}(z)$ belongs to $J(w)$ and thus wu is in $A^*w \cap \mathcal{L}(T)$. Moreover, u does not have a proper prefix ending with w, and thus u is in $\mathcal{R}_X(w)$.

As a consequence, one has $T^n(z)$ in $J(w)$ if and only if $S^n\Sigma_T^+(z)$ is in $[w]$ and thus $X(T')$ is the derivative of X with respect to $[w]$. ∎

A dual statement holds for the transformation T' induced on the semi-interval $I(w)$. One has $u \in \mathcal{R}'_X(w)$ if and only if $\Sigma_T^-(z) = \Sigma_T^-(T'(z))u$ for some $z \in I(w)$.

Example 8.1.12 Consider the 2-interval exchange of Figure 8.5 on the left (it is a rotation of angle α). The transformation induced on $J(b)$ is the 2-interval exchange shown on the right. We have $\mathcal{R}_X(b) = \{b, ab\}$, and the morphism associated with the induction is $a \mapsto ab, b \mapsto b$.

8.1.5 Planar Dendric Shifts

A shift space is called *planar dendric* if it is dendric and if there are two orders $\leq_1$ and $\leq_2$ on the alphabet A such that for every $w \in \mathcal{L}(X)$, the tree $\mathcal{E}(w)$ is compatible with these orders, that is, for every $(a, b), (c, d) \in \mathcal{E}(w)$, one has $a \leq_1 c$ if and only if $b \leq_2 d$.

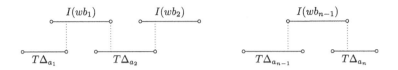

Figure 8.6 A path from a_1 to a_n in $\mathcal{E}(w)$.

Thus, placing the vertices of $L(w)$ on a vertical line in the order given by $\leq_1$ and those of $R(w)$ on a parallel line in the order given by $\leq_2$, the tree $\mathcal{E}(w)$ becomes planar. Note that the orders $\leq_1$ and $\leq_2$ do not depend on w.

Proposition 8.1.13 *A regular interval exchange transformation shift is a minimal and planar dendric shift.*

Proof Assume that T is a regular interval exchange transformation relative to the intervals $(\Delta_a)_{a \in A}$. Let X be the natural coding of T. We consider, on the alphabet A, the two orders defined by

1. $a <_{\text{top}} b$ if Δ_a is to the left of Δ_b,
2. $a <_{\text{bottom}} b$ if $T\Delta_a$ is to the left of $T\Delta_b$.

Note that, if wb, wc are in $\mathcal{L}(X)$, then $b <_{\text{top}} c$ if and only if $I(wb)$ is to the left of $I(wc)$. Indeed, $I(wb), I(wc) \subset I(w)$ and $T^{|w|}$ is a translation on $I(w)$.

For $a, a' \in L(w)$ with $a <_{\text{bottom}} a'$, there is a unique reduced path in $\mathcal{E}(w)$ from a to a'. Indeed, there is a unique sequence $a_1, b_1, \ldots a_n$ with $a_1 = a$ and $a_n = a'$ with $a_1 <_{\text{bottom}} a_2 <_{\text{bottom}} \cdots <_{\text{bottom}} a_n$, $b_1 <_{\text{top}} b_2 <_{\text{top}} \cdots <_{\text{top}} b_{n-1}$ and $T\Delta_{a_i} \cap I_{wb_i} \neq \emptyset$, $T\Delta_{a_{i+1}} \cap I_{wb_i} \neq \emptyset$ for $1 \leq i \leq n - 1$ (see Figure 8.6). Such a sequence is a path in $\mathcal{E}(w)$ and conversely.

Note that the hypothesis that T is regular is needed here since otherwise the right boundary of $T\Delta_{a_i}$ could be the left boundary of $I(wb_i)$. Thus, $\mathcal{E}(w)$ is a tree. It is compatible with the orders $<_{\text{bottom}}, <_{\text{top}}$ since the above shows that $a <_{\text{bottom}} a'$ implies that the letters b_1, b_{n-1} such that $(a, b_1), (a', b_{n-1}) \in E(w)$ satisfy $b_1 \leq_{\text{top}} b_{n-1}$. ∎

The converse of Proposition 8.1.13 is also true (see Section 8.6).

Example 8.1.14 Let X be the squared Fibonacci shift (Example 8.1.7). The extension graph of ε is represented in Figure 8.7. The orders $<_{top}$ and $<_{bottom}$ are

$$a <_{top} b <_{top} c \quad \text{and} \quad b <_{bottom} c <_{bottom} a.$$

Figure 8.7 The extension graph $\mathcal{E}(\varepsilon)$.

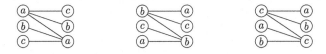

Figure 8.8 The graphs $\mathcal{E}(\varepsilon)$, $\mathcal{E}(a)$ and $\mathcal{E}(aba)$ in the Tribonacci set.

The following example shows that the Tribonacci shift is not a planar dendric shift and thus, that the Tribonacci shift is not a regular interval exchange shift (this can also be proved directly; see Exercise 8.1).

Example 8.1.15 Let X be the Tribonacci shift (see Example 1.5.3). The words a, aba and $abacaba$ are bispecial. Thus, the words ba, $caba$ are right-special and the words ab, $abac$ are left-special. The graphs $\mathcal{E}(\varepsilon)$, $\mathcal{E}(a)$ and $\mathcal{E}(aba)$ are shown in Figure 8.8. One sees easily that it is not possible to find two orders on A making the representation of the three graphs simultaneously planar.

Since, by Proposition 8.1.13, a regular interval exchange shift is dendric, we derive from Proposition 7.1.6 the following consequence.

Corollary 8.1.16 *The factor complexity of a regular k-interval exchange shift X is $p_n(X) = (k - 1)n + 1$.*

8.1.6 Rauzy Induction

Since the natural coding of a minimal interval exchange T is dendric, it has by Theorem 7.1.40 a primitive $\mathcal{S}_e$-adic representation with a directive sequence $\tau = (\tau_n)$ formed of elementary morphisms $\tau_n \in \mathcal{S}_e$. We will now give a method, called *Rauzy induction*, which allows one to build directly this representation from the interval exchange T.

Proposition 8.1.17 *Let T be the k-interval exchange transformation on $[0, 1)$ with the intervals Δ_i of lengths λ_i and with the permutation π. Let $r = 1 - \min$*

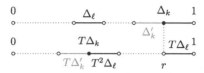

Figure 8.9 The Case 0 of Rauzy induction.

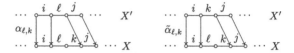

Figure 8.10 The natural coding of T' (case 0 and case 1).

$\{\lambda_k, \lambda_{\pi(k)}\}$. *If T is minimal, the transformation T' induced by T on the interval $[0, r)$ is again a minimal k-interval exchange transformation.*

Proof Set $\ell = \pi(k)$. Assume first that $\lambda_\ell < \lambda_k$ (Case 0, see Figure 8.9). Set $\Delta'_k = \Delta_k \setminus T\Delta_\ell$. The transformation T' induced by T on $[0, r)$ is such that for every $z \in [0, r)$,

$$T'z = \begin{cases} Tz & \text{if } x \notin \Delta_\ell \\ T^2z & \text{otherwise.} \end{cases}$$

Thus, T' is the same as T, except for the points of Δ_ℓ, which are sent to $T^2\Delta_\ell$. This shows that T' is an interval exchange on the k intervals:

$$\Delta_1, \ldots, \Delta_{k-1}, \Delta'_k.$$

In the case $\lambda_k < \lambda_{\pi(k)}$ (Case 1), we replace T by T^{-1} and we find the first case again. The case $\lambda_{\pi(k)} = \lambda_k$ cannot occur because T is minimal. ∎

The natural coding $X = X(T)$ of T is the image of the natural coding $X' = X(T')$ of T' by a morphism that is an elementary automorphism of the free group (see Section 7.1.5) and called the automorphism *associated* with the Rauzy induction, which is defined as follows.

In Case 0 ($\lambda_{\pi(k)} < \lambda_k$), we have $X = \alpha_{\ell,k}(X')$ where $\ell = \pi(k)$ and $\alpha_{\ell,k}$ is the elementary automorphism

$$\alpha_{\ell,k} : \ell \mapsto \ell k, \ i \mapsto i \text{ for } i \neq \ell, \tag{8.8}$$

which places a k after each ℓ. Indeed, in $X(T)$, every symbol ℓ is followed by k, and in the coding of T' every ℓk is replaced by ℓ (see Figure 8.10 on the left).

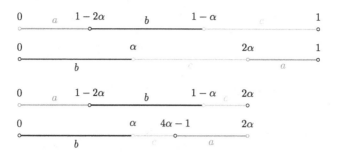

Figure 8.11 The effect of Rauzy induction on the rotation of angle 2α.

Figure 8.12 Case 0 of Rauzy induction.

In Case 1 ($\lambda_k < \lambda_{\pi(k)}$), we have $X = \tilde{\alpha}_{k,\ell}(X')$ where

$$\tilde{\alpha}_{k,\ell} : k \mapsto \ell k, \ i \mapsto i \text{ for } i \neq k \qquad (8.9)$$

is the morphism which places an ℓ before each k (see the right-hand side of Figure 8.10).

Example 8.1.18 Consider again the squared Fibonacci shift. It is the 3-interval exchange of Example 8.1.7, which is a rotation of angle 2α (reproduced for readability in Figure 8.11 top). The result of Rauzy induction is represented in Figure 8.11 bottom. The associated automorphism is $\alpha_{a,c} : a \to ac, b \to b, c \to c$ (indeed, in the coding of T, each a is followed by c and we obtain the coding of T' by replacing ac by a. The automorphism works the inverse way, replacing a by ac).

An important point is the relation of Rauzy induction with continued fractions. Consider indeed the particular case where T a rotation of angle α with $0 < \alpha < 1$. Then $X(T)$ is the Sturmian shift of slope α and, by Theorem 1.5.4, the characteristic sequence $c_\alpha = \gamma_\alpha(\alpha)$ is of the form $c_\alpha = \text{Pal}(x)$ for some

$$x = 0^{d_1} 1^{d_2} 0^{d_3} \cdots$$

with $d_1 \geq 0$ and $d_n \geq 1$ for $n \geq 2$. By Proposition 1.5.12, the continued fraction expansion of α is

$$\alpha = [0; 1 + d_1, d_2, \ldots].$$

Figure 8.13 Case 1 of Rauzy induction.

Case 0 of Rauzy induction corresponds to $\alpha > 1 - \alpha$ or, equivalently, $\alpha > 1/2$. This is also equivalent to $d_1 = 0$ since it means that c_α begins with 1. In this case, the Rauzy induction gives after normalization the rotation of angle $\beta = (2\alpha - 1)/\alpha$ (see Figure 8.12). Accordingly, we have

$$\beta = \begin{cases} [0; 1, d_2 - 1, d_3, \ldots] & \text{if } d_2 > 1 \\ [0; d_3 + 1, \ldots] & \text{otherwise,} \end{cases}$$

which is the slope of the characteristic word $c_\beta = \mathrm{Pal}(1^{d_2}0^{d_3} \cdots)$.

In case 1, the induction gives, after normalization, a rotation of angle $\beta = \alpha/(1 - \alpha)$ (see Figure 8.13). We have in this case $\beta = [0; d_1 - 1, d_2, \ldots]$.

Thus, there is a simple correspondence between the continued fraction expansion of α and the sequence of angles of rotation obtained by iterating the Rauzy induction.

In Case 1, the elementary automorphism $\tilde{\alpha}_{1,0}$ associated with the induction is the Rauzy automorphism $L_0 \colon 0 \to 0, 1 \mapsto 01$. In case 0, it is the automorphism $\alpha_{0,1} \colon 0 \mapsto 01, 1 \mapsto 1$, which is conjugate to the Rauzy automorphism $L_1 \colon 0 \mapsto 10, 1 \mapsto 1$ since $\alpha_{0,1}(i) = 1^{-1}L_1(i)1$ for $i = 0, 1$.

8.1.7 Admissible Semi-intervals

Let T be a minimal interval exchange transformation on $I = [\ell, r)$. Let $J \subset [\ell, r)$ be a semi-interval. Since T is minimal, for each $z \in [\ell, r)$, there is an integer $n > 0$ such that $T^n(z) \in J$.

As seen before, the transformation induced by T on J is the transformation $S \colon J \to J$ defined for $z \in J$ by $S(z) = T^n(z)$ with $n = \min\{i > 0 \mid T^i(z) \in J\}$. We also say that S is the *first return map* (of T) on J. The semi-interval J is called the *domain* of S, denoted $D(S)$.

Example 8.1.19 Let T be the transformation of Example 8.1.7. Let $J = [0, 2\alpha)$. The transformation induced by T on J is

$$S(z) = \begin{cases} T^2(z) & \text{if } 0 \le z < 1 - 2\alpha \\ T(z) & \text{otherwise.} \end{cases}$$

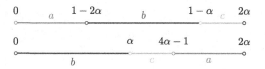

Figure 8.14 The transformation induced on I.

Let T be an interval exchange transformation relative to $(\Delta_a)_{a \in A}$. Denote by d_a for $a \in A$ the separation points. In order to fomulate our next result, we introduce the following definition. For $\ell < t < r$, the semi-interval $[\ell, t)$ is *right admissible* for T if there is an $n \in \mathbb{Z}$ such that $t = T^n(d_a)$ for some $a \in A$ and

(i) if $n > 0$, then $t < T^h(d_a)$ for all h such that $0 < h < n$,
(ii) if $n \leq 0$, then $t < T^h(d_a)$ for all h such that $n < h \leq 0$.

We also say that t itself is right admissible. Note that all semi-intervals $[\ell, d_a)$ with $\ell < d_a$ are right admissible. Similarly, all semi-intervals $[\ell, T d_a)$ with $\ell < T d_a$ are right admissible.

Example 8.1.20 Let T be the interval exchange transformation of Example 8.1.7. The semi-interval $[0, t)$ for $t = 1 - 2\alpha$ or $t = 1 - \alpha$ is right admissible since $1 - 2\alpha = d_b$ and $1 - \alpha = d_c$. On the contrary, for $t = 2 - 3\alpha$, it is not right admissible because $t = T^{-1}(d_c)$ but $d_c < t$, contradicting (ii).

The following result characterizes the semi-intervals obtained by iterating Rauzy induction.

Theorem 8.1.21 (Rauzy) *Let T be a regular k-interval exchange transformation and let J be a right admissible semi-interval for T. The transformation induced by T on J is a regular k-interval exchange transformation, which can be obtained by a sequence of Rauzy inductions.*

We shall obtain Theorem 8.1.21 as a corollary of a more general result describing the two-sided version of Rauzy induction (Theorem 8.1.25).

Note that, for a k-interval exchange transformation T on $[\ell, r)$, the transformation induced by T on any semi-interval included in $[\ell, r)$ is always an interval exchange transformation on at most $k + 2$ intervals by Theorem 8.1.3.

Example 8.1.22 Consider again the transformation of Example 8.1.7. The transformation induced by T on the semi-interval $I = [0, 2\alpha)$ is the 3-interval exchange transformation represented in Figure 8.14.

The notion of left admissible interval is symmetrical to that of right admissible. For $\ell < t < r$, the semi-interval $[t, r)$ is *left admissible* for T if there is a $k \in \mathbb{Z}$ such that $t = T^k(d_a)$ for some $a \in A$ and

(i) if $k > 0$, then $T^h(d_a) < t$ for all h such that $0 < h < k$,
(ii) if $k \leq 0$, then $T^h(d_a) < t$ for all h such that $k < h \leq 0$.

We also say that t itself is left admissible. Note that, as for right induction, the semi-intervals $[d_a, r)$ and $[T d_a, r)$ are left admissible. The symmetrical statements of Theorem 8.1.21 also hold for left admissible intervals.

Let us now generalize the notion of admissibility to a two-sided version. For a semi-interval $J = [u, v) \subset [\ell, r)$, we define the following functions on $[\ell, r)$:

$$\rho_{J,T}^+(z) = \min\{n > 0 \mid T^n(z) \in (u, v)\},$$
$$\rho_{J,T}^-(z) = \min\{n \geq 0 \mid T^{-n}(z) \in (u, v)\}.$$

We then define for every $z \in [\ell, r)$ three sets. First, let $E_{J,T}(z)$ be the following set of indices:

$$E_{J,T}(z) = \{k \mid -\rho_{J,T}^-(z) \leq k < \rho_{J,T}^+(z)\}.$$

Next, the set of *neighbors* of z with respect to J and T is

$$N_{J,T}(z) = \{T^k(z) \mid k \in E_{J,T}(z)\}.$$

Finally, the set of *division points* of I with respect to T is the finite set

$$\mathrm{Div}(J, T) = \bigcup_{i=1}^{k} N_{J,T}(d_i).$$

We now formulate the following definition. For $\ell \leq u < v \leq r$, we say that the semi-interval $J = [u, v)$ is *admissible* for T if $u, v \in \mathrm{Div}(J, T) \cup \{r\}$.

Note that a semi-interval $[\ell, v)$ is right admissible if and only if it is admissible and that a semi-interval $[u, r)$ is left admissible if and only if it is admissible. Note also that $[\ell, r)$ is admissible.

Note also that for a regular interval exchange transformation relative to a partition $(\Delta_a)_{a \in A}$, each of the semi-intervals Δ_a (or $T\Delta_a$) is admissible although only the first one is right admissible (and the last one is left admissible). Actually, we will prove that for every word w, the semi-intervals $I(w)$ and $J(w)$ are admissible. In order to do that, we need the following lemma.

Lemma 8.1.23 *Let T be a regular k-interval exchange transformation on the semi-interval $[\ell, r)$. For every $n \geq 1$, the set $P_n = \{T^h(d_i) \mid 1 \leq i \leq k, 1 \leq$*

$h \leq n\}$ is the set of $(k-1)n+1$ *left boundaries of the semi-intervals* $J(w)$ *for all words* $w \in \mathcal{L}_n(T)$.

Proof Let Q_n be the set of left boundaries of the intervals $J(w)$ for $w \in \mathcal{L}_n(T)$. Since $X(T)$ is dendric, we have $\mathrm{Card}(\mathcal{L}_n(T)) = (k-1)n+1$ by proposition 7.1.6 and thus $\mathrm{Card}(Q_n) = (k-1)n+1$. Since T is regular, the set $R_n = \{T^h(d_i) \mid 2 \leq i \leq k, \ 1 \leq h \leq n\}$ is made of $(k-1)n$ distinct points. Moreover, since

$$d_1 = T(d_{\pi(1)}), \ T(d_1) = T^2(d_{\pi(1)}), \dots, T^{n-1}(d_1) = T^n(d_{\pi(1)}),$$

we have $P_n = R_n \cup \{T^n(d_1)\}$. This implies $\mathrm{Card}(P_n) \leq (k-1)n+1$. On the other hand, if $w = b_0 \cdots b_{n-1}$, then $J(w) = \cap_{i=0}^{n-1} T^{n-i} I(b_i)$. Thus, the left boundary of each $J(w)$ is the left boundary of some $T^h I(a)$ for some h with $1 \leq h \leq n$ and some $a \in A$. Consequently $Q_n \subset P_n$. This proves that $\mathrm{Card}(P_n) = (k-1)n+1$ and that consequently $P_n = Q_n$. ∎

A dual statement holds for the semi-intervals $I(w)$.

Proposition 8.1.24 *Let T be a regular k-interval exchange transformation on the semi-interval $[\ell, r)$. For every $w \in \mathcal{L}(T)$, the semi-interval $J(w)$ is admissible.*

Proof Set $|w| = n$ and $J(w) = [u, v)$. By Lemma 8.1.23, we have $u = T^g(d_i)$ for $1 \leq i \leq k$ and $1 \leq g \leq n$. Similarly, we have $v = r$ or $v = T^d(d_j)$ for $1 \leq j \leq k$ and $1 \leq d \leq n$.

For $1 < h < g$, the point $T^h(d_i)$ is the left boundary of some semi-interval $J(y)$ with $|y| = n$ and thus $T^h(d_i) \notin J(w)$. This shows that $g \in E_{J(w),T}(d_i)$ and thus that $u \in \mathrm{Div}(J(w), T)$.

If $v = r$, then $v \in \mathrm{Div}(J(w), T)$. Otherwise, one shows in the same way as above that $v \in \mathrm{Div}(J(w), T)$. Thus, $J(w)$ is admissible. ∎

Note that the same statement holds for the semi-intervals $I(w)$ instead of the semi-intervals $J(w)$ (using the dual statement of Lemma 8.1.23).

It can be useful to reformulate the definition of a division point and of an admissible semi-interval using the terminology of graphs. Consider the (infinite) graph with vertex set $[\ell, r)$ and edges the pairs $(z, T(z))$ for $z \in [\ell, r[$. Then, if T is minimal and J is a semi-interval, for every $z \in [\ell, r)$, there is a path $P_{J,T}(z)$ such that its origin x and its end y are in J, z is on the path, $z \neq y$ and no vertices of the path except x, y are in J (actually $x = T^{-n}(z)$ with $n = \rho^-_{J,T}(z)$ and $y = T^m(z)$ with $m = \rho^+_{J,T}(z)$). Then the division points of J are the vertices that are on a path $P_{J,T}(d_i)$ but not at its end (see Figure 8.15).

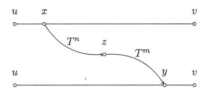

Figure 8.15 The neighbors of z with respect to $J = [u, v[$.

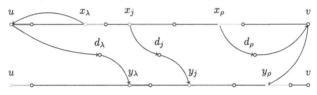

Figure 8.16 The transformation induced on $[u, v)$.

The following is a generalization of Theorem 8.1.21. Recall that Sep(T) denotes the set of separation points of T, that is, the points $d_1 = 0, d_2, \ldots, d_k$ (which are the left boundaries of the semi-intervals $\Delta_1, \ldots, \Delta_k$).

Theorem 8.1.25 *Let T be a regular k-interval exchange transformation on $[\ell, r)$. For every admissible semi-interval $I = [u, v)$, the transformation S induced by T on I is a regular k-interval exchange transformation with separation points* Sep(S) = Div(I, T) $\cap I$.

Proof Since T is regular, it is minimal. Thus, for each $i \in \{2, \ldots, k\}$ there are points $x_i, y_i \in (u, v)$ such that there is a path from x_i to y_i passing by d_i but not containing any point of I except at its origin and its end. Since T is regular, the x_i are all distinct and the y_i are all distinct.

Since I is admissible, there exist $\lambda, \rho \in \{1, \ldots, k\}$ such that $u \in N_{I,T}(d_\lambda)$ and $v \in N_{I,T}(d_\rho)$. Moreover, since u is a neighbor of d_λ with respect to I, u is on the path from x_λ to y_λ (it can be either before or after d_λ). Similarly, v is on the path from x_ρ to y_ρ (see Figure 8.16 where u is before d_λ and v is after d_ρ).

Set $x_1 = y_1 = u$. Let $(\Delta_j)_{1 \le j \le k}$ be the partition of I in semi-intervals such that x_j is the left boundary of Δ_j for $1 \le j \le k$. Let (J_j) be the partition of I such that y_j is the left boundary of J_j for $1 \le j \le k$. We will prove that

$$
S(\Delta_j) = \begin{cases} J_j & \text{if } j \neq 1, \lambda \\ J_1 & \text{if } j = \lambda \\ J_\lambda & \text{if } j = 1 \end{cases}
$$

and that the restriction of S to Δ_j is a translation.

Assume first that $j \neq 1, \lambda$. Then $S(x_j) = y_j$. Let n be such that $y_j = T^n(x_j)$ and denote $I'_j = \Delta_j \setminus x_j$. We will prove by induction on h that for $0 \leq h \leq n - 1$, the set $T^h(I'_j)$ does not contain u, v or any x_i. It is true for $h = 0$. Assume that it holds up to $h < n - 1$.

For any h' with $0 \leq h' \leq h$, the set $T^{h'}(\Delta'_j)$ does not contain any d_i. Indeed, otherwise there would exist h'' with $0 \leq h'' \leq h'$ such that $x_i \in T^{h''}(I'_j)$, a contradiction. Thus, T is a translation on $T^{h'}(\Delta_j)$. This implies that T^h is a translation on Δ_j. Note also that $T^h(I'_j) \cap I = \emptyset$. Assume the contrary. We first observe that we cannot have $T^h(x_j) \in I$. Indeed, $h < n$ implies that $T^h(x_j) \notin (u, v)$. And we cannot have $T^h(x_j) = u$ since $j \neq \lambda$. Thus, $T^h(I'_j) \cap I \neq \emptyset$ implies that $u \in T^h(I'_j)$, a contradiction.

Suppose that $u = T^{h+1}(z)$ for some $z \in I'_j$. Since u is on the path from x_λ to y_λ, it implies that for some h' with $0 \leq h' \leq h$ we have $x_\lambda = T^{h'}(z)$, a contradiction with the induction hypothesis. A similar proof (using the fact that v is on the path from x_ρ to y_ρ) shows that $T^{h+1}(I'_j)$ does not contain v. Finally suppose that some x_i is in $T^{h+1}(I'_j)$. Since the restriction of T^h to Δ_j is a translation, $T^h(\Delta_j)$ is a semi-interval. Since $T^{h+1}(x_j)$ is not in I the fact that $T^{h+1}(\Delta_j) \cap I$ is not empty implies that $u \in T^h(\Delta_j)$, a contradiction.

This shows that T^n is continuous at each point of I'_j and that $S = T^n(x)$ for all $x \in \Delta_j$. This implies that the restriction of S to Δ_j is a translation into J_j.

If $j = 1$, then $S(x_1) = S(u) = y_\lambda$. The same argument as above proves that the restriction of S to Δ_1 is a translation form Δ_1 into J_λ. Finally if $j = \lambda$, then $S(x_\lambda) = x_1 = u$ and, similarly, we obtain that the restriction of S to Δ_λ is a translation into Δ_1.

Since S is the transformation induced by the transformation T that is one-to-one, it is also one-to-one. This implies that the restriction of S to each of the semi-intervals Δ_j is a bijection onto the corresponding interval J_j, J_1 or J_λ according to the value of j.

This shows that S is an s-interval exchange transformation. Since the orbits of the points $x_2, \cdots, x_k$ relative to S are included in the orbits of $d_2, \ldots, d_k$, they are infinite and disjoint. Thus, S is regular.

Let us finally show that $\mathrm{Sep}(S) = \mathrm{Div}(I, T) \cap I$. We have $\mathrm{Sep}(S) = \{x_1, x_2, \ldots, x_k\}$ and $x_i \in N_{I,T}(d_i)$. Thus, $\mathrm{Sep}(S) \subset \mathrm{Div}(I, T) \cap I$. Conversely, let $x \in \mathrm{Div}(I, T) \cap I$. Then $x \in N_{I,T}(d_i) \cap I$ for some $1 \leq i \leq k$. If $i \neq 1, \lambda$, then $x = x_i$. If $i = 1$, then either $x = u$ (if $u = \ell$) or $x = x_{\pi(1)}$ since $d_1 = T(d_{\pi(1)})$. Finally, if $i = \lambda$, then $x = u$ or $x = x_\lambda$. Thus, $x \in \mathrm{Sep}(S)$ in all cases. ∎

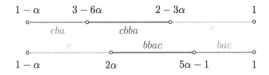

Figure 8.17 The transformation induced on Δ_c.

8.1.8 A Closure Property

We will prove that the family of regular interval exchange shifts is closed under derivation. The same property holds for minimal dendric shifts by Theorem 7.1.24. Thus, regular interval exchange shifts form a subfamily closed under derivation of minimal dendric shifts.

Theorem 8.1.26 *Every derivative shift of a regular k-interval exchange shift X with respect to some $w \in \mathcal{L}(X)$ is a regular k-interval exchange shift.*

Proof Let T be a regular s-interval exchange transformation and let X be its natural coding.

Let $w \in \mathcal{L}(X)$. Since the semi-interval J_w is admissible according to Proposition 8.1.24, the transformation T' induced by T on J_w is, by Theorem 8.1.25, a k-interval exchange transformation. By Proposition 8.1.11, the corresponding partition of $J(w)$ is the family $(J(wx))_{x \in \mathcal{R}_X(w)}$.

Using Proposition 8.1.11, it is clear that $\Sigma_T(z) = \varphi(\Sigma_{T'}(z))$, where z is a point of $J(w)$ and $\varphi \colon B^* \to \mathcal{R}_X(w)^*$ is a coding morphism for $\mathcal{R}_X(w)$.

Set $x = \Sigma_T^+(T^{-|w|}(z))$ and $y = \Sigma_T^+(z)$. Then $x = wy$ and thus $\Sigma_{T'}^+(z) = \mathcal{D}_w(x)$. This shows that the derivative shift of X with respect to w is the natural coding of T'. ∎

Example 8.1.27 Consider the squared Fibonacci shift X of Example 8.1.7. The set of return words to c is $\mathcal{R}_X(c) = \{bac, bbac, c\}$. The natural coding of transformation induced on the interval Δ_c is the interval exchange represented in Figure 8.17. The intervals of the upper level are labeled by $\mathcal{R}'_X(c)$ and those of the lower level by $\mathcal{R}_X(c)$ (we shall describe this transformation in more detail in Examples 8.2.7 and 8.2.16).

8.1.9 BV-Representation of Interval Exchange

Given a regular interval exchange, the successive inductions on the first semi-interval allow us to obtain a BV-representation of $X(T)$ described as follows.

Theorem 8.1.28 *The natural coding $X = X(T)$ of a regular k-interval exchange T has a BV-representation (X_E, T_E) where $(V, E, \leq)$ is a telescoping of a diagram with simple hat such that $\mathrm{Card}(V(n)) = k$ for all $n \geq 1$, and the adjacency matrix $M(n)$ for $n \geq 2$ has the form*

$$
M(n) = \begin{bmatrix}
1 & 0 & \cdots & 0 & 0 & s_1 \\
0 & 1 & \cdots & 0 & 0 & s_2 \\
\vdots & \vdots & \ddots & & \vdots & \vdots \\
0 & 0 & & 1 & 0 & s_{k-2} \\
0 & 0 & & 0 & 1 & s_{k-1} \\
0 & 0 & \cdots & 0 & 1 & s_k
\end{bmatrix},
$$

where $s_i \in \{0, m, m+1\}$, $s_{k-1} = m$ and $s_k = m + 1$ for some $m \geq 0$.

Proof Consider the sequence $(I_n, T_n)_{n \geq 0}$ of regular k-interval exchanges obtained as follows. Start with $I_0 = I$ and $T_0 = T$. Suppose that (I_n, T_n) is already defined. Let I_{n+1} be the first of the k intervals that partition I_n and let T_{n+1} be the transformation induced by T_n on I_{n+1}. Then, as one may verify easily by induction on n, $I_n = I(w_n)$, where $w_1, w_2, \ldots$ is the list of words of $\mathcal{L}(T)$ in lexicographic order. Since $I(w_n)$ is admissible (by the dual of Proposition 8.1.24), the induction T_n on I_{n+1} is obtained by a sequence of Rauzy inductions. Let θ_n be the morphism associated with the induction of $X(T)$ on $X(T_n)$. Then, by the dual of Proposition 8.1.11, θ_n is a coding morphism for $\mathcal{R}'_X(w_n)$. Thus, by the dual of Lemma 6.4.11, the sequence (θ_n) is a primitive left proper $\mathcal{S}$-adic representation of X. Changing each θ_n to some conjugate as in Lemma 6.4.9, we obtain a Bratteli diagram associated with a sequence of morphisms (θ'_n) with the same matrices as the (θ_n), which is a BV-representation of X.

Consider now the sequence (J_n, S_n) of regular k-interval exchanges obtained as follows. Start with $J_0 = I$ and $S_0 = T$. If (J_n, S_n) is already defined, let J_{n+1} be the union of the first $k - 1$ of the k semi-intervals partitioning J_n. The induction of S_n on J_{n+1} is obtained by a sequence of Rauzy inductions described in more detail below. This implies that the sequence (I_n, T_n) is a subsequence of the sequence (J_n, S_n). Thus, the matrices $M(\theta_n)$ are products of the matrices $M(\tau_n)$ where τ_n is the morphism associated with the induction of $X(S_n)$ on $X(S_{n+1})$.

There remains to show that the matrices $M(n) = M(\tau_n)$ have the form indicated.

Consider first the case where $\lambda_k < \lambda_{\pi(k)}$ (Case 0). Then (I_1, T_1) is obtained by one step of Rauzy induction and τ_0 is the morphism placing ℓ before k,

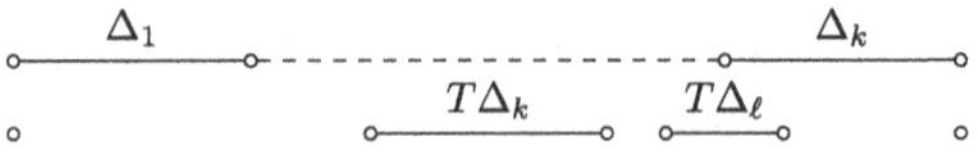

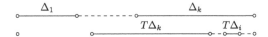

Figure 8.18 The second case.

Figure 8.19 The third case.

where ℓ is defined by $T\Delta_\ell = \Delta_{\pi(k)}$. Then $M(2)$ has the form (placing the indices ℓ, k at the end)

$$
M(2) = \begin{bmatrix} 1 & & & 0 \\ 0 & \ddots & 0 \\ 0 & & 1 & 0 \\ 0 & & 1 & 1 \end{bmatrix},
$$

which is of the form shown in Theorem 8.1.28 with $m = 0$ and $s_i = 0$ for $i \neq k$.

Consider now the case where $\lambda_{\pi(k)} < \lambda_k$ (Case 1) and assume first that Δ_k does not overlap with $T\Delta_k$. Let ℓ be such that $T\Delta_\ell$ contains the left boundary of Δ_k (see Figure 8.18). A sequence of Rauzy inductions produces (I_1, T_1), and the matrix $M(1)$ (placing the indices ℓ, k at the end) is

$$
M(2) = \begin{bmatrix} 1 & & & & & 0 \\ & \ddots & & & & \\ 0 & & 1 & & 0 & 1 \\ 0 & & & \ddots & 0 & 1 \\ 0 & & & & 1 & 0 \\ 0 & & 0 & & 1 & 1 \end{bmatrix},
$$

which is of the form shows in Theorem 8.1.28 with $m = 0$.

Assume finally that $\lambda_{\pi(k)} < \lambda_k$ (Case 1) but that Δ_k overlaps $T\Delta_k$ (see Figure 8.19). Let C be the set of indices i such that $T\Delta_i$ is contained in Δ_k and set $\mu = \sum_{i \in C} \lambda_i$. A first round of $\mathrm{Card}(C)$ Rauzy inductions cuts the right part of length μ of Δ_k. Continuing in this way, a sequence of $m = \lfloor \lambda_k / \mu \rfloor$ rounds of Rauzy inductions produces an interval exchange (I', T') such that a subset C'' of indices i in C are still such that $T'\Delta_i'$ is contained in Δ_k' (see Example 2.5.2 for an illustration). A last round of $\mathrm{Card}(C'')$ Rauzy inductions

Figure 8.20 The last case of the proof of Theorem 8.1.28.

Figure 8.21 The result of a first round of four iterations of Rauzy induction.

produces the interval exchange (I_1, T_1) and the matrix $M(2)$ (placing at the end before k the indices $i \in C$ and with $C' = C \setminus C''$)

$$
M(2) = \begin{array}{c} \\ \\ C'' \\ \\ C' \\ k \end{array}
\begin{bmatrix}
1 & 0 & & & & & & 0 \\
0 & \ddots & 0 & & & & & 0 \\
0 & & 1 & & & & & 0 \\
& & & 1 & 0 & & & m+1 \\
& & & 0 & \ddots & 0 & & m+1 \\
& & & 0 & & 1 & & m+1 \\
& & & & & 1 & 0 & m \\
& & & & & 0 & \ddots & 0 & m \\
& & & & & 0 & & 1 & m \\
& & 0 & & & & 1 & 0 & m+1
\end{bmatrix},
$$

which is of the form shown in Theorem 8.1.28. ∎

We develop in the next example the argument at the end of the proof.

Example 8.1.29 Assume that $k = 5$ and $\pi = (15)(24)$. Suppose that $\mu = \lambda_1 + \lambda_2 + \lambda_3 + \lambda_4$ is such that $\mu < \lambda_5 < \mu + \lambda_1 + \lambda_2$ (see Figure 8.20).

Thus, $C = \{1, 2, 3, 4\}$. A first round of Card$(C) = 4$ iterations of Rauzy induction produces the interval exchange of Figure 8.21.

Thus, $m = 1$ and $C'' = \{1, 2\}$. A round of Card$(C'') = 2$ Rauzy inductions produces the interval exchange of Figure 8.22 on the left and a final induction gives the interval exchange on the right.

The automorphism corresponding to this sequence of inductions is

$$1 \to 155, \ 2 \to 255, \ 3 \to 35, \ 4 \to 45, \ 5 \to 355$$

with incidence matrix

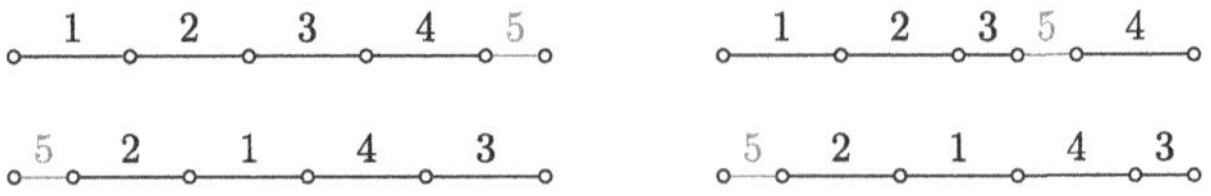

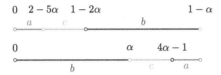

Figure 8.22 The last three iterations of Rauzy induction.

Figure 8.23 A second step of Rauzy induction on the rotation of angle 2α.

$$M = \begin{bmatrix} 1 & 0 & 0 & 0 & 2 \\ 0 & 1 & 0 & 0 & 2 \\ 0 & 0 & 1 & 0 & 1 \\ 0 & 0 & 0 & 1 & 1 \\ 0 & 0 & 1 & 0 & 2 \end{bmatrix}.$$

Example 8.1.30 Let us consider once again the squared Fibonacci shift, which is the rotation of angle 2α of Example 8.1.7. A first step of Rauzy induction gives the interval exchange of Example 8.1.18. A second step gives the interval exchange represented in Figure 8.23. The combination of the two steps corresponds to the morphism $\varphi \colon a \to ac, b \to b, c \to acc$. The incidence matrix of this morphism is of the form given in Theorem 8.1.28 with the order $b < a < c$. Indeed, with this order, the incidence matrix of φ is

$$M = \begin{bmatrix} 1 & 0 & 0 \\ 0 & 1 & 1 \\ 0 & 1 & 2 \end{bmatrix},$$

which is of form given in Theorem 8.1.28 with $m = 1$.

Actually, it is easy to compute a BV-representation of the squared Fibonacci shift directly. Indeed, we have seen in Example 8.1.7 that it is the shift generated by the morphism $\psi \colon a \mapsto baccb, b \mapsto bacc, c \mapsto bacb$. This morphism is conjugate to $\psi' \colon a \mapsto accbb, b \mapsto accb, c \mapsto acbb$, which is primitive and proper. Thus, we obtain a BV-representation of the squared Fibonacci shift using a stationary diagram with matrix

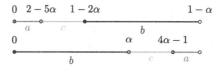

Figure 8.24 The transformation $\psi^2(T)$.

$$M = \begin{bmatrix} 1 & 2 & 2 \\ 1 & 1 & 2 \\ 1 & 2 & 1 \end{bmatrix}.$$

Its eigenvalues are $-1, 2 + \sqrt{5}, 2 - \sqrt{5}$. A left eigenvector for the maximal eigenvalue $2 + \sqrt{5}$ is $\begin{bmatrix} 2 & 1 + \sqrt{5} & 1 + \sqrt{5} \end{bmatrix}$. Thus, the dimension group is isomorphic $\mathbb{Z}^3$ with positive cone $\{(x, y, z) \in \mathbb{Z}^3 \mid x\sqrt{5} + y + z \geq 0\}$ and unit $(1, 1, 1)$.

8.2 Branching Rauzy Induction

In this section we introduce a branching version of Rauzy induction and we characterize in Theorem 8.2.2 the admissible semi-intervals for an interval exchange transformation.

8.2.1 Branching Induction

Let $T = T_{\lambda,\pi}$ be a regular k-interval exchange transformation on $[\ell, r)$. Set $Z(T) = [\ell, r')$ with $r' = 1 - \min\{\lambda_k, \lambda_{\pi(k)}\}$. Note that $Z(T)$ is the largest semi-interval that is right-admissible for T. We denote by $\psi(T)$ the transformation induced by T on $Z(T)$. The map $T \mapsto \psi(T)$ is called the *right Rauzy induction*.

Example 8.2.1 Consider again the transformation T of Example 8.1.7. Since $Z(T) = [0, 2\alpha)$, the transformation $\psi(T)$ is represented in Figure 8.14. The transformation $\psi^2(T)$ is represented in Figure 8.24.

The symmetrical notion of *left Rauzy induction* is defined similarly as follows. Let $T = T_{\lambda,\pi}$ be a regular s-interval exchange transformation on $[\ell, r)$. Set $Y(T) = [\min\{\lambda_1, \lambda_{\pi(1)}\}, r)$. We denote by $\varphi(T)$ the transformation induced by T on $Y(T)$. The map $T \mapsto \varphi(T)$ is called the *left Rauzy induction*.

The following is an addition to Theorem 8.1.21, in which we only considered right admissible semi-intervals.

Theorem 8.2.2 *Let T be a regular k-interval exchange transformation on $[\ell, r)$. A semi-interval I is admissible for T if and only if there is a sequence $\chi \in \{\varphi, \psi\}^*$ such that I is the domain of $\chi(T)$. In this case, the transformation induced by T on I is $\chi(T)$.*

We first prove the following lemmas, in which we assume that T is a regular s-interval exchange transformation on $[\ell, r)$. Recall that $Y(T), Z(T)$ are the domains of $\varphi(T), \psi(T)$ respectively.

Lemma 8.2.3 *If a semi-interval I strictly included in $[\ell, r)$ is admissible for T, then either $I \subset Y(T)$ or $I \subset Z(T)$.*

Proof Set $I = [u, v)$. Since I is strictly included in $[\ell, r)$, we have either $\ell < u$ or $v < r$. Set $Y(T) = [y, r)$ and $Z(T) = [\ell, z)$.

Assume that $v < r$. If $y \leq u$, then $I \subset Y(T)$. Otherwise, let us show that $v \leq z$. Assume the contrary. Since I is admissible, we have $v = T^j(d_i)$ with $j \in E_{I,T}(d_i)$ for some i with $1 \leq i \leq k$. But $j > 0$ is impossible since $u < T(d_i) < v$ implies $T(d_i) \in (u, v)$, in contradiction with the fact that $j < \rho_I^+(d_i)$. Similarly, $j \leq 0$ is impossible since $u < d_i < v$ implies $d_i \in \,]u, v)$. Thus, $I \subset Z(T)$.

The proof in the case $\ell < u$ is symmetric. ∎

Lemma 8.2.4 *Let T be a regular k-interval exchange transformation on $[\ell, r)$. Let J be an admissible semi-interval for T and let S be the transformation induced by T on J. A semi-interval $I \subset J$ is admissible for T if and only if it is admissible for S. Moreover $\mathrm{Div}(J, T) \subset \mathrm{Div}(I, T)$.*

Proof Set $J = [t, w)$ and $I = [u, v)$. Since J is admissible for T, the transformation S is a regular k-interval exchange transformation by Theorem 8.1.25.

Suppose first that I is admissible for T. Then $u = T^g(d_i)$ with $g \in E_{I,T}(d_i)$ for some $1 \leq i \leq k$, and $v = T^d(d_j)$ with $d \in E_{I,T}(d_j)$ for some $1 \leq j \leq k$ or $v = r$.

Since S is the transformation induced by T on J, there is a separation point x of S of the form $x = T^m(d_i)$ with $m = -\rho_{J,T}^-(d_i)$ and thus $m \in E_{J,T}(d_i)$. Thus, $u = T^{g-m}(x)$.

Assume first that $g - m > 0$. Since $u, x \in J$, there is an integer n with $0 < n \leq g - m$ such that $u = S^n(x)$.

Let us show that $n \in E_{I,S}(x)$. Assume by contradiction that $\rho_{I,S}^+(x) \leq n$. Then there is some j with $0 < j \leq n$ such that $S^j(x) \in (u, v)$. But we cannot have $j = n$ since $u \notin (u, v)$. Thus, $j < n$.

Next, there is h with $0 < h < g - m$ such that $T^h(x) = S^j(x)$. Indeed, setting $y = S^j(x)$, we have $u = T^{g-m-h}(y) = S^{n-j}(y)$ and thus $h < g-m$. If $0 < h \leq -m$, then $T^h(x) = T^{m+h}(d_i) \in I \subset J$, contradicting the hypothesis that $m \in E_{J,T}(d_i)$. If $-m < h < g - m$, then $T^h(x) = T^{m+h}(d_i) \in I$, contradicting the fact that $g \in E_{I,T}(d_i)$. This shows that $n \in E_{I,S}(x)$ and thus that $u \in \mathrm{Div}(I, S)$.

Assume next that $g - m \leq 0$. There is an integer n with $g - m \leq n \leq 0$ such that $u = S^n(x)$. Let us show that $n \in E_{I,S}(x)$. Assume by contradiction that $n < -\rho_{I,S}^-(x)$. Then there is some j with $n < j < 0$ such that $S^j(x) = T^h(x)$. Then $T^h(x) = T^{h+m}(d_i) \in I$ with $g < h + m < m$, in contradiction with the hypothesis that $m \in E_{I,T}(d_i)$.

We have proved that $u \in \mathrm{Div}(I, S)$. If $v = r$, the proof that I is admissible for S is complete. Otherwise, the proof that $v \in \mathrm{Div}(I, S)$ is similar to the proof for u.

Conversely, if I is admissible for S, there is some $x \in \mathrm{Sep}(S)$ and $g \in E_{I,S}(x)$ such that $u = S^g(x)$. But $x = T^m(d_i)$, and since $u, x \in J$ there is some n such that $u = T^n(d_i)$.

Assume for instance that $n > 0$ and suppose that there exists k with $0 < k < n$ such that $T^k(d_i) \in (u, v)$. Then, since $I \subset J$, $T^k(d_i)$ is of the form $S^h(x)$ with $0 < h < g$, which contradicts the fact that $g \in E_{I,S}(x)$. Thus, $n \in E_{I,T}(d_i)$ and $u \in \mathrm{Div}(I, T)$.

The proof is similar in the case $n \leq 0$.

If $v = r$, we have proved that I is admissible for T. Otherwise, the proof that $v \in \mathrm{Div}(I, T)$ is similar.

Finally, assume that I is admissible for T (and thus for S). For any $d_i \in \mathrm{Sep}(T)$, one has

$$\rho_{I,T}^-(d_i) \geq \rho_{J,T}^-(d_i) \quad \text{and} \quad \rho_{I,T}^+(d_i) \geq \rho_{J,T}^+(d_i),$$

showing that $\mathrm{Div}(J, T) \subset \mathrm{Div}(I, T)$. ∎

The last lemma is the key argument to prove Theorem 8.2.2.

Lemma 8.2.5 *For every admissible semi-interval $I \subset [\ell, r)$, the set $\mathcal{F}$ of sequences $\chi \in \{\varphi, \psi\}^*$ such that $I \subset D(\chi(T))$ is finite.*

Proof The set $\mathcal{F}$ is suffix-closed. Indeed it contains the empty word because $[\ell, r)$ is admissible. Moreover, for any $\xi, \chi \in \{\varphi, \psi\}^*$, one has $D(\xi \chi(T)) \subset D(\chi(T))$ and thus $\xi \chi \in \mathcal{F}$ implies $\chi \in \mathcal{F}$.

The set $\mathcal{F}$ is finite. Indeed, by Lemma 8.2.4, applied to $J = D(\chi(T))$, for any $\chi \in \mathcal{F}$, one has $\mathrm{Div}(D(\chi(T)), T) \subset \mathrm{Div}(I, T)$. In particular, the boundaries of $D(\chi(T))$ belong to $\mathrm{Div}(I, T)$. Since $\mathrm{Div}(I, T)$ is a finite set, this implies that there is a finite number of possible semi-intervals $D(\chi(T))$. Thus, there is no infinite word with all its suffixes in $\mathcal{F}$. Since the sequences χ are binary, this implies that $\mathcal{F}$ is finite.   ∎

Proof of Theorem 8.2.2 We consider χ as a word on the alphabet φ, ψ. We first prove by induction on the length of χ that the domain I of $\chi(T)$ is admissible and that the transformation induced by T on I is $\chi(T)$. It is true for $|\chi| = 0$ since $[\ell, r)$ is admissible and $\chi(T) = T$. Next, assume that $J = D(\chi(T))$ is admissible and that the transformation induced by T on J is $\chi(T)$. Then $D(\varphi\chi(T))$ is admissible for $\chi(T)$ since $D(\varphi\chi(T)) = Y(\chi(T))$. Thus, $I = D(\varphi\chi(T))$ is admissible for T by Lemma 8.2.4, and the transformation induced by T on I is $\varphi\chi(T)$. The same proof holds for $\psi\chi$.

Conversely, assume that I is admissible. By Lemma 8.2.5, the set $\mathcal{F}$ of sequences $\chi \in \{\varphi, \psi\}^*$ such that $I \subset D(\chi(T))$ is finite.

Thus, there is some $\chi \in \mathcal{F}$ such that $\varphi\chi, \psi\chi \notin \mathcal{F}$. If I is strictly included in $D(\chi(T))$, then by Lemma 8.2.3 applied to $\chi(T)$, we have $I \subset Y(\chi(T)) = D(\varphi\chi(T))$ or $I \subset Z(\chi(T)) = D(\psi\chi(T))$, a contradiction. Thus, $I = D(\chi(T))$.   ∎

We close this subsection with a result concerning the dynamics of the branching induction.

Theorem 8.2.6 *For every sequence $(T_n)_{n\geq 0}$ of regular interval exchange transformations such that $T_{n+1} = \varphi(T_n)$ or $T_{n+1} = \psi(T_n)$ for all $n \geq 0$, the length of the domain of T_n tends to 0 when $n \to \infty$.*

Proof Assume the contrary and let I be an open interval included in the domain of T_n for all $n \geq 0$. The set $\mathrm{Div}(I, T) \cap I$ is formed of s points. For any pair u, v of consecutive elements of this set, the semi-interval $[u, v)$ is admissible. By Lemma 8.2.5, there is an integer n such that the domain of T_n does not contain $[u, v)$, a contradiction.   ∎

8.2.2 Equivalent Transformations

Let $[\ell_1, r_1), [\ell_2, r_2)$ be two semi-intervals of the real line. Let $T_1 = T_{\lambda_1, \pi_1}$ be a k-interval exchange transformation relative to a partition of $[\ell_1, r_1)$ and $T_2 = T_{\lambda_2, \pi_2}$ another k-interval exchange transformation relative to $[\ell_2, r_2)$. We say that T_1 and T_2 are *equivalent* if $\pi_1 = \pi_2$ and $\lambda_1 = c\lambda_2$ for some $c > 0$. Thus, two interval exchange transformations are equivalent if we can obtain the second from the first by a rescaling following by a translation. We denote by $[T_{\lambda, \pi}]$ the equivalence class of $T_{\lambda, \pi}$.

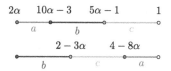

Figure 8.25 The transformation S.

Example 8.2.7 Let $S = T_{\mu,\pi}$ be the 3-interval exchange transformation on a partition of the semi-interval $[2\alpha, 1)$, with $\alpha = (3 - \sqrt{5})/2$, represented in Figure 8.25. S is equivalent to the transformation $T = T_{\lambda,\pi}$ of Example 8.1.7, with length vector $\lambda = (1 - 2\alpha, \alpha, \alpha)$ and permutation the cycle $\pi = (132)$. Indeed the length vector $\mu = (8\alpha - 3, 2 - 5\alpha, 2 - 5\alpha)$ satisfies $\mu = \frac{2-5\alpha}{\alpha}\lambda$.

Note that if T is a minimal (resp. regular) interval exchange transformation and $[S] = [T]$, then S is also minimal (resp. regular).

For an interval exchange transformation T we consider the directed labeled graph $\mathcal{G}(T)$, called the *induction graph* of T, defined as follows. The vertices are the equivalence classes of transformations obtained starting from T and applying all possible $\chi \in \{\psi, \varphi\}^*$. There is an edge labeled ψ (resp. φ) from a vertex $[S]$ to a vertex $[U]$ if and only if $U = \psi(S)$ (resp $\varphi(S)$) for two transformations $S \in [S]$ and $U \in [U]$.

Example 8.2.8 Let $\alpha = \frac{3-\sqrt{5}}{2}$ and R be a rotation of angle α. It is a 2-interval exchange transformation on $[0, 1)$ relative to the partition $[0, 1-\alpha), [1-\alpha, 1)$. The induction graph $\mathcal{G}(R)$ of the transformation is represented in the left of Figure 8.27.

Note that for a 2-interval exchange transformation T, one has $[\psi(T)] = [\varphi(T)]$, whereas in general the two transformations are not equivalent.

The induction graph of an interval exchange transformation can be infinite. A sufficient condition for the induction graph to be finite is given in Section 8.3.

Let us now introduce a variant of this equivalence relation (and of the related graph).

For a k-interval exchange transformation $T = T_{\lambda,\pi}$, with length vector $\lambda = (\lambda_1, \lambda_2, \ldots, \lambda_k)$, we define the *mirror transformation* $\widetilde{T} = T_{\widetilde{\lambda}, \tau \circ \pi}$ of T, where $\widetilde{\lambda} = (\lambda_k, \lambda_{k-1}, \ldots, \lambda_1)$ and $\tau : i \mapsto (k - i + 1)$ is the permutation that reverses the names of the semi-intervals.

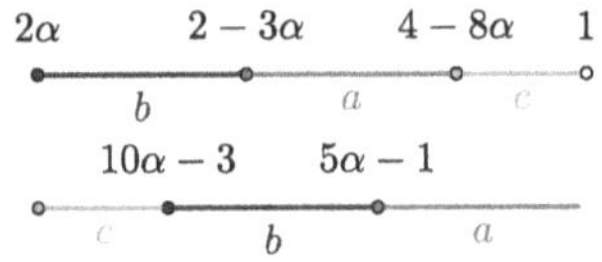

Figure 8.26 The transformation U.

Given two interval exchange transformations T_1 and T_2 on the same alphabet relative to two partitions of two semi-intervals $[\ell_1, r_1)$ and $[\ell_2, r_2)$ respectively, we say that T_1 and T_2 are *similar* either if $[T_1] = [T_2]$ or $[T_1] = [\widetilde{T_2}]$. Clearly, similarity is also an equivalent relation. We denote by $\langle T \rangle$ the class of transformations similar to T.

Example 8.2.9 Let T be the interval exchange transformation of Example 8.1.7. The transformation $U = \varphi^6(T)$ is represented in Figure 8.26 (see also Example 8.2.17). It is easy to verify that U is similar to the transformation S of Example 8.2.7. Indeed, we can obtain the second transformation (up to the separation points and the end points) by taking the mirror image of the domain.

Note that the order of the labels, that is, the order of the letters of the alphabet, may be different from the order of the original transformation.

As the equivalence relation, similarity also preserves minimality and regularity.

Let T be an interval exchange transformation. We denote by

$$\mathcal{O}(T) = \bigcup_{n \in \mathbb{Z}} T^n \left(\mathrm{Sep}(T) \right)$$

the union of the orbits of the separation points. Let S be an interval exchange transformation similar to T. Thus, there exists a bijection $f : D(T) \setminus \mathcal{O}(T) \to D(S) \setminus \mathcal{S}(S)$. This bijection is given by an affine transformation, namely a rescaling followed by a translation if T and S are equivalent and a rescaling followed by a translation and a reflection otherwise. By the previous remark, if T is a minimal interval exchange transformation and S is similar to T, then the two interval exchange sets $\mathcal{L}(T)$ and $\mathcal{L}(S)$ are equal up to permutation, that is, there exists a permutation π such that for every $w = a_0 a_1 \cdots a_{n-1} \in \mathcal{L}(T)$ there exists a unique word $v = b_0 b_1 \cdots b_{n-1} \in \mathcal{L}(S)$ such that $b_i = \pi(a_i)$ for all $i = 1, 2, \ldots, n - 1$.

In a similar way as before, we can use the similarity in order to construct a graph. For an interval exchange transformation T we define $\widetilde{\mathcal{G}}(T)$ the *modified induction graph* of T as the directed (unlabeled) graph with vertices the

Figure 8.27 Induction graph and modified induction graph of the rotation R of angle $\alpha = (3 - \sqrt{5})/2$.

similarity classes of transformations obtained starting from T and applying all possible $\chi \in \{\psi, \varphi\}^*$ and an edge from $\langle S \rangle$ to $\langle U \rangle$ if $U = \psi(S)$ or $U = \varphi(S)$ for two transformations $S \in \langle S \rangle$ and $U \in \langle U \rangle$.

Note that this notion appears naturally when considering Rauzy induction of a 2-interval exchange transformation as a continued fraction expansion (see Section 8.1.6). There exists a natural bijection between the closed interval $[0, 1]$ of the real line and the set of 2-interval exchange transformations on $[0, 1)$ given by the map $x \mapsto T_{\lambda, \pi}$

In this context, Rauzy induction corresponds to a step of the Euclidean algorithm, that is, the map $\mathcal{E} : \mathbb{R}_+^2 \to \mathbb{R}_+^2$ given by

$$\mathcal{E}(\lambda_1, \lambda_2) = \begin{cases} (\lambda_1 - \lambda_2, \lambda_2) & \text{if } \lambda_1 \geq \lambda_2 \\ (\lambda_1, \lambda_2 - \lambda_2) & \text{otherwise.} \end{cases}$$

Applying iteratively the Rauzy induction starting from T corresponds then to the continued fraction expansion of x, as already seen in Section 8.1.6.

Example 8.2.10 Let α and R be as in Example 8.2.8. The modified induction graph $\widetilde{\mathcal{G}}(R)$ of the transformation is represented on the right of Figure 8.27. Note that the ratio of the two lengths of the semi-intervals exchanged by T is

$$\frac{1 - \alpha}{\alpha} = \frac{1 + \sqrt{5}}{2} = \phi = 1 + \frac{1}{1 + \frac{1}{1 + \cdots}}.$$

8.2.3 Induction and Automorphisms

Let $T = T_{\lambda, \pi}$ be a regular interval exchange on $[\ell, r)$ relative to $(I(a))_{a \in A}$. Set $A = \{a_1, \ldots, a_k\}$. Recall now from Subsection 8.1.2 that for any $z \in [\ell, r)$, the natural coding of T relative to z is the sequence $\Sigma_T(z) = (b_n)_{n \in \mathbb{Z}}$ on the alphabet A with $b_n \in A$ defined for $n \in \mathbb{Z}$ by $b_n = a$ if $T^n(z) \in \Delta_a$.

Denote by θ_1, θ_2 the morphisms from A^* into itself defined by

$$\theta_1(a) = \begin{cases} a_{\pi(k)} a_k & \text{if } a = a_{\pi(k)} \\ a & \text{otherwise,} \end{cases} \qquad \theta_2(a) = \begin{cases} a_{\pi(k)} a_k & \text{if } a = a_k \\ a & \text{otherwise.} \end{cases}$$

The morphisms θ_1, θ_2 extend to automorphisms of the free group on A.

Proposition 8.2.11 *Let T be a regular interval exchange transformation on the alphabet A and let $S = \psi(T)$, $I = Z(T)$. There exists an automorphism θ of the free group on A such that $\Sigma_T(z) = \theta(\Sigma_S(z))$ for any $z \in I$.*

Proof Assume first that $d_k < T d_{\pi(k)}$ (Case 0). We have $Z(T) = [\ell, T d_{\pi(k)})$ and for any $x \in Z(T)$,

$$S(z) = \begin{cases} T^2(z) & \text{if } z \in K(a_{\pi(k)}) = I(a_{\pi(k)}) \\ T(z) & \text{otherwise.} \end{cases}$$

We will prove by induction on the length of w that for any $z \in I$, $\Sigma_S^+(z) \in wA^*$ if and only if $\Sigma_T^+(z) \in \theta_1(w)A^{\mathbb{N}}$. The property is true if w is the empty word. Assume next that $w = av$ with $a \in A$ and thus that $z \in I(a)$. If $a \neq a_{\pi(k)}$, then $\theta_1(a) = a$, $S(z) = T(z)$ and

$$\Sigma_S^+(z) \in avA^{\mathbb{N}} \Leftrightarrow \Sigma_S^+(S(z)) \in vA^{\mathbb{N}} \Leftrightarrow \Sigma_T^+(T(z)) \in \theta_1(v)A^{\mathbb{N}}$$
$$\Leftrightarrow \Sigma_T^+(z) \in \theta_1(w)A^{\mathbb{N}}.$$

Otherwise, $\theta_1(a) = a_{\pi(k)}a_k$, $S(z) = T^2(z)$. Moreover, by (8.2), $\Sigma_T^+(z) = a_{\pi(k)}a_k\Sigma_T^+(T^2(z))$ and thus

$$\Sigma_S^+(z) \in avA^{\mathbb{N}} \Leftrightarrow \Sigma_S^+(S(z)) \in vA^{\mathbb{N}} \Leftrightarrow \Sigma_T^+(T^2(z)) \in \theta_1(v)A^{\mathbb{N}}$$
$$\Leftrightarrow \Sigma_T^+(z) \in \theta_1(w)A^{\mathbb{N}}.$$

If $T(d_{\pi(k)}) < d_k$ (Case 1), we have $Z(T) = [\ell, d_k)$ and for any $z \in Z(T)$,

$$S(z) = \begin{cases} T^2(z) & \text{if } z \in K(a_k) = T^{-1}I(a_k) \\ T(z) & \text{otherwise.} \end{cases}$$

As in Case 0, we will prove by induction on the length of w that for any $z \in I$, $\Sigma_S^+(z) \in wA^{\mathbb{N}}$ if and only if $\Sigma_T^+(z) \in \theta_2(w)A^{\mathbb{N}}$.

The property is true if w is empty. Assume next that $w = av$ with $a \in A$. If $a \neq a_k$, then $\theta_2(a) = a$, $S(z) = T(z)$ and $z \in K(a) = I(a)$. Thus,

$$\Sigma_S^+(z) \in avA^{\mathbb{N}} \Leftrightarrow \Sigma_S^+(S(z)) \in vA^{\mathbb{N}} \Leftrightarrow \Sigma_T^+(T(z)) \in \theta_2(v)A^{\mathbb{N}}$$
$$\Leftrightarrow \Sigma_T^+(z) \in \theta_2(w)A^{\mathbb{N}}.$$

Next, if $a = a_k$, then $\theta_2(a) = a_{\pi(k)}a_k$, $S(z) = T^2(z)$ and $z \in K_{a_k} = T^{-1}(\Delta_{a_k}) \subset I(a_{\pi(k)})$. Thus,

$$\Sigma_S^+(z) \in avA^{\mathbb{N}} \Leftrightarrow \Sigma_S^+(S(z)) \in vA^{\mathbb{N}} \Leftrightarrow \Sigma_T^+(T^2(z)) \in \theta_2(v)A^{\mathbb{N}}$$
$$\Leftrightarrow \Sigma_T^+(z) \in \theta_2(w)A^{\mathbb{N}},$$

where the last equivalence results from the fact that $\Sigma_T^+(z) \in a_{\pi(k)} a_k A^{\mathbb{N}}$. This proves that $\Sigma_T(z) = \theta_2(\Sigma_S(z))$. ∎

Example 8.2.12 Let T be the transformation of Example 8.1.7. The automorphism θ_1 is defined by

$$\theta_1(a) = ac, \quad \theta_1(b) = b, \quad \theta_1(c) = c.$$

The right Rauzy induction gives the transformation $S = \psi(T)$ computed in Example 8.1.22. One has $\Sigma_S^+(\alpha) = bacba \cdots$ and $\Sigma_T^+(\alpha) = baccbac \cdots = \theta_1(\Sigma_S(\alpha))$.

We state the symmetric version of Proposition 8.2.11 for left Rauzy induction. The proof is analogous.

Proposition 8.2.13 *Let T be a regular interval exchange transformation on the alphabet A and let $S = \varphi(T)$, $I = Y(T)$. There exists an automorphism θ of the free group on A such that $\Sigma_T(z) = \theta(\Sigma_S(z))$ for any $z \in I$.*

Combining Propositions 8.2.11 and 8.2.13, we obtain the following statement.

Theorem 8.2.14 *Let T be a regular interval exchange transformation. For $\chi \in \{\varphi, \psi\}^*$, let $S = \chi(T)$ and let I be the domain of S. There exists an automorphism θ of the free group on A such that $\Sigma_T(z) = \theta(\Sigma_S(z))$ for all $z \in I$.*

Proof The proof follows easily by induction on the length of χ using Propositions 8.2.11 and 8.2.13. ∎

Note that if the transformations T and $S = \chi(T)$, with $\chi \in \{\psi, \varphi\}^*$, are equivalent, then there exists a point $z_0 \in D(S) \subseteq D(T)$ such that z_0 is a fixed point of the homothety that transforms $D(S)$ into $D(T)$ (if χ is different from the identity map, this point is unique). In that case one has $\Sigma_S(z_0) = \Sigma_T(z_0) = \theta(\Sigma_S(z_0))$ for an appropriate automorphism θ (i.e., $\Sigma_T(z_0)$ is a fixed point of an appropriate automorphism).

We now prove the following statement, which gives for regular interval exchanges a direct proof of the Return Theorem (Theorem 7.1.15).

Corollary 8.2.15 *Let T be a regular interval exchange transformation and let $X = X(T)$. For $w \in \mathcal{L}(T)$, the set $\mathcal{R}_X(w)$ is a basis of the free group on A.*

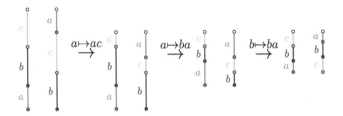

Figure 8.28   The sequence $\chi \in \{\varphi, \psi\}^*$.

Proof   By Proposition 8.1.24, the semi-interval $J(w)$ is admissible. By Theorem 8.2.2 there is a sequence $\chi \in \{\varphi, \psi\}^*$ such that $D(\chi(T)) = J(w)$. Moreover, the transformation $S = \chi(T)$ is the transformation induced by T on $J(w)$. By Theorem 8.2.14 there is an automorphism θ of the free group on A such that $\Sigma_T(z) = \theta(\Sigma_S(z))$ for any $z \in J(w)$.

By Lemma 8.1.11, we have $x \in \mathcal{R}_X(w)$ if and only if $\Sigma_T^+(z) = x \Sigma_T^+(S(z))$ for some $z \in J(w)$. This implies that $\mathcal{R}_X(w) = \theta(A)$. Indeed, for any $z \in J(w)$, let a is the first letter of $\Sigma_S^+(z)$. Then

$$\Sigma_T^+(z) = \theta(\Sigma_S^+(z)) = \theta(a\Sigma_S^+(S(z))) = \theta(a)\theta(\Sigma_S^+(Sz)) = \theta(a)\Sigma_T^+(S(z)).$$

Thus, $x \in \mathcal{R}_X(w)$ if and only if there is $a \in A$ such that $x = \theta(a)$. This proves that the set $\mathcal{R}_X(w)$ is a basis of the free group on A.   ∎

We illustrate this result with the following examples.

Example 8.2.16   We consider again the transformation T of Example 8.1.7 and $X = X(T)$. We have $\mathcal{R}_X(c) = \{bac, bbac, c\}$ (see Example 8.1.7). We represent in Figure 8.28 the sequence χ of Rauzy inductions such that $J(c)$ is the domain of $\chi(T)$.

The sequence is composed of a right induction followed by two left inductions. We have indicated on each edge the associated automorphism (indicating only the image of the letter which is modified). We have $\chi = \varphi^2\psi$, and the resulting composition θ of automorphisms gives

$$\theta(a) = bac, \quad \theta(b) = bbac, \quad \theta(c) = c.$$

Thus, $\mathcal{R}_X(c) = \theta(A)$.

Example 8.2.17   Let T and X be as in the preceding example. Let U be the transformation induced by T on J_a. We have $U = \varphi^6(T)$ and a computation shows that for any $z \in J_a$, $\Sigma_T(z) = \theta(\Sigma_U(z))$ where θ is the automorphism of the free group on $A = \{a, b, c\}$, which is the coding morphism for $\mathcal{R}_X(a)$ defined by

$$\theta(a) = ccba, \quad \theta(b) = cbba, \quad \theta(c) = ccbba.$$

One can verify that $\mathcal{L}(U) = \mathcal{L}(S)$, where S is the transformation obtained from T by permuting the labels of the intervals according to the permutation $\pi = (acb)$.

Note that $\mathcal{L}(U) = \mathcal{L}(S)$ although S and U are not identical, even up to rescaling the intervals. Actually, the rescaling of U to a transformation on $[0, 1)$ corresponds to the mirror image of S, obtained by taking the image of the intervals by a symmetry centered at $1/2$.

Note that in the above examples, all lengths of the intervals belong to the quadratic number field $\mathbb{Q}[\sqrt{5}]$.

In the next section we will see that if a regular interval exchange transformation T is defined over a quadratic field, then the family of transformations obtained from T by the Rauzy inductions contains finitely many distinct transformations up to rescaling.

8.3 Interval Exchange over a Quadratic Field

An interval exchange transformation is said to be defined over a number field $K \subset \mathbb{R}$ if the lengths of all exchanged semi-intervals belong to K. Let T be a minimal interval exchange transformation defined over a quadratic number field. Let $(T_n)_{n \geq 0}$ be a sequence of interval exchange transformation such that $T_0 = T$ and T_{n+1} is the transformation induced by T_n on one of its exchanged semi-intervals I_n. We state without proof the following result, which is a generalization of Lagrange Theorem on continued fraction expansions of quadratic numbers.

Theorem 8.3.1 (Boshernitzan, Carroll) *If T is defined over a quadratic number field, up to rescaling all semi-intervals I_n to the same length, the sequence (T_n) contains finitely many distinct transformations.*

It is possible to generalize this result and prove that, under the above hypothesis on the lengths of the semi-intervals and up to rescaling and translation, there are finitely many transformations obtained by the branching Rauzy induction defined in Section 8.2.

Theorem 8.3.2 *Let T be a regular interval exchange transformation defined over a quadratic field. The family of all induced transformation of T over an admissible semi-interval contains finitely many distinct transformations up to equivalence.*

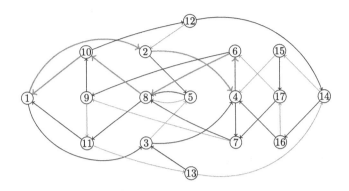

Figure 8.29 Modified induction graph of the transformation T.

An immediate corollary of Theorem 8.3.2 is the following.

Corollary 8.3.3 *Let T be a regular interval exchange transformation defined over a quadratic field. Then the induction graph $\mathcal{G}(T)$ and the modified induction graph $\widetilde{\mathcal{G}}(T)$ are finite.*

Example 8.3.4 Let T be the regular interval exchange transformation of Example 8.1.7. The modified induction graph $\widetilde{\mathcal{G}}(T)$ is represented in Figure 8.29. There are 17 similarity classes numbered from 1 to 17. The transformation T belongs to the similarity class 1 as well as transformations S of Example 8.2.7 and U of Example 8.2.9. The transformations $\psi(T)$ and $\psi^2(T)$ of Example 8.2.1 belong respectively to classes 3 and 4, while the two last transformations of Figure 8.28, namely $\varphi\psi(T)$ and $\varphi^2\psi(T)$, belong respectively to 5 and 8. Finally, the left Rauzy induction sequence from T to $U = \varphi^6(T)$ corresponds to the loop $1, 2, 4, 6, 8, 10$ in $\widetilde{\mathcal{G}}(T)$ (indicated by thick edges in Figure 8.29).

8.3.1 Primitive Substitutive Interval Exchange Shifts

In this section we prove an important property of interval exchange transformations defined over a quadratic field, namely that the corresponding interval exchange shifts are primitive substitutive shifts.

Theorem 8.3.5 *Let T be a regular interval exchange transformation defined over a quadratic field. The interval exchange shift $X(T)$ is a primitive substitutive shift.*

Example 8.3.6 Let $T = T_{\lambda,\pi}$ be the transformation of Example 8.1.7 (see also 8.1.7). The shift $X(T)$ is a primitive substitution shift, as we have seen. This can be obtained as a consequence of Theorem 8.3.5. Indeed the transformation T is regular and the length vector $\lambda = (1 - 2\alpha, \alpha, \alpha)$ belongs to $\mathbb{Q}\left[\sqrt{5}\right]^3$.

In order to prove Theorem 8.3.5, we need two preliminary results.

We close this subsection with a lemma that will be useful in Section 8.3.1.

Lemma 8.3.7 *Let T be a minimal interval exchange transformation on I. For every $N > 0$, there exists an $\varepsilon > 0$ such that for every $z \in I$ and for every $n > 0$, one has*

$$\left| T^n(z) - z \right| < \varepsilon \quad \Longrightarrow \quad n \geq N.$$

Proof Let $\alpha_1, \alpha_2, \ldots, \alpha_k$ be the translation values of T. For every $N > 0$, it is sufficient to choose

$$\varepsilon = \min \left\{ \left| \sum_{i_j=1}^{M} \alpha_{i_j} \right| i_1 \cdots i_M \in \mathcal{L}(T) \text{ and } M \leq N \right\}. \qquad \blacksquare$$

Proposition 8.3.8 *Let T, $\chi(T)$ be two equivalent regular interval exchange transformations with $\chi \in \{\varphi, \psi\}^*$. There exist a primitive morphism θ and a point $z \in D(T)$ such that the natural coding of T relative to z is a fixed point of θ.*

Proof Since T is regular, it is minimal, and thus the set $\mathcal{L}(T)$ is uniformly recurrent. Thus, there exists a positive integer N such that every letter of the alphabet appears in every word of length N of $\mathcal{L}(T)$. Moreover, by Theorem 8.2.6, applying iteratively the Rauzy induction, the length of the domains tends to zero.

Consider $T' = \chi^m(T)$, for a positive integer m, such that $D(T') < \varepsilon$, where ε is the positive real number for which, by Lemma 8.3.7, the first return time for every point of the domain is larger than N, that is, $T'(z) = T^{n(z)}(z)$, with $n(z) \geq N$, for every $z \in D(T')$.

By Theorem 8.2.14 and the remark following it, there exists an automorphism θ of the free group and a point $z \in D(T') \subseteq D(T)$ such that the natural coding of T relative to z is a fixed point of θ, that is, $\Sigma_T(z) = \theta(\Sigma_T(z))$.

By the previous argument, the image of every letter by θ is longer than N, hence it contains every letter of the alphabet as a factor. Therefore θ is a primitive morphism. ∎

Using the previous results, we can finally prove Theorem 8.3.5.

Proof of Theorem 8.3.5 By Theorem 8.3.2 there exists a regular interval transformation S such that we can find in the induction graph $\mathcal{G}(T)$ a path from $[T]$ to $[S]$ followed by a cycle on $[S]$. Thus, by Theorem 8.2.14 there exist a point $z \in D(S)$ and two automorphisms θ, η of the free group such that $\Sigma_T(z) = \theta\,(\Sigma_S(z))$, with $\Sigma_S(z)$ a fixed point of η.

By Proposition 8.3.8 we can suppose, without loss of generality, that η is primitive. Therefore, by Rauzy's Lemma (Proposition 6.2.10), $X(T)$ is a primitive substitutive shift. ∎

8.4 Linear Involutions

Let A be an alphabet of cardinality k with an involution θ and the corresponding specular group G_θ (see Section 7.3). Note that we allow θ to have fixed points. Recall that, in the group G_θ, we have $\theta(a) = a^{-1}$. Thus, when θ has no fixed points, the alphabet A can be identified with $B \cup B^{-1}$ in such a way that G_θ is the free group on B.

We consider two copies $I \times \{0\}$ and $I \times \{1\}$ of an open interval I of the real line and denote $\hat{I} = I \times \{0, 1\}$. We call the sets $I \times \{0\}$ and $I \times \{1\}$ the two *components* of $\hat{I}$. We consider each component as an open interval.

A *generalized permutation* on A of type (ℓ, m), with $\ell + m = k$, is a bijection $\pi : \{1, 2, \ldots, k\} \to A$. We represent it by a two-row array:

$$\pi = \begin{pmatrix} \pi(1)\,\pi(2)\,\ldots\pi(\ell) \\ \pi(\ell+1)\,\ldots\pi(\ell+m) \end{pmatrix}.$$

A *length data* associated with (ℓ, m, π) is a nonnegative vector $\lambda \in \mathbb{R}_+^A = \mathbb{R}_+^k$ such that

$$\lambda_{\pi(1)} + \cdots + \lambda_{\pi(\ell)} = \lambda_{\pi(\ell+1)} + \cdots + \lambda_{\pi(k)} \text{ and } \lambda_a = \lambda_{a^{-1}} \text{ for all } a \in A.$$

We consider a partition of $I \times \{0\}$ (minus $\ell - 1$ points) in ℓ open intervals $\Delta_{\pi(1)}, \ldots, \Delta_{\pi(\ell)}$ of lengths $\lambda_{\pi(1)}, \ldots, \lambda_{\pi(\ell)}$ and a partition of $I \times \{1\}$ (minus $m - 1$ points) in m open intervals $\Delta_{\pi(\ell+1)}, \ldots, \Delta_{\pi(\ell+m)}$ of lengths $\lambda_{\pi(\ell+1)}, \ldots, \lambda_{\pi(\ell+m)}$. Let Σ be the set of $k - 2$ *division points* separating the intervals Δ_a for $a \in A$.

The *linear involution* on I relative to these data is the map $T = \sigma_2 \circ \sigma_1$ defined on the set $\hat{I} \setminus \Sigma$, formed of $\hat{I}$ minus the $k - 2$ division points, and which is the composition of two involutions defined as follows.

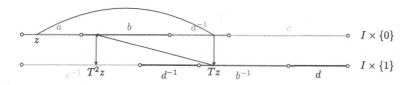

Figure 8.30 A linear involution.

(i) The first involution σ_1 is defined on $\hat{I} \setminus \Sigma$. It is such that for each $a \in A$, its restriction to Δ_a is either a translation or a symmetry from Δ_a onto $\Delta_{a^{-1}}$.

(ii) The second involution σ_2 exchanges the two components of $\hat{I}$. It is defined, for $(x, \delta) \in \hat{I}$, by $\sigma_2(x, \delta) = (x, 1 - \delta)$. The image of z by σ_2 is called the *mirror image* of z.

We also say that T is a linear involution on I and relative to the alphabet A or that it is a k-linear involution to express the fact that the alphabet A has k elements.

Example 8.4.1 Let $A = \{a, b, c, d, a^{-1}, b^{-1}, c^{-1}, d^{-1}\}$ and

$$\pi = \begin{pmatrix} a & b & a^{-1} & c \\ c^{-1} & d^{-1} & b^{-1} & d \end{pmatrix}.$$

Let T be the 8-linear involution corresponding to the length data represented in Figure 8.30 (we represent $I \times \{0\}$ above $I \times \{1\}$) with the assumption that the restriction of σ_1 to Δ_a and Δ_d is a symmetry while its restriction to Δ_b, Δ_c is a translation.

We indicate on the figure the effect of the transformation T on a point z located in the left part of the interval Δ_a. The point $\sigma_1(z)$ is located in the right part of $\Delta_{a^{-1}}$ and the point $T(z) = \sigma_2\sigma_1(z)$ is just below on the left of $\Delta_{b^{-1}}$. Next, the point $\sigma_1 T(z)$ is located on the left part of Δ_b and the point $T^2(z)$ just below.

Thus, the notion of linear involution is an extension of the notion of interval exchange transformation in the following sense. Assume that

(i) $\ell = m$,
(ii) for each letter $a \in A$, the interval Δ_a belongs to $I \times \{0\}$ if and only if $\Delta_{a^{-1}}$ belongs to $I \times \{1\}$,
(iii) the restriction of σ_1 to each subinterval is a translation.

Then, the restriction of T to $I \times \{0\}$ is an interval exchange (and so is its restriction to $I \times \{1\}$, which is the inverse of the first one). Thus, in this case, T is a pair of mutually inverse interval exchange transformations.

Note that we consider here interval exchange transformations defined by a partition of an open interval minus $\ell - 1$ points in ℓ open intervals. The usual notion of interval exchange transformation uses a partition of a semi-interval in a finite number of semi-intervals. One recovers the usual notion of interval exchange transformation on a semi-interval by attaching to each open interval its left endpoint.

A linear involution T is a bijection from $\hat{I} \setminus \Sigma$ onto $\hat{I} \setminus \sigma_2(\Sigma)$. Since σ_1, σ_2 are involutions and $T = \sigma_2 \circ \sigma_1$, the inverse of T is $T^{-1} = \sigma_1 \circ \sigma_2$.

The set Σ of division points is also the set of singular points of T, and their mirror images are the singular points of T^{-1} (which are the points where T (resp. T^{-1}) is not defined). Note that these singular points z may be "false" singularities, in the sense that T can have a continuous extension to an open neighborhood of z.

Two particular cases of linear involutions deserve attention.

A linear involution T on the alphabet A relative to a generalized permutation π of type (ℓ, m) is said to be *nonorientable* if there are indices $i, j \leq \ell$ such that $\pi(i) = \pi(j)^{-1}$ (and thus indices $i, j \geq \ell + 1$ such that $\pi(i) = \pi(j)^{-1}$). In other words, there is some $a \in A$ for which Δ_a and $\Delta_{a^{-1}}$ belong to the same component of $\hat{I}$. Otherwise T is said to be *orientable*.

A linear involution $T = \sigma_2 \circ \sigma_1$ on I relative to the alphabet A is said to be *coherent* if, for each $a \in A$, the restriction of σ_1 to Δ_a is a translation if and only if Δ_a and $\Delta_{a^{-1}}$ belong to distinct components of $\hat{I}$.

Example 8.4.2 The linear involution of Example 8.4.1 is coherent.

Linear involutions that are orientable and coherent correspond to interval exchange transformations, whereas orientable but noncoherent linear involutions are called *interval exchanges with flip*.

A *connection* of a linear involution T is a triple (x, y, n) where x is a singularity of T^{-1}, y is a singularity of T, $n \geq 0$ and $T^n x = y$.

Example 8.4.3 Let us consider the linear involution T, which is the same as in Example 8.4.1 but such that the restriction of σ_1 to Δ_c is a symmetry. Thus, T is not coherent. We assume that $I =]0, 1[$, that $\lambda_a = \lambda_d$. Let $x = (1 - \lambda_d, 0)$ and $y = (\lambda_a, 0)$.

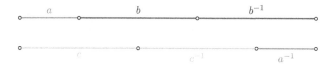

Figure 8.31 A linear involution without connections.

Then x is a singularity of T^{-1} ($\sigma_2(x)$ is the left endpoint of Δ_d), y is a singularity of T (it is the right endpoint of Δ_a) and $T(x) = y$. Thus, $(x, 1, y)$ is a connection.

Example 8.4.4 Let T be the linear involution on $I =]0, 1[$ represented in Figure 8.31. We assume that the restriction of σ_1 to Δ_a is a translation, whereas the restriction to Δ_b and Δ_c is a symmetry. We choose $(3 - \sqrt{5})/2$ for the length of the interval Δ_c (or Δ_b). With this choice, T has no connections.

Let T be a linear involution without connections. Let

$$O = \bigcup_{n \geq 0} T^{-n}(\Sigma) \quad \text{and} \quad \hat{O} = O \cup \sigma_2(O) \tag{8.10}$$

be respectively the negative orbit of the singular points and its closure under mirror image. Then T is a bijection from $\hat{I} \setminus \hat{O}$ onto itself. Indeed, assume that $T(z) \in \hat{O}$. If $T(z) \in O$, then $z \in O$. Next, if $T(z) \in \sigma_2(O)$, then $T(z) \in \sigma_2(T^{-n}(\Sigma)) = T^n(\sigma_2(\Sigma))$ for some $n \geq 0$. We cannot have $n = 0$ since $\sigma_2(\Sigma)$ is not in the image of T. Thus, $z \in T^{n-1}(\sigma_2(\Sigma)) = \sigma_2(T^{-n+1}(\Sigma)) \subset \sigma_2(O)$. Therefore, in both cases $z \in \hat{O}$. The converse implication is proved in the same way.

8.4.1 Minimal Linear Involutions

A linear involution T on I without connections is *minimal* if for any point $z \in \hat{I} \setminus \hat{O}$ the nonnegative orbit of z is dense in $\hat{I}$.

Note that when a linear involution is orientable, that is, when it is a pair of interval exchange transformations (with or without flips), the interval exchange transformations can be minimal, although the linear involution is not since each component of $\hat{I}$ is stable by the action of T.

Example 8.4.5 Let us consider the noncoherent linear involution T, which is the same as in Example 8.4.1 but such that the restriction of σ_1 to Δ_c is a symmetry, as in Example 8.4.3. We assume that $I =]0, 1[$, that $\lambda_a = \lambda_d$, that $1/4 < \lambda_c < 1/2$ and that $\lambda_a + \lambda_b < 1/2$. Let $z = 1/2 + \lambda_c$ (see Figure 8.32). We have then $T^3(z) = z$, showing that T is not minimal. Indeed, since $z \in \Delta_c$,

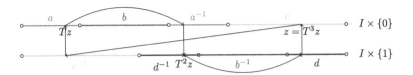

Figure 8.32 A noncoherent linear involution.

we have $T(z) = 1 - z = 1/2 - \lambda_c$. Since $T(z) \in \Delta_a$, we have $T^2(z) = (\lambda_a + \lambda_b) + (\lambda_a - 1 + z) = z - \lambda_c = 1/2$. Finally, since $T^2(z) \in \Delta_{d^{-1}}$, we obtain $1 - T^3(z) = T^2(z) - \lambda_c = 1 - z$ and thus $T^3(z) = z$.

We quote without proof the following result, analogous to Keane Theorem (Theorem 8.1.2) for interval exchange transformations.

Proposition 8.4.6 *Let T be a linear involution without connections on I. If T is nonorientable, it is minimal. Otherwise, its restriction to each component of $\hat{I}$ is minimal.*

8.4.2 Natural Coding

Let T be a linear involution on I, let $\hat{I} = I \times \{0, 1\}$ and let $\hat{O}$ be the set defined by Eq. (8.10).

Given $z \in \hat{I} \setminus \hat{O}$, the *natural coding* of T relative to z is the infinite sequence $\Sigma_T(z) = (a_n)_{n \in \mathbb{Z}}$ on the alphabet A defined by

$$a_n = a \quad \text{if} \quad T^n(z) \in \Delta_a.$$

We first observe that the factors of $\Sigma_T(z)$ are reduced. Indeed, assume that $a_n = a$ and $a_{n+1} = a^{-1}$ with $a \in A$. Set $x = T^n(z)$ and $y = T(x) = T^{n+1}(z)$. Then $x \in \Delta_a$ and $y \in \Delta_{a^{-1}}$. But $y = \sigma_2(u)$ with $u = \sigma_1(x)$. Since $x \in \Delta_a$, we have $u \in \Delta_{a^{-1}}$. This implies that $y = \sigma_2(u)$ and u belong to the same component of $\hat{I}$, a contradiction.

We denote by $X(T)$ the set of natural codings of T. We say that $X(T)$ is the *natural coding* of T. We also denote $\mathcal{L}(T) = \mathcal{L}(X(T))$. We say that $X(T)$ is a *linear involution shift*.

Example 8.4.7 Let T be the linear involution of Example 8.4.4. The words of length at most 3 of $\mathcal{L}(T)$ are represented in Figure 8.33.

The set $\mathcal{L}(T)$ can actually be defined directly as the set $\mathcal{L}(f)$ for the substitution

$$\tau : a \mapsto cb^{-1}, \quad b \mapsto c, \quad c \mapsto ab^{-1},$$

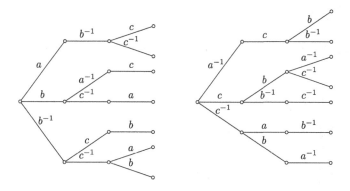

Figure 8.33 The words of length at most 3 of $\mathcal{L}(T)$.

which extends to an automorphism of the free group on $\{a, b, c\}$. Indeed, applying twice Rauzy induction on T gives a linear involution, which is the same as T (with the two copies of $[0, 1]$ interchanged). This gives the explanation of why the substitution shift of Example 7.3.30 is specular.

Define, as for an interval exchange transformation, for a word $w = a_0 a_1 \cdots a_n$ with $a_i \in A \cup A^{-1}$, $I(w) = \Delta_{a_0} \cap T^{-1}(\Delta_{a_1}) \cap \cdots \cap T^{-n}(\Delta_{a_n})$. It is clear that

$$w \in \mathcal{L}(T) \Leftrightarrow I(w) \neq \emptyset. \tag{8.11}$$

Proposition 8.4.8 *Let T be a linear involution. The set $\mathcal{L}(T)$ is a laminary set.*

Proof Set $T = \sigma_2 \circ \sigma_1$. We claim that for any nonempty word $u \in \mathcal{L}(T)$, one has $I(u^{-1}) = \sigma_1 T^{|u|-1}(I(u))$.

To prove the claim, we use an induction on the length of u. The property holds for $|u| = 1$ by definition of σ_1. Next, consider $u \in \mathcal{L}(T)$ and $a \in A \cup A^{-1}$ such that $ua \in \mathcal{L}(T)$.

Since $T^{-1} = \sigma_1 \circ \sigma_2$, we have, using the induction hypothesis,

$$\sigma_1 T^{|u|}(I(ua)) = \sigma_1 T^{|u|}(I(u) \cap T^{-|u|}(\Delta_a)) = \sigma_1 T^{|u|}(I(u)) \cap \sigma_1(\Delta_a)$$
$$= \sigma_1 \sigma_2 \sigma_1 T^{|u|-1}(I(u)) \cap \sigma_1(\Delta_a) = \sigma_1 \sigma_2(I(u^{-1})) \cap \Delta_{a^{-1}}$$
$$= I(a^{-1} u^{-1}),$$

where the last equality results from $I(a^{-1}u^{-1}) = T^{-1}I(u^{-1}) \cap \Delta_{a^{-1}}$.

We easily deduce from the claim that the set $\mathcal{L}(T)$ is closed under taking inverses. Furthermore it is a factorial subset of the group G_θ. It is thus a laminary set. ∎

We prove the following result.

Theorem 8.4.9 *The natural coding of a linear involution without connections is a specular shift.*

We first prove the following lemma.

Lemma 8.4.10 *Let T be a linear involution. For every nonempty word w and letter $a \in A$, one has*

(i) $a \in L(w) \Leftrightarrow \sigma_2(\Delta_{a^{-1}}) \cap I(w) \neq \emptyset$,
(ii) $a \in R(w) \Leftrightarrow \sigma_2(\Delta_a) \cap I(w^{-1}) \neq \emptyset$.

Proof By (8.11), we have $a \in L(w)$ if and only if $I(aw) \neq \emptyset$, which is also equivalent to $T(I(aw)) \neq \emptyset$. By definition of $I(aw)$, we have $T(I(aw)) = T(\Delta_a) \cap I(w)$. Since $T = \sigma_2 \circ \sigma_1$ and since $\sigma_1(\Delta_a) = \Delta_{a^{-1}}$, we have $a \in L(w)$ if and only if $\sigma_2(\Delta_{a^{-1}}) \cap I(w) \neq \emptyset$. Next, since $\mathcal{L}(T)$ is closed under taking inverses by Proposition 8.4.8, we have $aw \in \mathcal{L}(T)$ if and only if $w^{-1}a^{-1} \in \mathcal{L}(T)$. Thus, $a \in R(w)$ if and only if $a^{-1} \in L(w^{-1})$, whence the second equivalence. ∎

Given a linear involution T on I, we introduce two orders on $\mathcal{L}(T)$ as follows. For any $u, v \in \mathcal{L}(T)$, one has

(i) $u <_R v$ if and only if $I(u) < I(v)$,
(ii) $u <_L v$ if and only if $I(u^{-1}) < I(v^{-1})$.

Lemma 8.4.11 *Let T be a linear involution on I without connexion. Let $w \in \mathcal{L}(T)$ and $a, a' \in L(w)$ (resp. $b, b' \in R(w)$). Then $1 \otimes a$, $1 \otimes a'$ (resp. $b \otimes 1$, $b' \otimes 1$) are in the same connected component of $\mathcal{E}(w)$ if and only if $\Delta_{a^{-1}}$, $\Delta_{a'^{-1}}$ (resp. Δ_b, Δ_b) are in the same component of $\hat{I}$.*

Proof If $(1 \otimes a, b \otimes 1) \in \mathcal{E}(w)$, then $\sigma_2(\Delta_{a^{-1}}) \cap I(wb) \neq \emptyset$. Thus, $\Delta_{a^{-1}}$ and $I(wb)$ belong to distinct components of $\hat{I}$. Consequently, if $a, a' \in L(w)$ (resp. $R(w)$) belong to the same connected component of $\mathcal{E}(w)$, then $\Delta_{a^{-1}}$, $\Delta_{a'^{-1}}$ (resp. $I(wa)$, $I(wa')$) belong to the same component of $\hat{I}$. Conversely, let $a, a' \in L(w)$ be such that a, a' belong to the same component of $\hat{I}$. We may assume that $a <_L a'$. There is a reduced path, that is, it does not use twice consecutively the same edge) in $\mathcal{E}(w)$ from a to a', which is the sequence $a_1, b_1, \ldots, b_{n-1}, a_n$ with $a_1 = a$ and $a_n = a'$ with $a_1 <_L a_2 <_L \cdots <_L a_n$, $wb_1 <_R wb_2 <_R \cdots <_R wb_{n-1}$ and $\sigma_2(\Delta_{a_i^{-1}}) \cap$

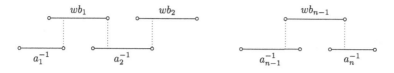

Figure 8.34 A path from a_1 to a_n in $\mathcal{E}(w)$.

$I(wb_i) \neq \emptyset$, $\sigma_2(\Delta_{a_{i+1}^{-1}}) \cap I(wb_i) \neq \emptyset$ for $1 \leq i \leq n-1$ (see Figure 8.34 for an illustration).

Note that the hypothesis that T is without connection is needed since otherwise the right boundary of $\sigma_2(\Delta_{a^{-1}})$ could be the left boundary of $I(wb_i)$. The assertion concerning $b, b' \in R(w)$ is a consequence of the first one since $b, b' \in R(w)$ if and only if $b^{-1}, b'^{-1} \in L(w^{-1})$. ∎

Proof of Theorem 8.4.9 Let T be a linear involution without connections. By Proposition 8.4.8, the set $\mathcal{L}(T)$ is symmetric. Since it is by definition extendable and formed of reduced words, it is a laminary set.

There remains to show that $X(T)$ is dendric of characteristic 2. Let us first prove that for any $w \in \mathcal{L}(T)$, the graph $\mathcal{E}(w)$ is acyclic. Assume that $(1 \otimes a_1, b_1 \otimes 1, \ldots, 1 \otimes a_n, b_n \otimes 1)$ is a path in $\mathcal{E}(w)$ with $(a_1, \ldots, a_n) \in L(w)$ and $b_1, \ldots, b_n \in R(w)$. We may assume that the path is reduced, that $n \geq 2$ and also that $a_1 <_L a_2$. It follows that $a_1 <_L \cdots <_L a_n$ and $wb_1 <_R \cdots <_R wb_n$ (see Figure 8.34). Thus, it is not possible to have an edge (a_1, b_n), which shows that $\mathcal{E}(w)$ is acyclic.

Let $a, a' \in A$. If $\Delta_{a^{-1}}$ and $\Delta_{a'^{-1}}$ are in the same component of $\hat{I}$, then $1 \otimes a$ and $1 \otimes a'$ are in the same connected component of $\mathcal{E}(\varepsilon)$. Thus, $\mathcal{E}(\varepsilon)$ is a union of two trees.

Next, if $w \in \mathcal{L}(T)$ is nonempty and $1 \otimes a, 1 \otimes a' \in L(w)$, then $\Delta_{a^{-1}}$ and $\Delta_{a'^{-1}}$ are in the same component of $\hat{I}$ (by Lemma 8.4.10), and thus $1 \otimes a, 1 \otimes a'$ are in the same connected component of $\mathcal{E}(w)$. Thus, $\mathcal{E}(w)$ is a tree. ∎

We now present an example of a linear involution on an alphabet A where the involution θ has fixed points.

Example 8.4.12 Let $A = \{a, b, c, d\}$ be as in Example 7.3.1 (in particular, $d = b^{-1}$, $a = a^{-1}$, $c = c^{-1}$). Let T be the linear involution represented in Figure 8.35 with σ_1 being a translation on Δ_b and a symmetry on Δ_a, Δ_c. Choosing $(3 - \sqrt{5})/2$ for the length of Δ_b, the involution is without connections. Thus, $\mathcal{L}(T)$ is a specular set.

Figure 8.35 A linear involution on $A = \{a, b, c, d\}$.

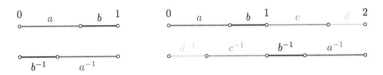

Figure 8.36 Interval exchanges U and V for the Fibonacci set and its doubling.

Note that the natural coding of the linear involution T is equal to the set of factors of the shift of Example 7.3.14. Indeed, consider the interval exchange V on the interval $Y =]0, 1[$ represented on the right-hand side of Figure 8.36, which is obtained by using two copies of the interval exchange U defining the Fibonacci set (see the left-hand side of Figure 8.36). Let $X =]0, 1[\times \{0, 1\}$ and let $\alpha \colon Y \to X$ be the map defined by

$$\alpha(z) = \begin{cases} (z, 0) & \text{if } z \in]0, 1[\\ (2 - z, 1) & \text{otherwise.} \end{cases}$$

Then $\alpha \circ V = T \circ \alpha$ and thus $\mathcal{L}(V) = \mathcal{L}(T)$.

8.5 Exercises

Section 8.1

8.1 Show that an Arnoux–Rauzy shift that is not a Sturmian shift is not an interval exchange transformation.

8.2 Let $T = T_{\lambda,\pi}$ be a k-interval exchange transformation on $I = [0, r)$ exchanging the intervals $\Delta_1, \ldots, \Delta_k$. Let μ be an invariant measure on (I, T) and set $\mu_i = \mu(\Delta_i)$ for $1 \le i \le k$. Let $\varphi \colon I \to I$ be the map defined by

$$\varphi(x) = \mu([0, x)).$$

Show that if T is minimal, then φ is a continuous isomorphism of dynamical systems from (I, T) onto $(I, T_{\mu,\pi})$, where $T_{\mu,\pi}$ is the interval exchange transformation on $[0, \mu(I))$ defined by the vector $\mu = (\mu_1, \ldots, \mu_k)$ and the permutation π.

8.3 Let $T = T_{\lambda, \pi}$ be as above. Set $T(x) = x + \alpha_i$ for $x \in \Delta_i$. Show that there is an antisymmetric matrix M_π such that the column vector α with coordinates α_i is defined by

$$\alpha = M_\pi \lambda. \tag{8.12}$$

Compute the matrix M_π and $M_{\pi'}$ for $\pi = (132)$ and $\pi' = (13)$.

8.4 Prove Theorem 8.1.5.

8.5 Let M_π be the antisymmetric matrix of Exercise 8.3. Show that if μ, ν are invariant measures on (I, T), then

$$\mu^t M_\pi \nu = 0,$$

where μ, ν denote the vectors $(\mu(\Delta_i)), (\nu(\Delta_i))$.

8.6 Let $T = T_{\lambda, \pi}$ be minimal and such that π has the property that $\pi(j) \neq \pi(j+1)$ for all j or, equivalently, that $T_{\lambda, \pi}$ has exactly $k - 1$ points of discontinuity. Let $V_{\lambda, \pi}$ be the vector space generated by the cone $\mathcal{I}(I, T)$ of invariant measures on (I, T). Show that $V_{\lambda, \pi} \cap \ker M_\pi = \{0\}$.

8.7 Show that if $T = T_{\lambda, \mu}$ is minimal, then $V_{\lambda, \mu}$ has dimension at most $1/2 \operatorname{rank}(M_\pi)$ and consequently that (I, T) admits at most $1/2 \operatorname{rank}(M_\pi)$ ergodic measures.

8.8 Define for every word $w \in A^*$ a subset $I(w)$ of I as follows. Set $I(\varepsilon) = I$ and $I(au) = \Delta_a \cap T^{-1}(I(u))$ for $u \in A^*$ and $a \in A$. Show that every nonempty $I(w)$ is a semi-interval.

8.9 Let $J(w)$ be defined by $J(\varepsilon) = I$ and by

$$J(ua) = T(J(u)) \cap T(\Delta_a) \tag{8.13}$$

for $a \in A$ and $u \in A^*$. Show that the nonempty sets $J(w)$ are semi-intervals.

8.10 Let (I, T) be an interval exchange on the semi-intervals $(\Delta_a)_{a \in A}$ and let G be a finite permutation group on a finite set Q. Let $a \mapsto g_a$ be a map from A into G. Show that the skew product $(I \times Q, S)$ of (I, T) with (Q, G) defined by

$$S(z, q) = (Tz, g_a(q)) \text{ if } z \in \Delta_a$$

is an interval exchange. Use this to show that the decoding of an interval exchange shift X by a finite X-maximal bifix code is an interval exchange shift. Hint: use Exercise 7.13.

8.11 Show that the dimension group of the squared Fibonacci shift is $\mathbb{Z}[\frac{1+\sqrt{5}}{2}] \times \mathbb{Z}$ with positive cone $\mathbb{Z}_+[\frac{1+\sqrt{5}}{2}] \times \mathbb{Z}$ and unit $(1, 0)$. Hint: use the fact that $X = X(\psi)$ with ψ given in Example 8.1.7.

8.6 Notes

8.6.1 Interval Exchange Transformations

Interval exchange transformations were introduced by Keane (1975), who proved Theorem 8.1.2. The condition defining regular interval exchange transformations is also called the *infinite disjoint orbit condition* or *idoc* .

If (I, T) is an interval exchange transformation and μ is any invariant Borel probability on I, then T is not mixing (see Cornfeld et al. (1982)). However, it was shown, solving a longstanding problem, by Avila and Forni (2007) that an interval exchange defined by an irreducible permutation, which is not a rotation, is almost surely weakly mixing.

The idea of Rauzy induction and Theorem 8.1.3 are from Rauzy (1979). For more details on interval exchange transformations we refer to Cornfeld et al. (1982), which we follow closely for the proof of Theorems 8.1.3 and 8.1.2. See also the book by Viana (2008) entirely devoted to interval exchanges.

The Cantor version of interval exchange transformations was introduced by Keane (1975).

The notion of planar dendric shifts and Proposition 8.1.13 are from Berthé et al. (2015d). The converse of Proposition 8.1.13, characterizing the languages of regular interval exchange transformations, is proved in Ferenczi and Zamboni (2008).

Theorem 8.1.28 is due to Gjerde and Johansen (2002). Their statement includes the case of a non-regular interval exchange. They also showed that there are BV-dynamical systems satisfying the hypothesis of the theorem that are not isomorphic to an interval exchange shift.

Theorem 8.1.21 is Theorem 14 in Rauzy (1979), while the one-sided version of Theorem 8.2.2 is Theorem 23. Lemma 8.2.4 is the two-sided version of Lemma 22 in Rauzy (1979).

We have noted that for any s-interval exchange transformation on $[\ell, r)$ and any semi-interval I of $[\ell, r[$, the transformation S induced by T on I is an interval exchange transformation on at most $s + 2$-intervals (Lemma 8.1.3). Actually, it follows from the proof of Lemma 2, page 128 in Cornfeld et al. (1982) that, if T is regular and S is an s-interval exchange transformation with separation points $\mathrm{Sep}(S) = \mathrm{Div}(I, T) \cap I$, then I is admissible. Thus, the converse of Theorem 8.1.25 is also true.

Branching Rauzy induction is introduced in Dolce and Perrin (2017b) where Theorem 8.2.2 appears. Actually, left Rauzy induction is already considered in Veech (1990), and Theorem 8.2.2 appears independently in Fickenscher (2017).

Proposition 8.2.11 appears in Jullian (2013).

On the relation between Rauzy induction and continued fractions, see Miernowski and Nogueira (2013) for more details.

Theorem 8.3.1 is from Boshernitzan and Carroll (1997). In the same paper, an extension to right Rauzy induction is suggested (but not completely developed). Theorem 8.3.2 is from Dolce and Perrin (2017b).

It was conjectured by Keane (1975) that a minimal interval exchange transformation is uniquely ergodic. This was disproved by Keynes and Newton (1976) with a 5-interval exchange transformation. Later Keane (1977) found examples with $k = 4$ and conjectured that almost all minimal interval exchange transformations are uniquely ergodic. The conjecture was proved independently by Veech (1982) and by Masur (1982). Exercises 8.2 to 8.7 are from Veech (1978).

Exercise 8.10 is from Berthé et al. (2015d). The skew product of an interval exchange with a finite group has been called *interval exchange transformation on a stack* in Boshernitzan and Carroll (1997) (see also Veech (1975)).

8.6.2 Linear Involutions

Linear involutions were introduced by Danthony and Nogueira (1990). It is also an extension of the notion of interval exchange with flip (Nogueira (1989); Nogueira et al. (2013)). Our definition is somewhat more general than the one used in Danthony and Nogueira (1990) and also that of Berthé et al. (2017b). Orientable linear involutions correspond to orientable laminations, whereas coherent linear involutions correspond to orientable surfaces. Thus, coherent nonorientable involutions correspond to nonorientable laminations on orientable surfaces.

Contrary to what happens with interval exchanges, it is shown in Danthony and Nogueira (1990) that noncoherent linear involutions are almost surely not minimal.

Proposition 8.4.6 (already proved in Boissy and Lanneau (2009, Proposition 4.2) for the class of coherent involutions) is Berthé et al. (2017b, Proposition 3.7). The proof uses Keane's theorem, proving that an interval exchange transformation without connections is minimal (Keane, 1975)).

The interval exchange U built in Example 8.4.12 is actually the orientation covering of the linear involution T (see Berthé et al. (2017b)).

9

Bratteli Diagrams and C^*-Algebras

In this chapter, we give a short introduction to the connection between Bratteli diagrams and C^*-algebras. These notions are closely related to our subject. Indeed, one may associate to every ordered Bratteli diagram both a C^*-algebra and a Cantor system. Minimal Cantor systems correspond to the so-called properly ordered Bratteli diagrams, and the dimension groups of the Cantor system and of the algebra are the same.

We define approximately finite dimensional algebras (AF algebras) as follows. A C^*-algebra $\mathfrak{A}$ is an AF algebra if it is the closure of an increasing sequence of finite dimensional subalgebras $(\mathfrak{A}_k)_{k \geq 0}$. Let t_k be the dimension of $\mathfrak{A}_k$. To such an algebra $\mathfrak{A}$ we associate a dimension group as a direct limit of abelian groups $\mathbb{Z}^{t_k}$.

The main object of this chapter is to prove the theorem of Elliott (Theorem 9.3.21). It asserts that two unital approximately finite algebras are $*$-isomorphic if and only if their dimension groups are isomorphic. In this way the difficult problem of isomorphism of C^*-algebras is, in the case of AF algebras, reduced to the relatively easier problem of isomorphism of dimension groups.

In the Section 9.1, we show how Bratteli diagrams can describe embeddings of sequences of finite dimensional algebras. In the next section (Section 9.2) we give a brief introduction to C^*-algebras. In Section 9.2.3 we introduce enveloping C^*-algebras and use them to define direct limits of C^*-algebras. Elliott's Theorem (Theorem 9.3.21) is presented in Section 9.3.

9.1 Bratteli Diagrams

Let us first give a modified version of Bratteli diagrams, which is well adapted to the purpose of this chapter and essentially differs by the addition of integer

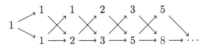

Figure 9.1 A Bratteli diagram.

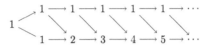

Figure 9.2 A second Bratteli diagram.

labels to the vertices. A Bratteli diagram, as defined in Chapter 5, is an infinite labeled multigraph. Its set of nodes of level k consists of pairs (k, j) with $p \geq 0$ and $1 \leq j \leq t(k)$ with $t(0) = 1$ (there is only one vertex at level 0). There are $a_{ij}^{(k)} \geq 0$ arrows from (k, j) to $(k + 1, i)$. We denote by M_k the $t(k+1) \times t(k)$-matrix with coefficients $a_{ij}^{(k)}$. Thus, M_k is the adjacency matrix $M(k + 1)$.

We label the vertex $(k + 1, i)$ of the diagram by the integer

$$n(k + 1, i) = \sum_{j=1}^{t(p)} a_{ij}^{(k)} n(k, j), \tag{9.1}$$

with $n(0, 1) = 1$. In this way, $n(k, j)$ is the number of paths from $V(0)$ to the vertex (k, j).

Conversely, any sequence $(M_k)_{k \geq 1}$ of nonnegative integer matrices defines a Bratteli diagram as above. By convention, we assume that $M_0 = \begin{bmatrix} 1 & 1 & \dots & 1 \end{bmatrix}^t$ and that $n(0, 1) = 1$. Thus, $n(1, i) = 1$ for $1 \leq i \leq t(1)$. This means that we are considering diagrams with a simple hat.

Example 9.1.1 The graph represented in Figure 9.1 with all matrices M_k for $k \geq 1$ equal to the matrix $\begin{bmatrix} 1 & 1 \\ 1 & 0 \end{bmatrix}$ is a Bratteli diagram. The levels $k = 0, 1, 2, \dots$ are growing horizontally from left to right, and on each level the vertices are numbered vertically from bottom to top. The integer labeling each node (k, j) is $n(k, j)$.

Example 9.1.2 The graph represented in Figure 9.2 is a Bratteli diagram that corresponds to all matrices M_n equal to $\begin{bmatrix} 1 & 1 \\ 0 & 1 \end{bmatrix}$.

We will now explain how a Bratteli diagram describes a sequence,

$$\mathfrak{A}_1 \xrightarrow{\pi_1} \mathfrak{A}_2 \xrightarrow{\pi_2} \mathfrak{A}_3 \xrightarrow{\pi_3} \cdots,$$

of algebras $\mathfrak{A}_k$ and morphisms $\pi_k : \mathfrak{A}_k \to \mathfrak{A}_{k+1}$.

Let M be an $s \times t$-matrix with coefficients $a_{ij} \geq 0$, let $(m_1, \ldots, m_t)$ be integers, and let $(n_1, \ldots, n_s)$ be such that

$$n_i = \sum_{i=1}^{t} a_{ij} m_j \quad (1 \leq i \leq s).$$

Let $\mathfrak{A}_1 = \mathcal{M}_{m_1} \oplus \cdots \oplus \mathcal{M}_{m_t}$ and $\mathfrak{A}_2 = \mathcal{M}_{n_1} \oplus \cdots \oplus \mathcal{M}_{n_s}$, where $\mathcal{M}_n$ denotes the algebra of $n \times n$-matrices with complex coefficients. We associate to an $s \times t$-matrix M with integer nonnegative coefficients a_{ij} the morphism $\varphi : \mathfrak{A}_1 \to \mathfrak{A}_2$ defined as $\varphi = \varphi_1 \oplus \cdots \oplus \varphi_s$ with

$$\varphi_i = \mathrm{id}_{m_1}^{(a_{i1})} \oplus \cdots \oplus \mathrm{id}_{m_t}^{(a_{it})}, \tag{9.2}$$

where id_n is the identity matrix of $\mathcal{M}_n$ and $\mathrm{id}_n^{(e)} : \mathcal{M}_n \to \mathcal{M}_{en}$ is the morphism $x \mapsto (x, \ldots, x)$ (e times). The matrix M is called the matrix of *partial multiplicities* associated with φ, and φ is the morphism determined by M.

We associate to a Bratteli diagram the sequence

$$\mathfrak{A}_1 \xrightarrow{\pi_1} \mathfrak{A}_2 \xrightarrow{\pi_2} \mathfrak{A}_3 \xrightarrow{\pi_3} \cdots \tag{9.3}$$

of algebras $\mathfrak{A}_k$ with algebra morphisms $\pi_k : \mathfrak{A}_k \to \mathfrak{A}_{k+1}$, where

$$\mathfrak{A}_k = \mathcal{M}_{n(k,1)} \oplus \cdots \oplus \mathcal{M}_{n(k,t(k))} \tag{9.4}$$

and where $\pi_k : \mathfrak{A}_k \to \mathfrak{A}_{k+1}$ is associated with the matrix M_k as above.

Example 9.1.3 The sequence of algebras $\mathfrak{A}_k$ associated with the Bratteli diagram of Example 9.1.1 is $\mathfrak{A}_k = \mathcal{M}_{n_{k+1}} \oplus \mathcal{M}_{n_k}$, where (n_k) is the Fibonacci sequence of numbers defined by $n_1 = n_2 = 1$ and $n_{k+1} = n_k + n_{k-1}$ for $k \geq 2$. The morphism π_k is defined by $\pi_k(x, y) = (x \oplus y, x)$.

Example 9.1.4 The sequence of algebras $\mathfrak{A}_k$ associated with the Bratteli diagram of Figure 9.2 is $\mathfrak{A}_k = \mathcal{M}_k \oplus \mathbb{C}$. The corresponding morphism $\pi_k : \mathfrak{A}_k \to \mathfrak{A}_{k+1}$ is defined by $\pi_k(x, \lambda) = (x \oplus \lambda, \lambda)$.

As a more general notion of Bratteli diagram, one may consider *partial diagrams* in which the labels $n(p, i)$ satisfy the inequalities

$$n(k+1, j) \geq \sum_{i=1}^{t_p} a_{ji}^{(k)} n(k, i) \tag{9.5}$$

instead of an equality.

$$1 \longrightarrow 2 \longrightarrow 3 \longrightarrow 4 \longrightarrow \cdots$$

Figure 9.3 A partial Bratteli diagram.

Example 9.1.5 The diagram represented in Figure 9.3 with all matrices M_k equal to [1] is a partial Bratteli diagram.

A partial Bratteli diagram represents as above a sequence of $*$-morphisms between finite dimensional algebras, but the morphisms are this time not unital.

Example 9.1.6 The Bratteli diagram of Figure 9.3 corresponds to the sequence of algebras $(\mathcal{M}_k)_{k \geq 1}$ with the morphisms $\pi_k : \mathcal{M}_k \to \mathcal{M}_{k+1}$ defined by $\pi_k(x) = x \oplus 0$.

9.2 C^*-**Algebras**

A $*$-*algebra* $\mathfrak{A}$ is a complex algebra with an idempotent map (called the *adjoint*) $M \mapsto M^*$ related to the algebra structure by

$$(M + N)^* = M^* + N^*, \quad (MN)^* = N^*M^*, \text{ and } (\lambda M)^* = \bar{\lambda} M^* \quad (9.6)$$

for every $M, N \in \mathfrak{A}$ and $\lambda \in \mathbb{C}$. It is usual to denote with capitals the elements of a C^*-algebra.

A C^*-*algebra* $\mathfrak{A}$ is a Banach $*$-algebra with a norm $\| \ \|$ such that for all $M \in \mathfrak{A}$, the identity

$$\|M^*M\| = \|M\|^2 \quad (9.7)$$

is satisfied. Equation (9.7) is called the C^*-*identity*. It implies that $\|M\| = \|M^*\|$, that is, the adjoint map $M \mapsto M^*$ is an isometry (see Exercise 9.1). Thus, (9.7) implies also

$$\|M^*M\| = \|M\|\|M^*\|. \quad (9.8)$$

A *morphism* from a C^*-algebra $\mathfrak{A}$ to a C^*-algebra $\mathfrak{B}$ (also called a $*$-*morphism*) is an algebra morphism π such that $\pi(M^*) = \pi(M)^*$ for all $M \in \mathfrak{A}$. It can be shown that this implies $\|\pi(M)\| \leq \|M\|$ for every $M \in \mathfrak{A}$ (Exercise 9.3).

A C^*-algebra is *unital* if it has an identity element I. It follows from (9.6) that $I^* = I$ and from (9.7) that $\|I\| = 1$.

Each algebra $\mathcal{M}_k$ is a C^*-algebra, using the usual conjugate transpose M^* of M as adjoint of M and using the matrix norm induced by the Hermitian norm. Indeed, Eq. (9.7) is a consequence of the Cauchy–Schwarz inequality $|\langle x, y \rangle| \leq \|x\|\|y\|$, which implies

$$\|M^*M\| = \sup_{\|x\|=\|y\|=1} \langle M^*Mx, y\rangle = \sup_{\|x\|=\|y\|=1} \langle Mx, My\rangle = \|M\|^2.$$

More generally, the algebra $\mathfrak{B}(\mathcal{H})$ of bounded linear operators on a Hilbert space $\mathcal{H}$ is a C^*-algebra. The adjoint of M is defined by $\langle M^*x, y\rangle = \langle x, My\rangle$ for all $x, y \in \mathcal{H}$. As a subalgebra of $\mathfrak{B}(\mathcal{H})$, the algebra $\mathfrak{K}(\mathcal{H})$ of *compact operators* on $\mathcal{H}$ is a C^*-algebra. When $\mathcal{H}$ is the Hilbert space $\ell^2(\mathbb{C})$ of sequences x of complex numbers such that $\|x\|_2 < \infty$, we denote it by $\mathfrak{K}$.

As a second fundamental example, the space $C(X)$ of continuous complex valued functions on a compact space X is a C^*-algebra with complex conjugation as adjoint operation. We have indeed

$$\|\overline{f}f\| = \sup_{x \in X} |\overline{f(x)}f(x)| = \sup_{x \in X} |f(x)|^2 = \|f\|^2.$$

An element M of a C^*-algebra is *self-adjoint* if $M = M^*$. It is *normal* if $MM^* = M^*M$. It is a *projection* if it is self-adjoint and $M^2 = M$.

Recall that $\mathrm{spr}(M)$ denotes the spectral radius of M (see Appendix B.4). The following property is clear when $\mathfrak{A} = \mathcal{M}_k$ (as well as the necessity of the hypothesis that M is normal).

Proposition 9.2.1 *In a C^*-algebra, one has $\|M\| = \mathrm{spr}(M)$ for every normal element of M.*

Proof Let first M be self-adjoint. Then by repeated use of Eq. (9.7), we have

$$\mathrm{spr}(M) = \lim \|M^{2^n}\|^{2^{-n}} = \|M\|. \tag{9.9}$$

When M is normal, using the preceding case, we have

$$\mathrm{spr}(M)^2 \leq \|M\|^2 = \|M^*M\| = \lim \|(M^*M)^n\|^{1/n}$$
$$\leq \lim(\|(M^*)^n\|\|(M^n)\|)^{1/n} = \mathrm{spr}(M)^2. \qquad \blacksquare$$

One can deduce from this property that the norm is unique in a C^*-algebra (Exercise 9.2).

A subalgebra $\mathfrak{B}$ of a C^*-algebra $\mathfrak{A}$ is *self-adjoint* if M^* is in $\mathfrak{B}$ for every $M \in \mathfrak{B}$. It is easy to verify that a closed self-adjoint subalgebra of a C^*-algebra is again a C^*-algebra.

The direct sum $\mathfrak{A} \oplus \mathfrak{B}$ of two C^*-algebras $\mathfrak{A}, \mathfrak{B}$ is itself a C^*-algebra, using on the set $\mathfrak{A} \times \mathfrak{B}$ the componentwise sum, product and $*$, and defining $\|(M, N)\| = \max(\|M\|, \|N\|)$.

9.2.1 Ideals in C^*-Algebras

An *ideal* in a C^*-algebra $\mathfrak{A}$ is a norm-closed, two-sided ideal of the algebra $\mathfrak{A}$. As for ordinary algebras, a C^*-algebra is *simple* if it has no nontrivial ideals. As is well known, every full matrix algebra is simple.

It can be shown that every ideal $\mathcal{J}$ is self-adjoint, that is, such that J^* is in $\mathcal{J}$ for every $J \in \mathcal{J}$.

The quotient $\mathfrak{A}/\mathcal{J}$ of a C^*-algebra $\mathfrak{A}$ by an ideal $\mathcal{J}$ is the set of cosets $M + \mathcal{J} = \{M + J \mid J \in \mathcal{J}\}$ with the adjoint defined by $(M + \mathcal{J})^* = M^* + \mathcal{J}$ and the norm defined by $\|M + \mathcal{J}\| = \inf_{J \in \mathcal{J}} \|M + J\|$.

Since $\mathcal{J}$ is self-adjoint, we have $\|(M + \mathcal{J})^*\| = \|M + \mathcal{J}\|$. Actually, the C^* identity is also satisfied, and thus one has the following statement.

Theorem 9.2.2 *For every ideal $\mathcal{J}$ in a C^*-algebra $\mathfrak{A}$, the quotient algebra $\mathfrak{A}/\mathcal{J}$ is a C^*-algebra.*

The following now shows that the basic isomorphism theorem for algebras still holds for C^*-algebras.

Theorem 9.2.3 *Let $\mathcal{J}$ be an ideal in C^*-algebra $\mathfrak{A}$ and let $\mathfrak{B}$ be a C^*-subalgebra of $\mathfrak{A}$. Then $\mathfrak{B} + \mathcal{J}$ is a C^*-algebra and*

$$\mathfrak{B}/(\mathfrak{B} \cap \mathcal{J}) \simeq (\mathfrak{B} + \mathcal{J})/\mathcal{J}.$$

A C^*-algebra is *semisimple* if it is a direct sum of simple algebras. The following result is classical.

Theorem 9.2.4 *Every C^*-algebra is semisimple.*

We give the proof for a finite dimensional C^*-algebra. Consider a C^*-algebra $\mathfrak{A}$ of matrices contained in $\mathcal{M}_n$. Suppose that I is a nilpotent ideal of $\mathfrak{A}$. Up to a change of basis, all matrices in I are upper diagonal with zeroes on the diagonal. But since I is self-adjoint, this forces $I = \{0\}$.

9.2.2 Finite Dimensional C^*-Algebras

As a direct sum of C^*-algebras, any algebra of the form (9.4) is itself a C^*-algebra. Actually, any C^*-algebra that is finite dimensional (as a vector space) is of this form.

Theorem 9.2.5 *Every finite dimensional C^*-algebra is $*$-isomorphic to a direct sum of full matrix algebras*

$$\mathfrak{A} = \mathcal{M}_{n_1} \oplus \mathcal{M}_{n_2} \oplus \cdots \oplus \mathcal{M}_{n_k}.$$

Proof Since a C^*-algebra is semisimple by Theorem 9.2.4, the result follows from the Wedderburn Theorem (see Appendix B.2). ∎

An element U of a C^*-algebra is *unitary* if $UU^* = U^*U = I$, that is, if $U^* = U^{-1}$. Two elements M, N are *unitarily equivalent* if $M = UNU^*$ where U is a unitary element. Two morphisms $\varphi, \psi \colon \mathfrak{A} \to \mathfrak{B}$ are unitarily equivalent if there is a unitary element U in $\mathfrak{B}$ such that $\varphi = \mathrm{Ad}(U) \circ \psi$ where $\mathrm{Ad}(U)(N) = UNU^{-1}$ for every $N \in \mathfrak{B}$.

The following result shows that the theory of finite dimensional C^*-algebras is similar to that of ordinary semisimple algebras.

Proposition 9.2.6 *Suppose that φ is a unital $*$-morphism of finite dimensional C^*-algebras from $\mathfrak{A} = \mathcal{M}_{m_1} \oplus \cdots \oplus \mathcal{M}_{m_t}$ into $\mathfrak{B} = \mathcal{M}_{n_1} \oplus \cdots \oplus \mathcal{M}_{n_s}$. Then φ is determined up to unitary equivalence by its $s \times t$-matrix of partial multiplicities.*

Proof By the analysis made in Section B.2.6, there is an invertible matrix U such that $\mathrm{Ad}(U) \circ \varphi = \varphi_1 \oplus \cdots \oplus \varphi_s$ is of the form $\varphi_i = \mathrm{id}_{m_1}^{(a_{i1})} \oplus \cdots \oplus \mathrm{id}_{n_t}^{(a_{it})}$ for $1 \leq i \leq s$.

There remains to show that if $\varphi \colon M \mapsto UMU^{-1}$ is a C^*-algebra morphism from $\mathcal{M}_n$ onto itself, then U is unitary. Consider the elementary matrices $E_{ij} \in \mathcal{M}_n$ having all entries equal to 0 except the entry i, j, which is 1. We have

$$\varphi(E_{ij}) = \ell_i r_j, \tag{9.10}$$

where ℓ_i is the column of index i of U and r_j is the row of index j of U^{-1}. Such a decomposition, with vectors r_i, ℓ_i such that $r_i \ell_i = 1$, is unique. Since $E_{ij}^* = E_{ji}$ and since $\varphi(E_{ij})^* = E_{ij}^*$, we have $r_j^* = \ell_j$ and $\ell_i^* = r_i$. Thus, $U^{-1} = U^*$ and U is unitary. ∎

Thus, every sequence $\mathfrak{A}_1 \overset{\varphi_1}{\to} \mathfrak{A}_2 \overset{\varphi_1}{\to} \mathfrak{A}_3 \ldots$ of $*$-morphisms between finite dimensional C^*-algebras is determined by a Bratteli diagram. This fact plays a fundamental role in the sequel.

9.2.3 Direct Limits of C^*-Algebras

Let $\mathfrak{A}$ be a $*$-algebra. A C^*-*seminorm* on $\mathfrak{A}$ is a seminorm ρ on $\mathfrak{A}$ such that for all $M, N \in \mathfrak{A}$,

$$\rho(MN) \leq \rho(M)\rho(N), \quad \rho(M^*) = \rho(M), \quad \text{and } \rho(M^*M) = \rho(M)^2.$$

If, additionally, ρ is in fact a norm, it is called a C^*-*norm*.

If $\varphi \colon \mathfrak{A} \to \mathfrak{B}$ is a $*$-morphism, the map from $\mathfrak{A}$ into $\mathbb{R}_+$ defined by $\rho(M) = \|\varphi(M)\|$ is a $*$-seminorm on $\mathfrak{A}$.

If ρ is a C^*-seminorm on a $*$-algebra $\mathfrak{A}$, the set $\mathcal{N} = \rho^{-1}(0)$ is self-adjoint ideal of $\mathfrak{A}$ and we get a C^*-norm on the quotient $*$-algebra $\mathfrak{A}/\mathcal{N}$ by setting $\|M + \mathcal{N}\| = \rho(M)$. If $\mathfrak{B}$ denotes the Banach space completion of $\mathfrak{A}/\mathcal{N}$ with this norm, it can be checked that the operations on $\mathfrak{A}/\mathcal{N}$ extend uniquely to $\mathfrak{B}$ to make a C^*-algebra, called the *enveloping C^*-algebra* of $\mathfrak{A}$ with respect to ρ.

Now let $(\mathfrak{A}_k)$ be a sequence of C^*-algebras with morphisms $\varphi_{k,k+1} \colon \mathfrak{A}_k \to \mathfrak{A}_{k+1}$. The set

$$\mathfrak{B} = \{(M_k)_{k \geq 0} \mid M_k \in \mathfrak{A}_k, \, M_{k+1} = \varphi_{k,k+1}(M_k) \text{ for every } k \text{ large enough}\}$$

is a $*$-subalgebra of the direct product $\prod_{k \geq 0} \mathfrak{A}_k$. Since the $\varphi_{k,k+1}$ are C^*-algebra morphisms, they are norm decreasing and we can define a semi-norm on $\mathfrak{B}$ by setting $\rho(M) = \lim \|M_n\|$ for $M = (M_n)_{n \geq 0}$. Note that $\rho(M) = 0$ for every $M = (M_n)_{n \geq 0}$ such that $M_n = 0$ for n large enough.

The enveloping C^*-algebra $\mathfrak{A}$ of $\mathfrak{B}$ with respect to ρ is called the *direct limit* of the sequence $(\mathfrak{A}_n)$, denoted $\varinjlim \mathfrak{A}_k$.

As in the case of a direct limit of groups, there is a natural morphism φ_k from each $\mathfrak{A}_k$ into $\mathfrak{A}$ that sends $M \in \mathfrak{A}_k$ to the class of any sequence $(M_\ell)_{\ell \geq 0}$ such that $M_k = M$ and $M_{\ell+1} = \varphi_{\ell,\ell+1}(M_\ell)$ for all $\ell \geq k$.

We will see examples of direct limits of C^*-algebras in the next section.

9.2.4　Positive Elements

An element M of a C^*-algebra $\mathfrak{A}$ is *positive* if $M = M^*$ (that is, M is self-adjoint) and its spectrum $\sigma(M)$ is contained in $\mathbb{R}_+$.

A useful property of positive elements is the existence, for every positive element $M \in \mathfrak{A}$, of a unique positive square root denoted $M^{1/2}$.

It can be shown that if $M, N \in \mathfrak{A}$ are positive, then $M + N$ is positive. As a consequence, the positive elements determine an order on the set of self-adjoint elements by $M \leq N$ if $N - M$ is positive. Indeed, if $M \leq N$ and $N \leq P$, then $P - M = (P - N) + (N - M)$, which is positive and thus $M \leq P$.

Theorem 9.2.7 *For every $M \in \mathfrak{A}$, the element M^*M is positive.*

A *positive linear functional* on C^*-algebra is a linear functional $f \colon \mathfrak{A} \to \mathbb{C}$ such that $f(M) \geq 0$ whenever $M \geq 0$. A *state* is a positive linear functional of norm 1.

For example, let $\mathcal{H}$ be a Hilbert space and let π be a $*$-morphism from $\mathfrak{A}$ into $\mathcal{B}(\mathcal{H})$. Then

$$f(M) = \langle \pi(M)x, x \rangle$$

is a positive linear functional. Indeed, if $M \geq 0$, using the square root $M^{1/2}$ of M, we have

$$f(M) = \langle \pi(M^{1/2})^2 x, x \rangle = \| \pi(M^{1/2})x \|^2 \geq 0.$$

It is a state if $\mathfrak{A}$ is unital and $\|x\| = 1$.

9.3 Approximately Finite Algebras

A C^*-algebra $\mathfrak{A}$ is *approximately finite dimensional* (or an *AF* algebra) if it is the closure in $\mathfrak{A}$ of an increasing union of finite dimensional subalgebras $(\mathfrak{A}_k)_{k \geq 0}$. When $\mathfrak{A}$ is unital, we further stipulate that $\mathfrak{A}_0$ consists of scalar multiples of the identity element 1.

As an equivalent definition, an *AF* algebra is a direct limit of finite dimensional C^*-algebras. Indeed, if $\varphi_{k,k+1} \colon \mathfrak{A}_k \to \mathfrak{A}_{k+1}$ are morphisms, then $\mathfrak{A} = \varinjlim \mathfrak{A}_k$ is an *AF* algebra since it is the closure of the images $\varphi_k(\mathfrak{A}_k)$ of the finite dimensional C^*-algebras $\mathfrak{A}_k$ by the natural morphisms φ_k.

Let $\mathfrak{A} = \overline{\cup_{k \geq 1} \mathfrak{A}_k}$ be an AF algebra with embeddings α_k from $\mathfrak{A}_k$ into $\mathfrak{A}_{k+1}$ We associate to the sequence $(\mathfrak{A}_k)$ the Bratteli diagram defined by the sequence of matrices of partial multiplicities of the morphisms α_k.

An AF algebra is separable. Indeed, every finite dimensional C^*-algebra is separable. The following statement, which we will admit, gives an equivalent definition of AF algebras that does not depend on a sequence of subalgebras.

Theorem 9.3.1 *A C^*-algebra is AF if and only if it is separable and for every $n \geq 1$, every $M_1, \ldots, M_n$ in $\mathfrak{A}$ and every $\varepsilon > 0$ there is a finite dimensional C^*-subalgebra $\mathfrak{B}$ of $\mathfrak{A}$ and $N_1, \ldots, N_n \in \mathfrak{B}$ such that $\|M_i - N_i\| \leq \varepsilon$ for $1 \leq i \leq n$.*

As a simple example of a C^*-algebra that is not an AF algebra, the algebra $C([0, 1])$ is not an AF algebra (Exercise 9.6).

Example 9.3.2 Consider again the sequence of algebras $\mathfrak{A}_k = \mathcal{M}_{n_k} \oplus \mathcal{M}_{n_{k-1}}$ with the morphisms $\pi_k \colon \mathfrak{A}_k \to \mathfrak{A}_{k+1}$ where n_k is the Fibonacci sequence (Example 9.1.3). The direct limit of the sequence of C^*-algebras $\mathfrak{A}_k$ is the *Fibonacci algebra*.

As a generalization, consider an irrational number $\alpha > 0$, its continued fraction expansion $\alpha = [a_0; a_1, a_2, \ldots]$ and the corresponding partial quotients p_k, q_k (see Appendix B.1). The integers q_k are defined by $q_0 = 1$, $q_1 = a_1$ and $q_{k+1} = q_k + q_{k-1}$.

We define an AF algebra $\mathfrak{A}_\alpha$, called a *Sturmian algebra*, as the limit of the algebras $\mathfrak{A}_k = \mathcal{M}_{q_{k+1}} \oplus \mathcal{M}_{q_k}$ with embeddings given by the matrix of multiplicities

$$M_k = \begin{bmatrix} a_k & 1 \\ 1 & 0 \end{bmatrix}.$$

When $\alpha = (1 + \sqrt{5})/2$, we have $\alpha = [1; 1, 1, \ldots]$, and thus the Fibonacci algebra is the algebra $\mathfrak{A}_{\frac{1+\sqrt{5}}{2}}$.

Example 9.3.3 Consider the sequence $\mathcal{M}_k \oplus \mathbb{C}$ of algebras with the embeddings $\pi_k(x, \lambda) = (x \oplus \lambda, \lambda)$, that is,

$$\pi_k(M, \lambda) = \left(\begin{bmatrix} M & 0 \\ 0 & \lambda \end{bmatrix}, \lambda \right)$$

of Example 9.1.4. The corresponding AF algebra is the C^*-algebra $\mathfrak{K} + \mathbb{C}I$, where $\mathfrak{K}$ is the C^*-algebra of compact operators on the Hilbert space $\ell^2(\mathbb{C})$. Indeed, let P_k be the orthogonal projection on the space E_k generated by the first k unit vectors and $P_k^\perp$ be the projection on $E_k^\perp$. Then $\mathfrak{A}_k = \mathbb{C}P_k^\perp + P_k \mathfrak{K} P_k \simeq \mathbb{C} \oplus \mathcal{M}_k$. Since every compact operator on a Hilbert space is a limit of operators of finite rank, $\cup_k \mathfrak{A}_k$ is dense in $\mathfrak{K}$, this proves the claim. Thus, $\mathfrak{K} \oplus \mathbb{C}I$ is a unital AF algebra. The C^*-algebra $\mathfrak{K}$ itself (which is not unital) is obtained as the closure of the sequence of subalgebras $\mathcal{M}_k$ as in Example 9.1.6.

Example 9.3.4 Consider $\mathfrak{A}_k = \mathcal{M}_{2^k}$ with the embedding of $\mathfrak{A}_k$ into $\mathfrak{A}_{k+1}$ being

$$\varphi_k(M) = \begin{bmatrix} M & 0 \\ 0 & M \end{bmatrix}.$$

The corresponding AF algebra is called the *CAR algebra*. The corresponding Bratteli diagram is represented in Figure 9.4. We recognize, of course, the BV-representation of the (2^n) odometer. Let $\mathfrak{B}_k = \mathcal{M}_{2^k} \oplus \mathcal{M}_{2^k}$. Since

$$\varphi_{k+1}\left(\begin{bmatrix} N_1 & 0 \\ 0 & N_2 \end{bmatrix} \right) = \begin{bmatrix} N_1 & 0 & 0 & 0 \\ 0 & N_2 & 0 & 0 \\ 0 & 0 & N_1 & 0 \\ 0 & 0 & 0 & N_2 \end{bmatrix},$$

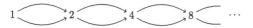

Figure 9.4 A Bratteli diagram for the CAR algebra.

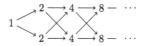

Figure 9.5 Another Bratteli diagram for the CAR algebra.

the morphism φ_{k+1} embeds $\mathfrak{B}_k$ into $\mathfrak{B}_{k+1}$. This shows that $\mathfrak{A}$ is also the algebra defined by the diagram of Figure 9.5.

We now give two examples of AF-algebras with nonstationary diagram.

Example 9.3.5 Let $X = \{0, 1\}^{\mathbb{N}}$ be the one-sided full shift on $\{0, 1\}$. The space $\mathfrak{A} = C(X)$ of continuous complex-valued functions on X is a commutative C^*-algebra. Let $\mathfrak{A}_k$ be the subalgebra of functions that are constant on each cylinder $[w]$ with w of length k. Then $\mathfrak{A} = \overline{\cup \mathfrak{A}_k}$ and thus $\mathfrak{A}$ is an AF algebra. A Bratteli diagram for $\mathfrak{A}$ is represented in Figure 9.6.

Example 9.3.6 Consider the Bratteli diagram represented in Figure 9.7. The corresponding AF algebra $\mathfrak{A}$ is called the *GICAR* algebra. We have $\mathfrak{A} = \varinjlim \mathfrak{A}_k$ where $\mathfrak{A}_k$ is

$$\mathfrak{A}_k = \mathcal{M}_{\binom{k}{0}} \oplus \mathcal{M}_{\binom{k}{1}} \oplus \cdots \oplus \mathcal{M}_{\binom{k}{k}},$$

where

$$\binom{k}{p} = \frac{k!}{(k-p)!p!}$$

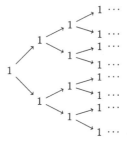

Figure 9.6 A diagram for the algebra $C(\{0, 1\}^{\mathbb{N}})$.

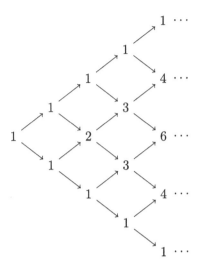

Figure 9.7 The Pascal triangle.

is the binomial coefficient. The embedding of $\mathfrak{A}_k$ into $\mathfrak{A}_{k+1}$ is

$$\varphi_k(M_0, M_1, \ldots, M_k) = (M_0, M_0 + M_1, \ldots, M_{k-1} + M_k, M_k).$$

The following result, which we will admit, is of fundamental importance.

Theorem 9.3.7 (**Bratteli**) *Let* $\mathfrak{A} = \overline{\cup \mathfrak{A}_m} = \overline{\cup \mathfrak{B}_n}$ *be an AF algebra obtained from two chains of finite dimensional C*-algebras* $\mathfrak{A}_m, \mathfrak{B}_n$. *There exist subsequences* m_k, n_k *and an automorphism* α *of* $\mathfrak{A}$ *such that*

$$\mathfrak{A}_{m_k} \subset \alpha(\mathfrak{B}_{n_k}) \subset \mathfrak{A}_{m_{k+1}}$$

for all $k \geq 1$.

We deduce the following characterization of Bratteli diagrams defining isomorphic AF algebras. It uses the notion of intertwining of diagrams introduced in Chapter 5 and playing an essential role in the Strong Orbit Equivalence Theorem (Theorem 5.5.1).

Corollary 9.3.8 *Two Bratteli diagrams define isomorphic AF algebras if and only if they have a common intertwining.*

Proof This is a direct consequence of Theorem 9.3.7. Indeed, the Bratteli diagram corresponding to the sequence

$$\mathfrak{A}_{m_1} \subset \alpha(\mathfrak{B}_{n_1}) \subset \mathfrak{A}_{m_2} \subset \cdots$$

is an intertwining of the diagrams corresponding to the sequences $(\mathfrak{A}_m)$ and $(\alpha(\mathfrak{B}_n))$, the latter being the same as the diagram corresponding to the sequence $(\mathfrak{B}_n)$. ∎

9.3.1 AF Algebras and Cantor Systems

Let (X, T) be a minimal invertible Cantor system. By the BV-representation theorem (Theorem 5.3.3), there exists a properly ordered Bratteli diagram $(V, E, \leq)$ such that (X, T) is isomorphic to (X_E, T_E). Let $M(k)$ be the adjacency matrix corresponding to $E(k)$. Set $V(k) = \{v_1, \ldots, v_{t(k)}\}$. Let

$$n(k, j) = (M(k) \cdots M(1))_{1,j}$$

be the number of paths from $V(0) = \{1\}$ to the vertex $v_j \in V(k)$ and let

$$\mathfrak{A}_k = \mathcal{M}_{n(k,1)} \oplus \mathcal{M}_{n(k,2)} \oplus \cdots \oplus \mathcal{M}_{n(k,t(k))}.$$

Let $\varphi_{k,k+1} \colon \mathfrak{A}_k \to \mathfrak{A}_{k+1}$ be the morphism determined by the adjacency matrix $M(k)$ as in (9.2). The direct limit of the sequence

$$\mathfrak{A}_1 \stackrel{\varphi_{1,2}}{\to} \mathfrak{A}_2 \stackrel{\varphi_{2,3}}{\to} \mathfrak{A}_3 \cdots$$

is an AF algebra $\mathfrak{A}$ called the *C^*-algebra associated with* the Cantor system (X, T). The following result is a complement to the Strong Orbit Equivalence Theorem (Theorem 5.5.1) It shows in particular that the algebra associated with (X, T) does not depend on a particular BV-representation of (X, T).

We can now state the following complement to the Strong Orbit Equivalence Theorem (Theorem 5.5.1).

Theorem 9.3.9 *Let (X, T) and (X', T') be two minimal invertible Cantor systems. The AF algebras associated with (X, T) and (X', T') are isomorphic if and only if the systems are strong orbit equivalent.*

Proof By Theorem 5.5.1, the two systems are strong orbit equivalent if and only if they have BV representations $(V, E, \leq)$ and $(V', E', \leq')$, which have a common intertwining. By Corollary 9.3.8, this is equivalent to the fact that the associated algebras are isomorphic. ∎

We illustrate these notions on two familiar Cantor systems.

Example 9.3.10 We have seen (Example 6.2.16) that the Fibonacci shift has a BV-representation as in Figure 9.8 (with the values of the labels indicated on the vertices). Since this diagram is telescoping equivalent to the diagram of

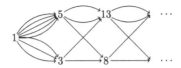

Figure 9.8 A BV-representation of the Fibonacci shift.

Figure 9.1, the associated AF algebra is the Fibonacci algebra $\mathfrak{A}_\alpha$ with $\alpha = (1 + \sqrt{5})/2$.

Example 9.3.11 As already noted in Example 9.3.4, Figure 9.4 is a BV representation of the (2^n) odometer. Thus, the AF algebra associated with the (2^n) odometer is the CAR algebra.

9.3.2 Simple AF Algebras

The following result explains why the term of simple Bratteli diagram was chosen.

Theorem 9.3.12 *The AF algebra defined by a Bratteli diagram is simple if and only if the diagram is simple.*

We first prove the following lemma.

Lemma 9.3.13 *If $\mathcal{J}$ is an ideal of an algebra $\mathfrak{A} = \overline{\cup_{k \geq 1} \mathfrak{A}_k}$, then*

$$\mathcal{J} = \overline{\cup_{k \geq 1}(\mathcal{J} \cap \mathfrak{A}_k)}. \tag{9.11}$$

In particular, if $\mathfrak{A}$ is an AF algebra, $\mathcal{J}$ is an AF algebra.

Proof Consider the commutative diagram

$$
\begin{array}{ccccc}
\mathcal{J} \cap \mathfrak{A}_n & \longrightarrow & \mathfrak{A}_n & \xrightarrow{\ \pi\ } & \mathfrak{A}_n/(\mathcal{J} \cap \mathfrak{A}_n) \\
\downarrow & & \downarrow & & \downarrow{\scriptstyle \alpha} \\
\mathcal{J} & \longrightarrow & \mathfrak{A}_n + \mathcal{J} & \xrightarrow{\ \pi\ } & (\mathfrak{A}_n + \mathcal{J})/\mathcal{J}
\end{array}
$$

where the unlabeled arrows are the canonical injections and π is the canonical quotient map from $\mathfrak{A}$ to $\mathfrak{A}/\mathcal{J}$. The map α is an isomorphism by Theorem 9.2.3. If $J \in \mathcal{J}$, since $\mathfrak{A} = \overline{\cup_{k \geq 1} \mathfrak{A}_k}$, there is for every $\varepsilon > 0$ an element M of some $\mathfrak{A}_k$ such that $\|J - M\| < \varepsilon$. Then, by definition of the norm in the quotient, $\|M + \mathcal{J}\| < \varepsilon$. Since α is an isomorphism, we have also $\|M + (\mathcal{J} \cap \mathfrak{A}_k)\| < \varepsilon$, and thus there is $J' \in \mathcal{J} \cap \mathfrak{A}_k$ such that $\|M - J'\| < \varepsilon$. Since $\|J - J'\| < 2\varepsilon$, we conclude that $\mathcal{J} = \overline{\cup_{k \geq 1}(\mathcal{J} \cap \mathfrak{A}_k)}$.

If $\mathfrak{A}$ is an AF algebra, we have $\mathfrak{A} = \overline{\cup_{k \geq 1} \mathfrak{A}_k}$ with finite dimensional algebras $\mathfrak{A}_k$. Thus, J is an AF algebra by (9.11). ∎

We can now give the proof of Theorem 9.3.12.

Proof of Theorem 9.3.12 Let (V, E) be a Bratteli diagram and let $\mathfrak{A}$ be the corresponding AF algebra. We shall establish a one-to-one correspondence between the ideals of $\mathfrak{A}$ and the sets W of vertices, which are both directed and hereditary.

By Proposition 5.1.1, (V, E) is simple if and only if there is no nontrivial directed and hereditary subset of V. Thus, the above correspondence will prove the theorem.

Let first $\mathcal{J}$ be an ideal of $\mathfrak{A}$. For each $p \geq 1$, the set $\mathcal{J}_p = \mathcal{J} \cap \mathfrak{A}_p$ is an ideal of $\mathfrak{A}_p = \mathcal{M}_{(p,1)} \oplus \cdots \oplus \mathcal{M}_{(p,k)}$. Since every summand $\mathcal{M}_{(p,i)}$ is simple, the ideal $\mathcal{J}_p$ corresponds to a set W_p of vertices (p, i). Let $W = \cup_{p \geq 1} W_p$.

Suppose that $v = (p, i)$ is in W_p and that there is an edge $(p, i) \to (p + 1, j)$. Then, denoting α_p the injection of $\mathfrak{A}_p$ into $\mathfrak{A}_{p+1}$, we have

$$\mathcal{J} \cap \mathcal{M}_{(p+1,j)} \supset \alpha_p(\mathcal{M}_{(p,i)}) \cap \mathcal{M}_{(p+1,j)} \neq \emptyset. \tag{9.12}$$

Hence $\mathcal{J}$ contains $\mathcal{M}_{(p+1,j)}$ and thus $(p + 1, j)$ is in W_{p+1}. This shows that W is directed.

Next, let (p, i) be a vertex and let $J = \{j \mid (p, i) \to (p + 1, j)\}$. Assume that W contains all $(p + 1, j)$ such that $\in J$. Then

$$\alpha_p(\mathcal{M}_{(p,i)}) \subset \sum_{j \in J} \mathcal{M}_{(p+1,j)} \subset \mathcal{J}, \tag{9.13}$$

which implies that $(p, i) \in W$. Thus, W is also hereditary. This allows us to associate to every ideal of $\mathcal{J}$ a directed and hereditary set $W = \alpha(\mathcal{J})$ of vertices of the diagram. By Lemma 9.3.13, the ideal $\mathcal{J}$ can be recovered from W and thus the map α is injective.

Conversely, let W be a directed and hereditary subset of W. Let $\mathcal{J}_p$ be the ideal of $\mathfrak{A}_p$ corresponding to the vertices of W at level p. Since W is directed, the sequence $\mathcal{J}_p$ is increasing. The closure of the union is an ideal $\mathcal{J}$ of $\mathfrak{A}$. Since W is hereditary, we have $\mathcal{J}_p = \mathfrak{A}_p \cap \mathcal{J}$. Indeed, if $\mathcal{M}_{(p,i)}$ is in $\mathcal{J}$, there is a $q \geq p$ such that $\mathcal{M}_{(p,i)} \in \mathfrak{A}_p$. All descendants of (p, i) at level q are then in W_q, and thus (p, i) is in W because W is hereditary. Thus, $W = \alpha(\mathcal{J})$, and this completes the proof. ∎

Example 9.3.14 The algebra $\mathbb{C}I + \mathfrak{K}$ of Example 9.3.3 is not simple since $\mathcal{G}_k$ is a proper ideal. Accordingly, the Bratteli diagram of Figure 9.2 is not simple. The vertices of the lower level form a directed hereditary set of vertices. The restriction of the diagram to this set of vertices is the partial Bratteli diagram of Figure 9.3, which represents the C^*-algebra $\mathfrak{K}$.

9.3.3 Dimension Groups of AF Algebras

We define the dimension group of an AF algebra as follows. Let $\mathfrak{A} = \overline{\cup_{n \geq 1} \mathfrak{A}_n}$ be an AF algebra where the sequence $(\mathfrak{A}_n)_{n \geq 1}$ is defined by the sequence of $k_{n+1} \times k_n$-matrices M_n. The dimension group $K_0(\mathfrak{A})$ of $\mathfrak{A}$ is the direct limit of the sequence

$$\mathbb{Z}^{k_1} \overset{M_1}{\to} \mathbb{Z}^{k_2} \overset{M_2}{\to} \mathbb{Z}^{k_3} \overset{M_3}{\to} \cdots .$$

Thus, the dimension group of an AF algebra is the dimension group of its Bratteli diagram (although the diagram is not unique, the group is well defined; see below).

Note that if $\mathfrak{B} = \mathcal{M}_{n_1} \oplus \cdots \oplus \mathcal{M}_{n_k}$, we can set

$$K_0(\mathfrak{B}) = (\mathbb{Z}^k, \mathbb{Z}^k_+, (n_1, \ldots, n_k)^t).$$

In this way, for $\mathfrak{A} = \overline{\cup_{n \geq 1} \mathfrak{A}_n}$, one can also write

$$K_0(\mathfrak{A}) = \lim_{\to} K_0(\mathfrak{A}_n)$$

with the connecting morphisms given by the matrices M_n.

It follows from Corollary 9.3.8 that the definition of the dimension group is independent of the sequence $\mathfrak{A}_n$ such that $\mathfrak{A} = \overline{\cup \mathfrak{A}_n}$. Indeed, by Theorem 5.1.5, the dimension groups of Bratteli diagrams equivalent by intertwining are isomorphic.

Note that if (X, T) is a minimal invertible Cantor system, the dimension group $K^0(X, T)$ is isomorphic to the group $K_0(\mathfrak{A})$, where $\mathfrak{A}$ is the AF algebra associated with (X, T). Indeed, by Theorem 5.3.6, the group $K^0(X, T)$ is the dimension group of a BV representation of (X, T).

Proposition 9.3.15 *For every irrational number $\alpha > 0$, the dimension group of the Sturmian algebra $\mathfrak{A}_\alpha$ is $\mathbb{Z} + \alpha \mathbb{Z}$.*

Proof The dimension group of $\mathfrak{A}_\alpha$ can identified with $\mathbb{Z}^2$ with positive cone formed of the pairs $(x, y) \in \mathbb{Z}^2$ such that

$$\begin{bmatrix} p_n & q_n \\ p_{n-1} & q_{n-1} \end{bmatrix} \begin{bmatrix} x \\ y \end{bmatrix} > 0$$

for some $n \geq 0$, which is equivalent to $\alpha x + y > 0$ since p_n/q_n converges to α.   ∎

Example 9.3.16 Consider the Fibonacci algebra $\mathfrak{A}$ of Example 9.3.2. The dimension group $K_0(\mathfrak{A})$ is the group $\mathbb{Z}[\alpha]$ where $\alpha = (1 + \sqrt{5})/2$. Indeed, it is the direct limit $\mathbb{Z}^2 \xrightarrow{M} \mathbb{Z}^2 \xrightarrow{M} \mathbb{Z}^2 \cdots$ with $M = \begin{bmatrix} 1 & 1 \\ 1 & 0 \end{bmatrix}$ and thus the assertion results from Example 2.3.7.

Example 9.3.17 The dimension group of the algebra $\mathbb{C}I \oplus \mathfrak{K}$ (Example 9.3.3) is the ordered group $\mathbb{Z}^2$ with positive cone $\{(x, y) \mid y > 0\} \cup \{(x, 0) \mid x \geq 0\}$ (see Example 2.3.8).

Example 9.3.18 The dimension group of the CAR algebra is $\mathbb{Z}[1/2]$ (see Example 2.3.1).

The next example introduces an ordered group not seen before.

Example 9.3.19 Consider the GICAR algebra (Example 9.3.6). We have $K_0(\mathfrak{A}_n) = \mathbb{Z}^{n+1}$ with morphisms

$$\varphi_n(a_0, a_1, \ldots, a_n) = (a_0, a_0 + a_1, \ldots, a_n).$$

Let us represent the group $K_0(\mathfrak{A}_n)$ as the set of polynomials with integer coefficients of degree at most n though the map

$$(a_0, a_1, \ldots, a_n) \mapsto a_0 + a_1 x + \cdots + a_n x^n.$$

Since

$$(1+x)(a_0+a_1x+\cdots+a_nx^n) = a_0+(a_0+a_1)x+\cdots+(a_{n-1}+a_nx^n+a_nx^{n+1},$$

the morphisms φ_n are now replaced by the multiplication by $1 + x$. Thus, the dimension group of the GICAR algebra is the group

$$G = \{(1 + x)^{-n}p(x) \mid p \in \mathbb{Z}[x] \text{ of degree at most } n, n \geq 0\}.$$

The positive cone G_+ corresponds to the polynomials p such that $(1 + x)^N p(x) > 0$ for some $N \geq 1$. One can show that a polynomial is of this form if and only if $p(x) > 0$ for $x \in]0, \infty[$ (Exercise 9.7).

A *trace* in a C^*-algebra $\mathfrak{A}$ is a state τ such that $\tau(MN) = \tau(NM)$ for all $M, N \in \mathfrak{A}$. When $\mathfrak{A} = \mathcal{M}_n$, there is a unique trace on $\mathfrak{A}$ that is

$$\tau(M) = \frac{1}{n} \text{Tr}(M),$$

where $\text{Tr}(M)$ is the usual trace of the matrix M. More generally, when $\mathfrak{A} = \mathcal{M}_{n_1} \oplus \cdots \oplus \mathcal{M}_{n_k}$ is a finite dimensional C^*-algebra, the traces on $\mathfrak{A}$ are the maps τ such that

$$\tau(M_1 \oplus \cdots \oplus M_k) = \frac{t_1}{n_1} \operatorname{Tr}(M_1) + \cdots + \frac{t_k}{n_k} \operatorname{Tr}(M_k), \qquad (9.14)$$

where $t_1, \ldots, t_k \geq 0$ are such that $\sum_{j=1}^{k} t_k = 1$.

Theorem 9.3.20 *The traces on an AF algebra $\mathfrak{A}$ are in one-to-one correspondence with the states on $K_0(\mathfrak{A})$.*

Proof Let first τ be a trace on the AF algebra $\mathfrak{A} = \overline{\cup \mathfrak{A}_n}$ where $\mathfrak{A}_n = \mathcal{M}_{n_1} \oplus \cdots \oplus \mathcal{M}_{n_{k_n}}$. The restriction of τ to $\mathfrak{A}_n$ is a trace on $\mathfrak{A}_n$, and thus it is given by Eq. (9.14). Set $t_j = \tau(I_j)$, where I_j is the identity of the $\mathcal{M}_{n_j}$. Let τ_n be the map on $K_0(\mathfrak{A}_n)$ defined by

$$\tau_n(x_1, \ldots, x_{k_n}) = t_1 x_1 + \cdots + t_{k_n} x_{k_n}.$$

Then τ_n is a state on $(\mathbb{Z}^{k_n}, \mathbb{Z}_+^{k_n}, \begin{bmatrix} n_1 \\ \vdots \\ n_{k_n} \end{bmatrix})$.

Conversely, for every state σ on $K_0(\mathfrak{A})$, we define a state on $\mathfrak{A}_n = \mathcal{M}_{n_1} \oplus \cdots \oplus \mathcal{M}_{n_{k_n}}$ by (9.14) with $t_i = \sigma(e_i)$ and e_i the ith basis vector of $\mathbb{Z}^{k_n}$. ∎

9.3.4 Elliott Theorem

The following classification result is due to Elliott. We may consider it as a complement to the Strong Orbit Equivalence Theorem (Theorem 5.5.1) and also to Theorem 9.3.9 since it adds one more equivalent condition to the strong orbit equivalence of minimal invertible Cantor systems: the isomorphism of the associated AF algebras. Although a proof using Theorem 5.5.1 is possible (see the forthcoming remark), we find it useful to give a direct proof.

Theorem 9.3.21 (Elliott) *Two unital AF algebras $\mathfrak{A}$ and $\mathfrak{B}$ are $*$-isomorphic if and only if their dimension groups are isomorphic.*

The condition is certainly necessary. To prove that it is sufficient, we first prove two lemmas that treat particular cases.

Let φ be unital $*$-morphism from $\mathfrak{A} = \mathcal{M}_{n_1} \oplus \cdots \oplus \mathcal{M}_{n_k}$ into $\mathfrak{B} = \mathcal{M}_{m_1} \oplus \cdots \oplus \mathcal{M}_{m_\ell}$ as in Proposition 9.2.6. We denote by φ_* the morphism from $K_0(\mathfrak{A}) = (\mathbb{Z}^k, \mathbb{Z}_+^k, (n_1, \ldots, n_k)^t)$ to $K_0(\mathfrak{B}) = (\mathbb{Z}^\ell, \mathbb{Z}_+^\ell, (m_1, \ldots, m_\ell)^t)$, which corresponds to the multiplication by the matrix of partial multiplicities M of φ. Since M has nonnegative coefficients, the morphism φ_* is positive, and since M satisfies (9.1), it is unital. Thus, φ_* is a morphism of unital ordered groups.

The next result shows how to recover morphisms of finite dimensional AF algebras from morphisms of their dimension groups.

Lemma 9.3.22 *Suppose that* $\mathfrak{A}, \mathfrak{B}$ *are finite dimensional* C^*-algebras and that ψ is a unital morphism of ordered groups from $K_0(\mathfrak{A})$ into $K_0(\mathfrak{B})$. Then there is a unital $*$-morphism $\varphi \colon \mathfrak{A} \to \mathfrak{B}$ such that $\varphi_* = \psi$ and φ is unique up to unitary equivalence.

Proof Set $\mathfrak{A} = \mathcal{M}_{n_1} \oplus \cdots \oplus \mathcal{M}_{n_k}$ and $\mathfrak{B} = \mathcal{M}_{m_1} \oplus \cdots \oplus \mathcal{M}_{m_\ell}$. Then $K_0(\mathfrak{A}) = (\mathbb{Z}^k, \mathbb{Z}^k_+, (n_1, \ldots, n_k)^t)$ and $K_0(\mathfrak{B}) = (\mathbb{Z}^\ell, \mathbb{Z}^\ell_+, (m_1, \ldots, m_\ell)^t)$. Let M be the matrix of the morphism ψ, which is such that $\psi(v) = Mv$ for every $v \in \mathbb{Z}^k$. Since ψ is positive, the coefficients of M are nonnegative. And since ψ is unital, (9.1) is satisfied. Thus, the result follows from Proposition 9.2.6. ∎

We now improve the last result by replacing $\mathfrak{B}$ by an AF algebra. For this, we introduce a notation. Let $\mathfrak{B} = \overline{\cup_{m \geq 1} \mathfrak{B}_m}$ be an AF algebra with embeddings $\beta_m \colon \mathfrak{B}_m \to \mathfrak{B}$. Then $K_0(\mathfrak{B}) = \varinjlim K_0(\mathfrak{B}_m)$ with connecting morphisms $\beta_{m,n}$. We denote by β_{m*} the natural morphism from $K_0(\mathfrak{B}_m)$ into $K_0(\mathfrak{B})$, which corresponds to the connecting morphisms β_{mn*}. Thus $\beta_{m*}(x)$ is, for $x \in \mathfrak{B}_m$, the class of sequences $(x_k) \in \prod_{k \geq 1} K_0(\mathfrak{B}_k)$ such that $x_n = x$ and $x_{m+1} = \beta_{m+1,m*}(x_m)$ for all $m \geq n$.

Lemma 9.3.23 *Let* $\mathfrak{A}$ *be a finite dimensional* C^*-algebra. Let $\mathfrak{B} = \overline{\cup_{m \geq 1} \mathfrak{B}_m}$ be an AF algebra with the embeddings $\beta_m \colon \mathfrak{B}_m \to \mathfrak{B}$. Let $\psi \colon K_0(\mathfrak{A}) \to K_0(\mathfrak{B})$ be a unital morphism of ordered groups. Then there is an integer m and a $*$-morphism φ from $\mathfrak{A}$ into $\mathfrak{B}_m$ such that $\beta_{m*}\varphi_* = \psi$. Moreover, φ is unique up to unitary equivalence.

Proof Set $\mathfrak{A} = \mathcal{M}_{n_1} \oplus \cdots \oplus \mathcal{M}_{n_k}$. Then, as we have seen, $K_0(\mathfrak{A})$ is the unital ordered group $(\mathbb{Z}^k, \mathbb{Z}^k_+, (n_1, \ldots, n_k)^t)$. Let $e_j \in \mathbb{Z}^k$ be the jth basis vector of $\mathbb{Z}^k$. For each $j = 1, \ldots, k$, since $K_0(\mathfrak{B})$ is the direct limit of the ordered groups $K_0(\mathfrak{B}_m)$ with natural morphisms β_{m*}, there is, by definition of the direct limit, an integer m such that $\psi(e_j)$ is in $\beta_{m*}(K_0^+(\mathfrak{B}_m))$. Taking the maximum of these integers, we may assume that this holds with the same m for all $j = 1, \ldots, k$. Set $\psi(e_j) = \beta_{m*}(v_j)$ for some $v_j \in K_0^+(\mathfrak{B}_m)$.

We define a morphism $\rho \colon \mathbb{Z}^k \to K_0(\mathfrak{B}_m)$ by $\rho(e_j) = v_j$ for $j = 1, \ldots, k$. Then $\psi = \beta_{m*}\rho$ and we have the commutative diagram below:

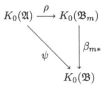

The morphism ρ is actually a morphism of unital ordered groups. Indeed, ρ is positive since $\rho(e_j)$ belongs to $K_0^+(\mathfrak{B}_m)$, and ρ is unital since

$$\rho(n_1, \ldots, n_k)^t = \sum_{j=1}^k n_j \rho(e_j) = \sum_{j=1}^k n_j \psi(e_j)$$

$$= \psi(n_1, \ldots, n_k)^t = 1_{K_0(\mathfrak{B})}.$$

We now apply Lemma 9.3.22 to obtain a unital $*$-morphism $\varphi \colon \mathfrak{A} \to \mathfrak{B}_m$ such that $\varphi_* = \rho$.

Assume that $\varphi' \colon \mathfrak{A} \to \mathfrak{B}_{m'}$ is another map with the same properties. Taking the maximum of m, m', we may assume that $m = m'$. Again by definition of the direct limit, there is an integer $p \geq m$ such that $\beta_{p,m*}(\varphi_*) = \beta_{p,m*}(\varphi'_*)$. We replace φ, φ' by $\beta_{p,m}(\varphi)$ and $\beta_{p,m}\varphi'$. Since $\beta_{p,m}(\varphi_*) = \beta_{p,m*}(\varphi_*)$, applying Lemma 9.3.22, we obtain that φ, φ' are unitarily equivalent. ∎

Proof of Theorem 9.3.21 We will prove that given a unital isomorphism $\rho \colon K_0(\mathfrak{A}) \to K_0(\mathfrak{B})$, there is a $*$-isomorphism $\varphi \colon \mathfrak{A} \to \mathfrak{B}$ such that $\varphi_* = \rho$.

Let $\mathfrak{A} = \lim \mathfrak{A}_m$ and $\mathfrak{B} = \lim \mathfrak{B}_n$ with natural maps α_m and β_n respectively. We will build a commutative diagram with increasing sequences $m_1 < m_2 < \cdots$ and $n_1 < n_2 < \cdots$ as represented below so that $\varphi_* = \rho$ and $\psi_* = \rho^{-1}$.

We start with $m_1 = 1$. Apply Lemma 9.3.23 to the map $\rho \circ \alpha_{m_1*} \colon K_0(\mathfrak{A}_{m_1}) \to K_0(\mathfrak{B})$ to obtain an integer n_1 and a morphism $\varphi_1 \colon \mathfrak{A}_{m_1} \to \mathfrak{B}_{n_1}$ such that $\beta_{n_1*}\varphi_{1*} = \rho\alpha_{m_1*}$ (see the left-hand side of Figure 9.10).

Now apply Lemma 9.3.23 to the map $\rho^{-1} \circ \beta_{n_1*} \colon K_0(\mathfrak{B}_{n_1}) \to K_0(\mathfrak{A})$ to obtain a positive integer m and a morphism ψ of $\mathfrak{B}_{n_1}$ into $\mathfrak{A}_m$ such that $\beta_{m*} \circ \psi_* = \rho^{-1} \circ \beta_{m_1*}$ (see the right-hand side of Figure 9.10). Then, since

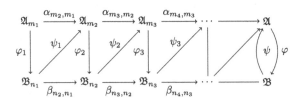

Figure 9.9 The sequences $m_1, m_2 \ldots$ and $n_1, n_2, \ldots$.

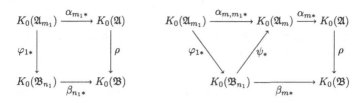

Figure 9.10 The construction of φ_1 and ψ_1.

$$\alpha_{m*} \circ \psi_* \circ \varphi_{1*} = \alpha_{m*} \circ \alpha_{m_1,m*},$$

there is an integer $m_2 \geq m$ such that

$$\alpha_{m_2,m*} \circ \psi_* \circ \varphi_{1*} = \alpha_{m_2,m*} \circ \alpha_{m_1 m*}.$$

By Lemma 9.3.22, this implies that $\alpha_{m_2m} \psi \varphi_1$ and α_{m_2m} are unitarily equivalent and thus that there is a unitary element $U \in \mathfrak{A}_{m_2}$ such that $\alpha_{m_2m} = \mathrm{Ad}(U) \circ \alpha_{m_2m} \circ \psi \circ \varphi_1$. We set $\psi_1 = \mathrm{Ad}(U) \circ \alpha_{m_2m} \circ \psi$. In this way, we have completed the upper left triangle of Figure 9.9.

We proceed in the same way to obtain the sequence $m_1 < m_2 < \cdots$, the sequence $n_1 < n_2 < \cdots$ and the morphisms φ_k, ψ_k as in Figure 9.9.

Since $\mathfrak{A}$ is the closure of $\cup_{m \geq 1} \mathfrak{A}_m$ and similarly for $\mathfrak{B}$, there is a unique continuous map $\varphi \colon \mathfrak{A} \to \mathfrak{B}$ such that its restriction to $\mathfrak{A}_k$ is equal to φ_k. Moreover, φ is a unital $*$-morphism. Similarly, there is a unique continuous map $\psi \colon \mathfrak{B} \to \mathfrak{A}$ whose restriction to $\mathfrak{B}_k$ is ψ_k. Since $\psi_k \circ \varphi_k$ is the embedding α_{m_{k+1},m_k} of $\mathfrak{A}_k$ into $\mathfrak{A}_{k+1}$, its restriction to $\mathfrak{A}_k$ is the identity. This implies that $\psi \circ \varphi = \mathrm{id}_{\mathfrak{A}}$. Similarly $\varphi \circ \psi = \mathrm{id}_{\mathfrak{B}}$.

Thus, φ, ψ are mutually inverse isomorphisms from $\mathfrak{A}$ to $\mathfrak{B}$. Finally, in view of Figure 9.10, we have the commutative diagrams below and thus $\varphi_* = \rho$ as asserted.

$$\begin{array}{ccc} K_0(\mathfrak{A}_{m_k}) & \xrightarrow{\alpha_{m_k*}} & K_0(\mathfrak{A}) \\ \downarrow{\scriptstyle \varphi_{k*}} & & \downarrow{\scriptstyle \varphi_*} \\ K_0(\mathfrak{B}_{n_k}) & \xrightarrow{\beta_{n_k*}} & K_0(\mathfrak{B}) \end{array} \quad \text{and} \quad \begin{array}{ccc} K_0(\mathfrak{A}_{m_k}) & \xrightarrow{\alpha_{m_k*}} & K_0(\mathfrak{A}) \\ \downarrow{\scriptstyle \varphi_{k*}} & & \downarrow{\scriptstyle \rho} \\ K_0(\mathfrak{B}_{n_k}) & \xrightarrow{\beta_{n_k*}} & K_0(\mathfrak{B}) \end{array} \qquad \blacksquare$$

Another proof of Theorem 9.3.21 is as follows. Assume that $\mathfrak{A}$ and $\mathfrak{A}'$ are AF algebras such that $K_0(\mathfrak{A}) = K_0(\mathfrak{A}')$. Let $\mathcal{B} = (V, E)$ and $\mathcal{B}' = (V', E')$ be Bratteli diagrams defining $\mathfrak{A}$ and $\mathfrak{A}'$. Since the dimension groups of $\mathcal{B}$

and $\mathcal{B}'$ are isomorphic to $K_0(\mathfrak{A})$ and $K_0(\mathfrak{A}')$, they are isomorphic. By Theorem 5.5.1, this implies that $\mathcal{B}$ and $\mathcal{B}'$ have a common intertwining and thus, by Corollary 9.3.8, that $\mathfrak{A}$ and $\mathfrak{A}'$ are isomorphic.

9.3.5 Morita Equivalence

The counterpart in terms of algebras of the Kakutani Strong Orbit Equivalence Theorem (Theorem 5.5.5) uses the notion of strong Morita equivalence of C^*-algebras. Two C^*-algebras $\mathfrak{B}$ and $\mathfrak{C}$ are *strong Morita equivalent* if there exist a simple unital and separable C^*-algebra $\mathfrak{A}$ and projections P, Q in $\mathfrak{A}$ such that

$$\mathfrak{B} = P\mathfrak{A}P, \quad \mathfrak{C} = Q\mathfrak{A}Q.$$

Theorem 9.3.24 *Two minimal invertible Cantor systems are Kakutani strong orbit equivalent if and only if the associated AF algebras are strong Morita equivalent.*

Actually, it can be shown that two simple unital AF algebras $\mathfrak{B}$ and $\mathfrak{C}$ are strong Morita equivalent if and only if the dimension groups $K_0(\mathfrak{B})$ and $K_0(\mathfrak{C})$ are isomorphic as ordered groups (but possibly with distinct units). Thus, taking this for granted, Theorem 9.3.24 is a reformulation of Theorem 5.5.5.

9.4 Exercises

Section 9.2

9.1 Show that the C^*-identity (9.7) implies (9.8) $\|M\| = \|M^*\|$.

9.2 Show that the norm in a C^*-algebra is unique.

9.3 Show that if $\pi : \mathfrak{A} \to \mathfrak{B}$ is a $*$-morphism, then $\|\pi(M)\| \leq \|M\|$ for all $M \in \mathfrak{A}$.

Section 9.3

9.4 Show that if $\mathfrak{A} = \overline{\cup_{n \geq 1} \mathfrak{A}_n}$ and $\mathfrak{B} = \overline{\cup_{n \geq 1} \mathfrak{B}_n}$ have the same Bratteli diagram, they are isomorphic.

9.5 A C^*-algebra is called *uniformly hyperfinite* of UHF if it is the increasing union of full matrix algebras $\mathcal{M}_{k_n}$ with $k_1|k_2|\dots$. The *supernatural number* associated with such a C^*-algebra $\mathfrak{A}$ is $\delta(\mathfrak{A}) = \prod_p p^{n_p}$, where the product is over all prime numbers p and $n_p \in \mathbb{N} \cup \infty$, is the supremum of the exponents of powers of p that divide k_n. Show that two UHF algebras are isomorphic if and only if $\delta(\mathfrak{A}) = \delta(\mathfrak{B})$.

9.6 Show that $C([0, 1])$ is not an AF algebra.

9.7 Show that for every $(a, b) \in \mathbb{R}^2$ with $b \neq 0$, there is an $N \geq 1$ such that $(1 + x)^N (x^2 - 2ax + a^2 + b^2)$ has positive coefficients. Conclude that a polynomial p is such that $(1 + x)^N p(x)$ has positive coefficients for some $N \geq 1$ if and only if $p(x) > 0$ for every $x \in]0, \infty[$.

9.5 Notes

The reader is referred for a more detailed presentation to the numerous monographs on the subject, including Davidson (1996), which we follow here, and also the book (Pedersen, 2018) edited by Soren Eilers and Dorte Oleson, which we occasionally follow.

The proof that for every ideal $\mathcal{J}$ in a C^*-algebra $\mathfrak{A}$, the quotient algebra $\mathfrak{A}/\mathcal{J}$ is a C^*-algebra (Theorem 9.2.2) is in Davidson (1996, Theorem I.5.4).

For a proof of Proposition 9.2.6, see Davidson (1996, Corollary III.2.1).

Theorem 9.3.1 appears in Bratteli (1972, Theorem 2.2) and actually earlier in Glimm (1960, Theorem 1.13).

The definition of the dimension group of an AF algebra given here is not the usual one. The standard presentation involves a development of the K-theory of AF algebras. Indeed, K_0 is a functor that assigns an ordered abelian group to each ring based on the structure of idempotents in the matrix algebra over the ring. It occurs that for AF algebras, this group coincides with the dimension group of a Bratteli diagram defining the algebra, a result due to Elliott (1976).

The CAR algebra (Example 9.3.4) comes from quantum mechanics (see Bratteli and Robinson (1987)). It is named for the *Canonical Anticommutation Relations algebra*. If V is a vector space with a nonsingular symmetric bilinear form, the unital $*$-algebra generated by the elements of V subject to the relations

$$fg + gf = \langle f, g \rangle$$
$$f^* = f$$

for every $f, g \in V$ is the CAR algebra on V. It can be shown that the CAR algebra, as defined in Example 9.3.4, is isomorphic with the CAR algebra on a separable Hilbert space (see Davidson (1996)).

The name of the GICAR algebra (Example 9.3.6) stands for the *Gauge Invariant* CAR and is also called the *current algebra* (see Davidson (1996)). On positive polynomials (Exercise 9.7), see Handelman (1985). Note that Figure 9.7 is the diagram defining the Pascal adic system (Exercise 5.13).

Theorem 9.3.7, originally from Bratteli (1972), is Theorem III.3.5 in Davidson (1996).

The way in which we define the C^*-algebra associated with a Cantor minimal system is not the standard one, which uses a C^*-*crossed product*. If (X, T) is a minimal Cantor system, the C^*-crossed product $C(X) \times_T \mathbb{Z}$ is the universal C^*-algebra generated by $C(X)$ and a unitary element U such that $UfU^* = f \circ T$ for every $f \in C(X)$. This algebra is not approximately finite but, as shown in Giordano et al. (1995), two Cantor minimal systems are strong orbit equivalent if and only if the associated crossed product C^*-algebras are isomorphic.

Elliott's theorem (Theorem 9.3.21) appeared in Elliott (1976). The proof follows that of Davidson (1996, Theorem IV.5.3).

The notion of strong Morita equivalence of C^*-algebras is introduced in Rieffel (1982). It is a version of Morita equivalence used for rings (see Appendix B.2) (which do not have the additional $*$ operation). Theorem 9.3.24 is from Giordano et al. (1995).

Let us finally mention perspectives on aspects that have not been treated here. We have been able to associate to every minimal Cantor system a BV-representation (Theorem 5.3.3) and thus a C^*-algebra (by the results of this chapter) in a way that conjugate systems define isomorphic algebras. This has been pursued using different constructions for general shift spaces as an extension of the early constructions of Krieger for shifts of finite type (see Cuntz and Krieger (1980), Matsumoto (1999) and Krieger and Matsumoto (2002)). The link of this approach with substitutive shifts was described in Carlsen and Eilers (2004).

The question of the computability of a Bratteli diagram for an AF-algebra has been investigated by Mundici (2020). In particular, the diagram is computable for AF-algebras $\mathfrak{A}$ such that the dimension group $K^0(\mathfrak{A})$ is lattice ordered (Mundici, 2020).

Appendix A

Solutions to Exercises

A.1 Chapter 1

Section 1.1

1.1 Let F be a closed subset of X. Then, it is compact. Thus, the set $(\phi^{-1})^{-1}(F) = \phi(F)$ is compact and hence closed, proving that ϕ^{-1} is continuous.

1.2 Assume that $U \cap T^{-n}V$ is nonempty. Let x in $U \cap T^{-n}V \neq \emptyset$. Then $T^n x$ belongs to $T^n U \cap V$. Conversely, if y is in $T^n U \cap V$, there is some $x \in U$ such that $T^n(x) = y$. Thus, x is an element of $U \cap T^{-n}V$.

1.3 If $T^n x$ is in U, then x belongs to $U \cap T^{-n}U$. Since the latter is an open set, there is an $m \geq 0$ such that $T^m x$ belongs to $U \cap T^{-n}U$ and thus $T^{n+m}x$ is in U. The same argument can be repeated to obtain the conclusion.

1.4 (i)$\Rightarrow$(ii) Let $(U_n)_{n \geq 0}$ be a countable basis of open sets (this exists for any metric space). For every $n \geq 0$, the open set $V_n = \cup_{m \geq 0} T^{-m}U_n$ is dense since (X, T) is recurrent. The set $V = \cap_{n \geq 0} V_n$ is formed of the points with a dense positive orbit. By the Baire Category Theorem (Theorem B.4.2), the set V is dense in X.

(ii)$\Rightarrow$ (i) Let $x_0 \in X$ be a point with a dense positive orbit. For every pair U, V of nonempty open sets, there are, by Exercise 1.3, arbitrary large integers n, m such that $T^n x$ is in U and $T^m x$ in V. Choosing $n < m$, we obtain that the set $U \cap T^{m-n}V$ is nonempty.

1.5 Let $\phi : (X, T) \rightarrow (X', T')$ be a factor map from a minimal system (X, T) to (X', T'). Let Y' be a closed stable nonempty subset of X'. Then

435

$Y = f^{-1}(Y')$ is nonempty and closed. It is also stable because for $y \in Y$, we have $\phi(T(y)) = T'(\phi(y)) \in T'Y' \subset Y'$ and thus $T(y)$ is in Y. Hence $Y = X$, which implies $Y' = X'$.

1.6 Let $\mathcal{F}$ be the family of nonempty closed subsets of X that are stable. Every descending chain $Y_0 \supset Y_1 \supset \cdots$ of elements of $\mathcal{F}$ has a lower bound $\cap Y_n$ in $\mathcal{F}$. Thus, by Zorn's Lemma, $\mathcal{F}$ has a minimal element, which is a nonempty closed invariant subset of X.

1.7 Assume first that α is irrational. Fix $\varepsilon > 0$ and set $T = T_\alpha$. If all points of the orbit of x are at distance at least ε on $\mathbb{T}$, there can be only a finite number of points in the orbit, a contradiction. Let $m < n$ be such that $T^m(x), T^n(x)$ are at distance $\delta < \varepsilon$. Then, since T is an isometry, for every point $y = T^p(x)$ of the orbit of x,

$$d(y, T^{n-m}(y)) = d(T^p(x), T^{m-n+p}(x)) = d(T^m(x), T^n(x)) = \delta.$$

This shows that every interval of length ε contains an element of the orbit of x and thus the orbit of x is dense.

Suppose now α is rational. Then the orbit of 0 in $\mathbb{T}$ will visit finitely many points and thus cannot be dense in $\mathbb{T}$.

1.8 If T is invertible, the inverse of T^h is the map that sends (x, i) to $(x, i-1)$ if $i \geq 1$ and to $(T^{-1}x, h(T^{-1}x) - 1)$ otherwise. Conversely if T^h is invertible the inverse of T is the map that sends x to the unique y such that $T^h(y, h(y) - 1) = (x, 0)$.

If (X, T) is minimal, the orbit of every $(x, i) \in X^h$ is clearly dense in X^h. Thus, (X^h, T^h) is minimal. Conversely, if (X^h, T^h) is minimal then each orbit is dense in $X \times \{0\}$ and thus (X, T) is minimal.

The last assertion result from the fact that $f(x)$ is the return time to $U = X \times \{0\}$ and thus

$$(T^h)_U(x, 0) = (T^h)^{h(x)}(x, 0) = (Tx, 0).$$

1.9 The map ϕ is clearly continuous and satisfies $\phi \circ T = S \circ \phi$.

1.10 Let (X, T) be an infinite recurrent symbolic system. Since X is a closed subset of $A^{\mathbb{Z}}$, it is compact and totally disconnected. Suppose x is an isolated point: There is a open set U such that $U \cap X = \{x\}$. Since X is recurrent, there is some point $y \in X$ such that the orbit $(T^n(y))_n$ is dense. Then there is some $n < m$ such that $T^n(y)$ and $T^m(y)$ belong to U and consequently

$T^n(y) = T^m(y) = x$. Thus, the orbit of x should be $\{x, T(x), \ldots, T^{m-n-1}(x)\}$. This would imply $X = \{x, T(x), \ldots, T^{m-n-1}(x)\}$, a contradiction.

1.11 The map $n \mapsto T^n$ defines an $\mathbb{N}$-dynamical system that is a $\mathbb{Z}$-dynamical system if (X, T) is invertible.

1.12 The distance $d(Tx, Tx')$ can be made arbitrary small. The same is true of the distance $d(T_{h_x}y, T_{h_{x'}}y')$ since, by hypothesis, we have $T_{h_x} = T_{h_{x'}}$ for x, x' close enough.

Assume now that (X, T) is a shift space on the alphabet A and Y is finite. Up to a conjugacy, we may assume that $x \mapsto h_x$ depends only on x_0, where $x = (x_n)_n$. Set $\alpha_{x_0} = T_{h_x}$. Then the map $(x, y) \mapsto (x_n, \alpha_{x_n}y)_{n \in \mathbb{Z}}$ is an injective morphism from $(X \times Y, U)$ into the full shift $(A \times Y)^{\mathbb{Z}}$.

Section 1.2

1.13 Assume that w is a primitive word of length n and that $w = uv = vu$. Then for every $k \geq 1$ we have $w^k = u^k v^k$, as shown easily by induction on k. This implies that $w^\omega = u^\omega$ and thus $w = u$.

1.14 Assume $p < q$. We use induction on $p+q$. The result is trivial if $p = d$. Otherwise, let u be the prefix of length $q - d$ of w. Then, for $1 \leq i \leq p - d$, we have $u_i = u_{i+p} = u_{i+p-q}$. This shows that u has period $q - p$ and p. By induction hypothesis, u has period d. Since $|u| \geq p$, the word w has also period d.

1.15 Let $u = a_0a_1 \cdots a_{n+c-1}$ be a factor of x of length $n + c$ whose factors of length n are all conservative. Set $x = vuy$ and for $0 \leq i \leq c$, set $p_i = a_ia_{i+1} \cdots a_{i+n-1}$. Since there are only c conservative factors of x of length n, there are indices i, j with $0 \leq i < j \leq c$ such that $p_i = p_j$. Then $x = va_0 \cdots a_{i-1}(p_ia_{i+n} \cdots a_{j+n-1})^\omega$.

1.16 Let X be the set of $x \in A^{\mathbb{Z}}$ with all its factors in L. Clearly X is the largest shift space such that $\mathcal{L}(X) \subset L$. For $u \in L$, the language L being extendable, there are sequences $(a_n)_n \in A^{\mathbb{N}}$ and $(b_n)_n \in A^{\mathbb{N}}$ such that $a_n \cdots a_1ub_1b_2 \cdots b_n \in L$ for every $n \geq 0$. Then $\cdots a_2a_1 \cdot ub_1b_2 \cdots$ is in X and thus $u \in \mathcal{L}(X)$.

1.17 Let $\varphi: a \mapsto ab, b \mapsto a$ be the Fibonacci morphism and let x be the Fibonacci word. Let F_n be the Fibonacci sequence defined by $F_0 = 0$, $F_1 = 1$

and $F_{n+1} = F_n + F_{n-1}$ for $n \geq 1$. Note that $F_n \leq F_{n+1}$ implies $F_{n+1} \leq 2F_n$ and $F_{n+2} \leq 3F_n$ for $n \geq 1$.

For every $n \geq 1$, we have $\varphi^{n+1}(a) = \varphi^n(a)\varphi^n(b) = \varphi^n(a)\varphi^{n-1}(a)$. Thus, $|\varphi^n(a)| = F_{n+2}$ and $|\varphi^n(b)| = F_{n+1}$.

Let $w \in \mathcal{L}(x)$ and let n be the least integer such that $|w| < F_{n+1}$. By the choice of n, we have $F_n \leq |w|$.

Since w has no factor in $\varphi^n(A)$, it is a factor of $\varphi^n(A^2)$. But the largest difference between the occurrences of a word of length 2 in the Fibonacci word is 5 (this bound is reached by aa in $aababaa$). Thus, two occurrences of w in x are separated by at most $5F_{n+2} \leq 15F_n \leq 15|w|$. This shows that x is linearly recurrent.

1.18 Let $f : \mathcal{L}_{k+\ell+1}(X) \to \mathcal{L}_1(X')$ be a block map that defines a factor map from X onto X'. Then f defines for all $n \geq 1$ a surjective map from $\mathcal{L}_{n+k+\ell}(X)$ onto $\mathcal{L}_n(X')$ and thus $p_n(X') \leq p_{n+k+\ell}(X)$, whence the result with $t = k + \ell$. If $p_n(X) \leq rn + s$, then $p_n(X') \leq r(n + t) + s = rn + s'$ with $s' = t + s$.

1.19 Assume that $(T^i x)^+ = (T^j y)^+$ and that $(T^k y)^+ = (T^\ell z)^+$. If $j \leq k$, we have $(T^{i+k-j}x)^+ = (T^\ell z)^+$ and otherwise we have $(T^i x)^+ = (T^{\ell+j-k}z)^+$. Thus, the relation is transitive.

Let $(x_1, y_1), \ldots, (x_\ell, y_\ell)$ be ℓ pairs of distinct elements of X belonging to right-asymptotic components $C_1, \ldots, C_\ell$ such that for all $1 \leq i \leq \ell$, one has $x_i^+ = y_i^+$ and $(x_i)_{-1} \neq (y_i)_{-1}$. For n large enough, the prefixes of length n of the x_i^+ are ℓ distinct left-special words and thus $\ell \leq s_n(X)$ since by Proposition 1.2.17 the number of left-special words is bounded by $s_n(X)$. This shows that the number of right-asymptotic components is finite and bounded by k.

Finally, a conjugacy preserves asymptotic equivalence and thus the number of nontrivial components is an invariant.

Section 1.3

1.20 Assume first that (X, S) is a shift of finite type defined by a finite set I of forbidden blocks. Let n be the maximal length of the words of I. It is clear that every $v \in \mathcal{L}_n(X)$ satisfies (1.41).

Conversely, consider the Rauzy graph $G = \Gamma_n(X)$. By condition (1.41), the label of every every infinite path in G is in X. Thus, the edge shift on G can be identified with (X, S).

1.21 Let $\varphi \colon X \to Y$ be a sliding block code from (X, S) to a shift of finite type (Y, S), which is a conjugacy. We may suppose that φ, φ^{-1} are defined by block maps f, g with memory and anticipation ℓ. Let k be the integer such that every word in $\mathcal{L}_k(Y)$ satisfies (1.41). Set $m = k + 4\ell$. Then, one may verify that every word in $\mathcal{L}_m(X)$ satisfies (1.41). Thus, (X, S) is a shift of finite type.

1.22 Let $I \subset A^*$ be a finite set of forbidden blocks and let (X, S) be the shift of finite type corresponding to I. Let n be the maximal length of the words in I and consider the Rauzy graph $\Gamma_n(X)$. Then the edge shift on $\Gamma_n(X)$ is the nth higher block presentation of (X, S). If (X, S) is recurrent, then $\Gamma_n(X)$ is strongly connected.

1.23 Suppose first that X is sofic and let $G = (V, E)$ be a finite labeled graph such that $X = X_G$. The one-block map assigning to an edge its label is a factor map from the edge shift on G onto X. Thus, X is a factor of a shift of finite type.

Conversely, let Y be a shift of finite type, which may be assumed to be an edge shift. Let X be the image of Y by a sliding block code $f \colon \mathcal{L}_{m+n-1}(X)$ with memory m and anticipation n. Consider the $n + m - 1$th higher block presentation Z of Y. Then Z is conjugate to Y and thus is a shift of finite type by Exercise 1.21. Let G be the Rauzy graph $\Gamma_{m+n-1}(Y)$. Then Y can be identified with the edge shift on G. Let H be the graph that is the same as G but with the labeling defined by the block map f. Clearly $X = X_H$ and thus X is a sofic shift.

1.24 Let $X = X_G$ be a sofic shift, where $G = (V, E)$ is a finite graph with labels in A. Let $H = (W, F)$ be the following graph. The set W is the set $\mathfrak{P}(V)$ of subsets of V. For $P, Q \subset V$, there is an edge from P to Q labeled a if $Q = \{q \in V \mid$ there is an edge $p \overset{a}{\to} q$ with $p \in P\}$. Then H is right-resolving and it is easy to see that $X = X_H$.

Assume now that X is an irreducible sofic shift. Any minimal right-resolving presentation of X is clearly strongly connected.

Set $L = \mathcal{L}(X)$. Let M be the following labeled graph. Its vertices are the *follower sets*

$$F(u) = \{v \in L \mid uv \in L\}$$

for $u \in L$. There is an edge from p to q labeled a if $p = F(u)$ and $q = F(ua)$. Let us show that any strongly connected right-resolving minimal presentation $G = (Q, E)$ of X can be identified with M. For every $p \in Q$, let $F(p)$ be the set of all $u \in L$ such that there is a path with label u starting at p. Define an

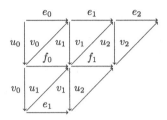

Figure A.1 The conjugacies σ and τ.

equivalence on Q by $p \sim q$ if $F(p) = F(q)$. The quotient $Q/\sim$ is clearly again a right-resolving presentation of X. Since G is minimal, the equivalence $\sim$ is the equality.

Consider, for $u \in L$, the set $I(u)$ of all $q \in Q$ such that there is a path in G with label u ending at q. Clearly, if $I(u) = I(v)$, then $F(u) = F(v)$. Take $u \in L$ such that $I(u)$ is of minimal cardinality. For every $p, q \in I(u)$, we have $p \sim q$. Thus, $I(u)$ has only one element. Since G is strongly connected, every element of Q appears in this way. This allows us to associate to every $p \in Q$ the follower set $F(u)$, where $u \in L$ is such that $I(u) = \{p\}$. This identifies G with M.

1.25 It is enough to consider the case of two elementary equivalent matrices $M = UV$ and $N = VU$. Let $G_M = (V_M, E_M)$ and $G_N = (V_N, E_N)$ be graphs with matrices M, N respectively. Let also G_U be the graph on $V_M \cup V_N$ having U_{xy} edges from $x \in V_M$ to $y \in V_N$ and similarly for G_V.

Since $M = UV$, there is a bijection $e \mapsto (u(e), v(e))$ from E_M onto the paths of length 2 made of an edge of G_U followed by an edge of G_V. We denote $e(u, v)$ the inverse map. Similarly, we have a bijection $f \mapsto (v(f), u(f))$ from E_N onto the paths formed of an edge of G_V followed by an edge of G_U with an inverse map denoted $f(v, u)$.

We define a 2-block map s (with memory 0) from $\mathcal{L}_2(X_M)$ to E_N by $s(e_0 e_1) = f(v(e_0)u(e_1))$. Let $\sigma : X_M \to X_N$ be the sliding block code defined by s. Similarly, let t be the 2-block map (with memory 0) from $\mathcal{L}_2(X_N)$ to E_M defined by $t(f_0 f_1) = e(u(f_0)v(f_1))$. Let $\tau : X_N \to X_M$ be the corresponding sliding block code. We have (see Figure A.1) $\tau \circ \sigma = S_M$, where S_M is the shift transformation on X_M.

Section 1.4

1.26 (i) $\Rightarrow$ (ii) is clear since $\mathcal{L}(\sigma) = \mathcal{L}(X)$ implies that $\mathcal{L}(\sigma)$ is extendable. (ii) $\Rightarrow$ (iii) is obvious.

(iii) $\Rightarrow$ (ii). Consider $u \in \mathcal{L}(\sigma)$. Let $a \in A$ and $n \geq 1$ be such that u is a factor of $\sigma^n(a)$. Set $\sigma^n(a) = pus$ with $p, s \in A^*$. Let $b, c \in A$ be such that $bac \in \mathcal{L}(X)$ and let $m \geq 1$, $d \in A$ be such that bac is a factor of $\sigma^m(d)$. Then $\sigma^n(b)pus\sigma^n(c)$ is a factor of $\sigma^{n+m}(d)$. Since σ is nonerasing, we have $euf \in \mathcal{L}(\sigma)$, where e is the last letter of $\sigma^n(b)p$ and f is the first letter of $s\sigma^n(c)$. This shows that $\mathcal{L}(\sigma)$ is extendable.

(ii) $\Rightarrow$ (i). Let $u \in \mathcal{L}(\sigma)$. An induction on n proves that there exists a sequence (a_n, b_n) of letters such that $a_n \cdots a_0 u b_0 \cdots b_n \in \mathcal{L}(\sigma)$ for every $n \geq 0$. Then $x = \cdots a_n \cdots a_0 u b_0 \cdots b_n \cdots$ is in X and thus $\mathcal{L}(\sigma) \subset \mathcal{L}(X)$. The other inclusion holds for any morphism.

1.27 Assume that σ is primitive. Let $m \geq 1$ be such that $|\sigma^m(b)|_a > 0$ for every $a, b \in A$. Consider $w \in \mathcal{L}(\sigma)$. Let $p \geq 1$ and $a \in A$ be such that w is a factor of $\sigma^p(a)$. Let $q \geq m$ be such that $p + q$ is a multiple of n. Then w is a factor of $\sigma^{p+q}(a)$ and thus w is in $\mathcal{L}(\sigma^n)$.

Let $\sigma: a \mapsto bb, b \mapsto c, c \mapsto c$. Then bb is in $\mathcal{L}(\sigma)$ although it is not in $\mathcal{L}(\sigma^2)$.

1.28 Let $\tau: 0 \mapsto 01, 1 \mapsto 10$ be the Thue–Morse morphism. The bispecial words of length $n \geq 2$ are of length 2^k or $2^k + 2^{k-1}$. Indeed, there are two bispecial words of length 2^k that are $\tau^k(01)$ and $\tau^k(10)$ and two bispecial words of length $2^k + 2^{k-1}$ that are $\tau^k(010)$ and $\tau^k(101)$. Their multiplicities are $m(\tau^k(01)) = m(\tau^k(10)) = 1$ while $m(\tau^k(010)) = m(\tau^k(101)) = -1$. Thus, by Proposition 1.2.17, the sequence $b_n(X)$ defined by (1.6) is given for $n \geq 2$ by

$$b_n(X) = \begin{cases} 2 & \text{if } n = 2^k \\ -2 & \text{if } n = 2^k + 2^{k-1} \\ 0 & \text{otherwise.} \end{cases}$$

This shows that for $n \geq 2$,

$$s_{n+1}(X) = \begin{cases} 4 & \text{if } 2^k \leq n < 2^k + 2^{k-1} \\ 2 & \text{otherwise,} \end{cases}$$

whence the formula for $p_{n+1}(X)$.

1.29 Let $\bar{a} = b$ and $\bar{b} = a$. With this notation, the Thue–Morse morphism is defined by $\tau(x) = x\bar{x}$ for every $x \in \{a, b\}$. The property is true for $n = 0, 1$. Next, since $\tau(x) = x$, and $\tau(x_0 \cdots x_{n-1}) = x_0 \cdots x_{2n-1}$, we

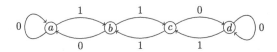

Figure A.2 The automaton of the Rudin–Shapiro sequence.

have $\tau(x_n) = x_{2n}x_{2n+1}$. Thus, $x_{2n} = x_n$ and $x_{2n+1} = \bar{x}_n$. This proves the property by induction on n.

1.30 Let $\tau: a \mapsto ab, b \mapsto ba$ be the Thue–Morse morphism. The left-special words in $\mathcal{L}(\tau)$ are the prefixes of $\tau^n(a)$ and $\tau^n(b)$. Thus, there are two right-asymptotic components. The first one is formed of the pairs (x, y), where x is in the orbit of $\tau^{2\omega}(a \cdots a)$ and y in the orbit of $\tau^{2\omega}(a \cdots a)$. The second one of the pairs (z, t), where z is in the orbit of $\tau^{2\omega}(a \cdots b)$ and y in the orbit of $\tau^{2\omega}(b \cdots b)$.

1.31 Consider the labeled graph represented in Figure A.2.

Let q_n be the vertex reached from a following a path labeled by the binary representation $b(n)$ of n. It is easy to verify by induction on the length of $b(n)$ that one has $x_n = q_n$ and

$$\varepsilon_n = \begin{cases} 1 & \text{if } q_n \in \{a, b\} \\ -1 & \text{otherwise.} \end{cases}$$

Thus, we conclude that $\varepsilon_n = \phi(x_n)$.

1.32 Set $X = X(\sigma)$. We have the initial values

n	0	1	2	3	4	5
$p_n(X)$	1	4	8	16	24	32
$s_n(X)$	3	4	8	8	8	
$b_n(X)$	1	4	0	0	0	

The eight words of $\mathcal{L}_2(X)$ are bispecial, four of them with multiplicity 1 and the others with multiplicity -1. All bispecial words for $n \geq 4$ are the images by the powers of σ of these eight words and thus $b_n(X) = 0$ for $n \geq 3$. This implies that $s_n(X) = 8$ for $n \geq 2$ and thus the result for the complexity of X. The morphism $\phi: a \mapsto 1, b \mapsto 1, c \mapsto -1, d \mapsto -1$ is injective on $\mathcal{L}_n(X)$ for $n \geq 8$ and thus the complexity of the binary Rudin–Shapiro sequence x is the same for $n \geq 8$.

1.33 Since $000, 010 \in \mathcal{L}(\sigma)$, the letters $0, 1$ are extendable in $\mathcal{L}(\sigma)$. Thus, it follows from Exercise 1.26 that σ is a substitution. Set $v_n = \sigma^n(0)$. Then $v_{n+1} = v_n v_n 1 v_n$. Any $u \in \mathcal{L}(\sigma)$ is a factor of some v_n. Thus, for every $n \geq 1$ there exists N such that every word of $\mathcal{L}_n(\sigma)$ is a factor of v_N and thus an integer M such that every word in $\mathcal{L}_n(\sigma)$ is a factor of every word in $\mathcal{L}_M(\sigma)$.

1.34 Since σ is growing, some power σ^n of σ is prolongable on some $a \in A$. Set $x = \sigma^{n\omega}(a)$. Since $X = X(\sigma)$ is minimal, $X(\sigma)$ is generated by x. Since σ is a substitution, $\mathcal{L}(\sigma) = \mathcal{L}(X) = \mathcal{L}(x)$. Thus, every $b \in A$ appears in x. Since X is minimal, x is uniformly recurrent. Since σ is growing, for every $b \in A$ there is some $n \geq 1$ such that a appears in $\sigma^n(b)$. Since σ is prolongable on a, the letter a also appears in all $\sigma^m(b)$ for $m \geq n$. Thus, there is an N such that a appears in all $\sigma^N(b)$ for $b \in A$. By minimality again, there is a $k \geq 1$ such that every $c \in A$ appears in $\sigma^k(a)$. Then every c appears in every $\sigma^{N+k}(b)$.

1.35 Set $M = M(\sigma)$. Since $\begin{bmatrix} 1 & 1 & \cdots & 1 \end{bmatrix} M = n \begin{bmatrix} 1 & 1 & \cdots & 1 \end{bmatrix}$, and since λ_M is, by the Perron–Frobenius Theorem (Theorem B.3.1), the only eigenvalue with a nonnegative eigenvector, we have $\lambda_M = n$.

1.36 If $M(\sigma)$ is reducible, it has, up to a permutation of the indices, a block triangular decomposition and its characteristic polynomial is reducible. Since the characteristic polynomial of $M(\sigma)$ is irreducible, the matrix $M(\sigma)$ is irreducible. Next, since the dominant eigenvalue is a Pisot number, there are no other eigenvalues of the same modulus. This implies, by the Perron–Frobenius Theorem (Theorem B.3.1) that $M(\sigma)$ is primitive.

1.37 We use an induction on $\ell(\sigma)$. If $\ell(\sigma) = 0$, set $B = \sigma(A)$. Let α be the identity on B and let $\beta = \sigma$. All conditions are clearly satisfied.

Assume now the statement true for $\ell < \ell(\sigma)$. Since σ is not injective on $A^{\mathbb{N}}$, we have $\sigma(a_0 a_1 \cdots) = \sigma(a_0' a_1' \cdots)$ for some $a_i, a_i' \in A$ with $a_0 \neq a_0'$. We can assume that $\sigma(a_0)$ is a prefix of $\sigma(a_0')$. Set $\sigma(a_0') = \sigma(a_0)v$. If v is empty, set $B = A \setminus \{a_0\}$. Let α be the restriction of σ to B and let β be defined by $\beta(a_0') = a_0$ and $\beta(a) = a$ for $a \neq a_0'$. Clearly $\sigma = \alpha \circ \beta$ and all conditions are satisfied.

Finally, assume the v is not empty. Define $\alpha_1 : A^* \to C^*$ by $\alpha_1(a_0') = v$ and $\alpha_1(a) = \sigma(a)$ for $a \neq a_0'$. Next, define $\beta_1 : A^* \to A^*$ by $\beta_1(a_0') = a_0 a_0'$ and $\beta_1(a) = a$ for $a \neq a_0'$. Then $\sigma = \alpha_1 \circ \beta_1$ since

$$\alpha_1 \circ \beta_1(a_0') = \alpha_1(a_0 a_0') = \sigma(a_0)v = \sigma(a_0')$$

and $\alpha_1 \circ \beta_1(a) = \alpha_1(a) = \sigma(a)$ if $a \neq a_0'$. The morphism β_1 is injective on $A^{\mathbb{N}}$ because no word in $\beta_1(A)$ begins with a_0'. Thus, α_1 is not injective on $A^{\mathbb{N}}$. By Proposition 1.4.1, we have $\ell(\alpha_1) < \ell(\sigma)$. By induction hypothesis, we have a decomposition $\alpha_1 = \alpha_2 \circ \beta_2$ for $\beta_2 \colon A^* \to B^*$ and $\alpha_2 \colon B^* \to C^*$ with $\mathrm{Card}(B) < \mathrm{Card}(A)$, the morphism α_2 being injective on $B^{\mathbb{N}}$ and every letter $b \in B$ appearing as initial letter in the word $\beta_2(a)$ for some $a \in A$.

Set $\beta = \beta_2 \circ \beta_1$. Then $\sigma = \alpha_1 \circ \beta_1 = \sigma_2 \circ \beta_2 \circ \beta_1 = \alpha_2 \circ \beta$. The decomposition $\sigma = \alpha_2 \circ \beta$ satisfies all the required conditions. Indeed, every $b \in B$ appears as the first letter of some $\beta_2(a)$ with $a \in A$. If $a \neq a_0'$, we have $\beta_1(a) = a$ and thus b is the first letter of $\beta(a)$. Suppose next that $a = a_0'$. Since $\sigma(a_0 a_1 \cdots) = \sigma(a_0' a_1' \cdots)$ and since α_2 is injective on $B^{\mathbb{N}}$, we have $\beta(a_0 a_1 \cdots) = \beta(a_0' a_1' \cdots)$. Since $\beta_1(a_0) = a_0$ and $\beta_1(a_0') = a_0 a_0'$, we obtain $\beta_2(a_0)\beta(a_1 \cdots) = \beta_2(a_0 a_0')\beta(a_1' \cdots)$ and thus

$$\beta(a_1 \cdots) = \beta_2(a_0')\beta(a_1' \cdots)$$

showing that b is the initial letter of $\beta(a_1)$.

1.38 Let $p(n)$ be the number of factors of length n of x. Let us show that if $p(n) < p(n + 1)$, then there is an $m > n$ such that $p(m) < p(m + 1)$. Indeed if $p(n) < p(n + 1)$, there is a factor u of length n of x that is right-special, that is, there are two distinct letters a, b such that $ua, ub \in \mathcal{L}(x)$. Since φ is indecomposable, it is injective on $A^{\mathbb{N}}$ (Exercise 1.37). Thus, there are some v, w such that $\varphi(av) \neq \varphi(bw)$. If $|\varphi(u)| > |u|$ or if $\varphi(a), \varphi(b)$ begin by the same letter, the longest common prefix of $\varphi(uav), \varphi(ubw)$ is a right-special word of length $m > n$. Otherwise we may replace u by $\varphi(u)$. Since φ is primitive, the second case can happen only a bounded number of times. If x is periodic, then $p(n)$ is bounded and by the preceding argument, no letter can be right-special, which implies that the period of x is at most $\mathrm{Card}(A)$.

1.39 We use an induction on $\mathrm{Card}(A)$. If $\mathrm{Card}(A) = 1$, the result is true since the period is 1. Otherwise, either φ is indecomposable and thus injective by Exercise 1.37. Then, by Exercise 1.38, the period of x is at most equal to $\mathrm{Card}(A)$. Finally, assume that $\varphi = \alpha \circ \beta$ with $\beta \colon A^* \to B^*$ and $\alpha \colon B^* \to A^*$ and $\mathrm{Card}(B) < \mathrm{Card}(A)$. Set $y = \beta(x)$ and $\psi = \beta \circ \alpha$. Then $\psi(y) = \beta \circ \alpha \circ \beta(x) = \beta(x) = y$. Thus, y is a fixed point of ψ. Since ψ is primitive, we may apply the induction hypothesis. Since the period of y is at most $|\psi|^{\mathrm{Card}(B)-1}$, the period of x is at most $|\alpha| \, |\psi|^{\mathrm{Card}(B)-1} \leq |\varphi|^{\mathrm{Card}(A)-1}$.

1.40 Assume first that $M = U^*$, where U is a code. Then, the unique factorization of $u(vw) = (uv)w$ forces $v \in M$. Conversely, let U be the set of nonempty words in M that cannot be written as a product of strictly shorter words in M. Take $x_1 x_2 \cdots x_n = y_1 y_2 \cdots y_m$ with $x_i, y_j \in U$, and $n, m \geq 0$ minimal. By definition of U, we have $n, m \geq 1$. Assume $|x_1| \leq |y_1|$ and set $y_1 = x_1 v$. Then, $x_2 \cdots x_n = v y_1 \cdots y_m$. By (1.42), we have $v \in M$, which forces $v = \varepsilon$ and $x_1 = y_1$, a contradiction with the minimality of n, m.

1.41 The first assertion is clear since, by construction of $\mathcal{A}(U)$, the nonempty simple paths from ω to ω are in bijection with U.

1. Suppose first that U is a code. By definition there can be at most one path from ω to ω labeled w since otherwise w would have two factorizations in words of U. Next, if there are two distinct paths labeled w from p to q, there would be two distinct paths labeled uwv from ω to ω for words u, v labeling paths from ω to p and q to ω respectively, a contradiction. The converse is obvious.

2. The number of paths labeled w from (u, v) to itself is the number of factorizations $w = vxu$ with $x \in U^*$. Assume that U is circular. There cannot be cycles labeled w from ω to ω and also from (u, v) to (u, v) by definition of a circular code. Next, suppose that there are cycles labeled w starting at (u, v) and starting at (u', v'). Set $w = vxu = v'x'u'$. Assuming $|v| < |v'|$, set $v' = vz$ and $t = x'u'$. Then $xu = zt$ implies $ztv = xuv$, which shows that $ztv \in U^*$. But $vzt, ztv \in U^*$ implies $zt, v \in U^*$ by (1.20). Thus, $w = vxu = vzt$ is in U^*, a contradiction. The converse is obvious.

1.42 The first assertion results from the definition of a synchronizing pair. The second one is clear since $\mu(xy)$ is the product of the column of index ω of $\mu(x)$ by the row of index ω of $\mu(y)$.

If $\mu(x), \mu(y)$ have rank one, consider $u, v \in A^*$ such that $uxyv \in U^*$. Since $\mu(x)$ has rank one, it follows from $uxyv, x \in U^*$ that $ux, xyv \in U^*$. Similarly, $uxy, yv \in U^*$. This shows that (x, y) is synchronizing.

1.43 (i) $\Rightarrow$ (ii). Let $\mathcal{A}(U) = (Q, E)$ be the flower automaton of U and let μ be as in Exercise 1.42. Let S be the finite semigroup $\mu(A^+)$. For $m \in S$ write $p \xrightarrow{m} q$ for $m_{p,q} = 1$. Since S is finite, there is for every $m \in S$ an integer $k \geq 1$ such that m^k is idempotent, that is $m^{2k} = m^k$. Let e be such an idempotent. For every $p, q \in Q$ there is, since $e = e^3$ some $r, s \in Q$ such that $p \xrightarrow{e} r \xrightarrow{e} s \xrightarrow{e} q$. By unambiguity, $r = s$, that is, r is a fixed point of m. But an element of S cannot have more than one fixed point by Exercise 1.41 and thus e has rank one.

Let J be the set of elements of S of rank one. Since all idempotents of S are in J, we have $S^n \subset J$ for $n = \mathrm{Card}(S) + 1$. Indeed, if $m \in S^n$ with

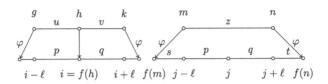

Figure A.3 Alternative definition of recognizability.

$n \geq \operatorname{Card}(S) + 1$, there is a factorization $m = uvw$ with $uv = u$. Then $uv^k = u$ for all $k \geq 1$. When v^k is idempotent, we have $v^k \in J$ and thus $m \in J$. This shows that for every word x of length n, $\mu(x)$ has rank one. We conclude using Exercise 1.42 that U has finite synchronization delay.

(ii) $\Rightarrow$ (iii) is clear by definition of a synchronizing pair.

(iii) $\Rightarrow$ (i). Suppose that $p, q \in A^*$ are such that $pq, qp \in U^*$. Then the sequence $\cdots qp \cdot pqpq \cdots$ has two factorizations unless $p, q \in U^*$. Thus, U is circular.

1.44 As seen in Section 1.4.4, we may consider X^φ as a shift on the alphabet $A^\varphi = \{(a, i) \mid 0 \leq i < |\varphi(a)|\}$. Assume first that φ is injective. For every $y \in \hat{\varphi}(X^\varphi)$, the set $\hat{\varphi}^{-1}(y)$ has at most $(\operatorname{Card}(A^\varphi))^2$ elements. Indeed, assume that $\hat{\varphi}(x) = y$. For $n \geq 1$, set $x_{-n} = (a, i)$ and $x_n = (b, j)$. Then, there are unique $k \geq 0$ and $a_1, \ldots, a_k \in A$ such that

$$\varphi(aa_1 \ldots a_k b) = y_{-n-i} \cdots y_{n+|\varphi(b)|-j}.$$

Thus, the symbols x_{-n} and x_n and the sequence y determine the symbols x_i for $-n \leq i \leq n$. Conversely, if there are distinct w, w' such that $\varphi(w) = \varphi(w')$, then the set $\hat{\varphi}^{-1}(\varphi(w)^\infty)$ is infinite.

1.45 Assume first that φ is recognizable and suppose that $x_{[i-\ell,i+\ell)} = x_{[j-\ell,j+\ell)}$. Set $p = x_{[i-\ell,i)}$ and $q = x_{[i,i+\ell)}$. If $i \in f(\mathbb{N})$, the pair (p, q) is parsable. For ℓ large enough, it is synchronizing. Let $r < s$ be such that $f(r) \leq j - \ell < j + \ell \leq f(s)$. Then pq is a factor of $\varphi(x_{[r,s)})$. Since (p, q) is synchronizing, we have $j \in f(\mathbb{N})$ and $x_{f^{-1}(i)} = x_{f^{-1}(j)}$.

Conversely, let $p, q \in \mathcal{L}_\ell(X)$ be such that the pair (p, q) is (u, v)-parsable with $uv \in \mathcal{L}(X)$. Let $g < h < k$ be such that $u = x_{[g,h)}$ and $v = x_{[h,k)}$ (see Figure A.3 on the left). We have $pq = x_{[i-\ell,i+\ell)}$ and $i = f(h)$. Let $z \in \mathcal{L}(X)$ be such that pq is a factor of $\varphi(z)$. Set $\varphi(z) = spqt$ and $z = x_{[m,n)}$ (see Figure A.3 on the right). Then $pq = x_{[j-\ell,j+\ell)}$ with $j = f(m) + |s| + \ell$. By the hypothesis, we have $j \in f(\mathbb{N})$ and $x_h = x_{f^{-1}(j)}$, which implies that (p, q) is synchronizing.

Section 1.5

1.46 The words $F_n = \varphi^n(a)$ are left-special. Indeed, this is true for $n = 0$ since $aa, ba \in \mathcal{L}(\varphi)$ and $aF_{n+1} = \varphi(bF_n)$, $abF_{n+1} = \varphi(aF_n)$ shows the claim by induction on n. It is easy to see (again by induction) that conversely every left-special word is a prefix of some F_n. This implies that there is exactly one left-special word of each length that is moreover extendable by every letter and is additionally a prefix of x and thus the conclusion that x is standard episturmian.

1.47 Set $p_n(U) = \mathrm{Card}(U \cap \{0, 1\}^n)$. Assume by contradiction that $p_n(U) \geq n + 2$ for some $n \geq 1$. If n is chosen minimal, we have $p_{n-1}(U) \leq n$. Thus, there exist two distinct left-special words u, v of length $n - 1$. Since $u \neq v$, there is a word w such that $w0$ is a prefix of, say u, and $w1$ a prefix of v. Then $0w0, 1w1$ are in U, a contradiction.

1.48 Let u, v be words of minimal length such that $\delta(u, v) \geq 2$. The first letters of u, v are distinct. Assuming that u starts with 0, we have factorizations $u = 0wau'$ and $v = 1wbv'$, where a, b are distinct letters. If $a = 1$ and $b = 0$, then $\delta(u', v') = \delta(u, v)$, a contradiction with the minimality of n. Thus, $a = 0$ and $b = 1$. By minimality again, we found $u = 0w0$ and $v = 1w1$. If w is not palindrome, there is a word z such that za is a prefix of w and $b\tilde{z}$ is a suffix of w with a, b distinct letters. If $a = 0$, then $0z0$ and $1\tilde{z}1$ are in $\mathcal{L}(s)$, a contradiction. Thus, $a = 1$ and $b = 0$. Set $u = 0z1u''$ and $v = v''1\tilde{z}0$. Then $\delta(u'', v'') = \delta(u, v)$, a contradiction again. We conclude that w is a palindrome.

1.49 Let $p_n(x) = \mathrm{Card}(\mathcal{L}_n(x))$ be the complexity of x. If x is aperiodic, then $p_n(x) \geq n + 1$ by the variant of Theorem 1.2.4 for one-sided sequences. Since x is balanced, $p_n(x) \leq n + 1$ by Exercise 1.47. Thus, $p_n(x) = n + 1$ showing that x is Sturmian.

Conversely, assume that x is Sturmian and unbalanced. We prove that x is eventually periodic. By Exercise 1.48, there is a palindrome w such that $0w0, 1w1$ are factors of x. Thus, w is right-special. Set $n = |w| + 1$. Since x is Sturmian there is a unique right-special factor of length $n + 1$, which must be either $0w$ or $1w$. Assume that $0w$ is right-special. Let v be a word of length $n - 1$ such that $u = 1w1v$ is a factor of x. In view of using Exercise 1.15, let us show that all factors of length n of u are conservative. It is enough to show that $0w$ is not a factor of u. Assume the contrary. Then we have $w = s0t$, $v = yz$ and $w = t1y$ (see Figure A.4). Since w is a palindrome, the first equality implies $w = \tilde{t}0\tilde{s}$. The words t and $\tilde{t}$ having the same length,

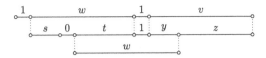

Figure A.4 The words u, v and w.

this implies $t1 = \tilde{t}0$, a contradiction. We conclude that u is a factor of x of length $2n$ in which all factors of length n are conservative. Since x is Sturmian, there are n conservative factors of length n. Thus, x is eventually periodic by Exercise 1.15.

1.50 Let x be a Sturmian sequence. Set $U = \mathcal{L}(x) \cup \widetilde{\mathcal{L}(x)}$. Since U is balanced, we have $\operatorname{Card}(U \cap \{0, 1\}^n) \leq n + 1$ by Exercise 1.47. Since $\operatorname{Card}(\mathcal{L}(x) \cap \{0, 1\}^n) = n + 1$, this forces $\mathcal{L}(x) = \widetilde{\mathcal{L}(x)}$.

1.51 Let X be the shift generated by φ. Let us first verify that $\mathcal{L}(X)$ is closed under reversal. Denote $\rho(w) = \tilde{w}$ the reversal of w. Let $\psi: A^* \to A^*$ be defined by $\psi(w) = \varphi(w)a$. Observe that the map $\tilde{\varphi}$ defined by $\tilde{\varphi}(w) = a^{-1}\varphi(w)a$ is a morphism and that $\rho(\varphi(w)) = \tilde{\varphi}(\tilde{w})$. This implies that $\psi^n(a)$ is a palindrome for every $n \geq 1$ because

$$\rho(\psi^{n+1}(a)) = \rho(\varphi(\psi^n(a)a) = a\tilde{\varphi}(\rho(\psi^n(a)))$$
$$= a\tilde{\varphi}(\psi^n(a)) = \varphi(\psi^n(a))a = \psi^{n+1}(a).$$

Now every $\varphi^n(a)$ is a prefix of $\psi^n(a)$ because $\psi^n(a) = \varphi^n(a)\varphi^{n-1}(a)\ldots\varphi(a)a$ and every $\psi^n(a)$ is a prefix of $\varphi^{n+2}(a)$, as one proves easily by induction. Thus, the set of factors of $\varphi^n(a)$ is also closed by reversal.

Next, we claim that the prefixes of the Tribonacci word are the left special words of $\mathcal{L}(X)$ and that they have three extensions. Indeed, let us verify by induction on n that the words $T_n = \varphi^n(a)$ are left-special with 3 extensions. This is true for $n = 0$ since $aa, ba, ca \in \mathcal{L}(X)$. Next, the equalities

$$aT_{n+1} = \varphi(cT_n), \quad abT_{n+1} = \varphi(aT_n), \quad acT_{n+1} = \varphi(bT_n)$$

prove the property by induction. Conversely, any left-special word is a prefix of some T_n, whence the assertion.

1.52 By Proposition 1.5.1, it is equivalent to prove that s is balanced and aperiodic. Let $u = s_{[n,n+p)}$ and set $h(u) = |u|_1$. Then $h(u) = \lfloor \alpha(n + p) + \rho \rfloor - \lfloor \alpha n + \rho \rfloor$. Thus,

$$\alpha|u| < h(u) < \alpha|u| + 1. \tag{A.1}$$

This implies that $\lfloor \alpha |u| \rfloor \le h(u) \le \lfloor \alpha |u| \rfloor + 1$ and shows that $h(u)$ takes only two consecutive values when u range over the factors of fixed length of s. Thus, s is balanced. Moreover, by (A.1), the real number $\pi(u) = h(u)/|u|$ satisfies

$$|\pi(u) - \alpha| < \frac{1}{|u|}.$$

This proves that $\pi(u)$ tends to α when $|u| \to \infty$ (and thus the last statement) and that s is aperiodic, since otherwise the limit of $\pi(u)$ would be rational.

1.53 Any word $uv^{-1}\tilde{u}$ is palindrome and has u as a prefix. Let us show that the palindromic closure is of this form. Set $u^{(+)} = ur = \tilde{r}\tilde{u}$. We have $|r| < |u|$ since we can choose the last letter of u as palindrome suffix. Set $u = \tilde{r}s$. Then $ur = \tilde{r}\tilde{u}$ implies $s = \tilde{s}$. This proves that $u^{(+)}$ is of the form above and thus corresponds to the longest possible v.

1.54 The formula is clear when $|w|_a = 0$. Otherwise, the longest palindromic suffix of $\mathrm{Pal}(w)a$ is $a\,\mathrm{Pal}(w_1)a$. By Exercise 1.53, we have

$$\mathrm{Pal}(wa) = (\mathrm{Pal}(w)a)^{(+)} = \mathrm{Pal}(w)a(a\,\mathrm{Pal}(w_1)a)^{-1}a\,\mathrm{Pal}(w)$$
$$= \mathrm{Pal}(w)\,\mathrm{Pal}(w_1)^{-1}\,\mathrm{Pal}(w).$$

1.55 This will follow from the fact that for every word u and every letter a, one has the formula $L_a(\tilde{u})a = \widetilde{L_a(u)a}$ easily established by induction on the length of u.

1.56 We use induction on the length of w. The formula is true if w is empty. Otherwise, set $w = vb$. Assume first that $|v|_b = 0$. Then $\mathrm{Pal}(w) = \mathrm{Pal}(v)b\,\mathrm{Pal}(v)$. Next, using the induction hypothesis,

$$L_a(\mathrm{Pal}(w))a = L_a(\mathrm{Pal}(v))b\,\mathrm{Pal}(v))a = L_a(\mathrm{Pal}(v))aa^{-1}L_a(b)L_a(\mathrm{Pal}(v))a$$
$$= \mathrm{Pal}(av)a^{-1}L_a(b)\,\mathrm{Pal}(av)$$
$$= \begin{cases} \mathrm{Pal}(av)b\,\mathrm{Pal}(av) = \mathrm{Pal}(avb) = \mathrm{Pal}(aw) & \text{if } a \neq b \\ \mathrm{Pal}(av)\,\mathrm{Pal}(av) = \mathrm{Pal}(ava) = \mathrm{Pal}(aw) & \text{otherwise.} \end{cases}$$

Now, if $|v|_b > 0$, set $v = v_1 b v_2$ with $|v_2|_b = 0$. By Exercise 1.54, we have $\mathrm{Pal}(w) = \mathrm{Pal}(v)\,\mathrm{Pal}(v_1)^{-1}\,\mathrm{Pal}(v)$. Thus,

$$L_a(\mathrm{Pal}(w))a = L_a(\mathrm{Pal}(v))aa^{-1}L_a(\mathrm{Pal}(v_1))^{-1}L_a(\mathrm{Pal}(v))a$$
$$= \mathrm{Pal}(av)\,\mathrm{Pal}(av_1)^{-1}\,\mathrm{Pal}(av) = \mathrm{Pal}(aw).$$

1.57 We use induction on the length of v. The formula is true for $|v| = 0$. Otherwise, set $v = av_1$. Using Exercise 1.56 and the induction hypothesis, we have

$$\mathrm{Pal}(vw) = \mathrm{Pal}(av_1w) = L_a(\mathrm{Pal}(v_1w))a = L_a(L_{v_1}(\mathrm{Pal}(w))\,\mathrm{Pal}(v_1))a$$
$$= L_{av_1}(\mathrm{Pal}(w))L_a(\mathrm{Pal}(v_1))a = L_v(\mathrm{Pal}(w))\,\mathrm{Pal}(v).$$

1.58 We first observe that for every $a \in A$, there is at most one $w \in \mathcal{R}'_X(u_n)$ that ends with a. Indeed, assume that w, w' are distinct words in $\mathcal{R}'_X(u_n)$ ending with the same letter. Set $w = w_1w_2$ and $w' = w'_1w_2$ with w'_1, w_1 ending with different letters. Then w_2u_n is left-special and thus begins with u_n, a contradiction.

Consider now $w \in \mathcal{R}'_X(u_n)$. Let a be the last letter of w. Since u_n can be followed by a, there is an $m \geq 0$ such that $x_{n+m} = a$. Then $u_{n+m+1} = (u_{n+m}a)^{(+)}$ has a factor $(u_na)^{(+)} = \mathrm{Pal}(x_0x_1 \cdots x_{n-1}a)$. By Justin's Formula (1.31), we have for every $a \in A$,

$$L_{x_0x_1\cdots x_{n-1}}(a)u_n = \mathrm{Pal}(x_0x_1 \cdots x_{n-1}a).$$

Since $\mathrm{Pal}(x_0x_1 \cdots x_{n-1}a) = (u_na)^{(+)}$ is the shortest palindrome with prefix u_na, it begins and ends with u_n with no other occurrence of u_n. Thus, the left return word to u_n ending with a is $L_{x_0x_1\cdots x_{n-1}}(a)$. This proves the inclusion $\mathcal{R}'_X(u_n) \subset \{L_{x_0x_1\cdots x_n}(a) \mid a \in A\}$.

Assume now that s is strict and consider $a \in A$. Since every letter appears infinitely often in x, there is an $m \geq 0$ such that $x_{n+m} = a$. Then $u_{n+m+1} = (u_{n+m}a)^{(+)}$ has a factor $(u_na)^{(+)} = \mathrm{Pal}(x_0x_1 \cdots x_{n-1}a)$. This word is clearly a left return word to u_n. This shows that $L_{x_0x_1\cdots x_{n-1}}(a)$ is in $\mathcal{R}'_X(u_n)$.

The statement concerning $\mathcal{R}'_X(w)$ is clear since u_n is the shortest left special word having w as a suffix.

1.59 The condition is necessary since a strict episturmian shift is closed under reversal. Conversely, let X be a minimal shift space satisfying the condition. Since X is minimal, every $u \in \mathcal{L}(X)$ is a prefix of a right-special word. This implies that every word in $\mathcal{L}(X)$ is a factor of some bispecial word. But the condition implies that bispecial words are palindrome. Indeed, if w is bispecial, then $\tilde{w}$ is also bispecial and thus $w = \tilde{w}$. This shows that $\mathcal{L}(X)$ is closed by reversal.

1.60 1. Set $u_{n+1} = yuy'$ with $|y|$ minimal. Since $u_{n+1} = u_nx_n\tilde{t}$, and since $|yu| \geq |u_nx_n|$, we have $|y'| \leq |t|$. Similarly, since $\tilde{u}$ is not a factor of u_n, we have $|y'u| \geq |u_nx_n|$ and thus $|y| \leq |t|$. Set $t = yz$ and $\tilde{t} = z'y'$. Then

$$u_{n+1} = yuy' = yzwz'y'.$$

2. We have $u = avb = zwz'$. If z, z' are nonempty, then w is a factor of v. Since v is a palindrome prefix of s shorter than u_{n+1}, it is a prefix of u_n and thus w is a factor of u_n. This contradicts the fact that w is unioccurrent in $u_n x_n$.

1.61 By Exercise 1.53, we have $u_{n+1} = u a_n v a_n \tilde{u}$, where $a_n v a_n$ is the longest palindrome suffix of $u_n a_n$ and $u_n = u a_n v$. By definition of u_n, we have $|v| \leq |u_{n-1}|$ with strict inequality if $a_n \neq a_{n-1}$. Thus,

$$|u_{n+1}| = 2|u| + 2 + |v| = 2|u_n| - |v| \geq 2|u_n| - |u_{n-1}|$$

with strict inequality if $a_{n-1} \neq a_n$. Since the sequence $(|u_n|)$ is increasing and its second difference is also nonnegative and not eventually zero, this proves the claim.

1.62 We first prove that the words s_n are primitive. For two words x, y on the alphabet $\{0, 1\}$, denote

$$M(x, y) = \begin{bmatrix} |x|_0 & |x|_1 \\ |y|_0 & |y|_1 \end{bmatrix}.$$

We prove by induction on $n \geq 1$ that $\det M(s_n, s_{n-1}) = 1$. This implies that $|s_n|_0, |s_n|_1$ are relatively prime and thus every s_n is primitive. The equality is true for $n = 1$. Next, for $i \geq 1$, we have

$$M(s_n, s_{n-1}) = \begin{bmatrix} d_n & 1 \\ 1 & 0 \end{bmatrix} M(s_{n-1}, s_{n-2})$$

whence the conclusion.

Let us now show that that every s_n is a prefix of c_α. For this, let

$$h_n = \begin{cases} L_{0^{d_1} 1^{d_2} \dots 1^{d_n}} & \text{if } n \text{ is even} \\ L_{0^{d_1} 1^{d_2} \dots 0^{d_n}} & \text{if } n \text{ is odd.} \end{cases}$$

One can easily verify by induction on $n \geq 1$ that

$$h_{2n-1}(1) = s_{2n-1} = h_{2n}(1),$$
$$h_{2n}(0) = s_{2n} = h_{2n+1}(0).$$

By Justin's Formula, this implies that s_n is a prefix of c_α.

1.63 We show that for $n \geq 3$, one has

$$s_{n-1} s_n = s_n t_{n-1} \text{ with } t_n = s_{n-1}^{d_{n-1}} s_{n-2} s_{n-1}.$$

Indeed,

$$s_{n-1}s_n = s_{n-1}s_{n-1}^{d_n}s_{n-2} = s_{n-1}^{d_n}s_{n-2}^{d_{n-1}}s_{n-3}s_{n-2}$$
$$= s_{n-1}^{d_n}s_{n-2}s_{n-2}^{d_{n-1}-1}s_{n-3}s_{n-2} = s_n t_{n-1}.$$

Observe that t_{n-1} is not a prefix of s_n since otherwise $s_n = t_{n-1}u$ for some word u and $s_{n-1}s_n u = s_n^2$, a contradiction since s_n is primitive by Exercise 1.62. Clearly $s_{n+1}s_n$ is a prefix of the characteristic sequence c_α. Since

$$s_{n+1}s_n = s_n^{d_{n+1}}s_{n-1}s_n = s_n^{d_{n+1}}s_n t_{n-1} = s_n^{d_{n+1}+1}t_{n-1},$$

the word $s_n^{d_{n+1}+1}$ is a prefix of c_α and since t_{n-1} is not a prefix of s_n, the word $s_n^{d_{n+1}+2}$ is not a prefix of c_α.

1.64 It suffices to consider the characteristic Sturmian sequence c_α with $\alpha = [0; 1 + d_1, d_2, \ldots]$. If the sequence of a_i is unbounded, then $s_n^{d_{n+1}}$ is a prefix of c_α and consequently c_α is not d-power free for any d.

1.65 Let x be a two-sided Surmian sequence of slope $\alpha = [a_0; a_1, \ldots]$. Assume that x is linearly recurrent with constant K. Since x is Sturmian it is not periodic. Thus, by Proposition 1.2.9, it is $(K + 1)$-power free. By Exercise 1.64, the coefficients a_i are bounded.

1.66 We have $t' = \lim_{z \to 1_-} \gamma_\alpha(z)$. Thus, $Tt' = \lim_{z \to \alpha_-} \gamma_\alpha(z)$ and thus $t_n = t'_n$ for all $n \geq 1$. Similarly, we have $T^{-1}t' = \lim_{z \to (1-\alpha)_-} \gamma_\alpha(z)$ and thus $t_n = t'_n$ for all $n \leq -2$.

1.67 The result follows from the observation that if the composition matrix of L_u is

$$M_u = \begin{bmatrix} n & n' \\ m & m' \end{bmatrix},$$

the labels of the ancestors of u are m/n and m'/n'. Indeed, for the left son of u, we have

$$M_{u0} = \begin{bmatrix} n & n' \\ m & m' \end{bmatrix}\begin{bmatrix} 1 & 1 \\ 0 & 1 \end{bmatrix} = \begin{bmatrix} n & n+n' \\ m & m+m' \end{bmatrix}$$

and the rule is satisfied since the left ancestors of u and $u0$ are the same while u is the right ancestor of $u0$. The argument for $u1$ is similar.

Section 1.6

1.68 Let X be a nonperiodic Toeplitz shift. Since X is infinite, it contains arbitrary long left-special words (see Section 1.2.3). Let $x, y \in X$ be such that $x_0 \neq y_0$ and $x_n = y_n$ for all $n \geq 1$. If x, y are both Toeplitz sequences, there are p, q such that $x_{kp} = x_0$ and $y_{kq} = y_0$ for all $k \geq 0$. But then $x_{pq} \neq y_{pq}$, a contradiction.

1.69 The formula is true for $n = 0$. Next, $v_2(2n + 2) \equiv v_2(n + 1) + 1 \bmod 2$ and $v_2(2n + 1) = 0$ for all $n \geq 0$.

1.70 Let $\tau : 0 \mapsto 01, 1 \mapsto 10$ be the Thue–Morse morphism and let $x = \tau^\omega(0)$ be the Thue–Morse sequence. Let $\sigma : 0 \mapsto 01, 1 \mapsto 00$ be the period-doubling morphism and let $y = \sigma^\omega(0)$ be the period-doubling sequence. We have to verify that $\gamma(x) = y$ or, equivalently, that $\gamma(x_n x_{n+1}) = y_n$ for every $n \geq 0$. Assume that $x_n = x_{n+1} = 0$. Since x_n is the parity of the number of 1 in the binary expansion $v_2(n)$ of the integer n (Exercise 1.29), we have $v_2(n) = u01^{2i+1}$, where u has an odd number of 1 and $i \geq 0$. Then $v_2(n+1) = u10^{2i+1}$ ends with an odd number of 0 and thus $y(n) = 1 = \gamma(00)$. The verification in the other cases is left to the reader.

1.71 We treat the case of the one-sided shift X generated by σ. The case of the two-sided shift is similar. Observe that the powers of σ also have coincidences. Thus, we can suppose σ has an admissible one-sided fixed point x.

There is a letter a such that, for all letters b, one has $\sigma(b) = u(b)av(b)$, where $b \mapsto |u(b)|$ is constant equal to k. Let $k(N) = \sum_{j=0}^{N} kn^j$. The sequence x being a concatenation of images of σ one has $x_k = x_{k+in}$ for all i. As a consequence, for all N, the word $\sigma^N(a)$ appears in x at the occurrence $k(N) + n^{N+1}i$ for all $i \geq 0$. One can check that $\sigma^N(a)$ is a prefix of $S^{k(N)}x$. Let y be an accumulation point of $(S^{k(N)}x)_N$. From the observation above it is a Toeplitz sequence and thus X is a Toeplitz shift.

1.72 Let $\sigma : 0 \mapsto 01, 1 \mapsto 00$ be the period-doubling morphism and $X = X(\sigma)$. Define $\alpha(u) = \sigma(u)0$ for every word u on 0, 1. The nonempty bispecial words are of the form $\alpha^n(0)$ or $\alpha^n(00)$ for $n \geq 1$. Thus,

$$b_n(X) = \begin{cases} 1 & \text{if } n = 2^k - 1 \\ -1 & \text{if } n = 2^k + 2^{k-1} - 1 \\ 0 & \text{otherwise.} \end{cases}$$

In the first case, the multiplicity is 1 and it is -1 in the second case. This implies that $1 \leq s_n(X) \leq 2$ whence the conclusion.

1.73 Let $\sigma : 0 \mapsto 01, 1 \mapsto 00$ be the period-doubling morphism and let $X = X(\sigma)$. Since σ is recognizable on X, there is a unique map $\pi : X \to \mathbb{Z}_2$ defined for $x \in X$ by $\pi(x) = a + 2\pi(y)$ if $x = S^a \sigma(y)$ with $0 \leq a \leq 1$ and $y \in X$. It is easy to verify that π is a factor map from X onto the odometer $(\mathbb{Z}_2, +1)$.

A.2 Chapter 2

Section 2.1

2.1 The set $S - S$ contains 0 and is closed under addition because $(s - t) + (s' - t') = (s + s') - (t + t')$. It is also closed by taking inverses because $-(s - t) = t - s$. Thus, it is a subgroup of G. Since any subgroup containing S contains $S - S$, the statement is proved.

2.2 If G is simple, consider $u \in G^+$. By Proposition 2.1.7, the set $J = [u] - [u]$ is an ideal such that $J \cap G^+ = [u]$. Thus, $J = G$ and $[u] = G^+$, which implies that u is an order unit. Conversely, if every nonzero element of G^+ is an order unit, consider an ideal J of G not reduced to 0. Let $u \in J^+$ with $u \neq 0$. Since u is an order unit, we have $[u] = G^+$ and thus $J = J^+ - J^+ = G^+ - G^+ = G$.

2.3 We use an induction on n. For $n = 1$, the result is true since a subgroup of $\mathbb{Z}$ is cyclic. Next, assume the result true for $n - 1$ and consider a subgroup H of $\mathbb{Z}^n$. Let $\pi : \mathbb{Z}^n \to \mathbb{Z}^{n-1}$ be the projection on the first $n-1$ components. Thus, $\pi(x_1, x_2, \ldots, x_n) = (x_1, x_2, \ldots, x_{n-1})$. By induction hypothesis, the group $\pi(H)$ is generated by k elements $\pi(h_1), \ldots, \pi(h_k)$ with $k \leq n - 1$. On the other hand, $\ker(\pi)$ is isomorphic to $\mathbb{Z}$ and thus $H \cap \ker(\pi)$ is cyclic. Let h_{k+1} be a generator of $\ker(\pi)$. For every $h \in H$, we have $\pi(h) = \sum_{i=1}^{k} n_i \pi(h_i)$ for some $n_i \in \mathbb{Z}$. Since $\pi(h - \sum_{i=1}^{k} n_i h_i) = 0$, we have $h - \sum_{i=1}^{k} n_i h_i = n_{k+1} h_{k+1}$ and thus $h = \sum_{i=1}^{k+1} n_i h_i$. This shows that H is generated by $h_1, h_2, \ldots, h_{k+1}$.

Section 2.2

2.4 Let g be an infinitesimal element. Let $v \in G^+$ be another order unit. By definition, there is an integer m such that $u \leq mv$. Since g is infinitesimal, we have $mng \leq u$ for any $n \in \mathbb{Z}$ and thus $mng \leq mv$, which implies $ng \leq v$.

2.5 Assume first that g is infinitesimal and consider $\varepsilon \in \mathbb{Q}_+$. Set $\varepsilon = p/q$ with $p, q \geq 0$. We have $qg \leq u \leq pu$ and thus $g \leq \varepsilon u$. Similarly, $qg \leq u$ implies $-pu \leq -u \leq -qg$ and thus $-\varepsilon u \leq g$.

Conversely if $n \geq 0$, $g \leq (1/n)u$ implies $ng \leq u$ and if $n \leq 0$, $-(1/n)u \leq g$ implies also $ng \leq u$.

2.6 Set $H = G/\operatorname{Inf}(G)$ and $H^+ = \{\dot{g} \mid g + h \geq 0 \text{ for some } h \in \operatorname{Inf}(G)\}$. If $\dot{g} \in H^+ \cap (-H^+)$ we have $g + h \geq 0$ and $-g - h' \geq 0$ for some $h, h' \in \operatorname{Inf}(G)$. Then $h - h' \geq 0$. Since $\mathcal{G}$ is simple, $h - h'$ is an order unit although an infinitesimal and thus $h = h'$. We conclude that $g + h = 0$ and thus that $\dot{g} = 0$.

If $g + h > 0$, there is since $\mathcal{G}$ is simple an integer $n > 0$ such that $u \leq n(g + h)$. Then $\dot{u} \leq n\dot{g}$ since $nh \in \operatorname{Inf}(G)$. Thus, (H, H^+) is simple.

Let $g \in G$ be such that $\dot{g} \in \operatorname{Inf}(H)$. Then for every $\varepsilon \in \mathbb{Q}_+$, we have $\dot{g} \leq \varepsilon\dot{u}$ and thus $g + h \leq \varepsilon u$ for some $h \in \operatorname{Inf}(G)$. Since $g - \varepsilon u \leq g + h \leq \varepsilon u$ it follows that $g \leq 2\varepsilon u$. A similar argument shows that $-2\varepsilon u \leq g$. Thus, $g \in \operatorname{Inf}(G)$ and finally $\dot{g} = 0$.

2.7 Note first that since $0u \leq 1g$, we have $f_*(g) \geq 0$. Next, for every $n \geq 0$ and $m > 0$ such that $nu \leq mg$, we have $n/m = p(nu)/m \leq \alpha(g)$ showing that $f_*(g) \leq \alpha(g)$. Next, consider $x \in H$ and $m > 0$ such that $x \leq mg$. Set $x = nu$ with $n \in \mathbb{Z}$. If $n < 0$, then $p(x)/m = n/m < 0 \leq f_*(g)$. Next if $n \geq 0$, we have $p(x)/m = n/m \leq f_*(g)$. Thus, $\alpha(g) \leq f_*(g)$. We conclude that $\alpha(g) = f_*(g)$. The proof that $f^*(g) = \beta(g)$ is similar.

2.8 By Exercise 2.7, we have $f_*(g) = \alpha(g)$ and $f^*(g) = \beta(g)$, where α, β are as in Lemma 2.2.6 with $H = \mathbb{Z}u$. By Lemma 2.2.6, this shows that

$$0 \leq f_*(g) \leq f^*(g) < \infty$$

and that for every state p on $H + \mathbb{Z}g$, one has $f_*(g) \leq p(g) \leq f^*(g)$. By Lemma 2.2.7, this inequality holds for every state p on $\mathcal{G}$.

We finally claim that, if $f_*(g) \leq \gamma \leq f^*(g)$, there is a state $p \in S(\mathcal{G})$ such that $p(g) = \gamma$. Indeed, by Lemma 2.2.6 (3), there is a state r on $H + \mathbb{Z}g$ such that $r(g) = \gamma$. By Lemma 2.2.7, r extends to a state p on $\mathcal{G}$. This proves the claim.

This shows that $\inf\{p(g) \mid p \in S(\mathcal{G})\} = f_*(g)$.

2.9 Let $H = \mathbb{Z}u$. By Lemma 2.2.6, there is a state q on $H + \mathbb{Z}g$ such that $q(g) = \beta(g)$. By Exercise 2.7, we have $\beta(g) = f^*(g) = \inf\{\varepsilon \in \mathbb{Q}_+ \mid g \leq \varepsilon u\}$. By Lemma 2.2.7, there is a state p of $\mathcal{G}$ that extends q. Thus, p is a state of $\mathcal{G}$ such that $p(g) = \inf\{\varepsilon \in \mathbb{Q}_+ \mid g \leq \varepsilon u\}$.

2.10 Since G is directed, there is $j > 0$ such that $z \leq ju$. Set $t = ju - z$ and $f^*(t) = \inf\{\varepsilon \in \mathbb{Q}_+ \mid t \leq \varepsilon u\}$. Since $t \in G^+$, by Exercise 2.9, there is a state p on G such that $p(t) = f^*(t)$. Then $j - f^*(t) = p(ju - t) = p(z) \geq 0$ implies $f^*(t) \leq j$. But since $f^*(t) < j + 1$, we can find $k, n > 0$ such that $nt \leq ku$ with $k/n < j + 1$ and thus $kt \leq (j + 1)nt$. Finally, we obtain

$$kju = kz + kt \leq kz + (j + 1)nt \leq kz + (j + 1)ku,$$

whence $k(z + u) \geq 0$ and the conclusion $z + u \geq 0$ since G is unperforated.

Section 2.3

2.11 It is easy to verify that G is a group with neutral element $[0]$ since the operation is the unique operation on G that extends the operations of the G_n. Consider the map $\pi \colon \cup_{n \geq 0} G_n \to \Delta$, which sends $g \in G_n$ to $\pi(g) = (0, \ldots, 0, g, i_{n+1,n}(g), \ldots)$. Since $\pi^{-1}(\Delta^0) = [0]$, the map π it induces an isomorphism from G onto $\varinjlim G_n$.

2.12 Let (G_n) be a sequence of abelian groups with connecting morphisms $i_{n+1,n} \colon G_n \to G_{n+1}$ and let $\alpha_n \colon G_n \to H$ be for each $n \geq 0$ a morphism such that $\alpha_n = \alpha_{n+1} \circ i_{n+1,n}$. Let $(g_n) \in \Delta$ be such that $g_{m+1} = i_{m+1,m}(g_m)$ for every $m \geq n$. Then $\alpha_{m+1}(g_{m+1}) = \alpha_{m+1} \circ i_{m+1,m}(g_m) = \alpha_n(g_n)$. Thus, there is a morphism $h \colon \Delta \to H$ such that $h(g_0, g_1, \ldots) = \alpha_m(g_m)$ for all $m \geq n$. Since $\Delta^0 \subset \ker h$, the morphism h induces a morphism $\varphi \colon G \to H$ such that $\alpha_n = \varphi \circ i_n$.

2.13 We have seen in Exercise 2.11 that the map π, which sends $g \in G_n$ to $(0, \ldots, 0, g, i_{n+1,n}(g), \ldots)$, induces an isomorphism from G onto the direct limit of the G_n. Since $\pi(g) \in \Delta^+$ if and only if $g \in \cup G_n^+$ and since $\pi(1_0) = (1_n)_{n \geq 0}$, the triple $(G, G^+, \mathbf{1})$ is a unital ordered group isomorphic to the direct limit of the $\mathcal{G}_n$.

2.14 This follows from

$$\mathcal{R}_M = \left\{ \begin{bmatrix} x \\ x \end{bmatrix} \mid x \in \mathbb{R} \right\}.$$

2.15 We have to prove the transitivity. Assume that $(R, S) \colon M \sim_\mathbb{Z} N$ (lag ℓ) and $(T, U) \colon N \sim_\mathbb{Z} P$ (lag k). Then $(RT, US) \colon M \sim_\mathbb{Z} P$ (lag $k + \ell$). Indeed, $MRT = RNT = RTP$, $USM = UNS = PUS$ and

$$M^{\ell+k} = RSM^k = RN^k S = RTUS, \quad P^{\ell+k} = P^\ell UT = UN^\ell T = USRT.$$

2.16 Assume that $(R, S)\colon M \sim_{\mathbb{Z}} N$ (lag ℓ). Let λ be a nonzero eigenvalue of M and let $v \neq 0$ be a corresponding eigenvector. Set $w = Sv$. We have

$$Rw = RSv = M^{\ell} v = \lambda^{\ell} v$$

and thus $w \neq 0$. Next,

$$Nw = NSv = SMv = \lambda Sv = \lambda w.$$

Thus, λ is an eigenvalue of N. The proof that every nonzero eigenvalue of N is an eigenvalue of M is similar.

2.17 First suppose that $(R, S)\colon M \sim_{\mathbb{Z}} N$ (lag ℓ). Denote by m, n the sizes of M, N. Let $\tilde{R}$ and $\tilde{S}$ be the maps defined respectively on $\mathcal{R}_N$ and $\mathcal{R}_M$ by $\tilde{R}(v) = Rv$ and $\tilde{S}(w) = Sw$. It follows from (2.11) and (2.12) that $\tilde{R}$ and $\tilde{S}$ are mutually inverse linear isomorphisms between $\mathcal{R}_M$ and $\mathcal{R}_N$. Suppose that $w \in \Delta_N$ and let $k \geq 1$ be such that $N^k w \in \mathbb{Z}^n$. Then, since R is integral,

$$M^k(Rw) = RN^k w \in \mathbb{Z}^m$$

showing that $\tilde{R}(w) \in \Delta_M$. Thus, $\tilde{R}(\Delta_N) \subset \Delta_M$. Similarly $\tilde{S}(\Delta_M) \subset \Delta_N$. By (2.12), we have the commutative diagrams below and this shows that $(\Delta_M, \delta_M) \simeq (\Delta_N, \delta_N)$.

$$
\begin{array}{ccccc}
\Delta_N & \xrightarrow{\tilde{R}} & \Delta_M & \xrightarrow{\tilde{S}} & \Delta_N \\
\downarrow{\scriptstyle \delta_N} & & \downarrow{\scriptstyle \delta_M} & & \downarrow{\scriptstyle \delta_N} \\
\Delta_N & \xrightarrow{\tilde{R}} & \Delta_M & \xrightarrow{\tilde{S}} & \Delta_N
\end{array}
\qquad \text{(A.2)}
$$

Conversely, suppose that $\theta\colon \Delta_M \to \Delta_N$ is a linear isomorphism such that $\delta_N \circ \theta = \theta \circ \delta_M$. Since $\theta(\Delta_M) \subset \Delta_N$, there is, for every $v \in \mathbb{Z}^m$, a $k \geq 1$ such that $N^k \theta(M^m v) \in \mathbb{Z}^m$. Let S be the matrix of the linear map $v \mapsto N^k \theta(M^m v)$ from $\mathbb{R}^m$ to $\mathbb{R}^n$. Then for every $v \in \mathbb{R}^m$,

$$SMv = N^k \theta(M^{m+1} v) = N^{k+1} \theta(M^m v) = NSv$$

and thus $SM = NS$. Similarly, there is an ℓ such that $M^{\ell} \theta^{-1}(N^n w) \in \mathbb{Z}^n$ for every $w \in \mathbb{Z}^n$. Let R be the matrix of the linear map $w \mapsto M^{\ell} \theta^{-1}(N^n w)$. Then we have as above $MR = RN$. Since finally $RS = M^{\ell} \theta^{-1} N^n N^k \theta M^m = M^{k+\ell+m+n}$, we conclude that M and N and shift equivalent over $\mathbb{Z}$.

2.18 Let $(R, S)\colon M \sim N$ (lag ℓ) with R, S nonnegative. Then the isomorphism $\tilde{S}\colon \Delta_M \to \Delta_N$ defined in the solution of Exercise 2.17 maps Δ_M^+ into Δ_n^+. and similarly $\tilde{R}(\Delta_N^+) \subset \Delta_M^+$. But $\tilde{R} \circ \tilde{S} = \delta_M^{\ell}$ maps Δ_M^+ onto itself, so that $\tilde{S}$ maps Δ_M^+ onto Δ_N^+. Hence $(\Delta_M, \Delta_M^+, \delta_M) \simeq (\Delta_N, \Delta_N^+, \delta_N)$.

Conversely, if $(\Delta_M, \Delta_M^+, \delta_M) \simeq (\Delta_N, \Delta_N^+, \delta_N)$, it is easy to verify that the matrices R, S defined in the solution of Exercise 2.17 are nonnegative.

Section 2.4

2.19 Assume first that (G, G^+) is isomorphic to $\mathbb{Z}^n$ with the usual order. Let $\alpha: \mathbb{Z}^n \to G$ be an isomorphism such that $\alpha(\mathbb{Z}_+^n) = G^+$. Then G^+ is generated by the images of the elementary basis vectors, which form an independent set.

Conversely, assume that $S \subset G^+$ is a finite independent set that generates G^+ as a semigroup. Set $S = \{s_1, s_2, \ldots, s_n\}$ and let $\alpha: \mathbb{Z}^n \to S$ be the linear map sending the ith elementary basis vector to s_i. Since $G = G^+ - G^+$, the map α is surjective. It is injective since S is independent. Finally, $\alpha(\mathbb{Z}_+^n) = G^+$ and thus (G, G^+) is isomorphic to $(\mathbb{Z}^n, \mathbb{Z}_+^n)$.

2.20 Assume first that G satisfies the Riesz interpolation property and consider $x_1, x_2, y_1, y_2 \geq 0$ such that $x_1 + x_2 = y_1 + y_2$. Since $0 \leq x_1 \leq y_1 + y_2$, we have $0, x_1 - y_2 \leq x_1, y_1$ and thus by the interpolation property, there is some z_{11} such that $0, x_1 - y_2 \leq z_{11} \leq x_1, y_1$. Set

$$z_{12} = x_1 - z_{11}, \ z_{21} = y_1 - z_{11}, \ z_{22} = y_2 - z_{12}.$$

These elements are all positive and $x_1 = z_{11} + z_{12}$, $y_1 = z_{11} + z_{21}$, $y_2 = z_{12} + z_{22}$. Finally $z_{21} + z_{22} = y_1 - z_{11} + y_2 - z_{12} = y_1 + y_2 - x_1 = x_2$.

Conversely, assume that G satisfies the decomposition property. By substraction, it is enough to prove the interpolation property for $0, x \leq y_1, y_2$. We then have $0 \leq y_1 \leq y_1 + (y_2 - x) = y_2 + (y_1 - x)$. Let $z \geq 0$ be such that $y_1 + z = y_2 + (y_1 - x)$. By the decomposition property, there are $z_{ij} \geq 0$ such that $y_1 = z_{11} + z_{12}$, $y_2 = z_{11} + z_{21}$ and $y_1 - x = z_{12} + z_{22}$. Then $x = z_{11} - z_{22}$ and thus $0, x \leq z_{11} \leq y_1, y_2$.

2.21 We use an induction on k. The property holds trivially for $k = 1$. Next, assume that $x, y_1, y_2, \ldots, y_{k+1} \in G^+$ satisfy the hypothesis. Since $0, x - y_{k+1} \leq x, y_1 + \cdots + y_k$, there is by Riesz interpolation some $z \in G$ such that $0, x - y_{k+1} \leq z \leq x, y_1 + \cdots + y_k$. By induction hypothesis, there are $x_1, \ldots, x_k$ such that $z = x_1 + \cdots + x_k$ with $x_i \leq y_i$ for $1 \leq i \leq k$. But then $x_1, \ldots, x_k, x_{k+1}$ with $x_{k+1} = x - z$ are a solution.

2.22 Indeed, we have $\beta_i(\eta_i(b_{i-1})) = \alpha_i(b_{i-1}) = \frac{1}{2^{i-1}}$, $\beta_i(\eta_i(a_i)) = \alpha_i(a_i) = \frac{1}{2^i}$ and $\ker(\eta_i) = \ker(\alpha_i) = \mathbb{Z}(b_{i-1} - 2a_i)$.

A.3 Chapter 3

Section 3.1

3.1 We have, extending the composition with T to rational fractions in T,

$$f^{(n)} = f \circ \frac{1 - T^n}{1 - T}.$$

This holds also for $n < 0$ since

$$f \circ \frac{1 - T^{-n}}{1 - T} = -f \circ T^{-n} \circ \frac{1 - T^n}{1 - T} = -f^{(n)}(T^{-n}).$$

The cohomological equation is then easy to verify since

$$f^{(n)} + f^{(m)} \circ T^n = f \circ \frac{1 - T^n}{1 - T} + f \circ \frac{1 - T^m}{1 - T} \circ T^n$$

$$= f \circ \frac{1 - T^n + T^n - T^{n+m}}{1 - T} = f \circ \frac{1 - T^{n+m}}{1 - T}$$

$$= f^{(n+m)}.$$

3.2 We have to verify the transitivity. Assume that f_0, f_1 are connected by f_t and that $g_0 = f_1, g_1$ are connected by g_t. Then f_0, g_1 are connected by h_t, where

$$h_t(x) = \begin{cases} f_{2t}(x) & \text{if } 0 \le t < 1/2 \\ g_{2t-1}(x) & \text{if } 1/2 < t \le 1. \end{cases}$$

Let $f_0, f_1 \colon X \to Y$ be connected by a homotopy f_t. For $g \colon Y \to Z$, the maps $h_0 = g \circ f_0$ and $h_1 = g \circ f_1$ are connected by $h_t = g \circ f_t$. Similarly, if $g_0, g_1 \colon X \to Z$ are connected by g_t, then for $f \colon X \to Y$, the maps $g_0 \circ f$ and $g_1 \circ f$ are connected by $g_t \circ f$. This proves the compatibility with composition.

3.3 Set $f_t(x) = ta + (1 - t)f(x)$.

3.4 First $\tilde{X}$ is compact because it can be identified with the quotient of $X \times [0, 1]$ by the equivalence that identifies $(x, 1)$ and $(Tx, 0)$ for all $x \in X$. As a continuous image of a compact metric space, it is metrizable (see Willard (2004)). The map $(y, t) \mapsto T_t(y)$ is well defined and continuous since $(y, s) \mapsto (y, s + t)$ is continuous from $X \times \mathbb{R}$ to itself.

3.5 Set $X = \{0, 1, \ldots, n - 1\}$ with $T(i) = i + 1 \bmod n$. Then $\tilde{X}$ can be identified with the torus $\mathbb{T}$ by the map $(i, t) \in X \times [0, 1[\mapsto (i + t)/n$.

3.6 Assume that $\varphi\colon (X, S) \to (Y, T)$ is a conjugacy. Then the map $(x, t) \to (\varphi(x), t)$ induces an equivalence from the suspension flow over (X, S) onto the suspension flow over (Y, T). All periodic systems are flow equivalent by Exercise 3.5. Thus, flow equivalence is weaker than conjugacy.

3.7 Consider the map $\pi\colon C(X, \mathbb{Z}) \to C(\tilde{X}, \mathbb{T})$, which associates to $f \in C(X, \mathbb{Z})$ the map $\pi(f) \in C(\tilde{X}, \mathbb{T})$ defined by $\pi(f)(x, s)) = \tau(f(x)s)$.

We first show that π induces a surjective map from $C(X, \mathbb{Z})$ to $H^1(\tilde{X}, \mathbb{Z})$. Let $\tilde{f}\colon \tilde{X} \to \mathbb{T}$ be continuous. By Lemma 3.3.7 there is a continuous function $f\colon X \to \mathbb{R}$ such that $\tilde{f}(x, 0) = \tau(f(x))$. Set for $0 \le s < 1$,

$$h(x, s) = \tau(f(x)(1 - s) + f(Tx)s).$$

Then h is nullhomotopic (by Exercise 3.3 because $f(x)(1 - s) + f(Tx)s$ extends to a continuous map from $X \times [0, 1]$ to $\mathbb{R}$) and $h(x, 0) = \tilde{f}(x, 0)$. Therefore the map $g_1(x, s) = \tilde{f}(x, s)/h(x, s)$ is homotopic to $\tilde{f}$ and $g_1(x, 0) = g_1(Tx, 1) = 0$ for all $x \in X$. For $x \in X$, let $r(x)$ be the number of times the loop $g_1(x, s)$ wraps around $\mathbb{T}$ as s increases from 0 to 1. Then r is continuous and, for each $x \in X$, $g(x, s) = \tau(r(x)s)$ wraps around $\mathbb{T}$ the same number of times as $g_1(x, s)$. Hence g_1 and g are holomorphic. This shows that π induces a surjective map.

Next, let $f, g \in C(X, \mathbb{Z})$ be such that $\pi(f)$ and $\pi(g)$ are in the same homotopy class. We may write for $0 \le s < 1$, $\tau((f(x) - g(x))s) = \tau(r(x, s))$ for some continuous function r from $\tilde{X}$ to $\mathbb{R}$ (because $\tau(r(x, s))$ is nullhomotopic). Then $(f(x) - g(x))s = r(x, s) + P(x, s)$, where $P(x, s) \in \mathbb{Z}$. But since $P(x, s)$ is continuous in s, this forces $P(x, s) = P(x, 0)$ for all s. Set $p(x) = P(x, 0)$. Then

$$p(Tx) = -r(Tx, 0) = -r(x, 1) = p(x) - (f(x) - g(x)),$$

which shows that $f(x) - g(x)$ is a coboundary. Thus, π induces an injective map from $H(X, T, \mathbb{Z})$ to $H^1(\tilde{X}, \mathbb{Z})$.

Section 3.2

3.8 Let g, g' be two solutions. Let x be a recurrent point in X. Then for any $y = T^n x$, we have by (3.2) $g(y) = f^{(n)}(x) - f(x) + g(x)$ and thus $g(y) - g'(y) = g(x) - g'(x)$. Since the positive orbit of x is dense, this shows that g, g' differ by a constant.

3.9 Since $f^{(n+1)}(x) = f(x) + f^{(n)}(Tx)$ for all $n \in \mathbb{Z}$, by Exercise 3.1, we have

$$
\begin{aligned}
g(x) = \sup_{n \in \mathbb{Z}} f^{(n)}(x) &= \sup_{n \in \mathbb{Z}} f^{((n+1))}(x) = \sup_{n \in \mathbb{Z}}(f(x) + f^{(n)}(Tx)) \\
&= f(x) + g(Tx)
\end{aligned}
$$

whence the result.

3.10 We first note that $|f^{(n)}(x)|$ is bounded for all $x \in X$. Indeed, set $M = \sup_{n \geq 0} |f^{(n)}(x_0)|$. If $|f^{(n)}(y)| > 2M$, the same inequality holds for any z sufficiently close to y, in particular for some iterate $T^m(x_0)$. But then $2M < |f^{(n)}(T^m x_0)| \leq |f^{(n+m)}(x_0)| + |f^{(m)}(x_0)|$ contrary to the definition of M.

Thus, $f^{(n)}(x)$ is bounded for all $n \in \mathbb{Z}$ and, by Exercise 3.9, the map $g(x) = \sup_{n \in \mathbb{Z}} f^{(n)}(x)$ is such that $\partial g = f$.

Define the *oscillation* of a real-valued function $h \colon X \to \mathbb{R}$ defined on a metric space X at a point x as

$$
\operatorname{Osc}_h(x) = \lim_{\delta \to 0}(\sup\{h(y) \mid d(x, y) < \delta\} - \inf\{h(y) \mid d(x, y) < \delta\}).
$$

Note that the oscillation of a function h at x vanishes if and only if h is continuous at x. Since f is continuous, we have $\operatorname{Osc}_h(x) = 0$ for $h = \partial g$. This implies $\operatorname{Osc}_{g \circ T}(x) = \operatorname{Osc}_g(x)$ or $\operatorname{Osc}_g \circ T = \operatorname{Osc}_g$ and thus that the function $x \mapsto \operatorname{Osc}_g(x)$ is invariant. For $\varepsilon > 0$, let $O_{\varepsilon,n}$ be the set of $x \in X$ such that $f(x) - f^{(n)}(x) \leq \varepsilon/2$. Since $O_{\varepsilon,n}$ is closed, the set

$$
\{x \in O_{\varepsilon,n} \mid \operatorname{Osc}_g(x) \leq \varepsilon\}
$$

is closed invariant and nonempty it is equal to X. Thus, g is continuous.

3.11 1. Follows from the hypothesis that the sequence $f^{(n)}(x_0)$ is bounded.

2. Assume that $(x, u), (x, v) \in K$ for some $x \in X$ and $u \neq v$. The map S commutes with the vertical translations $T_u \colon (x, t) \mapsto (x, t + u)$. Thus, $T_{u-v} K = \{(x, t + u - v) \mid (x, t) \in K\}$ is also an S-invariant and minimal compact set. It intersects K since $T_{u-v}(x, v) = (x, u)$. By minimality of K, this implies $T_{u-v} K = K$, which contradicts the fact that K is bounded. This shows that K is the graph of a function g. Its domain is X since it is closed and T-invariant. The function g is continuous since K is compact.

3. Since K is S-invariant, we have $S(x, g(x)) = (Tx, g(x) + f(x)) \in K$ for every $x \in X$. Thus, $g(Tx) = g(x) + f(x)$.

Section 3.5

3.12 Suppose that $\chi_{[ab]} = \partial\phi$ with $\phi \in C(X, \mathbb{Z})$. Let n be such that $\phi(x)$ depends only on $x_{[-n,n)}$. Let α be the value of ϕ on $[a^n \cdot a^n]$. We proceed in four steps.

1. For $2 \leq i \leq n$, we have $\phi \circ T(x) = \phi(x)$ and thus ϕ is constant with value α on $[a^n \cdot a]$.
2. Next, since $1 = \phi \circ T(x) - \phi(x)$ for every $x \in [a^n \cdot ab]$, ϕ is constant with value $\alpha + 1$ on $[a^n \cdot b]$.
3. Since $\phi \circ T(x) = \phi(x)$ for every $x \in [a^n \cdot ac]$, we have that ϕ is constant with value α on $[a^n \cdot c]$.
4. Finally, we have $\phi \circ T(x) = \phi(x)$ for every $x \in [a^n \cdot bc]$, which implies that ϕ is constant with value $\alpha + 1$ on $[a^n \cdot c]$, a contradiction.

3.13 To prove that $G_n^+(X) \cap (-G_n^+(X)) = \{0\}$, it is enough to prove that if $\phi \in Z_{n-1}(X)$ is such that $\partial_{n-1}\phi \in Z_n^+(X)$, then $\partial_{n-1}\phi = 0$. Let $u, v \in \mathcal{L}_{n-1}(X)$. Since X is recurrent, there exists $m \geq n$ and $w \in L_m(X)$ such that u is a prefix of w and v is a suffix of w. Then, since $u = w_{[1,n-1]}$ and $v = w_{[m-n+2,m]}$, we have, because the first sum is telescopic,

$$\phi(v) - \phi(u) = \sum_{i=1}^{m-n+1} (\phi(w_{[i+1,i+n-1]}) - \phi(w_{[i,i+n-2]}))$$
$$= \sum_{i=1}^{m-n+1} (\partial_{n-1}\phi)(w_{[i,i+n-1]}) \geq 0.$$

Since this is true for every u, v, it implies that ϕ is constant and thus that $\partial_{n-1}\phi = 0$.

3.14 If $X = \{x, Tx, \ldots, T^{n-1}x\}$ is periodic of period n, then $\chi_{\{x\}} \sim \chi_{\{Tx\}}$ and $\chi_X = n\chi_{\{x\}}$. Thus, $K^0(X, T) = (\mathbb{Z}, \mathbb{Z}_+, n)$. Conversely, let $U_1, \ldots, U_m$ be a partition of X in $m \geq n$ nonempty clopen sets. If the class of χ_{U_i} is identified with the integer α_i, we have $\sum \alpha_i = n$. Thus, $m = n$ and each U_i is reduced to one point, which implies that X is periodic of period n.

Section 3.8

3.15 Let (X, T) be a topological dynamical system. The set $\mathcal{M}_X$ is convex and compact by Theorem B.5.5. The map $T^*: \mathcal{M}_X \to \mathcal{M}_X$ defined by $T^*\mu =$

$\mu \circ T^{-1}$ is a continuous linear map. Thus, by the Markov–Kakutani Theorem, it has a fixed point μ, which is obviously an invariant measure.

3.16 Assume that μ is ergodic. Let U be a Borel set such that $T^{-1}(U) = U \bmod \mu$. Set

$$V = \cap_{n \geq 0} \cup_{i \geq n} T^{-i}(U).$$

It is clear that $T^{-1}V = V$. We claim that $U = V \bmod \mu$.

To prove the claim, note first that for two sets U, V, one has $U = V \bmod \mu$ if and only if $\mu(U \triangle V) = 0$, where $U \triangle V = (U \cup V) \backslash (U \cap V)$ is the *symmetric difference* of U, V. We will use the fact that

$$U \triangle W \subset (U \triangle V) \cup (V \triangle W). \tag{A.3}$$

We first observe that

$$\mu(U \triangle (\cup_{i \leq n-1} T^{-i} U)) = 0. \tag{A.4}$$

Indeed, using (A.3) to telescope the union on the right-hand side, we have

$$T^{-n} U \triangle U \subset \cup_{i=0}^{n-1} (T^{-i-1} U \triangle T^{-i} U) = \cup_{i=0}^{n-1} T^{-i} (T^{-1} U \triangle U)$$

and thus $\mu(T^{-n} U \triangle U) \leq n\mu(T^{-1} U \triangle U) = 0$. This implies that

$$\mu(U \triangle (\cup_{i \geq n} T^{-i} U)) \leq \sum_{i \geq n} \mu(U \triangle T^{-i} U) = 0,$$

which proves (A.4). Now (A.4) implies that $\mu(U \triangle V) = 0$ and thus $\mu(U) = \mu(V)$, proving the claim.

Since μ is ergodic, we have $\mu(V) = 0$ or 1. This shows that (i) implies (ii). The converse is obvious.

3.17 Denote $S(U, V) = \frac{1}{n} \sum_{i=0}^{n-1} \mu(U \cap T^{-i} V)$. Suppose first that (3.26) holds for all clopen sets U, V. Then it holds for all Borel sets U, V. Indeed, for $\varepsilon > 0$, let U_0, V_0 be clopen sets such that $\mu(U \triangle U_0) < \varepsilon$ and $\mu(V \triangle V_0) < \varepsilon$ (they exist by Theorem B.5.1). Then $|S(U_0, V_0) - \mu(U_0)\mu(V_0)| < \varepsilon$ for n large enough. Since $(U \cap T^{-i} V) \triangle (U_0 \cap T^{-i} V_0) \subset (U \triangle U_0) \cup (T^{-i} V \triangle V_0)$, we have $\mu((U \cap T^{-i} V) \triangle (U_0 \cap T^{-i} V_0)) < 2\varepsilon$. Thus, we obtain

$$\begin{aligned}
|S(U, V) - \mu(U)\mu(V)| \leq {} & |S(U, V) - S(U_0, V_0)| + |S(U_0, V_0) - \mu(U_0)\mu(V_0)| \\
& + |\mu(U_0)\mu(V_0) - \mu(U)\mu(V_0)| \\
& + |\mu(U)\mu(V_0) - \mu(U)\mu(V)| \leq 5\varepsilon.
\end{aligned}$$

Then for every invariant Borel set U, we have $\mu(U) = \frac{1}{n} \sum_{i=0}^{n-1} \mu(U) = \frac{1}{n} \sum_{i=0}^{n-1} \mu(U \cap T^{-i} U) = \mu(U)^2$ and thus $\mu(U)$ is 0 or 1.

Conversely, by the ergodic Theorem (Theorem 3.8.5) applied to the characteristic function χ_V of the Borel set V, we have

$$\lim_{n \to \infty} \frac{1}{n} \chi_V^{(n)}(x) = \int \chi_V d\mu = \mu(V)$$

almost everywhere. Multiplying both sides by $\chi_U(x)$ and integrating, we obtain (3.26).

3.18 It is clear that for all Borel sets U, V, if (3.27) holds, then (3.26) also holds. Thus, mixing implies ergodic. To show that it is enough for μ to be mixing that (3.27) holds for every clopen sets U, V, we use the same argument as in Exercise 3.17.

If μ is a Bernoulli measure, then (3.27) holds for all clopen sets U, V and thus μ is mixing, and therefore, a fortiori, ergodic.

3.19 Assume first that P is not irreducible. Let $B \subset A$ be a subset of the alphabet corresponding to a strongly connected component of the graph associated with P (for $a, b \in A$, there is an edge from a to b if and only if $P_{a,b} > 0$). Then $U = B^{\mathbb{Z}}$ is an invariant subset of $A^{\mathbb{Z}}$ such that $0 < \mu(U) < 1$.

Conversely, by Exercise 3.17, it is enough to prove that $\mu(U \cap T^{-n}(V))$ converges in mean to $\mu(U)\mu(V)$ for U, V clopen. Since the family of sets U, V for which this holds is closed under union and translation, it is enough to consider $U = [u]$ and $V = [w]$. Set $u = a_1 \cdots a_s$ and $w = b_1 \cdots b_t$. Then

$$\mu([u] \cap T^{-n}[w]) = v_{a_1} P_{a_1,a_2} \cdots P_{a_{s-1},a_s} (P^{n-s})_{a_s,b_1} P_{b_1,b_2} \cdots P_{b_{t-1},b_t}.$$

By Theorem B.3.1, the sequence P^n converges in mean to the matrix with all rows equal to v. Thus, $\mu([u] \cap T^{-n}[w])$ converges in mean to

$$v_{a_1} P_{a_1,a_2} \cdots P_{a_{s-1},a_s} v_{b_1} P_{b_1,b_2} \cdots P_{b_{t-1},b_t} = \mu([u])\mu([w]).$$

Finally, by Exercise 3.18, μ is mixing if and only if $\mu(U \cap T^{-n}V)$ converges to $\mu(U)\mu(V)$ for all U, V clopen. Again it is enough to consider $U = [u]$ and $V = [v]$ as above. By Theorem B.3.1, the matrix P^n converges to the matrix with all rows equal to v if and only if P is primitive.

3.20 It is clear that for every Borel set U, V, (3.27) $\Rightarrow$ (3.28) $\Rightarrow$ (3.26). Thus, mixing $\Rightarrow$ weakly mixing $\Rightarrow$ ergodic.

3.21 Let $U, V \subset X$ be nonempty open sets. Since μ is mixing, $\mu(U \cap T^{-n}V) \to \mu(U)\mu(V) > 0$. This implies that there is some N such that $\mu(U \cap T^{-n}V) > 0$ for all $n \geq N$.

3.22 Let R be the set of recurrent points in (X, T) (see Exercise 1.3). We show that $\mu(R) = 1$. This proves that (X, T) is recurrent by Exercise 1.4.

Let $(U_n)_{n \geq 1}$ be a countable basis of open sets of X (this exists because X is a metric space). Set

$$V_n = \cap_{k \geq 0} T^{-k}(X \setminus U_n).$$

We have $X \setminus R = \cup_{n \geq 1} V_n$. Since $T^{-1}V_n \subset V_n$ and $\mu(T^{-1}V_n) = \mu(V_n)$, it follows that $T^{-1}V_n = V_n$ mod μ and thus $\mu(V_n) = 0$ or 1 by Proposition 3.8.6 since μ is ergodic. But $U_n \subset X \setminus V_n$ implies $\mu(U_n) \leq \mu(X \setminus V_n)$. By our hypothesis, this implies $\mu(X \setminus V_n) > 0$ and thus $\mu(V_n) = 0$. This gives finally $\mu(R) = 1$.

3.23 By Weierstrass Theorem, the linear combinations of the functions $\chi_m(z) = z^m$ are dense in $C(S^1, \mathbb{C})$. Thus, by Theorem 3.8.12, it is enough to prove that for every $m \geq 0$, the averages $\chi_m^{(n)}$ converge uniformly to a constant. This is trivially true if $m = 0$. Otherwise, the rotation $R_\alpha(z) = \lambda z$ with $\lambda = e^{2i\pi\alpha}$ satisfies

$$\left| \frac{1}{n} \sum_{k=0}^{n-1} \chi_m(R_\alpha^k(z)) \right| = \left| \frac{1}{n} \sum_{k=0}^{n-1} e^{2i\pi km\alpha} \right|$$

$$= \frac{|1 - e^{2i\pi mn\alpha}|}{n|1 - e^{2i\pi m\alpha}|} \leq \frac{2}{n|1 - e^{2i\pi m\alpha}|} \to 0.$$

Every Sturmian shift (X, S) of slope α is, by Proposition 1.5.12, the image by the natural coding γ_α of the rotation $(\mathbb{T}, T_\alpha)$ of angle α. Since γ_α is one-to-one except on a denumerable set (the orbit of 0 under R_α), the invariant Borel probability measures on $(\mathbb{T}, T_\alpha)$ and on (X, S) are exchanged by γ_α. Thus, (X, S) is uniquely ergodic and its unique invariant measure μ is such that $\mu([0]) = 1 - \alpha$, $\mu([1]) = \alpha$.

3.24 Let $U \subset [0, 1]$ be an interval of length $1/4$. There is an infinity of n such that $T^{-n}U \cap U = \emptyset$ and thus $\mu(T^{-n}U \cap U)$ cannot converge to $1/16$.

3.25 It follows from the definition that for $1 \leq j \leq i$, the number of elements of E_j in the interval $0 < n \leq k_i$ is exactly $(k_i/k_j)(2k_{j-1} + 1)$. Hence an upper bound to the number of elements of $E_1 \cup E_2 \cup \cdots \cup E_i$ in the interval $0 < n \leq k_i$ is

$$\sum_{j=1}^{i} \frac{3k_i k_{j-1}}{k_j} \leq 3k_i \sum_{j \geq 1} \frac{k_{j-1}}{k_j} < \frac{1}{4}k_i.$$

It follows that $p(n) = i + 1$ for at least 3/4 of the numbers n in the interval $0 < n \leq k_i$. This implies that

$$\left| \frac{1}{k_i} \sum_{n=1}^{k_i} x(n) - \frac{1}{k_{i+1}} \sum_{n=1}^{k_{i+1}} x(n) \right| \geq \frac{1}{2},$$

and thus the frequency of 1 is not defined.

3.26 Since T is continuous, it is measurable and since μ is invariant, we have $\mu(T^{-1}V) = \mu(V)$ for every borel subset V of X. Thus, T preserves μ.

3.27 Set $\langle f, g \rangle = \int fg d\mu$. We have, by change of variable (Eq. (B.13)), for every $f, g \in L^2(X)$,

$$\langle U^*Uf, g \rangle = \langle Uf, Ug \rangle = \int (Uf)(Ug)d\mu = \int U(fg)d\mu = \int fg d\mu = \langle f, g \rangle.$$

Thus, $U^*U = I$. Since T is invertible, U is surjective. Let V be its right inverse, that is, such that $UV = I$. Then $V = U^*UV = U^*$. Thus, $UU^* = I$. This shows that U is unitary.

3.28 If $f = Ug - g$ with $g \in H$, then $||f^{(n)}|| = ||U^n g - g|| \leq 2||g||$ and thus the sequence is bounded.

Conversely, assume that $||f^{(n)}|| \leq k$ for $n \geq 1$. Let S be the set of all convex linear combinations of $U^n f$ and let $\bar{S}$ be its closure in the weak topology of H. Then $\bar{S}$ is convex and as it is contained in the weakly compact set $\{h \in H \mid ||h|| \leq k\}$ it is weakly compact. Moreover $\bar{S}$ is invariant under the continuous affine map $h \mapsto f + Uh$. By the Schauder–Tychonoff Theorem this map has a fixed point in $\bar{S}$, that is, there is $g \in \bar{S}$ such that $g = f + Ug$ and thus $f = Ug - g$ is the coboundary of $g \in H$.

3.29 By the choice of p, the value of φ_k^p is determined by the two first letters $a_0 a_1$ of $x = a_0 a_1 \cdots a_{k-1}$. Let $\alpha: \mathcal{L}_2(X)^* \to \mathcal{L}_k(X)^*$ be the morphism defined by $\alpha(a_0 a_1) = \varphi_k^p(x)$. Then U is the incidence matrix of α. Since obviously $\varphi_k \circ \alpha = \alpha \circ \varphi_2$, we obtain $U M_k = M_2 U$. Since $v_k M_k = v_2 U M_k = v_2 M_2 U = \rho v_2 U = \rho v_k$ the last assertion follows.

3.30 We have $\mu^3(aa) = abbabaab \cdot abbabaab$, whence the value of the first row of U and similarly for the others.

3.31 The equalities

$$M_2 U = U M_k, \quad V M_2 = M_k V, \quad \text{and } M_2^p = UV, \quad M_k^p = VU$$

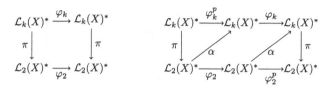

Figure A.5 A commutative diagram.

follow from the commutative diagram of Figure A.5. The diagram on the left gives the equality $V M_2 = M_k V$ (see diagram (1.26)). The other three equalities follow from the diagram on the right.

Section 3.9

3.32 Let us show that there is no decomposition $\chi_{[b]} \simeq \phi_1 + \phi_2$ such that $\phi_1 \leq \chi_{[d]}$ and $\phi_2 \leq \chi_{[e]}$. This will show that $H(X, S, \mathbb{Z})$ is not a Riesz group and thus not a dimension group. Assume the contrary. Then there is an integer $n \geq 1$ such that ϕ_1, ϕ_2 can be defined in $Z_n(X)$. Since $\chi_{[a^{n-1}b]} \leq \chi_{[b]}$, we must have $\chi_{[a^{n-1}b]} \leq \chi_{[d]}$ or $\chi_{[a^{n-1}b]} \leq \chi_{[e]}$. It is easy to see that this is not possible by assigning real values to a, b, c, d, e, f on the graph of Figure 3.4 defining an invariant Borel probability measure on X. For example, using the stationary Markov chain defined by

$$v = \begin{bmatrix} a & b & c & d & e & f \\ \frac{1}{4} & \frac{1}{4} & 0 & 0 & \frac{1}{4} & \frac{1}{4} \end{bmatrix}, \quad P = \begin{array}{c} \\ a \\ b \\ c \\ d \\ e \\ f \end{array}\begin{bmatrix} a & b & c & d & e & f \\ \frac{1}{2} & \frac{1}{2} & 0 & 0 & 0 & 0 \\ 0 & 0 & 0 & 0 & \frac{1}{2} & \frac{1}{2} \\ 0 & 0 & 1 & 0 & 0 & 0 \\ 0 & 0 & 0 & 1 & 0 & 0 \\ \frac{1}{2} & \frac{1}{2} & 0 & 0 & 0 & 0 \\ 0 & 0 & 0 & 0 & \frac{1}{2} & \frac{1}{2} \end{bmatrix},$$

we obtain an invariant probability measure μ on X (in fact, the nonzero entries of P are all edges of the Rauzy graph $\Gamma_2(X)$).

By Theorem 3.9.3, we have for every cylinder $[u], [v] \in X$, the implication $\chi_{[u]} \leq \chi_{[v]} \Rightarrow \mu([u]) \leq \mu([v])$.

This shows that $\chi_{[a^{n-1}b]} \leq \chi_{[d]}$ is impossible since the measure of the cylinder $[a^{n-1}b]$ is $1/2^{n+1}$ while the second one has measure 0. A similar argument shows that $\chi_{[a^{n-1}b]} \leq \chi_{[e]}$ is also impossible and we therefore obtain a contradiction.

3.33 Formula 3.30 defines clearly a measure on X. Since $T^n(X_n) = \cup_{m \geq n} U \cap X_m$, one has $\hat{v}(X) = \frac{1}{\lambda} \sum_{n \geq 1} v(T^n(X_n)) = 1$ and thus $\hat{v}$ is a probability measure. Next, for $V \subset U$, one has with $W_n = V \cap X_n$,

$$\hat{v}(V) = \frac{1}{\lambda} \sum_{n \geq 1} v(T^n W_n) = \frac{1}{\lambda} \sum_{n \geq 1} v(T_U W_n) = \frac{1}{\lambda} \sum_{n \geq 1} v(W_n) = \frac{1}{\lambda} v(V)$$

showing (together wih $\hat{v}(U)\lambda = 1$) that $\hat{v}$ determines v. Finally, let $V \subset X_n$. If $n > 1$, one has $TV \subset X_{n-1}$ and thus $\hat{v}(TV) = \hat{v}(V)$ by definition. If $n = 1$, then since $TV \subset U$, we have $\hat{v}(TV) = \frac{1}{\lambda} v(TV) = \hat{v}(V)$. This shows that $\hat{v}$ is invariant.

3.34 Let $\hat{v}$ be defined as in Exercise 3.33. Then $\mu = \hat{v} \circ \hat{\varphi}^{-1}$ and thus μ is an invariant Borel probability measure.

A.4 Chapter 4

Section 4.1

4.1 Let B, B' be the bases of $\mathfrak{P}$, $\mathfrak{P}'$. Since $\mathfrak{P}$ is nested in $\mathfrak{P}'$, we have $B \subset B'$. If $h > j$, then $T^{h-j} B'_k$ contains an element of $B \subset B'$, a contradiction. Thus, $0 \leq j - h$. Next, set $\ell = h_i - j$. Then $T^{h+\ell} B'_k \subset B'$ implies $h + \ell \geq h'_k$. Thus, $j - h \leq j - (h'_k - \ell) = h_i - h'_k$.

4.2 Since σ is primitive and proper and $X(\sigma)$ is not periodic, the sequence $(\mathfrak{P}(n))$ is a refining sequence of partitions in towers by Lemma 6.2.4 (to be proved in Chapter 6). The sequence $\mathfrak{P}'(n)$ satisfies (KR1) (since $B'(n) = B(n)$) and (KR2) but it cannot satisfy (KR3). Indeed, otherwise, the shift space $X(\sigma)$ would have a BV-representation with one vertex at each level and would be an odometer (by Theorem 6.1.1).

4.3 Each $\mathfrak{P}(n)$ is a partition in towers since σ is primitive and $X(\sigma)$ is infinite. The sequence satisfies (KR2) as any sequence of partitions built in this way. It satisfies (KR3) because $\sigma^n[a]$ and $\sigma^n[c]$ tend both to $\sigma^\omega(b \cdot a)$ while $\sigma^n[b]$ tends to $\sigma^\omega(b \cdot b)$. But condition (KR1) is not satisfied since there are two admissible fixed points.

Section 4.4

4.4 Assume that a is the first index and that $d = 3$ for simplicity. The extreme points of S are the vectors with one coefficient zero and all others equal to $1/(d-1)$. The image of the extreme points of S are the vectors

$$\begin{bmatrix} 0 & \frac{1}{2} & \frac{1}{2} \end{bmatrix}, \begin{bmatrix} \frac{1}{2} & \frac{1}{2} & 1 \end{bmatrix}, \begin{bmatrix} \frac{1}{2} & 1 & \frac{1}{2} \end{bmatrix}.$$

After normalization to sum 1, we obtain

$$\begin{bmatrix} 0 & \frac{1}{2} & \frac{1}{2} \end{bmatrix}, \begin{bmatrix} \frac{1}{4} & \frac{1}{4} & \frac{1}{2} \end{bmatrix}, \begin{bmatrix} \frac{1}{4} & \frac{1}{2} & \frac{1}{4} \end{bmatrix},$$

which all belong to S (and actually to the triangle of Figure 4.17).

4.5 It is clear that $\delta(x, y) \geq 0$ and, since the vectors have sum 1, with equality if and only if $x = y$. We have to check the triangular inequality. Set $Q(x, y) = \max x_i/y_i$ and $q(x, y) = \min x_i/y_i$. For every triple of positive vectors of sum 1, we have

$$x_i \leq Q(x, y)y_i \leq Q(x, y)Q(y, z)z_i$$

and thus $x_i/z_i \leq Q(x, y)Q(y, z)$, which implies $Q(x, z) \leq Q(x, y)Q(y, z)$. Similarly $q(x, z) \geq q(x, y)q(y, z)$. Thus, $\delta(x, z) \leq \delta(x, y) + \delta(y, z)$.

4.6 Set $\hat{x} = Mx$ and $\hat{y} = My$. Since M is positive, $\hat{x}$ and $\hat{y}$ are positive. We have

$$\frac{\hat{x}_i}{\hat{y}_i} = \frac{a_{i1}x_1 + \cdots + a_{in}x_n}{a_{i1}y_1 + \cdots + a_{in}y_n}.$$

Using the inequality

$$\min \frac{r_i}{s_i} \leq \frac{r_1 + \cdots + r_n}{s_1 + \cdots + s_n} \leq \max \frac{r_i}{s_i},$$

we obtain $q(x, y) \leq \hat{x}_i/\hat{y}_i \leq Q(x, y)$ and thus the inequalities $Q(\hat{x}, \hat{y}) \leq Q(x, y)$ and $q(\hat{x}, \hat{y}) \geq q(x, y)$ whence the inequality $d(Mx, My) \leq d(x, y)$.

4.7 1. Assume the statement is true for $m = 2$. We have

$$d(Mx, My) = \frac{\langle m_i, x \rangle \langle m_i, y \rangle^{-1}}{\langle m_j, x \rangle \langle m_j, y \rangle^{-1}}$$

for some indices i, j. Let M' be the $2 \times n$-matrix with rows m_i, m_j. Then $d(M'x, M'y) = d(Mx, My) > 1$ and therefore $1 < d(M')$. Then, by the hypothesis,

$$d(Mx, My) = d(M'x, M'y) < d(x, y)^{\tau(d(M'))} = d(x, y)^{\tau(d(m_i, m_j))}$$
$$< d(x, y)^{\tau(M)},$$

where the last inequality holds because τ is increasing.

2. Consider a $2 \times n$-matrix M with rows a, b. We must prove that

$$\frac{\langle a, x \rangle \langle a, y \rangle^{(-1)}}{\langle b, x \rangle \langle b, y \rangle^{(-1)}} < d(x, y)^{\tau(d(a,b))}$$

or equivalently

$$\frac{\langle a, b^{-1}x\rangle\langle a, b^{-1}y\rangle^{(-1)}}{\langle b, b^{-1}x\rangle\langle b, b^{-1}y\rangle^{(-1)}} < d(b^{-1}x, b^{-1}y)^{\tau(d(a,b))}.$$

Set $r = ab^{-1}$ and $s = xy^{-1}$. Then we may rewrite the left-hand side above as

$$\begin{aligned}\frac{\langle ab^{-1}, x\rangle\langle ab^{-1}, y\rangle^{(-1)}}{\langle bb^{-1}, x\rangle\langle bb^{-1}, y\rangle^{(-1)}} &= \frac{\langle ab^{-1}, x\rangle\langle \mathbf{1}, y\rangle}{\langle ab^{-1}, y\rangle\langle \mathbf{1}, x\rangle}\\ &= \frac{\langle r, sy\rangle\langle \mathbf{1}, y\rangle}{\langle r, y\rangle\langle \mathbf{1}, sy\rangle} = \frac{\langle rs, y\rangle\langle \mathbf{1}, y\rangle}{\langle r, y\rangle\langle s, y\rangle} = F(r, s, y),\end{aligned}$$

whence the desired formula.

3. By scaling r and s suitably, we may assume that $\langle r, y\rangle = 1$ and $\langle s - \mathbf{1}, y\rangle = 0$. Then $F(r, s, y) = \langle rs, y\rangle$ is linear in y. By the Fundamental Theorem of Linear Programming, there is a vector y maximizing $\langle rs, y\rangle$ under the constraints $\langle r, y\rangle = 1$ and $\langle s - \mathbf{1}, y\rangle = 0$ with at most two nonzero coordinates. Therefore, it is enough to prove (4.26) for $r, s, y \in \mathbb{R}^2$.

4. We may assume that r, s, y have the form $(1, r), (1, s), (1, y)$ for positive real numbers r, s, y with $r > 1$ and $s \neq 1$. Then an easy computation gives

$$F(r, s, y) = \frac{1 + rsy^2 + (1 + rs)y}{1 + rsy^2 + (r + s)y}.$$

If $s < 1$, then $0 > (r - 1)(s - 1) = (1 + rs) - (r + s)$ so that $F(r, s, y) < 1$. We may thus assume $s > 1$. Since $1 + rsy^2 \geq 2y\sqrt{rs}$, we have $F(r, s, y) \leq \left(\frac{1+\sqrt{rs}}{\sqrt{r}+\sqrt{s}}\right)^2$, and it is sufficient to prove that for $r, s > 1$,

$$\frac{1 + \sqrt{rs}}{\sqrt{r} + \sqrt{s}} < (\sqrt{s})^{\tau(r)}.$$

Replacing $\sqrt{r}, \sqrt{s}$ by r, s, we are left with proving that $(1+rs)/(r+s) < s^{\frac{r-1}{r+1}}$. Since both sides take the same value for $s = 1$, it is enough to verify the corresponding inequality for their partial derivatives with respect to s, that is,

$$\frac{r^2 - 1}{(r + s)^2} < \frac{r - 1}{r + 1}s^{\frac{-2}{r+1}}$$

or equivalently

$$\left(\frac{r + 1}{r + s}\right)^2 < s^{\frac{-2}{r+1}}.$$

The last inequality is equivalent to

$$\left(\frac{r + s}{r + 1}\right) > s^{\frac{1}{r+1}}.$$

Again both sides take the same value for $s = 1$. Taking again the partial deva-tives of both sides, we are left with the inequality $1/(r + 1) > s^{\frac{-r}{r+1}}/(r + 1)$, which is clearly true for $r > 0$ and $s > 1$.

Section 4.5

4.8 Set $\mathrm{Card}(V) = n$ and $\mathrm{Card}(E) = m$. Clearly, κ is injective and $\mathbb{Z}(\Gamma) = \mathrm{Im}(\kappa) \subset \ker(\beta)$. Next $\mathrm{Im}(\beta) = \ker(\gamma)$ and $\dim(\ker(\gamma)) = n - 1$. Thus,

$$m = \dim(\mathbb{Z}(E) = \dim(\ker(\beta)) + \dim(\mathrm{Im}(\beta)) = \dim(\ker(\beta)) + n - 1.$$

This implies that the dimension of $\ker(\beta)$ is $m - n + 1$. On the other hand, since G is strongly connected, any covering tree T of G has $n - 1$ elements. Since Γ has a basis of $\mathrm{Card}(E) - \mathrm{Card}(T) = m - n + 1$ elements (see Appendix B.2), this implies that the dimension of $\mathbb{Z}(\Gamma)$ is $m - n + 1$. We conclude that $\mathbb{Z}(\Gamma) = \ker(\beta)$.

Section 4.6

4.9 Let $X = X(\sigma)$ be the Chacon binary shift. We have $\mathcal{L}_2(X) = \{00, 01, 10\}$ and the Rauzy graph $\Gamma_2(X)$ is represented in Figure A.6. Let $\varphi \colon a \mapsto 00, b \mapsto 01, c \mapsto 10$. The 2-block presentation of σ is $\sigma_2 \colon a \mapsto abca, b \mapsto abcb, c \mapsto c$. Thus, the matrix M_2, P and N_2 are

$$M_2 = \begin{bmatrix} 2 & 1 & 1 \\ 1 & 2 & 1 \\ 0 & 0 & 1 \end{bmatrix}, \quad P = \begin{bmatrix} 1 & 0 & 0 \\ 0 & 1 & 1 \end{bmatrix}, \quad N_2 = \begin{bmatrix} 2 & 1 \\ 1 & 2 \end{bmatrix}.$$

The matrix N_2 has two eigenvalues 3 and 1 and corresponding eigenvectors $u = [1\ 1]^t, v = [1\ -1]^t$. Thus, the dimension group is isomorphic to $\mathbb{Z}[1/3] \times \mathbb{Z}$ with positive cone $\mathbb{Z}_+[1/3]$ and order unit $(3, -1)$, which can be normalized to $(1, 1)$.

4.10 Let $A_2 = \{x, y, z, t, u\}$ be an alphabet in order preserving bijection with $\mathcal{L}_2(X) = \{aa, ab, ac, ba, ca\}$. The morphism $\varphi_2 \colon A_2^* \to A_2^*$ is $x \mapsto yt, y \mapsto yt, z \mapsto yt, t \to zu, u \mapsto x$. The matrices M and M_2 are

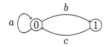

Figure A.6 The Rauzy graph $\Gamma_2(X)$.

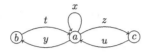

Figure A.7 The Rauzy graph $\Gamma_2(X)$ for the Tribonacci shift.

$$M = \begin{bmatrix} 1 & 1 & 0 \\ 1 & 0 & 1 \\ 1 & 0 & 0 \end{bmatrix}, \quad M_2 = \begin{bmatrix} 0 & 1 & 0 & 1 & 0 \\ 0 & 1 & 0 & 1 & 0 \\ 0 & 1 & 0 & 1 & 0 \\ 0 & 0 & 1 & 0 & 1 \\ 1 & 0 & 0 & 0 & 0 \end{bmatrix}.$$

The Rauzy graph $\Gamma_2(X)$ is represented in Figure A.7.

Thus, the matrix P is

$$P = \begin{bmatrix} 0 & 1 & 0 & 1 & 0 \\ 0 & 0 & 1 & 0 & 1 \\ 1 & 0 & 0 & 0 & 0 \end{bmatrix},$$

and $N_2 = M$. The matrix M is invertible and its dominant eigenvalue is the positive real number λ such that $\lambda^3 = \lambda^2 + \lambda + 1$. A corresponding row eigenvector is $[\lambda^2, \lambda, 1]$. Thus, the dimension group is $\mathbb{Z}[\lambda]$.

4.11 Let $X = X(\sigma)$, where $\sigma: a \mapsto ab, b \mapsto ac, c \mapsto db, d \mapsto dc$ is the Rudin–Shapiro morphism. Let $f: A_2 \to \mathcal{L}_2(X)$ be the map sending $A_2 = \{s, t, u, v, w, x, y, z, t\}$ to $\mathcal{L}_2(X) = \{ab, ac, ba, bd, ca, cd, d, dc\}$ in alphabetical order. The set of return words to a is

$$\mathcal{R}_X(a) = \{ba, bdba, bdbdcdba, bdbdcdbbdca, ca,$$
$$cdca, cdcdba, cdcdbdca\} \tag{A.5}$$

and the Rauzy graph $\Gamma_2(X)$ is represented in Figure A.8. The matrix P with rows the compositions of the elements of $f(a\mathcal{R}_X(a))$ and the matrix N_2 such that $PM_2 = N_2P$ are

$$P = \begin{bmatrix} 1 & 0 & 1 & 0 & 0 & 0 & 0 & 0 \\ 1 & 0 & 1 & 1 & 0 & 0 & 1 & 0 \\ 1 & 0 & 1 & 2 & 0 & 1 & 2 & 1 \\ 1 & 0 & 0 & 3 & 1 & 1 & 2 & 2 \\ 0 & 1 & 0 & 0 & 1 & 0 & 0 & 0 \\ 0 & 1 & 0 & 0 & 1 & 1 & 0 & 1 \\ 0 & 1 & 1 & 0 & 0 & 2 & 1 & 1 \\ 0 & 1 & 0 & 1 & 1 & 2 & 1 & 2 \end{bmatrix}, \quad N_2 = \begin{bmatrix} 1 & 0 & 0 & 0 & 1 & 0 & 0 & 0 \\ 1 & 0 & 0 & 0 & 1 & 1 & 0 & 0 \\ 1 & 0 & 0 & 0 & 1 & 1 & 0 & 1 \\ 1 & 0 & 0 & 0 & 0 & 1 & 1 & 1 \\ 0 & 1 & 0 & 0 & 0 & 0 & 0 & 0 \\ 0 & 0 & 1 & 0 & 0 & 0 & 0 & 0 \\ 0 & 0 & 0 & 1 & 1 & 0 & 0 & 0 \\ 0 & 0 & 0 & 1 & 0 & 0 & 1 & 0 \end{bmatrix}.$$

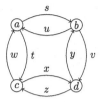

Figure A.8 The Rauzy graph $\Gamma_2(X)$ for the Rudin–Shapiro shift.

The eigenvalues of N_2 are $2, -1, \sqrt{2}, -\sqrt{2}$ and 0 with multiplicity 4. Thus, the dimension group is $\mathbb{Z}[1/2]^3 \times \mathbb{Z}$ with positive cone $\mathbb{Z}_+[1/2] \times \mathbb{Z}[1/2]^2 \times \mathbb{Z}^3$ and unit $(1, 0, 0, 0)$.

A.5 Chapter 5

Section 5.1

5.1 (i)$\Rightarrow$(ii) is clear since every $V(m)$ is finite. (ii)$\Rightarrow$(iii) Denote $f : \mathbb{N} \to \mathbb{N}$ the function defined by $fm) = n$ if $n > m$ is the least integer such that (ii) holds. Then the telescoping of (V, E) with respect to this sequence has the desired property. (iii)$\Rightarrow$(i) is clear.

5.2 We have to show that the relation is transitive. For this, consider Bratteli diagrams B_1, B_2, B_3 such that T_1 is an intertwining of B_1, B_2 and T_2 is an intertwining of B_2, B_3. We will build an intertwining T_3 of B_1, B_3. This will prove that the relation is transitive.

Denote by $B_1(n, m)$ the matrix of B_1 between levels n and m and similarly for B_2, B_3, T_1, T_2. For every $n \geq 1$, there is an integer $\ell_1(n)$ such that

$$T_1(n, 0) = \begin{cases} B_1(\ell_1(n), 0) & \text{if } n \text{ is odd} \\ B_2(\ell_1(n), 0) & \text{if } n \text{ is even,} \end{cases}$$

and there is a similar integer $\ell_2(n)$ for T_2 related to B_2 and B_3. We start with $T_3(1, 0) = T_1(1, 0)$. Next, by telescoping T_2, we can obtain $\ell_2(1) \geq \ell_1(2)$. We define

$$T_3(2, 1) = T_2(2, 1)B_2(\ell_2(1), \ell_1(2))T_1(2, 1).$$

This corresponds to the path from level 2 in T_2 to level 1 in T_1 indicated in Figure A.9. We verify that, with this choice, $T_3(2, 0) = B_3(\ell_2(2), 0)$. This follows by inspection of Figure A.9 or by a patient verification as below.

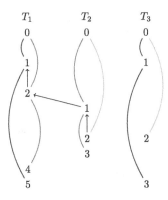

Figure A.9 The construction of the intertwining T_3.

$$T_3(2, 0) = T_3(2, 1)T_3(1, 0)$$
$$= (T_2(2, 1)B_2(\ell_2(1), \ell_1(2))T_1(2, 1))T_3(1, 0)$$
$$= (T_2(2, 1)B_2(\ell_2(1), \ell_1(2))T_1(2, 1))T_1(1, 0)$$
$$= T_2(2, 1)B_2(\ell_2(1), \ell_1(2))T_1(2, 0) = T_2(2, 1)B_2(\ell_2(1), 0)$$
$$= T_2(2, 0) = B_3(\ell_2(2), 0).$$

We perform one more step to convince everybody that the construction can continue in the same way forever. We may assume, again by telescoping T_2 if necessary, that $\ell_1(4) \geq \ell_2(3)$. Then we define

$$T_3(3, 2) = T_1(5, 4)B_2(\ell_1(4), \ell_2(3))T_2(3, 2)$$

and the reader may verify as above that $T_3(3, 1) = B_1(\ell_1(5), \ell_1(1))$.

5.3 The fact that C^*-equivalence is the same as the existence of an intertwining of the stationary Bratteli diagrams with matrices M, N appears clearly in Figure A.10. As a consequence, the C^*-equivalence is an equivalence relation since the existence of an intertwining is an equivalence (Exercise 5.2).

5.4 The matrices M, N given by Eq. (5.21) have the same maximal eigenvalue 2 and the same left and right eigenvectors $\begin{bmatrix} 1 & 1 & 1 & 1 & 1 \end{bmatrix}$. We may conjugate to write

$$M = \begin{bmatrix} 2 & 0 \\ 0 & M_1 \end{bmatrix}, \quad N = \begin{bmatrix} 2 & 0 \\ 0 & N_1 \end{bmatrix}.$$

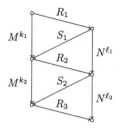

Figure A.10 Intertwining of stationary Bratteli digrams.

Let λ be the maximal modulus of the eigenvalues of M_1, M_2 and let μ be the maximal modulus of eigenvalues of their inverses. For every $k \geq 1$, we have

$$N^{-k}M^{ck} = \begin{bmatrix} 2^{(c-1)k} & 0 \\ 0 & 0 \end{bmatrix} + \begin{bmatrix} 0 & 0 \\ 0 & N_1^{-k}M_1^{ck} \end{bmatrix}.$$

After we conjugate back, the entries of the first term will be at least $c_1 2^{(c-1)k}$ and those of the second term will be at most $c_2 \lambda^{ck} \mu^k$. Thus, if we choose c such that $2^{(c-1)k} > \lambda^c \mu$, all entries of $N^{-k}M^{ck}$ will be positive. This shows that for all large enough k, the matrix $N^{-k}M^n$ is positive for all n large enough. It is also integral because M, N have both determinant equal to 2.

Consider now the construction of sequences (R_n, S_n) and (k_n, ℓ_n) such that Eq. (5.20) holds for all $n \geq 1$. Start with $S_1 = I$. Eq. (5.20) are equivalent to the equations

$$R_1 = M^{k_1}, \quad R_2 = M^{-k_1}N^{\ell_1}, \quad S_2 = N^{-\ell_1}M^{k_1+k_2}, \quad R_3 = M^{-k_1-k_2}N^{\ell_1+\ell_2}$$

and so on. By the preceding remark, we can successively choose $k_1, \ell_1, k_2, \ell_2, \ldots$ in such a way that all R_n, S_n are nonnegative and integral. Thus, M, N are C^*-equivalent.

Section 5.2

5.5 The space X_E is compact. Assume that it is infinite. If (V', E') is obtained by telescoping (V, E), the induced map from X_E to $X_{E'}$ is a homeomorphism. Thus, we can assume that there is an edge from every vertex in $V(n-1)$ to every vertex in $V(n)$. Assume that $x = (e_1, e_2, \ldots)$ is an isolated point. Then $[e_1, \ldots, e_n] = \{x\}$ for some $n \geq 1$. But then $E(m)$ has to be reduced to $\{e_m\}$ for all $m > n$, a contradiction with our hypothesis that X_E is infinite.

5.6 Assume that (V, E) is simple. Let $e = (e_n)_{n \geq 1}$ and $f = (f_n)_{n \geq 1}$ be elements of X_E. We show that f belongs to the closure of the class of e. Since

(V, E) is simple, for every $n \geq 1$, there is by Exercise 5.1, an integer $m > n + 1$ such that there is a path $(g_{n+1}, \ldots, g_{m-1})$ from $r(f_n)$ to $s(e_m)$. Then $h^{(n)} = (f_1, \ldots, f_n, g_{n+1}, \ldots, g_{m-1}, e_m, e_{m+1}, \ldots)$ is a path in X_E that is in the cofinality class of e. Since the sequence $h^{(n)}$ tends to f when $n \to \infty$, the claim is proved.

5.7 For each $n \geq 1$, set $F_n = \{[e_1, \ldots, e_n] \in E_{1,n} \mid \text{every } e_i \text{ is maximal}\}$.

An easy induction on n shows that for every vertex $v \in V(n + 1)$ there is an element $[e_1, \ldots, e_n]$ in F_n such that $r(e_n) = v$. This proves that the set of maximal edges forms a spanning tree. Next F_n is closed. Thus, the set $X_E^{\max}$ is the intersection of the nonempty closed sets F_n. Since X_E is compact, it is nonempty. The argument for $X_E^{\min}$ is similar.

5.8 Let (V, E) be a simple Bratteli diagram. Fix an order on $V(n)$ for each $n \geq 1$. We order the edges $e \in E(n)$ in such a way that, for two edges e, f with the same range, one has $s(e) < s(f) \Rightarrow e < f$. Then any long enough path made of minimal edges leads to the minimal vertex.

5.9 The inverse of T_E can be described as follows. First $T_E^{(-1)}(x_{\min}) = x_{\max}$. Next, for $e = (e_n)_{n \geq 1}$ distinct of $x_{\min}$, let $k \geq 1$ be the minimal index such that e_k is not a minimal edge. Then $T_E^{-1}(e) = (f_1, \ldots, f_{k-1}, f_k, e_{k+1}, e_{k+2}, \ldots)$, where f_k is the antecedent of e_k and $(f_1, \ldots, f_{k-1})$ is the maximal path from $v(0)$ to $s(f_k)$.

5.10 The map T_E is invertible by Exercise 5.9. We claim that if $e_m = f_m$ for $m > n$, and if $e_n < f_n$, then there is a $k \geq 0$ such that $T_E^k(e) = f$. This proves that the classes of cofinality are contained in the orbits of T_E and thus implies that (X_E, T_E) is minimal since the classes of R_E are dense by Exercise 5.6.

One proves the claim by induction on $n \geq 1$. It is clearly true for $n = 1$. Next, let $(g_1, \ldots, g_{n-1})$ be a path of maximal edges from $v(0)$ to $s(e_n)$. Then there is an integer $k \geq 1$ and a path $(h_1, \ldots, h_{n-1})$ of minimal edges such that $T_E^k(g_1, \ldots, g_{n-1}, e_n, e_{n+1}, \ldots) = (h_1, \ldots, h_{n-1}, f_n, f_{n+1}, \ldots)$. By induction hypothesis, there is an $\ell \geq 0$ such that $T_E^\ell(h_1, \ldots, h_{n-1}, f_n, f_{n+1}, \ldots) = f$. This proves the claim.

5.11 Let $m_0 = 0 < m_1 < m_2 < \cdots$ be the sequence defining the telescoping from $(V, E, \leq)$ to $(V', E', \leq')$. Let $\varphi \colon X_E \to X_{E'}$ be the map defined by $y = \varphi(x)$ if $x = (e_1, e_2, \ldots)$ and $y = (f_1, f_2, \ldots)$ with $f_n = (e_{m_n+1}, \ldots, e_{m_{n+1}})$. The map φ is clearly a homeomorphism from X_E onto $X_{E'}$. To show that it is a conjugacy, we have to show that $\varphi(T_E x) = T_{E'} \varphi(x)$. Consider first the case of $x = x_{max}$. Then $\varphi(T_E x) = \varphi(x_{min}) = y_{min}$ while $T_{E'}\varphi(x) = T_{E'}y_{max} = y_{min}$. Next, let $x = (e_1, e_2, \ldots) \neq x_{max}$ and $y = (e'_1, e'_2, \ldots) = \varphi(x)$. Let k be the least index n such that e_n is not maximal and let f_k be the successor of e_k. Let n be such that $m_n < k \leq m_{n+1}$. Then

$$\varphi(T_E x) = \varphi(e_1, \ldots, f_k, \ldots) = (e'_1, \ldots, f'_k, \ldots)$$

with $f'_k = (e_{m_n+1}, \ldots, f_k, \ldots, e_{m_{n+1}})$ while

$$T_{E'}\varphi(x) = T_{E'}(e'_1, e'_2, \ldots) = (e'_1, \ldots, f'_k, \ldots)$$

since f'_k is the successor of e'_k in E'_k. Thus, φ is a conjugacy.

5.12 By Proposition 2.5.1, the dimension group $DG(V, E)$ of the stationary system (V, E) has a unique state. Now, By Theorem 5.3.6, the dimension group $K^0(X_E, T_E)$ is isomorphic to $DG(V, E)$ and thus (X_E, T_E) has a unique invariant Borel probability measure by Theorem 3.9.3. Let $\mathfrak{P}(n) = \{[e_1, e_2, \ldots, e_n] \mid (e_1, e_2, \ldots, e_n) \in E_{1n}\}$ be the natural sequence of partitions associated with (V, E).

By Lemma 5.3.7, the group $K^0(X_E, T_E)$ is the direct limit of the groups $(G(n), G^+(n), \mathbf{1}_M)$ with connecting morphisms given by M. The group $G(n)$ is identified with $\mathbb{Z}^t$ and the connecting morphism is $x \mapsto Mx$. The sequence of maps $p(n)$ defined on $G(n)$ by $p(n)(x) = \langle \alpha, x \rangle / \lambda^{n-1}$ is a sequence of states on $G(n)$ compatible with the connecting morphisms since

$$p(n)(Mx) = \langle \alpha, Mx \rangle / \lambda^{n-1} = \langle \alpha, x \rangle / \lambda^{n-2} = p(n-1)(x).$$

Thus, it defines a state p on $K^0(X_E, T_E)$. The corresponding invariant measure μ is given by

$$\mu([u]) = p(n)(i_{v(u)}) = \alpha_{v(u)} / \lambda^{n-1}$$

for a path u of length n from $V(0)$ to $r(u)$, where $i_{r(u)}$ denotes the unit vector in $\mathbb{Z}^t$ with 1 in position $r(u)$.

5.13 A path $x \in X_E$ is in X_E^{max} if all its edges are maximal. Thus, either they are all 0, or after some initial path 0^n, it uses only edges 1. The argument for X_E^{min} is symmetrical. The transformation T is clearly measurable. It is measure-preserving because $T(0^n 1^m 10y) = 1^m 0^n 01y$ for $n, m \geq 0$.

Section 5.3

5.14 Let $\psi : X \to X_E$ be defined as follows. For every $x \in X$ and $n \geq 1$, since $\mathfrak{P}(n)$ is a partition of X, there is a unique pair (k_n, t_n) with $1 \leq k_n \leq$

$h_{t_n}(n) - 1$ and $1 \le t_n \le t(n)$ such that $x \in T^{k_n} B_{t_n}(n)$. We set $k_0 = 0$ and $t_0 = 1$. Then $k_n \ge k_{n-1}$ for all $n \ge 1$. The map $\psi(x) = (n, t_{n-1}, t_n, k_n - k_{n-1})_{n \ge 1}$ is well defined and continuous. Since ψ is obviously the inverse of ϕ, the result follows.

5.15 By definition of (V, E), we have $e_i = (i, t_{i-1}, t_i, j_i)$ for $1 \le i \le n$. Since each e_i is minimal, we have $j_i = 0$. Thus, by definition of ϕ,

$$\phi([e_1, \ldots, e_n]) = T^{\sum_{i=1}^n j_i} B_{t_n}(n) = B_{t_n}(n).$$

Section 5.4

5.16 Let first (Z, U) be a common primitive of (X, S) and (Y, T). Thus, there are nonempty clopen sets $A, B \subset Z$ such that $(X, S) \simeq (A, U_A)$ and $(Y, T) \simeq (B, U_B)$. For every $n \ge 0$, (A, U_A) is isomorphic to $(U^n A, U_{U^n A})$. Since U is minimal, we may assume that $C = U^n A \cap B \ne \emptyset$. Thus, (C, U_C) is a common derivative of (X, S) and (Y, T).

Conversely, let (Z, U) be a common derivative of (X, S) and (Y, T). Then, using the tower construction of Exercise 1.8, there exist continuous maps $g, h \colon Z \to \mathbb{N}$ such that (X, S) is isomorphic to (Z^g, U^g) and (X, T) is isomorphic to (Z^h, U^h). Let (W, R) be the result of the tower construction corresponding to $g + h$. Then (W, R) is isomorphic to (X^k, T^k) with k defined by

$$k(z, i) = \begin{cases} 1 & \text{if } i < g(z) \\ h(z) + 1 & \text{if } i = g(z) \end{cases}$$

through the map

$$\varphi(z, \ell) = \begin{cases} (z, \ell, 1) & \text{if } \ell < g(z) \\ (z, g(z), \ell + 1) & \text{otherwise.} \end{cases}$$

Thus, (W, R) is a primitive of (X, S). In the same way, (W, R) is a primitive of (Y, T).

5.17 We have to show the transitivity of the relation $(X, S) \sim (Y, T)$ if (X, S) and (Y, T) are Kakutani equivalent. Suppose that $(X, S) \sim (Y, T)$ and $(Y, T) \sim (Z, U)$. Let (X_1, S_1) be a common derivative of (X, S) and (Y, T) and let (X_2, S_2) be a common derivative of (Y, T) and (Z, U) (see Figure A.11). Since (X_1, S_1) and (Z_1, U_1) have a common primitive, namely (Y, T) they have by Exercise 5.16 a common derivative (X_2, S_2). Thus, $(X, S) \sim (Z, U)$.

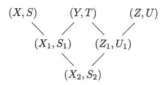

Figure A.11 Transitivity of Kakutani equivalence.

5.18 Let (V, E) be a stationary Bratteli diagram. By Theorem 5.4.1, a system (X, T) Kakutani equivalent to (X_E, T_E) is isomorphic to a BV-system $(X_{E'}, V_{E'})$, where (V', E') is obtained from (V, E) by a finite number of changes. Then a telescoping of (V', E') gives again a stationary Bratteli diagram. Thus, (X, T) is conjugate to a stationary BV-system.

5.19 Let $U \subset X$ be a clopen set, let (U, S_U) be the shift induced by X on U and let $\varphi \colon (Y, T) \to (X, S)$ be the morphism associated with the induction. The morphism φ sends right-asymptotic pairs of Y to right-asymptotic pairs of X and the induced map on the classes is onto.

Section 5.5

5.20 1. Set $\tilde{S} = \phi^{-1} \circ S \circ \phi$. Then $T(x) = \tilde{S}^{\alpha(x)}(x)$. Thus, replacing S by $\tilde{S}$, we may assume $X = Y$. We have then $f_k(x) = \alpha^{(k)}(x)$ and thus each f_k is continuous (concerning the definition of $\alpha^{(k)}$ for $k < 0$, see Exercise 3.1). Since (X, T) is a minimal Cantor system, there are no periodic points. Thus, $k \mapsto f_k(x)$ is a bijection. It satisfies (5.24) since $T^{k+j}x = S^{f_k(T^j x)}(T^j x) = S^{f_k(T^j x) + f_j(x)}x$.

2. Follows from compactness since the f_k are continuous.

3. Set $n_0 = \sup_{x \in X} |\alpha(x)|$ and choose K such that $[-n_0, n_0] \subset \{f_i(x) \mid -K \le i \le K\}$ for all $x \in X$. Then $X = A_K \cup B_K$. Since A_K, B_K are closed and invariant, the conclusion follows.

4. Since $f_0(x) = 0$, we have $N_M(x) + P_M(x) = 2m$ and thus $N_N(x), P_M(x)$ have the same parity. The equality $a(Sx) = a(x) - j + 1$ follows by a counting argument since $f_{j+k}(x) = f_k(Sx) + 1$ for every k by (5.24).

5. We have $g \circ S(x) = T^{a(Sx)}(Sx) = T^{a(x) - j + 1}(T^j(x)) = T^{a(x) + 1}(x) = T(T^{a(x)}(x)) = T \circ g(x)$. Thus, g is a conjugacy from (X, T) onto (Y, S). The case $B = B_K$ leads similarly to the conclusion that (X, T) is conjugate to (Y, S^{-1}).

Section 5.6

5.21 The equivalence R is not minimal because the closure of the set $\{x+2n \mid n \in \mathbb{Z}\}$ is the clopen set $[0]$ if x begins with 0 and $[1]$ otherwise. However, there are no nontrivial closed R-invariant sets. Indeed, the orbit of any $y \neq x$ is dense and the closure of each of the sets $\{x + 2n \mid n \in \mathbb{Z}\}$ or $\{x + 2n + 1 \mid n \in \mathbb{Z}\}$ contains points that are not in the orbit of x and whose orbit is dense.

Section 5.7

5.22 Set $\alpha = \inf u_n/n$. We show that for every $\varepsilon > 0$, we have $(u_n/n) - \alpha \leq \varepsilon$ for all large enough n. Let k be such that $u_k/k < \alpha + \varepsilon/2$. Then, for $0 \leq j < k$ and $m \geq 1$, we have

$$\frac{u_{mk+j}}{mk+j} \leq \frac{u_{mk}}{mk+j} + \frac{u_j}{mk+j} \leq \frac{u_{mk}}{mk} + \frac{u_j}{mk}$$
$$\leq \frac{mu_k}{mk} + \frac{ju_1}{mk} \leq \frac{u_k}{k} + \frac{u_1}{m} \leq \alpha + \frac{\varepsilon}{2} + \frac{u_1}{m}.$$

Hence, if $nmk + j$ is large enough so that $u_1/m < \varepsilon/2$, then $u_n/n < \alpha + \varepsilon$.

Set $v_n = p_n(X)$. Then $v_{n+m} \leq v_n v_m$ since a word of length $n + m$ is determined by its prefix of length n and its suffix of length m. Thus, the sequence $u_n = \log v_n$ is subadditive.

5.23   Let us show that if $\varphi \colon X \to Y$ is a surjective morphism from (X, T) onto (Y, S), then $h(Y) \leq h(X)$. By Theorem 1.2.12, φ is the sliding block code associated with some block map $f \colon \mathcal{L}_{n+m+1} \to B$, extended to a map from $\mathcal{L}_{n+m+k}$ to $\mathcal{L}_k(Y)$ for every $k \geq 1$. Since φ is surjective, f is surjective. Thus, $p_k(Y) \leq p_{n+m+k}(X)$ for every $k \geq 1$, which implies

$$\frac{p_k(Y)}{k} \leq \frac{p_k(X)}{k} + \frac{p_n(X) + p_m(X)}{k}.$$

The second term of the right-hand side tends to 0 when $k \to \infty$ and thus $h(Y) \leq h(X)$.

5.24   It is easy to show that $c\lambda^n \leq \sum_{i,j} M_{i,j}^n \leq d\lambda^n$ for some constants c, d. Since $p_n(X) = \sum_{i,j} M_{i,j}^n$, the result follows.

5.25   Since $m_1(k) = 1$ and $m_2(k) = k$, we have $\lambda_j(1) = 0$. Next, $m_j(2) = 2^{j-1}$ implies $\lambda_j(2) = 2^{-j+1} \log 2$. Thus, $\lambda(2) = 0$.

Next, for $k \geq 3$, let us first show by induction on j that

$$e^{m_j \lambda_j} > m_j + 1. \tag{A.6}$$

Indeed, it is true for $j = 1$ since $e^{m_1 \lambda_1} = k > 2$. Next, we have

$$e^{m_{j+1} \lambda_{j+1}} = (e^{m_j \lambda_j})! > (e^{m_j \lambda_j} - 2)e^{m_j \lambda_j} + 1$$

since $n! > (n - 2) + 1$ for $n \geq 3$. Thus, using the induction hypothesis,

$$e^{m_{j+1} \lambda_{j+1}} > m_j e^{m_j \lambda_j} + 1 = m_{j+1} + 1,$$

which proves (A.6).

Since $n! < n^n$, we have $e^{m_{j+1} \lambda_{j+1}} < e^{m_j \lambda_j} e^{m_j \lambda_j} = e^{m_{j+1} \lambda_{j+1}}$ showing that the sequence $(\lambda_j(k))_j$ is decreasing.

Using now Stirling Formula with $n = e^{m_j \lambda_j}$ and taking the logarithm, we estimate

$$0 \leq m_{j+1} \lambda_{j+1} - \frac{1}{2}(\log 2\pi + m_j \lambda_j) - m_{j+1} \lambda_j + e^{m_j \lambda_j} \leq \frac{1}{12} e^{-m_j \lambda_j} \leq \frac{1}{36},$$

where the last inequality follows from $e^{m_j \lambda_j} = k!^{(j-1)} \geq 3$. Dividing the leftmost and rightmost sides of this inequality by m_{j+1}, we obtain

$$0 \leq \lambda_{j+1} - \lambda_j + \frac{1}{m_j} \leq \frac{1}{2m_{j+1}}(\frac{1}{18} + \log 2\pi + m_j \lambda_j).$$

The right-hand side can be bounded by $(2/m_{j+1})m_j \lambda_j < 2 \log k e^{-m_j \lambda_j}$. Summing up on j, we obtain

$$0 \leq \lambda(k) - \log k + \sum_{j \geq 1} \frac{1}{m_j} \leq 2 \log k \sum_{j \geq 1} e^{-m_j \lambda_j}.$$

Now $1 < \sum_{j \geq 1} \frac{1}{m_j} < \sum_{j \geq 0} 1/k^j = k/(k - 1)$ and thus $\lim_{k \to \infty} \sum_{j \geq 1} \frac{1}{m_j} = 1$. Similarly, $\sum_{j \geq 1} e^{-m_j \lambda_j} < \sum_{j \geq 0} k^{-j-1} = 1/(k - 1)$ and thus

$$0 < \lim_{k \to \infty} (\lambda(k) - \log k - 1) < \lim_{k \to \infty} 2 \frac{\log k}{k - 1} = 0.$$

5.26 We have $\text{Card}(U_j) = e^{m_j \lambda_j}$ and all words in U_j have length m_j.

To show that x is uniformly recurrent, consider $u \in \mathcal{L}(x)$. Then u appears in some $\ell_j r_j$ with $\ell_j \neq r_j$. Then $\ell_j r_j$ appears in some $v \in U_{j+1}$, which appears itself in every $w \in U_{j+2}$. Thus, u occurs in every $x_{[tm_j,(t+1)m_j)}$, which shows that x is uniformly recurrent.

There remains to show that $h(x) = \lambda(k)$. Set $c_j = \text{Card}(U_j) = e^{m_j \lambda_j}$. Since $p_{m_j}(x) \geq \text{Card}(U_j)$, we have $h(x) \geq \lambda(k)$. Next, consider $u \in \mathcal{L}_{m_{j+1}}(x)$. There exist $u_1, \ldots, u_{c_j+1} \in U_j$ such that u is a factor of $u_1 \cdots u_{c_j+1}$. Thus,

$$p_{m_{j+1}}(x) \leq m_j c_j^{c_j+1} = m_{j+1} c_j^{c_j}.$$

This implies

$$h(x) \leq \frac{1}{m_{j+1}} \log p_{m_{j+1}}(x) \leq \frac{\log m_{j+1}}{m_{j+1}} + \frac{c_j}{m_{j+1}} \log c_j \leq \frac{\log m_{j+1}}{m_{j+1}} + \lambda_j$$

whence the conclusion since the right-hand side tends to $\lambda(k)$ when $j \to \infty$.

For $k = 2$, we have $U_1 = \{0, 1\}$ and $U_{j+1} = \{\tau^j(0), \tau^j(1)\}$, where $\tau: 0 \mapsto 01, 1 \mapsto 10$ is the Thue–Morse morphism. Thus, with $\ell_1 = r_1 = 0$, we obtain $x = \tau^\omega(0 \cdot 0)$.

A.6 Chapter 6

Section 6.1

6.1 Each $\mathbb{Z}/p_n\mathbb{Z}$ is a compact topological ring for the discrete topology. The direct product of compact topological rings is a topological ring that is compact for the product topology by Tychonov's theorem. Since $\mathbb{Z}_{(p_n)}$ is a closed subring of the direct product of the $\mathbb{Z}/p_n\mathbb{Z}$, the result follows.

Let G be the set of $x = (x_n) \in \mathbb{Z}_{(p_n)}$ such that x_n is eventually constant. The map $\varphi: G \to \mathbb{Z}$ such that $\varphi(x)$ is the value of all x_n for n large enough is an injective morphism from G into $\mathbb{Z}$. Since the sequence (p_n) is strictly increasing, it is onto. It is clear that G is dense in $\mathbb{Z}_{(p_n)}$.

The representation of $\mathbb{Z}_{(p_n)}$ by (p_n)-adic expansions shows that it is a Cantor space.

6.2 Assume first that $\varphi: X \to X'$ is a topological conjugacy from (X, T) onto (X', T'). Set $\alpha(x) = \varphi(x) - \varphi(0)$. Then α is another conjugacy from (X, T) onto (X', T'). It satisfies $\alpha(0) = 0$ and $\alpha(1) = \alpha(T0) = T'\alpha(0) = T'0 = 1$. Since $\mathbb{Z}$ is dense in X and X', it is an isomorphism. Conversely, assume that $\varphi: X \to X'$ is a group isomorphism. Since $\varphi(1)$ generates $\mathbb{Z}$, we have $\varphi(1) = 1$ or -1. In the first case, φ is a conjugacy. In the second case, the map $\psi(x) = -\varphi(x)$ is a conjugacy.

6.3 The map φ is a morphism because $y_0 + y_1q_1 + y_2q_1q_2 + \cdots = y_0 + p_1(y_1 + p_2(y_2 + \cdots))$. The inverse of φ is the map $x \mapsto y$ from $\mathbb{Z}_{(p_n)}$ defined by $y_0 = x_1$ and $y_n = (x_{n+1} - x_n)/p_n$.

6.4 Set $x_n = x \bmod (n + 1)!$ with $x_0 = 0$. Then $c_n = (x_{n+1} - x_n)/(n + 1)!$ with $n \geq 0$ gives the unique factorial expansion of x.

6.5 This holds because $1 + 2.2! + \cdots + n.n! = (n+1)! - 1$, as one may verify by induction on n.

6.6 The verification is easy.

6.7 Condition (KR1) is satisfied since $\cap_{n\geq 1} B(n) = \{0\}$. Condition (KR2) is also satisfied because $B(n+1) \subset B(n)$ and for every $j \leq p_{n+1}$,

$$B^j(n+1) = \cup_{k \equiv j \bmod p_n} B^k B(n).$$

Finally, since $\cap_{n\geq 1} T^{x_n} B(n) = \{x\}$, condition (KR3) is also satisfied.

6.8 If (X, T) and (X', T') are conjugate, then $\sigma(X) = \sigma(X')$. Indeed, for every prime p, p^n divides some p_n if and only if the group X has elements of order p^n.

Conversely, let $\sigma = \prod_p p^{n_p}$ be a supernatural number. Consider the odometer (Y, S) corresponding to the sequence $(q_1, q_2, \ldots)$ defined by

$$q_n = \prod_{n_p = \infty, p \leq n} p^n \times \prod_{n_p < \infty, p \leq n} p^{n_p}.$$

By construction, $\sigma(Y) = \sigma$.

Now let (X, T) be an arbitrary odometer corresponding to $(p_1, p_2, \ldots)$ such that $\sigma(X) = \sigma$. Note that for each $x \in X$ and each integer $q \geq 1$ the sequence $(x_n \bmod q)_{n\geq 1}$ is nondecreasing and eventually equal to an integer denoted $\max(x \bmod q)$.

Consider the continuous map $y \mapsto x$ from Y to X defined by

$$x_n = \max(y \bmod p_n).$$

Its inverse is the map $x \mapsto y$ from X to Y, where y_n is the unique integer $< q_n$ such that

$$y_n \equiv \max x \bmod p^m$$

for all $p < n$ with $m = n$ if $n_p = \infty$ and $m = n_p$ otherwise. The existence and uniqueness of y_n are guaranteed by the Chinese Remainder Theorem. This shows that (X, T) and (Y, S) are topologically conjugate. Thus, all odometers such that $\sigma(X) = \sigma(Y)$ are conjugate.

6.9 We have seen in Exercise 6.6 that the odometer (X, T) associated with the sequence $(p_1, p_2, \ldots)$ is the inverse limit of the finite groups $\mathbb{Z}/p_n\mathbb{Z}$.

6.10 Let $\varphi \colon \mathbb{Z}/p_1\mathbb{Z} \to \mathbb{Z}$ be defined by $\varphi(x) = 0$ if $x \neq -1$ and $\varphi(-1) = 1$. Then

$$T(x, y) = (x + 1, T^{\varphi(x)}y).$$

6.11 Arguing by contradiction, assume that there is an $\varepsilon_0 > 0$, a strictly increasing sequence n_i and a sequence x_i, y_i of points such that $d(x_i, y_i) \geq \varepsilon_0$ and x_i, $y_i \in \vee_{j=-n_i}^{j=n} T^{-j}C_{i,j}$, where $C_{i,j} \in \gamma$. Choose a subsequence i_k such that $x_{i_k} \to x$ and $y_{i_k} \to y$. Then $d(x, y) \geq \varepsilon_0$. For every $j \in \mathbb{Z}$, there is an ℓ_j such that x_{i_k}, $y_{i_k} \in T^{-j}C_{\ell_j}$ for an infinity of k. Thus, $x, y \in T^{-j}\overline{C_{\ell_j}}$ and therefore $d(x, y) \leq \varepsilon$. This implies $x = y$, a contradiction.

6.12 1. It is clear that γ is a closed cover of X by sets with radius at most ε. Next, for $i < j$, we have $C_i \cap C_j = C_i \cap \partial C_j$ because $\mathrm{int}(C_j) = B_j \setminus \overline{(B_0 \cup \cdots \cup B_{j-1})}$ and thus $C_i \cap C_j = C_i \cap \partial C_j = \partial C_i \cap \partial C_j$ because $\partial C_i \cap \mathrm{int}(C_j) \subset B_i \setminus (B_0 \cup \cdots \cup B_{j-1}) = \emptyset$.

By the Baire Category Theorem, the set $X \setminus D_\infty$ is dense. The map $\psi \colon X \setminus D_\infty$ is injective. Indeed, suppose that $\psi(x) = \psi(y)$. For every $\alpha > 0$, by Exercise 6.11, we can choose N such that $\mathrm{diam}(\vee_{|n| \leq N} T^n\gamma) < \alpha$. Then $\psi(x)_n = \psi(y)_n$ implies that x, y are in the same element of $\vee_{|n| \leq N} T^n\gamma$ and thus $d(x, y) < \alpha$. This shows that $x = y$. The extension of ψ to X is then a conjugacy from X to the closure of $\psi(X)$, which is a subshift of $A^\mathbb{Z}$.

6.13 The map ϕ is one-to-one out of the countable set of points that represent ordinary integers. It sends the Haar measure on $\mathbb{Z}_2$ to the Lebesgue measure on $[0, 1]$. Since moreover $\varphi(x + 1) = R(\varphi(x) + 1)$ for every $x \in \mathbb{Z}_2$ it is an isomorphism.

Section 6.2

6.14 The diagram of Figure 6.3 is stationary. The associated proper morphism $\tau \colon 0 \mapsto 1000, 1 \mapsto 1010$ is conjugate to $\sigma^2 \colon 0 \to 0100, 1 \to 0101$. The matrix $M = M(\tau)$ is

$$M = \begin{bmatrix} 3 & 1 \\ 2 & 2 \end{bmatrix}.$$

Its eigenvalues are 4, 1 with left eigenvectors $u = [1 \ 1]^t$ and $v = [1 \ -2]$. Thus, the dimension group is formed of the $\alpha u + \beta v$ such that $3\alpha \in \mathbb{Z}[1/2]$ and $3\beta \in \mathbb{Z}$, whence the result.

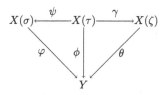

Figure A.12 The conjugacy from $X(\zeta)$ onto Y.

6.15 Let $\sigma: A^* \to A^*$ be a primitive morphism and let $X = X(\sigma)$. Let $\varphi: A^* \to B^*$ be a nonerasing morphism and let $Y = \varphi(X)$. Replacing if necessary σ by one of its powers, let $x = \sigma^\omega(r \cdot \ell)$ be an admissible fixed point of σ. Let $\psi: B \to \mathcal{R}_X(r \cdot \ell)$ be a coding morphism and let $\tau: B^* \to B^*$ be the morphism such that $\psi \circ \tau = \sigma \circ \psi$. By Proposition 6.2.12, the morphism τ is primitive. Thus, by Rauzy's Lemma (Proposition 6.2.10), applied to τ and $\phi = \varphi \circ \psi$, the shift space Y is conjugate to a primitive substitution shift $X(\zeta)$ (see Figure A.12).

6.16 Set $w_n = 0t_n$ and thus $w'_n = 2t_n$ for $n \geq 0$. We have then

$$w_{n+1} = \tau(w_n) = 0012\tau(t_n) = 0\tau(2t_n) = 0\tau(w'_n)$$

showing that $t_{n+1} = \tau(w'_n)$ for $n \geq 0$. Thus,

$$w_{n+1} = \tau(w_{n-1}w_{n-1}1w'_{n-1}) = w_n w_n 12\tau(w'_{n-1})$$
$$= w_n w_n 12t_n = w_n w_n 1w'_n.$$

The map θ sends the infinite word $\tau^\omega(0)$ to $\sigma^\omega(0)$ and thus maps $X(\tau)$ to $X(\sigma)$. Its inverse is the map that replaces 0 by 2 when there is a 1 just before. Thus, θ is a conjugacy.

6.17 The bispecial words of length at most 3 are ε, 0, 012 and 120. Their multiplicities are $m(\varepsilon) = m(0) = 0$, $m(012) = 1$, $m(120) = -1$.

Now let α be the map defined by $\alpha(x) = 012\tau(x)$. A bispecial word y of length at least 4 begins and ends with 012 (see the trees of left and right special words in Figure A.13).

Thus, $y = 012\tau(x) = \alpha(x)$, where x is a bispecial word.

Set $E_X(w) = \{(a, b) \in A \times A \mid awb \in \mathcal{L}(X)\}$. Let us verify that if $E_X(x) = E_X(012)$, the same holds for $E_X(y)$ with $y = \alpha(x)$. Indeed, since $0x0 \in \mathcal{L}(X)$, the word $\tau(0x0) = 0012\tau(x)0012 = 0y0012$ is also in $\mathcal{L}(X)$ and thus $(0, 0) \in \mathcal{E}(y)$. Since $2x0 \in \mathcal{L}(X)$ and since a letter 2 is always preceded by a letter 1, we have $bcxa \in \mathcal{L}(X)$. Thus, $\tau(12x0) = 12y0012 \in \mathcal{L}(X)$ and

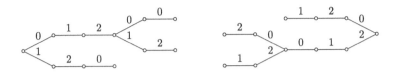

Figure A.13 The trees of left and right special words.

thus $(2, 0) \in E_X(y)$. The proof of the other cases is similar. The same property holds for a word x such that $E_X(x) = E_X(120)$.

We conclude that $b_n(X) = 0$ for every $n \geq 0$. Indeed, this is true for $n = 0, 1$ since $m(\varepsilon) = m(0) = 0$ and for $n = 3$ since $m(012) = 1$, $m(120) = -1$. Let $n \geq 4$ be such that there are bispecial words of length n. These bispecial words are of the form $\alpha^k(012)$, $\alpha^k(120)$ (note that these two words have the same length) and thus $b_n = m(\alpha^k(012)) + m(\alpha^k(120)) = m(012) + m(120) = 1 - 1 = 0$.

6.18   Let M be the composition matrix of τ and let v be the vector $(|\phi(b)|)_{b \in B}$. Let $B = (V, E, \leq)$ be the stationary diagram with matrices $(v, M, M, \ldots)$. Let P, Q be the nonnegative matrices, with P a partition matrix, such that $M = PQ$ defined by Proposition 6.2.8. Set $M' = QP$ and $w = \begin{bmatrix} 1 & 1 & \ldots & 1 \end{bmatrix}^t$. Let $B' = (V', E', \leq')$ be the ordered Bratteli diagram with matrices $(w, M', M', \ldots)$ with the order induced from $(V, E, \leq)$.

Let C be the index of the columns of P identified with $\{b_p \mid 1 \leq p \leq |\phi(b)|\}$. Let ζ be the morphism read on B'. Let $\gamma : B^* \to C^*$ be the morphism defined by $\gamma(b) = b_1 \cdots b_{|\phi(b)|}$. Let finally $\theta : C^* \to A^*$ be the letter-to-letter morphism defined by $\theta(b_p) = \phi(b)_p$.

It is clear that $\zeta \circ \gamma = \gamma \circ \tau$ since both are equal to the morphism read on PQP (with the order induced by B). The equality $\phi = \theta \circ \gamma$ holds by definition of γ and θ. This implies Assertion 1.

Assertion 2 is clear since ϕ is injective from $X(\tau)$ to X.

Assume finally that τ is eventually proper. We can also assume that $X(\tau)$ is aperiodic since otherwise the result is trivial. By Proposition 6.2.15 the system (X_E, T_E) is isomorphic to $(X(\tau), S)$. Since B is properly ordered, B' is also properly ordered and thus ζ is eventually proper by Proposition 6.2.2.

Consider $\tau : 0 \to 01, 1 \to 10$ and $\phi : 0 \to 01, 1 \to 0$. The matrices M, P, Q, M' are

$$
M = \begin{bmatrix} 1 & 1 \\ 1 & 1 \end{bmatrix}, \quad
P = \begin{bmatrix} 1 & 1 & 0 \\ 0 & 0 & 1 \end{bmatrix}, \quad
Q = \begin{bmatrix} 1 & 0 \\ 0 & 1 \\ 1 & 1 \end{bmatrix}, \quad
M' = \begin{bmatrix} 1 & 1 & 0 \\ 0 & 0 & 1 \\ 1 & 1 & 1 \end{bmatrix}.
$$

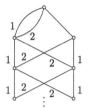 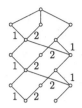

Figure A.14 The diagrams B and B'.

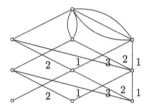

Figure A.15 A BV-representation of the Thue–Morse shift.

The morphisms γ, ζ and θ are

$$\gamma : 0 \to ab, 1 \to c \quad \zeta : a \to ab, b \to c, c \to cab$$
$$\theta : a \to a, b \to b, c \to a.$$

6.19 Set $\tau : u \to v, v \to wu, w \to wvu$. Then it is easy to verify that $\varphi_2 \circ \phi = \phi \circ \tau$ (note that the existence of such τ is guaranteed for every primitive morphism φ, see the proof of Proposition 4.6.3). The substitution τ is eventually proper and aperiodic and $\phi(B)$ is circular. Thus, by Proposition 6.2.15, a BV-representation of X is shown in Figure A.15. Note that the incidence matrix of τ is the matrix N_2 of Example 4.6.11.

6.20 Let σ be the Rudin–Shapiro morphism and let $X = X(\sigma)$. Consider the fixed point $\sigma^{2\omega}(b \cdot a)$. The set $\mathcal{R}(b \cdot a)$ of return words to $b \cdot a$ is formed of $w_1 = ab$, $w_2 = acabdb$, $w_3 = acdcacab$, $w_4 = acabdbdcdb$, $w_5 = abdb$, $w_6 = acdcacdcdbdcacab$, $w_7 = abdbdcdbdcacdcdb$, $w_8 = acdcacdcdbdcacdcdb$ and $w_9 = abdbdcdbdcacab$. Set $B = \{1, 2, 3, 4, 5, 6, 7, 8, 9\}$ and let $\phi : B^* \to \mathcal{R}(b \cdot a)^*$ be a coding morphism. The morphism τ such that $\phi \circ \tau = \sigma^2 \circ \phi$ is $\tau(1) = 12$, $\tau(2) = 1345$, $\tau(3) = 1634$, $\tau(4) = 13475$, $\tau(5) = 145$, $\tau(6) = 168932$, $\tau(7) = 147985$, $\tau(8) = 168985$, $\tau(9) = 147932$. Since τ is primitive and eventually proper and since ϕ is circular, we obtain a BV-representation of $X = X(\tau, \phi)$ using the stationary diagram with morphism τ.

6.21 Let $(V, E, \leq)$ be the stationary Bratteli diagram with matrix M. By Proposition 6.2.3, X is isomorphic to (X_E, V_E). By Exercise 5.12, the unique invariant Borel probability measure μ on (X_E, V_E) is given by (5.22).

6.22 Let n be the common length of the $\sigma(a)$ for $a \in A$. By Proposition 6.2.12, there is a primitive, eventually proper and aperiodic morphism $\tau \colon B^* \to B^*$ such that $X(\tau)$ is a derivative system of $X(\sigma)$ and moreover the spectral radiuses $\lambda(\sigma)$ and $\lambda(\tau)$ of the incidence matrices $M(\sigma)$ and $M(\tau)$ of σ, τ are equal. By Exercise 1.35, we have $\lambda(\sigma) = n$ and thus $\lambda(\tau) = n$. By Exercise 6.21, since the numbers α_b are rational, the image subgroup of $X(\tau)$ is a subgroup of $\mathbb{Q}$. Since $X(\tau)$ is an induced system of $X(\sigma)$, the image subgroup of $X(\sigma)$ is also a subgroup of $\mathbb{Q}$. By Corollary 6.1.4, this implies that $X(\sigma)$ is orbit equivalent to an odometer.

6.23 By Theorem 5.5.7, $X(\sigma)$ and $X(\sigma')$ are Kakutani orbit equivalent if and only if the image subgroups of $K^0(X(\sigma), S)$ and $K^0(X(\sigma'), S)$ are isomorphic as ordered groups. By Proposition 6.2.12, there are primitive eventually proper morphisms τ, τ' such that $X(\tau)$ is induced by $X(\sigma)$ and $X(\tau')$ by $X(\sigma')$ and the spectral radiuses satisfy $\lambda(\sigma) = \lambda(\tau)$ are $\lambda(\sigma') = \lambda(\tau')$. By Exercise 1.35, we have $\lambda(\sigma) = n$ and $\lambda(\sigma') = n'$. Now the subgroup generated by the α_a/n^k for some finite set of $\alpha_a \in \mathbb{Q}$ and all $k \geq 1$ is isomorphic, as ordered group, to $\mathbb{Z}[1/n]$. Finally $\mathbb{Z}[1/n]$ and $\mathbb{Z}[1/n']$ are isomorphic if and only if the prime factors of n, n' are the same.

6.24 The dimension group of the Rudin–Shapiro shift is $\mathbb{Z}[1/2]^3 \times \mathbb{Z}$ with positive cone $\mathbb{Z}_+[1/2] \times \mathbb{Z}[1/2]^2 \times \mathbb{Z}$ and unit $(1, 0, 0, 0)$ (Exercise 4.11). Thus, its image subgroup is $\mathbb{Z}[1/2]$ with unit 1. Finally, since the image subgroup is $\frac{1}{3}\mathbb{Z}[1/2]$ for the Thue–Morse shift, the last assertion follows.

6.25 Since σ differs from φ^5 by the order of the letters $\sigma(0)$, we have $M(\sigma) = M(\varphi)^5$. Since $\varphi(0)$ begins with 0^5, the language $\mathcal{L}(\sigma)$ contains all words $\varphi^n(0)^5$. On the contrary, $\mathcal{L}(\varphi)$ does not contain cubes w^3, as one may prove easily by induction on the length of w. This proves that $X(\sigma)$ and $X(\varphi)$ are not conjugate. Indeed, let $\gamma \colon X(\sigma) \to X(\tau)$ be a conjugacy defined by a block map f of window size s. Let w be a word of length exceeding s such that $w^5 \in \mathcal{L}(\sigma)$. Then $f(w^5)$ contains a word of exponent 3, a contradiction.

Section 6.3

6.26 The condition is obviously sufficient. Conversely, if $\mathcal{L}(X) \cap B^*$ is infinite, there are arbitrarily large integers n and $a \in A$ such that $\sigma^n(a) \in B^*$. If n is large enough, we can find i, j, k such that $i + j + k = n$ with

$$\sigma^i(a) = uby, \quad \sigma^j(b) = vbx, \quad \sigma^k(b) = w$$

with vx nonempty. In this way $\sigma^n(a) = \sigma^{j+k}(u)\sigma^k(vbx)\sigma^{j+k}(y)$ and thus $\sigma^k(vbx) \in B^*$. Thus, we find that $\sigma^j(b) = vbx$ and $\sigma^k(vbx)$ is in B^+.

6.27 Let $f: \mathcal{L}_k(X) \to A_k$ be a bijection extended as usual to a map $f: \mathcal{L}_{n+k-1}(X) \to \mathcal{L}_n(X^{(k)})$. For every $w \in \mathcal{L}_n(X^{(k)})$, we have

$$\mathcal{R}_{X^{(k)}}(w) = f(\mathcal{R}_X(u)),$$

where $w = f(u)$. Thus, we have for every $x \in \mathcal{R}_X(u)$,

$$|f(x)| = |x| - k + 1 \le K|u| - k + 1 \le K(n+k-1) - k + 1 \le K(k-1) + 1.$$

6.28 Let $x = \sigma^\omega(a)$. We have $x = a \prod_{i \ge 0} b^{2^i} dc^i$ as one may verify by computing $\sigma(x)$. For $0 \le i \le j \le (n-2)/2$, the word $b^i dc^j bw$ is in $\mathcal{L}(x)$ for some $w \in \{b, c, d\}^{n-i-j+2}$. This shows that $p_n(x)$ grows like n^2.

6.29 The necessity of the condition is proved with Lemma 6.3.10.

Let us show it is sufficient. Fix some n and let a be the maximum of the entries of $M(n)$. Let $h(n)$ be the vector with components $h_i(n)$ for $1 \le i \le t(n)$. By (5.13), we have $h(n) = M(n)h(n-1)$. Then,

$$a \min_i h_i(n) \le ||M(n)h(n)||_\infty = ||h(n+1)||_\infty \le L \min_i h_i(n)$$

and consequently a is less than L. This shows that the matrices $M(n)$ belong to a finite set.

6.30 Assume first that X is linearly recurrent with constant K. Let μ be an invariant probability measure on X. For $u \in \mathcal{L}_n(X)$, the family $\mathfrak{P} = \{S^j[vu] \mid v \in \mathcal{R}'_X(u), 0 \le j < |v|\}$ is a partition of X (see Proposition 4.1.1). Thus, we have

$$\sum_{v \in \mathcal{R}'_X(u)} |v|\mu([vu]) = 1.$$

Since μ is invariant, we have $\sum_{v \in \mathcal{R}'_X(u)} \mu([vu]) = \mu([u])$. Thus, we obtain

$$1 \le Kn \sum_{v \in \mathcal{R}'_X(u)} \mu([u]) = Kn\mu([u])$$

and finally $n\varepsilon_n(\mu) \ge 1/K$.

Conversely, assume $\inf n\varepsilon_n(\mu) > \varepsilon > 0$, where μ is some S-invariant measure on X. Let $u \in \mathcal{L}_n(X)$ and $w = w_1 w_2 \cdots w_N \in \mathcal{R}'_X(u)$ be of length N. We need to bound N/n. Consider, for $n \le k \le N$, the set W_k of words in $\mathcal{L}_N(X)$

that end with $w_1 \cdots w_k$. We set $U_k = \cup_{u \in W_k}[u]$. From the assumption we have $\mu(U_k) = \mu([w_1 \cdots w_k]) \geq \varepsilon/k$. Since the sets U_k are disjoint, we have

$$1 \geq \sum_{k=n}^{N} \frac{\varepsilon}{k} \geq \varepsilon \int_n^{N+1} \frac{dx}{x} \geq \varepsilon \log((N+1)/n)$$

and therefore we obtain the bound $N/n \leq \exp(1/\varepsilon)$. This shows the linear recurrence for the constant $K = \exp(1/\varepsilon)$.

Let $u \in \mathcal{L}_n(X)$. From Proposition 6.3.3 for all $v \in \mathcal{R}'_X(u)$ one has $(1/K)|u| \leq |v| \leq K|u|$. Consequently,

$$|u|\mu([u]) = |u| \sum_{v \in \mathcal{R}'_X(u)} \mu([vu]) \leq K \sum_{v \in \mathcal{R}'_X(u)} |v|\mu([vu]) = K.$$

Let μ, ν be two distinct ergodic measures on X. Since μ, ν are mutually singular (Theorem 3.8.11) the ratio $\nu([u])/\mu([u])$ should be unbounded when $|u|$ goes to infinity (indeed, otherwise, the ratio $\mu(U)/\nu(U)$ for U, V clopen would also be bounded, a contradiction with the fact that there exist clopen sets U_0, V_0 such that $\mu(U \Delta U_0), \nu(U \Delta U_0)$ are arbitrarily small). But one has

$$\frac{\nu([u])}{\mu([u])} = \frac{|u|\nu([u])}{|u|\mu([u])} \leq \frac{K}{\varepsilon}.$$

Thus, X is uniquely ergodic.

Section 6.4

6.31 For a primitive $\mathcal{S}$-adic shift, we have $\langle \tau[0, n) \rangle \to \infty$. Thus, (ii) and (iii) are equivalent with the same sequence τ.

(i) $\Rightarrow$ (iii) Let a_n be the first letter of $w^{(n)}$. Since $\tau_{[0,n)}(a_n)$ is a prefix of x with length that tends to ∞, this proves (iii).

(iii) $\Rightarrow$ (i) Set $w^{(0)} = x$. Let y_n be in $\tau_{[1,n)}(a_n)$ for each $n \geq 1$. Let y_{n_i} be a subsequence of the y_i converging to $y \in A_1^{\mathbb{N}}$. We replace τ by its telescoping with respect to the subsequence n_i and the sequence (a_n) by the sequence (a_{n_i}). Set $w^{(1)} = y$. Then $x = \tau_0(w^{(1)})$ and $\{w^{(1)}\} = \cap_{n \geq 1} \tau[1, n)(a_n)$. Continuing in this way, we build a sequence $w^{(n)}$ with the required properties.

6.32 Define σ_a by

$$\sigma_a(b) = \begin{cases} a\# & \text{if } b = a \\ b & \text{otherwise.} \end{cases}$$

6.33 The set $\mathcal{L}_2(X)$ is $\{aa, ab, ba, bb\}$ put in bijection with $\{x, y, z, t\}$. The morphism τ_2 is $x \to xyz$, $y \to xyt$, $z \to zx$, $t \to zy$. The composition matrices M and M_2 are

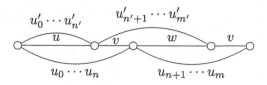

Figure A.16 The words u, v, w.

$$M = \begin{bmatrix} 2 & 1 \\ 1 & 1 \end{bmatrix}, \quad M_2 = \begin{bmatrix} 1 & 1 & 1 & 0 \\ 1 & 1 & 0 & 1 \\ 1 & 0 & 1 & 0 \\ 0 & 1 & 1 & 0 \end{bmatrix}.$$

The graph $\Gamma_2(X)$ is the complete graph on two vertices and thus a choice for the matrix P with rows a basis of its cycles is

$$P = \begin{bmatrix} 1 & 0 & 0 & 0 \\ 0 & 1 & 1 & 0 \\ 0 & 0 & 0 & 1 \end{bmatrix}, \quad N = \begin{bmatrix} 1 & 1 & 0 \\ 2 & 1 & 1 \\ 0 & 1 & 0 \end{bmatrix}.$$

The corresponding matrix N such that $PM_2 = NP$ is shown on the right. A left eigenvector for the maximal eigenvalue $\lambda^2 = (3 + \sqrt{5})/2$ is $\begin{bmatrix} 2\lambda & \lambda + 1 & 1 \end{bmatrix}$. We conclude by Proposition 4.6.6 that the dimension group has the form indicated (note that N is the matrix of Example 2.5.4).

6.34 By the hypothesis, there exist $u_0, u_1, \ldots$ and $u'_0, u'_1, \ldots$ with $u_i, u'_i \in U$ such that $u_0 u_1 \cdots = u'_0 u'_1 \cdots$. We may assume that $u_0 \neq u'_0$. For every $n \geq 0$, there is v_n and $n' \geq 0$ such that $u_0 \cdots u_n = u'_0 \cdots u'_{n'} v_n$. Since σ is injective, we have $v_n \notin U^*$. We may moreover assume that v_n is a prefix of $u'_{n'+1}$. Let $n < m$ be such that $v_n = v_m$. Set $u'_{n'+1} \cdots u'_{m'} = v_n w$ (see Figure A.16). Then $u = u'_0 \cdots u'_{n'}$, $v = v_n = v_m$ and w satisfy the required properties.

6.35 Set $U = \sigma(A)$. Since U is a prefix code, σ is injective on $A^{\mathbb{N}}$. If it is not injective on $A^{-\mathbb{N}}$, by the symmetric version of Exercise 6.34, there are words u, v, w such that $u, vu, vw, wv \in U^*$ but $v \notin U^*$. We may assume that u, v, w are chosen of minimal length. Since σ is left marked, either u is a prefix of w or w is a prefix of u. In the first case, set $w = uw'$. Then, since U is prefix, $u, wv = uw'v \in U^*$ imply $w'v \in U^*$. But then $(vu)(w'v)(u) = (vuw')(vu)$, a contradiction. Finally, if $u = wu'$, then $vw, vu = vwu' \in U^*$ imply $u' \in U^*$, and we may replace u, v, w by u', w, v, a contradiction again. Thus, σ is injective on $A^{-\mathbb{N}}$.

Assume now that $y \in B^{\mathbb{Z}}$ has two distinct representations, $y = S^k(\sigma(x)) = S^{k'}(\sigma(x'))$. We may assume $k = 0$. Since σ is injective on $A^{-\mathbb{N}}$, we have $k' \neq 0$. We will prove that y is periodic.

Let P be the set of proper prefixes of U. For $p \in P$ and $a \in A$, there is at most one $q \in P$ such that $p\sigma(a) \in U^*q$. We denote $q = p \cdot a$. Let $p_0 = y_{-k'} \cdots y_{-1}$. Since $y = \sigma(x) = S^{k'}(\sigma(x'))$, we have

$$\sigma(\cdots x'_{-2}x'_{-1})p_0 = \sigma(\cdots x_{-1}),$$
$$p_0\sigma(x_0x_1\cdots) = \sigma(x'_0x'_1\cdots).$$

As a consequence, there exists for each $n \in \mathbb{Z}$ a word $p_n \in P$ such that $p_n \cdot x_n = p_{n+1}$. Since σ is left marked, there is for every nonempty $p \in P$ at most one $a \in A$ such that $p \cdot a$ is in P. But all p_n are nonempty. This is clear if $n < 0$ since otherwise p_0 is also empty. For $n > 0$, we have

$$\sigma(\cdots x'_{i_n})p_n = \sigma(\cdots x_n),$$
$$p_n\sigma(x_{n+1}\cdots) = \sigma(x'_{i_n+1}\cdots)$$

and thus, since σ is injective on $A^{\mathbb{Z}}$, $p_n = \varepsilon$ implies that $x = x'$. Thus, x is periodic and y is also periodic.

Consider the labeled graph with P as set of vertices and edges (p, a, q) if $p \cdot a = q$. By what we have seen, the path $\cdots p_n \xrightarrow{x_n} p_{n+1} \cdots$ is a cycle and thus y is periodic.

Section 6.6

6.36 Set $x = 001^{\omega}$. Let $\varphi: 0 \to 01, 1 \to 1$ and $\phi: 0 \to 00, 1 \to 1$. Then $\varphi^{\omega}(0) = 01^{\omega}$ and $\phi \circ \varphi^{\omega}(0) = x$. Thus, 001^{ω} is substitutive. It cannot be purely substitutive because if $\psi(x) = x$, we have $\psi(0) = 001^n$ with $n \geq 0$ and thus $\psi(x)$ begins with 001^n001^n.

6.37 Let $\varphi: A^* \to A^*$ be a substitution and let y be a fixed point of φ. Let $\phi: A^* \to B^*$ be a letter-to-letter morphism and let $x = \phi(y)$. Let $X = X(\varphi)$ be the shift generated by y. For $k \geq 1$, let $f: \mathcal{L}_k(X) \to A_k$ be a bijection from $\mathcal{L}_k(X)$ onto an alphabet A_k and let $\gamma_k: X \to A_k^{\mathbb{Z}}$ be the corresponding higher block code. The shift $X^{(k)}$ is the substitution shift generated by the kth block presentation φ_k of φ and $z = \gamma_k(y)$ is a fixed point of φ_k. Define $\theta: A_k \to B$ by

$$\theta \circ f(y_n \cdots y_{n+k-1}) = \phi(y_{n+k-1}).$$

Then $\theta(z) = \phi(T^{k-1}x)$, which shows that the sequence $T^{k-1}x$ is substitutive.

6.38 We first prove that we can modify the pair (τ, ϕ) in such a way that

$$|\phi \circ \tau(b)| \geq |\phi(b)| \tag{A.7}$$

for every $b \in B$ and with strict inequality when $b = a$.

Since $\lim |\tau^n(a)| = \infty$, there are $1 \leq j < k$ such that $|\phi \circ \tau^j(b)| \leq |\phi \circ \tau^k(b)|$ for every $b \in B$ with strict inequality if $b = a$. Set $\tau' = \tau^{k-j}$, $\phi' = \phi \circ \tau^j$. Then

$$|\phi' \circ \tau'(b)| = |\phi \circ \tau^j \circ \tau^{k-j}(b)| = |\phi \circ \tau^k(b)|$$
$$\geq |\phi \circ \tau^j(b)| = |\phi'(b)|$$

for every $b \in B$ with strict inequality when $b = a$.

We now assume that (τ, ϕ) satisfies Eq. (A.7). Proceeding as in the proof of Rauzy's Lemma (Proposition 6.2.10), we define an alphabet $C = \{b_b \mid b \in B, 1 \leq p \leq \phi(b)\}$, a map $\theta : C \to A$ by $\theta(b_p) = (\phi(b))_p$ and a map $\gamma : B \to C^*$ by $\gamma(b) = b_1 b_2 \cdots b_{|\phi(b)|}$. In this way, we have $\theta \circ \gamma = \phi$.

Finally, we define the substitution ζ essentially as in the proof of Rauzy's Lemma (with the difference that this time the inequality $|\tau(b)| \geq |\phi(b)|$ is replaced by the weaker inequality (A.7)). For every $b \in B$, we have

$$|\gamma \circ \tau(b)| = |\phi \circ \tau(b)| \geq |\phi(b)|$$

by (A.7). Thus, we can define nonempty words $w_1, w_2, \ldots, w_{|\phi(b)|} \in C^*$ such that

$$\gamma \circ \tau(b) = w_1 w_2 \ldots w_{|\phi(b)|},$$

with $|w_1| > 1$ when $b = a$. Then we define the morphism $\zeta : C^* \to C^*$ by $\zeta(b_p) = w_p$. We have by construction $\zeta \circ \gamma = \gamma \circ \tau$ and thus $x = \theta \circ \zeta^\omega(a_1)$.

6.39 Let $\chi : 0 \to 0, 1 \to 1, 2 \to \varepsilon$ be the morphism erasing 0. Let $\mu : 0 \to 01, 1 \to 10$ be the Thue–Morse morphism and let $t = \mu^\omega(0)$. Since $\mu \circ \chi = \chi \circ \sigma$, we have $t = \chi(x)$.

Let τ be a nonerasing substitution such that $x = \tau^\omega(0)$. Then $\chi(\tau(2)^3) = \chi(\tau(2))^3$ is a factor of t that is a cube and thus $\chi(\tau(2)) = \varepsilon$. Since the factors of x in 2^* are $\varepsilon, 2, 22, 222$, this forces $\tau(2) = 2$. Now $\tau(0)$ is a prefix of x and thus $\tau(0) = 01u$ for some $u \in \{0, 1, 2\}^*$. Next, since $\tau(1222) = \tau(1)222$ cannot end with 2^4, we have $\tau(1) = ya$ with $y \in \{0, 1, 2\}^*$ and $a \in \{0, 1\}$. Finally, $\tau(10) = ya01u$ has a factor of length 3 in $\{0, 1\}^*$ while the factors of x in $\{0, 1\}^*$ are of length at most 2, a contradiction.

A.7 Chapter 7

Section 7.1

7.1 Let x be the fixed point $\sigma^\omega(a)$. Let π be the morphism from A^* onto $\{a, b\}^*$ defined by $\pi(a) = \pi(c) = a$ and $\pi(b) = \pi(d) = b$. The image of x by

Figure A.17 The inverse of the map π.

π is the Sturmian word y, which is the fixed point of the morphism $\tau: a \mapsto$ ab, $b \mapsto aba$. The word x can be obtained back from y by changing every other letter a into a c and any letter b after a c into a d (see Figure A.17). Thus, every word in $\mathcal{L}(y)$ gives rise to two words in $\mathcal{L}(x)$. In this way every bispecial word w of $\mathcal{L}(y)$ gives two bispecial words w', w'' of $\mathcal{L}(x)$ and their extension graphs are isomorphic to $\mathcal{E}(w)$. This proves the claim.

7.2 We have for $n \geq m$, $p_n(X) - p_m(X) = s_{n-1} + \cdots + s_m = (n - m)s_m$. Thus, $p_n(X) = ns_m + (p_m - ms_m)$.

7.3 This follows from the fact that, for every n, the word $\alpha^n(012)$ is bispecial not neutral by Exercise 6.17.

7.4 Let $g: \{a, c\}A^* \cap A^*\{a, c\} \to B^*$ be the map defined by

$$g(w) = \begin{cases} 3\tau(w) & \text{if } w \text{ begins and ends with } a \\ 3\tau(w)1 & \text{if } w \text{ begins with } a \text{ and ends with } c \\ 2\tau(w) & \text{if } w \text{ begins with } c \text{ and ends with } a \\ 2\tau(w)1 & \text{if } w \text{ begins with } c \text{ and ends with } c. \end{cases}$$

It can be verified that the set of bispecial words of $\mathcal{L}(Y)$ is the union of $\{\varepsilon, 2, 31\}$ and of the images by g of nonempty bispecial words of $\mathcal{L}(X)$ (described in the solution of Exercise 7.1). One may verify that these words are neutral. Since the words ε, 2, 31 are also neutral, the shift space X is neutral.

7.5 Consider a word $u \in \mathcal{L}(X)$, which is not a factor of a word in U. Since X is recurrent, there is a word v such that $uvu \in \mathcal{L}(X)$. Define a relation ρ on the set P of proper prefixes of u by $(p, q) \in \rho$ if one of the following conditions is satisfied (see Figure A.18).

(i) $q \in pU$.
(ii) $u = ps = qt$ and $svq = xyz$ with $x, z \in U$, $y \in U^*$, s a proper prefix of x, q a proper suffix of z.

Since U is bifix, the relation ρ is a partial bijection from P to itself. Assume first that U is X-maximal as a suffix code. Then the partial map ρ is onto. This

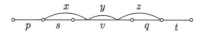

Figure A.18 The relation ρ.

implies that it is a bijection and thus u has a prefix in U. Since this is true for every long enough $u \in \mathcal{L}(X)$, it implies that U is X-maximal as a prefix code.

Assume now that U is neither X-maximal as a suffix code nor as a prefix code. Let $y, z \in \mathcal{L}(X)$ be such that $U \cup y$ is a prefix code and $U \cup z$ is a suffix code. Since X is recurrent, there is a v such that $yvz \in \mathcal{L}(X)$. Then $U \cup yvz$ is a bifix code, a contradiction.

7.6 This follows easily from the fact that T is also the set of suffixes of $\mathcal{R}'_X(u)u$ that are not suffixes of u.

7.7 One has clearly $U^* \cap \mathcal{L}(X) \subset \langle U \rangle \cap \mathcal{L}(X)$. Conversely, consider the coset graph C of U and let V be the set of labels of simple paths from ε to itself in C. By Proposition 7.1.33, we have $U \subset V$. Since C is Stallings reduced, V is a bifix code and since U is X-maximal, this implies $U = V \cap \mathcal{L}(X)$. Thus,

$$\langle U \rangle \cap \mathcal{L}(X) \subset \langle V \rangle \cap \mathcal{L}(X) = V^* \cap \mathcal{L}(X) = U^* \cap \mathcal{L}(X)$$

whence the conclusion.

7.8 If ua has a suffix in X, the number of parses of ua and u are the same. Otherwise, since U is a maximal suffix code, ua is a suffix of a word in U and thus ua has one more parse than u, namely $(ua, \varepsilon, \varepsilon)$. This proves (7.10).

Next $d_U(v) \le d_U(uvw)$ since U is X-maximal as a prefix code and as a suffix code. Indeed, every parse of v extends to a parse of uvw. Next, if v is not a factor of a word in U, let (s, x, p) be a parse of uvw. Since v is not a factor of a word of U, it cannot be a factor of any of s, x or p. Thus, there is a parse (q, y, r) of v and a factorization $x = zyt$ with $z, y, t \in U^*$ such that $sz = uq$ and $rw = zp$. This shows that every parse of uvw is an extension of a parse of v and thus that $d_U(uvw) = d_U(v)$. This proves (7.11).

7.9 Let $w \in \mathcal{L}(X)$ be such that $d_U(w) = d_U(X)$. Let $u \in \mathcal{L}(X)$ not a factor of a word in U. Since X is recurrent, there is a $v \in \mathcal{L}(X)$ such that $uvw \in \mathcal{L}(X)$. By Eq. (7.11), we have $d_U(u) = d_U(uvw) \ge d_U(w)$. Thus, $d_U(u) = d_U(X)$.

7.10 Every word of length at least n has clearly n parses.

7.11 For each i with $2 \leq i \leq d_U(X)$, let S_i be the set of proper suffixes s of U such that $d_U(s) = i$. Then S_i is a prefix code. Indeed, if $s, t \in S_i$ and if s is a proper prefix of t, then $d_U(s) = d_U(t)$ implies that t has a suffix in U by Eq. (7.10) of Exercise 7.8, a contradiction. Next, S_i is an X-maximal prefix code. Indeed, let $w \in \mathcal{L}(X)$ be long enough so that $d_U(w) = d$. Then w has nonempty prefixes $s_2, \ldots, s_d$ such that $s_i \in S_i$ for $2 \leq i \leq d$. We conclude that the set S of nonempty proper prefixes of U is a disjoint union of $d - 1$ X-maximal prefix codes.

7.12 Set $d = d_U(X)$. Let P be the set of proper prefixes of U. By the well-known formula relating the number of leaves of a tree to the number of children of its interior nodes, we have

$$\text{Card}(U) - 1 = \sum_{p \in P} (r_X(p) - 1). \tag{A.8}$$

By (the dual of) Exercise 7.11, the set $P \setminus \{\varepsilon\}$ is a disjoint union of $d - 1$ X-maximal suffix codes $V_1, \ldots, V_{d-1}$. Set $\rho(u) = r_X(u) - 1$ and, for $V \subset \mathcal{L}(X)$, denote $\rho(V) = \sum_{v \in V} \rho(v)$. By Lemma 7.1.18, we have $\rho(u) = \sum_{a \in L(u)} \rho(au)$. This implies that for any X-maximal suffix code V, one has

$$\rho(V) = \rho(\varepsilon) = \text{Card}(A) - 1,$$

where the last equality results of the hypothesis $A \subset \mathcal{L}(X)$. Thus, we have $\text{Card}(U) - 1 = \rho(P) = \rho(\varepsilon) + \sum_{i=1}^{d-1} \rho(V_i) = (\text{Card}(A) - 1)d$.

7.13 Suppose first that U is a finite X-maximal bifix code of X-degree $d = d_U(X)$ and let $H = \langle U \rangle$ be the subgroup generated by U. Let $w \in \mathcal{L}(X)$ be a word that is not a factor of a word in U. Let Q be the set of suffixes of w that are proper prefixes of U. Then, by Exercise 7.9, w has d parses and thus $\text{Card}(Q) = d$.

Moreover, we claim that it follows from Exercise 7.7 that the cosets Hq for $q \in Q$ are distinct. Indeed, let $p, q \in Q$ be such that $Hp = Hq$. Since p, q are suffixes of w, one is a suffix of the other. Assume that $q = tp$. Then $Hp = Htp$ implies $Ht = H$ and thus $t \in H$. By Exercise 7.7, this implies $t \in U^*$. Since t is a proper prefix of U, we conclude that $t = \varepsilon$ and thus $p = q$, which establishes the claim.

Consider the set $K = \{v \in F(A) \mid Qv \subset HQ\}$. It is a subgroup of $F(A)$. Indeed, by what precedes, the map $p \mapsto q$ if $pv \in Hq$ is a permutation of Q for every $v \in K$.

Next, we have $\mathcal{R}_X(w) \subset K$. In fact, consider $v \in \mathcal{R}_X(w)$. For every $p \in Q$, since U is X-complete, there is some $x \in U^*$ and some proper prefix q of U such that $pv = xq$. But since v is in $\mathcal{R}_X(w)$, pv ends with w and thus $q \in Q$.

Now, by Theorem 7.1.15, $\mathcal{R}_X(w)$ generates $F(A)$ and thus $K = F(A)$. We conclude that $F(A) \subset HQ$ and thus that Q is a set of representatives of the cosets of H. Thus, H has index d. Since U generates a subgroup of index d and since $\text{Card}(U) - 1 = d(\text{Card}(A) - 1)$, we conclude by Schreier's Formula that U is a basis of H.

Conversely, if the bifix code $U \subset \mathcal{L}(X)$ is a basis of a subgroup H of index d, let C be the coset graph of U. By Proposition 7.1.33, C is the Stallings graph of a subgroup of index d. Moreover U is contained in the set V of labels of simple paths from ε to ε. Then $W = V \cap \mathcal{L}(X)$ is an X-maximal bifix code of X-degree at most d. By Exercise 7.12, we have $\text{Card}(W) \leq d(\text{Card}(A) - 1)$ and thus, by Schreier's Formula $\text{Card}(W) \leq \text{Card}(U)$. This forces $U = W$ and concludes the proof.

7.14 Set $V = U \cap \mathcal{L}(X)$. Then V is a finite X-maximal bifix code of X-degree $e \leq d$. By Exercise 7.13, the set V is a basis of a subgroup of index e. Since $\langle V \rangle \subset \langle U \rangle$, the index e of $\langle V \rangle$ is a multiple of the index d of $\langle U \rangle$. Since $e \leq d$, this forces $d = e$.

7.15 Let U be the bifix code that generates the submonoid $\varphi^{-1}(1_G)$ and let $V = U \cap \mathcal{L}(X)$. By Exercise 7.14, the set U is a basis of a subgroup of index $\text{Card}(G)$ of $F(A)$.

Let $v \in V$ be of maximal length. Then v is not a proper factor of a word in V and thus it has d suffixes that are prefixes of V. Any two distinct such suffixes s, t have distinct images in G. Indeed, one of them, say s is a suffix of the other. But if $t = us$ with $\varphi(s) = \varphi(t)$, then $\varphi(u) = 1_G$ and thus $u \in V^*$, which forces $u = \varepsilon$. This forces the map $s \mapsto \varphi(s)$ to be surjective.

7.16 Let X be a minimal dendric shift and let U be a finite X-complete bifix code. Let $f: B^* \to A^*$ be a coding morphism for U. Set $Y = f^{-1}(X)$. By Exercise 7.13, the set U is a basis of a subgroup H of finite index of $F(A)$. Let $\varphi: F(A) \to G$ be the morphism from $F(A)$ onto the representation of $F(A)$ by permutations on the cosets of H.

Let $r, s \in \mathcal{L}(Y)$. Since X is recurrent, there is a word u such that $f(r)uf(s) \in \mathcal{L}(X)$. Set $t = f(r)uf(s)$. By Theorem 7.3.36, the set $\mathcal{R}_X(t)$ is a basis of $F(A)$. Let $\alpha: B^* \to A^*$ be a coding morphism for $\mathcal{R}_X(t)$ and let $\beta = \varphi \circ \alpha$. Since $\mathcal{R}_X(t)$ is a basis of $F(A)$, the morphism β is surjective. Now $Y = \alpha^{-1}(X)$ is, by Theorem 7.1.24, a minimal dendndric shift. By Exercise 7.15, this implies that the restriction of β to $\mathcal{L}(Y)$ is surjective.

Thus, there is some $v \in \mathcal{L}(X)$ such that tv is in $\mathcal{L}(X)$, tv ends with t and $\varphi(tv) = 1_G$. Set $tv = f(r)qf(s)$ (see Figure A.19). Now, since $f(r), f(r)qf(s)$ are in H, we have $qf(s) \in H$ and thus also $q \in H$. Let

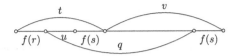

Figure A.19 Proving that Y is recurrent.

$w \in \mathcal{L}(Y)$ be such that $f(w) = q$. Then rws is in $\mathcal{L}(Y)$ and this shows that Y is recurrent. Finally, using Corollary 7.1.17, we obtain that Y is minimal.

7.17 The right-special words are the suffixes of the words $\sigma^n(c)$ for $n \geq 1$ and the left-special words are the prefixes of the words $\sigma^n(a)$ or $\sigma^n(c)$ for $n \geq 1$, as one may verify. Let us show by induction on the length of w that for any bispecial word $w \in \mathcal{L}(X)$, the graph $\mathcal{E}(w)$ is a tree. It is true for $w = c$ and $w = ac$. Assume that $|w| \geq 2$. Either w begins with a or with c. Assume the first case. Then w begins and ends with ac. We must have $w = ac\sigma(u)$, where u is a bispecial word beginning and ending with c. In the second case, w begins with $cbac$ and ends with ac. We must have $w = cbac\sigma(u)$, where u is a bispecial word beginning with a. In both cases, by induction hypothesis, $\mathcal{E}(u)$ is a tree and thus $\mathcal{E}(w)$ is a tree.

7.18 Denote by $LS(X)$ (resp. $LS_{\geq n}(X)$) the set of left-special words in $\mathcal{L}(X)$ (resp. $\mathcal{L}_{\geq n}(X)$).

Assume first that X is eventually dendric with threshold m. Then any word w in $LS_{\geq m}(X)$ has at least one right extension in $LS(X)$. Indeed, since $R_1(w)$ has at least two elements and since the graph $\mathcal{E}_1(w)$ is connected, there is at least one element r of $R_1(w)$ that is connected by an edge to more than one element of $L_1(w)$ and thus that $wr \in LS(X)$.

Next, the symmetric of Eq. (7.3) shows that for any $w \in LS_{\geq m}(X)$ that has more than one right extension in $LS(X)$, one has $\ell(wb) < \ell(w)$ for each such extension. Thus, the number of words in $LS_{\geq m}(X)$ that are prefixes of one another, and that have more than one right extension, is bounded by Card(A). This proves that there exists an $n \geq m$ such that for any $w \in LS_{\geq n}(X)$ there is exactly one $b \in A$ for which $wb \in LS(X)$ is left-special. Moreover, one has then $\ell(wb) = \ell(w)$ by the symmetric of Eq. (7.3). This proves the uniqueness of r. The proof for left extensions of right-special words is symmetric.

Conversely, assume that the regular bispecial condition is satisfied for some integer n. For any word w in $\mathcal{L}_{\geq n}(X)$, the graph $\mathcal{E}_1(w)$ is acyclic since all vertices in $R_1(w)$ except at most one have degree 1. Let $w \in LS_{\geq n}(X)$. If w is not bispecial, there is exactly one b such that $w \in \mathcal{L}(X)$ and then $wb \in LS(X)$.

Figure A.20 The extension graph of w.

If it is bispecial, it is regular and there is exactly one b such that $wb \in LS(X)$. Thus, in both cases, there is exactly one b such that $wb \in LS(X)$. One has then $L(wb) \subset L(w)$. Thus, there exists an $N \geq n$ such that for every $w \in LS_{\geq N}(X)$, one has $wb \in LS(X)$ and moreover $L(wb) = L(w)$. But for such a w, the extension graph $\mathcal{E}(w)$ is connected and thus it is a tree. This shows that X is eventually dendric.

7.19 1. We have $\sigma = \beta_{cb} \circ \beta_{bc} \circ \beta_{cb} \circ \beta_{ba} \circ \beta_{ac}$ since, for example,

$$a \xrightarrow{\beta_{ac}} a \xrightarrow{\beta_{ba}} ba \xrightarrow{\beta_{cb}} cba \xrightarrow{\beta_{bc}} bcba \xrightarrow{\beta_{cb}} cbccba.$$

The extension graph of the word $w = cbccb$ is shown in Figure A.20. Since this graph has a cycle, the shift $X(\sigma)$ is not dendric.

2. Let $\alpha = (\beta_{a_n b_n})$ be the directive sequence of morphisms defining the Brun shift X. Since α is primitive, there exists an increasing sequence (n_k) of integers such that the set $\{a_{n_i} \mid n_k \leq n_i < n_{k+1}\}$ is equal to A. Then all morphisms $\beta_{n_k} \circ \cdots \circ \beta_{n_{n+1}-1}$ are left proper. Indeed, let $\sigma = (\beta_{a_1 a_2})^{i_1} \circ (\beta_{a_2 a_3})^{i_2} \circ \cdots \circ (\beta_{a_n a_{n+1}})^{i_n}$ with $i_j = 1$ or 2. Then all words $\sigma(a_1), \ldots, \sigma(a_{n+1})$ begin with a_1. Using finally Lemma 6.4.9, we obtain the conclusion.

Section 7.2

7.20 The equivalence of (i) and (ii) is Corollary 7.2.7. Next, the implication (i) $\Rightarrow$ (iii) results from Proposition 1.2.9. Finally we have seen in Exercise 1.64 that (iii) $\Rightarrow$ (ii).

Section 7.3

7.21 This follows from the fact that every word of the even code U of length at least 2 is not an internal factor of U and has two parses.

7.22 Set $S = \mathcal{L}(X)$. Let P be the set of proper prefixes of $\mathcal{C}R_X(U)$. For $q \in P$, we define $\alpha(q) = \text{Card}\{a \in A \mid qa \in P \cup \mathcal{C}R_X(U)\} - 1$. For $P' \subset P$, we set $\alpha(P') = \sum_{p \in P'} \alpha(p)$. Since $\mathcal{C}R_X(U)$ is a finite nonempty prefix code, we have, by a well-known property of trees, $\mathrm{Card}(\mathcal{C}R_X(U)) = 1 + \alpha(P)$. Let P' be the set of words in P that are proper prefixes of U and let $Y = P \setminus P'$. Since P' is the set of proper prefixes of U, we have $\alpha(P') = \mathrm{Card}(U) - 1$.

For $u \in \mathcal{L}(X)$, set

$$\rho(u) = \begin{cases} r_X(u) - 1 & \text{if } u \neq \varepsilon \\ \mathrm{Card}(A) - 2 & \text{otherwise.} \end{cases}$$

In this way, we have $\rho(u) = \sum_{a \in L(u)} (au)$ for every $u \in \mathcal{L}(X)$. Thus, if Y is an X-maximal suffix code, we have $\sum_{y \in Y} \rho(y) = \rho(\varepsilon) = \mathrm{Card}(A) - 2$.

Since $P \cup \mathcal{C}R_X(U) \subset S$, one has $\alpha(q) \leq \rho(q)$ for any nonempty $q \in P$. Moreover, since X is recurrent, and since U has empty kernel, any word of S with a prefix in U is comparable for the prefix order with a word of $\mathcal{C}R_X(U)$. This implies that for any $q \in Y$ and any $b \in \mathcal{R}(q)$, one has $qb \in P \cup \mathcal{C}R(U)$. Consequently, we have $\alpha(q) = \rho(q)$ for any $q \in Y$. Thus, we have shown that $\mathrm{Card}(\mathcal{C}R_X(U)) = 1 + \alpha(P') + \rho(Y) = \mathrm{Card}(U) + \rho(Y)$.

Let us show that Y is an X-maximal suffix code. This will imply our conclusion. Suppose that $q, uq \in Y$ with u nonempty. Since q is in Y, it has a proper prefix in U. But this implies that uq has an internal factor in U, a contradiction. Thus, Y is a suffix code. Consider $w \in S$. Then, for any $x \in U$, there is some $u \in S$ such that $xuw \in S$. Let y be the shortest suffix of xuw that has a proper prefix in U. Then $y \in Y$. This shows that Y is an X-maximal suffix code.

A.8 Chapter 8

Section 8.1

8.1 Let (I, T) be an interval exchange transformation. Suppose that its natural coding $X = X(T)$ is an Arnoux-Rauzy shift. Since $X(T)$ is minimal, T is also minimal and thus the natural coding $\Sigma_T : I \to X$ is injective. For $w \in \mathcal{L}(X)$ long enough, the length of the interval $I(w)$ is arbitrary small and thus $\mathrm{Card}(R(w)) \leq 2$. This implies that $\mathrm{Card}(A) = 2$ and that X is a Sturmian shift.

8.2 If φ is not continuous, there is some $x \in [0, 1)$ such that $\mu(\{x\}) > 0$. This is not possible since T is minimal. For the same reason, φ is strictly increasing and thus bijective. Thus, φ is a homeomorphism.

Let us show that φ is a morphism of dynamical systems. For $1 \leq i \leq k$, set

$$\Delta_i = [\beta_{i-1}, \beta_i), \quad T\Delta_{\pi(i)} = [\gamma_{i-1}, \gamma_i)$$

and $\delta_i = \varphi(\beta_i), \varepsilon_j = \varphi(\gamma_j)$.

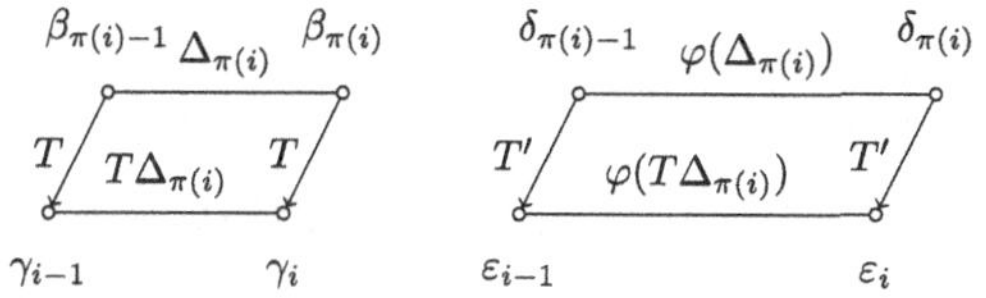

Figure A.21 The action of T and T'.

Let $T' = \varphi \circ T \circ \varphi^{-1}$. Since φ is increasing, we have $T'(\varphi(\Delta_{\pi(i)})) \subset [\varepsilon_{i-1}, \varepsilon_i)$ (see Figure A.21). Since μ is invariant, T' is a translation from $\varphi(\Delta_{\pi(i)})$ onto $[\varepsilon_{i-1}, \varepsilon_i)$. Thus, $T' = T_{\mu,\pi}$.

8.3 We have

$$\alpha_i = \sum_{\pi^{-1}(j) < \pi^{-1}(i)} \lambda_j - \sum_{j<i} \lambda_j. \tag{A.9}$$

Indeed, the first term of the right-hand side is the left boundary of $T\Delta_i$ and the second one is the left boundary of Δ_i. A pair (i, j) of integers with $1 \leq i < j \leq k$ is an *inversion* of a permutation $\sigma \in \mathfrak{S}_k$ if $\sigma(i) > \sigma(j)$. Setting

$$M_{ij} = \begin{cases} 1 & \text{if } (i, j) \text{ is an inversion of } \pi^{-1} \\ -1 & \text{if } (j, i) \text{ is an inversion of } \pi^{-1} \\ 0 & \text{otherwise,} \end{cases}$$

we obtain the desired antisymmetric matrix.

For $\pi = (123)$ and $\pi' = (13)$, the matrices are respectively

$$M_\pi = \begin{bmatrix} 0 & 1 & 1 \\ -1 & 0 & 0 \\ -1 & 0 & 0 \end{bmatrix}, \quad M_{\pi'} = \begin{bmatrix} 0 & 1 & 1 \\ -1 & 0 & 1 \\ -1 & -1 & 0 \end{bmatrix}.$$

8.4 Assume that (I, T) is not regular and thus that there is a connection. We will show that, assuming (I, T) irrational, we obtain a contradiction. Set $\Delta_i = [\beta_{i-1}, \beta_i)$ and $T\Delta_{\pi(i)} = [\gamma_{i-1}, \gamma_i)$ for $1 \leq i \leq k$. Let $1 \leq r, s \leq k - 1$ and $t \geq 0$ be such that $T^{t+1}\beta_r = \beta_s$ in such a way that $(r + 1, s + 1, t + 1)$ is a connection. For $1 \leq u \leq t$, let j_u with $0 \leq j_u \leq k - 1$ be such that $T^u\beta_r \in \Delta_{j_u+1}$ (note that we did not define j_0).

Define k_u by $k_u + 1 = \pi^{-1}(j_u + 1)$ for $1 \leq u \leq t$ and $k_0 + 1 = \pi^{-1}(r + 1)$. Then we have $T\beta_r = \gamma_{k_0}$ and for $1 \leq u \leq t$,

$$T^{u+1}\beta_r = T^u\beta_r + (\gamma_{k_u} - \beta_{j_u}). \tag{A.10}$$

Summing up these equalities gives

$$T^{t+1}\beta_r = \sum_{u=0}^{t} \gamma_{k_u} - \sum_{u=1}^{t} \beta_{j_u}. \tag{A.11}$$

Thus, setting now $j_0 = s$, the equality $T^{t+1}\beta_r = \beta_s$ gives $\sum_{u=0}^{t}(\gamma_{k_u} - \beta_{j_u}) = 0$ and thus

$$0 = \sum_{u=0}^{t} \left(\sum_{j=1}^{k_u} \lambda_\pi(j) - \sum_{j=1}^{j_u} \lambda_j \right) = \sum_{j=1}^{k}(c_j \lambda_{\pi(j)} - d_j \lambda_j), \tag{A.12}$$

where

$$c_j = \text{Card}\{u \mid 0 \le u \le t,\ k_u \ge j\},$$
$$d_j = \text{Card}\{u \mid 0 \le u \le t,\ j_u \ge j\}.$$

Since the numbers λ_j are independent over $\mathbb{Q}$, we obtain $c_j = d_{\pi(j)}$ for $1 \le j \le k$. We claim that, since π is irreducible and since $c_1 \ge c_2 \ge \cdots \ge c_k = 0$ and $d_1 \ge d_2 \ge \cdots \ge d_k = 0$, all the coefficients c_j, d_j are zero. Indeed, otherwise, let j be maximal such that $c_{j+1} = \cdots = c_n = d_{j+1} = \cdots = d_n$. Since π and π^{-1} are irreducible, there are $j + 1 \le i, i' \le k$ such that $\pi(i), \pi^{-1}(i') \le j$. Then $c_i = d_{i'} = 0$, which implies $c_j = d_j = 0$ since the sequences $(c_j), (d_j)$ are nonincreasing, and thus a contradiction. Finally, since all d_j are zero, we have $j_u = 0$ for all u, contradiction with the choice $j_0 = s$.

8.5 Let $\mathcal{I}(I, T)$ denote the family of invariant measures on (I, T). We have, by Exercise 8.2,

$$\mu \in \mathcal{I}(I, T_{\nu,\pi}) \Leftrightarrow \nu \in \mathcal{I}(I, T_{\mu,\pi}).$$

Thus, since M_π depends only on π, we only need to prove that, for $T = T_{\lambda,\pi}$ and $\mu \in \mathcal{I}(I, T)$, one has $\mu^t M_\pi \lambda = 0$.

We have indeed, since μ is invariant, $0 = \int (Tx - x) d\mu = \sum_{i=1}^{k}(M_\pi \lambda)_i \mu$ $(\Delta_i) = \mu^t M_\pi \lambda$.

8.6 Let λ', λ'' be invariant measures on (I, T) and suppose that $\lambda' - \lambda'' \in \ker M_\pi$. Since the interval exchange transformations $T_{\lambda',\pi}$ and $T_{\lambda'',\pi}$ are continuously isomorphic, we have for all $n \ge 1$ and $1 \le i \le k$,

$$\chi_{\Delta_i'}^{(n)}(0) = \chi_{\Delta_i''}^{(n)}(0), \tag{A.13}$$

where Δ_i', Δ_i'' are the intervals exchanged by $T_{\lambda',\pi}$ and $T_{\lambda'',\pi}$ respectively. Indeed, set $\lambda_t = t\lambda' + (1 - t)\lambda''$ for $0 \le t \le 1$ and let $\Delta_{t,i} = [\beta_{t,i-1}, \beta_{t,i}]$ be the intervals exchanged by $T_{\lambda_t,\pi}$. For all n, i either $T^n_{\lambda_t,\pi}0 > \beta_{t,i-1}$ for all t, or $T^n_{\lambda_t,\pi}0 = \beta_{t,i-1}$ for all t, or $T^n_{\lambda_t,\pi}0 < \beta_{t,i-1}$ for all t.

Now, using (A.13) and the fact that $M_\pi \lambda' = M_\pi \lambda''$, we obtain

$$T^n_{\lambda',\pi}0 = \sum_{i=0}^{k} \chi^{(n)}_{\Delta'_i}(0)(M_\pi\lambda')_i = \sum_{i=0}^{k} \chi^{(n)}_{\Delta''_i}(0)(M_\pi\lambda'')_i = T^n_{\lambda'',\pi}0.$$

Since $T_{\lambda',\pi}$ and $T_{\lambda'',\pi}$ are minimal, we conclude that $T_{\lambda',\pi} = T_{\lambda'',\pi}$ and thus that $\lambda' = \lambda''$.

8.7 We may assume that $T_{\lambda,\pi}$ has $k-1$ points of discontinuity (otherwise, we can reduce k without changing the rank of M_π). Let $W = \mathbb{R}^k/(V_{\lambda,\pi} + \ker(M_\pi))$ and let $\alpha: V_{\lambda,pi} \to W^*$ be the linear map defined for $v \in V_{\lambda,\pi}$ and $u \in \mathbb{R}^k$ by

$$(\alpha u)(v) = u^t M_\pi v.$$

Exercise 8.6 shows that this map is well defined and that it is injective. Since $V_{\lambda,\pi} \cap \ker(M_\pi) = \{0\}$, we have

$$\dim W = k - \dim V_{\lambda,\pi} - \dim \ker(M_\pi) = \operatorname{rank} M_\pi - \dim V_{\lambda,\pi}. \quad \text{(A.14)}$$

Since α is injective, we have also $\dim V_{\lambda,\pi} \leq \dim W$ and thus, using (A.14), $\dim V_{\lambda,\pi} \leq \operatorname{rank} M_\pi - \dim V_{\lambda,\pi}$ whence the conclusion $2\dim V_{\lambda,\pi} \leq \operatorname{rank} M_\pi$.

8.8 We use an induction on the length of w. The property is true if w is the empty word. Next, assume that $I(w)$ is a semi-interval and let a be a letter. Then $T(I(aw)) = T(\Delta_a) \cap I(w)$ is a semi-interval since $T(\Delta_a)$ is a semi-interval and also $I(w)$ by induction hypothesis. Since $I(aw) \subset \Delta_a$, the set $T(I(aw))$ is a translation of $I(aw)$, which is therefore also a semi-interval.

8.9 The proof is symmetrical to the proof for $I(w)$, using this time the fact that $T^{-1}J(wa) = J(w) \cap \Delta_a$ is a semi-interval and thus that $J(wa)$ is a semi-interval since $J(wa) \subset T\Delta(a)$.

8.10 We may consider the set $I \times Q$ for $Q = \{1, \ldots, n\}$ as a semi-interval by placing $I \times \{1\}, \ldots, I \times \{n\}$ successively on a line. Then S maps the semi-interval (Δ_a, i) onto $(T\Delta_a, g_a(i))$.

Let U be a finite X-maximal bifix code. By Exercise 7.13, U is a basis of a subgroup H of finite index n in $F(A)$. Let $Q = \{1, \ldots, n)\}$ be the set of cosets of H with 1 being H. We may identify the decoding of X by U with the system induced by the skew product S on $I \times \{1\}$, which is an interval exchange.

8.11 The morphism $\psi : a \mapsto baccb, b \mapsto bacc, c \mapsto bacb$ is conjugate to $a \mapsto accbb, b \mapsto accb, c \mapsto acbb$, which is primitive and proper. Thus, the dimension group of $X(\psi)$ is the group Δ_M of the incidence matrix

$$M = \begin{bmatrix} 1 & 2 & 2 \\ 1 & 1 & 2 \\ 1 & 2 & 1 \end{bmatrix}.$$

The eigenvalues of M are $-1, \lambda, 1/\lambda$, where $\lambda = 2 + \sqrt{5}$. Thus, the dimension group of X is isomorphic to $\mathbb{Z}[\frac{1+\sqrt{5}}{2}] \times \mathbb{Z}$ with positive cone $\mathbb{Z}_+[\frac{1+\sqrt{5}}{2}] \times \mathbb{Z}$ and unit $(1, 0)$.

A.9 Chapter 9

Section 9.2

9.1 Since $\|M\|^2 = \|M^*M\| \leq \|M^*\| \|M\|$, we have $\|M\| \leq \|M^*\| \leq \|M^{**}\| = \|M\|$ and thus $\|M\| = \|M^*\|$.

9.2 If M is self-adjoint, then $\|M\| = \mathrm{spr}(M)$ by Proposition 9.2.1. In the general case, we have

$$\|M\|^2 = \|M^*M\| = \mathrm{spr}(M^*M)$$

since M^*M is self-adjoint.

9.3 The spectrum of $\pi(M)$ is contained in the spectrum of M. Thus, if M is self-adjoint, $\pi(M)$ is also self-adjoint and we have by Proposition 9.2.1,

$$\|\pi(M)\| = \mathrm{spr}(\pi(M)) \leq spr(M) = \|M\|.$$

In the general case,

$$\|\pi(M)\|^2 = \|\pi(M)^*\pi(M)\| = \|\pi(M^*M)\| \leq \|M^*M\| = \|M\|^2.$$

Section 9.3

9.4 Denote α_n the embedding of $\mathfrak{A}_n$ into $\mathfrak{A}_{n+1}$ and by β_n the embedding of $\mathfrak{B}_n$ into $\mathfrak{B}_{n+1}$. Each $\mathcal{A}_n$ is isomorphic to $\mathcal{B}_n$, as one can prove easily by induction on n. Let $\varphi_n : \mathfrak{A}_n \to \mathfrak{B}_n$ be such isomorphism. Then $\varphi_{n+1} \circ \alpha_n$ and

$\beta_n \circ \varphi_n$ are embeddings of $\mathfrak{A}_n$ into $\mathfrak{B}_{n+1}$ with the same partial multiplicities. By Proposition 9.2.6, there is a unitary element U_{n+1} of $\mathfrak{B}_{n+1}$ such that

$$\beta_n \circ \varphi_n = \mathrm{Ad}(U_{n+1})(\varphi_{n+1} \circ \alpha_n).$$

Define recursively $V_n \in \mathfrak{B}_n$ and $\psi_n : \mathfrak{A}_n \to \mathfrak{B}_n$ by $\psi_1 = \varphi_1$ and $V_1 = I$ and

$$V_{n+1} = \beta_n(V_n)U_{n+1} \text{ and } \psi_{n+1} = \mathrm{Ad}(V_{n+1})\varphi_{n+1}$$

for $n \geq 1$. We then have

$$\beta_n \circ \psi_n = \beta_n \circ \mathrm{Ad}(V_n)\varphi_n = \mathrm{Ad}(\beta_n(V_n))\beta_n \circ \varphi_n$$
$$= \mathrm{Ad}(\beta_n(V_n))\,\mathrm{Ad}(U_{n+1})(\varphi_{n+1} \circ \alpha_n)$$
$$= \mathrm{Ad}(\beta_n(V_n)U_{n+1})\varphi_{n+1} \circ \alpha_n = \psi_{n+1} \circ \alpha_n,$$

where in the first line, we have used the identity $\beta \circ \mathrm{Ad}(V)\varphi = \mathrm{Ad}(\beta(V))\beta \circ \varphi$ for $*$-morphisms $\varphi : \mathfrak{A} \to \mathfrak{B}$ and $\varphi : \mathfrak{B} \to \mathfrak{B}'$. We therefore have the commutative diagram

$$
\begin{array}{ccccccc}
\mathfrak{A}_1 & \xrightarrow{\alpha_1} & \mathfrak{A}_2 & \xrightarrow{\alpha_2} & \mathfrak{A}_3 & \xrightarrow{\alpha_3} & \cdots \\
\downarrow{\psi_1} & & \downarrow{\psi_2} & & \downarrow{\psi_3} & & \\
\mathfrak{B}_1 & \xrightarrow{\beta_1} & \mathfrak{B}_2 & \xrightarrow{\beta_2} & \mathfrak{B}_3 & \xrightarrow{\beta_3} & \cdots
\end{array}
$$

showing that there is a map $\psi : \mathfrak{A} \to \mathfrak{B}$ that extends the maps ψ_n and is a $*$-isomorphism from $\cup_{n\geq 1}\mathfrak{A}_n$ to $\cup_{ge1}\mathfrak{B}_n$. Since ψ is an isometry, it extends to a $*$-isomorphism from $\mathfrak{A}$ onto $\mathfrak{B}$.

9.5 This is a consequence of Elliott's Theorem. In fact, two UHF algebras $\mathfrak{A}, \mathfrak{B}$ are such that $\delta(\mathfrak{A}) = \delta(\mathfrak{B})$ if and only if the odometers associated with the Bratteli diagrams of $\mathfrak{A}$ and $\mathfrak{B}$ have the same supernatural number and thus have isomorphic dimension groups (as seen in Exercise 6.8).

9.6 Set $\mathfrak{A} = C([0, 1])$. Assume that $\mathfrak{A} = \overline{\cup \mathfrak{A}_n}$ with $\mathfrak{A}_n$ finite dimensional. Let $\mathfrak{B}$ be the subalgebra of polynomials. Since $\mathfrak{B}$ is dense, we have $\mathfrak{A} = \overline{\cup \mathfrak{A}_n \cap \mathfrak{B}}$ with $\mathfrak{A}_n \cap \mathfrak{B}$ being a finite dimensional subalgebra of $\mathfrak{B}$. But the only finite dimensional subalgebra of $\mathfrak{B}$ is formed by the constant functions, a contradiction.

9.7 Set $\alpha = 2a$ and $\beta = a^2 + b^2$. We may assume that $\alpha > 0$. We have

$$(1 + x)^N(x^2 - \alpha x + \beta) = (x^2 - \alpha x + \beta) \sum_{k=0}^{N} \binom{N}{k} x^k$$

$$= \sum_{k=0}^{N+2} \left(\beta \binom{N}{k} - \alpha \binom{N}{k-1} + \binom{N}{k-2} \right) x^k$$

$$= \sum_{k=0}^{N+2} \frac{N!}{k!(N+2-k)!} a_{N,k} x^k,$$

where

$$a_{N,k} = \beta(N+2-k)(N+1-k) - \alpha k(N+2-k) + k(k-1)$$
$$= (1+\alpha+\beta)k^2 - (2\beta+\alpha)Nk - (3\beta+2\alpha+1)k + \beta(N^2+3N+2)$$
$$= (1+\alpha+\beta)\left(k - \frac{\beta+\alpha/2}{1+\alpha+\beta}N \right)^2 + \beta(3N - 3k + 2) - (2\alpha+1)k$$
$$+ (1+\alpha+\beta)^{-1}N^2((1+\alpha+\beta)\beta - (\beta+\alpha/2)^2).$$

Since $1 + \alpha + \beta > 0$ and $k \leq N$, we obtain

$$a_{N,k} \geq (1+\alpha+\beta)^{-1}N^2 \left(\beta - \frac{\alpha^2}{4} \right) + 2\beta - (2\alpha+1)N,$$

which is positive for N large enough, given that $4\beta > \alpha^2$.

The condition is clearly necessary. Conversely, a polynomial satisfying the condition has positive leading coefficient and no positive real root. Thus, it factors as

$$p(x) = c \prod_i (x + \lambda_i) \prod_j (x^2 - 2a_j x + a_j^2 + b_j^2),$$

with $\lambda_i \geq 0$ and $b_j > 0$. By what we have seen previously, there are integers N_j such that $(1+x)^{N_j}(x^2 - 2a_j x + a_j^2 + b_j^2)$ has positive coefficients. Thus, $N = \sum_j N_j$ is a solution.

Appendix B

Useful Definitions and Results

We have placed in this appendix a number of notions and results used in the book. It is intended to serve as a memento for notions that some of the readers may have missed in part or forgotten long ago.

B.1 Algebraic Number Theory

In this section, we give an introduction to the basic concepts and results of algebraic number theory. We assume the reader to know the basic concepts of algebra, as the notion of a *group*, of a *ring* or a *field*, of an *ideal* in a ring and of a *principal ring*. We also assume familiarity with the notions of *module* over a ring and of *vector space* over a field. A map $f : E \to F$ between modules E, F on a ring K is *linear* if $f(x + y) = f(x) + f(y)$ for all $x, y \in E$ and $f(ke) = kf(e)$ for all $k \in K$ and $e \in E$. It is injective if and only if its *kernel* $\ker(f) = f^{-1}(0)$ is reduced to $\{0\}$.

We denote as usual by $\mathbb{N} = \{0, 1, 2, \ldots\}$ the natural integers, by $\mathbb{Z}$ the ring of integers, by $\mathbb{Q}$ the field of rational numbers and by $\mathbb{C}$ the field of complex numbers.

B.1.1 Algebraic Numbers

An *algebraic number* is a complex number x solution of an equation

$$x^n + a_{n-1}x^n + \cdots + a_1 x + a_0 = 0 \tag{B.1}$$

with coefficients in $\mathbb{Q}$. It is an *algebraic integer* if the coefficients are in $\mathbb{Z}$. For $x \in \mathbb{C}$, we denote by $\mathbb{Q}[x]$ (resp. $\mathbb{Z}[x]$) the subring generated by $\mathbb{Q}$ (resp. $\mathbb{Z}$) and x.

Theorem B.1.1 *For $x \in \mathbb{C}$, the following conditions are equivalent.*

(i) *x is an algebraic number (resp. an algebraic integer).*
(ii) *The $\mathbb{Q}$-vector space $\mathbb{Q}[x]$ (resp. the $\mathbb{Z}$-module $\mathbb{Z}[x]$) is finitely generated.*

As a consequence, all elements of $\mathbb{Q}[x]$ (resp. $\mathbb{Z}[x]$) are algebraic numbers (resp. algebraic integers). Moreover, $\mathbb{Q}[x]$ is a field.

Let K be a field containing $\mathbb{Q}$. It is called an *algebraic extension* of $\mathbb{Q}$, or also a *number field*, if all its elements are algebraic over $\mathbb{Q}$.

When x is an algebraic number, the set of polynomials $p(X)$ such that $p(x) = 0$ is a nonzero ideal of the ring $\mathbb{Q}[X]$. Since $\mathbb{Q}[X]$ is a principal ring, this ideal is generated by a unique polynomial of the form (B.1) (that is, with leading coefficient 1). This polynomial is called the *minimal polynomial* of x. The roots of the minimal polynomial of x are called the *algebraic conjugates* of x.

A *Pisot number* is an algebraic number having all its algebraic conjugates of modulus strictly less than 1.

The *degree* over $\mathbb{Q}$ of an extension K, denoted $[K : \mathbb{Q}]$ is the dimension of the $\mathbb{Q}$-vector space K. By Theorem B.1.1 every extension of finite degree is algebraic (the converse is not true).

B.1.2 Quadratic Fields

A *quadratic field* is an extension of degree 2 of $\mathbb{Q}$. Every quadratic field is of the form $\mathbb{Q}[\sqrt{d}]$, where d is an integer without square factor.

Theorem B.1.2 *Let $K = \mathbb{Q}(\sqrt{d})$ be a quadratic field.*

1. *If $d \equiv 1, 3 \bmod 4$ the ring of integers of K is $\mathbb{Z} + \sqrt{d}\mathbb{Z}$.*
2. *If $d \equiv 1 \bmod 4$, the ring of integers of K is $\mathbb{Z} + \lambda\mathbb{Z}$ with $\lambda = (1 + \sqrt{d})/2$.*

Norm and Trace Let K be a number field. For $x \in K$, the multiplication by x in K is a $\mathbb{Q}$-linear map $\rho(x)$. The *norm* of x, denoted $N(x)$, is the determinant of $\rho(x)$. Its *trace*, denoted $\mathrm{Tr}(x)$ is the trace of $\rho(x)$.

If x is an algebraic integer, then $N(x)$ and $\mathrm{Tr}(x)$ are integers.

For example, if $K = \mathbb{Q}(\sqrt{d})$ and $x = a + b\sqrt{d}$, then $N(x) = a^2 - b^2 d$ and $\mathrm{Tr}(x) = 2a$.

Discriminant Let K be a number field of degree n over $\mathbb{Q}$. For $x_1, x_2, \ldots, x_n \in K$, the *discriminant* of $(x_1, x_2, \ldots, x_n)$ is

$$D(x_1, x_2, \ldots, x_n) = \det(\mathrm{Tr}(x_i x_j)).$$

One has $D(x_1, \ldots, x_n) \neq 0$ if and only if $x_1, x_2, \ldots, x_n$ is a basis of K.

If $y_1, y_2, \ldots, y_n$ is such that $y_i = \sum_{j=1}^{n} a_{ij} x_j$, then $D(y_1, y_2, \ldots, y_n) = \det(a_{ij})^2 D(x_1, \ldots, x_n)$. Thus, the set of discriminants of the bases of the $\mathbb{Z}$-module of algebraic integers of K is an ideal of $\mathbb{Z}$. The generator of this ideal is called the *discriminant* of K.

For example, the discriminant of $\mathbb{Q}(\sqrt{d})$ is $4d$ if $d \equiv 2, 3 \bmod 4$ and d if $d \equiv 1 \bmod 4$.

B.1.3 Classes of Ideals

Let K be a number field and let A be its ring of algebraic integers. Two ideals I, J of A are *equivalent* if there are nonzero $\alpha, \beta \in A$ such that $\alpha I = \beta J$.

The equivalence classes of ideals form a group with respect to the product, called the *class group* of F. The neutral element is the class of the ideal $I = A$, which can be shown to be formed of the principal ideals of F. Thus, when A is a principal ring, the class group has only one element.

The ring of integers of a number field may fail to be principal and thus to have unique factorization. For example, in $\mathbb{Q}[\sqrt{-5}]$, we have

$$(1 + \sqrt{-5})(1 - \sqrt{-5}) = 2 \cdot 3$$

although $1 + \sqrt{-5}$ has no nontrivial divisor. However, one has the following result.

Theorem B.1.3 (Dirichlet) *For every number field, the class group is finite.*

Units A *unit* in a number field K is an invertible element of the ring A of integers of K. The set of units forms a multiplicative group, called the *group of units* of A.

An algebraic integer in a number field K is a unit if and only if $N(x) = \pm 1$.

For example, the group of units in a real quadratic field $\mathbb{Q}[\sqrt{d}]$ with $d \geq 2$, is formed of the $a + b\sqrt{d}$ with $a, b \in \mathbb{Q}$ solution of

$$a^2 - db^2 = \pm 1, \tag{B.2}$$

which is known as *Pell's equation*.

The following result is known as *Dirichlet Unit Theorem*.

Theorem B.1.4 (Dirichlet) *The group of units of a number field K is of the form $\mathbb{Z}^r \times G$, where $r \geq 0$ and G is a finite cyclic group formed by the roots of unity contained in K.*

The integer r is $r = r_1 + r_2 - 1$, where r_1 is the number of real embeddings of K in $\mathbb{C}$ and $2r_2$ the number of complex embeddings. One has $n = r_1 + 2r_2$, where n is the degree of K over $\mathbb{Q}$.

For example, in a real quadratic field $\mathbb{Q}[\sqrt{d}]$, the group of units is $\mathbb{Z} \times \mathbb{Z}/2\mathbb{Z}$. Every unit is of the form $\pm u^n$, where u is a *fundamental unit*.

For $d = 2$, the positive fundamental unit is $1 + \sqrt{2}$. For $d = 5$, it is $(1 + \sqrt{5})/2$.

As another example, if $K = \mathbb{Q}[i]$, one has $r = 0$. The group of roots of unity is of order 4 and is generated by $-i$. The corresponding ring of integers is called the ring of *Gaussian integers*.

B.1.4 Continued Fractions

Every irrational real number $\alpha > 0$ has a unique *continued fraction* expansion

$$\alpha = a_0 + \cfrac{1}{a_1 + \cfrac{1}{a_2 + \cfrac{1}{\ddots}}} \tag{B.3}$$

where $a_0, a_1, \ldots$ are integers with $a_0 \geq 0$ and $a_n > 0$ for $n \geq 1$. Conversely, such an expression is an irrational number (rational numbers have a finite continued fraction expansion). We denote $\alpha = [a_0; a_1, \ldots]$. The integers a_i are the *coefficients* of the continued fraction.

Theorem B.1.5 (Lagrange) *An irrational number $\alpha > 0$ is quadratic if and only if its continued fraction expansion is eventually periodic.*

For example, we have $[0; 1, 1, 1, \ldots] = \frac{\sqrt{5}-1}{2}$ and $[0; 2, 2, \ldots] = \sqrt{2} - 1$.

One associates with the continued fraction expansion of α its sequence of *partial quotients* p_n/q_n defined as follows. Set $p_0 = a_0$, $q_0 = 1$, $p_1 = a_0 a_1 + 1$, $q_1 = a_1$ and inductively for $n \geq 2$,

$$\begin{bmatrix} p_n & q_n \\ p_{n-1} & q_{n-1} \end{bmatrix} = \begin{bmatrix} a_n & 1 \\ 1 & 0 \end{bmatrix} \begin{bmatrix} p_{n-1} & q_{n-1} \\ p_{n-2} & q_{n-2} \end{bmatrix}.$$

Then

$$\frac{p_n}{q_n} = a_0 + \cfrac{1}{a_1 + \cfrac{1}{\ddots + a_n}}$$

and the sequence (p_n/q_n) converges to α.

Notes For this brief introduction to algebraic number theory, we have followed Hardy and Wright (2008) and Samuel (1970) to which the reader is referred for a complete presentation.

B.2 Groups, Graphs and Algebras

We recall now some of the basic notions of algebra concerning free groups, fundamental graphs and simple algebras. We begin with some general notions on groups.

B.2.1 Groups

A *semigroup* is a nonempty set with just an associative operation. It is a *monoid* if it has a neutral element.

A subgroup H of a group G is *normal* if $gHg^{-1} = H$ for every $g \in G$. Two elements g, h of a group G are *conjugate* if $g = khk^{-1}$ for some $k \in G$. Two subgroups H, K of G are conjugate if $K = kHk^{-1}$ for some $k \in G$.

A *morphism* from a group G to a group H is a map $\varphi \colon G \to H$ such that $\varphi(g)\varphi(h) = \varphi(gh)$ for every $g, h \in G$. The *kernel* of a morphism $\varphi \colon G \to H$ is the set $\ker(\varphi) = \varphi^{-1}(1_H)$. It is a normal subgroup of G.

The set of morphisms from a group G into an abelian group H is a group for the product $\alpha\beta(g) = \alpha(g)\beta(g)$, which is denoted $\mathrm{Hom}(G, H)$.

Let H be a subgroup of a group G. A *right coset* of H in G is a set of the form $Hg = \{hg \mid h \in H\}$ for some $g \in G$. Two cosets are equal or disjoint. The *index* of H in G, denoted $[G \colon H]$, is the number of distinct cosets of H in G.

The *quotient* of G by a normal subgroup K, is a group denoted G/K. It is the set of cosets of K with the product $(Kg)(Kh) = (Kgh)$. If $\varphi \colon G \to H$ is a morphism from G onto H with kernel K, then H is isomorphic to G/K.

A *commutator* in a group G is an element of the form $[g, h] = ghg^{-1}h^{-1}$ for $g, h \in G$. The subgroup generated by the commutators, denoted $[G, G]$ is a normal subgroup called the *derived subgroup*. The quotient $G/[G, G]$ is an abelian group called the *abelianization* of the group G.

A *permutation* on a set Q is a bijection of Q onto Q. The set of all permutations on Q is a group called the *symmetric group*, denoted $\mathfrak{S}_Q$. A *permutation group* on a set Q is a subgroup of the symmetric group $\mathfrak{S}_Q$.

The *free abelian group* on a set A, denoted $\mathbb{Z}(A)$, is the additive group formed of linear combinations $\sum_{a \in A} n_a a$ with $n_a \in \mathbb{Z}$. When A is finite with n elements, it is isomorphic with $\mathbb{Z}^n$.

The following statement is known as the *Fundamental Theorem of abelian groups*. A group is *cyclic* if is generated by one element. A finite cyclic group is *primary* if its order is a power of a prime.

Theorem B.2.1 *Every finitely generated abelian group, is in a unique way a direct product of primary cyclic groups and infinite cyclic groups.*

Thus, every finitely generated abelian group G can be written uniquely as $G = \mathbb{Z}^n \times \mathbb{Z}/q_1\mathbb{Z} \times \cdots \times \mathbb{Z}/q_m\mathbb{Z}$ with $n, m \geq 0$ and all q_i powers of primes. The integer $n + m$ is called the *rank* of the group G, denoted rank(G).

As a closely related notion, the $\mathbb{Q}$-*rank* of G, denoted rank$_\mathbb{Q}(G)$, is the dimension over $\mathbb{Q}$ of the tensor product $\mathbb{Q} \otimes_\mathbb{Z} G$ called the *divisible hull* of G.

The divisible hull of G is the quotient of the set of pairs $(x, g) \in \mathbb{Q} \times G$ by the equivalence generated by $(xy, g) \sim (x, yg)$ for $x \in \mathbb{Q}$, $y \in \mathbb{Z}$ and $g \in G$. The class of (x, g) is denoted $x \otimes g$. It is a $\mathbb{Q}$-vector space for the sum $(p/q) \otimes g + (r/s) \otimes h = (1/sq) \otimes (psg + rqh)$ and the scalar product $(p/q)(x \otimes g) = (px/q) \otimes g$. One has rank$_\mathbb{Q}(G) \leq$ rank(G) with equality if rank(G) is finite.

B.2.2 Free Groups

Recall that the *free group* on the alphabet A is formed of the words on $A \cup A^{-1}$ which are *reduced* (that is, contain no aa^{-1} or $a^{-1}a$ for $a \in A$). The product of two reduced words u, v is the unique reduced word obtained by reduction of uv.

We denote by $F(A)$ the free group on A. For $U \subset F(A)$, we denote by $\langle U \rangle$ the subgroup generated by U. The abelianization of the free group $F(A)$ is the free abelian group $\mathbb{Z}(A)$.

The following result is known as the Nielsen–Schreier Theorem,

Theorem B.2.2 (Nielsen–Schreier) *Every subgroup of the free group is free.*

Thus, every subgroup H of the free group has a generating set U, called a *basis* of H, such that H is isomorphic to the free group on U. Two bases of a subgroup have the same number of elements, called the *rank* of the subgroup.

Let H be a subgroup of the free group $F(A)$. Let Q be a set of reduced words on A which is a prefix-closed set of representatives of the right cosets Hg of H. Such a set is traditionally called a *Schreier transversal* for H. Let

$$U = \{paq^{-1} \mid a \in A, p, q \in Q, pa \notin Q, pa \in Hq\}. \qquad \text{(B.4)}$$

Each word x of $U \cup U^{-1}$ has a unique factorization paq^{-1} with $p, q \in Q$ and $a \in A \cup A^{-1}$. The letter a is called the *central part* of x. The set U is a basis of H, called the *Schreier basis* relative to Q. It generates H because if $x = a_1 a_2 \cdots a_m \in H$ with $a_i \in A \cup A^{-1}$, then $x = (a_1 p_1^{-1})(p_1 a_2 p_2^{-1}) \cdots (p_{m-1} a_m)$ with $a_1 \cdots a_k \in Hp_k$ for $1 \le k \le m - 1$ is a factorization of x in elements of $U \cup U^{-1} \cup \{1\}$. Finally, if a product $x_1 x_2 \cdots x_m$ of elements of $U \cup U^{-1}$ is equal to 1, then $x_k x_{k+1} = 1$ for some index k since the central part a never cancels in a product of two elements of $U \cup U^{-1}$.

If A has k elements and H is a subgroup of finite index n in the free group on A, the rank r of H satisfies the equality

$$r - 1 = n(k - 1)$$

called *Schreier's Formula*.

A group G is called *residually finite* if for every element $g \ne 1$ of G, there is a morphism φ from G onto a finite group such that $\varphi(g) \ne 1$.

The following can be proved directly without much difficulty by associating to $g \ne 1$ a map φ from $F(A)$ into the symmetric group on $\{1, 2, \ldots, n\}$ such that $\varphi(g) \ne 1$.

Theorem B.2.3 *A finitely generated free group is residually finite.*

A group G is said to be *Hopfian* if any surjective morphism from G onto G is also injective.

Theorem B.2.4 (Malcev) *Every finitely generated, residually finite group is Hopfian.*

In particular, by Theorem B.2.3, a finitely generated free group is Hopfian. This can be proved directly as follows. Let $\alpha \colon F(A) \to F(A)$ be a surjective morphism. Since $\alpha(A)$ generates $F(A)$, it cannot have less than $\mathrm{Card}(A)$ elements. Thus, $\mathrm{Card}(\alpha(A)) = \mathrm{Card}(A)$ and consequently $\alpha(A)$ is a basis of $F(A)$. This implies that α is an automorphism.

B.2.3 Free Products

Given two groups G and H, the *free product* of G and H, denoted $G * H$ is the set of all $g_1 h_1 \cdots g_n h_n$ with $n \ge 1$, $g_i \in G$ and $h_i \in H$ for $1 \le i \le n$. Thus, the free group on A is the free product of the infinite cyclic groups generated by the $a \in A$.

The following result is known as the Kurosh Subgroup Theorem,

Theorem B.2.5 (Kurosh) *Any subgroup of a free product $G_1 * G_2 * \cdots * G_n$ is itself a free product of a free group and of groups conjugate to subgroups of the G_i.*

B.2.4 Graphs

Let $\Gamma = (V, E)$ be a graph with V as a set of *vertices* and E as a set of *edges*. Each edge $e \in E$ has an *origin* $s(e)$ and an *end* $r(e)$ (also called its source and range). We allow multiple edges and thus we are considering *directed multi-graphs*. One also considers *undirected* graphs in which an edge is an unordered pair of vertices.

The *degree* of v is the number of edges e such that v is the origin or the end of e (with the loops on v counted twice).

A *labeled graph* is a graph G with a label on every edge, which is a letter of some finite alphabet A. We usually denote $p \xrightarrow{a} q$ an edge from p to q labeled a. We assume that there are no *multiple labeled edges* in the sense that, given p, q, a, there is at most one edge from p to q with label a.

As a transition with what precedes, to every group G with a given set S of generators is associated a labeled graph called its *Cayley graph*. Its set of vertices is G and there is an edge from g to h labeled s whenever $gs = h$.

Two edges e and f are *consecutive* whenever $r(e) = s(f)$. A *path* in the graph is, as usual, a sequence $(e_1, \ldots, e_n)$ of consecutive edges. Its *origin* is $s(e_1)$, its *end* $r(e_n)$ and the integer n is its *length*. By convention, there is, for each vertex v, a path of length 0 of origin v and end v. The path is a *cycle* if its origin and end are equal. If $(e_1, \ldots, e_n)$ and $(f_1, \ldots, f_m)$ are two paths such that $r(e_n) = s(f_1)$, then the concatenation of these paths is the path $(e_1, \ldots, e_n, f_1, \ldots, f_m)$. A path is *simple* if it does not pass twice by the same vertex except possibly for its origin which may be equal to its end.

When G is a labeled graph, every path in G is also labeled. The label of the path $p_0 \xrightarrow{a_1} p_1 \xrightarrow{a_2} \cdots \xrightarrow{a_n} p_n$ is the word $a_1 a_2 \cdots a_n$.

A graph $G = (V, E)$ is *strongly connected* if for every pair of vertices u, v, there is a path from u to v.

These notions are the same for an undirected graph with the natural requirement for a path to be *reduced*, that is, it does not use twice consecutively the same edge. One says that it is *connected* if there is a path between every pair of vertices (the same term is used for a directed graph when the associated undirected graph is connected). An undirected graph is *acyclic* if there is no nontrivial cycle, that is, no reduced path is a cycle. An undirected graph is a *tree* if it is connected and acyclic.

A directed graph (V, E) with a distinguished vertex v_0 is a *tree* or *directed tree* with *root* v_0 if for every vertex v there is a unique path from v_0 to v. The corresponding undirected graph will then be a tree and, given v_0, there is a unique choice for the orientation of the edges which makes it a rooted tree with root v_0. The vertices of a directed tree are often called *nodes*. A node of a directed tree is a *leaf* if it has no successor. Otherwise, it is an *internal node*

An undirected graph (V, E) is *bipartite* if there is a partition $V = V_1 \cup V_2$ of the set of vertices such that the edges connect a vertex in E_1 to one in E_2 (or conversely).

For each edge $e \in E$ in a graph (V, E), we consider an inverse edge e^{-1} from $r(e)$ to $s(e)$. By convention, $(e^{-1})^{-1}$ is e. A *generalized path* in G is a sequence $(e_1, \ldots, e_n)$ of edges or their inverses such that, for all $i \in [1, n-1]$, the edges e_i and e_{i+1} are consecutive. The generalized path is *reduced* whenever e_i is different from e_{i+1}^{-1} for all i. Two paths are *equivalent* whenever they can be obtained one from an other by a sequence of insertions or deletions of a sequence (e, e^{-1}). Every generalized path is equivalent to a unique reduced generalized path. It is a *generalized cycle* if its origin and end are equal.

If the graph is labeled with labels in A, and $e\colon p \xrightarrow{a} q$ is an edge, the label of e^{-1} is the inverse a^{-1} of a in the free group. Thus, we have $e^{-1}\colon q \xrightarrow{a^{-1}} p$. The label of a generalized path is the product in the free group on A of the labels of the edges along the path.

Let $v \in V$ be a vertex of the graph $\Gamma = (V, E)$. The *fundamental group* $G(\Gamma, v)$ is the group formed by the generalized cycles from v to v. When G is connected, its isomorphism class does not depend on v and we denote $G(\Gamma)$ the fundamental group of Γ.

As is well known, the fundamental group of a connected graph G is a free group and a basis can be obtained as follows. Let T be a *spanning tree* of Γ rooted at v, that is, a set of edges (or inverse edges) such that for every $w \in V$, there is a unique reduced path p_w from v to w using the edges in T. Then, the set

$$\{p_{s(e)} e p_{r(e)}^{-1} \mid e \in E \setminus T\} \tag{B.5}$$

is a basis of $G(\Gamma, v)$.

Example B.2.6 Let Γ be the graph represented in Figure B.1 with $V = \{1, 2\}$ and $E = \{e, f, g, h\}$.

The set $T = \{f\}$ is a spanning tree rooted at 1. The corresponding basis of $G(\Gamma, 1)$ is $\{e, fg, fhf^{-1}\}$.

Figure B.1 A connected graph.

B.2.5 Stallings Graph

For a graph $\Gamma = (V, E)$ labeled by an alphabet A, the *group defined* by Γ with respect to a vertex v is the subgroup of the free group on A formed by the labels of generalized paths from v to itself. Thus, if all labels are distinct, the graph defined by G is the image of the fundamental graph of G under the map assigning to each generalized path its label.

A *Stallings folding* of a labeled graph is the following transformation. Suppose that two vertices p, p' of a graph G have edges (or inverse edges) going to q with the same label a. Then we change G to G' by merging p and p'. We call this transformation a Stallings folding at vertex q.

More precisely, we replace p, p' by a new vertex $\{p, p'\}$ which inherits the edges coming in p, p' or leaving p, p' without creating multiple labeled edges. Thus, if there were edges $p \xrightarrow{a} q$ and $p' \xrightarrow{a} q$ (and there was at least one such pair), we create only one edge from $\{p, p'\}$ to q. Similarly, if there were edges $q \xrightarrow{a} p$ and $q \xrightarrow{a} p'$, we create on only one edge $q \xrightarrow{a} \{p, p'\}$.

A Stallings folding does not change the subgroup defined by the graph. Indeed, any generalized path in G' can be obtained from a generalized path in G by insertion of paths of length 2 labeled aa^{-1}. A graph on which no Stallings folding can be performed is called *Stallings reduced*.

Given a finitely generated subgroup H of the free group, there is a unique Stallings reduced graph which defines H. This graph is called the *Stallings graph* of the subgroup H.

Example B.2.7 Let $A = \{a, b\}$ and let H be subgroup of $F(A)$ generated by $\{a, bab, bb\}$.

A graph defining H is represented in Figure B.2 on the left. The Stallings folding merging 2 and 3 gives the graph represented on the right, which is the Stallings graph of H.

B.2.6 Semisimple Algebras

An *algebra* over the field $\mathbb{C}$ of complex numbers is a vector space over $\mathbb{C}$ with a bilinear product (and thus making it a ring). A two-sided *ideal* of the algebra $\mathfrak{A}$ is a subspace which is a two-sided ideal of the ring $\mathfrak{A}$. The algebra is said

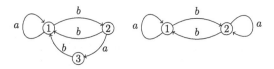

Figure B.2 A Stallings folding.

to be *simple* if it has no proper nonzero two-sided ideal. The algebra $\mathcal{M}_n$ of $n \times n$-matrices with coefficients in $\mathbb{C}$ is simple. Moreover, any automorphism of $\mathcal{M}_n$ is an inner automorphism, that is, of the form $M \mapsto AMA^{-1}$ for some invertible matrix $A \in \mathcal{M}_n$.

A direct sum $\mathfrak{A}_1 \oplus \cdots \oplus \mathfrak{A}_k$ of simple algebras $\mathfrak{A}_i$ is called a *semisimple* algebra. As an equivalent definition, an algebra is semisimple if does not contain nonzero nilpotent ideals.

Theorem B.2.8 (Wedderburn) *A finite dimensional algebra* $\mathfrak{A}$ *over* $\mathbb{C}$ *is semisimple if and only if* $\mathfrak{A} = \mathcal{M}_{n_1} \oplus \cdots \oplus \mathcal{M}_{n_t}$.

Moreover, if $\mathfrak{A} = \mathcal{M}_{n_1}^{(a_1)} \oplus \cdots \oplus \mathcal{M}_{n_k}^{(a_k)}$, where the n_i are distinct and each $\mathcal{M}_{n_i}$ is repeated a_i times, the n_i and the a_i are determined uniquely, up to a permutation. Consequently, every embedding of an algebra $\mathfrak{A}_1 = \mathcal{M}_{m_1} \oplus \cdots \oplus \mathcal{M}_{m_t}$ into an algebra $\mathfrak{A}_2 = \mathcal{M}_{n_1} \oplus \cdots \oplus \mathcal{M}_{n_s}$ is, up to conjugacy, of the type $\varphi = \varphi_1 \oplus \cdots \oplus \varphi_t$ with

$$\varphi_i = \mathrm{id}_{m_1}^{(a_{i1})} \oplus \cdots \oplus \mathrm{id}_{m_t}^{(a_{it})},$$

where id_m is the identity of $\mathcal{M}_m$ and $\mathrm{id}_m^{(a)} \colon \mathcal{M}_m \to \mathcal{M}_{am}$ is the morphism $x \mapsto (x, \ldots, x)$ (a times). In other words, each morphism $\varphi_i \colon \mathfrak{A}_1 \to \mathcal{M}_{n_i}$ has the form

$$\varphi_i(M_1, \ldots, M_s) = \begin{bmatrix} N_1 & & \\ & \ddots & \\ & & N_t \end{bmatrix}, \text{ with } N_j = \begin{bmatrix} M_j & & \\ & \ddots & \\ & & M_j \end{bmatrix} (a_{ij} \text{ times}).$$

B.2.7 Morita Equivalence

Two rings $\mathfrak{A}$, $\mathfrak{B}$ are *Morita equivalent* if the categories of $\mathfrak{A}$-modules and of $\mathfrak{B}$-modules are equivalent, which means that there are functors between the two categories inverse of each other. By Morita's Theorem, this is the case if and only if there is an integer $n \geq 1$ and an idempotent e in the ring $\mathcal{M}_n(\mathfrak{A})$ of $n \times n$-matrices with elements in $\mathfrak{A}$ such that $\mathcal{M}_n(\mathfrak{A}) e \mathcal{M}_n(\mathfrak{A}) = \mathcal{M}_n(\mathfrak{A})$ and

$$\mathfrak{B} \simeq e\mathcal{M}_n(\mathfrak{A})e. \tag{B.6}$$

Notes We have only briefly recalled the basic definitions and properties of free groups. For a systematic exposition (in particular of the Nielsen–Schreier Theorem), see Lyndon and Schupp (2002) or Magnus et al. (2004).

See Lyndon and Schupp (2001) (p.197) for a proof of Malcev Theorem B.2.4. The Stallings graphs are from Stallings (1983) (see also Kapovich and Myasnikov (2002)).

The elementary properties of semisimple algebras (serving as a preparation of the properties of finite dimensional C^*-algebras in Chapter 9), in particular, the Wedderburn Theorem, can be found in Lang (2002).

The notion of Morita equivalence and a proof of Morita's Theorem can be found in Rowen (1988). For an introduction to the terminology of categories, see Lang (2002).

B.3 Linear Algebra

In this section, we present some results concerning matrices playing an important role in the book.

B.3.1 Matrices

A $P \times Q$-matrix with coefficients in a ring K is a map $M : P \times Q \to K$ with P, Q two sets. The image of (p, q), denoted $M_{p,q}$, is called the coefficient of M at row p and column q. The sum of two $P \times Q$-matrices M, N is the $P \times Q$-matrix $M + N$ defined by $(M + N)_{p,q} = M_{p,q} + N_{p,q}$. The product of a $P \times Q$-matrix M and a $Q \times R$ matrix N is the $P \times R$-matrix MN defined by

$$(MN)_{p,r} = \sum_{q \in Q} M_{p,q} N_{q,r}.$$

With these operations, the set of $P \times P$-matrices over K forms a ring with identity $I_{p,q} = 1$ for $p = q$ and 0 otherwise.

Coming to a more familiar notation, an $m \times n$-matrix is a $P \times Q$-matrix with $P = \{1, 2, \ldots, m\}$ and $Q = \{1, 2, \ldots, n\}$.

We denote by M^t the *transpose* of M, which is defined by $M^t_{p,q} = M_{q,p}$. The matrix M is *symmetric* if $M^t = M$ and *antisymmetric* if $M^t = -M$.

The *trace* of a $P \times P$-matrix M is $\mathrm{Tr}(M) = \sum_{p \in P} M_{p,p}$. The trace is a linear map and moroever is such that

$$\mathrm{Tr}(MN) = \mathrm{Tr}(NM) \tag{B.7}$$

for all $P \times P$-matrices M, N.

Let E be a module over K and let $n \geq 1$. A map f from E^n to K is *multilinear* if for every i, given $x_1, \ldots, x_{i-1}, x_{i+1}, \ldots, x_n$, the map $x_i \mapsto f(x_1, \ldots, x_n)$ is linear. It is *alternating* if $f(x_1, \ldots, x_n) = 0$ whenever $x_i = x_{i+1}$ for some $i < n$.

There is a unique map $M \mapsto \det(M) \in K$ which, viewed as a function of the columns of the $n \times n$-matrix M, is multilinear, alternating and such that $\det(I) = 1$. The scalar $\det(M)$ is called the *determinant* of the matrix M.

The following are fundamental properties of the determinant. First, we have for every matrix M,

$$\det(M) = \det(M^t). \tag{B.8}$$

Next, for every $n \times n$-matrices M, N, we have

$$\det(MN) = \det(M)\det(N). \tag{B.9}$$

Finally, M is invertible in K if and only if $\det(M)$ is invertible.

B.3.2 Nonnegative Matrices

A matrix with real coefficients is *nonnegative* (resp. *positive*) if all its elements are nonnegative (resp. positive). The same terms are used for vectors.

A nonnegative square matrix M is *irreducible* if for every pair i, j of indices, there is an integer $n \geq 1$ such that $M_{i,j}^n > 0$. Otherwise, M is *reducible*. A matrix M is reducible if and only if, up to a permutation of the indices, it can be written

$$M = \begin{bmatrix} R & 0 \\ S & T \end{bmatrix}$$

for some matrices R, S, T with R, T being square matrices.

A nonnegative square matrix M is *primitive* if there is some $n \geq 1$ such that all entries of M^n are positive. A primitive matrix is irreducible but not conversely.

To every finite graph $G = (V, E)$, is associated a nonnegative matrix called its *adjacency matrix*. It is the $V \times V$-matrix with coefficients

$$M_{v,w} = \mathrm{Card}\{e \in E \mid s(e) = v, r(e) = w\}.$$

The matrix is irreducible if and only if G is strongly connected. It is primitive if and only if

(i) G is strongly connected.
(ii) For every $v \in V$ the greatest common divisor of the lengths of the cycles starting at v is 1.

The following result is well known.

Theorem B.3.1 (Perron, Frobenius) *Let* M *be a nonnegative real square matrix with* $M \neq 0$. *Then*

(i) *M has a positive eigenvalue λ_M such that $|\mu| \leq \lambda_M$ for every eigenvalue μ of M.*

(ii) *There corresponds to λ_M a nonnegative eigenvector v and λ_M is the only eigenvalue with a nonnegative eigenvector.*

(iii) *If M is irreducible (resp. primitive), the sequence (M^n/λ_M^n) converges in mean (resp. converges) at geometric rate to the matrix wv, where v, w are positive left and right eigenvectors such that $Mw = \lambda_M w$, $vM = \lambda_M v$ and $vw = 1$.*

If M is irreducible, then λ_M is simple and M is primitive if and only if $|\mu| < \lambda_M$ for every other eigenvalue μ of M.

The theorem expresses in particular that if a matrix M is primitive, its *spectral radius* $\rho(M) = \max\{|\lambda| \mid \lambda \in \mathrm{Spec}(M)\}$ is an eigenvalue of M which is algebraically simple. Furthermore, any eigenvalue of M different from $\rho(M)$ has modulus less than $\rho(M)$. We call $\rho(M)$ the *dominant eigenvalue of M*.

The term *convergent in mean*, used in Assertion (iii), means, for a sequence (x_n) of elements of a normed space, that $\frac{1}{n}\sum_{i=0}^{n-1} x_i$ converges.

The term *geometric rate* of convergence used in Assertion (iii) means that there is a constant $c > 0$ and a real number $r < 1$ such that for all $n \geq 0$,

$$\left\| \frac{1}{\lambda_M^n} M^n - wv \right\| \leq c\, r^n. \tag{B.10}$$

We can choose for r the quotient $r = \mu/\rho(M)$, where μ is the maximum of the $|\lambda|$ for λ an eigenvalue of M other than $\rho(M)$.

As an important particular case, a real nonnegative $A \times A$-matrix M is *stochastic* if its rows have sum 1, that is,

$$\sum_{b \in A} M_{a,b} = 1,$$

for every $a \in A$. It is easy to verify that the dominant eigenvalue of a stochastic matrix is 1. The column vector with all coefficients 1 is a corresponding eigenvector. When a stochastic matrix M is irreducible, by Theorem B.3.1, the sequence $\frac{1}{n}\sum_{i=0}^{n-1} M^i$ converges to the matrix with all rows equal to the row eigenvector v of sum 1 for the eigenvalue 1, that is, such that $vM = v$.

B.3.3 Linear Programming

A *linear program* is given by a triple (M, b, f), where $f \colon \mathbb{R}^n \to \mathbb{R}$ is a linear map, M is a real $m \times n$-matrix and $b \in \mathbb{R}^m$ is a vector. A *solution* is a vector $x \in \mathbb{R}^n_+$ such that $Mx = b$ and $f(x) \geq f(y)$ for all $y \in \mathbb{R}^n_+$ such that $My = b$. Thus, a solution maximizes f under the *constraint $Mx = b$*.

A solution $x = (x_1, \ldots, x_n)$ of the equation $Mx = b$ is a *basic solution* if $x_i = 0$ for a set of $n - m$ indices i.

The Fundamental Theorem of Linear Programming is the following result.

Theorem B.3.2 *If the linear program given by (M, b, f) with M of rank m has a solution, it has a basic solution.*

As a very simple example, if $M = \begin{bmatrix} 1 & 1 & 1 \end{bmatrix}$, $b = 1$ and $f(x) = x_1 + x_2 + x_3$, then $\begin{bmatrix} 1 & 0 & 0 \end{bmatrix}^t$ is a basic solution.

The proof is closely related to the *simplex algorithm* which consists in building a sequence of basic solutions of the equation $Mx = b$ increasing successively the values of the function f.

Notes The Perron–Frobenius Theorem is a classical and very useful result. A proof can be found in Queffélec (2010) or in the classical Gantmacher (1959).

See Chvátal (1983) or Luenberger and Ye (2016) for an introduction to Linear Programming and a proof of the Fundamental Theorem.

B.4 Topological, Metric and Normed Spaces

We give in this section a list of the main notions and results of general topology and elementary analysis used in the book.

B.4.1 Topological Spaces

A *topological space* is a set X with a family of subsets, called *open sets*, containing X and $\emptyset$ and closed by union and finite intersection. The complement of an open set is a *closed set*. The sets that are both open and closed are called *clopen sets*. An element of a topological space is commonly called a *point*.

The topology for which all subsets are open is called the *discrete topology*.

The *closure* of a subset S of a topological space X is the intersection of the closed sets containing S. It is also the smallest closed set containing S.

The *interior* of a set S is the union of all open sets contained in S.

A *neighborhood* of a point x is a set U containing an open set V which contains x. Thus, U is a neighborhood of x if x belongs to the interior of U.

Any subset S of a topological space inherits the topology of X by considering as open sets the family of $S \cap U$ for $U \subset X$ open. This is called the *induced topology* on S and U is called a *subspace* of S.

A topological space X is a *Hausdorff space* if for every distinct $x, y \in X$, there are disjoint open sets U, V such that $x \in U$ and $y \in V$.

An *isolated point* in a topological space is a point x such that $\{x\}$ is open.

A sequence (x_n) of points in X is *convergent* to a limit $x \in X$ if every open set U containing x contains all x_n for n large enough. When X is Hausdorff, the limit is unique. We denote $\lim x_n$ the limit.

A family $\mathcal{F}$ of subsets of a topological space X is a *basis* of the topology if every open set is a union of elements of $\mathcal{F}$. One also says that $\mathcal{F}$ *generates* the topology of X. For $\mathcal{F}$ to be a basis of some topology, it is necessary and sufficient that it satisfies the two following conditions: (i) Every point belongs to some $U \in \mathcal{F}$. (ii) For every $U, V \in \mathcal{F}$ with $x \in U \cap V$, there is some $W \subset U \cap V$ such that $W \in \mathcal{F}$ and $x \in W$.

For example, if X is a totally ordered set, the *order topology* is generated by the open intervals $(a, b) = \{x \in X \mid a < x < b\}$. It is the usual topology on $\mathbb{R}$.

Any direct product $\prod_{i \in I} X_i$ of topological spaces X_i is a topological space for the topology (called the *product topology*) with a basis of open sets formed by the sets $\prod_{i \in I} U_i$ with $U_i \subset X_i$ open sets such that $U_i = X_i$ for all but a finite number of indices $i \in I$.

As a particular case, the set Y^X of functions $f : X \to Y$ between topological spaces X, Y is a subset of the product $\prod_{x \in X} Y_x$. The topology induced by the product topology on Y^X is the topology of *pointwise convergence*. A sequence (f_n) converges in this topology to a function f if $\lim f_n(x) = f(x)$ for every $x \in X$. This notion is weaker than the notion of uniform convergence (see below).

A function $f : X \to Y$ between topological spaces X, Y is *continuous* if $f^{-1}(U)$ is open for every open set $U \subset Y$. A continuous invertible function with a continuous inverse is called a *homeomorphism*.

A topological space is *separable* if it has a countable dense subset. For example, the real line is separable with the usual topology but not separable with the discrete topology.

A *topological group* is a topological space G with a group operation that is continuous. Similarly, a *topological ring* is a topological space which is a ring such that the addition and the multiplication are both continuous.

B.4.2 Metric Spaces

A *distance* on a set X is a map $d: X \to \mathbb{R}_+$ such that

(i) $d(x, y) = d(y, x)$, for every $x, y \in X$,
(ii) $d(x, y) = 0$ if and only if $x = y$,
(iii) $d(x, z) \leq d(x, y) + d(y, z)$ for every $x, y, z \in X$ (*triangular inequality*).

A set X with a distance d is called a *metric space*.
  A distance satisfying the inequality

$$d(x, z) \leq \max\{d(x, y), d(y, z)\}$$

for all $x, y, z \in X$, stronger than (iii), is called *ultrametric*.
  For $x \in X$ and $\varepsilon > 0$, the *open ball* centered at x with radius ε is

$$B(x, \varepsilon) = \{y \in X \mid d(x, y) < \varepsilon\}.$$

The topology with basis of open sets the open balls is the *topology defined* by the distance d. In this topology, a sequence (x_n) converges to a limit x if for every $\varepsilon > 0$ there is an N such that $d(x_n, x) \leq \varepsilon$ for all $n \geq N$.
  A subset S of a metric space X is *bounded* if it is contained in some open ball or, equivalently if the set of distances $d(x, y)$ for $x, y \in S$ is bounded.
  The *diameter* of a set C in a metric space X, denoted diam(C), is the supremum of the distances of two points in the set.
  A topological space is *metrizable* if its topology can be defined by a distance. Not all topological spaces are metrizable, even if all topological spaces we meet in this book are metrizable. It is however useful to use the terminology of general topological spaces, even when working with metric spaces.
  A metric space is Hausdorff. Indeed, if $x \neq y$, the balls $B(x, \varepsilon)$ and $B(y, \varepsilon)$ are disjoint if $\varepsilon < d(x, y)/2$ by the triangular inequality.
  In a metric space, many notions become easier to handle. For example, a set U in a metric space X is closed if and only if U contains the limits of all converging sequences of elements of X.
  A sequence of functions $f_n: X \to Y$ between metric spaces X, Y is *uniformly convergent* to $f: X \to Y$ if for every $\epsilon > 0$ there is an $N \geq 1$ such that for all $n \geq N$, one has $d(f_n(x), f(x)) \leq \varepsilon$ for every $x \in X$. Uniform convergence implies pointwise convergence but the converse is false. The limit of a uniformly convergent sequence of continuous function is continuous.
  A sequence (x_n) in a metric space (X, d) is a *Cauchy sequence* if for every $\varepsilon > 0$ there is an integer $n \geq 1$ such that for all $i, j \geq n$, one has $d(x_i, x_j) \leq \varepsilon$. Every convergent sequence is a Cauchy sequence but the converse is not true in general.

A metric space (X, d) is *complete* if every Cauchy sequence converges. Every metric space X can be embedded in a complete metric space $\hat{X}$, called its *completion* such that X is dense in $\hat{X}$. The space $\hat{X}$ is built as the set of classes of Cauchy sequences which are equivalent, in the sense that $\lim d(x_n, y_n) = 0$. The distance on $\hat{X}$ is defined by continuity.

B.4.3 Normed Spaces

When X is a complex vector space, a *norm* on X is a map $x \to \|x\|$ from X to $\mathbb{R}_+$ such that

(i) $\|x\| = 0$ if and only if $x = 0$,
(ii) $\|\alpha x\| = |\alpha| \|x\|$ for $\alpha \in \mathbb{C}$ and $x \in X$, and
(iii) $\|x + y\| \le \|x\| + \|y\|$ for every $x, y \in X$.

The space X is called a *normed space*. If the separation property $\|x\| = 0 \Rightarrow x = 0$ is missing, one speaks of a *seminorm*. A norm defines a distance by $d(x, y) = \|x - y\|$. Indeed, using (iii), we have

$$d(x, y) + d(y, z) = \|x - y\| + \|y - z\| \ge \|x - z\| = d(x, z).$$

The *completion* of a normed space X is again a normed space. The addition $x + y$ is defined as follows. If (x_n) and (y_n) are Cauchy sequences in X, then $(x_n + y_n)$ is a Cauchy sequence and if (x'_n) is equivalent to (x_n), then $(x'_n + y_n)$ is equivalent to $(x_n + y_n)$. Thus, the addition is well defined on the classes.

A complex algebra $\mathfrak{A}$ is a *normed algebra* if it has a norm $\| \ \|$ such that $\|AB\| \le \|A\| \|B\|$ for all $A, B \in \mathfrak{A}$. The *completion* of a normed algebra X is again a normed algebra. One defines the product of two Cauchy sequences (x_n) and (y_n) as the sequence $(x_n y_n)$. The result is, as for the sum, compatible with the equivalence.

If X, Y are normed spaces, a linear map $f \colon X \to Y$ (also called a *linear operator*) is continuous if and only if it is *bounded* , that is, if there is a constant $c \ge 0$ such that $\|f(x)\| / \|x\| \le c$ for all $x \ne 0$. The least such c is called the *norm* of f *induced* by the norms of X, Y, denoted $\|f\|$. This turns the space $L(X, Y)$ of bounded linear maps from X to Y into a normed space and the algebra $\mathfrak{B}(X)$ of bounded linear maps from X into X into a normed algebra.

For every $p \ge 1$, the formula

$$\|x\|_p = \left(\sum_{i=1}^{n} |x_i|^p \right)^{1/p}$$

defines a norm on $\mathbb{R}^n$, called the *p*-norm or L^p-norm. For $p = 2$, it is called the *Euclidean norm* and the associated distance the *Euclidean distance*.

Additionally $\|x\|_\infty = \max\{|x_1|, |x_2|, \ldots, |x_n|\}$ is called the ∞-norm, or L^∞-norm. The matrix norm induced by the L^∞-norm is

$$\|M\|_\infty = \max \|M_i\|_1,$$

where M_i is the row of index i of M.

An *inner product* on a vector space X over $\mathbb{C}$ is map $(x, y) \in X \times X \mapsto \langle x, y \rangle \in \mathbb{C}$ such that

(i) $\langle x, y \rangle = \overline{\langle y, x \rangle}$ (conjugate symmetry),
(ii) $\langle x + y, z \rangle = \langle x, y \rangle + \langle y, z \rangle$ and $\langle \alpha x, y \rangle = \alpha \langle x, y \rangle$ (sesquilinearity),
(iii) $\langle x, x \rangle \geq 0$ and ($\langle x, x \rangle = 0$ only if $x = 0$) (definite positivity)

for every $x, y, z \in X$ and $\alpha \in \mathbb{C}$. By (iii), we can define a nonnegative real number $\|x\|$ by $\|x\|^2 = \langle x, x \rangle$. By the *Cauchy–Schwarz inequality*, we have

$$|\langle x, y \rangle| \leq \|x\| \|y\|.$$

It follows from this inequality that $\|x\|$ satisfies the triangular inequality and thus is a norm.

A *Hilbert space* is a complex vector space with an inner product which is complete for the topology induced by the norm.

The space $\mathbb{C}^n$ is a Hilbert space for the inner product

$$\langle x, y \rangle = \sum_{i=1}^n x_i \overline{y_i},$$

where $\bar{y}$ denotes the complex conjugate of y. The corresponding norm $\|x\|_2$ is called the *Hermitian norm*. This extends to the sequences $x = (x_i)$ such that $\|x\|_2 < \infty$, which form the Hilbert space $\ell_2(\mathbb{C})$.

A *Banach algebra* is a complete normed algebra. The algebra $\mathcal{M}_n$ of $n \times n$-matrices with complex elements is a Banach algebra (for the norm induced by some norm on $\mathbb{C}^n$).

If $\mathfrak{A}$ is a Banach algebra, the *spectrum* of $T \in \mathfrak{A}$ is $\sigma(T) = \{\lambda \in \mathbb{C} \mid \lambda I - T \text{ not invertible}\}$. It is a nonempty closed set. The *spectral radius* of $T \in \mathfrak{A}$ is $\mathrm{spr}(T) = \sup_{\lambda \in \sigma(M)} |\lambda|$. The spectral radius satisfies $\mathrm{spr}(T) \leq \|T\|$ and it is given by the formula

$$\mathrm{spr}(T) = \lim \|T^n\|^{1/n}. \tag{B.11}$$

The set of bounded linear operators from a Hilbert space $\mathcal{H}$ into itself, denoted $\mathfrak{B}(\mathcal{H})$, is a Banach algebra. The *adjoint* of $T \in \mathfrak{B}(\mathcal{H})$ is the unique operator T^* such that $\langle Tx, y \rangle = \langle x, T^*y \rangle$ for every $x, y \in \mathcal{H}$. For $\mathcal{H} = \mathbb{C}^n$, the matrix M of T^* is the *conjugate transpose* M^* of M defined by $M^*_{i,j} = \overline{M_{j,i}}$.

The operator $T \in \mathfrak{B}(\mathcal{H})$ is said to be *unitary* if $TT^* = T^*T = I$, that is, if T and T^* are mutually inverse.

B.4.4 Compact and Connected Spaces

A Hausdorff space X is *compact* if for every family $(U_i)_{i \in I}$ of open sets with union X, there is a finite subfamily with union X.

As a reformulation of the definition, say that a family $(U_i)_{i \in I}$ of subsets of X is a *cover* of X if $X = \cup_{i \in I} U_i$ and an *open cover* if, additionally, the U_i are open. Then X is compact if and only if, from every open cover, one may extract a finite cover.

Here are a few important properties of compact spaces.

1. A subset of a compact space is compact (for the induced topology) if and only if it is closed.
2. A metric space is compact if and only if every sequence has a converging subsequence.
3. The image of a compact space by a continuous map is compact.

It follows easily from the second property that every compact metric space is complete.

A subset of a topological space is *relatively compact* if its closure is compact. Thus, any subset of a compact space is relatively compact.

Theorem B.4.1 (Tychonov) *Any product of compact spaces is compact.*

This implies in particular that the set $A^{\mathbb{Z}}$ of sequences on a finite set A is compact. This particular case can of course be proved directly using *König's Lemma*, which states that every sequence of infinite words on a finite alphabet has a converging subsequence.

The proof in the general case uses the *axiom of choice*, in one of its equivalent forms, such as *Zorn's Lemma*, which states that if every descending chain $p_1 > p_2 > \cdots$ in a partially ordered nonempty set P has a lower bound, then P has a minimal element.

We state below (in one of its equivalent forms) the *Baire Category Theorem*.

Theorem B.4.2 (Baire) *In a compact metric space, a countable intersection of dense open sets is dense.*

In particular, in a nonempty compact metric space, a countable intersection of dense open sets is nonempty. Taking the complements, the space is not the countable union of closed sets with empty interior.

A function $f: X \to Y$ between metric spaces X, Y is said to be *uniformly continuous* if for every $\varepsilon > 0$, there exists $\delta > 0$ such that for all $x, y \in X$, $d(x, y) < \varepsilon$ implies $d(f(x), f(y)) < \delta$. Any continuous function on a compact metric space is uniformly continuous (Heine–Cantor Theorem).

The space $C(X, \mathbb{R})$ of continuous real-valued functions on a compact metric space X is a metric space for the norm $\|f\| = \sup_{x \in X} |f(x)|$. The corresponding topology is the topology of uniform convergence.

A linear map from a Banach space X to a Banach space Y is *compact* if the image of a bounded subset is relatively compact. A compact operator is continuous.

As a subalgebra of $\mathfrak{B}(\mathcal{H})$, the algebra $\mathfrak{K}(\mathcal{H})$ of *compact operators* on a Hilbert space $\mathcal{H}$ is a C^*-algebra. When $\mathcal{H}$ is the Hilbert space $\ell^2(\mathbb{C})$ of sequences x of complex numbers such that $\|x\|_2 < \infty$, we denote it by $\mathfrak{K}$. Every compact operator on a Hilbert space is a limit of operators of finite rank.

Connected Spaces A topological space is *disconnected* if it is a union of disjoint nonempty open sets. Otherwise, the space is *connected*. The following are important properties of connected sets. 1. A continuous image of a connected set is connected and 2. Any union of connected sets is connected.

Every topological space decomposes as a union of disjoint connected subsets, called its *connected components*. The connected component of x is the union of all connected sets containing x.

A topological space is *totally disconnected* if its connected components are singletons. Any product of totally disconnected spaces is totally disconnected and every subspace of a totally disconnected space is totally disconnected.

A topological space is *zero-dimensional* if it admits a basis consisting of clopen sets. A compact space is zero-dimensional if and only if it is totally disconnected

Notes These notes follow mostly Willard (2004) but a similar content can be found in most textbooks on topology. Most authors (not all of them, see for example Rudin (1987)) assume as we do a compact space to be Hausdorff.

B.5 Measure and Integration

We give below a short review of some definitions and concepts concerning measure theory.

B.5.1 Measures

A family $\mathcal{F}$ of subsets of a set X is a σ-*algebra* if it contains X and is closed under complement and countable union.

Let X be a topological space. The family of *Borel subsets* of X is the closure of the family of open sets under countable union and complement. It is thus the σ-algebra generated by the family of open sets.

A *measure* on a σ-algebra $\mathcal{F}$ is a map $\mu \colon \mathcal{F} \to \mathbb{R}_+ \cup \{\infty\}$ such that $\mu(\emptyset) = 0$ and which is countably additive, that is, such that $\mu(\cup_n X_n) = \sum_n \mu(X_n)$ for any family X_n of pairwise disjoint elements of $\mathcal{F}$ (note that we only consider measures with nonnegative values, also called *positive measures*). The measure μ is *finite* if $\mu(X) < \infty$.

The set X is then called a *measurable space* and the elements of $\mathcal{F}$ are the *measurable subsets* of X. We also say that $(X, \mathcal{F}, \mu)$ is a *measure space*.

A property of points in a measurable space is said to hold *almost everywhere* (often abbreviated a.e.) if the set of points for which it is false has measure 0.

A *probability measure* on X is a measure μ such that $\mu(X) = 1$.

Note the following important properties of measures.

1. A measure is *monotone*, that is, if $U \subset V$, then $\mu(U) \leq \mu(V)$.
2. A measure is *subadditive*, that is, $\mu(\cup_n U_n) \leq \sum_n \mu(U_n)$ for every family X_n of measurable sets.

A function $f \colon X \to Y$ between measurable spaces X, Y is *measurable* if $f^{-1}(U)$ is a measurable set for every measurable subset U in Y.

A Borel measure on a topological space X is a measure on the family of Borel sets of X.

As an example, in a Hausdorff space, for every $x \in X$, the *Dirac measure* δ_x, defined by $\delta_x(U) = 1$ if $x \in U$ and 0 otherwise, is a Borel probability measure on X.

The measure extending the Borel measure on $\mathbb{R}^n$ to sets differing of a Borel set by a subset of a Borel set of measure zero is called the *Lebesgue measure*.

The Carathéodory Extension Theorem below shows that one may extend measures defined on a Boolean algebra to one on a σ-algebra. Recall that a nonempty family R of subsets of a set X is a *Boolean algebra* if for every $U, V \in R$ one has $U \cup V, U \cap V, X \setminus U \in R$. A map $\mu \colon R \to [0, 1]$ is a probability measure on R if whenever the union of disjoint sets $U_n \in R$ is in R, then $\mu(\cup_n U_n) = \sum_n \mu(U_n)$.

Theorem B.5.1 (Carathéodory) *Let X be a topological space. Any probability measure μ on a Boolean algebra R has a unique extension μ^* to a* *probability measure on the σ-algebra $\mathcal{F}$ generated by R. One has for every set $U \in \mathcal{F}$,*

$$\mu^*(U) = \inf \sum_{n \geq 0} \mu(U_n) \tag{B.12}$$

for $U_n \in R$ such that $U \subset \cup_{n \geq 0} U_n$.

This result holds in particular when R is the Boolean algebra of clopen spaces, $\mathcal{F}$ the σ-algebra of Borel sets and X is a Cantor space. It shows in particular that any Borel measure on a Cantor space can be approximated by its value on clopen sets. Indeed, it implies that for every Borel set U and every $\varepsilon > 0$, there is a clopen set U_0 such that $\mu(U \Delta U_0) < \varepsilon$, where $U \Delta U_0 = (U \cup U_0) \backslash (U \cap U_0)$ is the *symmetric difference* of U, U_0.

B.5.2 Integration

Let μ be a positive measure on a measurable space X. For $A \subset X$, we denote by χ_A the *characteristic function* of A defined by $\chi_A(x) = 1$ if $x \in A$ and 0 otherwise. A *simple function* is of the form

$$s = \sum_{i=1}^{n} \alpha_i \chi_{A_i},$$

where the A_i are measurable sets, and $\alpha_i \in \mathbb{R}$. When s is a simple function, we define for a measurable set U,

$$\int_U s d\mu = \sum_{i=1}^{n} \alpha_i \mu(A_i \cap U),$$

with the convention $0 \cdot \infty = 0$.

If $f : X \to [0, \infty]$ is measurable, the *integral* (or *Lebesgue integral*) of f over U is defined as

$$\int_U f d\mu = \sup \int_U s d\mu,$$

the supremum being taken over the simple measurable functions s such that $0 \leq s \leq f$. For $U = X$, we omit the subscript U.

For an arbitrary measurable function $f : X \to \mathbb{R}$, we define

$$\int f d\mu = \int f^+ d\mu - \int f^- d\mu,$$

where $f^+ = \max\{f, 0\}$ and $f^- = -\min\{f, 0\}$. A measurable function $f : X \to \mathbb{R}$ is *integrable* if

$$\int |f| d\mu < \infty.$$

The set of integrable functions is denoted $L^1(X, \mu)$.

The following result is known as the *Monotone Convergence Theorem*.

Theorem B.5.2 *Let (f_n) be a sequence of measurable functions on X such that $0 \le f_1(x) \le f_2(x) \le \cdots$ and $f_n(x) \to f(x)$ for every $x \in X$. Then f is measurable and*

$$\int f_n d\mu \to \int f d\mu$$

as $n \to \infty$.

It follows from Theorem B.5.2 that if $f_n : X \to [0, \infty]$ is measurable for $n \ge 1$, then

$$\int \sum_{n \ge 1} f_n d\mu = \sum_{n \ge 1} \int f_n d\mu.$$

The next result is known as the *Dominated Convergence Theorem*.

Theorem B.5.3 (Lebesgue) *Suppose (f_n) is a sequence of measurable functions from X to $\mathbb{R}$ such that $f(x) = \lim_{n \to \infty} f_n(x)$ exists for every $x \in X$. If there is a function $g \in L^1(X, \mu)$ such that $|f_n(x)| \le g(x)$ for all $x \in X$, then*

$$\lim_{n \to \infty} \int f_n d\mu = \int f d\mu.$$

Let $(X, \mathcal{F}, \mu)$ be a measure space and let $T : X \to X$ be a measurable map. The map $\mu \circ T^{-1}$ is a measure on $\mathcal{F}$. The following formula is called the *change of variable formula* (its name relates it to the usual change of variable used to compute integrals of real functions of one variable). We have, for every real-valued measurable function f on X,

$$\int (f \circ T) d\mu = \int f d(\mu \circ T^{-1}). \tag{B.13}$$

Formula (B.13) is easy to verify in the case of a simple function $s = \sum_{i=1}^n \alpha_i \chi_{A_i}$. Indeed, we have $\int (s \circ T) d\mu = \sum_{i=1}^n \alpha_i \mu(T^{-1}A_i) = \int s d(\mu \circ T^{-1})$.

Every Borel measure μ on a topological space X defines a bounded linear map from $C(X, \mathbb{R})$ to $\mathbb{R}$ by $f \mapsto \int f d\mu$. By the *Riesz representation theorem*, the converse is true when X is a compact metric space. Thus, the space of measures on a compact metric space X can be identified with the (topological) dual of the space $C(X, \mathbb{R})$.

Let μ, ν be two Borel probability measures on X. Then ν is *absolutely continuous* with respect to μ, denoted $\nu \ll \mu$, if, for every Borel set $U \subset X$ such that $\mu(U) = 0$, one has $\nu(U) = 0$.

Theorem B.5.4 (Radon–Nikodym) *If $\nu \ll \mu$, there is a nonnegative μ-integrable function f such that $\nu(U) = \int_U f d\mu$ for every measurable set $U \subset X$.*

Two measures μ, ν on X are *mutually singular* if there is a measurable set U such that $\mu(U) = \nu(X \setminus U) = 0$.

A subset S of a real vector space is *convex* if for every $x, y \in S$ and $t \in [0, 1]$, one has $tx + (1 - t)y \in S$. The *convex hull* of S is the intersection of all convex sets containing S. The convex hull of a finite set of points in $\mathbb{R}^d$ is called a *simplex*.

The set $\mathcal{M}(X)$ of Borel probability measures on a compact metric space X is a convex subspace of the dual space of $C(X, \mathbb{R})$. The *weak-$*$ topology* on $\mathcal{M}(X)$ is the topology for which a sequence (μ_n) converges to $\mu \in \mathcal{M}(X)$ if for all $f \in C(X, \mathbb{R})$,

$$\int f d\mu_n \to \int f d\mu.$$

Theorem B.5.5 (Banach–Alaoglu) *For every compact metric space X, the space $\mathcal{M}(X)$ is metrizable and compact for the weak-$*$ topology.*

An *extreme point* of a convex set K is a point which does not belong to any open line segment in K. The closed convex hull of a set K is the closure of the intersection of all closed convex subsets containing K.

Theorem B.5.6 (Krein–Milman) *A compact convex set in a normed space is the closed convex hull of its extreme points.*

Notes We have mostly followed the classical Rudin (1987) for this summary of notions on measures and integration. We restricted the presentation to positive measures to simplify the picture. For the Carathéodory Extension Theorem (also known as the Kolmogorov extension Theorem), see Halmos (1974). On the weak-star topology (or weak-$*$ topology) and the Banach–Alaoglu Theorem, see Rudin (1991). The Krein–Milman Theorem is proved in Rudin (1991, Theorem 3.23).

B.6 Topological Entropy

We give a short introduction to the notion of topological entropy.

B.6.1 Entropy and Covers

Let (X, T) be a topological dynamical system. If α, β are covers of X, the *join* of α, β, denoted $\alpha \vee \beta$, is the cover formed of the $A \cap B$ for $A \in \alpha$ and $B \in \beta$. This extends to the join $\vee_{i=0}^{n-1} \alpha_i$ of n covers α_i for $0 \le i \le n - 1$. We also denote $T^{-j}\alpha$ the cover formed by the $T^{-j}A$ for $A \in \alpha$.

If α is an open cover of X, define the *entropy* of α by $H(\alpha) = \log N(\alpha)$, where $N(\alpha)$ is the number of elements of a finite subcover of α with minimal cardinality.

The *entropy of T relative to* α is the possibly infinite real number

$$h(T, \alpha) = \lim_{n \to \infty} \frac{1}{n} H(\vee_{i=0}^{n-1} T^{-i}\alpha)$$

(the limit can be shown to exist).

The *topological entropy* of T is

$$h(T) = \sup_{\alpha} h(T, \alpha),$$

where α ranges over all open covers of X. It can be shown that entropy is an invariant under conjugacy.

B.6.2 Computation of the Entropy

The following result is useful to compute the topological entropy. We denote by $\mathrm{diam}(\alpha) = \sup_{A \in \alpha} \mathrm{diam}(A)$ the diameter of an open cover α, where $\mathrm{diam}(A)$ is the diameter of the set A.

Theorem B.6.1 *If $(\alpha_n)_{n \ge 0}$ is a sequence of open covers such that* $\mathrm{diam}(\alpha_n) \to 0$, *then $h(T) = \lim_{n \to \infty} h(T, \alpha_n)$.*

The topological entropy of a shift space X has a simple expression in terms of its factor complexity $p_n(X)$.

Theorem B.6.2 *The topological entropy of a shift space (X, S) is $h(S) = \lim_{n \to \infty} \frac{1}{n} \log p_n(X)$.*

Actually, the topological entropy of a shift space on the alphabet A is the entropy of S relative to the natural partition $\{[a] \mid a \in A\}$.

Notes Topological entropy is a counterpart for topological dynamical systems of the notion of entropy for measure-theoretic system. Our presentation follows Walters (1975). Theorem B.6.1 is Theorem 7.6 and Theorem B.6.2 is Theorem 7.13.

Appendix C
Summary of Examples

We have summarized here a number of examples presented in the various chapters. The columns of Table C.1 give the values associated with various shifts (or families of shifts) which appear as the rows. For each row a comment is made after the array. The columns give successively, for every row (with more information in the comment below the table), the following:

1. the factor complexity, denoted p_n (except for the odometer for which the factor complexity is not defined),
2. the dimension group (with the positive cone and the unit given below),
3. the image subgroup,
4. the topological rank,
5. and finally, when available, the corresponding C^*-algebra.

Fibonacci shift. The Fibonacci shift is generated by the morphism $\varphi\colon a \mapsto ab, b \mapsto a$. It is the Sturmian shift of slope $\alpha = \frac{1+\sqrt{5}}{2}$ (Example 1.5.15). Thus, its dimension group is $\mathbb{Z}[\alpha]$ with the order induced by the reals and unit 1 (Example 3.9.7). It has a stationary BV-representation of rank 2 (Example 6.2.16). The Fibonacci algebra $\mathfrak{A}_\alpha$ corresponds to the Bratteli diagram of Figure C.1.

Sturmian shifts. All Sturmian shifts have, by definition, complexity $n + 1$. The dimension group of a Sturmian shift of slope α is $\mathbb{Z} + \alpha\mathbb{Z}$ (Theorem 3.9.6).

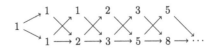

Figure C.1 The Bratteli diagram for the Fibonacci algebra $\mathfrak{A}_\alpha$.

Table C.1 *Summary of the examples*

name	complexity	K^0	Image	rank	C^*-algebra
Fibonacci	$n+1$	$\mathbb{Z}[\alpha], \alpha = \frac{1+\sqrt{5}}{2}$	$\mathbb{Z}[\alpha]$	2	$\mathfrak{A}_\alpha$
Sturmian	$n+1$	$\mathbb{Z} + \alpha\mathbb{Z}$	$\mathbb{Z} + \alpha\mathbb{Z}$	2	$\mathfrak{A}_\alpha$
Thue–Morse	$3n \le p_{n+1} < 4n$	$\mathbb{Z}[1/2] \times \mathbb{Z}$	$\frac{1}{3}\mathbb{Z}[1/2]$	3	
2-odometer		$\mathbb{Z}[1/2]$	$\mathbb{Z}[1/2]$	1	CAR
Chacon	$2n+1$	$\mathbb{Z}[1/3] \times \mathbb{Z}$	$\mathbb{Z}[1/3]$	2	
period-doubling	$n \le p_n \le 2n$	$\mathbb{Z}[1/2] \times \mathbb{Z}$	$\frac{1}{3}\mathbb{Z}[1/2]$	2	
Rudin–Shapiro	$8(n-1)$	$\mathbb{Z}[1/2]^3 \times \mathbb{Z}$	$\mathbb{Z}[1/2]$	≤ 9	
Tribonacci	$2n+1$	$\mathbb{Z}[\lambda]$	$\mathbb{Z}[\lambda]$	3	
Arnoux–Rauzy	$(d-1)n+1$	$\mathbb{Z}^d$		d	
dendric	$(d-1)n+1$	$\mathbb{Z}^d$		d	
squared Fibonacci	$2n+1$	$\mathbb{Z}^3$	$\mathbb{Z}[\sqrt{5}]$	3	
Cassaigne shift	$2n+2$	$\mathbb{Z}^3$	$\frac{1}{2}\mathbb{Z}[\sqrt{2}]$	3	
specular	$(d-2)n+2$	$\mathbb{Z}^{d-1}$		$d-1$	

It has rank 2 by Corollary 7.2.4. Its associated C^*-algebra is the Sturmian algebra $\mathfrak{A}_\alpha$.

Thue–Morse shift. The complexity of the Thue–Morse shift is given in Exercise 1.28. Its dimension group is computed in Example 4.6.11 as being isomorphic to $\mathbb{Z}[1/2] \times \mathbb{Z}$ with positive cone $\{(\alpha, \beta) \mid 3\alpha + \beta > 0\} \cup \{0\}$ and unit $(1, 0)$. The image subgroup is $(1/3)\mathbb{Z}[1/2]$. Its topological rank is ≤ 3 (Exercise 6.19) but cannot be 2 (see Example 6.3.8).

2-odometer. The dimension group of the 2-odometer is $\mathbb{Z}[1/2]$ (see Proposition 6.1.2). Its topological rank is 1 and the associated AF algebra is the CAR algebra (Example 9.3.4). More generally, the dimension group of the (p_n)-odometer is $\mathbb{Z}[(p_n)]$ by Proposition 6.1.2. Its AF algebra is a UHF algebra by Exercise 9.5.

Chacon shift The ternary Chacon shift has complexity $2n+1$ (Exercise 6.17). It is conjugate to the binary Chacon shift (Exercise 6.16) which has a dimension group isomorphic to $\mathbb{Z}[1/3] \times \mathbb{Z}$ with positive cone $\mathbb{Z}_+[1/3]$ and unit $(1, 1)$ (Exercise 4.9). It has a BV-representation with rank 2 by (6.10) since there are two return words to 0 in the binary Chacon shift.

Period-doubling. The period-doubling shift, generated by $0 \mapsto 01, 1 \mapsto 00$ is a Toeplitz shift. Its complexity is described in Exercise 1.72. A BV-representation of rank 2 and its dimension group are computed in Example 6.1.8 (and Exercise 6.14).

Rudin–Shapiro shift. The complexity of the Rudin–Shapiro shift is $8(n - 1)$ for $n \geq 2$ (Exercise 1.32). Its dimension group is $\mathbb{Z}[1/2]^3 \times \mathbb{Z}$ with positive cone $\mathbb{Z}_+[1/2] \times \mathbb{Z}[1/2]^2 \times \mathbb{Z}$ and unit $(1, 0, 0, 0)$ (Exercise 4.11). A BV-representation with 9 vertices is computed in Exercise 6.20.

Tribonacci. The Tribonacci shift is an Arnoux–Rauzy shift (Example 1.5.3) and thus a dendric shift. Thus, its complexity is $2n + 1$, as a dendric shift on 3 letters. Its dimension group is $\mathbb{Z}[\lambda]$ where $\lambda^3 = \lambda^2 + \lambda + 1$ (Example 4.4.6). It has a BV-representation with 3 vertices (Example 6.2.5). Thus, its topological rank is 3 as its dimension group has rank 3.

Arnoux–Rauzy. The complexity of an Arnoux–Rauzy shift X on an alphabet A with d symbols is $(d-1)n+1$ since it is a dendric shift (Proposition 7.1.2). Its dimension group is $\mathbb{Z}^A$ with positive cone $\{x \in \mathbb{Z}^A \mid \langle x, \boldsymbol{\mu} \rangle > 0\} \cup \{0\}$ where μ is the unique ergodic measure on X and $\boldsymbol{\mu} = (\mu(a))_{a \in A}$ (Theorem 4.4.7). It has a BV-representation with d vertices at each level, as all dendric shifts. The rank is equal to d by (6.9).

Dendric shifts. The complexity of a dendric shift on $d \geq 2$ letters is $(d-1)n + 1$. Its dimension group is $\mathbb{Z}^d$ with positive cone given by the ergodic measures on X (their number is bounded by $d - 1$, see Theorem 3.9.5 and actually by $d/2$, see the notes of Chapter 7). It has a BV-representation with d vertices at each level (Theorem 7.1.43) and rank d, as for Arnoux–Rauzy shifts.

Squared Fibonacci. The squared Fibonacci shift is obtained as a 3-interval exchange corresponding to a rotation of angle $3 - \sqrt{5}$ (Example 8.1.7). As a regular interval exchange shift, it is dendric and has complexity $2n + 1$. Its dimension group is isomorphic to $\mathbb{Z}^3$ with positive cone $\{(x, y, z) \mid \sqrt{5}x + y + z \geq 0\}$ and unit $(1, 1, 1)$ (see Example 8.1.30). It has a BV-representation with 3 vertices at each level, as every dendric shift on 3 symbols and its rank is 3 because the $\mathbb{Q}$-rank of its dimension group is 3.

Cassaigne. The Cassaigne shift is generated by the morphism $a \mapsto ab, b \mapsto cda, c \mapsto cd, d \mapsto abc$. It is a specular shift with complexity $2n + 2$ for $n \geq 1$ (Exercise 7.2). Its dimension group is $\mathbb{Z}^3$ with positive cone $\{(x, y, z) \in \mathbb{Z}^3 \mid x + y\sqrt{2} + z > 0\}$ and unit $(3, 4, 3)$. The image subgroup is isomorphic to $\frac{1}{2}\mathbb{Z}[\sqrt{2}]$ (Example 7.3.41). The rank is 3 by Eq. (6.10) since every set of return words has 3 elements.

Specular shifts. The complexity of a specular shift X on d letters is $p_n(X) = (d - 2)n + 2$ for $n \geq 1$ (Proposition 7.3.7). It has an induced system which is dendric on $d - 1$ symbols by Theorem 7.3.40. Thus, its dimension group is isomorphic to $\mathbb{Z}^{d-1}$ and has a BV-representation with $d - 1$ vertices and thus rank $d - 1$.

Appendix D

Equivalent Definitions of Sturmian Shifts

The following statement, included for the convenience of the reader, lists the numerous equivalent definitions of Sturmian sequences and Sturmian shifts. Recall first that a one-sided sequence s is Sturmian if its complexity is $n + 1$. Its slope $\alpha = \lim_{n \to \infty} f_n$ where $f_n = |s_{[0,n)}|_1 / n$ is the frequency of 1 in s.

Theorem D.0.1 *The following conditions are equivalent for a shift X on the alphabet $\{0, 1\}$ and an irrational number $0 < \alpha < 1$.*

(i) *X is Sturmian of slope α.*

(ii) *$\mathcal{L}(X)$ is the language of a one-sided Sturmian sequence s of slope α.*

(iii) *X is minimal of complexity $n + 1$ and α is the frequency of 1 in every $x \in X$.*

(iv) *X is aperiodic and $\mathcal{L}(X)$ is balanced.*

(v) *X is the natural coding of a rotation of angle α.*

(vi) *X is generated by a mechanical sequence of slope α.*

(vii) *X is generated by a cutting sequence of slope $\alpha/(1 - \alpha)$.*

(viii) *X is strict episturmian with directive sequence $x = 0^{d_1} 1^{d_2} \cdots$ and $\alpha = [0; 1 + d_1, d_2, \ldots]$.*

(ix) *X is aperiodic and generated by a palindrome closed one-sided sequence.*

(x) *X has an S-adic representation of the form $(L_{x_n})_{n \geq 0}$ with $x = x_0 x_1 \cdots$ such that $x = 0^{d_1} 1^{d_2} \cdots$ and $\alpha = [0; 1 + d_1, d_2, \ldots]$.*

Let us give references for these conditions.

(ii) was taken as the definition of a Sturmian shift. (iii) is easily proved equivalent to (ii). The equivalence with (iv) is Proposition 1.5.1. The equivalence with (v) or (vi) is Exercise 1.52 in one direction and Proposition 1.5.12 in the other.

The equivalence with (vii) is, in one direction, a consequence of the fact that the language of a Sturmian shift is closed under reversal (Exercise 1.50) and, in the other direction, of the fact that, by Eq. (1.34), an episturmian shift is minimal. The link of the slope α with the directive word x is Proposition 1.5.12.

(viii) The relation of the slope with the directive sequence is given by Proposition 1.5.12.

(ix) The condition is necessary since a sequence of the form $s = \text{Pal}(x)$ is a palindrome closed by Proposition 1.5.8. Conversely, by Proposition 1.5.10, a palindrome closed sequence is of this form. If 0 and 1 do not appear infinitely often in x, the sequence x is eventually periodic and thus also $\text{Pal}(x)$.

(x) The $\mathcal{S}$-adic representation is given by Justin's Formula (1.32). This gives the equivalence with (vi).

Appendix E

Open Problems

Substitutions

1. **Pisot conjecture for substitutions.** Let σ be a primitive substitution on d letters whose dominant eigenvalue α is a Pisot number. Suppose that the characteristic polynomial of the incidence matrix of σ is irreducible. Let X be the shift generated by σ. Is it true that there exists a factor map $f \colon (X, S) \to (\mathbb{T}^d, T_\alpha)$, where $T_\alpha \colon z \mapsto z + \alpha$ is the rotation of angle $\alpha \in \mathbb{T}^d$, which is a measure-theoretic isomorphism? It is known to be true when the alphabet has two letters (see Hollander and Solomyak (2003) and Barge and Diamond (2002)).

2. **Computation of the topological rank.** Find an algorithm that computes the topological rank of a minimal substitution shift (bounds are given by Inequalities (6.9) and (6.10)).

3. **Topological mixing.** Find an algorithm to decide whether a minimal substitution shift is topologically mixing (a partial characterization is given in Kenyon et al. (2005)).

4. **Stabilizer of a substitution fixed point.** The stabilizer of a one-sided infinite sequence $x \in A^{\mathbb{N}}$, denoted by $\mathrm{Stab}(x)$, is the monoid of endomorphisms σ defined on the alphabet A that satisfy $\sigma(x) = x$. Infinite sequences that have a cyclic stabilizer are called *rigid*.

   Can we characterize rigid sequences? Can we describe $\mathrm{Stab}(x)$?

   There exist numerous results on the two-letter case concerning rigidity (see Séébold (1998); Richomme and Séébold (2012) and also Berthé et al. (2012)). The situation is more contrasted as soon as the size of the alphabet increases. For instance, over a ternary alphabet, the stabilizer can be infinitely generated, even when the word is generated by iterating an invertible primitive morphism (see Krieger (2008); Diekert and Krieger (2009)). But for dendric fixed points of substitutions, more is known (see Berthé et al. (2018)).

5. **Equality problem of fixed points.** Given two morphic sequences x and y, is it decidable whether $x = y$ (Equality problem for morphic sequences)? For purely morphic sequences, this problem is called the *D0L ω-equivalence problem* and was solved in 1984 by Karel Culik and Tero Harju (1984). In Durand (2012), it is shown that it is true for primitive morphic sequences and that more can be said of the D0L ω-equivalence problem in this case: the two (underlying) primitive substitutions have some nontrivial powers that coincide on some cylinder whenever $x = y$.

   The general case of the Equality problem remains open.

6. **Decidability of language inclusion.** Let σ, τ be two endomorphisms. Is it decidable whether $X(\sigma)$ is included in $X(\tau)$ or, equivalently, whether $\mathcal{L}(\sigma) \subset \mathcal{L}(\tau)$ (Inclusion problem for substitution shifts or languages) or $X(\sigma) = X(\tau)$ (Equality problem for substitution shifts)?

   In Fagnot (1997a) a positive answer is given to the Inclusion problem for some families of purely morphic sequences and, in Fagnot (1997b), it is solved for automatic sequences.

   In Durand and Leroy (2020), it is shown that the Inclusion problem, and thus the Equality problem, is decidable for minimal substitution shifts. Both problems are open if we do not assume uniform recurrence.

7. **Topological factors.** Find an algorithm that computes the set of (non-isomorphic) topological factors of minimal substitution shift.

   Let σ be a primitive substitution. We know from Durand (2000) that the number of (non-isomorphic) aperiodic shift factors of $X(\sigma)$ is finite and that all such factors are conjugate to a substitution shift. We can thus formulate a more precise question, as follows.

   Can we find an algorithm that computes a finite set of primitive substitutions $\{\sigma_1, \ldots, \sigma_N\}$ such that any aperiodic shift factor of $X(\sigma)$ is isomorphic to some $X(\sigma_n)$?

   Periodic factors can easily be determined using Section 6 in Durand (2000). For constant-length substitutions the problem is solved. Algorithms are given in Coven et al. (2017) and Durand and Leroy (2020) to find constant-length substitution factors. In Müllner and Yassawi (2021), it is shown that factors of primitive constant-length shifts are isomorphic to constant-length substitution shifts.

   The non–constant-length case is open.

8. **Arithmetic subsequences and eigenvalues.** Let $x = (x_n)$ be a fixed point of a substitution σ. Let $p \in \mathbb{N}$, $k \in \mathbb{Z}$. The sequence $(x_{pn+k})_n$ is substitutive with respect to some (τ, ϕ) (Durand (1996); Durand and Goyheneche (2019)).) It can be shown that the eigenvalues of $M(\sigma)$, with the

possible exception of 0 and roots of the unity, are eigenvalues of $M(\tau)$. But there are examples where some eigenvalues, different from 0 and roots of unity, of $M(\tau)$ that are not eigenvalues of $M(\sigma)$.

What are the eigenvalues of $M(\tau)$?

Finite Rank

1. **Characterization of the dimension groups of unimodular S-adic shifts.**
   Consider the dimension group $\mathbb{Z}^d$, with positive cone $\{x \in \mathbb{Z}^d \mid \langle x, \mu_i \rangle > 0,$
   $1 \leq i \leq d'\} \cup \mathbf{0}$ and unit $\mathbf{1}$, where $\mathbf{1}$ is the vector with all components equal
   to 1 and the μ_i are some vectors with $d' \leq d$. All dimension groups of
   primitive unimodular proper S-adic shifts are of this type with $d' \leq d - 1$
   (Effros et al. (1980) or Corollary 6.5.2). The converse is not true. According to Riedel (1981b), not every strong orbit equivalence class represented
   by such a dimension group contains a primitive unimodular proper S-adic
   shift. But when the dimension group has a unique state, that is, when the
   systems it represents are uniquely ergodic, then the class contains such a
   primitive unimodular S-adic shift (Riedel, 1981a).

   Does there exist a complete characterization of the vectors μ_i, $1 \leq i \leq d-1$
   defining a strong orbit equivalence class containing a primitive unimodular
   proper S-adic shift?

   Of course, when it exists, there is a kind of continued fraction algorithm,
   given by the unimodular matrices, that approximates simultaneously the
   vectors μ_i, (Riedel, 1981b). This suggests the following questions.

2. Is there a cone property of being "badly approximable" that would be
   equivalent to the absence of a primitive unimodular proper S-adic shift?
   In such a situation, does the strong orbit equivalence class contain a finite
   rank system?

3. Primitive unimodular proper S-adic shifts on d letters have at most $d - 1$
   ergodic measures as recalled above. Is it a sharp bound?

Bratteli Diagrams

1. **Weak mixing of the Pascal adic transformation.** Show that the Pascal adic transformation is weakly mixing (see Exercise 5.13, Méla and
   Petersen (2003) and Janvresse and de la Rue (2004)).

2. **Expansiveness in strong orbit equivalence classes.** Is it possible to
   characterize expansiveness within a given strong orbit equivalence class?
   Observe (it is not difficult to prove) that there always exists a minimal shift
   in a strong orbit equivalence class of a minimal Cantor system.

Recognizability

1. **Number of periodic orbits.** Let $\varphi \colon A^* \to B^*$ be an indecomposable morphism. Show that the number of orbits of periodic sequences with more than one factorization is at most $\mathrm{Card}(A)$ (see Karhumäki et al. (2003)).
2. **Disjoint factorizations.** Let $\varphi \colon A^* \to B^*$ be a morphism. If there is a sequence $x \in B^{\mathbb{Z}}$ with k disjoint factorizations, then $\varphi = \alpha \circ \beta$ with $A^* \xrightarrow{\beta} C^* \xrightarrow{\alpha} B^*$ and $\mathrm{Card}(C) \leq \mathrm{Card}(A) - k + 1$ (see Karhumäki and Maňuch (2002)).

Complexity and $\mathcal{S}$-adic Sequences

1. **Unimodular BV-representation of unimodular $\mathcal{S}$-adic shifts.** Is it true that every primitive unimodular $\mathcal{S}$-adic shift has a primitive unimodular BV-representation? Even in the substitutive case the question is open.
2. **Polynomial complexity and finite rank systems.** Let X be an aperiodic minimal $\mathcal{S}$-adic shift generated by a primitive directive sequence $\tau = (\tau_n \colon \mathcal{A}_{n+1}^* \to \mathcal{A}_n^*)_{n \geq 0}$ with $\liminf |\mathcal{A}_n| \leq r$ for some $r \geq 1$. Is there an integer $d \geq 1$ such that $\liminf_{n \to \infty} p_n(X)/n^d = 0$?
3. **$\mathcal{S}$-adic conjecture.** Does there exist a property (P) defined on sequences of morphisms belonging to a finite set such that a sequence x has subaffine complexity if and only if x is $\mathcal{S}$-adic with the property (P)? Can this property be formulated as a Büchi condition or other effectively defined condition? (On this conjecture, see Ferenczi (1996) and Durand et al. (2013)).

Dendric Shifts

1. **Converse of the Return Theorem.** Is it true that a minimal shift X on the alphabet A such that $A \subset \mathcal{L}(X)$ is dendric if and only if every set of return words $\mathcal{R}_X(w)$ for $w \in \mathcal{L}(X)$ is a basis of the free group on A? (The converse holds by Theorem 7.1.15.)
2. **Substitution dendric shifts.**
   Is there a characterization of substitution shifts that are dendric?
3. **Factors of eventually dendric shifts.** Is every factor of an eventually dendric shift eventually dendric? (The class is closed under conjugacy; see Dolce and Perrin (2017a).)
4. **$\mathcal{S}$-adic representation of dendric shifts.** Is there a characterization of eventually dendric shifts in terms of an $\mathcal{S}$-adic representation, with $\mathcal{S}$ being the set of elementary automorphisms?

References

Aho, Alfred V., Hopcroft, John E., and Ullman, Jeffrey D. *The Design and Analysis of Computer Algorithms*. Addison-Wesley, Boston, 1974.

Allauzen, Cyril. Une caractérisation simple des nombres de Sturm. *J. Théor. Nombres Bordeaux*, 10(2):237–241, 1998.

Allouche, Jean-Paul and Shallit, Jeffrey. *Automatic Sequences*. Cambridge University Press, Cambridge, 2003.

Almeida, Jorge, Costa, Alfredo, Kyriakoglou, Revekka, and Perrin, Dominique. *Profinite Semigroups and Symbolic Dynamics*. Springer, Cham, 2020.

Araújo, Isabel M. and Bruyère, Véronique. Words derived from Sturmian words. *Theor. Comput. Sci.*, 340(1):204–219, 2005.

Arnoux, Pierre and Rauzy, Gérard. Représentation géométrique de suites de complexité $2n + 1$. *Bull. Soc. Math. France*, 119(2):199–215, 1991.

Arnoux, Pierre, Mizutani, Masahiro, and Sellami, Tarek. Random product of substitutions with the same incidence matrix. *Theor. Comput. Sci.*, 543:68–78, 2014.

Ashley, Jonathan, Marcus, Brian, Perrin, Dominique, and Tuncel, Selim. Surjective extensions of sliding-block codes. *SIAM J. Discrete Math.*, 6(4):582–611, 1993.

Atiyah, Michael. *K-Theory*. Benjamin, 1967. (New edition, CRC Press, Boca Raton, FL, 2018).

Avila, Artur and Delecroix, Vincent. Pisot property for Brun and fully subtractive algorithm. 2013. arXiv:1506.03692.

Avila, Artur and Forni, Giovanni. Weak mixing for interval exchange transformations and translation flows. *Ann. Math.*, 165(2):637–664, 2007.

Balková, L'ubomíra, Pelantová, Edita, and Steiner, Wolfgang. Sequences with constant number of return words. *Monatsh. Math.*, 155(3–4):251–263, 2008.

Barge, Marcy and Diamond, Beverly. Coincidence for substitutions of Pisot type. *Bull. Soc. Math. France*, 130(4):619–626, 2002.

Berstel, Jean. *Transductions and Context-Free Languages*, vol. 38 of *Leitfäden der Angewandten Mathematik und Mechanik [Guides to Applied Mathematics and Mechanics]*. Teubner, Stuttgart, 1979.

Berstel, Jean. On the index of Sturmian words. In Juhani Karhumaki, Hermann Maurer, Gheorghe Paun, and Gregorz Rozenberg, eds., *Jewels Are Forever*. Springer-Verlag, Berlin and Heidelberg, 1999.

Berstel, Jean, Perrin, Dominique, and Reutenauer, Christophe. *Codes and Automata*. Cambridge University Press, New York, 2009.

Berstel, Jean, Perrin, Dominique, Perrot, Jean François, and Restivo, Antonio. Sur le théoréme du défaut. *Journal of Algebra*, 60(1):169–180, 1979.

Berthé, Valérie and Delecroix, Vincent. Beyond substitutive dynamical systems: S-adic expansions. In *Numeration and Substitution 2012*, RIMS Kôkyûroku Bessatsu, B46, pages 81–123. Research Institute for Mathematical Sciences (RIMS), Kyoto, 2014.

Berthé, Valérie and Rigo, Michel, eds. *Combinatorics, Automata and Number Theory*, vol. 135 of *Encyclopedia of Mathematics and Its Applications*. Cambridge University Press, Cambridge, 2010.

Berthé, Valérie, Frettlöh, Dirk, and Sirvent, Victor. Selfdual substitutions in dimension one. *European J. Combin.*, 33(6):981–1000, 2012. DOI:https://doi.org/10.1016/j.ejc.2012.01.001.

Berthé, Valérie, De Felice, Clelia, Dolce, Francesco, Leroy, Julien, Perrin, Dominique, Reutenauer, Christophe, and Rindone, Giuseppina. Acyclic, connected and tree sets. *Monatsh. Math.*, 176(4):521–550, 2015a.

Berthé, Valérie, De Felice, Clelia, Dolce, Francesco, Leroy, Julien, Perrin, Dominique, Reutenauer, Christophe, and Rindone, Giuseppina. Maximal bifix decoding. *Discrete Math.*, 338(5):725–742, 2015b.

Berthé, Valérie, De Felice, Clelia, Dolce, Francesco, Leroy, Julien, Perrin, Dominique, Reutenauer, Christophe, and Rindone, Giuseppina. The finite index basis property. *J. Pure Appl. Algebra*, 219(7):2521–2537, 2015c.

Berthé, Valérie, De Felice, Clelia, Dolce, Francesco, Leroy, Julien, Perrin, Dominique, Reutenauer, Christophe, and Rindone, Giuseppina. Bifix codes and interval exchanges. *J. Pure Appl. Algebra*, 219(7):2781–2798, 2015d.

Berthé, Valérie, De Felice, Clelia, Delecroix, Vincent, Dolce, Francesco, Leroy, Julien, Perrin, Dominique, Reutenauer, Christophe, and Rindone, Giuseppina. Specular sets. *Theoret. Comput. Sci.*, 684:3–28, 2017a.

Berthé, Valérie, Delecroix, Vincent, Dolce, Francesco, Perrin, Dominique, Reutenauer, Christophe, and Rindone, Giuseppina. Return words of linear involutions and fundamental groups. *Ergod. Theory Dyn. Syst.*, 37(3):693–715, 2017b.

Berthé, Valérie, Dolce, Francesco, Durand, Fabien, Leroy, Julien, and Perrin, Dominique. Rigidity and substitutive dendric words. *Internat. J. Found. Comput. Sci.*, 29(5):705–720, 2018.

Berthé, Valérie, Steiner, Wolfgang, and Thuswaldner, Jörg M. Geometry, dynamics, and arithmetic of S-adic shifts. *Ann. Inst. Fourier (Grenoble)*, 69(3):1347–1409, 2019a.

Berthé, Valérie, Steiner, Wolfgang, Thuswaldner, Jörg M., and Yassawi, Reem. Recognizability for sequences of morphisms. *Ergod. Theory Dyn. Syst.*, 39(11):2896–2931, 2019b.

Berthé, Valérie, Cecchi Bernales, Paulina, Durand, Fabien, Leroy, Julien, Perrin, Dominique, and Petite, Samuel. On the dimension group of unimodular S-adic subshifts. *Monatsh. Math.*, 194(4):687–717, 2021.

Besbes, Adnene, Boshernitzan, Michael, and Lenz, Daniel. Delone sets with finite local complexity: linear repetitivity versus positivity of weights. *Discrete Comput. Geom.*, 49(2):335–347, 2013.

Bezuglyi, Sergey, Kwiatkowski, Jan, and Medynets, Konstantin. Aperiodic substitution systems and their Bratteli diagrams. *Ergod. Theory Dyn. Syst.*, 29(1):37–72, 2009.

Bezuglyi, Sergey, Kwiatkowski, Jan, Medynets, Konstantin, and Solomyak, Boris. Finite rank Bratteli diagrams: structure of invariant measures. *Trans. Amer. Math. Soc.*, 365(5):2637–2679, 2013.

Bezuglyi, Sergey and Karpel, Olena. Orbit equivalent substitution dynamical systems and complexity. *Proc. Amer. Math. Soc.*, 142(12):4155–4169, 2014.

Birkhoff, Garrett. Extensions of Jentzsch's theorem. *Trans. Amer. Math. Soc.*, 85:219–227, 1957.

Birkhoff, Garrett. *Lattice Theory*. 3rd ed. American Mathematical Society Colloquium Publications, vol. XXV. American Mathematical Society, Providence, RI, 1967.

Blanchard, François and Perrin, Dominique. Relèvement d'une mesure ergodique par un codage. *Z. Wahrsch. Verw. Gebiete*, 54(3):303–311, 1980.

Boissy, Corentin and Lanneau, Erwan. Dynamics and geometry of the Rauzy-Veech induction for quadratic differentials. *Ergod. Theory Dyn. Syst.*, 29(3):767–816, 2009.

Boshernitzan, Michael. A unique ergodicity of minimal symbolic flows with linear block growth. *J. Anal. Math.*, 44:77–96, 1984.

Boshernitzan, Michael D. A condition for unique ergodicity of minimal symbolic flows. *Ergod. Theory Dyn. Syst.*, 12(3):425–428, 1992.

Boshernitzan, Michael D. and Carroll, C. R. An extension of Lagrange's theorem to interval exchange transformations over quadratic fields. *J. Anal. Math.*, 72:21–44, 1997.

Boyle, Mike. Topological Orbit Equivalence and Factor Maps in Symbolic Dynamics. PhD thesis, University of Washington, 1983.

Boyle, Mike. Open problems in symbolic dynamics. In *Geometric and Probabilistic Structures in Dynamics*, vol. 469 of *Contemp. Math.*, pages 69–118. American Mathematical Society, Providence, RI, 2008.

Boyle, Mike and Handelman, David. Entropy versus orbit equivalence for minimal homeomorphisms. *Pacific J. Math.*, 164(1):1–13, 1994.

Boyle, Mike and Handelman, David. Orbit equivalence, flow equivalence and ordered cohomology. *Isr. J. Math.*, 95:169–210, 1996.

Boyle, Mike and Krieger, Wolfgang. Periodic points and automorphisms of the shift. *Trans. Amer. Math. Soc.*, 302(1):125–149, 1987.

Boyle, Mike and Tomiyama, Jun. Bounded topological orbit equivalence and C^*-algebras. *J. Math. Soc. Japan*, 50(2):317–329, 1998.

Bratteli, Ola. Inductive limits of finite dimensional C^*-algebras. *Trans. Amer. Math. Soc.*, 171:195–234, 1972.

Bratteli, Ola and Robinson, Derek W. *Operator Algebras and Quantum Statistical Mechanics. 1*. Texts and Monographs in Physics. Springer, New York, 2nd ed., 1987.

Bratteli, Ola, Jørgensen, Palle E. T., Kim, Ki Hang, and Roush, Fred. Non-stationarity of isomorphism between AF algebras defined by stationary Bratteli diagrams. *Ergod. Theory Dyn. Syst.*, 20(6):1639–1656, 2000.

Bratteli, Ola, Jørgensen, Palle E. T., Kim, Ki Hang, and Roush, Fred. Decidability of the isomorphism problem for stationary AF-algebras and the associated ordered simple dimension groups. *Ergod. Theory Dyn. Syst.*, 21(6):1625–1655, 2001.

Brocot, Achille. Calcul des rouages par approximation, nouvelle méthode. *Revue Chronométrique*, 6:186–194, 1860.

Brown, James R. *Ergodic Theory and Topological Dynamics*. Academic Press, Waltham, MA, 1976.

Brun, Viggo. Algorithmes euclidiens pour trois et quatre nombres. In *Treizième congrès des mathèmaticiens scandinaves, tenu à Helsinki 18–23 août 1957*, pages 45–64. Mercators Tryckeri, Helsinki, 1958.

Canterini, Vincent and Siegel, Anne. Geometric representation of substitutions of Pisot type. *Trans. Amer. Math. Soc.*, 353(12):5121–5144, 2001.

Carlsen, Toke M. and Eilers, Søren. Augmenting dimension group invariants for substitution dynamics. *Ergod. Theory Dyn. Syst.*, 24(4):1015–1039, 2004.

Carroll, Joseph E. Birkhoff's contraction coefficient. *Linear Algebra Appl.*, 389:227–234, 2004.

Cassaigne, Julien. An algorithm to test if a given circular HD0L-language avoids a pattern. In Information Processing 1994, *IFIP Transactions A: Computer Science and Technology*, A-51, pages 459–464. North-Holland, Amsterdam, 1994.

Cassaigne, Julien. Special factors of sequences with linear subword complexity. In *Developments in Language Theory II*, pages 25–34. World Scientific, Hackensack, NJ, 1996.

Cassaigne, Julien. Complexité et facteurs spéciaux. *Bull. Belg. Math. Soc. Simon Stevin*, 4(1):67–88, 1997.

Cassaigne, Julien. Personal communication, 2015.

Cassaigne, Julien and Nicolas, François. Quelques propriétés des mots substitutifs. *Bull. Belg. Math. Soc. Simon Stevin*, 10(suppl.):661–676, 2003.

Champernowne, David G. The construction of decimals normal in the scale of ten. *J. Lond. Math. Soc.*, 8(4):254–260, 1933.

Chvátal, Vašek. *Linear Programming*. A Series of Books in the Mathematical Sciences. W. H. Freeman and Company, New York, 1983.

Cobham, Alan. On the Hartmanis-Stearns problem for a class of tag machines. In *SWAT '68: Proceedings of the 9th Annual Symposium on Switching and Automata Theory (swat 1968)*, pages 51–60, IEEE Computer Society, Washington, DC, 1968.

Cobham, Alan. Uniform tag sequences. *Mathematical Systems Theory*, 6:164–192, 1972.

Cohn, Paul M. *Free Rings and Their Relations*, vol. 19 of London Mathematical Society Monographs. Academic Press, Inc. [Harcourt Brace Jovanovich, Publishers], London, 2nd ed., 1985.

Cornfeld, Isaac P., Fomin, Sergei V., and Sinai, Yakov G. *Ergodic Theory*, vol. 245 of *Grundlehren der Mathematischen Wissenschaften [Fundamental Principles of Mathematical Sciences]*. Springer, New York, 1982. Translated from the Russian by A. B. Sosinskii.

Cortez, María Isabel, Durand, Fabien, Host, Bernard, and Maass, Alejandro. Continuous and measurable eigenfunctions of linearly recurrent dynamical cantor systems. *J. Lond. Math. Soc.*, 67(3):790–804, 2003.

Coulbois, Thierry, Hilion, Arnaud, and Lustig, Martin. $\mathbb{R}$-trees and laminations for free groups. I. Algebraic laminations. *J. Lond. Math. Soc. (2)*, 78(3):723–736, 2008.

Coven, Ethan, Dekking, Michel, and Keane, Michael. Topological conjugacy of constant length substitution dynamical systems. *Indag. Math. (N.S.)*, 28(1):91–107, 2017.

Coven, Ethan M. and Hedlund, Gustav A. Sequences with minimal block growth. *Math. Systems Theory*, 7:138–153, 1973.

Crochemore, Maxime, Lecroq, Thierry, and Rytter, Wojciech. *125 Problems in Text Algorithms*. Cambridge University Press, 2021.

Culik, Karel and Harju, Tero. The ω-sequence equivalence problem for D0L systems is decidable. *J. Assoc. Comput. Mach.*, 31(2):282–298, 1984.

Cuntz, Joachim and Krieger, Wolfgang. Topological Markov chains with dicyclic dimension groups. *J. Reine Angew. Math.*, 320:44–51, 1980.

Damanik, David and Lenz, Daniel. The index of Sturmian sequences. *European J. Combin.*, 23(1):23–29, 2002.

Damanik, David and Lenz, Daniel. Powers in Sturmian sequences. *European J. Combin.*, 24(4):377–390, 2003.

Damanik, David and Lenz, Daniel. Substitution dynamical systems: Characterization of linear repetitivity and applications. *J. Math. Anal. Appl.*, 321(2):766–780, 2006.

Damron, Michael and Fickenscher, Jon. The number of ergodic measures for transitive subshifts under the regular bispecial condition. *Ergod. Theory Dyn. Syst.*, pages 1–55, 2020. doi: https://doi.org/10.1017/etds.2020.134

Danthony, Claude and Nogueira, Arnaldo. Measured foliations on nonorientable surfaces. *Ann. Sci. École Norm. Sup. (4)*, 23(3):469–494, 1990.

Dartnell, Pablo R., Durand, Fabien, and Maass, Alejandro. Orbit equivalence and Kakutani equivalence with Sturmian subshifts. *Studia Math.*, 142(1):25–45, 2000.

Davidson, Kenneth R. *C*-Algebras by Example*, vol. 6 of *Fields Institute Monographs*. American Mathematical Society, Providence, RI, 1996.

de Luca, Aldo. Sturmian words: Structure, combinatorics, and their arithmetics. *Theoret. Comput. Sci.*, 183(1):45–82, 1997.

Dekking, F. Michel. The spectrum of dynamical systems arising from substitutions of constant length. *Z. Wahrscheinlichkeitstheorie und Verw. Gebiete*, 41(3):221–239, 1977/78.

Delecroix, Vincent. Interval exchange transformations. CIMPA Research School Cantor-Salta 2015: Dynamics on Cantor Sets. November 2–13, 2015, Universidad Nacional de Salta, Argentina. Accessed at https://cantorsalta2015.sciencesconf.org.

Delecroix, Vincent, Hejda, Tomáš, and Steiner, Wolfgang. Balancedness of Arnoux-Rauzy and Brun words. In *Combinatorics on Words*, vol. 8079 of *Lect. Notes Comput. Sci.*, pages 119–131. Springer-Verlag, Heidelberg, 2013.

Diekert, Volker and Krieger, Dalia. Some remarks about stabilizers. *Theoret. Comput. Sci.*, 410(30–32):2935–2946, 2009.

Dolce, Francesco and Perrin, Dominique. Neutral and tree sets of arbitrary characteristic. *Theoret. Comput. Sci.*, 658(part A):159–174, 2017a.

Dolce, Francesco and Perrin, Dominique. Interval exchanges, admissibility and branching Rauzy induction. *RAIRO Theor. Inform. Appl.*, 51(3):141–166, 2017b.

Dolce, Francesco and Perrin, Dominique. Eventually dendric shift spaces. *Ergod. Theory Dyn. Syst.*, pages 1–26, 2020. doi: https://doi.org/10.1017/etds.2020.35.

Donoso, Sebastian, Durand, Fabien, Maass, Alejandro, and Petite, Samuel. Interplay between finite topological rank minimal cantor systems, S-adic subshifts and their complexity. *Trans. Amer. Math. Soc.*, 374(5):3453–3489, 2021.

Downarowicz, Tomasz. Survey of odometers and Toeplitz flows. *Contemp. Math.*, 385:7–37, 2005.

Downarowicz, Tomasz and Maass, Alejandro. Finite-rank Bratteli-Vershik diagrams are expansive. *Ergod. Theory Dyn. Syst.*, 28(3):739–747, 2008.

Droubay, Xavier, Justin, Jacques, and Pirillo, Giuseppe. Episturmian words and some constructions of de Luca and Rauzy. *Theoret. Comput. Sci.*, 255(1–2):539–553, 2001.

Du, Chen Fei, Mousavi, Hamoon, Schaeffer, Luke, and Shallit, Jeffrey O. Decision algorithms for Fibonacci-automatic words, with applications to pattern avoidance. *CoRR*, 2014. arXiv:1406.0670.

Dumont, Jean-Marie and Thomas, Alain. Systemes de numeration et fonctions fractales relatifs aux substitutions. *Theoret. Comput. Sci.*, 65(2):153–169, 1989.

Dunford, Nelson and Schwartz, Jacob T. *Linear Operators, Part I, General Theory.* Wiley, New York, 1988.

Durand, Fabien. Contributions à l'étude des suites et systèmes dynamiques substitutifs. PhD thesis, Université de la Méditerranée, 1996.

Durand, Fabien. A characterization of substitutive sequences using return words. *Discrete Math.*, 179(1–3):89–101, 1998.

Durand, Fabien. Linearly recurrent subshifts have a finite number of non-periodic subshift factors. *Ergod. Theory Dyn. Syst.*, 20(4):1061–1078, 2000.

Durand, Fabien. Corrigendum and addendum to: "Linearly recurrent subshifts have a finite number of non-periodic subshift factors." *Ergod. Theory Dyn. Syst.*, 23(2):663–669, 2003.

Durand, Fabien. Combinatorics on Bratteli diagrams and dynamical systems. In *Combinatorics, Automata and Number Theory*, vol. 135 of *Encyclopedia Math. Appl.*, pages 324–372. Cambridge University Press, Cambridge, 2010.

Durand, Fabien. H0L ω-equivalence and periodicity problems in the primitive case (dedicated to the memory of G. Rauzy). *J. Unif. Distrib. Theory*, 7(1):199–215, 2012.

Durand, Fabien. Decidability of the HD0L ultimate periodicity problem. *RAIRO Theor. Inform. Appl.*, 47(2):201–214, 2013a.

Durand, Fabien. Decidability of uniform recurrence of morphic sequences. *Int. J. Found. Comput. Sci.*, 24(1):123–146, 2013b.

Durand, Fabien and Goyheneche, Valérie. Decidability, arithmetic subsequences and eigenvalues of morphic subshifts. *Bull. Belg. Math. Soc. Simon Stevin*, 26(4):591–618, 2019.

Durand, Fabien and Leroy, Julien. S-adic conjecture and Bratteli diagrams. *C. R. Math. Acad. Sci. Paris*, 350(21–22):979–983, 2012.

Durand, Fabien and Leroy, Julien. Decidability of the isomorphism and the factorization between minimal substitution subshifts. 2020. arXiv:1806.04891.

Durand, Fabien, Host, Bernard, and Skau, Christian. Substitutional dynamical systems, Bratteli diagrams and dimension groups. *Ergod. Theory Dyn. Syst.*, 19(4):953–993, 1999.

Durand, Fabien, Leroy, Julien, and Richomme, Gwenael. Do the properties of an S-adic representation determine factor complexity? *J. Integer Seq.*, 16(2), Article 13.2.6, 2013.

Effros, Edward G. *Dimensions and C*-Algebras*, vol. 46 of CBMS Regional Conference Series in Mathematics. Conference Board of the Mathematical Sciences, Washington, DC, 1981.

Effros, Edward G. and Shen, Chao-Liang. Dimension groups and finite difference equations. *J. Oper. Theory*, 2(2):215–231, 1979.

Effros, Edward G., Handelman, David E., and Shen, Chao-Liang Dimension groups and their affine representations. *Amer. J. Math.*, 102(2):385–407, 1980.

Ehrenfeucht, Andrew and Rozenberg, Gregorz. Elementary homomorphisms and a solution of the D0L sequence equivalence problem. *Theoret. Comput. Sci.*, 7(2): 169–183, 1978.

Ehrenfeucht, Andrzej, Lee, Kwok Pun, and Rozenberg, Grzegorz. Subword complexities of various classes of deterministic developmental languages without interactions. *Theor. Comput. Sci.*, 1(1):59–75, 1975.

Eilenberg, Samuel. *Automata, Languages and Machines*, vol. B. Academic Press, Waltham, MA, 1976.

Elliott, George A. On the classification of inductive limits of sequences of semisimple finite-dimensional algebras. *J. Algebra*, 38(1):29–44, 1976.

Elliott, George A. On totally ordered groups, and K_0. In Ring Theory (Proc. Conf., Univ. Waterloo, Waterloo, 1978), vol. 734 of Lecture Notes in Mathematics, pages 1–49. Springer-Verlag, Berlin, 1979.

Espinoza, Bastián and Maass, Alejandro. On the automorphism group of minimal S-adic subshifts of finite alphabet rank. 2020. arXiv:2008.05996.

Fagnot, Isabelle. Sur les facteurs des mots automatiques. *Theor. Comput. Sci.*, 172(1–2):67–89, 1997a.

Fagnot, Isabelle. On the subword equivalence problem for morphic words. *Discrete Appl. Math.*, 75(3):231–253, 1997b.

Fekete, Michael. Über die Verteilung der Wurzeln bei gewissen algebraischen Gleichungen mit ganzzahligen Koeffizienten. *Math. Z.*, 17(1):228–249, 1923.

Ferenczi, Sébastien. Les transformations de Chacon: combinatoire, structure géométrique, lien avec les systèmes de complexité $2n + 1$. *Bull. Soc. Math. France*, 123(2):271–292, 1995.

Ferenczi, Sébastien. Rank and symbolic complexity. *Ergod. Theory Dyn. Syst.*, 16(4):663–682, 1996.

Ferenczi, Sébastien and Zamboni, Luca Q. Languages of k-interval exchange transformations. *Bull. Lond. Math. Soc.*, 40(4):705–714, 2008.

Fickenscher, Jon. Decoding Rauzy induction: An answer to Bufetov's general question. *Bull. Soc. Math. France*, 145(4):603–621, 2017.

Fine, Nathan J. and Wilf, Herbert S. Uniqueness theorems for periodic functions. *Proc. Amer. Math. Soc.*, 16:109–114, 1965.

Fischer, Roland. Sofic systems and graphs. *Monatsh. Math.*, 80(3):179–186, 1975.

Fogg, N. Pytheas. *Substitutions in Dynamics, Arithmetics and Combinatorics*, vol. 1794 of *Lecture Notes in Mathematics*. Springer-Verlag, Berlin, 2002. Edited by V. Berthé, S. Ferenczi, C. Mauduit and A. Siegel.

Forrest, Alan H. K-groups associated with substitution minimal systems. *Isr. J. Math.*, 98:101–139, 1997.

Fuchs, Laszlo. *Partially Ordered Algebraic Systems*. Dover, Mineola, NY, 1963.

Fuchs, Laszlo. Riesz groups. *Ann. Scuola Norm. Sup. Pisa (3)*, 19:1–34, 1965.

Furstenberg, Harry. *Recurrence in Ergodic Theory and Combinatorial Number Theory*. Princeton University Press, 1981.

Gantmacher, Felix R. *The Theory of Matrices*, 2 vols. Chelsea, New York, 1959. Translated from the Russian original.

Giordano, Thierry, Putnam, Ian F., and Skau, Christian F. Topological orbit equivalence and C^*-crossed products. *J. Reine Angew. Math.*, 469:51–111, 1995.

Giordano, Thierry, Handelman, David, and Hosseini, Maryam. Orbit equivalence of Cantor minimal systems and their continuous spectra. *Math. Z.*, 289(3–4):1199–1218, 2018.

Gjerde, Richard and Johansen, Orjan. Bratteli–Vershik models for Cantor minimal systems: Applications to Toeplitz flows. *Ergod. Theory Dyn. Syst.*, 20(6):1687–1710, 2000.

Gjerde, Richard and Johansen, Orjan. Bratteli–Vershik models for Cantor minimal systems associated to interval exchange transformations. *Math. Scand.*, 90(1):87–100, 2002.

Glasner, Eli and Weiss, Benjamin. Weak orbit equivalence of Cantor minimal systems. *Internat. J. Math.*, 6(4):559–579, 1995.

Glass, Andrew M. W. and Madden, James J. The word problem versus the isomorphism problem. *J. Lond. Math. Soc. (2)*, 30(1):53–61, 1984.

Glimm, James G. On a certain class of operator algebras. *Trans. Amer. Math. Soc.*, 95:318–340, 1960.

Goodearl, Kenneth R. *Partially Ordered Abelian Groups with Interpolation*, vol. 20 of Mathematical Surveys and Monographs. American Mathematical Society, Providence, RI, 1986.

Goodearl, Kenneth R. and Handelman, David. Rank functions and K_0 of regular rings. *J. Pure Appl. Algebra*, 7(2):195–216, 1976.

Gottschalk, Walter Helbig and Hedlund, Gustav Arnold. *Topological Dynamics*. American Mathematical Society Colloquium Publications, vol. 36. American Mathematical Society, Providence, RI, 1955.

Goulet-Ouellet, Herman. Suffix-connected languages, 2021. arXiv:2106.00452.

Graham, Ronald L., Knuth, Donald E., and Patashnik, Oren. *Concrete Mathematics: A Foundation for Computer Science*. Addison-Wesley, Boston, 1989.

Grillenberger, Christian. Constructions of strictly ergodic systems. I. Given entropy. *Z. Wahrscheinlichkeitstheorie und Verw. Gebiete*, 25:323–334, 1972/73.

Halmos, Paul R. *Measure Theory*, vol. 18 in Graduate Texts in Mathematics. Springer-Verlag, Berlin, 1974.

Handelman, David. *Positive Polynomials and Product Type Actions of Compact Groups*. Memoirs of the American Mathematical Society, vol. 54 no. 320. American Mathematical Society, Providence, RI, 1985.

Hansel, Georges and Perrin, Dominique. Codes and Bernoulli partitions. *Math. Syst. Theory*, 16(2):133–157, 1983.

Hardy, Godfrey H. and Wright, Edward M.. *An Introduction to the Theory of Numbers*. Oxford University Press, Oxford, 6th ed., 2008. Revised by D. R. Heath-Brown and J. H. Silverman, with a foreword by Andrew Wiles.

Harju, Tero and Linna, Matti. On the periodicity of morphisms on free monoids. *RAIRO Inform. Théor. Appl.*, 20(1):47–54, 1986.

Harpe, Pierre de la. *Topics in Geometric Group Theory*. Chicago Lectures in Mathematics. University of Chicago Press, Chicago, 2000.

Hartfiel, Darald J. *Nonhomogeneous Matrix Products*. World Scientific Publishing Co., River Edge, NJ, 2002.

Hatcher, Allen. *Algebraic Topology.* Cambridge University Press, 2001.

Hedlund, George A. Endomorphisms and automorphisms of the shift dynamical system. *Math. Systems Theory,* 3:320–375, 1969.

Herman, Richard H., Putnam, Ian F., and Skau, Christian F. Ordered Bratteli diagrams, dimension groups and topological dynamics. *Internat. J. Math.,* 3(6):827–864, 1992.

Hollander, Michael and Solomyak, Boris. Two-symbol Pisot substitutions have pure discrete spectrum. *Ergod. Theory Dyn. Syst.,* 23(2):533–540, 2003.

Holton, Charles and Zamboni, Luca Q. Descendants of primitive substitutions. *Theory Comput. Syst.,* 32(2):133–157, 1999.

Honkala, Juha. On the simplification of infinite morphic words. *Theor. Comput. Sci.,* 410(8):997–1000, 2009.

Host, Bernard. Dimension groups and substitution dynamical systems. Laboratoire de Mathématiques discrètes, Marseille, 1995.

Host, Bernard. Substitution subshifts and Bratteli diagrams. In *Topics in Symbolic Dynamics and Applications (Temuco, 1997),* vol. 279 of *London Mathematical Society Lecture Note Series,* pages 35–55. Cambridge University Press, Cambridge, 2000.

Jacobs, Konrad and Keane, Michael. 0-1-sequences of Toeplitz type. *Z. Wahrscheinlichkeitstheorie Verw. Gebiete,* 13:123–131, 1969.

Janvresse, Élise and de la Rue, Thierry. The Pascal adic transformation is loosely Bernoulli. *Annales de l'Institut Henri Poincaré (B) Probability and Statistics,* 40(2):133–139, 2004.

Jullian, Yann. An algorithm to identify automorphisms which arise from self-induced interval exchange transformations. *Math. Z.,* 274(1–2):33–55, 2013.

Justin, Jacques and Vuillon, Laurent. Return words in Sturmian and episturmian words. *Theor. Inform. Appl.,* 34(5):343–356, 2000.

Kakutani, Shizuo. Induced measure preserving transformations. *Proc. Imp. Acad. Tokyo,* 19:635–641, 1943.

Kapovich, Ilya and Myasnikov, Alexei. Stallings foldings and subgroups of free groups. *J. Algebra,* 248(2):608–668, 2002.

Karhumäki, Juhani and Maňuch, Ján. Multiple factorizations of words and defect effect. *Theor. Comput. Sci.,* 273(1):81–97, 2002.

Karhumäki, Juhani, Maňuch, Ján, and Plandowski, Wojciech. A defect theorem for bi-infinite words. *Theor. Comput. Sci.,* 292(1):237–243, 2003.

Katok, Anatole and Hasselblatt, Boris. *Introduction to the Modern Theory of Dynamical Systems,* vol. 54 of *Encyclopedia of Mathematics and its Applications.* Cambridge University Press, Cambridge, 1995. With a supplementary chapter by Katok and Leonardo Mendoza.

Katok, Anatole B. Invariant measures of flows on orientable surfaces. *Dokl. Akad. Nauk SSSR,* 211:775–778, 1973.

Keane, Michael. Interval exchange transformations. *Math. Z.,* 141:25–31, 1975.

Keane, Michael. Non-ergodic interval exchange transformations. *Isr. J. Math.,* 26(2):188–196, 1977.

Kenyon, Richard, Sadun, Lorenzo, and Solomyak, Boris. Topological mixing for substitutions on two letters. *Ergod. Theory Dyn. Syst.,* 25(6):1919–34, 2005.

Kerov, Serguei. PhD thesis, Leningrad State University.

Keynes, Harvey B. and Newton, Dan. A "minimal," non-uniquely ergodic interval exchange transformation. *Math. Z.*, 148(2):101–106, 1976.

Kim, Ki Hang and Roush, Fred W. Some results on decidability of shift equivalence. *J. Combin. Inform. System Sci.*, 4(2):123–146, 1979.

Kim, Ki Hang and Roush, Fred W. Decidability of shift equivalence. In *Dynamical systems (College Park, MD, 1986–87)*, vol. 1342 of *Lecture Notes in Math.*, pages 374–424. Springer-Verlag, Berlin, 1988.

Kim, Ki Hang and Roush, Fred W. Williams's conjecture is false for reducible subshifts. *J. Amer. Math. Soc.*, 5(1):213–215, 1992.

Kim, Ki Hang and Roush, Fred W. The Williams conjecture is false for irreducible subshifts. *Electron. Res. Announc. Amer. Math. Soc.*, 3:105–109, 1997.

Kim, Ki Hang, Roush, Fred W., and Williams, Susan G. Duality and its consequences for ordered cohomology of finite type subshifts. In *Combinatorial & computational mathematics (Pohang, 2000)*, pages 243–265. World Scientific Publishing, River Edge, NJ, 2001.

Klouda, Karel and Starosta, Stepán. Characterization of circular D0L-systems. *Theor. Comput. Sci.*, 790:131–137, 2019.

Knuth, Donald E. *The Art of Computer Programming. Vol. 2.* Seminumerical algorithms. Addison-Wesley, Reading, MA, 1998.

Koblitz, Neal. *p-adic Numbers, p-adic Analysis, and Zeta-Functions*, vol. 58 of Graduate Texts in Mathematics. Springer, New York, 2nd ed., 1984.

Krieger, Dalia. On stabilizers of infinite words. *Theoret. Comput. Sci.*, 400(1–3):169–181, 2008.

Krieger, Wolfgang. On topological Markov chains. In *Dynamical Systems, Vol. II—Warsaw*, pages 193–196. Astérisque, No. 50. 1977.

Krieger, Wolfgang. On dimension functions and topological Markov chains. *Invent. Math.*, 56(3):239–250, 1980a.

Krieger, Wolfgang. On a dimension for a class of homeomorphism groups. *Math. Ann.*, 252(2):87–95, 1980b.

Krieger, Wolfgang and Matsumoto, Kengo. Shannon graphs, subshifts and lambda-graph systems. *J. Math. Soc. Japan*, 54(4):877–899, 2002.

Krylov, Nicolas and Bogolioubov, Nicolas. La théorie générale de la mesure dans son application à l'étude des systèmes dynamiques de la mécanique non linéaire. *Ann. Math.*, 38(1):65–113, 1937.

Kurka, Petr. *Topological and Symbolic Dynamics*, vol. 11 of *Cours Spécialisés [Specialized Courses]*. Société Mathématique de France, Paris, 2003.

Kyriakoglou, Revekka. Recognizable Substitutions. PhD thesis, Université Paris-Est, 2019.

Labbé, Sébastien and Leroy, Julien. Bispecial factors in the Brun S-adic system. In *Developments in Language Theory*, vol. 9840 of *Lecture Notes in Comput. Sci.*, pages 280–292. Springer-Verlag, Berlin, 2016.

Lang, Serge. *Algebra*, vol. 211 of Graduate Texts in Mathematics. Springer, New York, 3rd ed., 2002.

Leroy, Julien. An S-adic characterization of minimal subshifts with first difference of complexity $1 \leq p(n+1) - p(n) \leq 2$. *Discrete Math. Theor. Comput. Sci.*, 16(1):233–286, 2014a.

Leroy, Julien. An S-adic characterization of ternary tree sets. 2014b. in preparation.

Lind, Douglas and Marcus, Brian. *An Introduction to Symbolic Dynamics and Coding*. Cambridge University Press, Cambridge, 1995. (2nd edition 2021).

Lindenmayer, Aristid. Mathematical models for cellular interaction in development. *J. Theoret. Biology*, 18:280–315, 1968.

Linna, Matti. The decidability of the D0L prefix problem. *International Journal of Computer Mathematics*, 6(2):127–142, 1977.

Lopez, Luis-Miguel and Narbel, Philippe. Lamination languages. *Ergod. Theory Dyn. Syst.*, 33(6):1813–1863, 2013.

Lothaire, M. *Combinatorics on Words*. Cambridge University Press, 2nd ed., 1997.

Lothaire, M. *Algebraic Combinatorics on Words*. Cambridge University Press, 2002.

Luenberger, David G. and Ye, Yinyu. *Linear and Nonlinear Programming*, vol. 228 of *International Series in Operations Research & Management Science*. Springer, Cham, 4th ed., 2016.

Lyndon, Roger C. and Schupp, Paul E. *Combinatorial Group Theory*. Classics in Mathematics. Springer-Verlag, Berlin, 2001. Reprint of the 1977 edition.

Magnus, Wilhelm, Karrass, Abraham, and Solitar, Donald. *Combinatorial Group Theory*. Dover Publications, Inc., Mineola, NY, 2nd ed., 2004.

Maloney, Gregory R. and Rust, Dan. Beyond primitivity for one-dimensional substitution subshifts and tiling spaces. *Ergod. Theory Dyn. Syst.*, 38(3):1086–1117, 2018.

Markov, Alexander A. The insolubility of the problem of homeomorphy. *Dokl. Akad. Nauk SSSR*, 121:218–220, 1958.

Martin, John C. Minimal flows arising from substitutions of non-constant length. *Math. Systems Theory*, 7:72–82, 1973.

Masur, Howard. Interval exchange transformations and measured foliations. *Ann. Math.*, 115(1):169–200, 1982.

Matsumoto, Kengo. A simple C^*-algebra arising from a certain subshift. *J. Oper. Theory*, 42(2):351–370, 1999. ISSN 0379-4024.

Medynets, Konstantin. Cantor aperiodic systems and Bratteli diagrams. *C. R. Math. Acad. Sci. Paris*, 342(1):43–46, 2006.

Méla, Xavier and Petersen, Karl. Dynamical properties of the pascal adic transformation. *Ergod. Theory Dyn. Syst.*, 25(1):227–256, 2003.

Michel, Pierre. Stricte ergodicité d'ensembles minimaux de substitution. *C. R. Acad. Sci. Paris Sér. A*, 278:811–813, 1974.

Michel, Pierre. Stricte ergodicite d'ensembles minimaux de substitution. In Jean-Pierre Conze and Michael S. Keane, eds., *Théorie Ergodique*, pages 189–201, Springer-Verlag, Berlin and Heidelberg, 1976.

Miernowski, Tomasz and Nogueira, Arnaldo. Exactness of the Euclidean algorithm and of the Rauzy induction on the space of interval exchange transformations. *Ergod. Theory Dyn. Syst.*, 33(1):221–246, 2013.

Mignosi, Filippo. On the number of factors of Sturmian words. *Theoret. Comput. Sci.*, 82(1):71–84, 1991.

Mignosi, Filippo and Séébold, Patrice. If a D0L language is k-power free then it is circular. In Andrzej Lingas, Rolf G. Karlsson, and Svante Carlsson, eds., *Automata, Languages and Programming, 20nd International Colloquium, ICALP93, Lund, Sweden, July 5–9, 1993, Proceedings*, vol. 700 of *Lecture Notes in Computer Science*, pages 507–518. Springer-Verlag, Berlin, 1993.

Monteil, Thierry. An upper bound for the number of ergodic invariant measures of a minimal subshift with linear complexity. 2009. Preprint.

Morse, Marston and Hedlund, Gustav A. Symbolic dynamics. *Amer. J. Math.*, 60(4):815–866, 1938.

Morse, Marston and Hedlund, Gustav A. Symbolic dynamics II. Sturmian trajectories. *Amer. J. Math.*, 62:1–42, 1940.

Mossé, Brigitte. Puissances de mots et reconnaissabilité des points fixes d'une substitution. *Theoret. Comput. Sci.*, 99(2):327–334, 1992.

Mossé, Brigitte. Reconnaissabilité des substitutions et complexité des suites automatiques. *Bull. Soc. Math. France*, 124(2):329–346, 1996.

Mousavi, Hamoon. Automatic theorem proving in Walnut, 2016. `arXiv:1603.06017`.

Muller, David E. and Schupp, Paul E. Groups, the theory of ends, and context-free languages. *J. Comput. System Sci.*, 26(3):295–310, 1983.

Müllner, Clemens and Yassawi, Reem. Automorphisms of automatic shifts. *Ergod. Theory Dyn. Syst.*, 41(5):1530–1559, 2021.

Mundici, Daniele. Simple Bratteli diagrams with a Gödel-incomplete C^*-equivalence problem. *Trans. Amer. Math. Soc.*, 356(5):1937–1955, 2004.

Mundici, Daniele. Bratteli diagrams via the De Concini-Procesi Teorem. *Commun. Contemp. Math.*, 2020. doi: https://doi.org/10.1142/S021919972050073X

Mundici, Daniele and Panti, Giovanni. The equivalence problem for Bratteli diagrams. Technical Report 259, University of Siena (Italy), 1993.

Nogueira, Arnaldo. Almost all interval exchange transformations with flips are nonergodic. *Ergod. Theory Dyn. Syst.*, 9(3):515–525, 1989.

Nogueira, Arnaldo, Pires, Benito, and Troubetzkoy, Serge. Orbit structure of interval exchange transformations with flip. *Nonlinearity*, 26(2):525–537, 2013.

Ormes, Nicholas S. Real coboundaries for minimal Cantor systems. *Pacific J. Math.*, 195(2):453–476, 2000.

Ornstein, Donald S., Rudolph, Daniel J., and Weiss, Benjamin. Equivalence of measure preserving transformations. *Mem. Amer. Math. Soc.*, 37(262):xii+116, 1982.

Oxtoby, John C. Ergodic sets. *Bull. Amer. Math. Soc.*, 58:116–136, 1952.

Pansiot, Jean-Jacques. Hiérarchie et fermeture de certaines classes de tag-systèmes. *Acta Informatica*, 20(2):179–196, 1983.

Pansiot, Jean-Jacques. Complexité des facteurs des mots infinis engendrés par morphismes itérés. In Jan Paredaens, ed., *Automata, Languages and Programming*, pages 380–389, Springer-Verlag, Berlin and Heidelberg, 1984.

Pansiot, Jean-Jacques. Decidability of periodicity for infinite words. *RAIRO Inform. Théor. Appl.*, 20(1):43–46, 1986.

Parry, William and Tuncel, Selim. *Classification Problems in Ergodic Theory*. Cambridge University Press, 1982.

Pedersen, Gert K. *C*-Algebras and Their Automorphism Groups*. Academic Press, Cambridge, MA, 2nd ed., 2018.

Perrin, Dominique and Pin, Jean-Eric. *Infinite Words: Automata, Semigroups, Logic and Games*. Elsevier, Amsterdam, 2004.

Perrin, Dominique and Restivo, Antonio. A note on Sturmian words. *Theor. Comput. Sci.*, 429:265–272, 2012.

Petersen, Karl. *Ergodic Theory*. Cambridge University Press, 1983.

Petite, Samuel. Personal communication, 2019.

Poincaré, Henry. Sur le problème des trois corps et les équations de la dynamique. *Acta Math*, 13(1–2):1–270, 1890.

Poon, Yiu Tung. A K-theoretic invariant for dynamical systems. *Trans. Amer. Math. Soc.*, 311(2):515–533, 1989.

Putnam, Ian F. The C^*-algebras associated with minimal homeomorphisms of the Cantor set. *Pacific J. Math.*, 136(2):329–353, 1989.

Putnam, Ian F. C^*-algebras arising from interval exchange transformations. *J. Oper. Theory*, 27(2):231–250, 1992.

Putnam, Ian F. Orbit equivalence of Cantor minimal systems: A survey and a new proof. *Expo. Math.*, 28(2):101–131, 2010.

Putnam, Ian F. A homology theory for Smale spaces. *Mem. Amer. Math. Soc.*, 232(1094):viii+122, 2014.

Putnam, Ian F. *Cantor Minimal Systems*. American Mathematical Society, Providence, RI, 2018.

Queffélec, Martine. *Substitution Dynamical Systems – Spectral Analysis*, vol. 1294 of *Lecture Notes in Mathematics*. Springer-Verlag, Berlin, 2nd ed., 2010.

Rauzy, Gérard. Échanges d'intervalles et transformations induites. *Acta Arith.*, 34(4):315–328, 1979.

Renault, Jean. *A Groupoid Approach to C*-Algebras*, vol. 793 of *Lecture Notes in Math.* Springer-Verlag, Berlin, 2014.

Restivo, Antonio. A combinatorial property of codes having finite synchronization delay. *Theoret. Comput. Sci.*, 1(2):95–101, 1975.

Reutenauer, Christophe. *From Christoffel Words to Markoff Numbers*. Oxford University Press, Oxford, 2019.

Richomme, Gwénaël and Séébold, Patrice. Completing a combinatorial proof of the rigidity of Sturmian words generated by morphisms. *Theoret. Comput. Sci.*, 428:92–97, 2012.

Riedel, Norbert. Classification of dimension groups and iterating systems. *Mathematica Scandinavica*, 48(2):226–234, 1981a.

Riedel, Norbert. A counterexample to the unimodular conjecture on finitely generated dimension groups. *Proc. Amer. Math. Soc.*, 83(1):11–15, 1981b.

Rieffel, Marc. Morita equivalence for operator algebras. *Proc. Sympos. Pure Math.*, 38:285–298, 1982. Accessed at http://math.berkeley.edu/~rieffel/papers/morita_equivalence.pdf

Rigo, Michel. *Formal Languages, Automata and Numeration Systems*. Wiley, Hoboken, NJ, 2014.

Robert, Alain M. *A Course in p-adic Analysis*, vol. 198 of Graduate Texts in Mathematics. Springer, New York, 2000.

Rokhlin, Vladimir. A "general" measure-preserving transformation is not mixing. *Doklady Akad. Nauk SSSR (N.S.)*, 60:349–351, 1948.

Rowen, Louis. *Ring Theory*. Elsevier, Amsterdam, 1988.

Rozenberg, Grzegorz and Salomaa, Arto. *The Mathematical Theory of L-Systems*. Academic Press, Cambridge, MA, 1980.

Rudin, Walter. Some theorems on Fourier coefficients. *Proc. Amer. Math. Soc.*, 10:855–859, 1959.

Rudin, Walter. *Real and Complex Analysis*. McGraw-Hill, New York, 3rd ed., 1987.

Rudin, Walter. *Functional Analysis*. International Series in Pure and Applied Mathematics. McGraw-Hill, New York, 2nd ed., 1991.

Rust, Dan and Balchin, Scott. Computations for symbolic substitutions. *J. Integer Seq.*, 20(4), 2017 arXiv:1705.11130.

SageMath. *The Sage Mathematics Software System (Version 9.2)*, 2020. www.sagemath.org.

Samuel, Pierre. *Algebraic Theory of Numbers*. Translated from the French by Allan J. Silberger. Houghton Mifflin Boston, 1970.

Schaeffer, Luke and Shallit, Jeffrey. The critical exponent is computable for automatic sequences. *Internat. J. Found. Comput. Sci.*, 23(8):1611–1626, 2012.

Schützenberger, Marcel-Paul. Une théorie algébrique du codage. In *Séminaire Dubreil-Pisot 1955–56*, page Exposé N°. 15, 1955.

Séébold, Patrice. On the conjugation of standard morphisms. *Theoret. Comput. Sci.*, 195(1):91–109, 1998.

Seneta, Eugene. *Non-negative Matrices and Markov Chains*. Springer Series in Statistics. Springer, New York, 2006. Revised reprint of the second (1981) edition.

Serre, Jean-Pierre. *Trees*. Springer Monographs in Mathematics. Springer-Verlag, Berlin, 2003. Translated from the French original by John Stillwell, Corrected 2nd printing of the 1980 English translation.

Shallit, Jeffrey O., 2020. Personal communication.

Shen, Chao Liang. On the classification of the ordered groups associated with the approximately finite-dimensional C^*-algebras. *Duke Math. J.*, 46(3):613–633, 1979.

Shimomura, Takashi. A simple approach to minimal substitution subshifts. *Topology and Its Applications*, 260:203–214, 2019.

Stallings, John R. Topology of finite graphs. *Invent. Math.*, 71(3):551–565, 1983.

Stern, Moritz. Ueber eine zahlentheoretische funktion. *Journal für die Reine und Angewandte Mathematik*, 55:193–220, 1858.

Sugisaki, Fumiaki. The relationship between entropy and strong orbit equivalence for the minimal homeomorphisms. I. *Internat. J. Math.*, 14(7):735–772, 2003.

Sugisaki, Fumiaki. On the subshift within a strong orbit equivalence class for minimal homeomorphisms. *Ergod. Theory Dyn. Syst.*, 27(3):971–990, 2007.

Tan, Bo, Wen, Zhi-Xiong, and Zhang, Yiping. The structure of invertible substitutions on a three-letter alphabet. *Adv. in Appl. Math.*, 32(4):736–753, 2004.

Thuswaldner, Jörg M. S-adic sequences, a bridge between dynamics, arithmetic and geometry. 2020. arXiv:1908.05954.

Toeplitz, Otto. Ein Beispiel zur Theorie der fastperiodischen Funktionen. *Math. Ann.*, 98(1):281–295, 1928.

Veech, William A. Finite group extensions of irrational rotations. *Israel J. Math.*, 21(2–3):240–259, 1975.

Veech, William A. Interval exchange transformations. *J. Analyse Math.*, 33:222–272, 1978.

Veech, William A. Gauss measures for transformations on the space of interval exchange maps. *Ann. Math.*, 115(2):201–242, 1982.

Veech, William A. Moduli spaces of quadratic differentials. *J. Analyse Math.*, 55:117–171, 1990.

Vershik, Anatol M. A theorem on periodical Markov approximation in ergodic theory. In *Ergodic Theory and Related Topics (Vitte, 1981)*, vol. 12 of *Math. Res.*, pages 195–206. Akademie-Verlag, Berlin, 1982.

Vershik, Anatol M. A theorem on the Markov periodical approximation in ergodic theory. *Journal Soviet. Math.*, 28:667–674, 1985.

Vershik, Anatol M. and Livshits, Anatol N. Adic models of ergodic transformations, spectral theory, substitutions, and related topics. In *Representation Theory and Dynamical Systems*, vol. 9 of Advances in Soviet Mathematics, pages 185–204. American Mathematical Society, Providence, RI, 1992.

Viana, Marcelo. *Dynamics of Interval Exchange Transformations and Teichmüller Flows*. 2008. http://w3.impa.br/~viana/.

Walters, Peter. *Ergodic Theory – Introductory Lectures.* Lecture Notes in Mathematics, vol. 458. Springer-Verlag, Berlin and New York, 1975.

Walters, Peter. *An Introduction to Ergodic Theory*, vol. 79 of Graduate Texts in Mathematics. Springer-Verlag, Berlin and New York, 1982.

Weiss, Benjamin. Subshifts of finite type and sofic systems. *Monatsh. Math.*, 77:462–474, 1973.

Willard, Stephen. *General Topology*. Dover Publications, Mineola, NY, 2004.

Williams, Robert F. Classification of subshifts of finite type. In *Recent Advances in Topological Dynamics*, pages 281–285. Lecture Notes in Mathematics, vol. 318, 1973.

Williams, Susan. Toeplitz minimal flows which are not uniquely ergodic. *Z. Wahrscheinlichkeitstheorie Verw. Gebiete*, 67(1):95–107, 1984.

Yuasa, Hisatoshi. On the topological orbit equivalence in a class of substitution minimal systems. *Tokyo J. Math.*, 25(2): 221–240, 2002.

Name Index

Index of Symbols

Subject Index

Printed in the United States
by Baker & Taylor Publisher Services